This volume includes all the published papers of Alfred Young
(1873-1940),
who made outstanding contributions to the
algebra of invariants and the theory of groups,
together with a biographical sketch published after Young's death
and a foreword by Professor G. de B. Robinson
of the University of Toronto.
It will be of interest to algebraists, combinatorialists,
theoretical physicists, and students of the
corresponding disciplines.

MATHEMATICAL EXPOSITIONS

VOLUMES PUBLISHED

1. *The Foundations of Geometry.* G. de B. Robinson
2. *Non-Euclidean Geometry.* H.S.M. Coxeter
3. *The Theory of Potential and Spherical Harmonics.* W.J. Sternberg and T.L. Smith
4. *The Variational Principles of Mechanics.* Cornelius Lanczos
5. *Tensor Calculus.* J.L. Synge and A.E. Schild (out of print)
6. *The Theory of Functions of a Real Variable.* R.L. Jeffery (out of print)
7. *General Topology.* Waclaw Sierpinski; translated by C. Cecilia Krieger (out of print)
8. *Bernstein Polynomials.* G.G. Lorentz (out of print)
9. *Partial Differential Equations.* G.F.D. Duff
10. *Variational Methods for Eigenvalue Problems.* S.H. Gould
11. *Differential Geometry.* Erwin Kreyszig (out of print)
12. *Representation Theory of the Symmetric Group.* G. de B. Robinson
13. *Geometry of Complex Numbers.* Hans Schwerdtfeger
14. *Rings and Radicals.* N.J. Divinsky
15. *Connectivity in Graphs.* W.T. Tutte
16. *Introduction to Differential Geometry and Riemannian Geometry.* Erwin Kreyszig
17. *Mathematical Theory of Dislocations and Fracture.* R.W. Lardner
18. *n-gons.* Friedrich Bachmann and Eckart Schmidt; translated by Cyril W.L. Garner
19. *Weighing Evidence in Language and Literature: A Statistical Approach.* Barron Brainerd
20. *Rudiments of Plane Affine Geometry.* P. Scherk and R. Lingenberg
21. *The Collected Papers of Alfred Young, 1873–1940*

MATHEMATICAL EXPOSITIONS NO. 21

The Collected Papers of Alfred Young

1873–1940

F.R.S., Sc.D., LL.D.

Fellow of Clare College, Cambridge

Canon of Chelmsford

Rural Dean of Belchamp

Rector of Birdbrook, Essex 1910–1940

University of Toronto Press

Toronto and Buffalo

Toronto and Buffalo
Printed in Canada
Reprinted in 2018

Canadian Cataloguing in Publication Data

Young, Alfred, 1873–1940.
The collected papers of Alfred Young, 1873–1940

(Mathematical expositions ; 21 ISSN 0076-5333)

ISBN 0-8020-2267-7
ISBN 978-1-4875-7272-3 (paper)
1. Mathematics – Collected works. I. Robinson, Gilbert de B., 1906– II. Series.

QA3.Y69 510'.8 C77-001375-9

This book has been published during the
Sesquicentennial year of the University of Toronto

Contents

*These papers involve group theory.

Foreword

Invariant theory and its applications were popular in the 19th century, but the work of Frobenius, Burnside, and Schur shifted the emphasis to group theory, in particular to the representation theory of symmetric groups, a subject which is of importance in quantum mechanics and modern combinatorics. Later developments have aroused so much interest that a symposium was held in Oberwolfach in 1975. At this symposium, on 'Combinatorics: Young Tableaux and Combinatorics in Symmetric Group Representation,' the need was expressed to go back to the pioneering work in the field, particularly that of Alfred Young. It was urged upon me, as Young's sole Ph.D. student in Cambridge, to bring out a volume of his collected papers - which are not now readily available in many libraries. It was suggested also that I might include some samples of his correspondence with me. I am grateful that the original publishers of the papers looked upon the suggestion with favour and granted the necessary permissions for reproduction.

Alfred Young's life and contribution to mathematical knowledge are well summarized in the biography by H.W. Turnbull published in the *Journal of the London Mathematical Society* in 1941 following his sudden death on 15 December 1940. It serves as an appropriate introduction to the present volume. A few months after Young's death, on 31 March 1941, his wife wrote me a charming letter, saying in part:

> I was so overwhelmed with the sudden passing of my dear one, he was out with me on the Wednesday afternoon visiting in the parish. He seemed as usual, when suddenly after tea he was taken with a pain in his side. The doctor had him rushed off to the hospital, operated upon him but he never really came around and passed away on Sunday morning. I know it was what he would have wished - to die at his post.
>
> Mr. Hall of King's College, Cambridge has kindly promised to look through his unpublished papers; he came over and took them all ...

Subsequently, H.W. Richmond, also of King's College, and Turnbull went through the papers after Philip Hall. The large box of papers was eventually sent to me at the University of Toronto in 1948, where they are now housed in the Robarts Library and are available for consultation by anyone interested. My introduction to the ninth paper on quantitative substitutional analysis (see pp. 650ff.) provides a summary of the contents noted in an examination by W.T. Sharp and me. It is worth mentioning here that Alfred Young appreciated the need to interpret his representation theory of S_n to apply to a subgroup $H \subset S_n$. Some day someone will continue where he left off and achieve the success which I still believe is possible.

In spite of his love of abstract ideas, Alfred Young was a very practical man. The rectory at Birdbrook was large and very beautiful. Turnbull refers to his small electrical engine used to pump water. During the war years 1914–18, Young worked on the paper published in 1920 (pp. 274–84), and in the box of his papers we found two patents. Following the 1976 Strasbourg sequel to the Oberwolfach symposium, I went to the Patent Office in London and obtained the specifications. Professor G.R. Slemon, Chairman of the Department of Electrical Engineering at the University of Toronto, examined these, commenting as follows:

Patent Specification 113,679 is entitled 'An Apparatus for the purpose of the Direct Conversion of the Energy of Motion of a Dielectric into Electric Energy, and Conversely for the Conversion of Electric Energy (of the right frequency) into the Motion of Mass.' This apparatus, patented in 1918, is similar in many ways to the magnetohydrodynamic generators currently under development for efficient conversion of thermo energy to electric energy. The invention converts mechanical energy in a moving fluid such as water or gas to high-frequency electric energy without the use of moving parts. Exploitation of the basic principle of this invention has had to wait on the development of superconducting coils to produce intense magnetic fields. While present-day magnetohydrodynamic generators produce either direct current or line-frequency alternating current, Young set out to produce high-frequency electric currents predominantly for use in the emerging technology of wireless telegraphy.

The Patent Specification 127,488 is entitled 'A Machine for the Generation of Electric Currents also Applicable as a Motor.' The patent, granted in 1919, is also for an apparatus to produce high-frequency alternating currents. It is one of a family of devices in which the mutual inductance between coils is varied by virtue of their relative motion and the varying inductance is used in a tuned circuit to produce high-frequency current. Similar devices existed at that time but were restricted to the production of low-frequency current by the inclusion of iron cores. The advantage claimed for this device was that it could operate at high frequencies because it used air-cored coils.

To the best of my knowledge, neither of these inventions was exploited during the period of validity of the patent. Both specifications display a sound knowledge of electrical principles. There is no question that the devices would work. Very few inventions, however, reach the stage of engineering practicality and these are among the many that have failed to achieve industrial application. Alfred Young appears to have been a genuine natural philosopher with added ingenuity.

After visiting the Patent Office, our son-in-law drove my wife and me up to Birdbrook, where I took the accompanying photographs. Mrs Young lived until 1950 and she and her husband are remembered in a stained glass window in the ancient church to which they both contributed so much over so many years.

Young's group theoretical ideas were developed in the long series of papers entitled 'On Quantitative Substitutional Analysis' (QSA), but his other applications of group theory to invariants are also of interest. Here we include all of Young's papers, whether on these or other topics, so it seemed desirable to distinguish those involving group theory by adding an asterisk before the title in the table of contents. Not included in the volume is Young's only book, *Algebra of Invariants* (Cambridge University Press, 1903), co-authored with J.H. Grace, which Turnbull describes ecstatically. The first six chapters are devoted to the 19th century approach using differential operators, determinants, and generating functions, and the remainder is largely of geometrical interest (Grace was a geometer) except for the last chapter in which Young describes his process of 'symmetrization.'

In QSA I (pp. 42–91) Young derives the fundamental equation $1 = \Sigma A_\alpha T_\alpha$ and applies his ideas to proving theorems of Gordan, Capelli, and Peano. In QSA II (pp. 92–128) he calculates A_α and studies the properties of the *PN*. The paper concludes with an application to binary forms. Burnside refereed QSA I and II, and Young told me that he advised him to read the works of Frobenius and Schur. But Young knew no German and it was not till QSA III (pp. 352–89) appeared in 1928 that he further pursued his group-theoretic ideas. In the introduction to this, one of his most important papers, he relates his work to that of Frobenius and proceeds to develop the idempotents and their properties in a masterly fashion. Young's work had attracted Weyl's attention (see *Gruppentheorie und Quantenmechaniks* [2nd ed., Hirzel, Leipzig, 1931]), and suddenly he was famous.

Church of St. Augustine of Canterbury, Birdbrook, Essex

Grave of Dr. and Mrs. Alfred Young, Birdbrook, Essex

In QSA IV, V, and VI (pp. 396–466) Young pursues and largely completes his contribution to group theory. He had been continuing his study of Frobenius and Schur and in QSA V he applied his method to obtain the representation theory of the hyperoctahedral group. In QSA VI he derives the fundamental rule for writing down the orthogonal representation of S_n, starting with the hook-representation which I had dealt with in 1931 in my thesis (see *Representation Theory of the Symmetric Group* [University of Toronto Press, 1961]).

In the remaining papers Young returns to his first love – invariant theory – and applies his symmetrizer to organize the field. In QSA VII and VIII (pp. 494–613) he comes to realize the significance of the sequence in which standard tableaux can be written. This is proving increasingly important in recent developments of the subject. In his 1935 paper he further relates his ideas to work of Frobenius and Schur. My introduction to QSA IX (pp. 650–84) concludes this summary of Young's work. As Turnbull so ably expresses it, Young's contributions had already been appreciated in 1941, and it is my hope that this volume will make them more accessible to another generation.

The papers are presented as they appeared originally, including the pagination of the journal at the top of the page for ease in following the many cross-references (the pagination of the present volume is at the foot of the pages). For the same reason the year of publication is placed in a prominent position in the table of contents. The papers are reproduced in facsimile, which precluded any adjustment of the text except for the correction of minor typographical errors noted on copies sent to me by Young. Unfortunately the reproduction presented a problem in places where the type was weak; I ask the reader to bear with these imperfections.

The following excerpts from two letters of Alfred Young relating to my first graduate student H.H. Ferns, whose thesis was published in the *Transactions of the Royal Society of Canada*, III (1934), 35–60, may be of interest.

June 19, 1933

I have been trying to get to grips with Ferns' dissertation and suddenly it seemed to me that a good deal of what he was after was very intimately connected with something I had included in my last paper Q.S.A. VIII which I sent in to the L.M.S. at the beginning of February last, and heard from Watson about a fortnight or so ago that it had been accepted for publication. As it is unlikely to get into print until some time next year, I thought I had better copy out the two sections of that paper from my MS, (I had it typed for the L.M.S. to have the MS by me), and let you have them, in case they may be of some assistance to you or Ferns. Of course the matter is approached from a different point of view; and with another purpose in prospect. But I think these two sections are understandable of themselves - they lead on from Q.S.A. VI (with probably the exception of pp. 4,5, i.e. Theorem XV in the enclosed MS - this has nothing to do with the matter at present in discussion of Ferns' work: - but of the greatest importance for my object - invariant algebra - it was put in here to avoid a gap in what I sent you) ...

On later thoughts I have also copied out and enclose Section VII of Q.S.A. VIII, which is closely connected with the subject; and is an illustration of Section VI. I should be immensely interested to see a geometrical interpretation of the double group matrix. There is a uniqueness about the irreducible representations, that might correspond to some very fundamental geometric property. There is probably some general simple form of the irreducible double group matrix applicable to all cases which it would be interesting to discover. - but I have had no time to go further into that than the discussion enclosed. I should imagine that Ferns is a man who should be encouraged as much as possible, he appears to possess that infinite capacity for taking pains which is the bedrock of success. And also to have the capacity of imagination which should lead him to turn his labours in the right direction.

August 26, 1933

... Many thanks for letting me see Ferns' work, which I have posted separately. His ideas should lead him on to quite a general theory, both interesting and useful. I hope he will not feel I have cut any ground from under his feet in that extract of Q.S.A. VIII I sent you; but there is ample room for him to develope his own way of looking at the algebra, and the geometry is a wide field which he has begun well to explore. I am in great hopes that my MS may help him to generalise his own work, and to cut down the mass of algebraic detail with which it is rather embarassed at present.

We have had the most wonderful summer I ever remember, the harvest is practically finished and appears to be excellent in every way. Many thanks for asking us to visit you in Canada, I should love to do so. But my wife is very averse to travelling, even to crossing the channel, so I much regret that it is a most unlikely event.

It is appropriate to conclude this Foreword by referring to the recent revival of interest in invariant theory (see C. Procesi, 'The Invariant Theory of $n \times n$ Matrices,' *Advances in Mathematics*, 19 [1976], 306–81). It is not so much the geometrical aspects which are being emphasized but the general algebraic structure which was so well described in the *Algebra of Invariants* and to which Young returned in his later years.

Since this Foreword was written the proceedings of the Strasbourg 'Table Ronde' have appeared in print (*Combinatoire et Représentations du Groupe Symétrique* [Springer-Verlag, Lecture Notes in Mathematics No. 579, 1977]), and it is interesting to read the many papers presented there in the present context. It is expected that the next volume in the Mathematical Expositions Series will be one by T.V. Narayana dealing with the application of Young's ideas in statistics.

G. de B. Robinson
Toronto, 1977

ALFRED YOUNG

1873–1940

H. W. Turnbull

Alfred Young was born at Birchfield, Farnworth, near Widnes, Lancashire, on 16 April 1873. He died after a short illness on Sunday, 15 December 1940. He was the youngest son of Edward Young, a prosperous Liverpool merchant and a Justice of the Peace for the county. His father married twice and had a large family, eleven living to grow up. The two youngest sons of the two branches of the family rose to scientific distinction: Sydney Young, of the elder family, became a distinguished chemist of Owen's College, Manchester, University College, Bristol, and finally, for many years, of Trinity College, Dublin. He was elected Fellow of the Royal Society in his thirty-sixth year and died in 1937. Alfred, who was fifteen years his junior, was elected Fellow in 1934, at the age of sixty, in recognition of his mathematical contributions to the algebra of invariants and the theory of groups, a work to which he had devoted over ten years of academic life followed by thirty years of leisure during his duties as rector of a country parish. Recognition of his remarkable powers came late but swiftly; he was admitted to the Fellowship in the year when his name first came up for election.

In 1879 the family moved to Bournemouth, and in due course the younger brothers went to school and later to a tutor, under whom Alfred suffered for his brain power, being the only boy considered worth keeping in. Next, he went to Monkton Combe School, near Bath, and there again his unusual mathematical ability was recognised. Thence he gained a scholarship at Clare College, Cambridge, where he matriculated in 1892. At Clare he formed his life-long friendship with G. H. A. Wilson, another distinguished mathematician, who eventually became Master of the College. Young was a good oarsman and rowed in the Junior Trial Eights as a freshman and in the Scratch Fours of 1893. His college friends still remember him as a shy, clever lad with a great humility of spirit which so marked him in his youth and indeed throughout his life. Early in his third year at college his interest in research began, and his enthusiasm doubtless diverted him from the subjects laid down for the Tripos examination, for which he was prepared by the celebrated coach, Webb of St. John's College. In 1895 he graduated as tenth Wrangler, his friend Wilson being placed fifth. It was a brilliant year; Bromwich was Senior Wrangler, Grace and Whittaker were bracketed second, and thereafter followed Hopkinson, Godfrey, and Maclaurin twelfth Wrangler, to whom one of the three Smith's Prizes was subsequently awarded. Young, who according to Grace was the most original man of his year, would probably have occupied a higher place in the list had he directed his attention to the examination schedule; but in turning to his research, as Wilson tells us, undoubtedly he chose the better path. In the following year Young was placed in the Second Class of the Mathematical Tripos, Part II. From 1901 to 1905 he lectured at Selwyn College, Cambridge, but resigned that appointment shortly after his election to a Fellowship at his own college, where he was also Bursar until 1910. His work received recognition when he was approved for the degree of Sc.D. at Cambridge in 1908.

Young had always intended to take Holy Orders, but it was not until 1908 that he was ordained, when he accepted a curacy at Christ Church, Blacklands, Hastings. Two years later he was presented by his College to the living of Birdbrook, a village of Essex about twenty-five miles east of Cambridge and on the borders of Suffolk. There he lived and worked for thirty years, quietly and faithfully performing the duties of a parish priest, beloved by his congregation, who respected and wondered at his great scholastic gifts so modestly set forth. He was always a welcome visitor in their homes, and he conducted the services in the parish church with dignity and sincerity, and was an excellent preacher. He readily undertook the responsibilities of his calling further afield, was appointed in 1923 to be Rural Dean of Belchamp, and in 1929 Chaplain to the Bishop of Colchester. As recently as November 1940 he was installed as Honorary Canon of Chelmsford Cathedral. In 1926 he accepted an invitation from the University of Cambridge to give a course of lectures on Higher Algebra, and this he continued to do for several years during the Lent, and occasionally the May, Term. In 1931 he was awarded the Honorary LL.D. Degree of the University of St. Andrews.

In 1907 Alfred Young married Edith Clara, daughter of Mr. Edward Wilson, of Sheffield, by whom he is survived. There were no children of the marriage. Their home was a typical country rectory, set in an old world garden full of colour and of great charm, where a warm welcome awaited a visitor from Cambridge or elsewhere, young or old, who sought out in this secluded corner of Essex a master of abstract algebra, and found more than a mathematician, a friend. After a thirty-mile bicycle ride (these were the days before the motor bus had become ubiquitous) the shade and peace of the rectory garden on a summer day were greatly refreshing.

"He was charming to us", an undergraduate wrote after such a visit, "and I remember how delighted I was with the Rectory, and how he told us of the excellence of the beer brewed with the water of his pond". He was a practical mechanic, and had a device by which all the water in the house was pumped up with the help of a little motor engine which was run round to the pump each day. He had also successfully set up a small electric light plant to supply the house. His expert knowledge of its working seemed odd amid those rustic surroundings, till one recalled his interests in the more practical mathematics besides the theory of groups, and the paper he had published on the electromagnetic properties of coils. He was very methodical in planning his duties and his leisure; he was a willing correspondent and would take great pains to answer the mathematical queries of his friends, sharing with them liberally his own abundant thoughts on algebra, invariants and geometry. He could take up or lay aside and again, after several weeks, resume a formidable piece of algebraic computation, without apparently losing the threads of the arguments and with the utmost composure. Every year he and his wife would regularly take their holiday shortly after Easter at a South Coast resort. and every Tuesday they would go over to Cambridge and thus keep in touch with their University friends, a practice which doubtless started when Young returned to the lecture room at the call of the University. His quiet determination and his unhurried devotion to the things of the mind and of the spirit were very impressive. Of such might Whittier have been thinking when he wrote: "And let our ordered lives confess, The beauty of Thy peace". Young was intellectually alert to the end of his life, and during his last year he was constantly working at his ninth memoir, to which he attached great importance. It was nearly finished and lay on his desk awaiting the final touches. During the last few days of illness he asked his doctor whether he could hope to live to finish his work.

My friend W. L. Edge has supplied a picture of the lecture course when Young resumed his teaching at Cambridge: "I remember (who could forget?) very well my experiences of attending his first lectures. This was only a course of one lecture a week for one term; you can see for yourself how much he got through. . . . Doubtless it is all standard work to you, but it will be interesting to see how the old warrior entered the lists again and what he considered should be given to his first hearers. I went along on 19 January 1926, in my third year, just two terms before my Tripos, to Clare . . . there were eleven of us and I was the only undergraduate who ventured. Others in the class were, I think, Cooper, now at Belfast; Broadbent, now at Greenwich; L. H. Thomas, who got a Smith's Prize and a Trinity Fellowship and went to America; Dirac certainly, and F. P. White, the only M.A. And I remember the tall clerical figure entering the room, and his surprise at so large an audience, and shaking hands with White with obvious pleasure. And so to linear transformations and Aronhold's symbolic notation. . . . At the end of his last lecture in March, Young said that he was so pleased that people had turned up that he would lecture again in the following term. And he and I were both surprised, and I very embarrassed, when no other member of the class but myself showed up in April. It was my Tripos term, but I was not going to miss his lectures ! . . . One lecture fell during the General Strike, and no preparation of room, blackboard, chalk or anything had been made by the college. So Young sat down beside me and wrote out the notes with his own hand".

Young had a quiet humour. "I remember an occasion", writes Wilson, "when he said to me with a grave face: I have lost all sense of personal security'. It appeared that the maidservant at his Cambridge lodgings had used some of his manuscripts to light the fire rather than waste clean paper for that purpose".

Young wrote his first mathematical paper in 1899 and continued to write and to publish for over forty years. With the exception of his work on electromagnetism in 1918, every paper was devoted to one theme, the algebra of groups. It began with the algebraic theory of invariants, a subject which was first explicitly started a hundred years ago, in 1841, by Boole, and then developed by Cayley, Salmon and Sylvester, and later by Macmahon and Elliott. It provided the analytical aspect of geometrical projection and of those properties of a figure which remain unchanged for any such projection. This led, first of all, to the discovery of algebraic forms which were invariant for the corresponding linear transformations, and then to the search for the basic set of forms, out of which all other invariants of a given system of ground forms could be constructed. Such a form is the binary n-ic

$$f = \sum_{r=0}^{n} \binom{n}{r} a_r x_1^{n-r} x_2^{r}$$

and the study resolved itself into the theory of annihilators, that is, of certain differential operators linearly composed of terms such as

$$a_{r-1} \frac{\partial}{\partial a_r}.$$

At an early stage it was supposed that, for binary forms higher than the quartic, the invariant theory was essentially different from that for lower forms. Indeed, in his second *Memoir on Quantics* (1855), Cayley had stated his conclusion that whereas the number of different irreducible invariants and covariants for a quartic was finite, this was no longer true of the quintic. But this surmise was upset in 1869 when Gordan startled the mathematical world by proving the finiteness of such systems for a binary form of any order. Gordan followed this up with the publication of his *Programm* at Erlangen in 1875 which widened the scope of the finiteness to all systems of binary forms. Finally, the theory was extended to all higher types of form by Hilbert in 1890. The influence of Gordan's work was apparent in the successful use of generating functions by Sylvester and MacMahon, who made important advances both in detailed systems and in the general theory. A friendly rivalry sprang up between the English mathematicians and the Continental algebraists, Gordan with his followers, the former following the non-symbolic method, as it is called, and the latter the symbolic. The work of the English school is ably expounded by Elliott in his *Algebra of Quantics*, which brings us to the beginning of the twentieth century.

The development by the symbolic method, which went on in Germany, grew out of certain hyperdeterminants invented by Cayley. In this branch of the theory the coefficient a_r in the form f was regarded as a numerical multiple of $\partial^n f/\partial x_1^{n-r}\partial x_2^r$ or, let us say, $\partial_1^{n-r}\partial_2^r f$, where $\partial_1 = \partial/\partial x_1$, $\partial_2 = \partial/\partial x_2$. All such forms together with all their invariants and covariants were expressed in terms of the ∂_1, ∂_2 belonging to the variables x_1, x_2, and of analogous symbols for any further cogredient variables. Clebsch and Aronhold perfected the technique of these symbols. Thereupon Clebsch and Gordan used them systematically for invariant theory with conspicuous success. They proved that every rational integral invariant of binary forms could be expressed as a polynomial aggregate of symbols $\partial^2/\partial x_1\partial y_2 - \partial^2/\partial x_2\partial y_1$ (the hyperdeterminants of Cayley) and of $y_1(\partial/\partial x_1)+y_2(\partial/\partial x_2)$, the polar operators (the First Fundamental Theorem); that all invariantive properties could be deduced by means of certain specified determinantal identities which left these characteristic hyperdeterminantal and polar structures unimpaired (the Second Fundamental Theorem); that every such invariant could be expressed rationally and integrally in terms of a *finite* number of invariants (Gordan's Theorem, 1869); and they established an important expansion (the Clebsch-Gordan series) which enables one to deal with forms involving many sets of variables by means of forms in fewer variables and their polars.

These symbolic advances, together with the generating functions and perpetuants of MacMahon, opened up a wide field of enquiry, and it was this into which Young entered, in company with his friend J. H. Grace, soon after and possibly even before they took their degrees. Their interest was first aroused in 1895 by reading Meyer's newly published *Bericht über den gegenwartigen Stand der Invariantentheorie*, which opened up a vast world of algebra and gave them their first ideas of modern mathematics. They came to grips with the symbolic method, which fascinated them, and at once began to make important contributions to a subject hitherto very little known in England. With the publication of their treatise, the *Algebra of Invariants*, in 1903, a new era dawned for the teaching and progress of Higher Algebra. This excellent book, with its fine display of algebraic technique and geometrical insight, had a considerable influence on the younger geometers and algebraists at Cambridge, particularly in the decade preceding the last war. Both authors were masters of their subject, Grace as a geometer and Young as an algebraist. The pages of the book have a deceptively simple appearance, owing to the extraordinary compactness of the symbolic notation, where, for example, an invariant $a_0 a_4 - 4a_1 a_3 + 3a_2^2$ of a binary quartic appears in the guise $\frac{1}{2}(ab)^4$. The book ends with four lively appendices, bringing the latest results to the notice of the reader—and one almost expects to see a stop press column on the last page!

From the outset Young's rapid and skilful handling of symbolic algebra bore all the signs of genius. Grace likens him to Ramanujan, not only for what each achieved but for what each ignored. Young at once found his own solutions for the complete systems of the binary octavic and septimic forms: and, of his ingenious device for treating the octavic as the symbolic square of a quartic, Grace says "I could not have thought of that in fifty years". Young began his research in algebra by solving the problem of binary quartic types—the invariant theory of any number of quartics; and it was through the practice of the symbolic methods upon such problems that he was led to his first great discovery, which he called *Quantitative Substitutional Analysis*. He was dealing with functions of a finite number of variables; and these variables always occurred in sets, let us say a, b, c, ..., each of which could be manipulated as a single vector, or as a column of a determinant. In the course of the work innumerable varieties of alternative expressions were produced, many of which only differed among themselves by sign, or else by derangements of these vectors, in much the same way as a three-rowed determinant, $|a_1 b_2 c_3| = \Delta$, assumes two values $\pm\Delta$ and six alphabetical forms, when the letters are permuted without disturbing the suffix sequence. His functions were, of course, usually much more complicated than single determinants; nevertheless the determinant provided the clue to a general theory which comprehends all the details of the algebra. Now functions such as these are evidently closely connected with the theory of finite groups, and it became clear to Young that sets of functions, which at first sight were quite distinct, but which on examination proved to belong to the same group, could be dealt with by a single prescription if only the properties of that particular group were thoroughly known. Young therefore set himself to extricate the group properties of these functions, and accordingly he expressed the functions as well as the relations between such functions by means of a new kind of symbolic operator depending at once on a substitution group.

This operator, which consisted of two main ingredients N_i and P_i, can best be explained by a simple example: thus, if $f(a, b, c)$ is a function of three variables, then

$$f(a, b, c)+f(b, a, c) = Pf(a, b, c),$$

$$f(a, b, c)-f(b, a, c) = Nf(a, b, c),$$

where $P = 1+(ab)$, $N = 1-(ab)$ and (ab) denotes the operation of interchanging a with b in the function $f(a, b, c)$. For n letters such a P has $n\,!$ positive terms, and forms the positive symmetric group, while N forms the negative symmetric group with the same terms, half of which have a negative sign. For n such elements $a_1, a_2, \ldots, a_n$, Young writes

$$P = \{a_1 a_2 \ldots a_n\}, \quad N = \{a_1 a_2 \ldots a_n\}'.$$

In the above two-letter illustration it will at once be seen that, when f is the determinant Δ, then $P\Delta = 0$, $N\Delta = 2\Delta$. Moreover, if *any* expression $\phi = \Sigma\,\lambda a_1 b_2 c_3$ is taken which involves each letter and each suffix once in each term, besides a numerical coefficient λ, then the effect of the three-letter operators P and N upon ϕ are $\mu\Delta_+$ and $\mu\Delta$, where $\mu = \Sigma\lambda$, and Δ_+, Δ are the *permanent* and the *determinant* respectively. These examples serve to explain how Young transferred the whole emphasis from the operand f to the substitutional operator, and very soon he had elaborated a considerable theory of these operators.

The main result was contained in *the method of the tableau* (1900). A tableau is an arrangement of the variable in rows and columns, equal or diminishing both downwards and from left to right. A function, depending on, let us say, five permutable variablés, possesses (among others) the tableau

$$\begin{matrix} \times\times\times \\ \times\times \end{matrix} \qquad \begin{matrix} abc \\ de \end{matrix}.$$

From this model Young constructs the operators

$$P_1 P_2 N_1 N_2 N_3 = \{abc\}\{de\}\{ad\}'\{be\}'\{c\}',$$

where each P refers to a particular row, and each N to a particular column. The sum of all 5! such expressions, due to the permutations of all five letters, Young calls $T_{3,2}$, the suffixes denoting the lengths of the rows of the tableau. Clearly there are as many such shapes (with the longest row and column always at the top and on the left) as there are partitions of n, the number of letters. For four letters there are five shapes

$$\times\times\times\times, \quad \begin{matrix} \times\times\times \\ \times \end{matrix}, \quad \begin{matrix} \times\times \\ \times\times \end{matrix}, \quad \begin{matrix} \times\times \\ \times \\ \times \end{matrix}, \quad \begin{matrix} \times \\ \times \\ \times \\ \times \end{matrix}, \qquad (1)$$

and therefore five operators T_4, $T_{3,1}$, $T_{2,2}$, $T_{2,1,1}$, $T_{1,1,1,1}$. Young found that, for all values of n, a certain positive linear combination of these T's was identically equal to unity, say

$$\sum_p A_{(p)} T_{(p)} = 1,$$

where each (p) denotes a different partition of n, and the coefficient $A_{(p)}$ is a non-zero perfect square rational number. In fact

$$A_{(p)} = \left(\prod_{r,s} (a_r - a_s - r + s)! \Big/ \prod_r (a_r + h - r)! \right)^2,$$

where a_r is the number of letters in the rth row of the corresponding tableau and h is the number of rows. This may well be called *Young's Theorem* (1900). From it he deduced the Clebsch-Gordan series and a host of other results. It acted as a powerful crystallising influence by turning an amorphous function f, depending on several sets of variables, into the highly organized but limited varieties of forms Tf. Moreover, many such forms Tf vanish identically (as in the example above), and so do many products $T_{(p)} T_{(q)}$. They certainly vanish if the partition (q) precedes (p) in the descending order as illustrated in (1). And, again, $T_{(p)} = A_{(p)} T^2_{(p)}$. It is remarkable that the whole of this theory was elaborated out of one simple basic fact—that it is impossible for a non-zero function to be simultaneously symmetric and skew symmetric in two of its variables. This fact is the unit out of which Young constructed his whole edifice. It was natural for the enquiry to be suggested, how to find the general solution for one or more substitutional equations such as were constantly occurring. The whole of Young's subsequent work provided a substantial answer to this.

These results were given in the early papers of 1900 and 1902 on Quantitative Substitutional Analysis, which at once attracted the notice of Burnside and Frobenius, the two greatest contemporary experts on the theory of groups. Indeed it had been at the request of Burnside, who refereed the papers, that they were thrown into a form which emphasized the underlying group theory. But Young published his work in ignorance of its close connection with that of Frobenius, who had begun to write long and deep memoirs (1896–1903) on the finite group, and had already used matrices and linear transformations to represent any such group. In particular, Frobenius had set himself the problem of finding, if possible, numerical coefficients such as should satisfy the equation $\Sigma\lambda A = (\Sigma\lambda A)^2$; that is to say, a certain linear combination of all the elements of the group was to be equal to its own square, with the significant proviso that elements belonging to any one and the same class should all have the same coefficient. The fundamental group property $AB = C$ affords an *a priori* reason for such a possibility. The successful answer to this question led Frobenius to study sets of coefficients involving *group characters*, whereby he opened up a wide field of research which is still far from exhausted.

When Young's work appeared Frobenius at once saw its close relation with his own, and explained the connection and adapted the tableau method to his own work in his "Die charakterischen Einheiten der symmetrischen Gruppe" (*Berliner Sitzungsberichte*, 1903, 349). Young first learnt of these memoirs through Burnside in 1906, but apparently it was not until after the last war—a period of inevitable mathematical inactivity for so many—that Young had fully mastered them. He had certainly started his research again in 1922 when he wrote his *Ternary Perpetuants*, and he had already been re-reading Frobenius for several months when I first met him in the summer of 1925. The reading was slow and painstaking for, as he remarked, the German was involved and he was no linguist. This opportunity to visit Young occurred when I sought his advice on the problem of extending Pascal's Theorem of the hexagram inscribed in a conic to the decagon inscribed in a quadric surface. Algebraically the hexagon satisfies the condition

$$(123)(156)(264)(345) - (456)(423)(531)(612) = 0,$$

where each expression (123) denotes the determinant of the coordinates for three of the points. The function on the left is skew symmetric in each pair of the six symbols; and a corresponding sum of terms containing five four-rowed determinants of ten points in space was known to exist. E. Study had conjectured that it would consist of four terms, but I had found it to have five, even in the special case when three of the ten points were in line. In reply to a query whether the five-term expression could be rendered quite general, Young, who was at once interested in the problem, suggested the trial of a certain simple operator, adding that it would produce either a zero result or else the desired form. Happily I was able to report to Young that it turned out to be non-zero; but, alas, the series had sixty times as many terms as Study had conjectured. This, however, did not daunt Young, who thought in terms of factorial n as easily as most of us with n. He at once went into the general theory of the symmetric group of ten letters, and in an amazingly short time produced a highly elaborate but complete account of the linear invariants of ten quadrics.

During my visit to Birdbrook he also told me that he was gradually mastering Frobenius, though the work was very abstract and he always preferred to embark upon a theory by way of a practical problem. Already he had learnt much from Frobenius and had greatly improved his own substitutional analysis by inventing *standard forms* and a new matrical representation of his T operators. These important results were published two years later (1927) in the third memoir of the series, in which he paid generous tribute to Frobenius. The original stages of Young's theory had suffered from a defect which is also inherent in the symbolic invariant theory; in it results would certainly be complete, but they might often be redundant. The discovery of standard forms removed the redundancy without impairing the completeness, and there were certain numerical checks, issuing from the theory of Frobenius, which would safeguard the accuracy of the results. It is indicative of the care and deliberation with which Young worked that he should have turned aside from this general theory to the problem of ten quadrics, which he treated as a challenge to his newer technique, before publishing the main result.

By standard forms he meant arrangements of the tableau such as the ten following, for the case of four letters,

$$abcd,\quad \begin{matrix}abc\\ d\end{matrix},\quad \begin{matrix}abd\\ c\end{matrix},\quad \begin{matrix}acd\\ b\end{matrix},\quad \begin{matrix}ab\\ cd\end{matrix},\quad \begin{matrix}ac\\ bd\end{matrix},\quad \begin{matrix}ab\\ c\\ d\end{matrix},\quad \begin{matrix}ac\\ b\\ d\end{matrix},\quad \begin{matrix}ad\\ b\\ c\end{matrix},\quad \begin{matrix}a\\ b\\ c\\ d\end{matrix}.$$

Here the alphabetical order is strictly preserved in each row and down each column. The 4! arrangements of each of the five shapes are now seen to be reduced to 1, 3, 2, 3, 1 arrangements respectively. Young found that all others can be expressed in terms of these standard arrangements, so that the series $\Sigma AT = 1$ can be greatly simplified. The resulting forms are also linearly independent and give exact information about the number of functions of a particular type. The most general function of the type $T_{(p)}f$ then has exactly q^2 arbitrary constants, where q is the number of standard forms of the partition (p). As an illustration of this remarkable result, which may be called Young's *Standard Theorem*, we may take the above case of four letters. In the list of ten standard forms the sum of the squares of the subsets $1^2+3^2+2^2+3^2+1^2$ must be $4\,!$; and in general $\Sigma q^2 = n!$. Also, in terms of the original coefficients $A_{(p)}$, $q = n!\,\sqrt{(A_{(p)})}$.

The original $T_{(p)}$ is now replaced by a modified form

$$T'_{(p)} = \Sigma \lambda_{rs} P_r \sigma_{rs} N_s$$

where each of r and s is summed from 1 to q, while λ_{rs} is numerical, and the P, σ, N are definite substitutional expressions arising from the tableau. This leads to a matrix $[\lambda_{rs}]$ of q rows and columns which completely specifies the function $T'_{(p)}f(x, y, \ldots, z)$, derived from any function f of n variables $x, y, \ldots, z$, which may undergo derangement. In fact the sum or product of two such modified T operators obeys the matrix law.

For the general function $f(x, \ldots)$ the numbers λ_{rs} in each matrix are quite arbitrary: so for four variables there are five matrices of orders 1×1, 3×3, 2×2, 3×3 and 1×1 respectively, thus possessing altogether 24 elements. Furthermore, any modification of the function due to symmetry, or to skew symmetry, or to any other such property of the variables, is at once visible in the matrices—blank spaces appear. For example, if $f(a, b, c, d)$ is symmetric in a, b, it is then capable of at most twelve values, by interchange of the variables in all possible ways. In this case Young found that half the rows of the matrices would be blank. Exactly which rows then survive provided a very interesting problem, and it was analysed directly from the standard forms; for the case in point Young proved that this amounted to rejecting each standard form wherein the symmetrical letters a, b occur in the same column. A glance shows that this leaves only 1, 2, 1, 1, 0 standard forms; and these tell us the numbers of non-zero rows, which have, of course, the same respective numbers of elements in them as before, namely, 1, 3, 2, 3, 1. Thence the full number of constants is found, by multiplying together respective pairs, to be $1+6+2+3+0 = 12$, correctly.

Young took a modest but wholehearted pleasure in these results, and his enthusiasm was infectious. "I am delighted", he wrote some months later (1926), "to find someone else really interested in the matter. The worst of modern mathematics is that it is now so extensive that one finds there is only about one person in the universe really interested in what you are". The tide turned in his favour with the appearance of the third memoir. Within a few years his method of the tableau and the standard theorem appeared in Weyl's *Theory of Groups and Quantum Mechanics* (2nd ed., 1930), as a means of elucidating the properties of quantum numbers, while the Clebsch-Gordan series, which had been largely responsible for substitutional analysis, was now found to be of fundamental importance for the whole of spectroscopy. Also during the last decade an interest in the theory of group characters has developed among several of the younger algebraists throughout the country.

During the last fourteen years of his life Young wrote a steady series of papers elaborating his theory and applying it to the problems of invariants and their generating functions. In a letter to a friend, he wrote (1930): "For the last two years I have been working at a paper on the application of substitutional analysis to invariants, but though I have obtained a good many interesting results that I think might be worth publishing, yet I feel that it is too scrappy as yet to write out. . . . My ambition is at the moment to present the complete system for a single cubic in any number of variables. Quite do-able if things turn out as I hope; but I am a confirmed optimist, and so suffer many defeats".

Something should perhaps be said in detail of this later work. The fourth memoir was mainly illustrative and technical, but it brought out the close relation between the matrices $[\lambda_{rs}]$ and the group matrices of Frobenius and Schur (1908). The fifth dealt with the group of rotations and reflexions of the hyperoctahedron in n dimensions. In the sixth a proof of Frobenius' generating function for characters of the symmetric group was obtained by the method of the tableaux, together with semi-normal (or triangular) group matrices. The seventh and eighth memoirs, together with the communication to the Royal Society (1935), were devoted to invariant theory; in the eighth was included an illustration from the invariants of a quaternary cubic, which has an irreducible system of six forms of degrees 8, 16, 24, 32, 40, 100. They had been discovered seventy years earlier by Clebsch and Salmon independently, but of this Young was unaware. The impression left by these examples, and by all the replies to his friends which he readily supplied on particular problems of geometry or invariants, was that in his hands the method of the tableau was irresistible. Known and unknown results alike were treated summarily and afresh; he merely reaffirmed (or corrected where they were wrong) the old and recorded the new.

The motif which ran through these last few memoirs related to certain well-known forms called gradients (homogeneous and isobaric polynomials in the coefficients of the ground forms). Gradients were fundamental in the nineteenth century progress of invariant theory and in all the work of Elliott. Their behaviour at bottom depended on additive properties of sets of the positive integers occurring among their indices and suffixes. But so, also, were the properties of the forms which the method of the tableau produced. When the semi-normal matrices were used, Young found a marked parallelism in these two methods, which he called respectively the method of leading gradients and that of irreductible forms. Indeed, for the binary case, parallelism became coincidence; and on this evidence, and his own instinct for algebraic truth, he surmised the following theorem:

In general the complete set of leading gradients is defined in the same way as the complete set of irreductible forms.

It is probable that his most recent and unfinished work deals with this, and in any case it is greatly to be hoped that the pages are complete enough to make their publication possible.

Young's work is never easy reading, for it lacks that quality which helps the reader to grasp the essential point at the right time. The very closest and constant attention is required to pick out some of the most fundamental results from a mass of detail. One could almost suppose that he camouflaged his principal theorems. His work resembles a noonday picture of a magnificent sunlit mountain scene rather than the same in high relief with all the light and shade of early morning or sunset. The craftsmanship is accurate and logical, and the ideas underlying many of the proofs are very beautiful. His powers of combining deep insight into abstract algebraic theory with an uncanny technical manipulative skill in all the practical applications give him a place unrivalled among his contemporaries. His humility, and perhaps his isolation and lack of teaching experience among undergraduates, prevented him from realising the importance of clarifying the crucial passage from abstract theory to detailed practice. It is hard to believe that processes which to the mind of Young were intuitively clear cannot yet be made part of our common mathematical heritage; for if they can, then there is a great future for algebra.

In drawing up this notice I have been very much indebted to Mrs. Gunnery, Mr. G. H. A. Wilson, Prof. E. T. Whittaker, Mr. J. H. Grace, Dr. A. C. Aitken and Mr. W. L. Edge for supplying family, academic or mathematical details.]

COLLECTED PAPERS

The Irreducible Concomitants of any Number of Binary Quartics.

By A. Young. Received and read February 9th, 1899.

The irreducible system is here arrived at by first finding the irreducible system of types and then the number of independent forms belonging to each type for a system of N quartics. Two concomitants are said to be of the same type when they can be obtained from the same form by polarization. For the purpose of discussing the system of types, a type is regarded as being of the first degree in the coefficients of each of the quantics concerned. The finiteness of the irreducible system of types has been established by Prof. Peano.* He proves that the complete system of concomitants for any number of binary n-ics may be obtained from the system for n n-ics by polarization alone; with the one possible exception of invariants of the type

$$\begin{vmatrix} A_0 & A_1 & \dots & A_n \\ B_0 & B_1 & \dots & B_n \\ \dots & \dots & & \dots \\ K_0 & K_1 & \dots & K_n \end{vmatrix}$$

In other words, every type of a binary n-ic which furnishes no irreducible form for n n-ics is reducible, with the possible exception just mentioned. It was with the help of this proposition that some of the reductions for the quartic were first arrived at; however, other

* *Atti di Torino*, t. XVII., p. 580.

methods have proved shorter. The latter part of his paper is devoted to the discovery of the cubic types. From the fact that

$$\begin{vmatrix} A_0 & A_1 & A_2 & A_3 \\ B_0 & B_1 & B_2 & B_3 \\ C_0 & C_1 & C_2 & C_3 \\ D_0 & D_1 & D_2 & D_3 \end{vmatrix}$$

is reducible, it is shown that all the types occur in the system for two cubics. His results are—

The irreducible system for N cubics belongs to 10 types, as follows:—

	Type.	Simplest Form.	Number of Forms.
I.	${}_3C_1$	One of the cubics	N
II.	${}_4C_2$	Jacobian of two cubics	$\binom{N}{2}$
III.	${}_2C_2$	Hessian of one cubic	$\binom{N+1}{2}$
IV.	I_2	Third transvectant of two cubics	$\binom{N}{2}$
V.	${}_3C_3$	Covariant order 3 of one cubic	$\binom{N+2}{3}$
VI.	${}_1C_3$	Second transvectant of I. and III.	$2\binom{N+1}{3}$
VII.	I_4	Discriminant of one cubic	$\binom{N+3}{4}$
VIII.	${}_2C_4$	Jacobian of two forms III.	$3\binom{N+2}{4}$
IX.	${}_1C_5$	First transvectant of III. and VI.	$4\binom{N+3}{5}$
X.	I_6	Resultant of two forms VI.	$\binom{N+4}{6}$

For the quartic, I have first expressed the types in symbols based on the quadratic. To do this, it is proved that the types of a binary mn-ic can be expressed in symbols based on the n-ic; the symbolical factors being of the form of n-ic types; just as, in ordinary symbolical algebra, the concomitants of the m-ic are expressed in symbols based on the linear form. It is easy then to show that there is only one type to be considered, of given degree and order. Writing this $(abc \dots k)$, the fundamental identities give relations of the form

$$(1+S_1+S_2+\dots+S_k)(abc \dots k) = R,$$

where R stands for reducible terms, and $S_1, S_2, \dots, S_k$ are certain substitutions.

The chief advantage obtained from quadratic symbols lies in the possibility of using symbolical operators, with the help of which relations between forms of one degree and order may be obtained from relations between forms of the same order but of one degree lower.

The invariant type of highest degree I_6 has been expressed in terms of determinants of five rows and columns; by means of this a number of syzygies may be at once written down, in fact $I_6 P$ equals a sum of products of forms, there being at least three forms in each product, where P is an irreducible form of any type except I_2 and I_3.

1. Consider any simultaneous system of binary nm-ics,

$$(A_0, A_1, \dots, A_{mn} \between x_1, x_2)^{mn} \equiv a^n_{x^m},$$

$$(B_0, B_1, \dots, B_{mn} \between x_1, x_2)^{mn} \equiv b^n_{x^m},$$

$$\dots \quad \dots \quad \dots \quad \dots \quad \dots$$

where $$a_{x^m} \equiv (a_0, a_1, \dots, a_m \between x_1, x_2)^m,$$

and the identities are taken to define the relations between the symbolical letters $a_0, a_1, \dots,$ and the actual coefficients. Let $f(A, B, \dots, K)$ be a type belonging to this system; writing in this for $A_0, \dots$ their values in terms of the symbolical letters, $f(A, B, \dots, K)$ takes the form $\phi(a, b, \dots, k)$, say. Now make any linear transformation, and denote by dashed letters the coefficients of the transformed quantics; then

$$f(A', B', \dots, K') = \mu f(A, B, \dots, K),$$

where μ is a power of the determinant of transformation; hence also

$$\phi(a', b', \dots, k') = \mu\phi(a, b, \dots, k).$$

Therefore ϕ is a concomitant of the m-ics $a_{x^m}, b_{x^m}, \dots, k_{x^m}$; and hence ϕ can be expressed as a sum of products each factor of which is of the same form as an irreducible type of the m-ic.

It is necessary now to show how to proceed from a symbolical product P to the form F in the actual coefficients, which it represents. Let

$$a_{x^m} \equiv \alpha_x^m, \quad b_{x^m} \equiv \beta_x^m, \; \dots;$$

then the same result will be obtained by writing in F the coefficients of the first quantic in terms of $a_0, a_1, \dots, a_m$, and then putting

$$a_r = \alpha_1^{m-r} \alpha^r$$

as by writing

$$A_r = \alpha_1^{mn-r} \alpha_2^r$$

directly in that type. Hence the result of writing in P

$$a_r = \alpha_1^{m-r} \alpha_2^r, \; \dots, \; b_r = \beta_1^{m-r} \beta_2^r, \; \dots$$

is the same mn-ic type expressed in linear symbols; the step from these to the actual coefficients presents no difficulty. As a matter of fact, an mn-ic type, when expressed in m-ic symbols, will rarely consist of a single symbolical product; still, given any symbolical product P—of the right degree in the symbols—a type F of the mn-ic may be arrived at, in general, with the help of linear symbols, which is such that when expressed in m-ic symbols it becomes $P+Q$, where the effect of substituting linear for m-ic symbols is to make Q vanish.

Hence products which vanish when the change is made to linear symbols may be ignored, and every other symbolical product of the right degree in the symbols may be regarded as giving a type of the mn-ic.

Let $(a, b, c, \dots)$ be an m-ic type; then among the factors which may occur are forms like $(a, a, c, \dots)$; a factor of this kind may be reducible, or when linear symbols are introduced it may vanish; in either case there is no need to consider it.

The lineo-linear type (a, b) is a case in point; a product with a factor (a, a) may always be ignored.

2. The types of a quadratic are—

$$a_0 b_2 + a_2 b_0 - 2a_1 b_1 \equiv [ab], \quad a_0 x_1^2 + 2a_1 x_1 x_2 + a_2 x_2^2 \equiv a_{x^2},$$

$$\begin{vmatrix} a_0 & a_1 & a_2 \\ b_0 & b_1 & b_2 \\ c_0 & c_1 & c_2 \end{vmatrix} \equiv |\, abc \,|, \quad \begin{vmatrix} a_0 & a_1 & a_2 \\ b_0 & b_1 & b_2 \\ x_2^2 & -x_1 x_2 & x_1^2 \end{vmatrix} \equiv |\, abx^2 \,|;$$

no symbolical factor in quadratic symbols need be considered in which the same letter occurs twice.

The fundamental identities are

$$2\,|abc|\,|def| = \begin{vmatrix} [ad] & [ae] & [af] \\ [bd] & [be] & [bf] \\ [cd] & [ce] & [cf] \end{vmatrix}, \qquad \text{I.}$$

$$[ae]\,|bcd| + [be]\,|cad| + [ce]\,|abd| = [de]\,|abc|, \qquad \text{II.}$$

$$\begin{vmatrix} [ab] & [ad] & [af] & [ah] \\ [cb] & [cd] & [cf] & [ch] \\ [eb] & [ed] & [ef] & [eh] \\ [gb] & [gd] & [gf] & [gh] \end{vmatrix} = \begin{vmatrix} a_0 & a_1 & a_2 & 0 \\ c_0 & c_1 & c_2 & 0 \\ e_0 & e_1 & e_2 & 0 \\ g_0 & g_1 & g_2 & 0 \end{vmatrix} \begin{vmatrix} b_2 & -2b_1 & b_0 & 0 \\ d_2 & -2d_1 & d_0 & 0 \\ f_2 & -2f_1 & f_0 & 0 \\ h_2 & -2h_1 & h_0 & 0 \end{vmatrix} = 0. \qquad \text{III.}$$

The identities for the forms a_{x^2}, $|abx^2|$ may be obtained at once from these, since $[ad]$, $|abd|$ become respectively a_{x^2}, $|abx^2|$ if $x_2^2, -x_1x_2, x_1^2$ are written for d_0, d_1, d_2.

From I. it follows that no product need be discussed in which more than one factor of either of the forms $|abc|$, $|abx^2|$ occurs. In quartic types each symbolical letter must occur twice in a product; hence, if there is a factor $|abc|$, there is also a factor e_{x^2}; but, from II.,

$$|abc|\,e_{x^2} = [ae]\,|bcx^2| + [be]\,|cax^2| + [ce]\,|abx^2|;$$

hence, for the quartic, factors $|abc|$ need not be considered. The invariant forms then are

$$[ab]^2,$$
$$[ab][bc][ca],$$
$$[ab][bc][cd][da],$$
$$\ldots \quad \ldots \quad \ldots \quad \ldots .$$

We shall find it convenient to use the notation

$$(abca) \equiv [ab][bc][ca],$$
$$(abcda) \equiv [ab][bc][cd][da],$$
$$\ldots \quad \ldots \quad \ldots \quad \ldots \quad \ldots \quad \ldots ;$$

then $$(abcd \ldots ka) \equiv (bcd \ldots kab) \equiv (ak \ldots dcba).$$

The general form of covariant type of order 2 is

$$[ab][bc] \dots [hk] \mid kax^2 \mid ,$$

which will be denoted by $(ab \dots hk)$, and here

$$(ab \dots hk) = -(kh \dots ba).$$

Covariants order 4 are of the form

$$[ab][bc] \dots [hk]\, h_x k_x \equiv (abc \dots hkx^2a)\,;$$

and so on.

3. *The Invariants.*—Multiply III. by $[bc][de][fg][ha]$, and expand; then

$$(abcdefgha)+(afgbcdeha)+(adefgbcha)+(abcfgdeha)+(afgdebcha) \\ +(adebcfgha) = R,$$

where R is used to express reducible terms, and terms which have factors of the form $[aa]$. The above may be conveniently written

$$[a,\, bc,\, de,\, fg,\, ha] = R.$$

From the identity

$$\begin{vmatrix} [ab] & [ac] & [ad] & [aa] \\ [bb] & [bc] & [bd] & [ba] \\ [cb] & [cc] & [cd] & [ca] \\ [db] & [dc] & [dd] & [da] \end{vmatrix} = 0,$$

we deduce

$$[a \dots a] \equiv (abcda)+(adbca)+(acdba)+(abdca)+(adcba)+(acbda) \\ = R.$$

The other relations obtainable may be written

$$[a \dots ea] = R, \quad [a,\, bc,\, d,\, e,\, a] = R, \quad [a,\, bc,\, de,\, f,\, a] = R, \text{ \&c.}$$

The types I_2 and I_3, viz., $[ab]^2$ and $(abca)$, are unaffected by the fundamental identity; hence for N quartics there are $\binom{N+1}{2}$ and $\binom{N+2}{3}$ independent irreducible invariants respectively of these types.

For I_4 there is one equation,

$$[a \dots a] = R,$$

or

$$2\,(abcda)+2\,(adbca)+2\,(acdba) = R.$$

There are then only two independent forms having the same letters. Let $(A^pB^q \dots)\, I_k$ denote an invariant of the type I_k which is of degree p in the coefficients of the quartic A, of degree q in the coefficients of the quartic B, and so on. Then there are two independent irreducible forms $(ABCD)\, I_4$, one form $(ABC^2)\, I_4$, and one form $(A^2B^2)\, I_4$; in other cases the invariant is reducible. Hence for N quartics there are $\binom{N}{2}+3\binom{N}{3}+2\binom{N}{4}=\binom{N+2}{4}+\binom{N+1}{4}$ invariants of the type I_4.

There are twelve possible forms $(ABCDE)\, I_5$, and ten relations between them, all of which are of the form

$$[ab \dots a] = R.$$

Simpler relations are obtained thus:—

$$[ab \dots a]+[ae \dots a]-[ac \dots a]-[ad \dots a]$$
$$\equiv 2\,(abcdea)+2\,(abdcea)-2\,(acbeda)-2\,(acebda) = R. \quad \text{IV.}$$

In connexion with this type consider the form

$$|\,ABCDE\,| \equiv \begin{vmatrix} A_0 & A_1 & A_2 & A_3 & A_4 \\ B_0 & B_1 & B_2 & B_3 & B_4 \\ C_0 & C_1 & C_2 & C_3 & C_4 \\ D_0 & D_1 & D_2 & D_3 & D_4 \\ E_0 & E_1 & E_2 & E_3 & E_4 \end{vmatrix}$$
$$= (\alpha\beta)(\alpha\gamma)(\beta\gamma)(\alpha\delta)(\beta\delta)(\gamma\delta)(\alpha\epsilon)(\beta\epsilon)(\gamma\epsilon)(\delta\epsilon).$$

where $\quad a_x^2 \equiv (A_0, A_1, A_2, A_3, A_4 \between x_1, x_2)^4 \equiv \alpha_x^4$, &c.

Using the identity

$$3\,(\alpha\beta)(\alpha\gamma)(\beta\gamma)(\alpha\delta)(\beta\delta)(\gamma\delta) = -(\alpha\beta)^3(\gamma\delta)^3-(\alpha\gamma)^3(\delta\beta)^3-(\alpha\delta)^3(\beta\gamma)^3,$$

we obtain $\quad 6\,|\,ABCDE\,|$

$$= -2\,(\alpha\beta)^3(\gamma\delta)^3(\alpha\epsilon)(\beta\epsilon)(\gamma\epsilon)(\delta\epsilon)-\dots$$
$$= -(\alpha\beta)^2(\gamma\delta)^2 \begin{vmatrix} (\alpha\gamma)^2 & (\alpha\delta)^2 & (\alpha\epsilon)^2 \\ (\beta\gamma)^2 & (\beta\delta)^2 & (\beta\epsilon)^2 \\ (\epsilon\gamma)^2 & (\epsilon\delta)^2 & 0 \end{vmatrix} -\dots$$
$$= (abedca)+(abdcea)+(acebda)+(acbdea)+(adecba)+(adcbea)$$
$$-(abecda)-(abcdea)-(acedba)-(acdbea)-(adebca)$$
$$-(adbcea)+R,$$

since $(abcdea)$ and $(\alpha\beta)^2(\beta\gamma^2)(\gamma\delta^2)(\delta\epsilon^2)(\epsilon\alpha)^2$ represent the same type.

Hence, from equation IV.,

$$(abdcea)-(acbeda)+R = (acebda)-(abcdea)+R$$
$$= (acbdea)-(adceba)+R = (adecba)-(acdbea)+R$$
$$= (adcbea)-(abdeca)+R = (abedca)-(adbcea)+R$$
$$= |\, ABCDE \,|$$
$$= \tfrac{1}{3}\left\{(abdcea)+(acebda)+(acbdea)+(adecba)+(adcbea) + (abedca)\right\}+R, \qquad \text{V.}$$

this last in virtue of

$$[ab \ldots a]+[ac \ldots a]+[ad \ldots a]+[ae \ldots a] = R.$$

These relations include all those from which we started; further they are independent; hence, there are six independent irreducible invariants $(ABCDE)\ I_5$. For N quartics the number of independent irreducible forms I_5 is

$$6\binom{N}{5}+8\binom{N}{4}+3\binom{N}{3} = 3\binom{N+2}{5}+2\binom{N+1}{5}+\binom{N}{5}.$$

For I_6 there are two systems of equations, viz.,

$$[abc \ldots a] = R \quad \text{and} \quad [a,\, bc,\, d,\, e,\, fa] = R.$$

A third equation, which, though not independent of these, is useful, may be found thus:

$$2\,|\,abc\,|\;|\,def\,| = \begin{vmatrix} [ad] & [ae] & [af] \\ [bd] & [be] & [bf] \\ [cd] & [ce] & [cf] \end{vmatrix};$$

expand the determinant and square both sides

$$4\,|\,abc\,|^2\;|\,def\,|^2 = R+(adbecfa)+(afbdcea)+(aebfcda) +(adbfcea)+(afbecda)+(aebdcfa)$$
$$= R+[a-b-c-a], \text{ say};$$

then $\qquad [a-b-c-a] = R.$

This last is the sum of two equations, for, if S be the sum of the sixty possible forms $(ABCDEF)\,I_6$,

$$[ad,\, be,\, c,\, f,\, a]+[af,\, bd,\, c,\, e,\, a]+[ae,\, bf,\, c,\, d,\, a]$$
$$+[a,\, d,\, be,\, cf,\, a]+[a,\, f,\, bd,\, ce,\, a]+[a,\, e,\, bf,\, cd,\, a]$$
$$+[ad,\, b,\, e,\, cf,\, a]+[af,\, b,\, d,\, ce,\, a]+[ae,\, b,\, f,\, cd,\, a]$$
$$+[abc\ldots a]+[acb\ldots a]+[ab\ldots ca]$$
$$-S+[a-b-c-a]$$
$$= 6\left[(adbecfa)+(afbdcea)+(aebfcda)\right] = R, \qquad \text{VI.}$$

since S is reducible, it being the sum of the ten expressions $[a-b-c-a]$.

The results of operating with $[ef]\left[f_0\frac{\partial}{\partial e_0}+f_1\frac{\partial}{\partial e_1}+f_2\frac{\partial}{\partial e_2}\right]$ on $[ea]$, $[ee]$ and $(abcdea)$ are $[ef][fa]$, $[ef]^2$, and $(abcdefa)+(abcdfea)$ respectively. Hence from any relation $\Sigma I_5 = R$ there may be at once deduced one of the form $\Sigma I_6 = R$. The above operation will be found equivalent to substituting the coefficients of that covariant of two quartics which is of the same type as the Hessian for the coefficients of a single quartic.

Applying this operator to V., we obtain

$$[ef]\left[f_0\frac{\partial}{\partial e_0}+f_1\frac{\partial}{\partial e_1}+f_2\frac{\partial}{\partial e_2}\right]\,|\,ABCDE\,|$$
$$= |\,ABCD,\, EF\,|$$
$$= (abdcefa)+(abdcfea)-(acbefda)-(acbfeda)+R$$
$$= (abefdca)+(abfedca)-(adbcefa)-(adbcfea)+R$$
$$= \tfrac{1}{2}\left\{-(afdbeca)-(aedbfca)+(afbdeca)+(aebdfca)\right\}+R,$$

using equations of the form VI.

Hence $\quad (abdcefa)+(abdcfea)+(adbcefa)+(adbcfea)+R$

$$= (abefdca)+(abfedca)+(acbefda)+(acbfeda)+R.$$

If b and e, f and d be interchanged in this, the first line is unaltered; it therefore

$$= (aebdfca)+(aedbfca)+(acebdfa)+(acedbfa)+R = R,$$

as is seen by adding the three lines and using equations of the form VI.; therefore

$$| ABCD, EF |$$
$$= \tfrac{1}{2}\{-(afdbeca)-(aedbfca)+(afbdeca)+(aebdfca)\}+R$$
$$= -(afdbeca)-(aedbfca)+R = (afbdeca)+(aebdfca)+R$$
$$= | ABCD, FE |.$$

It follows that the type I_6 may be expressed in the determinant forms $| ABCD, EF |$; for

$$(afdbeca)+(aedbfca) = | ABDC, EF | + R,$$
$$-(aedbfca)-(aecbfda) = | EFAB, DC | + R,$$
$$(aecbfda)+(afdbeca) = | DCEF, AB | + R.$$

Therefore

$$2(afdbeca) = | ABDC, EF | + | EFAB, DC | + | DCEF, AB | + R, \qquad \text{VII.}$$

There are fifteen possible forms $| ABCD, EF |$. Written in full,

$$| ABCD, EF |$$
$$= 2\begin{vmatrix} A_0 & B_0 & C_0 & D_0 & E_0F_2-2E_1F_1+E_2F_0 \\ A_1 & B_1 & C_1 & D_1 & \tfrac{1}{2}(E_0F_3-E_1F_2-E_2F_1+E_3F_0) \\ A_2 & B_2 & C_2 & D_2 & \tfrac{1}{6}(E_0F_4+2E_1F_3-6E_2F_2+2E_3F_1+E_4F_0) \\ A_3 & B_3 & C_3 & D_3 & \tfrac{1}{2}(E_1F_4-E_2F_3-E_3F_2+E_4F_1) \\ A_4 & B_4 & C_4 & D_4 & E_2F_4-2E_3F_3+E_4F_2 \end{vmatrix}$$

Amongst these forms one kind of equation exists, viz.,

$$| ABCD, EF | + | EABC, DF | + | DEAB, CF |$$
$$+ | CDEA, BF | + | BCDE, AF | = 0, \qquad \text{VIII.}$$

as may be verified by taking the coefficients of $F_0, F_1, \ldots, F_4$ in turn. When $(abcdefa)$ is expressed in the determinant forms, $[abc \ldots a] = R$ is identically satisfied, and $[ab, cd, e, f, a] = R$ becomes the sum of two equations VIII. Hence there are no relations between the forms $| ABCD, EF |$ beyond those included in VIII. There are six relations VIII., and five are independent. Hence there are ten independent irreducible forms $(ABCDEF)\ I_6$. For a system of N quartics there are

$$10\binom{N}{6}+20\binom{N}{5}+10\binom{N}{4}=10\binom{N+2}{6}$$

irreducible invariants of this type.

Invariant types of higher degree than 6 are reducible. For, using the operator $[fg]\left[g_0\frac{\partial}{\partial f_0}+g_1\frac{\partial}{\partial f_1}+g_2\frac{\partial}{\partial f_2}\right]$, we obtain from VII.

$$2\,(afgdbeca)+2\,(agfdbeca)$$

$$=|\,A, B, D, C, EFG\,|+|\,E, FG, A, B, DC\,|+|\,D, C, E, FG, AB\,|. \qquad \text{IX.}$$

Therefore

$$R=[a\ldots beca]$$

$$\begin{aligned}=\ &|\,A,B,D,C,EFG\,|+|\,E,FG,A,B,DC\,|+|\,D,C,E,FG,AB\,|\\ +\ &|\,A,B,G,C,EDF\,|+|\,E,DF,A,B,GC\,|+|\,G,C,E,DF,AB\,|\\ +\ &|\,A,B,F,C,EGD\,|+|\,E,GD,A,B,FC\,|+|\,F,C,E,GD,AB\,|.\end{aligned}$$

Interchange A and B, and add the result to the original equation

$$|\,D, C, E, FG, AB\,|+|\,G, C, E, DF, AB\,|+|\,F, C, E, GD, AB\,|=R. \qquad \text{X.}$$

Hence also

$$\begin{aligned}&|\,A, B, D, C, EFG\,|+|\,A, B, G, C, EDF\,|+|\,A, B, F, C, EGD\,|\\ +\ &|\,E, FG, A, B, DC\,|+|\,E, DF, A, B, GC\,|+|\,E, GD, A, B, FC\,|\\ &=R. \qquad \text{XI.}\end{aligned}$$

But, from VIII.,

$$\begin{aligned}&|\,A, B, D, C, EFG\,|+|\,FG, A, B, D, EC\,|+|\,C, FG, A, B, ED\,|\\ +\ &|\,D, C, FG, A, EB\,|+|\,B, D, C, FG, EA\,|=R.\end{aligned}$$

Hence, using equations of the form X.,

$$\begin{aligned}&|\,A, B, D, C, EFG\,|+|\,A, B, G, C, EDF\,|+|\,A, B, F, C, EGD\,|\\ =-&|\,C, FG, A, B, ED\,|-|\,C, DF, A, B, EG\,|-|\,C, DG, A, B, EF\,|\\ &+R;\end{aligned}$$

and therefore XI. becomes

$$| E, FG, A, B, DC | + | E, DF, A, B, GC | + | E, GD, A, B, FC |$$
$$= | C, FG, A, B, ED | + | C, DF, A, B, EG | + | C, DG, A, B, EF | + R$$
$$= | C, FG, E, B, AD | + | C, DF, E, B, AG | + | C, GD, E, B, AF | + R$$
$$= | A, FG, E, B, CD | + | A, DF, E, B, CG | + | A, GD, E, B, CF | + R$$
$$= R,$$

the last two equations being obtained from the first by substitutions. Interchange F and G, C and D in the last of these, and add the result to the original equation; then

$$2 \, | A, FG, E, B, CD | = R,$$

and, in virtue of the equation obtained from VIII.,

$$| A, B, C, D, EFG | = R.$$

Therefore IX. becomes

$$(afgdbeca) + (agfdbeca) = R.$$

From this it may at once be deduced that the sum of $(abcdefga)$ and any form obtained from it by a substitution formed of an odd number of transpositions is reducible.

Hence $(abcdefga) + (agfedcba) = R$ or $2\,(abcdefga) = R$,

and the type I_7 is reducible.

Operate on $(abcdefga)$ with $[gh]\left[h_0 \dfrac{\partial}{\partial g_0} + h_1 \dfrac{\partial}{\partial g_1} + h_2 \dfrac{\partial}{\partial g_2}\right]$; then

$$(abcdefgha) + (abcdefhga) = R. \qquad \text{XII.}$$

The equation $\qquad [ab, cd, ef, gh, a] = R$

has six terms, each obtainable from the first by means of a substitution formed of an even number of transpositions; therefore, by repeated use of XII., it gives

$$6\,(abcdefgha) = R.$$

The reduction of invariant types of higher degree follows in the same way.

4. The equations for covariants order 2 are:

(i.) Those obtained from the factors $[ab]$ only, viz.,

$$[a \ldots e] = R, \quad [a, bc, d, e, f] = R, \text{ \&c.}$$

(ii.) Those obtained from III. by writing in that identity $h_1x_1^2+h_2x_1x_2$ for h_2, $h_0x_1^2-h_2x_2^2$ for $2h_1$, and $-h_0x_1x_2-h_1x_2^2$ for h_0, and therefore $| ahx^2 |$ for $[ah]$; these are of the forms

$$[a\ldots]=R, \quad [a,\, bc,\, d,\, e,]=R, \ \&c.$$

(iii.) A system of equations obtained from II., thus:

$$[ef]\,|\,abc\,|\,d_{x^2} = d_{x^2}\,\{[af]\,|\,bce\,|+[bf]\,|\,cae\,|+[cf]\,|\,abe\,|\}$$
$$= [af]\{[bd]\,|\,cex^2\,|+[cd]\,|\,ebx^2\,|+[ed]\,|\,bcx^2\,|\}$$
$$+[bf]\{[cd]\,|\,aex^2\,|+[ad]\,|\,ecx^2\,|+[ed]\,|\,cax^2\,|\}$$
$$+[cf]\{[ad]\,|\,bex^2\,|+[bd]\,|\,eax^2\,|+[ed]\,|\,abx^2\,|\}.$$

Multiply this identity by $[ef][bd][ca]$; then

$$[bd,\, ca,\, fe] \equiv (bdcafe)+(cafedb)+(efbdca)+(cadbfe)+(efcadb)$$
$$+(bdefca)=R.$$

Similarly the following may be deduced:—

$$[bd,\, ca,\, e] \equiv (bdcae)+(caedb)+(ebdca)+(cadbe)+(ecadb)$$
$$+(bdeca) \quad =R,$$

$$[b,\, ca,\, e] \equiv (bcae)+(caeb)+(ebca)+(cabe)+(ecab)+(beca)=R.$$

For the types ${}_2C_2$, ${}_2C_3$ there are no equations; and for a system of N quartics there are $\binom{N}{2}$ and $2\binom{N+1}{3}+\binom{N}{3}$ irreducible covariants of these two types respectively.

There are six equations like

$$[b,\, ca,\, d]=R$$

amongst the twelve forms $(ABCD)\,{}_2C_4$, all of which are independent; the equations

$$[a\,\ldots]=R$$

are, however, deducible from these. Hence there are six independent irreducible forms $(ABCD)\,{}_2C_4$; and for N quartics there are $3\binom{N+2}{4}+3\binom{N+1}{4}$ irreducible covariants of this type.

As regards higher degrees, consider the form obtained by operating with $\epsilon_x\left[x_1\dfrac{\partial}{\partial\epsilon_2}-x_2\dfrac{\partial}{\partial\epsilon_1}\right]$ on $|\,ABCDE\,|$; where $\epsilon_x^4\equiv e_{x^2}^2$; from V.,

$$|\,ABCDE\,| = R+(\alpha\beta)^2(\beta\delta)^2(\delta\gamma)^2(\gamma\epsilon)^2(\epsilon\alpha)^2-(\alpha\gamma)^2(\gamma\beta)^2(\beta\epsilon)^2(\epsilon\delta)^2(\delta\alpha)^2;$$

therefore

$$\tfrac{1}{2}\epsilon_x\left[x_1\frac{\partial}{\partial\epsilon_1}-x_2\frac{\partial}{\partial\epsilon_1}\right]\mid ABCDE\mid\equiv\tfrac{1}{2}\mid ABCD,\,Ex^2\mid\equiv\tfrac{1}{2}\Delta_E\text{ (say)}$$

$$=R+(\alpha\beta)^2(\beta\delta)^2(\delta\gamma)^2(\gamma\epsilon)(\epsilon\alpha)\left\{\gamma_x(\epsilon\alpha)-\alpha_x(\gamma\epsilon)\right\}\epsilon_x$$

$$-(\alpha\gamma)^2(\gamma\beta)^2(\beta\epsilon)(\epsilon\delta)(\delta\alpha)^2\left\{\beta_x(\epsilon\delta)-\delta_x(\beta\epsilon)\right\}\epsilon_x$$

$$=R+(eabdc)-(abdce)-(edacb)+(dacbe)\equiv R+[eabdc]\text{ (say)}$$

$$=R+[ebdac]=R+[eacbd]=R+[ecbad]=R+[eadcb]$$

$$=R+[edcab].\qquad\text{XIII.}$$

Now $[dc,\,ba,\,e]+[ab,\,cd,\,e]+[a,\,bed,\,c]+[b,\,cea,\,d]$

$$=[bacde]-[dcabe]-[abdce]+[cdbae]+3\,(cbeda)+3\,(bcead).$$

Therefore $(cbeda)+(bcead)$

$$=\tfrac{1}{6}\left\{\mid BCDE,\,Ax^2\mid+\mid EABC,\,Dx^2\mid-\mid CDEA,\,Bx^2\mid-\mid DEAB,\,Cx^2\mid\right\}+R$$

$$=\tfrac{1}{6}\left\{\Delta_A-\Delta_B-\Delta_C+\Delta_D\right\}+R=\Sigma\Delta+R.\qquad\text{XIV.}$$

Again,

$$[cd,ab,e]-[b,aed,c]+[a,bed,c]-[cdabe]+[badce]-[abdce]+[cdbae]$$

$$=3\,(caedb)-3\,(cbeda)+(aedcb)+(cbaed)-(cabed)-(bedca)-(cebad)$$
$$-(aecdb)+(becda)+(ceabd)$$

$$=3\,(caedb)-3\,(cbeda)+(cbdae)+(edacb)-(edbca)-(cadbe)-(adbce)$$
$$-(dbcae)+(dacbe)+(bdace)+R+\Sigma\Delta\ \text{(by XIV.)}$$

$$=5\,(caedb)-5\,(cbeda)+[ad,\,cb,\,e]+[bc,\,da,\,e]-[ac,\,db,\,e]$$
$$-[bd,\,ca,\,e]+R+\Sigma\Delta.$$

Therefore $(caedb)-(cbeda)=R+\Sigma\Delta.$ XV.

Again, using XV.,

$$[acedb]+[bdeca]-[e\,\ldots\,a]$$
$$=(ecdab)+(edacb)-(edbca)-(ecdba)+R+\Sigma\Delta.$$

Subtract the sum of the results of interchanging c and a, c and b, respectively in this equation from its original form; then, with the help of XV., we obtain

$$(edacb)+(edcba)+(edbac)-(edcab)-(edabc)-(edbac)=R+\Sigma\Delta.$$

But $[ed\,\ldots]=R$;

therefore $(edacb)+(edcba)+(edbac)=R+\Sigma\Delta.$ XVI.

The equations XV. and XVI. contain all the equations for $(abcde)$. It follows from XV. and XVI. that any form

$$_2C_5 = \Sigma\,(a\ \dots) + \Sigma\Delta + R\,;$$

further they prove that forms having a definite letter b in the second or fifth place can be expressed in terms of those having b in the third or fourth place, and at the same time a in the first place; XV. leaves only six of these forms to be discussed, and amongst these we have one relation, viz.,

$$(acbde)+(acebd)+(aebcd)+(adcbe)+(adbec)+(aedbc) = R+\Sigma\Delta.$$

Hence there are five independent irreducible forms $(abcde)$, and five forms $|\ ABCD,\ Ex^2\ |$ which have yet to be discussed. These determinant forms prove to be reducible. Thus

$$[abcde]+[bcdea]+[cdeab]+[deabc]+[eabcd] \equiv 0\,;$$

therefore $$\Delta_A+\Delta_B+\Delta_C+\Delta_D+D_E = R. \qquad \text{XVII.}$$

Now, from XVI., XV., and $[e,\ da,\ c,\ b,] = R$ we obtain

$$\begin{aligned}(edacb)+(ecdab)+(ebcda) &= R-(edabc)-(ebdac)-(ecbda)\\ &= \Sigma\Delta+R = \lambda_1\Delta_A+\lambda_2\Delta_B+\lambda_3\Delta_C+\lambda_4\Delta_D+\lambda_5\Delta_E+R \text{ (say)}\\ &= \lambda_1\Delta_A+\lambda_3\Delta_B+\lambda_2\Delta_C+\lambda_4\Delta_D+\lambda_5\Delta_E+R,\end{aligned}$$

since the interchange of B and C merely changes the sign of each expression. Hence $\lambda_2=\lambda_3$, and, in virtue of XVII., they may be each taken to be zero. Interchange e and b, a and d in the above result; then

$$(badce)+(bcade)+(becad) = \lambda_4\Delta_A+\lambda_5\Delta_B+\lambda_1\Delta_D+R\,; \qquad \text{XVIII.}$$

adding,

$$\begin{aligned}(ebcda)+(becad) &= (\lambda_1+\lambda_4)(\Delta_A+\Delta_D)+\lambda_5(\Delta_B+\Delta_E)+R\\ &= \tfrac{1}{6}\left\{-\Delta_A-\Delta_D+\Delta_B+\Delta_E\right\}+R,\end{aligned}$$

in virtue of XIV. Therefore

$$\lambda_5 = \tfrac{1}{6} = -(\lambda_1+\lambda_4).$$

Substituting symmetrically the results of XVIII.,

$$\begin{aligned}9\Delta_E &= 3\,[eabdc]+3\,[ebdac]+3\,[eacbd]+3\,[ecbad]+3\,[eadcb]\\ &\qquad +3\,[edcab]+R\\ &= 2\left\{-12\lambda_5\Delta_E-3\,(\lambda_1+\lambda_4)(\Delta_A+\Delta_B+\Delta_C+\Delta_D)\right\}+R = -5\Delta_E+R.\end{aligned}$$

Therefore these determinant forms are reducible.

Hence for N quartics there are

$$5\binom{N}{5}+12\binom{N}{4}+9\binom{N}{3}+2\binom{N}{2}=2\binom{N+3}{5}+3\binom{N+2}{5}$$

independent irreducible covariants of the type ${}_2C_5$.

All covariants of order 2 and of degree higher than 5 are reducible; for

$$(abcdefa) = \Sigma \mid ABCD, EF \mid + R.$$

Operate with $a_x\left[x_1\frac{\partial}{\partial a_2}-x_2\frac{\partial}{\partial a_1}\right]$; then

$$(abcdef)-(bcdefa) = R+\Sigma\Delta.$$

The determinant $\mid Ax^2, B, C, D, EF \mid$ is obtainable from $\mid Ax^2, B, C, D, E \mid$ by an operation $[ef]\left[f_0\frac{\partial}{\partial e_0}+f_1\frac{\partial}{\partial e_1}+f_2\frac{\partial}{\partial e_2}\right]$, and is therefore reducible.

Further, $a_x\left[x_1\frac{\partial}{\partial a_2}-x_2\frac{\partial}{\partial a_1}\right]\mid BCDE, AF \mid$ is reducible owing to VIII. Therefore

$$(abcdef)-(bcdefa) = R.$$

Now $$[a, bcde, f] = R;$$

therefore $$3\,(abcdef)+3\,(fbcdea) = R.$$

Hence the sum of $(abcdef)$ and a form obtained from it by an odd substitution is reducible; therefore

$$(abcdef)+(bcdefa) = R,$$

and hence $(abcdef)$ is reducible.

The reducibility of forms of higher degree and of the second order may be obtained in exactly the same way as it was for invariants.

5. Covariants of order higher than 2 may be obtained from those of order 4 lower, by writing, in the quadratic symbolical expressions for these, for a_0, x_2^2; for a_1, $-x_1x_2$; for a_2, x_1^2. It is at once apparent that all covariants of order higher than 6 are reducible; and that the only forms which have yet to be discussed are ${}_4C_1$, ${}_4C_2$, ${}_4C_3$, ${}_4C_4$, ${}_4C_5$, ${}_6C_2$, ${}_6C_3$, ${}_6C_4$.

For the forms order 4, the only further reductions possible are those due to products of two covariants order 2. The first form to be affected by these is ${}_4C_4$.

Take then such a product

$$2\,[ab]\mid abx^2\mid[cd]\mid cdx^2\mid$$

$$= [ab][cd]\begin{vmatrix} [ac] & [ad] & a_{x^2} \\ [bc] & [bd] & b_{x^2} \\ c_{x^2} & d_{x^2} & 0 \end{vmatrix}$$

$$= -(abx^2dca)+(abx^2cda)+(abcdx^2a)-(abdcx^2a).$$

But, from V., $(abx^2dca) = (adbcx^2a)+\mid ABCDx^4\mid + R,$

$$(abdcx^2a) = (acbx^2da)+\mid ABCDx^4\mid + R.$$

Therefore

$$2\mid ABCDx^4\mid = (abx^2cda)+(abcdx^2a)-(adbcx^2a)-(acbx^2da)+R.$$

Perform the substitutions (bcd) and (bdc), and add the results; then

$$6\mid ABCDx^4\mid = R.$$

There are then only six forms to discuss, connected by the equations

$$(abx^2cda)+(abcdx^2a)+R$$
$$= (adbcx^2a)+(adx^2bca)+R = (acx^2dba)+(acdbx^2a)+R$$
$$= \tfrac{1}{3}\{(abx^2cda)+(abcdx^2a)+(adbcx^2a)+(adx^2bca)+(acx^2dba)+(acdbx^2a)\}+R$$
$$= R.$$

Hence there are three independent forms $(ABCD)\,{}_4C_4$, and for N quartics there are $3\binom{N+1}{4}$ independent irreducible covariants of the type ${}_4C_4$.

Since I_6 is expressible in the forms $\mid ABCD,\ EF\mid$, which are connected by equations VIII., it follows that ${}_4C_5$ may be expressed in terms of the determinants $\mid ABCx^4,\ EF\mid$; which are reducible, since $\mid ABCDx^4\mid$ is reducible.

Hence the type ${}_4C_5$ is reducible.

As regards the types degree 6, there are $\binom{N}{2}$ irreducible forms of the type ${}_6C_2$ for N quartics. For ${}_6C_3$ it is necessary to refer to ${}_2C_4$; the equations for ${}_2C_4$ give

$$(cax^2b)+(bx^2ca) = R \quad\text{or}\quad (bx^2ca) = R+(bx^2ac)$$

and $$(bcx^2a)+(cx^2ab)+(cx^2ba)+(acx^2b) = R.$$

Hence $$2\,(cx^2ab) = (ax^2cb)+(bx^2ca)+R,$$

and therefore the difference of any two forms $(ABC)\,{}_6C_3$ is reducible.

Hence the number of independent irreducible covariants of the type ${}_6C_3$ for N quartics is $\binom{N+2}{3}$.

Reference to the work for ${}_2C_5$ shows that forms ${}_6C_4 = \Sigma\,(x^2\,\ldots)+R$, but $(x^2\,\ldots) = 0$; hence the type ${}_6C_4$ is reducible.

6. The complete system of irreducible concomitants for a set of N quartics is as follows:—

Type.	Number of Independent Forms.
I_2	$\binom{N+1}{2}$
I_3	$\binom{N+2}{3}$
I_4	$\binom{N+1}{4}+\binom{N+2}{4}$
I_5	$\binom{N}{5}+2\binom{N+1}{5}+3\binom{N+2}{5}$
I_6	$10\binom{N+2}{6}$
${}_2C_2$	$\binom{N}{2}$
${}_2C_3$	$\binom{N}{3}+2\binom{N+1}{3}$
${}_2C_4$	$3\binom{N+1}{4}+3\binom{N+2}{4}$
${}_2C_5$	$3\binom{N+2}{5}+2\binom{N+3}{5}$
${}_4C_1$	N
${}_4C_2$	$\binom{N+1}{2}$
${}_4C_3$	$2\binom{N+1}{3}$
${}_4C_4$	$3\binom{N+1}{4}$
${}_6C_2$	$\binom{N}{2}$
${}_6C_3$	$\binom{N+2}{3}$.

The Invariant Syzygies of Lowest Degree for any number of Quartics. By A. YOUNG. Received and communicated June 14th, 1900.

In a previous paper* I found the irreducible system of types for any number of quartics. The reductions were effected by means of systems of equations between forms of the same degree and order; these were written $[abc...f] = R$, &c. If, instead of R, we write the sum of products of forms for which R stands, these systems of equations will give us the syzygies. This method has, as is shown in § 1, the merit of exhaustiveness.

So far as results are concernod, I have only dealt exhaustively with the invariant syzygies up to degree 7. There is no invariant syzygy of degree lower than 7; and all syzygies of this degree are linearly deducible from syzygies of a single form. From this syzygy it is possible to deduce, by means of differential operations, syzygies between covariants and syzygies of higher degree between invariants.

The system of syzygies for two quartics has been worked out by von Gall.† The syzygy degree 7 obtained here is identically zero when fewer than three quartics are under consideration, so that it does not appear in his paper. As some use is made of substitutions, it is convenient to use Roman letters for symbols indicating them.

The following abbreviations are used:—

{s} denotes the sum of powers of the substitution s.

{s_1, s_2} denotes the sum of the substitutions of the simplest group containing s_1 and s_2; and so on.

{a, b, c, ... } denotes the sum of the substitutions of the symmetric group of the letters $a, b, c, \ldots$.

{a, b, c, ... }′ denotes the sum of the substitutions of the alternating group, minus the sum of the substitutions which do not belong to the alternating group of the letters $a, b, c, \ldots$.

The invariant types in my paper referred to above were written (aba), $(abca)$; they are written here simply (ab), (abc), ...

* *Proc. Lond. Math. Soc.*, Vol. **xxx**., p. 290.
† *Math. Ann.*, Bd. xxxiii., p. 197, and Bd. xxxiv., p. 332.

1. All relations between quartic types, including, of course, syzygies, must depend ultimately on the fundamental identities found in § 2 of my previous paper. Hence, if we form from these all possible equations between quartic forms of a given degree and order, we shall obtain all the relations which exist between forms of that degree and order. To construct these equations it is only necessary to take those obtained by multiplying one or other of the identities by a single factor; all other equations being linearly dependent on these. Hence the relations $[abc\ldots g]=R$, &c., already used to obtain the irreducible system of types, must also give all the syzygies.

Having obtained a relation between types of a certain degree, it is possible, by a certain operation, to obtain from it a relation between types of a higher degree. Thus the result of operating with $(kb)\left[k_0\frac{\partial}{\partial b_0}+k_1\frac{\partial}{\partial b_1}+k_2\frac{\partial}{\partial b_2}\right]$ on the symbolical product $(abcd\ldots)$ is $(abkcd\ldots)+(akbcd\ldots)$. Before, however, using this operator, it is necessary to have the relation written out in full in quadratic symbols. Now we shall find it convenient to ignore symbolical factors of the form $[aa]$. When this is done, any single symbolical form represents a single invariant. In fact, the relations as written will be relations between invariants themselves, rather than between quadratic symbolical products. To pass from the actual coefficients of the quartics to quadratic symbols, and *vice versa*, we have the identity

$$(A_0,\ A_1,\ A_2,\ A_3,\ A_4\between x_1,\ x_2)^4\equiv a_x^2{}^2\equiv(a_0x_1^2+2a_1x_1x_2+a_2x_2^2)^2.$$

Therefore $\quad A_0=a_0^2,\quad A_1=a_0a_1,\quad A_3=a_1a_2,\quad A_4=a_2^2,$

$$6A_2=4a_1^2+2a_0a_2=6a_1^2+[aa]=6a_0a_2-2\,[aa].$$

Hence, if P be the type represented by the symbolical product $(abcd\ldots)$,

$$P=(abcd\ldots)+\Sigma\,[aa]\,P_a+\Sigma\,[aa][bb]\,P_{a,\,b}+\ldots.$$

To find P_b, consider the product

$$\begin{aligned}[ab][bc]\equiv B_0a_2c_2&-2B_1\,(a_2c_1+a_1c_2)+(B_2+\tfrac{1}{3}\,[bb])(a_2c_0+a_0c_2)\\ &+4\,(B_2-\tfrac{1}{6}\,[bb])\,a_1c_1-2B_3\,(a_1c_0+a_0c_1)+B_4a_0c_0\\ &=\text{a function of the coefficients }B+\tfrac{1}{3}\,[bb][ac].\end{aligned}$$

Therefore $P_b = -\frac{1}{3}(acd...)$.

The result, then, of operating with $[kb]\left[k_0\frac{\partial}{\partial b_0}+k_1\frac{\partial}{\partial b_1}+k_2\frac{\partial}{\partial b_2}\right]$ on the invariant type $(abcd...)$ is

$$(abkcd...)+(akbcd...)-\tfrac{2}{3}(bk)(acd...).$$

The effect of this operator, it is easy to see, is the same as that of substituting the coefficients of the covariant type $2[bk]\,b_{x^2}k_x$, for those of the quartic $b_{x^4}^2$.

2. Before proceeding to the discussion of syzygies by the above method, we shall establish certain syzygies which exist for the types expressible as determinants.

First,

$$96\,|ABCDE|\;|FGHKL| = \begin{vmatrix} (af) & (ag) & (ah) & (ak) & (al) \\ (bf) & (bg) & (bh) & (bk) & (bl) \\ (cf) & (cg) & (ch) & (ck) & (cl) \\ (df) & (dg) & (dh) & (dk) & (dl) \\ (ef) & (eg) & (eh) & (ek) & (el) \end{vmatrix}.$$

Similarly the product of $|ABCDE|$ and any form of the type I_6, as also that of two forms I_6, may be expressed as the sum of products of forms I_2 and I_3.

Next, any function homogeneous and linear in the coefficients of each of the five quartics $a_{x^2}^2$, $b_{x^2}^2$, $c_{x^2}^2$, $d_{x^2}^2$, $e_{x^2}^2$ is of the form

$$P = \Sigma\lambda A_{r_1}B_{r_2}C_{r_3}D_{r_4}E_{r_5},$$

where $r = 0, 1, 2, 3, 4$; and can therefore have only five distinct values. Hence

$$\{a, b, c, d, e\}'P = |ABCDE|\,Q. \qquad \text{(I.)}$$

Similarly, $\{a, b, c, d\}'P = |ABCDQ|$,

and, if P be also homogeneous and linear in the coefficients of f_{x^2},

$$\{a, b, c, d, e, f\}'P = 0. \qquad \text{(II.)}$$

In (I.) put $P = |ABCD, EF|$; the right-hand side must be zero, otherwise Q would be an invariant of the first degree of $f_{x^2}^2$; this gives the identity for the forms $|ABCD, EF|$ (found before).

In (II.) put $P=(fg)\,|ABCDE|$; the result is a relation which may be written

$$(fg)\,|ABCDE| = (ag)\,|FBCDE| + (bg)\,|AFCDE| + (cg)\,|ABFDE| + (dg)\,|ABCFE| + (eg)\,|ABCDF|. \quad \text{(III.)}$$

In this we may write for F_0, F_1, F_2, F_3, F_4 the coefficients of F_4, $-4F_3$, $6F_2$, $-4F_1$, F_0 in any type $(f\ldots)$, invariant or covariant. The result will be

$$(g\ldots)\,|ABCDE| = \Sigma I_2 J,$$

where J is reducible, except when $(g\ldots)$ belongs to one of the types I_2, I_3. Operate with $[eh]\left[h_0\frac{\partial}{\partial e_0}+h_1\frac{\partial}{\partial e_1}+h_2\frac{\partial}{\partial e_2}\right]$, and we obtain

$$J\,.\,I_6 = \Sigma P_3,$$

where ΣP_3 represents a sum of products of forms, there being at least three forms in each product, and J is any type except I_2 or I_3.

Operate on (III.) with

$$[eh]\left[h_0\frac{\partial}{\partial e_0}+h_1\frac{\partial}{\partial e_1}+h_2\frac{\partial}{\partial e_2}\right][fk]\left[k_0\frac{\partial}{\partial f_0}+k_1\frac{\partial}{\partial f_1}+k_2\frac{\partial}{\partial f_2}\right],$$

$$2\,(gfk)\,|ABCD,\,EH| = 2\,(geh)\,|ABCD,\,FK| + \Sigma P_3.$$

Similarly, $(geh)\,|ABCD,\,FK| = (fkh)\,|ABCD,\,GE| + \Sigma P_3$.

Therefore $(gfk)\,|ABCD,\,EH| = (hfk)\,|ABCD,\,EG| + \Sigma P_3$.

It is not possible to express $I_3 I_6$ in the form ΣP_3; for suppose

$$(abc)\,|DEFG,\,HK| = \Sigma P_3 = \Sigma\left\{I_5 I_2 I_2 + I_4 I_3 I_2 + I_3 I_3 I_3 + I_3 I_2 I_2 I_2\right\}.$$

Then the syzygy would still be true if we supposed the quartics represented by the letters a, b, c, h, k to be all the same—we may without confusion express this supposition by using only one letter for this quartic. Then

$$(aaa)\,|DEFG,\,AA| = \Sigma P_3 = \Sigma\left\{I_5 I_2 I_2 + \ldots\right\}.$$

Operate on this with $\{\mathrm{d, e, f, g}\}'$; a little consideration will show that the result is

$$24\,(aaa)\,|DEFG,\,AA| = \lambda\,(aa)(aa)\,|DEFGA|;$$

and therefore $\quad (aaa)\,|DEFA,\,AA| = 0.$

Hence $\quad (aaa) = 0 \quad \text{or} \quad |DEFA,\,AA| = 0;$

either result being absurd. In exactly the same way, it may be shown that there is no syzygy of the form

$$I_3 I_6 = \Sigma I_4 I_5 + \Sigma P_3$$

or of the form $$I_2 I_6 = \Sigma I_3 I_5 + \Sigma P_3.$$

3. For the sake of finding syzygies, the equations of my previous paper must be written out in full. As regards degree 4, there is only one equation,

$$2\,(abcd)+2\,(adbc)+2\,(acdb) = (ab)(cd)+(ad)(bc)+(ac)(db);$$

which is the expanded form of

$$\begin{vmatrix} [ab] & [ac] & [ad] & [aa] \\ [bb] & [bc] & [bd] & [ba] \\ [cb] & [cc] & [cd] & [ca] \\ [db] & [dc] & [dd] & [da] \end{vmatrix} = 0.$$

The only form of equation for degree 5 is

$$[ab...] = (abc)(de)+(abd)(ec)+(abe)(cd)+2\,(ab)(cde). \quad \text{(IV.)}$$

Now it was proved before that

$$60\,|ABCDE| = -\{\mathrm{a, b, c, d, e}\}'\,(abcde)+R.$$

In this case $R = \Sigma I_2 I_3$, but $\{\mathrm{a, b, c, d, e}\}' \Sigma I_2 I_3 = 0$; hence R must here be zero. It will be convenient to use the form

$$(abcde)' = (abcde)-\tfrac{1}{2}\{(ab)(cde)+(bc)(dea)+(cd)(eab) + (de)(abc)+(ea)(bcd)\}.$$

Then $$60\,|ABCDE| = -\{\mathrm{a, b, c, d, e}\}'\,(abcde)'.$$

Further $$[ab...]' = -\tfrac{1}{12}\{\mathrm{a, b, c, d, e}\}\,(ab)(cde).$$

Hence $[ab...]'+[ae...]'-[ac...]'-[ad...]'$

$$= 2\,(abcde)'+2\,(abdce)'-2\,(acbed)'-2\,(acebd)' = 0;$$

and therefore

$$\begin{aligned}(abdce)'-(acbed)' &= (acebd)'-(abcde)' = (adecb)'-(acdbe)' \\ &= \ldots = |ABCDE| \\ &= \tfrac{1}{3}\{(abdce)'+(acebd)'+(acbde)'+(adecb)'+(adcbe)'+(abedc)'\} \\ &\quad +\tfrac{1}{36}\{\mathrm{a, b, c, d, e}\}\,(ab)(cde). \quad \text{(V.)}\end{aligned}$$

The equations of degree 5 give nothing in the way of syzygies.

For degree 6 there are two kinds of equation to be discussed—

$$[abc...] = (abcd)(ef)+(abce)(fd)+(abcf)(de)+2(abc)(def) \quad \text{(VI.)}$$

and
$$[ab,\ cd,\ e,\ f] = (abcd)(ef)+(abef)(cd)+(abfe)(cd)+(ab)(cdef)$$
$$+(ab)(cdfe)+(abe)(cdf)+(abf)(cde)-(ab)(cd)(ef). \quad \text{(VII.)}$$

As in the former problem, we will use, in addition, the equation

$$4\,|abc|^2\,|def|^2$$
$$= 2[a-b-c-]+\{d,\ e,\ f\}\,[(ad)(be)(cf)-(ad)(becf)$$
$$-(bd)(aecf)-(cd)(aebf)]$$
$$= 4(abc)(def)$$

or
$$[a-b-c-]$$
$$= 2(abc)(def)-\tfrac{1}{2}\{d,\ e,\ f\}\Big[(ad)(be)(cf)-(ad)(becf)-(bd)(ceaf)$$
$$-(cd)(aebf)\Big]. \quad \text{(VIII.)}$$

Let S represent the sum of the sixty possible forms $(abcdef)$, and operate on (VIII.) with $\{a,\ b,\ c,\ d,\ e,\ f\}$;

$$72S = \{a,\ b,\ c,\ d,\ e,\ f\}\Big[2(abc)(def)-3(ad)(be)(cf)+9(ad)(becf)\Big]$$
$$= \{a,\ b,\ c,\ d,\ e,\ f\}\Big[2(abc)(def)+\tfrac{3}{2}(ad)(be)(cf)\Big]. \quad \text{(IX.)}$$

Now we proved before that

$$6\Big[(adbecf)+(afbdce)+(aebfcd)\Big]$$
$$= \{(abc),\ (def)\}\Big[[ad,\ be,\ c,\ f]+\tfrac{1}{3}[abc...]-\tfrac{1}{9}S+\tfrac{1}{9}[a-b-c-]\Big]$$
$$= \{(abc),\ (def)\}\Big[(cf)(adbe)+(ad)(becf)+(ad)(befc)+(be)(adcf)$$
$$+(be)(adfc)-(ad)(be)(cf)+(adc)(bef)+(adf)(bec)$$
$$+(ef)(abcd)+\tfrac{2}{3}(abc)(def)-\tfrac{2}{9}(abc)(def)$$
$$-2(abd)(cef)-\tfrac{1}{3}(ad)(be)(cf)-\tfrac{1}{3}(ad)(bf)(ce)$$
$$-(ab)(de)(cf)+\tfrac{2}{9}(abc)(def)-\tfrac{1}{6}(ad)(be)(cf)$$
$$-\tfrac{1}{6}(ad)(bf)(ce)+(ad)(becf)\Big]$$
$$= \{(abc),\ (def)\}\Big[\tfrac{2}{3}(abc)(def)+(ad)\{4(becf)+2(befc)\}$$
$$-\tfrac{1}{2}(ab)(de)(cf)-\tfrac{1}{2}(ad)(bf)(ce)$$
$$-\tfrac{3}{2}(ad)(be)(cf)\Big]. \quad \text{(X.)}$$

We proceed to find the exact expression for $(abcdef)$ in terms of the determinant forms degree 6 and products of types of lower degree. For this purpose, it is useful to note that $(abcdef)$ is unchanged by any substitution of the group $\{(\mathrm{abcdef}), (\mathrm{ac})(\mathrm{df})\}$. Operate, then, with the sum of the substitutions of this group on the expression given by (X.) for

$$(abcfed)+(afcdeb)+(adcbef).$$

The result is

$$\{(\mathrm{abcdef}), (\mathrm{ac})(\mathrm{df})\}\Big[3\,(abcfed)-(ace)(bdf)-(ab)\,\{2\,(cfed)+(cedf)\}$$
$$-(ad)\,\{2\,(cbef)+(cefb)\}-(af)\,\{2\,(cdeb)+(cebd)\}$$
$$+\tfrac{1}{4}\,(ab)(ce)(df)+\tfrac{1}{4}\,(ad)(ce)(bf)+\tfrac{1}{4}\,(af)(ce)(bd)$$
$$+\tfrac{1}{4}\,(ab)(cd)(ef)+\tfrac{1}{4}\,(ad)(cf)(eb)+\tfrac{1}{4}\,(af)(cb)(ed)$$
$$+\tfrac{3}{4}\,(ab)(cf)(de)+\tfrac{3}{4}\,(ad)(cb)(ef)+\tfrac{3}{4}\,(af)(cd)(eb)\Big]=0$$

or

$$\{(\mathrm{abcdef}), (\mathrm{ac})(\mathrm{df})\}\Big[3\,(abcfed)-(ace)(bdf)$$
$$-2\,(ab)\,\{2\,(cdef)+(cedf)\}-(ad)\,\{2\,(bcfe)+(bcef)\}$$
$$+\tfrac{1}{2}\,(ab)(ce)(df)+\tfrac{1}{2}\,(ab)(cd)(ef)+\tfrac{1}{4}\,(ad)(ce)(bf)$$
$$+\tfrac{1}{4}\,(ad)(cf)(be)+\tfrac{9}{4}\,(ab)(cf)(de)\Big]=0. \qquad \text{(XI.)}$$

Similarly, by operating on (X.) itself,

$$\{(\mathrm{abcdef}), (\mathrm{ac})(\mathrm{df})\}\Big[2\,(abecfd)+(abfdec)-(abc)(def)$$
$$-\tfrac{1}{3}\,(ab)\,\{4\,(cedf)+2\,(cefd)\}-\tfrac{1}{3}\,(ad)\,\{5\,(becf)+4\,(befc)\}$$
$$+\tfrac{1}{6}\,(ab)(cd)(ef)+\tfrac{5}{12}\,(ab)(de)(cf)+\tfrac{1}{2}\,(ab)(ce)(df)$$
$$-\tfrac{1}{12}\,(ad)(ce)(bf)+\tfrac{3}{4}\,(ad)(be)(cf)\Big]=0. \qquad \text{(XII.)}$$

Operate on equations (V.) with $[fc]\left[f_0\dfrac{\partial}{\partial c_0}+f_1\dfrac{\partial}{\partial c_1}+f_2\dfrac{\partial}{\partial c_2}\right]$; then

$$|DEAB,\,CF|$$
$$=(acfebd)+(afcebd)-(abcfde)-(abfcde)$$
$$-\tfrac{2}{3}\,(cf)\,\{(aebd)-(abde)\}-(acf)(ebd)-(cfe)(bda)+(bcf)(dea)$$
$$+(cfd)(eab)-\tfrac{1}{2}\,(eb)\,\{(dacf)+(dafc)\}-\tfrac{1}{2}\,(bd)\,\{(acfe)+(afce)\}$$
$$-\tfrac{1}{2}\,(da)\,\{(cfeb)+(fceb)\}+\tfrac{1}{2}\,(de)\,\{(abcf)+(abfc)\}$$
$$+\tfrac{1}{2}\,(ea)\,\{(bcfd)+(bfcd)\}+\tfrac{1}{2}\,(ab)\,\{(cfde)+(fcde)\}$$
$$+\tfrac{1}{3}\,(cf)\,\{2\,(eb)(da)-2\,(de)(ab)\}$$

$$= (adecfb) + (adefcb) - (acfdbe) - (afcdbe) - \tfrac{2}{3}(cf)\{(adeb) - (adbe)\}$$
$$-(ecf)(bad) - (cfb)(ade) + (acf)(dbe) + (cfd)(bea)$$
$$-\tfrac{1}{2}(ba)\{(decf) + (defc)\} - \tfrac{1}{2}(ad)\{(ecfb) + (efcb)\}$$
$$-\tfrac{1}{2}(de)\{(cfba) + (fcba)\} + \tfrac{1}{2}(db)\{(eacf) + (eafc)\}$$
$$+\tfrac{1}{2}(be)\{(acfd) + (afcd)\} + \tfrac{1}{2}(ea)\{(cfdb) + (fcdb)\}$$
$$+\tfrac{1}{3}(cf)\{2(ab)(de) - 2(bd)(ea)\}.$$

Operate on these with $\{(\mathrm{abcdef}), (\mathrm{ac})(\mathrm{df})\}$;

$$\{(\mathrm{abcdef}), (\mathrm{ac})(\mathrm{df})\}\,|DEAB, CF|$$
$$= \{(\mathrm{abcdef}), (\mathrm{ac})(\mathrm{df})\}\big[2(abecfd) - 2(abcefd)$$
$$+ (ab)\{(cdef) + (cedf)\} - \tfrac{1}{3}(ad)\{5(becf) + 3(befc) - 2(bcef)\}$$
$$+\tfrac{2}{3}(ad)(be)(cf) - \tfrac{2}{3}(ab)(de)(cf)\big]$$
$$= \{(\mathrm{abcdef}), (\mathrm{ac})(\mathrm{df})\}\big[(abcfed) - (acfdbe)$$
$$-(ab)\{(cdef) + (cedf)\} + (ac)\{(bedf) + (befd)\}$$
$$-\tfrac{2}{3}(ad)\{(befc) - (becf)\} + \tfrac{2}{3}(ab)(de)(cf) - \tfrac{2}{3}(ad)(ce)(bf)\big].$$

Hence, using $$[ad, be, c, f] = R,$$

the last expression for $\{(\mathrm{abcdef}), (\mathrm{ac})(\mathrm{df})\}\,|DEAB, CF|$ becomes

$$\{(\mathrm{abcdef}), (\mathrm{ac})(\mathrm{df})\}\big[2(abcfed) + 4(abecfd) - 2(abd)(cef)$$
$$-(ab)\{(cdef) + (cedf)\} + (ac)\{(bedf) + (befd)\}$$
$$-\tfrac{1}{3}(ad)\{7(becf) + 8(befc)\} + \tfrac{2}{3}(ab)(de)(cf)$$
$$-\tfrac{2}{3}(ad)(ce)(bf) + (ad)(be)(cf)\big]$$
$$= \{(\mathrm{abcdef}), (\mathrm{ac})(\mathrm{df})\}\big[4(abecfd) + \tfrac{2}{3}(ace)(bdf) - 2(abd)(cef)$$
$$+\tfrac{1}{3}(ab)\{5(cdef) + (cedf)\} + (ac)\{(bedf) + (befd)\}$$
$$-\tfrac{1}{3}(ad)\{7(becf) + 4(befc) - 2(bcef)\} - \tfrac{1}{3}(ab)(ce)(df)$$
$$-\tfrac{1}{3}(ab)(cd)(ef) - \tfrac{5}{6}(ab)(de)(cf) - \tfrac{5}{6}(ad)(ce)(bf)$$
$$+\tfrac{5}{6}(ad)(be)(cf)\big],$$

using (XI.).

From $[ab, de, c, f]$, we have

$$\{(\mathrm{abcdef}), (\mathrm{ac})(\mathrm{df})\}\big[(abcdef) + 4(abcefd) + (abfdec) - 2(abc)(def)$$
$$-2(ab)\{(decf) + (defc)\} - (ad)(bcef) + (ab)(de)(cf)\big] = 0.$$

Eliminate $(abfdec)$ between this equation and (XII.); then

$$\{(\text{abcdef}), (\text{ac})(\text{df})\} \Big[(abcdef)+4\,(abcefd)-2\,(abecfd)-(abc)(def)$$
$$-\tfrac{1}{3}\,(ab)\,\{6\,(cdef)+2\,(cedf)-2\,(cefd)\}$$
$$+\tfrac{1}{3}\,(ad)\,\{5\,(becf)+4\,(bcfe)-3\,(bcef)\}-\tfrac{1}{6}\,(ab)(cd)(ef)$$
$$+\tfrac{7}{12}\,(ab)(de)(cf)-\tfrac{1}{2}\,(ab)(ce)(df)+\tfrac{1}{12}\,(ad)(ce)(bf)$$
$$-\tfrac{3}{4}\,(ad)(be)(cf)\Big] = 0.$$

Eliminate $(abcefd)$ and $(abecfd)$ between this equation and the two expressions we have found for $\{(\text{abcdef}), (\text{ac})(\text{df})\}\,|\,DEAB,\ CF\,|$; then

$$\tfrac{3}{2}\,\{(\text{abcdef}), (\text{ac})(\text{df})\}\,|\,DEAB,\ CF\,|$$
$$= \{(\text{abcdef}), (\text{ac})(\text{df})\} \Big[(abcdef)-(abc)(def)-\tfrac{1}{3}\,(ace)(bdf)$$
$$+(abd)(cef)+\tfrac{1}{6}\,(ab)\,\{-5\,(cdef)+7\,(cedf)+4\,(cefd)\}$$
$$-\tfrac{1}{2}\,(ac)\,\{(bedf)+(befd)\}+\tfrac{1}{6}\,(ad)\,\{-3\,(becf)\}$$
$$-\tfrac{1}{3}(ab)(de)(cf)-\tfrac{1}{3}(ab)(ce)(df)+\tfrac{1}{2}(ad)(ce)(bf)+\tfrac{1}{6}(ad)(be)(cf)\Big].$$

Therefore

$$2\,(abcdef)$$
$$= |\,ADEB.\ CF\,| + |\,EBCF,\ AD\,| + |\,CFAD,\ EB\,|$$
$$+\{(\text{abcdef})\}\Big[\tfrac{1}{3}\,(abc)(def)+\tfrac{1}{9}\,(ace)(bdf)-\tfrac{1}{3}\,(abd)(cef)$$
$$+\tfrac{1}{2}\,(ab)(cdef)-\tfrac{1}{6}\,(ab)(cedf)-\tfrac{1}{6}\,(ac)(bdef)+\tfrac{1}{6}\,(ad)(becf)$$
$$-\tfrac{1}{9}\,(ab)(cd)(ef)+\tfrac{1}{6}\,(ab)(ce)(df)-\tfrac{1}{12}\,(ad)(ce)(bf)$$
$$-\tfrac{1}{18}\,(ad)(be)(cf)\Big]. \qquad \text{(XIII.)}$$

This expression for $(abcdef)$ satisfies the equations $[abc...] = R$ and $[ab,\ cd,\ e,\ f] = R$ identically; hence no syzygies amongst invariants degree 6 can exist.

4. *Degree* 7.—From the invariant identity degree 6, we have

$$\{\text{a, b, e}\}'\,|\,FBAD,\ EC\,| = 2\,|\,EBAD,\ FC\,| + 2\,|\,FBAE,\ CD\,|.$$

Hence

$$\{\text{a, b, e}\}'\Big[(afdbec)+(cfdeba)+2\,(efdcba)+(bfdace)\Big]$$
$$= \tfrac{1}{2}\,\{\text{a, b, e}\}'\Big[\,|\,ABEF,\ DC\,| + |\,DCAB,\ EF\,| + 3\,|\,ADFB,\ CE\,|$$
$$+|\,CEAD,\ FB\,| + |\,FBCE,\ AD\,|$$
$$+|\,ABFC,\ ED\,| + |\,EDAB,\ FC\,|\Big]+R$$
$$= 6\,|\,ABEF,DC\,| + \{\text{a,b,e}\}'\Big[(af)(dbec)+(bc)(aefd)-\tfrac{1}{2}(af)(bd)(ec)\Big].$$

Operate on this with $[gf]\left[g_0\frac{\partial}{\partial f_0}+g_1\frac{\partial}{\partial f_1}+g_2\frac{\partial}{\partial f_2}\right]$, and then with $\{g, d, f\}$; therefore, using the equations $[a...bec] = R$, &c.,

$$6\{g, d. f\}\,|A, B, E, FG, DC|$$

$$= \{g, d, f\}\{a, b, e\}'\left[\tfrac{1}{3}(fg)\{2(adbec)+(abcde)\}-2(afg)(dbec) + (afg)(bd)(ec)\right]$$

$$= 12|A, B, E, FG, DC| + 12|A, B, E, DF, GC| + 12|A, B, E, GD, FC|.$$

Hence, operating with $\frac{1}{2}\{(cd)\}$,

$$12|A, B, E, FG, DC|$$

$$= \{(cd), (fg)\}\{a, b, e\}'\left[\tfrac{1}{3}(dg)|AFBEC| + (dg)(afbec) -2(adg)(fbec)+(adg)(ec)(bf)\right],$$

and
$$12|A. B, C, E, DFG|$$

$$= -2\{a, b, c, e\}'|A, B, E, FG, DC|$$

$$= \{(fg)\}\{a, b, c, e\}'\left[\tfrac{1}{12}(dg)|AFBEC| + \tfrac{1}{12}(fg)|ADBEC| -(cg)(afbed)\right].$$

From these we proceed to find an expression for the reducible type $(abcdefg)$.

Operate then on (XIII.) with $[gf]\left[g_0\frac{\partial}{\partial f_0}+g_1\frac{\partial}{\partial f_1}+g_2\frac{\partial}{\partial f_2}\right]$, and we obtain, after multiplying by 12 and reducing,

$$24(abcdefg)+24(abcdegf)$$

$$\begin{aligned}= \{(fg), (ae)(bd)\}\big[&4(fg)(abcde)+6\{(ab)(cdefg)+(bc)(defga)\}\\
&-2\{(ab)(cedfg)+(bc)(dfgea)\}\\
&-\{2(ac)(bdefg)+(bd)(acefg)+(ea)(bcdfg)\}+4(ad)(becfg)\\
&+4(ace)(bdfg)+4(bcd)(aefg)+8(abc)(defg)\\
&-4\{(abd)(cefg)+(abe)(cdfg)+(acd)(befg)\}\\
&+(afg)\{6(bcde)-2(bdce)\}-2(bfg)(acde)+2(cfg)(adbe)\\
&+2(afg)\{(bd)(ce)-2(bc)(de)\}\\
&+2(bfg)\{(cd)(ae)+(ca)(ed)-(ad)(ce)\}\\
&-(cfg)\{2(ad)(be)+(ae)(bd)\}\end{aligned}$$

$-(fg)\{4(ab)(cde)+(ae)(bcd)+(bd)(ace)-2(ac)(bde)\}$

$+2(cg)\,|ABDEF|+2(ag)\{(bcdef)-(bdcef)\}$

$+2(bg)\{(adcef)+(dfeac)-(afcde)-(adfce)\}$

$-(aeg)\{2(bcdf)-2(bdcf)\}+(bdg)\{2(acef)-2(aecf)\}$

$-2(abg)\{(fced)-(fcde)\}-2(acg)\{(dbef)-(debf)\}$

$-2(bcg)\{(adef)-(daef)\}-(aeg)(bd)(cf)+(bdg)(ae)(cf)$

$-2(abg)(cd)(ef)+2(acg)(bd)(ef)+2(abe)(cf)(dg)$

$-2(acd)(bf)(eg)\big]$.

Now. $24(abcdefg)+24(adcbefg)$

$=-24(abdcefg)-24(adbcefg)-24(acbdefg)-24(acdbefg)$

$+24(bc)(adefg)+24(cd)(abefg)+24(db)(acefg)+48(bcd)(aefg)$

$=\{(\mathrm{bd}),\ (\mathrm{ae})(\mathrm{fg})\}\big[-2(bd)(acefg)-(ac)\{4(bdefg)-2(bdgef)\}$

$+2(fc)\{(abdeg)-2(abdge)\}$

$-(ef)\{6(bdcga)+6(bdgac)-4(bdacg)\}-4(fg)(bdcea)$

$-4(ae)(bdfcg)+2(eg)\{(bdcfa)+(bdcaf)\}-4(acf)(bdeg)$

$-4(cef)(bdag)+4(ace)(bdfg)+4(cfg)(bdae)-16(efg)(bdac)$

$+8(acg)(bdcf)-(abd)\{4(cefg)-2(cfeg)\}$

$-(fbd)\{4(aegc)-2(aceg)\}+(cbd)\{6(aefg)+2(aegf)\}$

$+2(abd)\{(ce)(fg)-(cg)(ef)\}+(fbd)\{6(ae)(cg)-2(ec)(ag)\}$

$+(cbd)\{4(ag)(ef)-2(af)(eg)\}$

$+(bd)\{6(ce)(afg)+4(ag)(cef)-2(cf)(aeg)\}$

$-4(fd)\,|ABCEG|+(cd)\{10(abefg)+2(abefg)\}$

$-2(fd)\{(cbeag)+(abceg)-(abgce)-(ebgac)\}$

$-2(ad)\{(ebcfg)+(fbecg)-(cbgef)-(gbfce)\}$

$+2(acd)\{(befg)-(begf)\}+2(egd)\{(acbf)-(abcf)\}$

$+2(fcd)\{(aegb)-(eagb)\}+2(agd)\{(bcfe)-(bcef)\}$

$+2(acd)\{(fb)(eg)+(eb)(fg)\}-2(egd)\{(fb)(ac)+(cb)(af)\}$

$-2(fcd)(ab)(eg)+2(agd)(cb)(ef)+2(aef)(cb)(gd)$

$-2(agc)(eb)(fd)-2(ace)(fb)(gd)+2(fgc)(ab)(ed)\big]$.

Similarly, $-24\,(adcbefg) - 24\,(edcbafg)$

$$= -24\,(ae)(dcbfg) - 24\,(fg)\,\{(adcbe) + (edcba)\}$$

$$-24\,(bcd)\,\{(efga) + (afge)\} - 24\,(dcba)(efg) - 24\,(dcbe)(afg)$$

$$+24\,(dcb)(ae)(fg) + 24\,(dcbeafg) + 24\,(dcbaefg)$$

$$+24\,(dcbfgea) + 24\,(dcbfgae)$$

$$= \{(ae), (bd)(fg)\}\,\big[2\,(ae)(dcbfg)$$

$$+(cd)\,\{4\,(fgbae) + 6\,(bfgae) - 2\,(gbfae)\}$$

$$+(fg)\,\{-6\,(dcbae) - 2\,(dbcae)\}$$

$$+(gd)\,\{6\,(fcbae) - 2\,(bfcae) - 2\,(cbfae)\} - 2\,(bd)(fgcae)$$

$$+(fd)\,\{4\,(cgbae) - 2\,(gcbae)\} + 2\,(gc)(bdfae) + 4\,(bdf)(cgae)$$

$$+4\,(fgd)(cbae) + 4\,(gdc)(bfae) - 4\,(bcd)(fgae) - 8\,(dcf)(gbae)$$

$$-4\,(gcf)(bdae) + (bae)\,\{6\,(fgdc) + 2\,(fdgc)\}$$

$$+(fae)\,\{4\,(dcbg) - 2\,(dcgb)\} - 2\,(cae)(dbfg)$$

$$-(bae)\,\{2\,(fd)(cg) + 2\,(fc)(gd) + 4\,(fg)(dc)\}$$

$$+(fae)\,\{2\,(gb)(cd) - 2\,(gd)(cb)\}$$

$$+(cae)\,\{2\,(bd)(fg) + 2\,(bf)(gd) - 2\,(bg)(df)\}$$

$$-(ae)\,\{4\,(cb)(fgd) - 2\,(fg)(bcd) + 2\,(bf)(cdg) + 2\,(cg)(bdf)$$
$$-2\,(bd)(cfg) - 2\,(df)(bcg)\}$$

$$+4\,(da)\,|CBEFG| + 2\,(ca)\,\{(gefbd) - (bedgf)\}$$

$$+2\,(ga)\,\{(befcd) + (cebfd) - (fedbc) - (decfb)\}$$

$$+2\,(ba)\,\{(cefgd) - (cefdg)\} + 4\,(bfa)\,\{(ecgd) - (ecdg)\}$$

$$-2\,(bca)\,\{2\,(edfg) - (edgf) - (egdf)\} - (fga)\,\{2\,(edbc) + 10\,(ebcd)\}$$

$$+2\,(gca)\,\{2\,(efdb) - (efbd) - (ebfd)\} - 2\,(bda)\,\{(egcf) - (ecfg)\}$$

$$-2\,(dfa)\,\{(ebgc) - (egbc)\} - 2\,(bfa)(cg)(de) - 2\,(bca)(gd)(fe)$$

$$-2\,(fga)(cd)(be) + 2\,(gca)(fb)(de) + 2\,(bda)(gc)(fe)$$

$$+2\,(dfa)(cg)(be) - 2\,(dcf)(ba)(ge) - 2\,(bdg)(ca)(fe)$$

$$+2\,(fbc)(da)(ge) + 2\,(fgb)(da)(ce)\big].$$

Add these three results, and we obtain, after some reduction,

$$
\begin{aligned}
&336\,(abcdefg)\\
&= \{(\mathrm{abcdefg}),\ (\mathrm{bg})(\mathrm{cf})(\mathrm{de})\}\,\big[4\,(ac)\,|BDEFG| + 16\,(ad)\,|BCEFG|\\
&\quad + (ab)\{40\,(cdefg) - 8\,(cdfge) - 8\,(cedfg)\}\\
&\quad + (ac)\{-8\,(bdefg) - 4\,(debfg) + 24\,(bdgef)\}\\
&\quad + (ad)\{-4\,(bcefg) + 16\,(cebfg)\} + (abc)\{36\,(defg) - 4\,(efdg)\}\\
&\quad + (abd)\{8\,(cfeg) - 16\,(cefg)\} + (abe)\{8\,(cfdg) - 24\,(cdfg)\}\\
&\quad + (ace)\{16\,(bdfg) - 20\,(bfdg)\}\\
&\quad + (abc)\{-4\,(ef)(dg) + 4\,(df)(eg) - 16\,(de)(fg)\}\\
&\quad + (abd)\{28\,(ce)(fg) - 12\,(cf)(eg)\}\\
&\quad + (abe)\{-2\,(cd)(fg) - 8\,(cg)(df) - 14\,(cf)(dg)\}\\
&\quad + (ace)\{-12\,(bd)(\ g) + 2\,(bg)(df) + 6\,(bf)(dg)\}\big].
\end{aligned}
$$

Now, if we write

$$
\begin{aligned}
(cdefg)'' = (cdefg) + \tfrac{1}{2}\,|CDEFG|&\\
+ \{(\mathrm{cdefg})\}\big[-\tfrac{1}{3}(cd)(efg) + \tfrac{1}{6}\,(ce)(dfg)\big]&,
\end{aligned}
$$

we shall have $\qquad (cdefg)'' = (cegdf)''$

and $\qquad \{\mathrm{e,\ f,\ g}\}\,(cdefg) = 0,$

so that the reductions will be much simplified. For the same reason we will use the form

$$(defg)' = (defg) - \tfrac{1}{6}\,\{(de)(fg) + (df)(ge) + (dg)(ef)\}.$$

With this notation we obtain

$$
\begin{aligned}
&336\,(abcdefg)\\
&= \{(\mathrm{abcdefg}),\ (\mathrm{bg})(\mathrm{cf})(\mathrm{de})\}\big[-20\,(ab)\,|CDEFG| - 2\,(ac)\,|BDEFG|\\
&\quad + 10\,(ad)\,|BCEFG| + (ab)\,\{40\,(cdefg)'' - 8\,(cdfge)'' - 8\,(cedfg)''\}\\
&\quad + (ac)\,\{-8\,(bdefg)'' - 4\,(debfg)'' + 24\,(bdgef)''\}\\
&\quad + (ad)\,\{-4\,(bcefg)'' + 16\,(cebfg)''\}\\
&\quad + (abc)\,\{36\,(defg)' - 4\,(efdg)'\} + (abd)\,\{8\,(cfeg)' - 16\,(cefg)'\}\\
&\quad + (abe)\,\{8\,(cfdg)' - 24\,(cdfg)'\} + (ace)\,\{16\,(bdfg)' - 20\,(bfdg)'\}\\
&\quad + \tfrac{28}{3}\,(abc)\,\{(de)(fg) + (df)(ge) + (dg)(ef)\}\\
&\quad - \tfrac{28}{3}\,(abd)\,\{(ce)(fg) + (cf)(ge) + (cg)(ef)\}\\
&\quad + \tfrac{14}{3}\,(abe)\,\{5\,(cd)(fg) - (cf)(gd) - (cg)(df)\}\\
&\quad + \tfrac{28}{3}\,(ace)\,\{-2\,(bd)(fg) + (bf)(dg) + (bg)(df)\}\big]. \qquad \text{(XIV.)}
\end{aligned}
$$

If we take this value of $(abcdefg)$ and substitute in the three identities degree 7, viz.,

$$[abcd...] = R, \quad [abc, de, f, g,] = R, \quad [ab, cd, ef, g,] = R,$$

we shall obtain all the syzygies of this degree. Owing to the form of the expression obtained for $(abcdefg)$, it is only necessary to take one equation of each type. Take, then, the equation

$$[abc, de, f, g,] = R\,;$$

this may be written

$$\begin{aligned}\{(\mathrm{fg}),\,(\mathrm{ac})(\mathrm{de})\}\Big[&2\,(abcdefg)+(abcfdeg)+\tfrac{1}{4}\,(fg)\,|ABCDE|\\ &-\tfrac{1}{2}\,(fg)(abcde)''-(de)(abcfg)''-(abc)(defg)'-(deg)(abcf)'\\ &-\tfrac{1}{6}\,(abc)\,\{(de)(fg)+2\,(df)(eg)\}\\ &-\tfrac{1}{12}\,(fg)\,\{4\,(abe)(cd)-2\,(abd)(ce)+4\,(ade)(bc)-2\,(acd)(be)\\ &\qquad\qquad-(bde)(ac)\}\\ &-\tfrac{1}{6}\,(de)\,\{2\,(abg)(cf)+4\,(afg)(bc)-2\,(acf)(bg)-(bfg)(ac)\}\\ &-\tfrac{1}{6}\,(deg)\,\{(ab)(cf)+(ac)(fb)+(af)(bc)\}\Big]=0.\end{aligned}$$

Hence, putting in the values of the first two forms,

$$\begin{aligned}\{(\mathrm{fg}),\,(\mathrm{ac})(\mathrm{de})\}\Big[&(fg)\,\{-3\,(abcde)''-(acdbe)''-8\,(acbed)''\}\\ &+(bf)\,\{8\,(cdaeg)''+6\,(daceg)''+4\,(adecg)''\}\\ &+(af)\,\{-12\,(ecbdg)''-4\,(bdecg)''+2\,(debcg)''+2\,(ebdcg)''\}\\ &+(ef)\,\{4\,(cbadg)''+6\,(cadbg)''+10\,(adcbg)''-8\,(bacdg)''\}\\ &-9\,(ac)(dbefg)''-9\,(de)(abcfg)''+18\,(cd)(abefg)''\\ &+(ab)\,\{8\,(cdefg)''-2\,(cedfg)''\}+(eb)\,\{8\,(acdfg)''-2\,(cadfg)''\}\\ &-3\,(bfg)(adce)'+6\,(afg)\,\{(cdbe)'-(cdeb)'\}\\ &+(efg)\,\{10\,(abcd)'+2\,(acdb)'\}-9\,(acf)(dbeg)'-9\,(def)(abcg)'\\ &+6\,(abf)(cdeg)'+2\,(ebf)\,\{(cadg)'-(cdag)'\}\\ &+4\,(aef)\,\{(cbdg)'-(cdbg)'\}+(adf)\,\{-14\,(bceg)'-10\,(becg)'\}\\ &+6\,(abc)(dfeg)'+6\,(dbe)(afcg)'-3\,(abe)(cfdg)'\\ &-9\,(abd)(cfeg)'\Big]=0. \qquad \text{(XV.)}\end{aligned}$$

This may be simplified thus:—Interchange a and e, c and d, and subtract the resulting from the original equation,

$$\begin{aligned}\{(\mathrm{fg}), (\mathrm{ac})(\mathrm{de})\}\big[&(fg)\{-2(abcde)''+2(acdbe)''\}\\ &+(bf)\{2(daceg)''-2(adecg)''\}+(af)\{4(becdg)''-4(debcg)''\}\\ &+(ef)\{4(cabdg)''-4(badcg)''\}+(afg)\{4(cbde)'-4(cdeb)'\}\\ &+(efg)\{4(dcab)'-4(dbca)'\}+(abf)\{4(cdeg)'-4(cedg)'\}\\ &+(ebf)\{4(dacg)'-4(dcag)'\}+4(aef)\{(dcbg)'-(cdbg)'\}\\ &+(adf)\{4(becg)'-4(bceg)'\}\big]=0.\end{aligned} \qquad \text{(XVI.)}$$

Now, let the expression

$$\begin{aligned}&(gf)(abcde)''+(af)(gbdce)''+(bf)(agced)''+(cf)(adgbe)''\\ &\quad+(df)(abegc)''+(ef)(acbdg)''-(agf)(bced)'-(bgf)(cdae)'\\ &\quad-(cgf)(deba)'-(dgf)(eacb)'-(egf)(abdc)'-(abf)(dcge)'\\ &\quad-(acf)(egdb)'-(adf)(cebg)'-(aef)(gbcd)'-(bcf)(edga)'\\ &\quad-(bdf)(agec)'-(bef)(dacg)'-(cdf)(gbae)'-(cef)(adbg)'\\ &\quad-(def)(gcba)'\equiv\phi(abcde, g, f);\end{aligned}$$

then $\qquad \phi(abcde, f, g)=\phi(bcdea, f, g)=\phi(edcba, f, g).$

Then equation (XVI.) is equivalent to

$$\phi(abcde, g, f)+\phi(abcde, f, g)=\phi(acdbe, g, f)+\phi(acdbe, f, g).$$

Operate on this equation with the substitutions (bcd), (abcde), $(\mathrm{abcde})^3$, $(\mathrm{abcde})^4$ respectively; and we obtain

$$\begin{aligned}\phi(abcde, g, f)+\phi(abcde, f, g)&=\phi(acdbe, g, f)+\phi(acdbe, f, g)\\ &=\phi(adbce, g, f)+\phi(adbce, f, g)\\ &=\phi(acedb, g, f)+\phi(acedb, f, g)\\ &=\phi(abecd, g, f)+\phi(abecd, f, g)\\ &=\phi(acbed, g, f)+\phi(acbed, f, g)\\ &=0,\end{aligned}$$

since the sum of the above six expressions is zero.

Consider then the expression $\phi(abcde, g, f)$; it may be written

$$\begin{aligned}\{(\mathrm{abcde})\}\big[\tfrac{1}{5}(fg)(abcde)''+(af)(gbdce)''-(agf)(bced)'\\ -(abf)(dcge)'-(acf)(egdb)'\big].\end{aligned}$$

But it might equally well be written

$$\{(\mathrm{gbdce})\}\left[\tfrac{1}{5}(af)(gbdce)''+(gf)(abcde)''-(gaf)(bced)'\right.$$
$$\left.-(gbf)(cdae)'-(gdf)(eacb)'\right].$$

Hence $\phi(abcde, g, f) = \phi(gbdce, a, f) = \phi(agced, b, f) = \dots$.

We have proved that

$$\{(\mathrm{fg})\}\,\phi(abcde, g, f) = 0;$$

hence also

$$\{(\mathrm{fa})\}\,\phi(abcde, g, f) = 0,\quad \{(\mathrm{fb})\}\,\phi(abcde, g, f) = 0,\quad \&c.$$

Suppose, now, we express $\phi(abcde, g, f)$ in linear symbols, the quartics being α_x, β_x^4, γ_x^4, δ_x^4, ϵ_x^4, ζ_x^4, η_x^4; then, if ζ and η be interchanged in ϕ and the result added to the original form, we obtain identically zero. Hence, if we put $\zeta_1/\zeta_2 = \eta_1/\eta_2$ in ϕ, the result is zero. Similarly, if we give ζ_1/ζ_2 any one of the six values α_1/α_2, β_1/β_2, &c., ϕ vanishes. Now, ϕ/ζ_2^4 is a rational integral function of ζ_1/ζ_2, of degree 4, and it vanishes for six values of the variable; hence $\phi \equiv 0$.

The syzygy may be written symmetrically thus

$$\{(\mathrm{abcde}),(\mathrm{gbdce})\}\left[(gf)(abcde)''-\tfrac{5}{2}(agf)(bced)'\right] = 0 \quad \text{(XVII.)}$$

or

$$\{(\mathrm{abcde}),(\mathrm{gbdce})\}\left[(gf)(abcde)-\tfrac{5}{2}(agf)(bced)-\tfrac{5}{3}(gf)(ab)(cde)\right.$$
$$+\tfrac{5}{6}(gf)(ac)(bde)+\tfrac{5}{6}(agf)(bc)(ed)$$
$$\left.+\tfrac{5}{12}(agf)(be)(cd)\right] = 0.$$

All other syzygies degree 7 are reducible to this one. In order to prove this, it is well to further abbreviate the notation. Thus, let

$$(gf)(abcde)_1 \equiv (gf)(abcde)''-\{(\mathrm{abcde})\}\left[\tfrac{1}{2}(agf)(bced)'\right].$$

Then the syzygy just obtained becomes

$$\{(\mathrm{abcde})\}\left[\tfrac{1}{5}(gf)(abcde)_1+(af)(gbdce)_1\right] = 0.$$

The syzygy

$$(fg)\{(becda)_1+(bedca)_1\}+(fe)\{(gbdac)_1+(gbdca)_1\}$$
$$+(fd)\{(gcbea)_1+(gebca)_1\}+(fc)\{(egdba)_1+(gedba)_1\}$$
$$+(fa)\{(gdecb)_1+(gedcb)_1\}+(eg)\{(fbcad)_1+(fbcda)_1\}$$
$$+(gc)\{(fdbea)_1+(febda)_1\}+(gd)\{(efcba)_1+(fecba)_1\}$$
$$+(ga)\{(fcedb)_1+(fecdb)_1\}+(ed)\{(gabfc)_1+(gfbac)_1\}$$
$$+(ea)\{(gfcbd)_1+(fgcbd)_1\}+(ec)\{(gdfab)_1+(gfdab)_1\}$$
$$+(dc)\{(gbfae)_1+(gbfea)_1\}+(da)\{(gfceb)_1+(gcfeb)_1\}$$
$$+(ac)\{(edgfb)_1+(egdfb)_1\} = 0 \quad \text{(XVIII.)}$$

is also required. It is obtained by subtracting

$$\{(\text{gedca})\}\left[\tfrac{1}{5}(bf)(gedca)_1+(bg)(fecda)_1\right]$$

from the sum of the syzygies

$$\{(\text{gedca})\}\left[\tfrac{1}{5}(bf)(gedca)_1+(fg)(becda)_1\right]=0,$$

$$\{(\text{fecda})\}\left[\tfrac{1}{5}(bg)(fecda)_1+(bf)(gedca)_1\right]=0, \text{ \&c.}$$

First consider equation (XV.). We obtained (XVI.) from it by interchanging a and e, c and d, and subtracting the resulting from the original equation. If we now add, instead of subtracting, and discuss the result, we shall have completely discussed (XV.). Hence we need only consider the result of operating with $\{(\text{ae})(\text{cd})\}$ on (XV.). *i.e.*,

$$\begin{aligned}\{(\text{fg}),\ (\text{ac})(\text{de}),\ (\text{ae})(\text{cd})\}\big[&(fg)\{-4(bcdea)_1-8(bedac)_1\}\\ &+(bf)\{8(cdaeg)_1+10(daceg)_1\}\\ &+(af)\{4(bcdeg)_1+8(ebdcg)_1+12(debcg)_1-20(ecbdg)_1\\ &\qquad -4(bdecg)_1\}\\ &+(ab)\{16(cdefg)_1-4(cedfg)_1\}-18(ac)(dbefg)_1\\ &+18(ae)(cbdfg)_1\big]=0. \qquad\text{(XIX.)}\end{aligned}$$

From $\{(\text{cdefg})\}\left[\tfrac{1}{5}(ab)(cdefg)_1+(bc)(adfeg)_1\right]=0,$

we obtain

$$\{(\text{fg}),\ (\text{ac})(\text{de}),\ (\text{ae})(\text{cd})\}\left[4(ab)(cdefg)_1+2(bf)(cdaeg)_1\right]=0.$$

Similarly,

$$\{(\text{fg}),\ (\text{ac})(\text{de}),\ (\text{ae})(\text{cd})\}\left[4(ab)(cedfg)_1+2(bf)(ceadg)_1\right]=0.$$

Also $\{(\text{fg}),\ (\text{ac})(\text{de}),\ (\text{ae})(\text{cd})\}\big[(bf)(daceg)_1+(fg)(bdcae)_1$

$$+(fa)\{2(edbcg)_1+2(bdceg)_1\}\big]=0$$

and $\{(\text{fg}),\ (\text{ac})(\text{de}),\ (\text{ae})(\text{cd})\}\big[(bf)(ceadg)_1+(fg)(deacb)_1$

$$+(fa)\{2(dbegc)_1+2(cedbg)_1\}\big]=0.$$

Using these four equations, (XIX.) becomes

$$\begin{aligned}\{(\text{fg}),\ (\text{ac})(\text{de}),\ (\text{ae})(\text{cd})\}\big[&(fg)\{-2(bedca)_1-14(bcdea)_1-8(bedca)_1\}\\ &+(af)\{4(ebdcg)_1+4(bcdeg)_1-8(cedbg)_1-8(debcg)_1\\ &\qquad -20(dcebg)_1-20(ecbdg)_1\}\\ &-18(ac)(dbefg)_1+18(ae)(cbdfg)_1\big]=0. \qquad\text{(XX.)}\end{aligned}$$

But, from (XVIII.),

$$\{(fg),\ (ac)(de),\ (ae)(cd)\}\left[2\,(fg)(bedca)_1-4(af)\{(becdg)_1+(dbceg)_1\}\right.$$
$$\left.-2\,(ac)(dbefg)_1-2\,(ad)(cbefg)_1+4\,(ae)(cbdfg)_1\right]=0.$$

Then (XX.) is merely the sum of numerical multiples of this, and two other equations similarly obtained, after interchanging c and e, and c and a, respectively in (XVIII.).

Take next the equation

$$[b,\ cd,\ ef,\ ga,]=R;$$

this may be written

$$\{(ce)(df),\ (ceg)(dfa)\}\left[6\,(abcdefg)-6\,(ef)(abcdg)-3\,(abg)(cdef)\right.$$
$$\left.+3\,(abg)(cd)(ef)\right]=0,$$

or, putting in the value of $(abcdefg)$,

$$\{(ce)(df),\ (ceg)(dfa),\ (cd)(ef)(ga)\}\left[(bc)\{20\,(defga)_1+16\,(deafg)_1\right.$$
$$+12\,(dfgea)_1\}$$
$$+(cd)\{-20(befga)_1+8\,(beafg)_1\}+(ce)\{24(bgfda)_1+12(bfdga)_1\}$$
$$\left.+(cf)\{4(bdega)_1+48\,(begda)_1-8\,(bgdea)_1+4\,(bdgae)_1\}\right]=0.$$

Eliminate the terms $(bc)(\ldots)_1$, from this by means of (XVII.), and the equation reduces to

$$\{(ce)(df),\ (ceg)(dfa),\ (cd)(ef)(ga)\}\left[20\,(cd)\{(begaf)_1-(befga)_1\}\right.$$
$$\left.+20\,(ce)\{2\,(begda)_1-(bgdea)_1-(bdgae)_1\}\right]=0.$$

And this equation is merely the result of operating on (XVIII.) with $\{(ce)(df),\ (ceg)(dfa),\ (cd)(ef)(ga)\}$. The equation

$$[b,\ cd,\ ef,\ ga,]=R,$$

then, gives no fresh syzygy.

We may proceed in exactly the same way with the equation

$$[abcd\ldots]=R,$$

with the same result. But the following is better. Suppose the quartics $e_{x^2}^2$, $f_{x^2}^2$, $g_{x^2}^2$, to be all the same; and in the invariant forms denote them all by the same letter e. (The form of the equation under consideration is practically equivalent to this supposition.)

Then $[abc,\ de,\ e,\ e,]=2\,(abcdeee)+2\,(abcedee)+2\,(abceede)$,

$$[abc,\ ed,\ e,\ e,]=2\,(abcedee)+2\,(abceede)+2\,(abceeed),$$
$$[abc,\ ee,\ d,\ e,]=2\,(abcdeee)+2\,(abceeed)$$
$$+(abcedee)+(abceede);$$

therefore $2\,[abc,\, ee,\, d,\, e,] + [abc,\, de,\, e,\, e,] - 2\,[abc,\, ed,\, e,\, e,]$

$$= 6\,(abcdeee) = [abcd\ldots].$$

If the values of these expressions in terms of irreducible types be substituted, the equation is still identically true. Hence the equation

$$[abcd\ldots] = R$$

can give us nothing more than is given by

$$[abc,\, de,\, f,\, g,] = R.$$

The above proof holds for types of any degree, and so we conclude that this form of equation never requires special consideration.

The expression for $(abcdefg)$ may be simplified with the help of (XVII.). Operating with $\{(\mathrm{abcdefg}),\, (\mathrm{bg})(\mathrm{cf})(\mathrm{de})\}$ on the various syzygies of the form (XVII.), we obtain five independent relations; by means of these (XIV.) reduces to

$$36\,(abcdefg)$$

$$\begin{aligned}
&= \{(\mathrm{abcdefg}),\, (\mathrm{bg})(\mathrm{cf})(\mathrm{de})\}\,\Big[6\,(ab)(cdefg)_1 - 3\,(ac)(bdefg)_1 \\
&\quad - 3\,(ad)(cbefg)_1 - \tfrac{15}{7}\,(ab)\,|CDEFG| - \tfrac{3}{14}\,(ac)\,|BDEFG| \\
&\quad + \tfrac{15}{14}\,(ad)\,|CBEFG| + (abc)\,\{6\,(defg)' - \tfrac{9}{2}\,(degf)'\} \\
&\quad + (abd)\,\{6\,(cfeg)' - 3\,(cefg)'\} - 3\,(abe)(cdfg)' \\
&\quad + (ace)\,\{3\,(bdfg)' - \tfrac{3}{2}\,(bfdg)'\} + (abc)\{(de)(fg) + (df)(ge) + (dg)(ef)\} \\
&\quad - (abd)\,\{(ce)(fg) + (cf)(ge) + (cg)(ef)\} \\
&\quad + (abe)\,\{\tfrac{5}{2}\,(cd)(fg) - \tfrac{1}{2}\,(cf)(dg) - \tfrac{1}{2}\,(cg)(df)\} \\
&\quad + (ace)\,\{-2\,(bd)(fg) + (bf)(gd) + (bg)(df)\}\Big] \\
&= \{(\mathrm{abcdefg}),\, (\mathrm{bg})(\mathrm{cf})(\mathrm{de})\}\,\Big[6\,(ab)(cdefg) - 3\,(ac)(bdefg) \\
&\quad - 3\,(ad)(cbefg) + 6\,(abc)(defg) - 6\,(abd)(cefg) - 3\,(abe)(cdgf) \\
&\quad - 3\,(ace)(bfdg) + (abc)\,\{-(dg)(ef) - 4\,(de)(fg) + 2\,(df)(eg)\} \\
&\quad + (abd)\,\{4\,(ce)(fg) + 4\,(cg)(ef) - 2\,(cf)(ge)\} \\
&\quad + (abe)\,\{-(cd)(fg) - (cg)(df) + 2\,(cf)(dg)\} \\
&\quad + \tfrac{1}{2}\,(ace)\,\{(bd)(fg) + (bf)(gd) + (bg)(df)\} + \tfrac{6}{7}\,(ab)\,|CDEFG| \\
&\quad - \tfrac{12}{7}\,(ac)\,|BDEFG| + \tfrac{18}{7}\,(ad)\,|BCEFG|\Big].
\end{aligned}$$

5. From (XVII.), syzygies of higher degree may be deduced. Thus, if we operate on (XVII.) with

$$[gh][hk][kl]\left[l_0\frac{\partial}{\partial g_0}+l_1\frac{\partial}{\partial g_1}+l_2\frac{\partial}{\partial g_2}\right],$$

we obtain at once $$(fghkl)(abcde)=\Sigma P_3,$$

where ΣP_3 is used, as before, to denote a sum of products of forms, there being at least three forms in each product. Again, by means of syzygies deduced from (XVII.) it is possible to show that $I_4I_5=\Sigma P_3$ Thus, operate on (XVII.) with $[hg]\left[h_0\frac{\partial}{\partial g_0}+h_1\frac{\partial}{\partial g_1}+h_2\frac{\partial}{\partial g_2}\right]$;

$$2\,(fgh)(abcde)+\{(\mathrm{abcde})\}\big[(af)\{(bdcegh)+(bdcehg)\}+(agfh)(bced)$$
$$-(abf)\{(edcgh)+(edchg)\}-(acf)\{(dbegh)+(dbehg)\}\big]=\Sigma P_3.$$

Operate on this with $\{\mathrm{f, c, d, e}\}'$; then

$$\{\mathrm{f,c,d,e}\}'\,\{\mathrm{g,h}\}\big[(fgh)(abcde)+\tfrac{1}{2}(af)\{|BEGD,\,CH|+|CHBE,GD|\}$$
$$+\tfrac{1}{2}(bf)\{|CAGE,\,DH|+|DHCA,\,GE|\}\big]=\Sigma P_3.$$

Hence

$$\{\mathrm{f, c, d, e}\}'\{\mathrm{a, b}\}'\{\mathrm{g, h}\}\big[(fgh)(abcde)$$
$$+\tfrac{2}{3}(af)\{|CEGD,\,BH|+|BECD,\,GH|\}\big]=\Sigma P_3;$$

and therefore

$$\{\mathrm{f, c, d, e}\}'\{\mathrm{a, b}\}'\{\mathrm{g, h}\}\big[(fgh)(abcde)+\tfrac{1}{6}(ag)\,|CEFD,\,BH|$$
$$+\tfrac{1}{3}(agh)\,|BECDF|\big]=\Sigma P_3.$$

Operate now with $[fk]\left[k_0\frac{\partial}{\partial f_0}+k_1\frac{\partial}{\partial f_1}+k_2\frac{\partial}{\partial f_2}\right]$, and we obtain

$$\{\mathrm{c, d, e}\}'\{\mathrm{a, b}\}'\{\mathrm{g, h}\}\big[-(fgkh)(abcde)$$
$$+(cgh)\{(abfked)+(abkfed)+(abdkfe)+(abdfke)+(abedfk)+(abedkf)\}$$
$$+\tfrac{4}{3}(agh)\,|BECD,\,FK|\big]=\Sigma P_3.$$

Or, since $(cgh)\,|AKEB,\,DF|-(dgh)\,|AKEB,\,CF|=\Sigma P_3$,

$$\{\mathrm{c, d, e}\}'\{\mathrm{a, b}\}'\{\mathrm{f, k}\}\big[-(fgkh)(abcde)+\tfrac{4}{3}(agh)\,|BECD,\,FK|$$
$$+(cgh)\{|BEDF,KA|+|BFED,KA|+|EDKA,BF|$$
$$+|EKAD,FB|\}\big]=\Sigma P_3;$$

and therefore

$$\{\mathrm{c, d, e}\}'\{\mathrm{a, b}\}'\{\mathrm{f, k}\}\big[-(fgkh)(abcde)\big]=\Sigma P_3.$$

Hence $(fgkh)\{(abcde)+(abdec)+(abecd)\} = \Sigma P_3$.

Similarly, $(fgkh)\{(abcde)+(cabde)+(bcade)\} = \Sigma P_3$;

therefore $(fgkh)\{(abecd)-(abdce)\} = \Sigma P_3$,

and hence $(fgkh)(abcde) = \Sigma P_3$

Thus the product of any pair of the invariant types I_4, I_5, I_6—with the exception of $I_4 . I_4$—is expressible in the form ΣP_3. But the product of I_2 or I_3 and any other invariant type is not so expressible.

Covariant syzygies may be obtained from (XVII.) by means of combinations of the operations just used with operations of the forms

$$\left[x_2^2 \frac{\partial}{\partial a_0} - x_1 x_2 \frac{\partial}{\partial a_1} + x_1^2 \frac{\partial}{\partial a_2}\right]^2$$

and $$\left[(-a_0 x_1 x_2 - a_1 x_2^2)\frac{\partial}{\partial b_0} + (a_0 x_1^2 - a_2 x_2^2)\frac{\partial}{\partial b_1} + (a_1 x_1^2 + a_2 x_1 x_2)\frac{\partial}{\partial b_2}\right]^2.$$

The search for covariant syzygies is, however, much simplified by noticing that each of the irreducible types, ${}_2C_4$, ${}_2C_5$, ${}_4C_3$, ${}_4C_4$, ${}_6C_2$, ${}_6C_3$, may be expressed in the form $R + \Sigma J$, where J denotes a Jacobian, and R a sum of products of concomitants; and therefore the product of any pair of these types may be expressed in the form ΣP_3.

On Quantitative Substitutional Analysis

By A. YOUNG

From any function P of n variables may be obtained $n!$ functions, not necessarily all different, by permuting the variables in P in all possible ways; or, what is the same thing, by operating on P with each of the $n!$ substitutions of the symmetric group of the variables. It frequently happens that between these functions linear relations with constant coefficients exist; such may be written

$$(\lambda_1 + \lambda_2 s_2 + \lambda_3 s_3 + \ldots)\, P = 0,$$

$\lambda_1, \lambda_2, \ldots$ being numbers positive or negative, integral or fractional, and $s_2, s_3, \ldots$ substitutions belonging to the symmetric group of the variables. The words "substitutional relation" will be used to denote a relation such as that just written down; and the expression "substitutional equation" will be used for the same relation when P is an unknown function for which this relation is true. The simplest form of such a relation is

$$(1-s)\,P = 0,$$

which merely implies that P is unaltered by the substitution s. This is dealt with in the theory of substitutions. The main object of the present paper is the discussion of single equations, such as that written down above, or of simultaneous systems of such equations, with a view to their solution; further, of the discussion of equations of the form

$$(\lambda_1 + \lambda_2 s_2 + \lambda_3 s_3 + \ldots)\,P = R,$$

where $\lambda_1, \lambda_2, \ldots, s_2, s_3, \ldots$ are defined as above, and R is a known function; these equations are also to be included in the term "substitutional equations." It will be seen, moreover, that the right-hand sides of such equations, when a single equation, or else a simultaneous system, is under consideration, are subject to restrictions, in that they have in general to satisfy certain substitutional relations.

The problem proposed is not a purely hypothetical one. In a paper on "The Irreducible Concomitants of any Number of Binary Quartics,"* I have shown that there is one type of concomitant to be discussed for each degree and order; and that such a type satisfies certain substitutional equations, the solution of which enables us to find how many concomitants of that type for a definite number of quartics are irreducible, and which these are. The equations were there discussed, and the irreducible system for any number of quartics was found. Thus, using the notation of that paper, the invariant type degree 6 may be written $(abcdef)$; it satisfies the equations

$$(abcdef) = (bcdefa) = (afedcb),$$

$$(abcdef) + (abcfde) + (abcefd) + (abcdfe) + (abcfed) + (abcedf) = R,$$

$$(abcdef) + (abfcde) + (abefcd) + (abcdfe) + (abecdf) + (abfecd) = R,$$

where R stands for certain reducible terms, with the form of which

* *Proc. Lond. Math. Soc.*, Vol. xxx., p. 290.

we are not concerned. The other equations satisfied by this type are a necessary consequence of the four written down.

Later, in a paper on "The Invariant Syzygies of Lowest Degree for any Number of Quartics,"* I proved that the substitutional equations satisfied by the quartic types gave all the syzygies between quartic concomitants; but here the form of the reducible terms on the right-hand sides of the equations had to be included in the discussion. The equations for invariant types up to and including degree 7 were discussed, with the result that no invariant syzygies existed of degree less than 7, and that the syzygies of degree 7 could all be obtained from one definite form. Incidentally, the method of discussing the equations with a view simply to finding the irreducible system was somewhat improved; and a theorem connected with substitutional analysis was proved, which has been generalized here, § 8.

The term "substitutional expression" is used to denote an expression of the form

$$\lambda_1+\lambda_2 s_2+\lambda_3 s_3+\ldots+\lambda_\rho s_\rho,$$

where λ_1, λ_2, ... are numerical constants (positive or negative) and s_2, s_3, ... are substitutions. It is shown, to start with, that the solution of substitutional equations, so far as rational integral algebraic functions are concerned, may be made to depend on the finding of substitutional expressions which satisfy the equations in virtue of the multiplication table of the group to which all the substitutions belong. The first seven paragraphs of this paper are concerned with substitutional equations; in § 9 some examples are given.

The second part of the paper has to do with two substitutional identities, one of which is proved in § 13, the other in § 15. By means of relations which are established between substitutional and polar operations on functions of a definite kind, from the first of these a proof of Gordon's series is obtained; from the second Capelli's extension of this series, a theorem due to Peano, and some corollaries concerning substitutional equations. An account of the paper which contains Capelli's theorem is also given, § 11, as there exists a fairly close connexion between the analysis of substitutional and polar operators. With this connexion § 12 has to do; it is somewhat further developed in that part of § 17 which has to do with Capelli's theorem.

For convenience, owing to the quantitative use of the symbols, the

---

* *Proc. Lond. Math. Soc.*, Vol. XXXII., p. 384.

substitutions next a function are regarded as operating on it before those further away, thus

$$s_1 s_2 P = s_1 (s_2 P).$$

To avoid confusion, as the symbol $(abc...)$ is used in two senses, viz., as a substitution and as a concomitant type of a quantic, Roman letters are used when it denotes a substitution, italics when it denotes a type. The usual symbols for a group are used in two senses: first, as a name for the group, and, secondly, to represent the sum of the substitutions of the group. The following notation is made use of:—

$\{s\}$ = the sum of the substitutions of the smallest group including s.

$\{s_1, s_2\}$ = the sum of the substitutions of the smallest group including s_1 and s_2.

$\{G_1, G_2\}$ = the sum of the substitutions of the smallest group having G_1 and G_2 for sub-groups.

{abc...} = the symmetric group of the letters $a, b, c, \ldots$.

{abc...}$'$ = the sum of the substitutions of the alternating group of the letters $a, b, c, \ldots$, minus the substitutions of these letters which do not belong to the alternating group.

The expression {abc...} is sometimes referred to as "the positive symmetric group"; while {abc...}$'$ is called "the negative symmetric group."

The paper has been rewritten and greatly enlarged at the request of the referees; my thanks are due to them—particularly to Prof. Burnside—for many valuable criticisms and suggestions.

1. Consider any rational integral algebraic function P of n variables $a_1, a_2, \ldots, a_n$; its terms may be arranged in sets $P_1, P_2, \ldots, P_m$, such that each set contains all those terms of P, and only those, which are obtainable from some particular term by means of substitutions and of positive or negative numerical factors. And P may be written

$$P = P_1 + P_2 + \ldots + P_m.$$

Now, consider any set P_1; let $A_1 a_1^{a_1} a_2^{a_2} \ldots a_n^{a_n}$ be any term of this set, A_1 being a positive or negative numerical coefficient; then

$$P_1 = (A_1 + A_2 s_2 + A_3 s_3 + \ldots)\, a_1^{a_1} a_2^{a_2} \ldots a_n^{a_n},$$

where $A_1, A_2, \ldots$ are numerical, and $s_2, s_3, \ldots$ substitutions belonging to the symmetric group of the n variables. The effects of substitutions on P_1, and consequently all substitutional properties of P_1, depend partly on the substitutional operator $(A_1+A_2s_2+A_3s_3+\dots)$, partly on the substitutional properties of the term $a_1^{a_1}a_2^{a_2}\dots a_n^{a_n}$. If in this term all the indices $a_1, a_2, \dots, a_n$ are different, we obtain by operating on it with the $n!$ substitutions of the symmetric group $\{a_1a_2\dots a_n\}$ of the variables $n!$ different terms which are connected by no linear relations with constant coefficients. In this case, then, $a_1^{a_1}a_2^{a_2}\dots a_n^{a_n}$ has no substitutional properties, and all the substitutional properties of P_1 are a consequence of the operator

$$(A_1+A_2s_2+A_3s_3+\dots).$$

Suppose next that $a_1=a_2=\dots=a_r=a$, and that $a, a_{r+1}, a_{r+2}, \dots, a_n$ are all different. The substitutional properties of the term

$$a_1^a a_2^a \dots a_r^a a_{r+1}^{a_{r+1}} a_{r+2}^{a_{r+2}} \dots a_n^{a_n}$$

consist solely of the fact that this term belongs to the group $\{a_1a_2\dots a_r\}$. For there result, by operating on it with the $n!$ substitutions of the group $\{a_1a_2\dots a_n\}$, $\frac{n!}{r!}$ different terms between which no linear relations with constant coefficients can exist. The substitutional properties of this term are then identical with those of

$$\{a_1a_2\dots a_r\}\, a_1^{a_1}a_2^{a_2}\dots a_n^{a_n},$$

where all the indices of the a's in $a_1^{a_1}a_2^{a_2}\dots a_n^{a_n}$ are different. Hence all the substitutional properties of P_1 are, in this case, a consequence of the operator when we write, as may be done,

$$P_1=\frac{1}{r!}\left[(A_1+A_2s_2+A_3s_3+\dots)\{a_1a_2\dots a_r\}\right]a_1^a a_2^a\dots a_r^a a_{r+1}^{a_{r+1}}a_{r+2}^{a_{r+2}}\dots a_n^{a_n}$$

$$=(B_1+B_2s_2+B_3s_3+\dots)\,a_1^a a_2^a\dots a_r^a a_{r+1}^{a_{r+1}}a_{r+2}^{a_{r+2}}\dots a_n^{a_n},$$

where

$$\frac{1}{r!}(A_1+A_2s_2+A_3s_3+\dots)\{a_1a_2\dots a_r\}=(B_1+B_2s_2+B_3s_3+\dots),$$

the B's being constants. In exactly the same way, whatever be the equalities amongst the indices in the term $a_1^{a_1}a_2^{a_2}\dots a_n^{a_n}$, a substitutional operator $(B_1+B_2s_2+B_3s_3+\dots)$ may be obtained, such that

$$P_1=(B_1+B_2s_2+B_3s_3+\dots)\,a_1^{a_1}a_2^{a_2}\dots a_n^{a_n},$$

all the substitutional properties of P_1 being a consequence of the operator alone.

Now, owing to the way in which the sets have been chosen, no substitution can change a term of one set into a term of a different set; and there can exist no substitutional relation between different ets. Hence, if P satisfy any substitutional equation

$$(\lambda_1+\lambda_2 s_2+\lambda_3 s_3+\ldots)P=0,$$

where $\lambda_1, \lambda_2, \ldots$ are constants, each set must independently satisfy this equation. And hence each set possesses all the substitutional properties of P.

Theorem.—Every rational integral algebraic function P of n variables may be written in the form $P=\sum_{i=1}^{i=m} P_i$, where P_i possesses all the substitutional properties of P, and possibly others as well. And P_i may be expressed in the form

$$P_i=(A_1^{(i)}+A_2^{(i)}s_2+A_3^{(i)}s_3+\ldots)F_i,$$

where $A_1^{(i)}, A_2^{(i)}, \ldots$ are positive or negative numerical coefficients, $s_2, s_3, \ldots$ are substitutions belonging to the symmetric group of the n variables, and F_i is a rational integral algebraic function of the variables. Further, the substitutional operator

$$(A_1^{(i)}+A_2^{(i)}s_2+A_3^{(i)}s_3+\ldots)$$

is such that all the substitutional properties of P_i are a direct consequence of it.

For example, take the form

$$P=\tfrac{1}{2}a_2-\tfrac{1}{2}a_3+3a_1^2a_2-\tfrac{1}{5}a_2^2a_3-3a_1^2a_3+\tfrac{1}{5}a_2a_3^2;$$

then
$$\begin{aligned}P_1&=\tfrac{1}{2}a_2-\tfrac{1}{2}a_3=\tfrac{1}{2}\{\mathrm{a}_2\mathrm{a}_3\}'a_2=\tfrac{1}{4}\{\mathrm{a}_2\mathrm{a}_3\}'\{\mathrm{a}_1\mathrm{a}_3\}\,a_2\\&=\tfrac{1}{4}\left[1-(\mathrm{a}_2\mathrm{a}_3)+(\mathrm{a}_1\mathrm{a}_3)-(\mathrm{a}_1\mathrm{a}_2\mathrm{a}_3)\right]a_2,\\P_2&=3a_1^2a_2-\tfrac{1}{5}a_2^2a_3-3a_1^2a_3+\tfrac{1}{5}a_2a_3^2\\&=\left[3-\tfrac{1}{5}(\mathrm{a}_1\mathrm{a}_2\mathrm{a}_3)-3(\mathrm{a}_2\mathrm{a}_3)+\tfrac{1}{5}(\mathrm{a}_1\mathrm{a}_3)\right]a_1^2a_2,\\P&=P_1+P_2.\end{aligned}$$

Here P, P_1, P_2 all satisfy the equation

$$\{\mathrm{a}_2\mathrm{a}_3\}P=0;$$

also P_1 satisfies the equation

$$\{a_1 a_2 a_3\}' P_1 = 0.$$

Again, if the substitutional properties of P are completely summed up by saying that P belongs to the group G of order ρ, it is sufficient and more convenient to write

$$P = \frac{1}{\rho} GP,$$

this being, as it is easy to verify, the necessary and sufficient condition that P may belong to the group G of order ρ.

Corollary.—Every rational integral algebraic solution P of a single equation

$$(\lambda_1 + \lambda_2 s_2 + \lambda_3 s_3 + \ldots) P = 0,$$

where $\lambda_1, \lambda_2, \ldots$ are constants, and $s_2, s_3, \ldots$ substitutions belonging to the symmetric group of the variables, of which P is supposed to be a function, or of a simultaneous system of such equations, may be obtained in the form

$$P = \sum_i (A_1^{(i)} + A_2^{(i)} s_2 + A_3^{(i)} s_3 + \ldots) F_i\,;$$

where $A_1^{(i)}, A^{(i)}, \ldots$ are constants, and F_i is a rational integral algebraic function of the variables, the substitutional operator of each term being such that

$$(\lambda_1 + \lambda_2 s_2 + \lambda_3 s_3 + \ldots)(A_1^{(i)} + A_2^{(i)} s_2 + A_3^{(i)} s_3 + \ldots) \equiv 0,$$

in virtue of the multiplication table of the group.

For P may be written in the form

$$P = \sum_i P_i,$$

where $$P_i = (A_1^{(i)} + A_2^{(i)} s_2 + \ldots) F_i,$$

all the substitutional properties of P_i being consequences of the operator, and, further, where P_i possesses all the substitutional properties of P, and hence is a solution of the equation, or system of equations, of which we are supposing P to be a solution. But, since every substitutional property of P_i is a consequence of the operator $(A_1^{(i)} + A_2^{(i)} s_2 + \ldots)$, it follows that

$$(\lambda_1 + \lambda_2 s_2 + \ldots)(A_1^{(i)} + A_2^{(i)} s_2 + \ldots) \equiv 0.$$

2. The applications of our theory at present required are entirely to functions rational integral algebraic in the variables. Consequently,

we may restrict ourselves to the discussion of such functions, and will throughout this paper tacitly assume that the functions considered are of this nature. Nevertheless, should the theorem of the preceding article be true for any kind of function—as seems to me probable—no restrictions as to the nature of the functions considered would be necessary.

In consequence of the corollary just proved, it follows that in order to obtain the solutions of a system of equations of the form

$$(\lambda_1+\lambda_2 s_2+\lambda_3 s_3+\ldots)\,P=0$$

it is only necessary to discover the substitutional expressions

$$(A_1+A_2 s_2+A_3 s_3\ldots)$$

which are such that

$$(\lambda_1+\lambda_2 s_2+\lambda_3 s_3+\ldots)(A_1+A_2 s_2+A_3 s_3+\ldots)\equiv 0,$$

in virtue of the multiplication table of the group. The solution is then a matter of relations between substitutional operators only. We may then proceed thus: Take the sum of all the substitutions of the group concerned with arbitrary coefficients; for brevity we write this S. Then expand the various expressions

$$(\lambda_1+\lambda_2 s_2+\ldots)\,S$$

obtained by substituting S for P in the various equations of the simultaneous system, and in the results equate the coefficient of each substitution to zero. A system of simultaneous linear equations is thus obtained for the arbitrary constants in S. As a rule, all the arbitrary constants cannot be definitely determined; but the result of solving these linear equations and substituting their values in S will be expressible in the form

$$\sum_{j=1}^{j=m} C_j S_j,$$

where C_j is an arbitrary constant and S_j is a substitutional expression containing no arbitrary constant, which is such that the result of substituting S_j for P in each of the substitutional equations is zero, in virtue of the multiplication table of the group. Every solution may then be written in the form

$$P=\sum_i\,[\sum_j C_{j,\,i}\,S_j]\,F_i=\sum_j S_j\,[\sum_i C_{j,\,i}\,F_i]=\sum_j S_j\Phi_j,$$

where $C_{j,\,i}$ is a definite constant

$$\Phi_j=\sum_i C_{j,\,i}\,F_i,$$

and F_i and P are functions of the nature under discussion.

An expression in terms of which every solution can be expressed, such as $\Sigma_j S\ \Phi_j$, we call the complete solution of the system of equations. It will be seen later on that this is not always unique.

It is well to remark that it is not necessary to take S equal to the sum of all the substitutions of the symmetric group of the variables with arbitrary coefficients. It is sufficient that S should contain all the substitutions of the smallest group G which contains all those substitutions which actually occur in the expressions of our equations. For, if $G = 1 + s_2 + \ldots + s_p$, it is well known that it is possible to obtain a table

$$\begin{array}{lllll} 1, & s_2, & s_3, & \ldots, & s_p, \\ \sigma_2, & s_2\sigma_2, & s_3\sigma_2, & \ldots, & s_p\sigma_2, \\ \ldots & \ldots & \ldots & & \ldots \\ \sigma_{p'}, & s_2\sigma_{p'}, & s_3\sigma_{p'}, & \ldots, & s_p\sigma_{p'}, \end{array}$$

such that every substitution of the symmetric group is contained once, and only once, in the table; and, further, that the result of multiplying on the left-hand side any substitution in this table by one of the substitutions in G changes it to another substitution in the same horizontal line. Hence, if S be the sum of all the substitutions of the symmetric group with arbitrary coefficients, the substitutional equations only give relations between the constants in the same horizontal line, and the relations for the various lines are the same.

As an example, consider the equation

$$\{(\mathrm{abcd})\}\, P = 0$$

$$S = A_1 + A_2\,(\mathrm{abcd}) + A_3\,(\mathrm{ac})(\mathrm{bd}) + A_4\,(\mathrm{adcb}).$$

Equating the coefficients in $\{(\mathrm{abcd})\}\, S$ to zero, we obtain

$$A_1 + A_2 + A_3 + A_4 = 0.$$

Hence

$$\begin{aligned} S &= -A_2 - A_3 - A_4 + A_2\,(\mathrm{abcd}) + A_3\,(\mathrm{abcd})^2 + A_4\,(\mathrm{abcd})^3 \\ &= \big[1 - (\mathrm{abcd})\big]\big[-A_2 - A_3\{1 + (\mathrm{abcd})\} - A_4\{1 + (\mathrm{abcd}) + (\mathrm{abcd})^2\}\big]. \end{aligned}$$

And the complete solution is

$$[1 - (\mathrm{abcd})]\, F.$$

3. Consider now a single equation, or a system of equations, of the form

$$[\lambda_1 + \lambda_2 s_2 + \lambda_3 s_3 + \ldots]\, P = R,$$

where, as before, λ_1, λ_2, ... are constants, s_2, s_3, ... substitutions, and R

is a given rational integral algebraic function of the variables. It is, in the first place, to be noticed that the above equation in general implies a restriction on R, viz., that R can be written in the form $[\lambda_1+\lambda_2 s_2+\dots]\,F$, and, as a consequence, satisfies certain substitutional equations. Thus, if $\lambda_1+\lambda_2 s_2+\dots = G$ the sum of the substitutions of a group, R must belong to the group G. Let P_1 be any solution of the equations; then, if P_2 be another solution,

$$[\lambda_1+\lambda_2 s_2+\dots]\,(P_1-P_2)=0.$$

Hence, as in linear differential equations, the work of solution may be divided into two parts. First, any particular solution P_1 is found; and then—what corresponds to the complementary function—the complete solution Q of the system

$$[\lambda_1+\lambda_2 s_2+\dots]\,Q=0.$$

The complete solution—that is, the solution in terms of which every other can be expressed—is then

$$P=P_1+Q.$$

It will be seen later on, in the applications made to the quadratic and quartic invariants, that, in general, R is subject to more conditions than that implied by

$$R=[\lambda_1+\lambda_2 s_2+\dots]\,F$$

when a simultaneous system of such equations is under discussion.

4. It may happen that the only solution of an equation

$$[\lambda_1+\lambda_2 s_2+\dots]\,P=0 \qquad \text{(I.)}$$

is $P=0$. Let G be the group of the substitutions which appear in this equation; then, if s be any substitution of G,

$$s\,[\lambda_1+\lambda_2 s_2+\dots]\,P=0.$$

Operating, then, on (I.) with each of the ρ substitutions of G, where ρ is the order of G, we obtain ρ linear equations with constant coefficients between the ρ quantities

$$P,\ s_2P,\ \dots$$

regarded as independent variables. The necessary and sufficient condition that there may be a solution other than zero is then expressed by the vanishing of a determinant of ρ columns and rows.

If $\qquad \lambda_1+\lambda_2 s_2+\dots = G = 1+s_2+s_3+\dots+s_\rho,$

the sum of the substitutions of a group G of order ρ, the complete solution of (1.) is of the form

$$P = \Sigma (A_1 + A_2 s_2 + \ldots + A_\rho s_\rho)\, F,$$

where
$$G\,(A_1 + A_2 s_2 + \ldots + A_\rho s_\rho) \equiv 0.$$

This gives
$$A_1 + A_2 + \ldots + A_\rho = 0.$$

Hence

$$A_1 + A_2 s_2 + \ldots + A_\rho s_\rho = A_2\,(s_2 - 1) + A_3\,(s_3 - 1) + \ldots + A_\rho\,(s_\rho - 1).$$

Now, let $\sigma_1, \sigma_2, \ldots, \sigma_m$ be any substitutions of G which are not all contained in one of its sub-groups, and hence are sufficient to generate G. Then every substitution s of G can be expressed in the form

$$s = \sigma_{r_1}^{\alpha_1} \sigma_{r_2}^{\alpha_2} \ldots \sigma_{r_k}^{\alpha_k},$$

where $r_1, r_2, \ldots, r_k$ are some of the numbers 1, 2, ..., m, not necessarily all different. But

$$s - 1 = \sigma_{r_1}^{\alpha_1} s' - 1 = (\sigma_{r_1}^{\alpha_1} - 1)\, s' + (s' - 1),$$

where
$$s' = \sigma_{r_2}^{\alpha_2} \ldots \sigma_{r_k}^{\alpha_k},$$

and hence

$$s - 1 = (\sigma_{r_1}^{\alpha_1} - 1)\, \sigma_{r_2}^{\alpha_2} \ldots \sigma_{r_k}^{\alpha_k} + (\sigma_{r_2}^{\alpha_2} - 1)\, \sigma_{r_3}^{\alpha_3} \ldots \sigma_{r_k}^{\alpha_k} + \ldots + (\sigma_{r_k}^{\alpha_k} - 1)$$
$$= (\sigma_1 - 1)\, S_1 + (\sigma_2 - 1)\, S_2 + \ldots + (\sigma_m - 1)\, S_m,$$

where $S_1, S_2, \ldots, S_m$ are substitutional expressions, some of which may be zero, or merely numerical.

Hence

$$A_1 + A_2 s_2 + \ldots + A_\rho s_\rho = (\sigma_1 - 1)\, T_1 + (\sigma_2 - 1)\, T_2 + \ldots + (\sigma_m - 1)\, T_m,$$

where $T_1, T_2, \ldots, T_m$ are substitutional expressions containing the arbitrary constants $A_2, A_3, \ldots, A_\rho$.

Moreover
$$G\,(\sigma - 1) = 0\,;$$

hence the complete solution of the equation

$$GP = 0$$

may be written

$$P = (\sigma_1 - 1)\, F_1 + (\sigma_2 - 1)\, F_2 \ldots + (\sigma_m - 1)\, F_m,$$

$F_1, F_2, \ldots, F_m$ being arbitrary functions.

Similarly, the complete solution of the equation

$$GP = R,$$

R necessarily belonging to the group G, is

$$P = \frac{R}{\rho} + (\sigma_1 - 1) F_1 + (\sigma_2 - 1) F_2 + \ldots + (\sigma_m - 1) F_m,$$

for

$$\frac{1}{\rho} GR = R,$$

and consequently $\frac{R}{\rho}$ is a particular solution.

If, for instance, $G = \{a_1 a_2 \ldots a_n\}$,

any one of the three following expressions may be taken as the complete solution :—

$$\{a_1 a_2\}' F_1 + \{a_1 a_3\}' F_2 + \ldots + \{a_1 a_n\}' F_n,$$

$$\{a_1 a_2\}' F_1 + \{a_2 a_3\}' F_2 + \ldots + \{a_{n-1} a_n\}' F_n,$$

$$\{a_1 a_2\}' F_1 + [1 - (a_1 a_2 a_3 \ldots a_n)] F_2 :$$

an illustration of the remark already made, that it would be found that the complete solution was not always unique.

It follows from the above that the solution of

$$\{G_1, G_2\} P = 0$$

may be written $P = P_1 + P_2$,

where $G_1 P_1 = 0$ and $G_2 P_2 = 0$.

For we may choose substitutions $\sigma_1, \sigma_2, \ldots, \sigma_h$ which generate G_1, and substitutions $\sigma_{h+1}, \sigma_{h+2}, \ldots, \sigma_m$ which generate G_2; these substitutions will then together generate $\{G_1, G_2\}$. The solution of

$$\{G_1, G_2\} P = 0$$

may then be written

$$P = (\sigma_1 - 1) F_1 + \ldots + (\sigma_h - 1) F_h + (\sigma_{h+1} - 1) F_{h+1} + \ldots + (\sigma_m - 1) F_m$$

$$= P_1 + P_2,$$

where $P_1 = (\sigma_1 - 1) F_1 + \ldots + (\sigma_h - 1) F_h$

and $P_2 = (\sigma_{h+1} - 1) F_{h+1} + \ldots + (\sigma_m - 1) F_m$,

and consequently $G_1 P_1 = 0$ and $G_2 P_2 = 0$.

5. When all the substitutions are powers of a single substitution the equations are easy to solve. Consider a single equation, the most general of its kind,

$$\phi(s) P = (A_0 + A_1 s + A_2 s^2 + \ldots + A_{n-1} s^{n-1}) P = 0,$$

where s is a substitution of order n.

We require to find the most general expression

$$\psi(s) = (B_0 + B_1 s + B_2 s^2 + \ldots + B_{n-1} s^{n-1}),$$

which is such that $\phi(s) \psi(s) \equiv 0.$

Now $\phi(x) \psi(x)$ only vanishes when $\phi(x) = 0$, or when $\psi(x) = 0$. Neither of these cases need be discussed here; then the product $\phi(s) \psi(s)$ must vanish solely in consequence of the equation

$$s^n = 1.$$

Hence $\phi(x) \psi(x) = (x^n - 1) \chi(x).$

To find ψ, we then obtain the H.C.F. of $x^n - 1$ and $\phi(x)$, say $\phi_1(x)$; then

$$\psi(x) = \frac{x^n - 1}{\phi_1(x)}.$$

Now, if a is not a root of $x^n - 1 = 0$,

any function Q may be written in the form

$$Q = (s^{n-1} + as^{n-2} + \ldots + a^{n-1}) F = \left[\frac{s^n - a^n}{s - a}\right] F = \left[\frac{1 - a^n}{s - a}\right] F,$$

for $$Q = \frac{s^n - a_n}{1 - a^n} Q = (s^{n-1} + as^{n-2} + \ldots + a^{n-1}) \left(\frac{s - a}{1 - a^n} Q\right).$$

Hence, if $\phi(x) = \phi_1(x)(x - a_1)(x - a_2) \ldots (x - a_r)$,

any solution P of the equation may be written

$$P = \psi(s) F = \left[\frac{s^n - 1}{\phi_1(s)}\right] F$$

$$= \left[\frac{s^n - 1}{\phi_1(s)}\right] \left[\frac{1}{(s - a_1)(s - a_2) \ldots (s - a_r)}\right] E$$

$$= \left[\frac{s^n - 1}{\phi(s)}\right] E,$$

where it is to be understood that the expression $\dfrac{1}{s - a_1}$ is equivalent to $\dfrac{s^{n-1} + as^{n-2} + \ldots + a^{n-1}}{1 - a_1^n}$, when a_1^n is not equal to unity.

It has been tacitly assumed that $\phi(s)$ has no squared factor which is also a factor of s^n-1; if such should occur, we may remove it by adding to $\phi(s)$ a multiple of s^n-1, which is in actual value zero, and then proceed as before. If $\phi(s)$ has no common factor with s^n-1, then $P=0$.

To find a particular solution of

$$\phi(s)\,P = R.$$

The restriction imposed on R by this equation is

$$\frac{s^n-1}{\phi_1(s)}\,R = 0.$$

Hence $$R = \phi_1(s)\,R'.$$

If there is no difficulty in finding R' from this, the particular solution

$$P = \frac{\phi_1(s)}{\phi(s)}\,R'$$

may be taken.

If the form of R' is not at once obvious, the particular solution may be found thus:—

$$\phi_1(s)\,\phi_1(s) = \phi_1(s)\,\phi_1(s) + \lambda\,(s^n-1) = \phi_1(s)\,\phi_2(s),$$

where s^n-1 and $\phi_2(s)$ have no common factor. Then

$$\phi_1(s)\,\phi(s)\,P = \phi_1(s)\,R = \phi_1(s)\left[\frac{\phi(s)}{\phi_1(s)}\,\phi_2(s)\right]P,$$

and $$\left[\frac{\phi_1(s)}{\phi(s)\,\phi_2(s)}\right]R$$

is a solution.

The extension to any set of simultaneous equations involving only powers of s is obvious.

Also it may be seen, in the same way, that the solution of any set of Abelian equations is a matter only of algebra.

Single equations which are not merely formed by the sum of the substitutions of a group, and in which the substitutions are not all contained in an Abelian group, may frequently be solved with the help of the solutions in these two cases. Thus, the solution of the equation $$\{ab\}\,[1+(abcd)+(abcd)^2]\,P = R$$

—which occurs in the reduction of the quartic invariant types—is

$$P = \tfrac{1}{3}\,[1-2\,(abcd)+(abcd)^2+(abcd)^3]\left[\frac{R}{2}+\{ab\}'\,F\right],$$

and R must satisfy the equation

$$\{ab\}' R = 0.$$

6. Consider now two simultaneous equations

$$\{s\}\, P = 0, \quad \{\sigma\}\, P = 0.$$

Then, if

$$s\sigma \equiv \tau_1, \quad \sigma s \equiv \tau_2,$$

$$\sigma\tau_1^a = \tau_2^a\sigma, \quad \tau_1^a = s\tau_2^{a-1}\sigma.$$

Hence, if m be the order of τ_1,

$$\tau_2^m = \sigma\tau_1^m\sigma^{-1} = 1,$$

and the orders of τ_1 and τ_2 must be identical.

Also the expression $(1-\sigma)\{\tau_1\} = (s-1)\{\tau_2\}\,\sigma$.

Hence

$$P = (1-\sigma)\,\{\tau_1\}\,F$$

is a solution of the equations.

Unfortunately this is not always the complete solution, for suppose that

$$s\sigma = \sigma s, \quad s^2 = 1, \quad \sigma^3 = 1\,;$$

then the complete solution may be written

$$P = (1-s)(1-\sigma)\,F;$$

but the expression $(1-\sigma)\{\tau_1\}$ vanishes identically, for $\{\tau_1\}$ is here equal to $\{s, \sigma\}$.

Again, whenever the substitutions s, σ are permutable, the solution

$$P = (1-\sigma)\,\{\tau_1\}\,F,$$

in addition to satisfying the two equations

$$\{s\}\,P = 0, \quad \{\sigma\}\,P = 0,$$

belongs to the group $\{\tau_1\}$, which is not in general the case with the complete solution

$$P = (1-\sigma)(1-s)\,F.$$

However, whenever

$$s^2 = 1 = \sigma^2,$$

the complete solution may be written

$$P = (1-\sigma)\,\{\tau_1\}\,F,$$

for

$$\{s, \sigma\} = \{\sigma, \tau_1\} = \{\sigma\}\,\{\tau_1\},$$

since

$$\tau_1\sigma = s = \sigma\tau_1^{-1}.$$

Hence we may write $S = \Sigma\,(1+A_u\sigma)\,B_u\tau_1^u$,

and find S, so that $\{s\}\,S = 0$ and $\{\sigma\}\,S = 0$.

The second equation gives $A_u = -1$.

Hence $S = (1-\sigma)(B_0 + B_1\tau_1 + \ldots + B_{m-1}\tau_1^{m-1})$

$$= \left[s\,(B_0\tau^{m-1} + B_1 + B_2\tau_2 + \ldots + B_{m-1}\tau_2^{m-2}) - (B_0 + B_1\tau_2 + \ldots + B_{m-1}\tau_2^{m-1})\right]\sigma.$$

The equation $\{s\}\,S = 0$

then gives $B_0 = B_1 = \ldots = B_{m-1}$.

Hence the complete solution is as stated.

A solution of any number of equations

$$\{s_1\}P = 0,\ \{s_2\}\,P = 0,\ \ldots,\ \{s_n\}\,P = 0$$

may then be seen to be

$$P = (1-s_1)\,\{s_2s_1,\ s_3s_1,\ \ldots,\ s_ns_1\}\,F.$$

If each of the substitutions $s_1, s_2, \ldots, s_n$ is of order 2, this is the complete solution. For it can be written in the form

$$P = (1-s_1)\,E,$$

where E is a rational integral algebraic function of the variables, since

$$\{s_1\}\,P = 0;$$

and by what we have seen above E must belong to the group $\{s_2s_1\}$, if

$$\{s_2\}\,P = 0.$$

Hence E must belong to the smallest group containing $s_2s_1, s_3s_1, \ldots, s_ns_1$.

7. It frequently happens that a function is given as belonging to a certain group, besides satisfying certain substitutional equations. Thus, the invariant type degree 5 of a quartic belongs to the group $\{(abcde), (ae)(bd)\}$, and satisfies the equation

$$\{abc\}\,I_5 = R,$$

the other equations which it satisfies being consequences of these facts. Further, in the case of irreducible invariants, we really only require to find the number of invariants of the form I_5 in terms of which the rest can be linearly expressed. In respect to this, we shall prove that:

If M be the number of arbitrary constants in the most general substitutional expression S_1, which may contain all the $n!$ substitutions of the symmetric group of the n variables under consideration, which satisfies the equations

$$G_1 S_1 \equiv 0, \quad G_2 S_1 \equiv r_2 S_1,$$

G_1 and G_2 being groups of orders r_1 and r_2 respectively, and if N be the number of arbitrary constants in the most general substitutional expression S_2 which satisfies the equations

$$G_2 S_2 \equiv 0, \quad G_1 S_2 \equiv r_1 S_2,$$

then
$$M - N = n! \left\{ \frac{1}{r_2} - \frac{1}{r_1} \right\}.$$

Consider S_1, and suppose that at first all the coefficients are arbitrary. Let A_s be the coefficient of s; then the equation

$$G_1 S_1 \equiv 0$$

gives $\frac{n!}{r_1}$ equations of the form

$$\Sigma A_s = 0, \qquad \text{(I.)}$$

and in no two of these equations does the same coefficient occur. Now, if σ be any substitution of G_2, it follows that, since S_1 has G_2 for a factor,

$$A_{\sigma s} = A_s.$$

Owing to this, there are only $\frac{n!}{r_2}$ different coefficients; and, if this be taken into account, the equations (I.) are not all independent. Let $T = 0$ be any relation between these equations written out in full; then this is an identity solely on account of the equations $A_{\sigma s} = A_s$. Hence, if substitutions applied to T be supposed to operate on the suffixes of the A's, we have the equation

$$G_2 T = 0.$$

And, further, from the form of equations (I.),

$$G_1 T = r_1 T,$$

for $T = 0$ is a relation between different equations (I.). If, then, T

be what T becomes when for each A_s we write s, T' will satisfy the equations for S_2. Hence, for every relation between the $\frac{n!}{r_1}$ equations to determine the $\frac{n!}{r_2}$ unknown constants in S_1, there is an expression of exactly the same form which satisfies the equations for S_2. Conversely, every solution of the equations for S_2 will give such a relation between the equations for the unknown constants in S_1. Hence the number of independent relations between the equations (I.) is N; consequently, the number of arbitrary constants left in S_1 when all the equations are satisfied is

$$M = \frac{n!}{r_2} - \left(\frac{n!}{r_1} - N\right);$$

and therefore $$M - N = n!\left(\frac{1}{r_2} - \frac{1}{r_1}\right).$$

Further, the number of those functions obtained from P by permuting the n variables, in terms of which the $n!$ possible functions thus obtained from P may be linearly expressed when P belongs to the group G_2 and satisfies

$$G_1 P = 0,$$

is equal to M, the number of arbitrary constants in the most general substitutional expression S_1 for which

$$G_1 S_1 \equiv 0, \quad G_2 S_1 \equiv r_2 S_1.$$

For, if P_s be the function obtained from P by operating on it with the substitution s, exactly the same linear equations exist between the functions P_s as between the coefficients A_s in S_1. Hence the number of linearly independent functions P_s is the same as the number of arbitrary coefficients in S_1.

8. If a function P satisfy each of the equations

$$\{a_1 a_2\} P = 0, \quad \{a_1 a_3\} P = 0, \quad \ldots, \quad \{a_1 a_n\} P = 0,$$

it is merely changed in sign when operated upon by any transposition of the letters $a_1, a_2, \ldots, a_n$. The complete solution of these equations is then

$$P = \{a_1 a_2 \ldots a_n\}' F.$$

The function P is an alternating function, and may be written, as is well known,

$$P = \sqrt{\Delta}\,\{\mathrm{a}_1 \mathrm{a}_2 \ldots \mathrm{a}_n\}\, F',$$

where Δ is the product of the squares of the differences of the letters $a_1, a_2, \ldots, a_n$.

Hence, if P is of degree less than $n-1$ in any one letter, it must be zero. Hence also, if Q be any rational integral function of degree $<n-1$ in each of its variables $a_2, a_2, \ldots, a_n$, it satisfies the equation

$$\{\mathrm{a}_1 \mathrm{a}_2 \ldots \mathrm{a}_n\}'\, Q = 0.$$

In this connection should be mentioned the following propositions already given for the quartic in my paper "On the Invariant Syzygies of Lowest Degree for any Number of Binary Quartics," viz.,

If P be a rational integral function homogeneous and linear in the coefficients of m binary n-ics,

$$(A_0^{(1)}, A_2^{(1)}, \ldots, A_n^{(1)} \between x_1, x_2)^n \ldots (A_0^{(m)}, A_1^{(m)}, \ldots, A_n^{(m)} \between x_1, x_2)^n,$$

m being greater than $n+1$, then

$$\{\mathrm{A}^{(1)} \mathrm{A}^{(2)} \ldots \mathrm{A}^{(n+2)}\}'\, P = 0, \qquad \text{(i.)}$$

$$\{\mathrm{A}^{(1)} \mathrm{A}^{(2)} \ldots \mathrm{A}^{(n+1)}\}'\, P = |\, A^{(1)} A^{(2)} \ldots A^{(n+1)} \,|\, P_1, \qquad \text{(ii.)}$$

$$\{\mathrm{A}^{(1)} \mathrm{A}^{(2)} \ldots \mathrm{A}^{(n)}\}'\, P = |\, A^{(1)} A^{(2)} \ldots A^{(n)} Q \,|\,, \qquad \text{(iii.)}$$

where a substitution $(\mathrm{A}^{(\alpha)} \mathrm{A}^{(\beta)})$ operating on P is regarded as interchanging (α) and (β) in all the indices in P; in fact it interchanges the positions held by the coefficients of the two quantics

$$(A_0^{(\alpha)}, A_1^{(\alpha)}, \ldots, A_n^{(\alpha)} \between x_1, x_2)^n, \quad (A_0^{(\beta)}, A_1^{(\beta)}, \ldots, A_n^{\beta} \between x_1, x_2)^n$$

in P; or else it may be regarded as an abbreviation for

$$(\mathrm{A}_0^{(\alpha)} \mathrm{A}_0^{(\beta)})(\mathrm{A}_1^{(\alpha)} \mathrm{A}_1^{(\beta)}) \ldots (\mathrm{A}_n^{(\alpha)} \mathrm{A}_n^{(\beta)}).$$

And $|\, A^{(1)} A^{(2)} \ldots A^{(n+1)} \,|$ is the determinant of $n+1$ rows and columns formed by the coefficients of the $n+1$ quantics concerned; $|\, A^{(1)} A^{(2)} \ldots A^{(n)} Q \,|$ is the same determinant with functions $Q_0, Q_1, \ldots, Q_n$ of the coefficients of the quantics represented by $A^{(n+1)}, A^{(n+2)}, \ldots, A^{(m)}$ of the same character as P, substituted for the coefficients $A_0^{(n+1)}$, $A_1^{(n+1)}, \ldots, A_n^{(n+1)}$; and P_1 is a function, having the same character as

P, of the coefficients of the quantics represented by $A^{(n+2)} \ldots A^{(m)}$. To prove (i.) we observe that P may be written in the form

$$P = \Sigma A^{(1)}_{r_1} A^{(2)}_{r_2} \ldots A^{(n+2)}_{r_{n+2}} P',$$

where each of the suffixes $r_1, r_2, \ldots, r_{n+2}$ is one of the $n+1$ numbers $0, 1, 2, \ldots, n$; hence in any case two suffixes must be equal, and consequently

$$\{\mathrm{A}^{(1)} \mathrm{A}^{(2)} \ldots \mathrm{A}^{(n+2)}\}' P = 0.$$

For (ii.) we write $$P = \Sigma A^{(1)}_{r_1} A^{(2)}_{r_2} \ldots A^{(n+1)}_{r_{n+1}} P',$$

and here it is possible for the suffixes to be all different; if this is so,

$$\{\mathrm{A}^{(1)} \mathrm{A}^{(2)} \ldots \mathrm{A}^{(n+1)}\}' A^{(1)}_{r_1} A^{(2)}_{r_2} \ldots A^{(n+1)}_{r_{n+1}}$$
$$= \pm \{\mathrm{A}^{(1)} \mathrm{A}^{(2)} \ldots \mathrm{A}^{(n+1)}\}' A^{(1)}_{0} A^{(2)}_{1} \ldots A^{(n+1)}_{n+1}$$
$$= \pm \mid A^{(1)} A^{(2)} \ldots A^{(n+1)} \mid ;$$

and therefore

$$\{\mathrm{A}^{(1)} \mathrm{A}^{(2)} \ldots \mathrm{A}^{(n+1)}\}' P = \mid A^{(1)} A^{(2)} \ldots A^{(n+1)} \mid P_1.$$

As regards (iii.) we write

$$P = \Sigma A^{(1)}_{r_1} A^{(2)}_{r_2} \ldots A^{(n)}_{r_n} P,$$

and distinguish the following cases:—first, terms R' in which two of the suffixes are equal; then terms R_0 in which the suffixes $r_1, r_2, \ldots, r_n$ are the numbers $1, 2, \ldots, n$ in some order; then terms R_1 in which the suffixes are the numbers $0, 2, 3, \ldots, n$ in some order, and so on; finally, terms R_n in which the suffixes are $0, 1, 2, \ldots, n-1$ in some order. Now operate with $\{\mathrm{A}^{(1)} \mathrm{A}^{(2)} \ldots \mathrm{A}^{(n)}\}'$; then

$$\{\mathrm{A}^{(1)} \mathrm{A}^{(2)} \ldots \mathrm{A}^{(n)}\}' R' = 0,$$

$$\{\mathrm{A}^{(1)} \mathrm{A}^{(2)} \ldots \mathrm{A}^{(n)}\}' R_0 = \begin{vmatrix} A^{(1)}_1 A^{(1)}_2 \ldots A^{(1)}_n \\ A^{(2)}_1 A^{(2)}_2 \ldots A^{(2)}_n \\ \ldots \quad \ldots \quad \ldots \\ A^{(n)}_1 A^{(n)}_2 \ldots A^{(n)}_n \end{vmatrix} Q_0.$$

The other terms are found in the same way; so that, taking the sum,

$$\{\mathrm{A}^{(1)} \mathrm{A}^{(2)} \ldots \mathrm{A}^{(n)}\}' P = \mid A^{(1)} A^{(2)} \ldots A^{(n)} Q \mid .$$

9. As an example, consider the invariants of any number of binary quadratics

$$a_x^2,\ b_x^2,\ \ldots .$$

The possible invariant forms are

$$(ab)^2,\ (ab)(bc)(ca),\ (ab)(bc)(cd)(da),\ \ldots;$$

then $\{\mathrm{bc}\}\,(ab)(bc)(cd) = (ab)(bc)(cd)-(ac)(bc)(bd) = -(bc)^2\,(ad)$;

so that, if b, c be any pair of consecutive letters in an invariant I, $\{\mathrm{bc}\}\ I$ is reducible.

Again,

$$\{\mathrm{bd}\}'(ab)(bc)(cd)(de) = (ab)(bc)(cd)(de)-(ad)(dc)(cb)(be)$$
$$= (bc)(cd)(db)(ae).$$

Similarly, any other interchange of letters may be dealt with. The number of irreducible invariants I of any degree n is equal to the number of linearly independent functions obtained from the function P by permuting the letters which it contains, when P satisfies the equations

$$\{\mathrm{ab}\}\,P = 0,\ \{\mathrm{bc}\}\,P = 0,\ \ldots,\ \{\mathrm{ac}\}'\,P = 0,\ \ldots,$$

and, in fact, all the equations which I satisfies, with the right-hand side of each replaced by zero [I being supposed $= (ab)(bc)(cd)\ldots(ha)$].

If n, the degree of I, be greater than 3, then by the last article

$$\{\mathrm{abcd}\}'\,I = 0.$$

Since $\qquad \{\mathrm{ab}\}\,P = 0,\ \{\mathrm{bc}\}\,P = 0,\ \ldots,\ \{\mathrm{ha}\}\,P = 0,$

$$P = \{\mathrm{abc}\ldots\mathrm{h}\}'\,F = \frac{1}{n!}\,\{\mathrm{abc}\ldots\mathrm{h}\}'\,P = 0,$$

and I is reducible when $n>3$. If the actual solution of the equations for I be carried out, it will be found that in general the expressions on the right-hand side have to satisfy relations; these relations will be the syzygies degree n for the quadratic invariant types. In regard to these equations, it should be noticed that in each separate equation for quadratic types, of the form

$$[\lambda_1+\lambda_2 s_2+\lambda_3 s_3+\ldots]\,I = R,$$

where R is a given reducible expression, it is obviously true that R possesses the substitutional properties involved in the operator on the left. The syzygies arise from the fact that I satisfies more than one equation of this kind. Hence R is subject, owing to the system of equations, to more conditions than those implied by the operator on the left-hand side. Exactly the same remark applies to the equations for quartic invariant types of degree greater than 6. The equations in their complete form for degree 7 are given in my paper, "On the Invariant Syzygies of Lowest Degree for any Number of Binary Quartics," already quoted.

As has been pointed out at the commencement of this paper, the invariants of any number of quartics give another illustration of substitutional equations. Thus, the invariant type $(abcde)$, degree 5, satisfies the equation

$$\{\mathrm{abc}\}\,(abcde) = R,$$

and is of group $\{(\mathrm{abcde}),\ (\mathrm{be})(\mathrm{cd})\}$. It has been shown that there are only six independent irreducible forms $(abcde)$. If, now, the theorem of § 7 be applied, we find that, if M be the number of the functions obtained from $[abcde]$ by interchanging the variables in terms of which all the functions obtained by every possible interchange can be linearly expressed, where $[abcde]$ is defined as being of group $\{\mathrm{abc}\}$ and as satisfying the equation

$$\{(\mathrm{abcde}),\ (\mathrm{bc})(\mathrm{cd})\}\,[abcde] = 0$$

then $$M-6 = 5!\,(\tfrac{1}{6}-\tfrac{1}{10}) = 8$$

and $$M = 14.$$

10. In what follows repeated use will be made of the symmetric group; it is convenient, then, to note that the sum of its substitutions may be factorized in a variety of ways. For instance,

$$\begin{aligned}\{\mathrm{a}_1\mathrm{a}_2 \dots \mathrm{a}_n\} &= \{(\mathrm{a}_1\mathrm{a}_2 \dots \mathrm{a}_n)\}\,\{\mathrm{a}_1\mathrm{a}_2 \dots \mathrm{a}_{n-1}\} \\ &= \left[1+(\mathrm{a}_1\mathrm{a}_n)+(\mathrm{a}_2\mathrm{a}_n)+\dots+(\mathrm{a}_{n-1}\mathrm{a}_n)\right]\{\mathrm{a}_1\mathrm{a}_2 \dots \mathrm{a}_{n-1}\} \\ &= \{\mathrm{a}_1\mathrm{a}_2\}\,G_n,\end{aligned}$$

where G_n is the alternating group of the n letters.

Now, any purely formal relation between functions of substitutions will still hold good if the sign of every transposition be changed, Hence the negative symmetric group may be factorized in the same way, thus

$$\{\mathrm{a}_1\mathrm{a}_2 \dots \mathrm{a}_n\}' = \left[1-(\mathrm{a}_1\mathrm{a}_n)-(\mathrm{a}_2\mathrm{a}_n)-\dots-(\mathrm{a}_{n-1}\mathrm{a}_n)\right]\{\mathrm{a}_1\mathrm{a}_n \dots \mathrm{a}_{n-1}\}'.$$

Again, the product of a group by itself is the group multiplied by a constant factor equal to its order. The product of a group by a sub-group is equal to the whole group multiplied by the order of the sub-group; for, if G be the whole group, and S a substitution belonging to the sub-group G_1, then

$$G.s = G.$$

Again, if $\{a_1 a_2 a_3 \ldots a_n\}$ be any positive symmetric group, and $\{a_1 a_2 b_3 \ldots b_m\}'$ a negative symmetric group,

$$\begin{aligned} &\{a_1 a_2 a_3 \ldots a_n\} \{a_1 a_2 b_3 \ldots b_m\}' \\ = \;& \{a_1 a_2 a_3 \ldots a_n\} (a_1 a_2) \left[-(a_1 a_2) \{a_1 a_2 b_3 \ldots b_m\}'\right] \\ = & -\{a_1 a_2 a_3 \ldots a_n\} \{a_1 a_2 b_3 \ldots b_m\}' = 0. \end{aligned}$$

Let $S[a_1 b_1 b_2 \ldots b_m]$ be any substitutional expression affecting the letters $a_1, b_1, b_2, \ldots, b_m$, and only these; then

$$\{a_2 a_3 \ldots a_n\} S[a_1 b_1 b_2 \ldots b_m] = S[a_1 b_1 b_2 \ldots b_m] \{a_2 a_3 \ldots a_n\}.$$

Hence $\qquad \{a_1 a_2 \ldots a_n\} S[a_1 b_1 b_2 \ldots b_m] \{a_1 a_2 \ldots a_n\}$

$$= [1+(a_1 a_2)+(a_1 a_3)+\ldots+(a_1 a_n)] \{a_2 a_3 \ldots a_n\} S[a_1 b_1 b_2 \ldots b_m\} \times \{a_1 a_2 \ldots a_n\}$$

$$= (n-1)!\, [1+(a_1 a_2)+(a_1 a_3)+\ldots+(a_1 a_n)] S[a_1 b_1 b_2 \ldots b_m] \{a_1 a_2 \ldots a_n\}$$

$$= (n-1)!\, \left[S[a_1 b_1 b_2 \ldots b_m]+S[a_2 b_1 b_2 \ldots b_m]+\ldots+S[a_n b_1 \ldots b_m]\right] \times \{a_1 a_2 \ldots a_n\};$$

or, as may be proved in the same way,

$$= (n-1)!\, \{a_1 a_2 \ldots a_n\} \left[S[a_1 b_1 \ldots b_m]+S[a_2 b_1 \ldots b_m]+\ldots \ldots+S[a_n b_1 \ldots b_m]\right].$$

11. As certain results, due in the first place to Capelli, are to be obtained in this paper by means of substitutional analysis, some account of the remarkable paper, "Sur les Opérations dans la Théorie des Formes Algébriques,"* in which they occur, is given here. In this paper Capelli considers functions rational, integral, algebraic,

---

* *Math. Ann.*, Bd. XXXVII., pp. 1–37.

of n sets of variables

$$\begin{matrix} x_1, & x_2, & \ldots, & x_m, \\ y_1, & y_2, & \ldots, & y_m, \\ \ldots & \ldots & & \ldots \\ u_1, & u_2, & \ldots, & u_m, \end{matrix}$$

there being m variables in each set, and homogeneous in the variables of each set. Such a function is written

$$f(x, y, \ldots, u).$$

He regards the polar operation

$$D_{xy} = y_1 \frac{\partial}{\partial x_1} + y_2 \frac{\partial}{\partial x_2} + \ldots + y_m \frac{\partial}{\partial x_m}$$

as fundamental, and proceeds in the first section to develop a theory of operations which can be expressed as rational integral functions with constant coefficients of operations of this kind, and proves that, if by Δ be understood some operation which can be thus expressed, every function $f(x, y, \ldots, u)$ of the above sets of variables which is homogeneous and of degree a_i in the variables whose index is i, for all values of i from 1 up to m, can be obtained in the form

$$f(x, y, \ldots, u) = \Delta x_1^{a_1} y_2^{a_2} \ldots v_m^{a_m},$$

there being the same number of sets expressed in the term on which Δ operates as there are variables in each set, Δ depending on the form of f.

His second section is devoted to the discussion of an operation H defined as follows :—

If $m = n$, $\quad H = | xy \ldots u | \left| \frac{\partial}{\partial x} \frac{\partial}{\partial y} \ldots \frac{\partial}{\partial u} \right|$,

if $m > n$, $\quad H = \sum_i | x_{i_1} y_{i_2} \ldots u_{i_n} | \left| \frac{\partial}{d x_{i_1}} \frac{\partial}{\partial x_{i_2}} \ldots \frac{\partial}{\partial u_{i_n}} \right|$,

is $m < n$, $\quad H = 0$,

where $| xy \ldots u |$ is the determinant formed by the variables, and $\left| \frac{\partial}{\partial x} \frac{\partial}{\partial y} \ldots \frac{\partial}{\partial u} \right|$, which is the determinant formed by the first differential operators with respect to the variables, is Cayley's operator Ω.

It is shown that H may be expressed in terms of the operators

D_{xy}, and the form of this expression is found; further, it is proved that H is commutative with all rational integral functions of the operators D_{xy}. It is then proved that, if a function $f(x, y, z, \ldots, t, u)$ of the kind considered, of n sets of variables, there being n variables in each set, is annihilated by each of $D_{xy}, D_{yz}, \ldots, D_{tu}$, it is equal to a power of $| xyz \ldots tu |$ multiplied by a function of the same nature of the sets $y, z, \ldots, t, u$, which is annihilated by $D_{yz}, \ldots, D_{tu}$.

In the third section it is proved that, if two functions of the same number of sets of variables, rational, integral, and homogeneous in the variables of each set, are obtainable from each other by means of a permutation of the sets, they are also obtainable from each other by means of the operators D_{xy}. In other words, an operator which is a rational, integral function of the operators D_{xy} may be always found which will have the same effect on $f(x, y, \ldots, u)$ as any given substitution operating on this function. In view of the importance of this theorem in connection with the present subject, I quote Capelli's illustration. Let $f(x, y, z)$ be any rational, integral function of the variables

$$\begin{array}{l} x_1,\ x_2,\ \ldots,\ x_m, \\ y_1,\ y_2,\ \ldots,\ y_m, \\ z_1,\ z_2,\ \ldots,\ z_m, \end{array}$$

homogeneous and of degrees λ, μ, ν respectively in the variables of the three sets. Let

$$\begin{array}{l} \xi_1,\ \xi_2,\ \ldots,\ \xi_m, \\ \eta_1,\ \eta_2,\ \ldots,\ \eta_m, \\ \zeta_1,\ \zeta_2,\ \ldots,\ \zeta_m \end{array}$$

be three new sets of variables, independent of each other and of the original sets; then

$$f(\xi, \eta, \zeta) = \frac{1}{\lambda!\,\mu!\,\nu!} D^{\lambda}_{x\xi} D^{\mu}_{y\eta} D^{\nu}_{z\zeta} f(x, y, z),$$

and
$$f(y, z, x) = \frac{1}{\lambda!\,\mu!\,\nu!} D^{\lambda}_{\xi y} D^{\mu}_{\eta z} D^{\nu}_{\zeta x} f(\xi, \eta, \zeta);$$

hence
$$f(y, z, x) = \left(\frac{1}{\lambda!\,\mu!\,\nu!}\right)^2 D^{\lambda}_{\xi y} D^{\mu}_{\eta z} D^{\nu}_{\zeta x} D^{\lambda}_{x\xi} D^{\mu}_{y\eta} D^{\nu}_{z\zeta} f(x, y, z).$$

By means of the methods laid down in the first section of Capelli's paper, it is possible to reduce this to the form $\Delta f(x, y, z)$, where the operators of which Δ is a function only affect the sets x, y, z.

In this section it is also proved that the condition that f should be expressible as a sum of terms each of which is derivable by operations of the kind considered from functions of a smaller number of sets of variables than that contained in f is

$$H.f = 0.$$

In § 4 the following important theorem is proved:

If $f(x, y, \ldots, u)$ is a rational, integral function of n sets of variables, there being n variables in each set, which is homogeneous in the variables of each set, then

$$f(x, y, \ldots, u) = \sum_{\mu, i} | xy \ldots u |^{\mu} . \Delta_i . \phi_i (y, z, \ldots, u),$$

where $\phi_i (y, z, \ldots, u) = D_{xy}^{\alpha_i} D_{xz}^{\beta_i} \ldots D_{xu}^{\lambda_i} . \Omega^{\mu} f$;

the Σ extending to all positive integral solutions of

$$\alpha_i + \beta_i + \ldots + \lambda_i + \mu = p,$$

where p is the degree of f in x, and where Δ_i is a rational integral function with constant coefficients of operators of the form D_{xy}, the form of which depends only on the degrees in which the variables occur in f; and, further, the coefficients of different powers of $| xy \ldots u |$ are unique. The last section is devoted to an extension of certain of the results to any analytic function.

12. In what follows substitutions are taken as the fundamental operators, instead of Capelli's operators D_{xy}. Functions $f(a, b, \ldots, k)$ are considered which are rational, integral, homogeneous, and linear in each of n sets of variables

$$\begin{array}{cccc} a_1, & a_2, & \ldots, & a_m, \\ b_1, & b_2, & \ldots, & b_m, \\ \ldots & \ldots & & \ldots \\ k_1, & k_2, & \ldots, & k_m, \end{array}$$

there being m variables in each set. The letters $a, b, \ldots, k$ are employed, as the applications considered are mainly to concomitant types of quantics. The restriction that f is to be linear in the variables of each set does not in reality restrict the generality of the results obtained; for, if $F(a, b, \ldots, k)$ be a function rational, integral, homogeneous, and of degrees $\alpha, \beta, \ldots, \kappa$ in the variables of the different sets, we may obtain a function f, such that

$$f(a^{(1)}, a^{(2)}, \ldots, a^{(\alpha)}, b^{(1)}, \ldots, b^{(\beta)}, \ldots, k^{(\kappa)})$$

$$= \frac{1}{\alpha!\,\beta!\ldots\kappa!} D_{aa^{(1)}} D_{aa^{(2)}} \ldots D_{aa^{(\alpha)}} D_{bb^{(1)}} \ldots D_{kk^{(\kappa)}} F(a, b, \ldots, k),$$

and consider, instead of F, the function

$$\frac{1}{\alpha!\,\beta!\ldots\kappa!} \{\mathrm{a}^{(1)}\mathrm{a}^{(2)} \ldots \mathrm{a}^{(\alpha)}\}\{\mathrm{b}^{(1)} \ldots \mathrm{b}^{(\beta)}\} \ldots \{\mathrm{k}^{(1)} \ldots \mathrm{k}^{(\kappa)}\} f.$$

For, if we write

$$a^{(1)} = a^{(2)} = \ldots = a, \quad b^{(1)} = \ldots = b, \quad \ldots, \quad k^{(1)} = \ldots = k,$$

this becomes F once more. There is a fairly close connexion between the theory of substitutional and of polar operators. Thus any function $f(a, b, \ldots, k)$ of n sets of variables, there being m variables in each set, which is homogeneous and linear in the variables of each set, and homogeneous and of degree α_i in the variables whose index is i, for all values of i from 1 up to m, may be expressed in the form

$$f(a, b, \ldots, k) = S a_1^{(1)} \ldots a_1^{(\alpha_1)} b_2^{(1)} \ldots b_2^{(\alpha_2)} \ldots k_m^{(\alpha_m)},$$

where S is a substitutional operator with constant coefficients. This follows at once from § 1; for there is only one kind of term which can occur here.

The operator H may be expressed as a substitutional operator thus:—We first suppose that H is to operate on a function homogeneous and linear in the variables of each of n sets, there being n variables in each set; then

$$H = |\, ab \ldots k \,| \left| \frac{\partial}{\partial a} \frac{\partial}{\partial b} \cdots \frac{\partial}{\partial k} \right|.$$

But in this case

$$|\, ab \ldots k \,| \left| \frac{\partial}{\partial a} \frac{\partial}{\partial b} \cdots \frac{\partial}{\partial k} \right| f = \{\mathrm{ab} \ldots \mathrm{k}\}' f.$$

For, if $A a_{i_1} b_{i_2} \ldots k_{i_m}$ be any term of f, the effect of both operators is zero, unless all the indices are different, and, if this is so, both operators give $A \,|\, ab \ldots k \,|$ as the result, the rule for determining the sign being the same in each case.

If f is still linear in the variables of each set, but the number of variables in a set is m, greater than the number n of sets, then H is still equivalent to $\{\mathrm{ab} \ldots \mathrm{k}\}'$, for, if $Aa_{i_1} b_{i_2} \ldots k_{i_m}$ be any term of f, then

$$\{\mathrm{ab} \ldots \mathrm{k}\}\, Aa_{i_1} b_{i_2} \ldots k_{i_m}$$

$$= | a_{i_1} b_{i_2} \ldots k_{i_m} | \left| \frac{\partial}{\partial a_{i_1}} \frac{\partial}{\partial b_{i_2}} \cdots \frac{\partial}{\partial k_{i_m}} \right| Aa_{i_2} \ldots k_{i_m}$$

$$= \left[\sum_j | a_{j_1} b \ \ldots k_{j_m} | \left| \frac{\partial}{\partial a_{j_1}} \frac{\partial}{\partial b_{j_2}} \cdots \frac{\partial}{\partial k_{j_m}} \right| \right] Aa_{i_1} \ldots k_{i_m};$$

for all terms of the Σ, except the one first quoted, give zero when operating on the term chosen.

If $m < n$, $H = 0$, and $\{\mathrm{ab} \ldots \mathrm{k}\}' f = 0$; for every term of f must contain at least two variables with the same indices.

Now consider any function F homogeneous but no longer linear in the variables of each set, having n sets and m variables in each set. Then we form from F a function f, as shown above, such that we may consider, instead of F,

$$P = \frac{1}{\alpha!\, \beta! \ldots \kappa!} \{\mathrm{a}^{(1)} \mathrm{a}^{(2)} \ldots \mathrm{a}^{(\alpha)}\} \{\mathrm{b}^{(1)} \ldots \mathrm{b}^{(\beta)}\} \ldots$$
$$\ldots \{\mathrm{k}^{(1)} \ldots \mathrm{k}^{(\kappa)}\} f(a^{(1)}, a^{(2)}, \ldots, a^{(\alpha)}, b^{(1)}, \ldots, k^{(\kappa)}).$$

Then, if $H_{a^{(1)} b^{(1)} \ldots k^{(1)}}$ be what H becomes when we write in it the sets $a^{(1)}, b^{(1)}, \ldots, k^{(1)}$ instead of the sets $a, b, \ldots, k$, HF becomes

$$\sum_{\alpha_i=1}^{\alpha_i=\alpha} \sum_{\beta_i=1}^{\beta_i=\beta} \cdots \sum_{\kappa_i=1}^{\kappa_i=\kappa} H_{a^{(\alpha_i)} b^{(\beta_i)} \ldots k^{(\kappa_i)}} P;$$

this last expression being $= HF$ when we write

$$a^{(1)} = a^{(2)} = \ldots = a^{(\alpha)} = a,$$
$$b^{(1)} = b^{(2)} = \ldots = b^{(\beta)} = b,$$
$$\ldots \quad \ldots \quad \ldots \quad \ldots$$
$$k^{(1)} = k^{(2)} = \ldots = k^{(\kappa)} = k.$$

But $$H_{a^{(\alpha_i)} b^{(\beta_i)} \ldots k^{(\kappa_i)}} P = \{\mathrm{a}^{(\alpha_i)} \mathrm{b}^{(\beta_i)} \ldots \mathrm{k}^{(\kappa_i)}\} P,$$

as we have already seen; hence in this case

$$\sum_{\alpha_i, \beta_i, \ldots, \kappa_i} \{\mathrm{a}^{(\alpha_i)} \mathrm{b}^{(\beta_i)} \ldots \mathrm{k}^{(\kappa_i)}\}'$$

is the equivalent of H.

Now, in the substitutional equivalent of H it is assumed that there is a substitutional operator

$$\{a^{(1)} a^{(2)} \dots a^{(\alpha)}\}\{b^{(1)} \dots b^{(\beta)}\} \dots \{k^{(1)} \dots k^{(\kappa)}\},$$

of definite form applied to the operand. The same operator may then be attached to this equivalent of H, without affecting the result except as regards a constant. Hence we may write

$$H = \frac{1}{\alpha!\,\beta!\,\dots\,\kappa!}\Sigma\,\{a^{(\alpha_i)} b^{(\beta_i)} \dots k^{(\kappa_i)}\}'\,\{a^{(1)} \dots a^{(\alpha)}\}\{b^{(1)} \dots b^{(\beta)}\} \dots \{k^{(1)} \dots k^{(\kappa)}\}$$

$$= \frac{1}{\alpha!\,\beta!\dots\kappa!\,(\alpha-1)!\,(\beta-1)!\dots(\kappa-1)!}$$

$$\times\{a^{(1)} \dots a^{(\alpha)}\}\{b^{(1)} \dots b^{(\beta)}\} \dots \{k^{(1)} \dots k^{(\kappa)}\}$$

$$\times\{a^{(1)}b^{(1)} \dots k^{(1)}\}'\{a^{(1)} \dots a^{(\alpha)}\}\{b^{(1)} \dots b^{(\beta)}\} \dots \{k^{(1)} \dots k^{(\kappa)}\}.$$

For $$\{a^{(1)} \dots a^{(\alpha)}\}\,\{a^{(1)} b^{(1)} \dots k^{(1)}\}'\,\{a^{(1)} \dots a^{(\alpha)}\}$$

$$= \left[1 + (a^{(1)} a^{(2)}) + (a^{(1)} a^{(3)}) + \dots + (a^{(1)} a^{(\alpha)})\right]$$

$$\times\{a^{(2)} \dots a^{(\alpha)}\}\,\{a^{(1)} b^{(1)} \dots k^{(1)}\}'\,\{a^{(1)} \dots a^{(\alpha)}\}$$

$$= (\alpha-1)!\left[1 + (a^{(1)}a^{(2)}) + \dots + (a^{(1)}a^{(\alpha)})\right]\{a^{(1)}b^{(1)} \dots k^{(1)}\}'\{a^{(1)} \dots a^{(\alpha)}\}$$

$$= (\alpha-1)!\sum_{\alpha_i=1}^{\alpha_i=\alpha}\{a^{(\alpha_i)}b^{(1)} \dots k^{(1)}\}\,\{a^{(1)} \dots a^{(\alpha)}\}.$$

Capelli has shown in the general case how a substitution may be expressed in terms of polar operators; in the case of functions homogeneous and linear in the variables of each set, the effect of a transposition may be obtained thus,

$$D_{ba} D_{ab} f(a, b, c, \dots) = D_{ba} f(b, b, c, \dots) = \{ab\} f(a, b, c, \dots);$$

hence $$(ab) f(a, b, c, \dots) = (D_{ba} D_{ab} - 1) f(a, b, c, \dots).$$

Any other substitution operating on f may be expressed as a product of transpositions, and so as a function of polar operators. The converse theorem is also true; for let D_{ab} be a polar operator, operating on a function F of degree α in the variables of the set a, and β in those of the set b; then we consider instead of F the function P

defined as above. The effect of the operator D_{ab} on F is the same as that of

$$\frac{1}{(\beta+1)!}\{b^{(1)}b^{(2)}\ldots b^{(\beta+1)}\}\left[D_{a^{(1)}b^{(\beta+1)}}+D_{a^{(2)}b^{(\beta+1)}}+\ldots+D_{a^{(\alpha)}b^{(\beta+1)}}\right]$$

on P. For each of the sets $a^{(1)}, a^{(2)}, \ldots, a^{(\alpha)}$ in P is in reality equivalent to a, and each of the sets $b^{(1)}, b^{(2)}, \ldots$ equivalent to b. Since P does not contain $b^{(\beta+1)}$,

$$D_{a^{(1)}b^{(\beta+1)}}P=(a^{(1)}b^{(\beta+1)})\,P,$$

the right-hand side being no longer a function of $a^{(1)}$.

Now, P is symmetric in the sets $a^{(1)}, \ldots, a^{(\alpha)}$; hence the function $(a^{(2)}b^{(\beta+1)})\,P$ is the same as $(a^{(1)}b^{(\beta+1)})\,P$, except that $a^{(1)}$ and $a^{(2)}$ are interchanged; hence the function $D_{ab}F$ is equivalent to

$$\frac{a}{(\beta+1)!}\{b^{(1)}b^{(2)}\ldots b^{(\beta+1)}\}\,(a^{(\alpha)}b^{(\beta+1)})\,P,$$

which does not contain the set $a^{(\alpha)}$. In this the new set $b^{(\beta+1)}$ may be replaced by the old set $a^{(\alpha)}$ by operating with $(a^{(\alpha)}b^{(\beta+1)})$, and the result becomes

$$\frac{a}{(\beta+1)!}\{b^{(1)}b^{(2)}\ldots b^{(\beta)}a^{(\alpha)}\}\,P,$$

where now $a^{(\alpha)}$ is to be regarded as equivalent to b.

13. If $T_{\alpha,0}\equiv S\{a_1b_1\}'\{a_2b_2\}'\ldots\{a_\alpha b_\alpha\}'\{b_1b_2\ldots b_m\}S$

and $\beta>0$, $T_{\alpha,\beta}\equiv S\{a_1b_1\}'\{a_2b_2\}'\ldots\{a_\alpha b_\alpha\}'\{a_{n-\beta+1}\ldots a_nb_1\ldots b_m\}S$,

where $S\equiv\{a_1a_2\ldots a_n\}\{b_1b_2\ldots b_m\}$,

then $T_{0,0}=A_{0,n}T_{0,n}+A_{1,n-1}T_{1,n-1}+\ldots+A_{n,0}T_{n,0}$,

if $m \nless n$; but, if $m<n$, the series must stop with $A_{m,n-m}T_{m,n-m}$, and the coefficients A are given by

$$A_{\alpha,\beta}=\binom{\alpha+\beta}{\beta}\frac{m!\,(m+1+\beta-\alpha)}{(m+\beta+1)!}.$$

The theorem to be established is purely formal, an identity between certain substitutional expressions.

When $\alpha<h<n-\beta+1$, the expression

$$S\{a_1b_1\}'\{a_2b_2\}'\ldots\{a_\alpha b_\alpha\}'(a_hb_1)\{a_{n-\beta+1}\ldots a_nb_1b_2\ldots b_m\}S$$
$$=S\,(a_hb_1)\{a_1a_h\}'\{a_2b_2\}'\ldots\{a_\alpha b_\alpha\}'\{a_{n-\beta+1}\ldots a_nb_1b_2\ldots b_m\}S,$$

for it is well known that, if s be any substitution, $(a_hb_1)\,s\,(a_hb_1)$ is the same as that substitution obtained from s by the interchange of a_h and b_1; and hence, if U be any substitutional expression,

$$(a_h b_1)\, U\, (a_h b_1) = U_1,$$

the expression obtained from U by the interchange of a_h and b_1; and hence

$$U\, (a_h b_1) = (a_h b_1)\, U_1.$$

Now, no one of the factors $\{a_2 b_2\}' \ldots \{a_a b_a\}' \{a_{n-\beta+1} \ldots a_n b_1 b_2 \ldots b_m\}$ contains either of the letters a_1 or a_h; hence

$$\begin{aligned} S\,(a_h b_1)\, \{a_1 a_h\}' \{a_2 b_2\}' \ldots \{a_a b_a\}' \{a_{n-\beta+1} \ldots a_n b_1 b_2 \ldots b_m\}\, S \\ = S\,(a_h b_1) \{a_2 b_2\}' \ldots \{a_a b_a\}' \{a_{n-\beta+1} \ldots a_n b_1 b_2 \ldots b_m\} \{a_1 a_h\}'\, S \\ = 0; \end{aligned}$$

for

$$\{a_1 a_h\}' \{a_1 a_2 \ldots a_n\} = 0;$$

and therefore

$$S\, \{a_1 b_1\}' \{a_2 b_2\}' \ldots \{a_a b_a\}' (a_h b_1) \{a_{n-\beta+1} \ldots a_n b_1 b_2 \ldots b_m\}\, S = 0.$$

Hence

$$\begin{aligned} T_{a,\beta} &= S\, \{a_1 b_1\}' \{a_2 b_2\}' \ldots \{a_a b_a\}' \{a_{n-\beta+1} \ldots a_n b_1 b_2 \ldots b_m\}\, S \\ &= S\, \{a_1 b_1\}' \ldots \{a_a b_a\}' \big[1 + (a_{n-\beta+1} a_{n-\beta+2}) + (a_{n-\beta+1} a_{n-\beta+3}) + \ldots \\ &\qquad + (a_{n-\beta+1} a_n) + (a_{n-\beta+1} b_1) + \ldots + (a_{n-\beta+1} b_m)\big] \\ &\qquad \times \{a_{n-\beta+2} \ldots a_n b_1 b_2 \ldots b_m\}\, S \\ &= S\, \{a_1 b_1\}' \ldots \{a_a b_a\}' \big[\beta + (a_{n-\beta+1} b_{a+1}) + \ldots + (a_{n-\beta+1} b_m)\big] \\ &\qquad \times \{a_{n-\beta+2} \ldots a_n b_1 b_2 \ldots b_m\}\, S \\ &= S\, \{a_1 b_1\}' \ldots \{a_a b_a\}' \big[-(m-a) \{a_{n-\beta+1} b_{a+1}\}' + (m+\beta-a)\big] \\ &\qquad \times \{a_{n-\beta+2} \ldots a_n b_1 b_2 \ldots b_m\}\, S \\ &= -(m-a)\, T_{a+1,\beta-1} + (m+\beta-a)\, T_{a,\beta-1}; \end{aligned}$$

therefore $$T_{a,\beta} = \frac{1}{m+\beta-a+1} T_{a,\beta+1} + \frac{m-a}{m+\beta-a+1} T_{a+1,\beta}.$$

By repeated application of this formula, we obtain

$$\begin{aligned} T_{00} &= \frac{1}{m+1} T_{0,1} + \frac{m}{m+1} T_{1,0} \\ &= \ldots \\ &\ldots \quad \ldots \quad \ldots \quad \ldots \\ &= A_{0,i}\, T_{0,i} + A_{1,i-1}\, T_{1,i-1} + \ldots + A_{i,0}\, T_{i,0}, \end{aligned}$$

except when $i > m$, in which case the series ends with $A_{m, i-m} T_{m, i-m}$; i being supposed to be not greater than n, and the A's being numerical coefficients.

To find these coefficients a recurrence formula is obtained by proceeding from the last line written down a step further. The coefficient of $T_{j, i-j+1}$ in this will be

$$A_{j, i-j+1} = \frac{m-j+1}{m+i-2j+3} A_{j-1, i-j+1} + \frac{1}{m+i-2j+1} A_{j, i-j}.$$

Hence $$A_{a, \beta} = \frac{m-a+1}{m+\beta-a+2} A_{a-1, \beta} + \frac{1}{m+\beta-a} A_{a, \beta-1}.$$

It follows from this that, if

$$A_{a, \beta} = \binom{a+\beta}{\beta} \frac{m!\,(m+1+\beta-a)}{(m+\beta+1)!},$$

when $a < a_1$, and also when $a = a_1$ so long as $\beta < \beta_1$, it is true when $a = a_1$ and $\beta = \beta_1$. Hence, on this hypothesis it is true whenever $a < a_1 + 1$. But this form of $A_{a, \beta}$ is correct, as it is easy to verify, when $a = 0$, and also when $\beta = 0$ and $a < n+1$. Hence it is true always when $a < n+1$. And the theorem is proved.

14. As has been pointed out, the theorem just proved is merely a substitutional identity. If the two sides of the identity be made to operate on the same function, the results must be equal. This operand may be taken to be any function of $m+n$ variables

$$a_1, a_2, \ldots, a_n, \quad b_1, b_2, \ldots, b_m;$$

or else any function of $m+n$ sets of variables

$$\begin{array}{llll} a_{1,1}, & a_{1,2}, & a_{1,3}, \ \ldots, & a_{1,p}, \\ a_{2,1}, & a_{2,2}, & \ldots\ldots\ldots, & a_{2,p}, \\ \ldots & \ldots & \ldots & \ldots \\ a_{n,1}, & a_{n,2}, & \ldots\ldots\ldots, & a_{n,p}, \\ b_{1,1}, & b_{1,2}, & \ldots\ldots\ldots, & b_{1,p}, \\ \ldots & \ldots & \ldots & \ldots \\ b_{m,1}, & b_{m,2}, & \ldots\ldots\ldots, & b_{m,p}, \end{array}$$

a substitution $(a_1 b_1)$ interchanging two sets, just as in §§ 10 and 11 a substitution on the functions there discussed interchanged two sets. In this case, as has been seen in § 11, when the operand F is linear and homogeneous in the variables of each set, the expression $\{\mathfrak{a}_1 \mathfrak{b}_1\}'$ is equivalent to H_{a_1, b_1}, where

$$H_{a_1, b_1} = \Sigma \mid a_{1, i_1} \; b_{1, i_2} \mid \left| \frac{\partial}{\partial a_{1, i_1}} \; \frac{\partial}{\partial b_{1, i_2}} \right| .$$

(1) Let us take for operand

$$F = a_{1_x} a_{2_x} \ldots a_{n_x} b_{1_y} b_{2_y} \ldots b_{m_y},$$

where the factors of F are binary symbolical factors, thus

$$a_{1_x} = a_{1,1} x_1 + a_{1,2} x_2,$$

$$b_{1_y} = b_{1,1} y_1 + b_{1,2} y_2.$$

Then $T_{0,0} F = \{\mathfrak{a}_1 \mathfrak{a}_2 \ldots \mathfrak{a}_n\} \{\mathfrak{b}_1 \mathfrak{b}_2 \ldots \mathfrak{b}_m\} \{\mathfrak{b}_1 \ldots \mathfrak{b}_m\} \{\mathfrak{a}_1 \ldots \mathfrak{a}_n\} \{\mathfrak{b}_1 \ldots \mathfrak{b}_m\} F$

$$= (n!)^2 (m!)^3 F.$$

Denote by D and Δ polar operators, such that, ϕ being homogeneous and of order n in x_1, x_2, and homogeneous and of order m in y_1, y_2, then

$$D\phi = \frac{1}{m} \left(x_1 \frac{\partial \phi}{\partial y_1} + x_2 \frac{\partial \phi}{\partial y_2} \right),$$

$$\Delta\phi = \frac{1}{n} \left(y_1 \frac{\partial \phi}{\partial x_1} + y_2 \frac{\partial \phi}{\partial x_2} \right),$$

also let

$$\Omega\phi = \frac{1}{mn} \left(\frac{\partial^2 \phi}{\partial x_1 \partial y_2} - \frac{\partial^2 \phi}{\partial x_2 \partial y_1} \right);$$

these being the operators used by Clebsch (*Binaren Formen*, pp. 13, 14, *et seq.*).

Then
$$D^m F = a_{1_x} a_{2_x} \ldots a_{n_x} b_{1_x} b_{2_x} \ldots b_{m_x},$$

and the effect of operating with Δ^m on this function is to change it to the sum of all possible terms obtained from $D^m F$ by writing y for x in m of its factors, divided by their number. But this is the same as

$$\frac{1}{(m+n)!} \{\mathfrak{a}_1 \mathfrak{a}_2 \ldots \mathfrak{a}_n \mathfrak{b}_1 \mathfrak{b}_2 \ldots \mathfrak{b}_m\} \, a_{1_x} a_{2_x} \ldots a_{n_x} b_{1_y} b_{2_y} \ldots b_{m_y}.$$

Hence

$$T_{0,n} F = \{\mathfrak{a}_1 \ldots \mathfrak{a}_n\} \{\mathfrak{b}_1 \ldots \mathfrak{b}_m\} \{\mathfrak{a}_1 \ldots \mathfrak{a}_n \mathfrak{b}_1 \ldots \mathfrak{b}_m\} \{\mathfrak{a}_1 \ldots \mathfrak{a}_n\} \{\mathfrak{b}_1 \ldots \mathfrak{b}_m\} F$$

$$= (n!)^2 (m!)^2 \{\mathfrak{a}_1 \ldots \mathfrak{a}_n \mathfrak{b}_1 \ldots \mathfrak{b}_m\} F$$

$$= (n!)^2 (m!)^2 (m+n)! \, \Delta^m D^m F.$$

Again,

$$\Omega F = \frac{1}{mn} \sum_{i=1}^{i=n} \sum_{j=1}^{j=m} (a_i b_j)\, a_{1_x} \dots a_{i-1_x} a_{i+1_x} \dots a_{n_x} b_{1_y} \dots b_{j-1_y} b_{j+1_y} \dots b_{m_y};$$

and $$D^{m-1} \Omega F = \frac{1}{mn} \sum_i \sum_j (a_i b_j)\, a_{1_x} \dots a_{i-1_x} a_{i+1_x} a_{n_x} b_{1_x} \dots b_{j-1_x} b_{j+1_x} \dots b_{m_x}.$$

And $\Delta^{m-1} D^{m-1} \Omega F$ is equal to the sum of all possible terms obtained by substituting y for x in $m-1$ of the factors of each term of $D^{m-1}\Omega F$ divided by their number

$$= \frac{1}{(m+n-2)!} \frac{1}{mn} \sum_{i,j} (a_i b_j) \{\mathrm{a}_1 \dots \mathrm{a}_{i-1} \mathrm{a}_{i+1} \dots \mathrm{a}_n \mathrm{b}_1 \dots \mathrm{b}_{j-1} \mathrm{b}_{j+1} \dots \mathrm{b}_m\}$$
$$\times a_{1_x} \dots a_{i-1_x} a_{i+1_x} \dots a_{n_x} b_{1_y} \dots b_{j-1_y} b_{j+1_y} \dots b_{m_y}$$

$$= \frac{1}{(m+n-2)!} \frac{1}{m!\, n!} \{\mathrm{a}_1 \mathrm{a}_2 \dots \mathrm{a}_n\} \{\mathrm{b}_1 \mathrm{b}_2 \dots \mathrm{b}_m\} (a_1 b_1) \{\mathrm{a}_2 \dots \mathrm{a}_n \mathrm{b}_2 \dots \mathrm{b}_m\}$$
$$\times a_2 \dots a_{n_x} b_{2_y} \dots b_{m_y}.$$

Hence

$$T_{1,n-1} F = S \{\mathrm{a}_1 \mathrm{b}_1\}' \{\mathrm{a}_2 \dots \mathrm{a}_n \mathrm{b}_1 \dots b_m\} SF$$
$$= n!\, m!\, S \{\mathrm{a}_1 \mathrm{b}_1\}' \{\mathrm{a}_2 \dots \mathrm{a}_n \mathrm{b}_1 \dots \mathrm{b}_m\} a_{1_x} \dots a_{n_x} b_{1_y} \dots b_{m_y}$$
$$= n!\, m!\, S \{\mathrm{a}_1 \mathrm{b}_1\}' \big[m a_{1_x} b_{1_y} \{\mathrm{a}_2 \dots \mathrm{a}_n \mathrm{b}_2 \dots \mathrm{b}_m\} a_{2_x} \dots a_{n_x} b_{2_y} \dots b_{m_y}$$
$$+ (n-1)\, a_1\, b_{1_x} \{\mathrm{a}_2 \dots \mathrm{a}_n \mathrm{b}_2 \dots \mathrm{b}_m\} a_{2_x} \dots a_{(n-1)_x} a_{n_y} b_{2_y} \dots b_{m_y} \big]$$
$$= n!\, m!\, m S (a_1 b_1)(xy) \{\mathrm{a}_2 \dots \mathrm{a}_n \mathrm{b}_2 \dots \mathrm{b}_m\} a_{2_x} \dots a_{n_x} b_2 \dots b_{m_y}$$
$$= (xy)(n!)^2 (m!)^2 (m+n-2)!\, m \Delta^{m-1} D^{m-1} \Omega F.$$

Proceeding in the same way,

$$\Omega^h F = \frac{(m-h)!\,(n-h)!}{m!\, n!} \sum (a_1 b_1)(a_2 b_2) \dots (a_h b_h)\, a_{h+1_x} \dots a_{n_x} b_{h+1_y} \dots b_{m_y},$$

$D^{m-h} \Omega^h F$

$$= \frac{(m-h)!\,(n-h)!}{m!\, n!} \sum (a_1 b_1)(a_2 b_2) \dots (a_h b_h)\, a_{h+1_x} \dots a_{n_x} b_{h+1_y} \dots b_{m_x},$$

$\Delta^{m-h} D^{m-h} \Omega^h F$

$$= \frac{1}{(m+n-2h)!} \frac{1}{m!\, n!} \{\mathrm{a}_1 \mathrm{a}_2 \dots \mathrm{a}_n\} \{\mathrm{b}_1 \dots \mathrm{b}_m\} (a_1 b_1)(a_2 b_2) \dots (a_h b_h)$$
$$\times \{\mathrm{a}_{h+1} \dots \mathrm{a}_n \mathrm{b}_{h+1} \dots \mathrm{b}_m\} a_{h+1_x} \dots a_{n_x} b_{h+1_y} \dots b_{m_y}.$$

And hence

$$T_{h,n-h}F = S\{\mathrm{a}_1\mathrm{b}_1\}'\{\mathrm{a}_2\mathrm{b}_2\}'\dots\{\mathrm{a}_h\mathrm{b}_h\}'\{\mathrm{a}_{h+1}\dots\mathrm{a}_n\,\mathrm{b}_1\dots\mathrm{b}_m\}SF$$

$$= n!\,m!\,S\{\mathrm{a}_1\mathrm{b}_1\}'\dots\{\mathrm{a}_h\mathrm{b}_h\}'\left[\frac{m!}{(m-h)!}a_{1_x}\dots a_{h_x}\,b_{1_y}\dots b_{h_y}\right.$$

$$\left.\times\{\mathrm{a}_{h+1}\dots\mathrm{a}_n\,\mathrm{b}_{h+1}\dots\mathrm{b}_m\}a_{h+1_x}\dots a_{n_x}\,b_{h+1_y}\dots b_{m_y}+P\right],$$

where P contains the product $a_{i_x}b_{i_x}$ for some value of i between 1 and h inclusive, and hence is annihilated by the product $\{\mathrm{a}_1\mathrm{b}_1\}'\dots\{\mathrm{a}_h\mathrm{b}_h\}'$; therefore

$$T_{h,n-h}F = (xy)^h\,n!\,m!\,\frac{m!}{(m-h)!}S(a_1b_1)\dots(a_hb_h)\{\mathrm{a}_{h+1}\dots\mathrm{a}_n\,\mathrm{b}_{h+1}\dots\mathrm{b}_m\}$$

$$\times a_{h+1_x}\dots a_{n_x}\,b_{h+1_y}\dots b_{m_y}$$

$$= (xy)^h(n!)^2(m!)^2(m+n-2h)!\,\frac{m!}{(m-h)!}\Delta^{m-h}D^{m-h}\Omega^hF;$$

and hence $$A_{h,n-h}T_{h,n-h}F$$

$$= (xy)^h\binom{n}{h}\frac{m!\,(m+1+n-2h)}{(m+n-h+1)!}(n!)^2(m!)^2(m+n-2h)!\,\frac{m!}{(m-h)!}$$

$$\times\Delta^{m-h}D^{m-h}\Omega^hF$$

$$= (n!)^2(m!)^3\,\frac{\binom{n}{h}\binom{m}{h}}{\binom{m+n-h+1}{h}}(xy)^h\Delta^{m-h}D^{m-h}\Omega^hF.$$

Hence we obtain Gordan's series

$$F = \sum_{h=0}^{h=m}\frac{\binom{n}{h}\binom{m}{h}}{\binom{m+n-h+1}{h}}(xy)^h\Delta^{m-h}D^{m-h}\Omega^hF,$$

if $n \geqslant m$; if $n<m$, the summation must be taken from $h=0$ to $h=n$.

(2) Let F as before $= a_{1_x}a_{2_x}\dots a_{n_x}b_{1_y}b_{2_y}\dots b_{m_y}$, where the factors of F are now ternary symbolical factors, thus

$$a_{1_x} = a_{1,1}x_1+a_{1,2}x_2+a_{1,3}x_3.$$

Then $$T_{0,0}F = (n!)^2(m!)^3F,$$

and $$T_{0,n}F = (n!)^2(m!)^3(m+n)!\,\Delta^mD^mF,$$

just as when the factors of F were binary; the definition of Δ and D being that, if ϕ is a function homogeneous and of order n in x_1, x_2, x_3, and homogeneous and of order m in y_1, y_2, y_3, then

$$D\phi = \frac{1}{m}\left(x_1\frac{\partial\phi}{\partial y_1} + x_2\frac{\partial\phi}{\partial y_2} + x_3\frac{\partial\phi}{\partial y_3}\right),$$

$$\Delta\phi = \frac{1}{n}\left(y_1\frac{\partial\phi}{\partial x_1} + y_2\frac{\partial\phi}{\partial x_2} + y_3\frac{\partial\phi}{\partial x_3}\right).$$

Let u_1, u_2, u_3 be three quantities defined by

$$u_1 = x_2y_3 - x_3y_2,$$

$$u_2 = x_3y_1 - x_1y_3,$$

$$u_3 = x_1y_2 - x_2y_1;$$

then $$a_xb_y - a_yb_x = (abu);$$

and, just as in the former case,

$$T_{h,\,n-h}F = n!\,m!\,S\{\mathrm{a}_1\mathrm{b}_1\}' \ldots \{\mathrm{a}_h\mathrm{b}_h\}' \frac{m!}{(m-h)!} a_{1_x}\ldots a_{h_x} b_{1_y}\ldots b_{h_y}$$

$$\times \{\mathrm{a}_{h+1}\ldots \mathrm{a}_n\, \mathrm{b}_{h+1}\ldots \mathrm{b}_m\}\, a_{h+1_x}\ldots a_{n_x} b_{h+1_y}\ldots b_{m_y}$$

$$= n!\,m!\,\frac{m!}{(m-h)!}\, S\,(a_1b_1u)\ldots(a_hb_hu)\,(m+n-2h)!\,\Delta^{m-h}D^{m-h}$$

$$\times\, a_{h+1_x}\ldots a_{n_x} b_{h+1_y}\ldots b_{m_y}.$$

Let us now suppose that $F = a_x^n b_y^m$, and that we may write

$$a_1 = a_2 = \ldots = a_n = a, \quad b_1 = b_2 = \ldots = b_m = b,$$

after all the substitutional operations have been performed on F; then

$$T_{h,\,n-h}F = (n!)^2\,(m!)^2\,\frac{m!}{(m-h)!}\,(m+n-2h)!\,(abu)^h\,\Delta^{m-h}D^{m-h}a_x^{n-h}b_y^{m-h},$$

and, as in the case of Gordan's series, we obtain

$$a_x^n b_y^m = \sum_{h=0}^{h=m} \frac{\binom{n}{h}\binom{m}{h}}{\binom{m+n-h+1}{h}}\,(abu)^h\,\Delta^{m-h}D^{m-h}a_x^{n-h}b_y^{m-h},$$

if $n \geqslant m$; if $n < m$, the summation must be taken from $h=0$ to $h=n$. The same series may be established in exactly the same way if a_x, b_y are p-ary symbolical factors, provided we write instead of (abu) the difference $a_xb_y - a_yb_x$.

(3) The series furnishes information concerning those functions F which satisfy the substitutional equations

$$\{a_1 b_1 b_2 \ldots b_m\} F = 0,$$
$$\{a_2 b_1 b_2 \ldots b_m\} F = 0,$$
$$\ldots \quad \ldots \quad \ldots \quad \ldots$$
$$\{a_n b_1 b_2 \ldots b_m\} F = 0.$$

For in this case $T_{\alpha,\beta} F = 0$, provided $\beta > 0$.

Hence

$$n!\,(m!)^2 \{a_1 a_2 \ldots a_n\}\{b_1 b_2 \ldots b_m\} F = T_{0,0} F = \frac{m+1-n}{m+1} T_{n,0} F \quad \text{or} = 0$$

according as $m+1$ is or is not greater than n. When n is greater than m,

$$\{a_1 a_2 \ldots a_{m+1}\}\{b_1 b_2 \ldots b_m\} F = 0.$$

15. Let the letters $a_1, a_2, \ldots, a_n$ be arranged in any manner in horizontal rows, so that each row has its first letter in the same vertical column, its second letter in a second vertical column, and so on, and so that no row contains more letters than any row above it; then form the substitutional expression

$$S = \Gamma_1' \Gamma_2' \ldots \Gamma_h' G_1 G_2 \ldots G_k,$$

such that Γ_1' is the negative symmetric group of the letters of the first row, Γ_2' that of the letters of the second row, and so on, Γ_h' being that of the letters of the last row; and that G_1 is the positive symmetric group of the letters of the first column, G_2 that of the letters of the second column, and so on, G_k being that of the letters of the last column (it being understood, in case a row or column contains only one letter, that the positive or negative symmetric group of a single letter is unity). Then, if $T_{a_1, a_2, \ldots, a_h}$ be the sum of all expressions S formed as above from all possible tabular arrangements of the letters, so that there are a_1 letters in the first row, a_2 in the second, and so on, the a's satisfying

$$a_1 + a_2 + \ldots + a_h = n,$$

and

$$a_1 \not< a_2 \not< a_3 \ldots \not< a_h,$$

it is possible to uniquely determine numerical coefficients $A_{a_1, a_2, \ldots, a_h}$ so that

$$1 = \Sigma A_{a_1, a_2, \ldots, a_h} T_{a_1, a_2, \ldots, a_h},$$

where the Σ extends to all possible positive integral values of $\alpha_1, \alpha_2, \ldots, \alpha_h$ which satisfy the two conditions just laid down, the number h of α's not being fixed.

Let us suppose the terms T to be arranged in order, so that $T_{\alpha_1, \alpha_2, \ldots, \alpha_h}$ will come before $T_{\beta_1, \beta_2, \ldots, \beta_{h'}}$, if $\alpha_1 < \beta_1$, or if $\alpha_1 = \beta_1$, but $\alpha_2 < \beta_2$, or if $\alpha_1 = \beta_1$, $\alpha_2 = \beta_2$, ..., $\alpha_{i-1} = \beta_{i-1}$, but $\alpha_i < \beta_i$.

Consider one of the expressions S of which $T_{\alpha_1, \alpha_2, \ldots, \alpha_h}$ is the sum, and the table of letters from which S is formed. Let N be the product of the negative symmetric groups of S, and P the product of its positive symmetric groups, so that

$$S = NP.$$

The degrees of the groups in N are $\alpha_1, \alpha_2, \ldots, \alpha_h$; the degrees of the groups in P depend solely on these numbers, as may be seen from the table, for these groups are formed by the vertical columns in the table. Thus there are only h rows, so that there cannot be more than h elements in any column; in the first α_h columns there are exactly h elements, since the number of letters α_h in the last row is not greater than that in any row above. Next, there are $\alpha_{h-1} - \alpha_h$ columns containing exactly $h-1$ elements, and so on. Hence in P there are first α_h groups of degree h, then $\alpha_{h-1} - \alpha_h$ groups of degree $h-1$, and so on, there being α_1 groups altogether.

Let Γ' be any negative symmetric group which contains a pair of letters out of some one column in the table for S; then $P\Gamma' = 0$, for P contains this pair of letters in a positive symmetric group; and always, as has been seen in § 10,

$$\{\text{abcd} \ldots\}\{\text{abc}'\text{d}' \ldots\} = 0.$$

Again, if Γ' be of degree greater than α_1, then it must contain a pair of letters out of some one column in the table for S, for there are only α_1 different columns. Hence, if the degree of Γ' is greater than α_1, $P\Gamma' = 0$.

Now, let S_1 be one of the expressions of which $T_{\beta_1, \beta_2, \ldots, \beta_{h'}}$ is the sum, where $T_{\beta_1, \beta_2, \ldots, \beta_{h'}}$ is a term which comes after the term $T_{\alpha_1, \alpha_2, \ldots, \alpha_h}$ when these terms are arranged in order; and let

$$S_1 = N_1 P_1,$$

where N_1 is the product of the negative symmetric groups of S_1, and P_1 that of the positive symmetric groups. Then

$$PN_1 = 0.$$

For, if $\beta_1 > \alpha_1$, N_1 contains a negative symmetric group Γ' of degree greater than α_1; and hence, as we have seen,

$$P\Gamma' = 0,$$

and therefore $$PN_1 = 0.$$

Now, since $T_{\beta_1, \beta_2, \ldots, \beta_{h'}}$ comes after $T_{\alpha_1, \alpha_2, \ldots, \alpha_h}$, then $\beta_1 > \alpha_1$; or $\beta_1 = \alpha_1$ and $\beta_2 > \alpha_2$; or $\beta_1 = \alpha_1$, $\beta_2 = \alpha_2$, ..., $\beta_{i-1} = \alpha_{i-1}$, but $\beta_i > \alpha_i$.

Let $$N_1 = \Gamma'_{\beta_1} \Gamma'_{\beta_2} \ldots \Gamma'_{\beta_{h'}},$$

the degrees of the different groups being equal to their suffixes. Suppose that $\beta_1 = \alpha_1$, and that Γ'_{β_1} contains no pair of letters which occur in any one column of the table for S (otherwise $PN_1 = 0$), and that $\beta_2 > \alpha_2$. Then Γ'_{β_1} contains one letter belonging to each of the columns, that is, one letter belonging to each of the $\beta_1 = \alpha_1$ groups of P. We will for the moment suppress all these letters belonging to Γ'_{β_1}. When this is done, let P become P', N_1 become N_1'; then P' and N_1' are related in exactly the same way as P and N_1 are. Thus there are only α_2 groups in P', and α_2 columns in the table which gives P', for one letter from each group or column has been suppressed, and thus $\alpha_1 - \alpha_2$ groups have gone altogether. But all the β_2 letters of Γ'_{β_2} occur in the table for P'; and, since $\beta_2 > \alpha_2$, some one of the α_2 columns of P' must contain more than one of the letters of Γ'_{β_2}; hence

$$P'\Gamma'_{\beta_2} = 0.$$

But P is obtained from P' by adding α_1 new letters to its groups; and hence, if one of the groups of P' has a pair of letters in common with Γ'_{β_2}, the same is true for P; and therefore

$$P\Gamma'_{\beta_2} = 0,$$

and $$PN_1 = 0.$$

The argument is exactly the same in the general case

$$\beta_1 = \alpha_1,\ \beta_2 = \alpha_2,\ \ldots,\ \beta_{i-1} = \alpha_{i-1},\ \beta_i > \alpha_i.$$

The letters of each of the groups $\Gamma'_{\beta_1}, \Gamma'_{\beta_2}, \ldots, \Gamma'_{\beta_{i-1}}$ are suppressed, it being supposed that

$$P\Gamma'_{\beta_1}\Gamma'_{\beta_2} \ldots \Gamma'_{\beta_{i-1}}$$

does not vanish. Then, if P and N_1 become P' and N_1', these products are related to each other in the same way as P and N_1, and the necessary consequence of $\beta_i > \alpha_i$ becomes

$$P'N_1' = 0,$$

for there is a group in N_1' which contains more letters than there are different columns in the table for P', and hence it must contain a pair of letters from the same column. Then P and N are obtained from P' and N_1' by adding new letters and new groups; but the letters in P' and N_1' are left undisturbed. Hence, if

$$P'N_1' = 0,$$

then

$$PN_1 = 0.$$

Hence, provided $T_{\beta_1, \beta_2, \ldots, \beta_{h'}}$ comes after $T_{a_1, a_2, \ldots, a_h}$, when the terms are arranged in order,

$$PN_1 = 0;$$

and hence

$$PS_1 = PN_1P_1 = 0,$$

and

$$PT_{\beta_1, \beta_2, \ldots, \beta_{h'}} = P.\Sigma S_1 = 0;$$

therefore

$$NP.T_{\beta_1, \beta_2, \ldots, \beta_{h'}} = 0,$$

and

$$T_{a_1, a_2, \ldots, a_h} T_{\beta_1, \beta_2, \ldots, \beta_{h'}} = (\Sigma NP)\, T_{\beta_1, \beta_2, \ldots, \beta_{h'}} = 0.$$

Let $t_{a_1, a_2, \ldots, a_h}$ represent the sum of all those substitutions of the group $\{a_1 a_2 \ldots a_n\}$ which are formed of h cycles of orders $a_1, a_2, \ldots, a_h$ respectively. Then, from the way in which $T_{a_1, a_2, \ldots, a_h}$ is formed, viz., as the sum of the expressions obtained when the letters in the table are permuted in any way, but so that the number of letters in any row or column is unchanged, it follows that $T_{a_1, a_2, \ldots, a_h}$ is a function of the expressions $t_{\beta_1, \beta_2, \ldots, \beta_{h'}}$ only. That is, if it contains any one substitution s multiplied by some constant, it contains every substitution similar to s multiplied by the same constant. Hence

$$T_{a_1, a_2, \ldots, a_h} = \Sigma \lambda_{\beta_1, \beta_2, \ldots, \beta_{h'}} t_{\beta_1, \beta_2, \ldots, \beta_{h'}},$$

where the λ's are constants.

Consider the coefficient of the identical substitution in the product

$$T_{a_1, a_2, \ldots, a_h} . T_{a_1, a_2, \ldots, a_h} \equiv T^2_{a_1, a_2, \ldots, a_h}.$$

To obtain it we have to multiply each term λs of the first T by the term $\lambda' s^{-1}$, involving the inverse substitution, in the second factor. But every substitution is similar to its own inverse, and therefore, if s is a term of $t_{\beta_1, \beta_2, \ldots, \beta_{h'}}$, s^{-1} is also a term of this expression. It follows from the form just found for $T_{a_1, a_2, \ldots, a_h}$ that the coefficients of s and s^{-1} are the same. Consequently the coefficient of the identical substitution in $T^2_{\alpha_1, \alpha_2, \ldots, \alpha_h}$ is

$$\Sigma \mu \lambda^2_{\beta_1, \beta_2, \ldots, \beta_{h'}},$$

where μ is the number of different substitutions in the sum

$$t_{\beta_1, \beta_2, \ldots, \beta_{h'}}.$$

Now, every term of $\Sigma \mu \lambda^2_{\beta_1, \beta_2, \ldots, \beta_{h'}}$ is essentially positive, for no unreal quantities can occur in the formation of T; this coefficient cannot then be zero. Consequently $T^2_{\alpha_1, \alpha_2, \ldots, \alpha_h}$ does not vanish identically.

We can now prove that no relation exists between the T's; for, suppose that one such exists, of which the first term when the T's are arranged according to their proper order is $\lambda T_{\alpha_1, \alpha_2, \ldots, \alpha_h}$. Multiply this equation by $T_{\alpha_1, \alpha_2, \ldots, \alpha_h}$; then every term but the first vanishes: for $TT' = 0$ if T' comes after T. Hence

$$\lambda T^2_{\alpha_1, \alpha_2, \ldots, \alpha_h} = 0;$$

and therefore, by what we have just proved, $\lambda = 0$. Hence $T_{\alpha_1, \alpha_2, \ldots, \alpha_h}$ cannot be the first term, and the relation is impossible.

The expression $t_{\alpha_1, \alpha_2, \ldots, \alpha_h}$ has been defined as the sum of all the substitutions of the group $\{a_1 a_2 \ldots a_n\}$ which are formed of cycles whose orders are $\alpha_1, \alpha_2, \ldots, \alpha_h$ respectively; if cycles order 1 are taken into consideration, the condition

$$\alpha_1 + \alpha_2 + \ldots + \alpha_h$$

may be introduced. Further, the order of the α's in the suffixes of $t_{\alpha_1, \alpha_2, \ldots, \alpha_h}$ is immaterial, so that they may be supposed to be in descending order of magnitude. Then $t_{\alpha_1, \alpha_2, \ldots, \alpha_h}$ thus defined depends on exactly the same numbers as $T_{\alpha_1, \alpha_2, \ldots, \alpha_h}$; hence there are the same number of expressions t as expressions T. Moreover, every T can be expressed in terms of the t's, and no relation can exist between the T's alone; so that we have the same number of independent linear equations as unknown quantities $t_{\alpha_1, \alpha_2, \ldots, \alpha_h}$. It is then possible to solve; hence in general

$$t_{\alpha_1, \alpha_2, \ldots, \alpha_h} = \Sigma \mu_{\beta_1, \beta_2, \ldots, \beta_{h'}} \cdot T_{\beta_1, \beta_2, \ldots, \beta_{h'}},$$

where μ is numerical; and therefore in particular

$$1 = t_{1, 1, 1, \ldots, 1} = \Sigma A_{\alpha_1, \alpha_2, \ldots, \alpha_h} \cdot T_{\alpha_1, \alpha_2, \ldots, \alpha_h}.$$

16. For $n = 2, 3, 4$, the work of finding the coefficients of the series by direct calculation is not too laborious: the results are

$$n = 2, \quad 1 = \tfrac{1}{2}\{a_1 a_2\} + \tfrac{1}{2}\{a_1 a_2\}';$$

$$n = 3, \quad 1 = \frac{1}{3!}\{a_1 a_2 a_3\} + \frac{1}{9}\Sigma\{a_1 a_2\}'\{a_1 a_3\} + \frac{1}{3!}\{a_1 a_2 a_3\}';$$

$$n = 4, \quad 1 = \frac{1}{4!}\{a_1 a_2 a_3 a_4\} + \frac{1}{32}\Sigma\{a_1 a_2\}'\{a_2 a_3 a_4\}$$

$$+ \frac{1}{36}\Sigma\{a_1 a_2\}'\{a_3 a_4\}'\{a_2 a_3\}\{a_1 a_4\}$$

$$+ \frac{1}{32}\Sigma\{a_1 a_2 a_3\}'\{a_3 a_4\} + \frac{1}{4!}\{a_1 a_2 a_3 a_4\}'.$$

It is worthy of remark too that, if N be the product of the negative and P that of the positive symmetric groups of one of the expressions of which T is the sum, then

$$T = \Sigma NP = \Sigma PN.$$

For
$$T = \Sigma NP = \lambda \Sigma PNP = \Sigma PN,$$

since PNP is equal to a numerical multiple of

$$(\Sigma N)\, P,$$

where ΣN is the sum of the different expressions obtained from N by operating on N with all the substitutions of P; for it was shown in § 10 that, if $S\,[a_1 b_1 b_2 \ldots b_m]$ is any substitutional expression affecting the letters $a_1, b_1, b_2, \ldots, b_m$,

$$\{a_1 a_2 \ldots a_n\}\, S\,[a_1 b_1 \ldots b_m]\,\{a_1 \ldots a_n\}$$
$$= (n-1)!\,\big[\Sigma S\,[a_1, b_1, \ldots b_m]\big]\,\{a_1 \ldots a_n\},$$

the result stated here being an extension of this. In the same way, PNP is the same multiple of

$$P\,(\Sigma N).$$

It is easy now to show that, if T and T' be any two different terms of the sum of § 15, then

$$T \,.\, T' = 0.$$

For, let
$$T = \Sigma NP, \quad T' = \Sigma N'P';$$

then, if T comes before T' in the series, it has been shown already that
$$T \,.\, T' = 0.$$

Suppose, then, that T comes after T'; then

$$T.T' = [\Sigma NP][\Sigma N'P']$$
$$= [\Sigma PN][\Sigma P'N'];$$

but in this case

$$NP' = 0;$$

hence
$$TT' = 0,$$

whenever T and T' are different.

Multiply now the series of § 15 by

$$T_{a_1, a_2, \ldots, a_h};$$

we then obtain $T_{a_1, a_2, \ldots, a_h} = A_{a_1, a_2, \ldots, a_h} \cdot T^2_{a_1, a_2, \ldots a_h}$.

17. The theorem of § 15, like that of § 13, is purely a substitutional identity; algebraic theorems may be deduced from it by suitably choosing the operand.

(1) *Capelli's Theorem.* — Let the operand be the function $f(a_1, a_2, \ldots, a_n)$ of the n sets of variables

$$\begin{array}{cccc} a_{1,1}, & a_{1,2}, & \ldots, & a_{1,m}, \\ a_{2,1}, & a_{2,2}, & \ldots, & a_{2,m}, \\ \ldots & \ldots & & \ldots \\ a_{n,1}, & a_{n,2}, & \ldots, & a_{n,m}, \end{array}$$

homogeneous and linear in the m variables of each set, such as was under discussion in § 12.

Let $\{a_1 a_2 \ldots a_\alpha\}$ be the positive symmetric group of α of the sets; then

$$\{a_1 a_2 \ldots a_\alpha\} f(a_1, a_2, \ldots, a_n)$$

may be obtained by means of polar operations only from the function

$$f(a_1, a_1, \ldots, a_1, a_{\alpha+1}, a_{\alpha+2}, \ldots, a_n).$$

For, if $\lambda a_{1,r_1} a_{2,r_2} \ldots a_{\alpha,r_\alpha} a_{\alpha+1,r_\alpha+1} \ldots a_{n,r_n}$ be any term of f, then

$$\{a_1 a_2 \ldots a_\alpha\} \lambda a_{1,r_1} a_{2,r_2} \ldots a_{\alpha,r_\alpha} \ldots a_{n,r_n}$$
$$= D_{a_1 a_2} D_{a_1 a_3} \ldots D_{a_1 a_\alpha} \lambda a_{1,r_1} a_{1,r_2} \ldots a_{1,r_\alpha} a_{\alpha+1,r_{\alpha+1}} \ldots a_{n,r_n},$$

where D_{xy} is Capelli's operator

$$y_1 \frac{\partial}{\partial x_1} + y_2 \frac{\partial}{\partial x_2} + \dots + y_m \frac{\partial}{\partial x_m}.$$

Hence $\{\mathfrak{a}_1 \mathfrak{a}_2 \dots \mathfrak{a}_u\} f = D_{a_1 a_2} D_{a_1 a_3} \dots D_{a_1 a_u} D_{a_2 a_1} D_{a_3 a_1} \dots D_{a_u a_1} f.$

And in the same way, if P be the product of β positive symmetric groups no two of which contain the same letter, and which between them contain all the letters $a_1, a_2, \dots, a_n$, groups of degree unity being taken into account, then Pf is a function which may be obtained by means of polar operations only from a function f_1 which contains only β variables, and f_1 is obtainable by means of polar operations only from f.

Again, it was shown in § 12 that

$$\{\mathfrak{a}_1 \mathfrak{a}_2 \dots \mathfrak{a}_u\}' f = H_{a_1 a_2 \dots a_u} f.$$

Hence
$$T_{\alpha_1, \alpha_2, \dots, \alpha_h} f = \Sigma H_{\alpha_1} H_{\alpha_2} \dots H_{\alpha_h} \Delta f$$
$$= \Sigma \Delta H_{\alpha_1} H_{\alpha_2} \dots H_{\alpha_h} f,$$

where, if $$T = \Sigma NP,$$

H_{α_1} is that H which affects the letters contained in the negative symmetric group degree α_1 of N, H_{α_2} that which affects the letters of the group degree α_2, and so on, and where Δ is the polar operation corresponding to P the form of which we have shown how to find.

If it is required to expand a function $F(x, y, \dots, u)$ of m sets of variables, there being m variables in each set, which is homogeneous but not linear in the variables of the different sets, we may obtain from this a function

$$f(a_1, a_2, \dots, a_n)$$

homogeneous and linear in the variables of each of n sets, there being m variables in each set, such that, when we put

$$a_1 = a_2 = \dots = a_{\rho_1} = x, \quad a_{\rho_1+1} = \dots = a_{\rho_2} = y, \quad \dots\dots, \quad a_n = u,$$

f becomes F; this was shown in § 12. Now, f may be expanded as we have just seen; in the result, the variables of F may be substituted for those of f, and the expansion becomes that for F. This expansion is the same as that obtained by Capelli, and quoted in § **11**.

For, if $\alpha_1 < m$, the function $T_{\alpha_1, \alpha_2, \dots, \alpha_h} f$ may be obtained from f_1 by means of polar operators only, where f_1 is a function of a_1 sets of variables, obtained from f by means of polar operators only. If $a_1 > m$, then

$$T_{a_1, a_2, \ldots, a_h} f = 0.$$

And, if $a_1 = a_2 = \ldots = a_i = m$, $a_{i+1} < m$, then $T_{a_1, a_2, \ldots, a_h} f$ gives rise to a term $| xy \ldots u |^i \phi$, where ϕ is a function obtained from a function of not more than $m-1$ variables by means of polar operations only, which is itself to be obtained by means of polar and Ω operations only from either f or F. For, in the expression P, where

$$T_{a_1, a_2, \ldots, a_h} f = \Sigma PNf$$

$$= \Sigma P . H_{a_1} H_{a_2} \ldots H_{a_h} f,$$

there are only a_{i+1} groups which affect the letters of

$$\Omega_{a_1} \Omega_{a_2} \ldots \Omega_{a_i} f,$$

where by Ω_a is understood the Ω operator which affects the letters contained in H_a.

The expansion might otherwise be obtained, viz., by considering the function

$$F = a_{1_x} a_{2_x} \ldots a_{p_{1_x}} a_{p_1+1_y} \ldots a_{n_u},$$

where the factors of f are m-ary symbolical factors, and then proceeding in a similar manner to that in which Gordan's series was obtained in § 14.

(2) *Peano's Theorem.**—Starting from Capelli's theorem, Peano has proved the following:—"The complete system of concomitants for any number of binary n-ics may be obtained from that for n n-ics by polarization alone; with the one possible exception of that invariant which is the determinant of $n+1$ rows formed by the coefficients of $n+1$ n-ics." He then deduced that the number of concomitant types of a binary n-ic is finite; and proceeded to find the types for a binary cubic, showing that they all give irreducible forms for two cubics because the invariant determinant type referred to above is reducible for the cubic. I have quoted his results for the cubic in my paper, already referred to, on "The Irreducible Concomitants of any Number of Binary Quartics." Peano's theorem may be deduced

* *Atti di Torino*, t. XVII., p. 580.

directly from that of § 15:—Let F be a type of a binary m-ic of degree n, linear in the coefficients of each of n m-ics; then

$$F = \Sigma A_{a_1, a_2, \ldots, a_h} \cdot T_{a_1, a_2, \ldots, a_h} F.$$

If $a_1 > m+1$, $T_{a_1, a_2, \ldots, a_h} F = 0$; if $a_1 = m+1$, $T_{a_1, a_2, \ldots, a_h} F$ is the sum of terms each one of which has for a factor the determinant of $m+1$ rows formed from the coefficients of $m+1$ of the m-ics, and is in consequence reducible. If $a_1 < m+1$, then

$$T_{a_1, a_2, \ldots, a_h} F = [\Sigma PN] F,$$

where P is the product of a_1 positive symmetric groups, no two of which contain a common element, and which between them contain all the letters $a_1, a_1, \ldots, a_n$; it being possible that one or more of these groups is of degree unity. In this case PNF is a function obtained by polarization from a function F_1 of only a_1 sets of variables, where F_1 is a function obtained by polarization from F, as has been proved already. Hence $T_{a_1, a_2, \ldots, a_h} F$ is reducible, unless F gives an irreducible concomitant for a_1 m-ics; for concomitants obtained by polarization from reducible concomitants are themselves reducible. Hence, if F is a type of a binary m-ic, which gives no irreducible concomitant for m m-ics, it is reducible, unless F is the determinant of $m+1$ rows formed by the coefficients of $m+1$ of the m-ics. Now, if $a_1 = m$, then

$$T_{a_1, a_2, \ldots, a_h} F = [\Sigma \Gamma'_{a_1} S_1] F,$$

where Γ'_{a_1} is the negative symmetric group degree a_1 in each of the expressions $\Gamma'_{a_1} S_1$ of which T is the sum. But it has been shown, § 8, that, if Φ be a concomitant type of a binary m-ic, and if $\{a_1, a_2, \ldots, a_m\}'$ be the negative symmetric group of the letters $a_1, a_2, \ldots, a_m$, each letter referring to a different quantic, then

$$\{a_1 a_2 \ldots a_m\}' \Phi = | \, a_1 a_2 \ldots a_m Q \, | \, ,$$

where Q refers to the coefficients of a concomitant type of order m, viz., $(Q_0, Q_1, \ldots, Q_m \between x_1, x_2)^m$. Hence, as Γ'_{a_1} is a negative symmetric group degree $a_1 = m$, in this case

$$T_{a_1, a_2, \ldots, a_h} F = \Sigma \, | \, a_1 a_2 \ldots a_m Q \, | \, .$$

And it follows that every rational integral concomitant of any number of m-ics can be expressed as a sum of terms each of which is a product of concomitants of types which give irreducible forms for $m-1$ m-ics, and of types of the form

$$|\, a_1 a_2 \ldots a_m Q \,| ,$$

where $(Q_0, Q_1, \ldots, Q_m \between x_1, x_2)^m$ is a covariant type order m. If $|\, a_1, a_2, \ldots, a_{m+1} \,|$ is reducible as in the case of the cubic, it follows at once that $|\, a_1 a_2 \ldots a_m Q \,|$ is reducible; and hence that all types which give no irreducible form for $m-1$ m-ics are reducible.

Similar results follow for ternary forms, and, in fact, for forms with any number of variables. Thus, for types of the ternary m-ic, we suppose, as before, that each letter refers to one m-ic, and that the coefficients of the m-ic a_1 are

$$a_{1,1}\ a_{1,2} \ldots a_{1,\frac{1}{2}(m+1)(m+2)}.$$

Thus we are dealing in reality with functions of n sets of variables, there being $\frac{1}{2}(m+1)(m+2)$ variables in each set. Every type which gives no irreducible concomitant for $\frac{1}{2}(m+1)(m+2)-1$ m-ics is reducible, with the single exception of the determinant of $\frac{1}{2}(m+1)(m+2)$ rows formed by the coefficients of this number of m-ics.

Moreover, the proof has nothing to do with the fact that the functions are invariant; except that none of the operations employed destroy the property of invariance. Similar results might be deduced for other kinds of algebraic functions.

Again, if $F=0$ be a syzygy between types of a binary m-ic, then every term of $\Gamma' F$ vanishes when Γ' is a negative symmetric group of degree greater than $m+1$. Hence, expanding F by the theorem of § 15, it follows that every syzygy between types must give at least one syzygy, when not more than $m+1$ m-ics are under discussion, which does not reduce to a mere identity; with the exception of syzygies which are wholly due to the fact that

$$\Gamma' Q = 0,$$

where Γ' is a negative symmetric group degree greater than $m+1$, and Q is any product of m-ic types. For, suppose that $F=0$ is a syzygy which always reduces to an identity when less than $m+2$ binary m-ics are under discussion; then, if $\alpha_1 < m+2$, each of the terms

$$T_{\alpha_1, \alpha_2, \ldots, \alpha} F$$

is identically zero. Further, if $\alpha_1 > m+1$, each of the terms

$$T_{\alpha_1, \alpha_2, \ldots, \alpha_h} F$$

is zero, being the sum of terms such as $\Gamma' Q$ mentioned above. Hence

F, which is $= \Sigma A_{a_1, \ldots, a_h} \cdot T_{a_1, \ldots, a_h} F$, is the sum of such terms, and $F = 0$ is a syzygy of that nature. As an example of a syzygy of this nature we have that between quadratic invariant types

$$[ab] = a_0 b_2 + a_2 b_0 - 2a_1 b_1,$$

viz., $$\{\mathrm{bdfh}\}' [ab][cd][ef][gh] = 0.$$

(3) To find the system of concomitants for r binary m-ics. Let F be any type, then, if Γ'_{r+1} be any negative symmetric group degree $r+1$, of the letters $a_1, a_2, \ldots, a_n$,

$$\Gamma'_{r+1} F = 0,$$

for there are not more than r different quantics represented by the letters, so that amongst $r+1$ letters at least two must refer to some one quantic. This is necessary; it is also sufficient, for

$$F = \Sigma A_{a_1, \ldots, a_h} \cdot T_{a_1, \ldots, a_h} F,$$

and, if $a_1 > r_1$, $$T_{a_1, \ldots, a_h} F = 0;$$

but, if a_1 is equal to or less than r, the term is obtainable by polarization from a concomitant of not more than r m-ics. Hence in this case we take the ordinary relations for the type F, coupled with all possible equations of the form

$$\Gamma'_{r+1} F = 0.$$

(4) The complete solution of the simultaneous system of equations

$$\Gamma'_{r+1} F = R,$$

where Γ'_{r+1} is any negative symmetric group of degree $r+1$, of the letters $a_1, a_2, \ldots, a_n$, and there is one equation for every combination of these letters $r+1$ at a time, is

$$F = \Sigma G_1 G_2 \ldots G_\nu f + R', \quad \nu < r+1,$$

where $G_1, G_2, \ldots, G_\nu$ are positive symmetric groups no two of which have a common letter, but which between them contain all the n letters, and R' is a function obtained from the R's by means of substitutions alone. This is evidently a solution, for, provided that R' is chosen so that

$$\Gamma'_{r+1} R' = R,$$

it satisfies each of the equations. Moreover

$$F = \Sigma A_{a_1, a_2, \ldots, a_h} \cdot T_{a_1, a_2, \ldots, a_h} F,$$

and $T_{a_1, a_2, \ldots, a_h} F = R_1$, if $a_1 > r$, where R_1 is obtained in a definite manner from the given functions R, since $T_{a_1, a_2, \ldots, a_h} = \Sigma P.N$, where N contains as a factor a negative symmetric group degree a_1. And further, if $a_1 \leqslant r_1$,

$$T_{a_1, a_2, \ldots, a_h} = \Sigma PN,$$

where

$$P = G_1 G_2 \ldots G_\nu,$$

ν being $< r+1$, and the groups $G_1, G_2, \ldots, G_\nu$ having the character laid down above. The solution is then the complete one, and we see further that, in each term of the sum $\Sigma G_1 G_2 \ldots G_\nu f$, f is such that it may be obtained from F by means of the operation of the product N of certain definite negative symmetric groups.

Conversely, the complete solution of all possible equations of the form

$$G_1 G_2 \ldots G_\nu F = R, \quad \nu < r+1,$$

where the groups $G_1, G_2, \ldots, G_\nu$ are positive symmetric groups, no two of which contain a common letter, and which between them contain all the letters $a_1, a_2, \ldots, a_n$, is

$$F = \Sigma \Gamma'_{r+1} f + R',$$

where Γ'_{r+1} is a negative symmetric group degree $r+1$, and R' a function obtained from the R's by means of substitutions alone; and, further, the f for each term may be obtained from F by means of substitutions alone.

(5) In exactly the same way it may be shown that, if G_{r+1} is a positive symmetric group degree $r+1$, the solution of all possible equations of the form

$$G_{r+1} F = R$$

is

$$F = \Sigma \Gamma'_1 \Gamma'_2 \ldots \Gamma'_\nu f + R', \quad \nu < r+1.$$

(6) Suppose that G_r is the alternating group of certain r letters, that $G_r^{(1)}$ is the positive symmetric group of the same letters, and that $G_r^{(2)}$ is the negative symmetric group. Then, if

$$G_r F = 0,$$

it follows that

$$G_r^{(1)} F = 0 \quad \text{and} \quad G_r^{(2)} F = 0.$$

For, if a and b are any two letters affected by G_r, then

$$G_r^{(1)} = \left[1 + (ab)\right] G_r$$

and

$$G_r^{(2)} = \left[1 - (ab)\right] G_r.$$

Consider, then, the simultaneous system of equations

$$G_r F = 0,$$

where the r letters affected by G_r are chosen in any manner from the letters $a_1, a_2, \ldots, a_n$, there being one equation for each combination of these letters r at a time; then

$$F = \Sigma A_{a_1, a_2, \ldots, a_h} \cdot T_{a_1, a_2, \ldots, a_h} F,$$

and all terms of this expansion vanish in which T possesses either positive or negative symmetric groups of degree $\geqslant r$. Hence, if

$$T_{a_1, a_2, \ldots, a_h} F$$

is not zero, $a_1 < r$ and $h < r$; for a_1 is the degree of the greatest negative symmetric group, and h that of the greatest positive symmetric group contained in T. Now

$$a_1 \not< a_2 \not< a_3 \not< \ldots \not< a_h;$$

and hence $$n = a_1 + a_2 + \ldots + a_h \not> h a_1;$$

and therefore, in order that both h and a_1 may be $< r$, we must have

$$n \not> (r-1) .$$

If therefore $n > (r-1)^2$, every term

$$T_{a_1, a_2, \ldots, a_h} F$$

is zero, and F itself is zero.

On Quantitative Substitutional Analysis (*Second Paper*)

By A. YOUNG

In § 16 of my first paper on the above subject* a series of substitutional expressions was obtained, which was written

$$1 = \Sigma A_{a_1, a_2, \ldots, a_h} T_{a_1, a_2, \ldots, a_h}.$$

The object of the first section of the present paper is to find the coefficient $A_{a_1, a_2, \ldots, a_h}$.

The second section deals with the relations between the forms PN of which one of the terms T of the above series is made up.

The latter part of the paper is devoted to the application of the theory already developed to modern algebra. By means of the above series it is shown that every integral function of the coefficients of any q-ary quantics may be expressed linearly in terms of coefficients of concomitants of these quantics.

The fifth section is devoted to the invariants of a single binary n-ic; it is shown that these may be expressed in terms of forms $f(\alpha_0, \alpha_1, \ldots, \alpha_n)$, where $\alpha_0, \alpha_1, \ldots, \alpha_n$ are certain numbers which completely define the invariant, and are such that

$$\alpha_0 + \alpha_1 + \ldots + \alpha_n = \text{the degree},$$

$$\alpha_1 + 2\alpha_2 + \ldots + n\alpha_n = \alpha_{n-1} + 2\alpha_{n-2} + \ldots + n\alpha_0$$
$$= \text{the weight}.$$

I. *The Coefficients* $A_{a_1, a_2, \ldots, a_h}$ *in the Substitutional Expansion* $1 = \Sigma A_{a_1, a_2, \ldots, a_h} T_{a_1, a_2, \ldots, a_h}$.

1. $T_{a_1, a_2, \ldots, a_h}$ is defined as follows :—

The letters $a_1, a_2, \ldots, a_n$ are arranged in any manner in h horizontal rows, so that each row has its first letter in the same vertical column, its second letter in a second vertical column, and

* *Proc. Lond. Math. Soc.*, Vol. XXXIII., p. 97.

so on; there being a_1 letters in the first row, a_2 in the second, &c., and finally, a_h in the last; the a's satisfying the relations

$$a_1+a_2+\ldots+a_h = n, \quad a_1 \not< a_2 \not< a_3 \ldots \not< a_h.$$

From this table an expression

$$S = \Gamma_1'\Gamma_2' \ldots \Gamma_h' G_1 G_2 \ldots G_k$$

is formed, such that Γ_1' is the negative symmetric group of the letters of the first row, Γ_2' that of the letters of the second row, and so on; G_1 is the positive symmetric group of the letters of the first column, G_2 of the second column, and so on (it being understood that the positive or negative symmetric group of a single letter is unity). Then $T_{a_1, a_2, \ldots, a_h}$ is the sum of the $n!$ expressions—not necessarily all different—obtained by permuting the letters in S in all possible ways.*

The product of the negative symmetric groups in S will be denoted by N, and that of the positive symmetric groups by P. It was proved that

$$T = \Sigma NP = \Sigma PN;$$

also that the product of two different T's is zero; and hence that

$$T_{a_1, a_2, \ldots, a_h} = A_{a_1, a_2, \ldots, a_h} T^2_{a_1, a_2, \ldots, a_h}.$$

2. Consider the expression PNP. If s be any substitution in one of the groups of P, $sNP = N'P$, where N' is the result of operating on N with the substitution s; hence

$$PNP = \Sigma N'P,$$

where the Σ contains $\beta_1!\,\beta_2!\,\ldots\,\beta_k!$ terms—the numbers $\beta_1, \beta_2, \ldots, \beta_k$ being the degrees of the different groups of P.

Hence, if $\lfloor S \rceil$ denotes a substitutional expression S which *operates on*, but does not multiply, the substitutions which follow it,

$$\lfloor\{a_1 a_2 \ldots a_n\}\rceil PNP = \beta_1!\,\beta_2!\,\ldots\,\beta_k!\,\lfloor\{a_1 a_2 \ldots a_n\}\rceil NP$$
$$= \beta_1!\,\beta!\,\ldots\,\beta_k!\,T_{a_1, a_2, \ldots, a_h}.$$

* In the paper referred to $T_{a_1, a_2, \ldots, a_h}$ was defined as the sum of all the *different* expressions S thus obtained. The above definition will be found more convenient.

Similarly it may be proved that

$$\lfloor\{a_1 a_2 \ldots a_n\}\rfloor NPN = a_1!\, a_2! \ldots a_h!\, T_{a_1, a_2, \ldots, a_h},$$

or, if Σ refer only to those terms which are different,

$$T_{a_1, a_2, \ldots, a_h} = \Sigma PNP = \Sigma NPN.$$

3. Consider now the expression

$$P'N'P' . NP.$$

If $P'N$ is not zero, no two letters belonging to any group of P' can occur in the same group of N. Thus each of the h groups of N must contain one letter of the largest group of P'—that order h. And by similar reasoning it follows that a table may be constructed of which the rows represent the groups of N and the columns those of P'. Now

$$P'N'P' = P'N''P',$$

where N'' is obtained from N' by any substitution of P'. Such a substitution is equivalent to a change in the table giving $N'P'$ which does not alter the column in which any letter lies. Thus no alterations in the arrangement of the letters in a column of this table can affect the value of $P'N'P'$. But by such a change we can obtain the table for NP'; hence

$$P'N'P' . NP = P'NP' . NP.$$

The same argument applies to the form

$$NP'N.$$

Hence $$P'NP' . NP = P'NP . NP.$$

But P' is transformed to P by the operation of a substitution s which is contained in N. Hence

$$P'N'P' . NP = P'NP . NP = \pm sPNP\,NP,$$

according as s is even or odd.

Now $$T = \Sigma P'N'P',$$

and therefore $$T . NP = (\Sigma P'N'P')\, NP,$$

where the Σ extends to those forms of P' for which $P'N \neq 0$.

But these forms of P' are those, and all those, which are obtained by operating on P with each of the substitutions of N. Hence

$$T . NP = (\Sigma P'N'P')\, NP = (\Sigma P'NP')\, NP$$
$$= (\Sigma \pm s)\, PNPNP = NPNPNP = (NP)^3.$$

Further $$T = \lfloor\{a_1 a_2 \ldots a_n\}\rceil NP;$$

therefore $$T^2 = \lfloor\{a_1 a_2 \ldots a_n\}\rceil T . NP$$

$$= \lfloor\{a_1 a_2 \ldots a_n\}\rceil (NP)^3.$$

We shall proceed to show that

$$(NP)^2 = \lambda NP;$$

and hence $$T^2 = \lambda^2 T,$$

from which we obtain the coefficient

$$A = \lambda^{-2}.$$

4. If S be any substitutional expression whatever, and G be a positive symmetric group which contains all the letters of one of the groups G_1 of P, and also one letter of a group G_2 of P which is not of higher degree, then

$$NSGP = 0.$$

For $$GP = P_1 S_1,$$

where the groups of P which contain no letters of G are groups of P_1, G replaces the group G_1, and the group G_2 is replaced by that group which contains all its letters except that one contained in G. Now, when the terms $T_{a_1, a_2, \ldots, a_h}$ are placed in order as explained before, it follows that P_1 belongs to an earlier T than P; and hence

$$NP_1 = 0,$$

whatever N is chosen from the terms of the particular T. Now, if s be any substitution

$$Ns = sN',$$

where N' is obtained from N by the interchange of certain of its letters, then, if

$$S = \Sigma \mu s,$$

$$NSGP = N(\Sigma\mu s)\, P_1 S_1 = \Sigma\mu s N' P_1 S_1 = 0.$$

5. Let the two groups of lowest order in P be

$$\{a_1 a_2 \ldots a_{\beta_1}\}\{b_1 b_2 \ldots b_{\beta_2}\} \quad (\beta_2 \not< \beta_1),$$

and let $$P = \{a_1 a_2 \ldots a_{\beta_1}\}\{b_1 b_2 \ldots b_{\beta_2}\}\, P_1.$$

Then $$NP . P = NP\left[1 - (1 + (a_1 b_1) + (a_1 b_2) + \ldots + (a_1 b_{\beta_2}))\right] P,$$

for $$(1 + (a_1 b_1) + (a_1 b_2) + \ldots (a_1 b_{\beta_2}))\{b_1 b_2 \ldots b_{\beta_2}\} = \{a_1 b_1 b_2 \ldots b_{\beta_2}\},$$

and, by what has just been proved,

$$NP\{a_1 b_1 b_2 \dots b_{\beta_2}\} P = 0.$$

Hence
$$NP.P = NP\left[-\beta_2 (a_1 b_1)\right] P$$
$$= NP\left[\beta_2 \{a_1 b_1\}' - \beta_2\right] P;$$

and therefore
$$NP.P = \frac{\beta_2}{\beta_2+1} NP\{a_1 b_1\}' P.$$

Let us assume then that

$$NP.P = \lambda_r NP\{a_1 b_1\}'\{a_2 b_2\}' \dots \{a_r b_r\}' P.$$

Then

$$0 = NP\{a_1 b_1\}'\{a_2 b_2\}' \dots \{a_r b_r\}' \big(1 + (b_1 a_{r+1}) + (b_2 a_{r+1}) + \dots$$
$$\dots + (b_{\beta_2} a_{r+1})\big) P$$
$$= NP\{a_1 b_1\}'\{a_2 b_2\}' \dots \{a_r b_r\}' \big(1 + r(b_1 a_{r+1}) + (\beta_2 - r)(a_{r+1} b_{r+1})\big) P$$
$$= NP\{a_1 b_1\}'\{a_2 b_2\}' \dots \{a_r b_r\}' \big(\beta_2 - r + 1 - (\beta_2 - r)\{a_{r+1} b_{r+1}\}'\big) P,$$

for $NP\{a_1 b_1\}'\{a_2 b_2\}' \dots \{a_r b_r\}' (b_1 a_{r+1}) P$

$$= NP(b_1 a_{r+1})\{a_{r+1} a_1\}'\{a_2 b_2\}' \dots \{a_r b_r\}' P = 0.$$

Hence
$$NP.P = \lambda_r NP\{a_1 b_1\}'\{a_2 b_2\}' \dots \{a_r b_r\}' P$$
$$= \lambda_r \frac{\beta_2 - r}{\beta_2 - r + 1} NP\{a_1 b_1\}' \dots \{a_{r+1} b_{r+1}\}' P;$$

and therefore

$$NP.P = \frac{\beta_2}{\beta_2+1} NP\{a_1 b_1\}' P = \dots$$
$$= \frac{\beta_2 - \beta_1 + 1}{\beta_2 + 1} NP\{a_1 b_1\}'\{a_2 b_2\}' \dots \{a_{\beta_1} b_{\beta_1}\}' P.$$

Let $\{c_1 c_2 \dots c_{\beta_3}\}$ be the group of next lowest order in P. Then

$$0 = NP\{a_1 b_1 c_1\}'\{a_2 b_2 c_2\}' \dots \{a_r b_r c_r\}'\{a_{r+1} b_{r+1}\}' \dots$$
$$\dots \{a_{\beta_1} b_{\beta_1}\}'\{a_{r+1} c_1 c_2 \dots c_{\beta_3}\} P$$
$$+ NP\{a_1 b_1 c_1\}'\{a_2 b_2 c_2\}' \dots \{a_r b_r c_r\}'\{a_{r+1} b_{r+1}\}' \dots$$
$$\dots \{a_{\beta_1} b_{\beta_1}\}'\{b_{r+1} c_1 c_2 \dots c_{\beta_3}\} P$$
$$= NP\{a_1 b_1 c_1\}' \dots \{a_r b_r c_r\}'\{a_{r+1} b_{r+1}\}' \dots \{a_{\beta_1} b_{\beta_1}\}'$$
$$\times \left[1 + r(a_{r+1} c_1) + (\beta_3 - r)(a_{r+1} c_{r+1}) + 1 + r(b_{r+1} c_1)\right.$$
$$\left. + (\beta_3 - r)(b_{r+1} c_{r+1})\right] P.\beta_3!$$
$$= NP\{a_1 b_1 c_1\}' \dots \{a_r b_r c_r\}'\{a_{r+1} b_{r+1}\}' \dots \{a_{\beta_1} b_{\beta_1}\}!$$
$$\times \left[2 + \beta_3 - r - (\beta_3 - r)\big(1 - (a_{r+1} c_{r+1}) - (b_{r+1} c_{r+1})\big)\right] P.\beta_3!\cdot$$

But $\{a_{r+1}b_{r+1}\}'\big(1-(a_{r+1}c_{r+1})-(b_{r+1}c_{r+1})\big)=\{a_{r+1}b_{r+1}c_{r+1}\}'$.

Hence

$$NP\,\{a_1b_1c_1\}'\,\{a_2b_2c_2\}'\,\ldots\,\{a_rb_rc_r\}'\,\{a_{r+1}b_{r+1}\}'\,\ldots\,\{a_{\beta_1}b_{\beta_1}\}'\,P$$

$$=\frac{\beta_3-r}{\beta_3-r+2}NP\,\{a_1b_1c_1\}'\,\{a_2b_2c_2\}'\,\ldots\,\{a_rb_rc_r\}'\,\{a_{r+1}b_{r+1}c_{r+1}\}'$$
$$\times\{a_{r+2}b_{r+2}\}'\,\ldots\,\{a_{\beta_1}b_{\beta_1}\}'\,P.$$

Hence

$$NP.P=\frac{\beta_2-\beta_1+1}{\beta_2+1}\,\frac{(\beta_3-\beta_1+1)(\beta_3-\beta_1+2)}{(\beta_3+1)(\beta_3+2)}NP\{a_1b_1c_1\}'\,\{a_2b_2c_2\}'\,\ldots$$
$$\ldots\{a_{\beta_1}b_{\beta_1}c_{\beta_1}\}'\,P$$

$$=\frac{\beta_2-\beta_1+1}{\beta_2+1}\,\frac{(\beta_3-\beta_1+1)(\beta_3-\beta_1+2)}{(\beta_3+1)(\beta_3+2)}\,\frac{\beta_3-\beta_2+1}{\beta_3-\beta_1+1},$$

$$NP\,\{a_1b_1c_1\}'\,\ldots\,\{a_{\beta_1}b_{\beta_1}c_{\beta_1}\}'\,\{b_{\beta_1+1}c_{\beta_1+1}\}'\,\ldots\,\{b_{\beta_2}c_{\beta_2}\}'\,P.$$

Thus it is easy to see that when the letters of the r-th group (of degree β_r) are introduced into the product of negative symmetric groups a factor must be introduced outside $=\lambda_r$, where

$$\lambda_r=\frac{(\beta_r-\beta_1+r-1)!\,\beta_r!}{(\beta_r-\beta_1)!\,(\beta_2+r-1)!}\,\frac{(\beta_r-\beta_2+r-2)!\,(\beta_r-\beta_1)!}{(\beta_r-\beta_2)!\,(\beta_r-\beta_1+r-2)!}\cdots$$
$$\cdots\frac{(\beta_r-\beta_s+r-s)!\,(\beta_r-\beta_{s-1})!}{(\beta_r-\beta_s)!\,(\beta_r-\beta_{s-1}+r-s)!}\cdots\frac{(\beta_r-\beta_{r-1}+1)!\,(\beta_r-\beta_{r-2})!}{(\beta_r-\beta_{r-1})!\,(\beta_r-\beta_{r-2}+1)!}$$

$$=\frac{(\beta_r-\beta_1+r-1)(\beta_r-\beta_2+r-2)\ldots(\beta_r-\beta_{r-1}+1)\,\beta_r!}{(\beta_r+r-1)!}.$$

Hence
$$NP=\frac{1}{\beta_1!\,\beta_2!\ldots\beta_k!}NP.P$$

$$=\frac{\lambda_1}{\beta_1!}\,\frac{\lambda_2}{\beta_2!}\cdots\frac{\lambda_k}{\beta_k!}(NP)^2$$

$$=\frac{\prod\limits_{\substack{r,s=1,\ldots,k\\ r>s}}(\beta_r-\beta_s+r-s)}{\prod\limits_{r=1}^{r=k}(\beta_r+r-1)!}(NP)^2.$$

Hence in the series

$$1=A'_{\beta_1,\beta_2,\ldots,\beta_k}T'_{\beta_1,\beta_2,\ldots,\beta_k},$$

where $\beta_1,\beta_2,\ldots,\beta_k$ are the degrees of the positive symmetric groups, such that

$$\beta_1 \not> \beta_2 \not> \beta_3 \not> \ldots \not> \beta_k,$$

the coefficient $$A'_{\beta_1, \beta_2, \ldots, \beta_k} = \left(\frac{\prod_{r, s} (\beta_r - \beta_s + r - s)}{\prod_r (\beta_r + r - 1)!} \right)^2.$$

6. The numbers $\beta_1, \beta_2, \ldots, \beta_k$ are here in ascending order of magnitude: we have, however, been taking the suffixes in descending order; then, if

$$\gamma_1 = \beta_k, \; \gamma_2 = \beta_{k-1}, \; \ldots, \; \gamma_k = \beta_1,$$

$$A''_{\gamma_1, \gamma_2, \ldots, \gamma_k} = A'_{\beta_1, \beta_2, \ldots, \beta_k} = \left(\frac{\prod_{r, s} (\gamma_r - \gamma_s - r + s)}{\prod_r (\gamma_r + k - r)!} \right)^2.$$

In this series $$1 = \Sigma A''_{\gamma_1, \gamma_2, \ldots, \gamma_k} T''_{\gamma_1, \gamma_2, \ldots, \gamma_k}.$$

Change the sign of every transposition; then

$$T''_{\gamma_1, \gamma_2, \ldots, \gamma_k}$$

becomes $$T_{\gamma_1, \gamma_2, \ldots, \gamma_k},$$

and, since the constants are unaffected by this change of sign, we obtain

$$A_{a_1, a_2, \ldots, a_h} = \left(\frac{\prod_{r, s} (a_r - a_s - r + s)}{\prod_r (a_r + h - r)!} \right)^2$$

for the value of the constants in the series

$$1 = \Sigma A_{a_1, a_2, \ldots, a_h} T_{a_1, a_2, \ldots, a_h}.$$

It is to be observed that $T_{a_1, \ldots, a_h}$ might be equally well defined by its positive symmetric groups: the value of the constant must be the same with this definition. Now the degrees of the positive symmetric groups of $T_{a_1, a_2, \ldots, a_h}$ are as follows: there are a_h groups of degree h; $a_{h-1} - a_h$ of degree $h-1$; $a_{h-2} - a_{h-1}$ of degree $h-2$, and so on; finally, there are $a_1 - a_2$ groups of degree unity. If these numbers be substituted for $a_1, a_2, \ldots, a_h$, it can be verified that the resulting coefficient $= A_{a_1, a_2, \ldots, a_h}$.

7. The following particular results of this investigation should be noticed.

If NP be a term of T, the coefficient in the series

$$1 = \Sigma A_{a_1, \ldots, a_h} T_{a_1, \ldots, a_h}$$

of T being A; then
$$(NP)^2 = A^{-1}NP,$$
$$T.NP = A^{-1}NP,$$
$$T.PN = A^{-1}PN.$$

Again,
$$\begin{aligned} TN &= (\Sigma P'N'P')\,N \\ &= \Sigma P'NP'N \\ &= NPNPN \\ &= A^{-1}NPN. \end{aligned}$$

Similarly,
$$T.P = A^{-1}PNP.$$

8. If P be expressed as a sum of substitutions, the coefficient of each is unity. Similarly, if N be expressed as a sum of substitutions, the coefficient of each is ± 1. Moreover, if two letters appear in the same cycle of one of the substitutions of P, they cannot appear in the same cycle of a substitution of N. Hence, if s be a substitution of P, s^{-1} cannot appear in N unless $s = 1$. The coefficient of the identical substitution in NP must then be 1.

Again, the coefficient of every substitution in NP is ± 1. For, let t be any substitution of this product, and let it arrive by taking the substitution s_1 of N with σ_1 of P, so that
$$s_1\sigma_1 = t.$$

Suppose that it occurs a second time in the sum, as, say, $s_2\sigma_2$. Then
$$s_1\sigma_1 = s_2\sigma_2,$$
and therefore
$$s_2^{-1}s_1\sigma_1\sigma_2^{-1} = 1.$$
But $s_2^{-1}s_1$ is a substitution of N, and $\sigma_1\sigma_2^{-1}$ is a substitution of P; hence
$$s_2^{-1}s_1 = 1 = \sigma_1\sigma_2^{-1},$$
and therefore
$$s_1 = s_2, \quad \sigma_1 = \sigma_2.$$
The substitution t then can only occur once in NP. The same remark applies to PN.

Now
$$T = \lfloor\{(a_1 a_2 \ldots a_n)\}\rfloor\, NP:$$
the coefficient of the identical substitution in T must then be $n!$. Hence we obtain the identity
$$\frac{1}{n!} = \Sigma A_{a_1, a_2, \ldots, a_n}$$
by equating the coefficients of the identical substitution. Remember-

ing that the operations are made in the order expressed by the equation

$$s_1 s_2 F = s_1 [s_2 F],$$

we observe that, if S be any substitutional expression,

$$\sigma S \sigma^{-1} = \lfloor \sigma \rfloor S.$$

Hence, if

$$S = \lfloor \{a_1 a_2 \ldots a_n\} \rfloor S',$$

$$\sigma S \sigma^{-1} = S$$

or

$$\sigma S = S \sigma.$$

It follows from this that any one of the expressions T is commutative with any single substitution; and therefore with any substitutional expression whatever which contains no letters which do not appear in T.

II. *Relations between different Forms NP.*

9. If s be any substitution, we obtain by means of the T series

$$s = \Sigma A_{a_1, a_2, \ldots, a_h} T_{a_1, a_2, \ldots, a_h} s\,;$$

thus every substitution can be expressed in terms of forms NPs (or PNs).

Between forms NP certain linear relations occur. To discuss them it is necessary to employ a notation which will completely define the form under discussion.

We will write

$$\left\{\begin{matrix} a_{1,1}\, a_{1,2} \cdots\cdots\cdots\cdots a_{1,\beta_1} \\ a_{2,1}\, a_{2,2} \cdots\cdots\cdots a_{2,\beta_2} \\ \cdots \quad \cdots \quad \cdots \quad \cdots \\ a_{k,1}\, a_{k,2} \cdots a_{k,\beta_k} \end{matrix}\right\} \equiv \{a_{1,1}\, a_{1,2} \ldots a_{1,\beta_1}\}\{a_{2,1} \ldots a_{2,\beta_2}\} \ldots$$

$$\ldots\{a_{k,1}\, a_{k,2} \ldots a_{k,\beta_k}\}\{a_{1,1}\, a_{2,1} \ldots a_{k,1}\}' \ldots$$

$$\ldots \{a_{1,\beta_1}\}'$$

$$= PN.$$

Here the rows are taken to define the positive symmetric groups, while the columns define the negative symmetric groups.

Again, we will write

$$\left\{\begin{matrix} a_{1,1}\, a_{1,2} \cdots\cdots\cdots a_{1,\alpha_1} \\ a_{2,1}\, a_{2,2} \cdots a_{2,\alpha_2} \\ \cdots \quad \cdots \quad \cdots \\ a_{h,1} \cdots a_{h,\alpha_h} \end{matrix}\right\}' \equiv \{a_{1,1}\, a_{1,2} \ldots a_{1,\alpha_1}\}' \{a_{2,1} \ldots a_{2,\alpha_2}\}' \ldots$$

$$\ldots \{a_{h,1} \ldots a_{h,\alpha_h}\}' \{a_{1,1}\, a_{2,1} \ldots a_{h,1}\} \ldots \{a_{1,\alpha_1}\}$$

$$= NP,$$

where the rows define the negative symmetric groups, the columns the positive symmetric groups. The notation used for a single positive or negative symmetric group is, it will be noticed, a particular case of that now introduced.

10. The product $(s > r)$

$$\{a_{r,1} \dots a_{r,\beta_r} a_{s,\lambda}\} \left\{ \begin{matrix} a_{1,1}\, a_{1,2} \, \dots\dots\dots \, a_{1,\beta_1} \\ a_{2,1} \, \dots\dots\dots\dots \\ \dots \quad \dots \quad \dots \\ a_{r,1} \, \dots\dots\dots \, a_{r,\beta_r} \\ \dots \quad \dots \quad \dots \\ a_{h,1} \dots a_{h,\beta_h} \end{matrix} \right\}$$

is zero. This follows at once from § 4.

Similarly, the product $(s > r)$

$$\left\{ \begin{matrix} a_{1,1}\, a_{1,2} \dots a_{1,r} \dots a_{1,\beta_1} \\ a_{2,1} \, \dots\dots \\ \dots\dots\dots\dots \, a_{a_n, r} \\ a_{h,1} \, \dots\dots \, a_{h,\beta_h} \end{matrix} \right\} \{a_{1,r}\, a_{2,r} \dots a_{a_r,r}\, a_{\lambda,s}\}' = 0.$$

All relations between forms PNS may be linearly obtained from relations of these two kinds and the obvious relations

$$\sigma PN\sigma^{-1} = \lfloor\sigma\rceil PN,$$

$$(a_{r,\lambda}\, a_{r,\mu})\, PN = PN,$$

$$PN\,(a_{\lambda,r}\, a_{\mu,r}) = -PN.$$

It is to be observed that, since

$$\{a_{r,1} \dots a_{r,\beta_r} a_{s,\lambda}\}$$
$$= \left[1 + (a_{s,\lambda}\, a_{r,1}) + (a_{s,\lambda}\, a_{r,2}) + \dots + (a_{s,\lambda}\, a_{r,\beta_r})\right] \{a_{r,1} \dots a_{r,\beta_r}\}$$

and

$$\{a_{1,r}\, a_{2,r} \dots a_{a_r,r}\, a_{\lambda,s}\}'$$
$$= \{a_{1,r}\, a_{2,r} \dots a_{a_r,r}\}\left[1 - (a_{1,r}\, a_{\lambda,s}) - \dots - (a_{a_r,r}\, a_{\lambda,s})\right],$$

the above identities may be written

$$\left[1+(a_{s,\lambda}\,a_{r,1})+(a_{s,\lambda}\,a_{r,2})+\ldots+(a_{s,\lambda}\,a_{r,\beta_r})\right]PN=0$$

and $$PN\left[1-(a_{1,r}\,a_{\lambda,s})-(a_{2,r}\,a_{\lambda,s})-\ldots-(a_{\alpha_r,r}\,a_{\lambda,s})\right]=0.$$

11. Before establishing the theorem just enunciated it is necessary to prove that, if s be any substitution,

$$PsPN = \lambda\, PN$$

and $$PN\,sN = \mu\, PN,$$

where λ and μ are numerical constants. Let us suppose s written as a product of cycles; then if any cycle σ contains two letters b_1, b_2 out of the same group of P,

$$P\sigma = P\sigma'\sigma'',$$

where $\sigma'\sigma''$ are two independent cycles all the letters of which appear in σ, but which are such that b_1 occurs in σ' and b_2 in σ''.

To prove this we observe that

$$P\sigma = P\,(b_1b_2)\,\sigma,$$

and consider the following cases.

(i.) Let $$\sigma = (ab_1\,b_2c\,\ldots);$$

then $$(b_1b_2)\,\sigma = (ab_2c\,\ldots) = \sigma'', \quad \sigma' = 1.$$

(ii.) Let $$\sigma = (ab_1c\,\ldots\,db_2ef\,\ldots);$$

then $$(b_1b_2)\,\sigma = (ab_2ef\,\ldots)(b_1c\,\ldots\,d).$$

Let the greatest group of P in which any of the letters of σ appear be $\{a_1a_2\,\ldots\,a_r\}$. Suppose that of these letters a_1, a_2, ..., a_h appear in s, necessarily all in different cycles. The cycle in which a_h appears we will call σ_1, and let

$$s = \sigma_1\sigma'.$$

Now σ_1 may be written in the form

$$\sigma_1'\,(a_hb),$$

where σ_1' is a cycle which contains the letter b, but not the letter a_h, and where b belongs to a group of P whose degree is $\not> r$.

Then

$$PsPN = P\sigma'\sigma_1'(a_h b)\,PN$$

$$= \frac{1}{r+1-h} P\sigma'\sigma_1'\left[(a_h b)+(a_{h+1} b)+\ldots+(a_r b)\right] PN$$

$$= \frac{1}{r+1-h} P\sigma'\sigma_1'\left[-\big(1+(a_1 b)+(a_2 b)+\ldots+(a_{h-1} b)\big) + \big(1+(a_1 b)+(a_2 b)+\ldots+(a_n b)\big)\right] PN$$

$$= \frac{-1}{r+1-h} P\sigma'\sigma_1'\left[1+(a_1 b)+(a_2 b)+\ldots+(a_{h-1} b)\right] PN,$$

owing to the relation satisfied by PN,

$$= \frac{-1}{r+1-h} P\,[\Sigma s']\,PN,$$

where s' affects fewer letters than s.

Proceeding thus step by step, we obtain

$$Ps\,PN = \lambda\,PN,$$

and similarly $$PNsN = \mu\,PN,$$

where λ and μ are numerical.

12. We proceed now to show that the relations of § 10 embrace all relations satisfied by PN. In the first place, it is to be noticed that the only relations used as yet are those of § 10; whether in proving that

$$TT' = 0, \quad (PN)^2 = A^{-1}PN, \quad TPN = A^{-1}PN,$$

or in the theorem of the last paragraph.

Let $$\Sigma\lambda_i P_i N_i s_i = 0 \qquad \text{(I.)}$$

be any such relation.

If all the forms $P_i N_i$ do not belong to the same member of the series $$1 = \Sigma AT,$$

let T be a member which contains some of them.

Then multiply the equation (I.) by T on the left-hand side. If $P_i N_i$ belongs to T, then

$$TP_i N_i s_i = A^{-1} P_i N_i s_i.$$

Otherwise $$TP_i N_i s = 0.$$

Hence we obtain $A^{-1}\{\Sigma\lambda_j P_j N_j s_j\} = 0,$

where the terms of this sum are just those terms of (I.) which belong to T. Hence no relation can connect forms belonging to different T's.

Consider the relation $\Sigma\lambda_j P_j N_j s_j = 0,$ (II.)

where each term belongs to T. Let PN be any definite term of T; then, by means of certain interchanges of the letters, we may change PN into $P_j N_j$; in other words,

$$P_j N_j = \sigma_j PN\sigma_j^{-1}.$$

The relation may then be written

$$\Sigma\lambda_j \sigma_j PN\sigma_j^{-1}s_j = 0.$$

Multiply this on the right-hand side by N; then

$$\Sigma\mu_j \sigma_j PN = 0, \quad \text{(III.)}$$

where $\lambda_j PN\sigma_j^{-1}N = \mu_j PN.$

Multiply (III.) on the left-hand side by P, and we obtain

$$\Sigma\nu_j PN = 0;$$

and hence $\Sigma\nu_j = 0,$

where $\mu_j P\sigma_j PN = \nu_j PN.$

Let us return to the original equation; then

$$\Sigma\lambda_j P_j N_j s_j AT = \Sigma\lambda_j P_j N_j s_j = 0;$$

but $T = \Sigma NPN.$

The left-hand side of this relation is then equal to a sum of expressions

$$\Sigma\lambda_j \sigma_j PN\sigma_j^{-1}s_j NPN,$$

each of which is of the form $\Sigma\mu_j\sigma_j PN,$

since $(PN)^2 = A^{-1}PN.$

But each of these expressions is separately zero by (III.). Hence the original equation is expressible in terms of relations (III.), by means of the relations of § 10.

Consider then the relation

$$AT\Sigma\mu_j\sigma_j PN = \Sigma\mu_j\sigma_j PN = 0.$$

Take any term $P'N'P'$ of T; then

$$P'N'P' = \sigma' PNP\sigma'^{-1}.$$

Hence this relation is equal to a sum of relations of the form

$$\sigma' PNP\sigma'^{-1} [\Sigma\mu_j\sigma_j] PN = 0,$$

and this reduces, by § 11, to

$$[\Sigma\nu_j]\, \sigma' PN = 0.$$

The original relation may then be reduced to a merely numerical identity, by the use of relations derived from those of § 10. Hence no relation can exist which is independent of these.

13. Consider the expression of any substitution σ in terms of the forms PNs. It is obtained by means of the identity

$$\sigma = \Sigma AT\sigma.$$

To simplify this expression it is necessary to consider the product of PN by any substitution.

Let $$PN = \begin{Bmatrix} a_{1,1}\, a_{1,2} \;\cdots\cdots\cdots\; a_{1,\beta_1} \\ a_{2,1}\, a_{2,2} \;\cdots\; a_{2,\beta_2} \\ \cdots \;\cdots\; \cdots \\ a_{k,1} \cdots a_{k,\beta_k} \end{Bmatrix}.$$

(i.) $\sigma = (a_{\lambda,s}\, a_{\mu,s})$; then

$$PN\sigma = -PN.$$

(ii.) $\sigma = (a_{r,\lambda}\, a_{r,\mu})$; then

$$PN\sigma = \lfloor\sigma\rceil PN.$$

(iii.) $\sigma = (a_{r,\lambda}\, a_{s,\mu})$; then

$$PN\sigma = P\sigma N',$$

where $$N' = \lfloor(a_{r,\lambda}\, a_{s,\mu})\rceil N.$$

Now $a_{r,\lambda}$, $a_{r,\mu}$ appear in the same group of N'; they also appear in the same group of P.

Hence $$PN(a_{r,\lambda}\, a_{s,\mu}) = P(a_{r,\lambda}\, a_{s,\mu})N' = -P(a_{r,\mu}\, a_{s,\mu})N'$$
$$= -PN''(a_{r,\mu}\, a_{s,\mu}),$$

where $$N'' = \lfloor(a_{r,\mu}\, a_{s,\mu})\rceil N'$$
$$= \lfloor(a_{r,\mu} a_{s,\mu})(a_{r,\lambda}\, a_{s,\mu})\rceil N$$
$$= \lfloor(a_{r,\mu}\, a_{r,\lambda})\rceil N,$$

since $$\lfloor (a_{r,\mu}\, a_{s,\mu}) \rceil N = N.$$

Hence $$\lfloor \{a_{r,\lambda}\, a_{r,\mu}\} \rceil PN\, (a_{r,\lambda}\, a_{s,\mu}) = 0.$$

Similarly, $$\lfloor \{a_{s,\lambda}\, a_{s,\mu}\} \rceil PN\, (a_{r,\lambda}\, a_{s,\mu}) = 0.$$

These identities presuppose that

$$\mu \not> \beta_r \quad \text{and} \quad \lambda \not> \beta_s;$$

since we know that $\mu \not> \beta_s$ and $\lambda \not> \beta_r$, one of these identities exists in every case.

14. It was proved in § 11 that, if s be any substitution,

$$Ns = \pm Ns',$$

where s' is a product of cycles no one of which contains a pair of letters from the same group of N. Hence, when considering a term $PN\sigma$, we may suppose σ to be a product of such cycles. In the same way we may break up the cycles of σ and obtain

$$PN\sigma = PN'\sigma',$$

where the cycles of σ' are such that no two letters either in the same row or the same column of the expression PN' occur in the same cycle.

The two equations just written down enable us to still further reduce σ.

15. If σ is not a substitution of PN, then

$$(\lfloor \sigma \rceil N)\, P = 0.$$

For $$N'P = 0,$$

if a group of N' contains two letters of one of the groups of P. But, if

$$N'P \neq 0,$$

we have seen that there is some substitution σ' of P such that

$$\lfloor \sigma' \rceil N = N'.$$

Hence, if $$\lfloor \sigma \rceil N = N',$$

$$\sigma = \sigma'\sigma'',$$

where σ'' is a substitution which leaves N unaltered, and hence belongs to N (unless N contains two or more equal groups, and these are interchanged *in toto;* this may be supposed done by a substitution of P and included in σ'). Therefore, if

$$(\lfloor\sigma\rceil N)\,P \neq 0,$$

σ must be a substitution of PN.

16. Consider $PN\sigma^{-1}$, where σ is not a substitution of PN,

$$PN\sigma^{-1} = P\sigma^{-1}\lfloor\sigma\rceil N.$$

Hence $$(PN\sigma^{-1})^2 = P\sigma^{-1}(\lfloor\sigma\rceil N)\,P\sigma^{-1}\lfloor\sigma\rceil N = 0.$$

Hence every such expression $PN\sigma^{-1}$ is a solution of the substitutional equation

$$S^2 = 0,$$

where S is an unknown substitutional expression.

If, however, σ is a substitution of PN, then

$$\sigma^{-1} = \sigma''^{-1}\sigma'^{-1},$$

where σ''^{-1} is a substitution of N; and σ'^{-1} is a substitution of P,

and therefore $$PN\sigma^{-1} = \pm PN'.$$

Any substitutional expression can therefore be linearly expressed in terms of forms PN, whose squares are the forms themselves multiplied by a constant; and of forms $PN\sigma$ whose squares are zero.

Consider the product of two different forms PN belonging to the same T. Let these be PN and $P'N'$.

If $N'P = 0$, then

$$P'N'PN = 0.$$

Suppose that $N'P$ is not zero. If $P'N \neq 0$, there is a substitution σ of N such that

$$\lfloor\sigma\rceil P' = P.$$

Hence $$P'N'PN = \pm P'N'\sigma P'N.$$

But, unless σ^{-1} is a substitution of $P'N'$, we must have

$$P'N'\sigma P' = 0.$$

Hence $\sigma = \sigma_1\sigma_2$, where σ_1 is a substitution of N', and σ_2 one of P'.

Therefore
$$\begin{aligned} P'N'PN &= \pm P'N'\sigma P'N \\ &= \pm P'N'\sigma_1\sigma_2 P'N \\ &= \pm P'N'P'N = \pm P'NP'N \\ &= \pm A^{-1}P'N. \end{aligned}$$

If $P'N = 0$, there exists a substitution σ such that

$$\ulcorner \sigma \lrcorner\, PN = P'N'.$$

Hence $$P'N'PN = \sigma PN\sigma^{-1}PN.$$

And here $PN\sigma^{-1}P = 0$, unless $\sigma = \sigma_1\sigma_2$, where σ_1 is a substitution of P, and σ_2 is a substitution of N; so that

$$\sigma^{-1} = \sigma_2^{-1}\sigma_1^{-1}.$$

Therefore
$$\begin{aligned} P'N'PN &= \sigma PN\sigma_2^{-1}\sigma_1^{-1}PN \\ &= \pm\sigma PNPN = \pm A^{-\frac{1}{2}}\sigma PN = \pm A^{-\frac{1}{2}}P'\sigma N. \end{aligned}$$

Similarly, $$(P'N's')(PNs) = 0$$

or
$$\begin{aligned} &= s'P''N''PNs \\ &= \pm A^{-\frac{1}{2}}s'P''\sigma Ns \\ &= \pm A^{-\frac{1}{2}}P's'\sigma Ns. \end{aligned}$$

17. To illustrate the last few paragraphs, we will consider one or two special forms PN.

Let $$PN = \left\{\begin{matrix} a_1 a_2 \\ a_3 \end{matrix}\right\};$$

then
$$\begin{aligned} \left\{\begin{matrix} a_1 a_2 \\ a_3 \end{matrix}\right\}(a_2a_3) &= \{a_1a_2\}\{a_1a_3\}'(a_2a_3) \\ &= \{a_1a_3\}(a_2a_3)\{a_1a_2\}' \\ &= -\{a_1a_2\}(a_1a_3)\{a_1a_2\}' \\ &= -\left\{\begin{matrix} a_2 a_1 \\ a_3 \end{matrix}\right\}(a_1a_3). \end{aligned}$$

Again, $$\left\{\begin{matrix} a_1 a_2 \\ a_3 \end{matrix}\right\}\left[1-(a_2a_1)-(a_2a_3)\right] = 0;$$

hence
$$\begin{aligned} \left\{\begin{matrix} a_1 a_2 \\ a_3 \end{matrix}\right\}(a_2a_3) &= \left\{\begin{matrix} a_1 a_2 \\ a_3 \end{matrix}\right\}\left[1-(a_1a_2)\right] \\ &= \left\{\begin{matrix} a_1 a_2 \\ a_3 \end{matrix}\right\} - \left\{\begin{matrix} a_2 a_1 \\ a_3 \end{matrix}\right\}. \end{aligned}$$

Again
$$\begin{aligned} 0 &= \left[1+(a_3a_1)+(a_3a_2)\right]\left\{\begin{matrix} a_1 a_2 \\ a_3 \end{matrix}\right\} \\ &= \left\{\begin{matrix} a_1 a_2 \\ a_3 \end{matrix}\right\} - \left\{\begin{matrix} a_3 a_2 \\ a_1 \end{matrix}\right\} + \left\{\begin{matrix} a_1 a_3 \\ a_2 \end{matrix}\right\}(a_2a_3) \\ &= \left\{\begin{matrix} a_1 a_2 \\ a_3 \end{matrix}\right\} - \left\{\begin{matrix} a_3 a_2 \\ a_1 \end{matrix}\right\} + \left\{\begin{matrix} a_1 a_3 \\ a_2 \end{matrix}\right\} - \left\{\begin{matrix} a_3 a_1 \\ a_2 \end{matrix}\right\}; \end{aligned}$$

hence $\left\{\begin{matrix} a_1 a_2 \\ a_3 \end{matrix}\right\} + \left\{\begin{matrix} a_1 a_3 \\ a_2 \end{matrix}\right\} = \left\{\begin{matrix} a_3 a_1 \\ a_2 \end{matrix}\right\} + \left\{\begin{matrix} a_3 a_2 \\ a_1 \end{matrix}\right\} = \left\{\begin{matrix} a_2 a_1 \\ a_3 \end{matrix}\right\} + \left\{\begin{matrix} a_2 a_3 \\ a_1 \end{matrix}\right\}$

$$= \tfrac{1}{3} T_{2,1}.$$

Let
$$PN = \left\{\begin{matrix} a_1 a_3 a_4 \dots a_{n-1} \\ a_2 a_n \end{matrix}\right\};$$

then
$$PN\left[1-(a_3 a_1)-(a_3 a_2)\right] = 0;$$

hence $\left\{\begin{matrix} a_1 a_3 a_4 \dots a_{n-1} \\ a_2 a_n \end{matrix}\right\} (a_3 a_2)$

$$= \left\{\begin{matrix} a_1 a_3 a_4 \dots a_{n-1} \\ a_2 a_n \end{matrix}\right\} - \left\{\begin{matrix} a_3 a_1 a_4 \dots a_{n-1} \\ a_2 a_n \end{matrix}\right\}.$$

Similarly $\left\{\begin{matrix} a_1 a_3 a_4 a_5 \dots a_{n-1} \\ a_2 a_n \end{matrix}\right\} (a_4 a_2)$

$$= \left\{\begin{matrix} a_1 a_3 a_4 a_5 \dots a_{n-1} \\ a_2 a_n \end{matrix}\right\} - \left\{\begin{matrix} a_4 a_3 a_1 a_5 \dots a_{n-1} \\ a_2 a_n \end{matrix}\right\}.$$

Again
$$\left[1+(a_2 a_1)+(a_2 a_3)+\dots+(a_2 a_{n-1})\right] PN = 0;$$

hence

$$\left\{\begin{matrix} a_1 a_3 a_4 a_5 \dots a_{n-1} \\ a_2 a_n \end{matrix}\right\} + \left\{\begin{matrix} a_1 a_2 a_4 a_5 \dots a_{n-1} \\ a_3 a_n \end{matrix}\right\} + \left\{\begin{matrix} a_1 a_3 a_2 a_5 \dots a_{n-1} \\ a_4 a_n \end{matrix}\right\} + \dots$$

$$= \left\{\begin{matrix} a_2 a_3 a_4 a_5 \dots a_{n-1} \\ a_1 a_n \end{matrix}\right\} + \left\{\begin{matrix} a_2 a_1 a_4 a_5 \dots a_{n-1} \\ a_3 a_n \end{matrix}\right\} + \left\{\begin{matrix} a_2 a_3 a_1 a_5 \dots a_{n-1} \\ a_4 a_n \end{matrix}\right\} + \dots;$$

that is

$$\left[\{a_1 a_2\}'\left[1+(a_2 a_3)+(a_2 a_4)+\dots+(a_2 a_{n-1})\right]\right] \left\{\begin{matrix} a_1 a_3 a_4 \dots a_{n-1} \\ a_2 a_n \end{matrix}\right\} = 0.$$

In general it is to be observed that, if a, b be two letters such that the groups of N in which they appear are each of degree <3, or else such that the groups of P in which they appear are each of degree <3, then
$$PN(ab) = \pm P'N' - P''N''.$$

III. *The Product of a Symmetric Group into PN.*

18. First consider the product

$$\{a_1 a_2\} \left\{\begin{matrix} a_1 b_1 c_1 \dots \\ a_2 b_2 \dots\dots \\ a_3 \dots \\ \dots \end{matrix}\right\}.$$

Since $$[1+(a_2a_1)+(a_2b_1)+(a_2c_1)+\dots]\begin{Bmatrix} a_1b_1c_1\dots \\ a_2b_2c_2\dots \\ \dots\ \dots\ \dots \end{Bmatrix}=0,$$

therefore

$$\{a_1a_2\}\begin{Bmatrix} a_1b_1\dots \\ a_2b_2\dots \\ \dots\ \ \dots \end{Bmatrix} = -\begin{Bmatrix} a_1a_2c_1\ \dots \\ b_1b_2c_2\ \dots \\ a_3b_3\dots\dots \\ \dots\ \dots\ \dots \end{Bmatrix}(a_2b_1) - \begin{Bmatrix} a_1b_1a_2\dots \\ c_1b_2c_2\ \dots \\ a_3b_3\dots\dots \\ \dots\ \dots\ \dots \end{Bmatrix}(a_2c_1),$$

every term of which is of the form $PN\sigma$, where $\{a_1a_2\}$ is a factor of one of the groups of P.

Similarly $$\{a_1b_2\}\begin{Bmatrix} a_1b_1c_1\dots \\ a_2b_2c_2\dots \\ \dots\ \dots\ \dots \end{Bmatrix} = -\Sigma PN\sigma,$$

where in each term $\{a_1b_2\}$ is a factor of P.

More generally the product

$$\{a_{1,1}a_{1,2}\dots a_{1,r}a_{2,s}\}\begin{Bmatrix} a_{1,1}a_{1,2}\dots a_{1,\beta_1} \\ a_{2,1}a_{2,2}\dots \\ a_{3,1}\dots \\ \dots\ \ \dots \end{Bmatrix}$$

$$= r!\,[1+(a_{2,s}a_{1,1})+(a_{2,s}a_{1,2})+\dots+(a_{2,s}a_{1,r})]\begin{Bmatrix} a_{1,1}a_{1,2}\dots a_{1,\beta_1} \\ a_{2,1}a_{2,2}\dots \\ a_{3,1}\dots \\ \dots\ \ \dots \end{Bmatrix}$$

$$= -r!\,[(a_{2,s}a_{1,r+1})+\dots+(a_{2,s}a_{1,\beta_1})]\begin{Bmatrix} a_{1,1}a_{1,2}\dots a_{1,\beta_1} \\ a_{2,1}a_{2,2}\dots \\ a_{3,1}\dots \\ \dots\ \ \dots \end{Bmatrix}$$

$$= -r!\,\Sigma P'N'\sigma,$$

in every term of which $\{a_{1,1}a_{1,2}\dots a_{1,r}a_{r,s}\}$ is a factor of P'. And, further, P' is only changed in that one of the letters $a_{1,t}\ (t>r)$, in the group $\{a_{1,1}a_{1,2}\dots a_{1,\beta_1}\}$ which contains $\{a_{1,1}\dots a_{1,r}\}$ as a factor, is interchanged with $a_{2,s}$; and this is done in all possible ways.

We have supposed that the letters $a_{1,1}, a_{1,2}, \dots, a_{1,r}$ lie in the greatest group of P; this is unnecessary, provided $a_{2,s}$ does not lie in a group of degree greater than that of the group which contains the letters $a_{1,1}, a_{1,2}, \dots$.

If, however, $\beta_1 \not> r$, then the above product is zero.

19. In the same way it may be shown that the product of a positive symmetric group $\{a_1 a_2 \dots a_r\}$ into any form PN may be expressed in the form $\Sigma\lambda P'N'\sigma$, where $\{a_1 a_2 \dots a_r\}$ is a factor of every P', and, moreover, is a factor of that group of P' which is of degree β, β being the degree of the greatest group of P which contains one of the letters $a_1, a_2, \dots, a_r$.

For, if a_1 is the letter which appears in the group of P degree β, then

$$\{a_1 a_2 \dots a_n\} PN$$
$$= [1+(a_n a_1)+(a_n a_2)+\dots+(a_n a_{n-1})][1+(a_{n-1} a_1)+\dots+(a_{n-1} a_{n-2})]\dots$$
$$\dots [1+(a_2 a_1)]\, PN.$$

We may proceed exactly as in the last paragraph at each step, and the result follows. We observe further that the expressions P' are just those expressions which may be obtained by filling up the places vacated by $a_2, \dots, a_r$ when these letters are moved into the group degree β in all possible ways by the remaining letters of this group.

IV. *Application to the Theory of Invariants.*

As already stated, the last three sections of this paper deal with an application of substitutional analysis to modern algebra. First, any homogeneous rational integral function F of the coefficients of certain binary quantics is considered. By the introduction of the ordinary symbolical notation this may be represented as a function $H_1\phi$; where ϕ is a function of certain sets of variables, there being q variables in each set, and H_1 is a substitutional expression denoting the fact that certain of the sets of variables refer to the same quantic. By polarization and the introduction of new sets of variables we may obtain from $H_1\phi$ a function $f = HGf_1$, which is linear in each of n sets of variables, there being q variables in each set, where G is a substitutional expression denoting the fact that certain of these sets are equivalent—*i.e.*, combine to replace a single set of $H_1\phi$—and H is the substitutional equivalent of H_1. This process was explained in § 12 of my former paper, and it was there shown that f is a function equivalent to $H_1\phi$.

Here it is proved that $HGPNf_1$ is a linear function of the coefficients of a certain concomitant of the quantics under consideration, and hence that the series $1 = \Sigma AT$ enables us to express f, and therefore F, as a linear function of coefficients of concomitants of the quantics; for
$$f = HGf_1 = HG\,(\Sigma AT) f_1.$$

Hence any rational integral function of the coefficients of certain q-ary quantics may be expressed linearly in terms of coefficients of concomitants of these quantics.

By the same process, taking F to be a concomitant, a proof of the fundamental theorem of symbolical algebra is obtained; viz., that F may be expressed as a sum of symbolical products, the factors being of certain definite forms.

Incidentally it appears that the function of the coefficients of transformation which appears as a factor of a concomitant after transformation must be a power of the modulus of transformation.

Further, if $HGPNf_1$ is a concomitant, then it is completely defined —except for a numerical factor—by the substitutional expression $HGPN$. Thus a certain set of concomitants is obtained in terms of which every concomitant can be linearly expressed. This set has the advantage that the linear relations between members of the set may be all simply obtained; they are given by the relations of § 10.

In Section V. the invariants of a single binary quantic are discussed from this point of view. It is shown that each of the members of the fundamental set is completely defined by certain $n+1$ numbers, where n is the order of the quantic considered. Calling these numbers $a_0, a_1, a_2, \ldots, a_n$, the invariant is written

$$f(a_0, a_1, \ldots, a_n).$$

If δ is the degree, and w the weight, of this invariant, then the relations

$$a_0+a_1+\ldots+a_n = \delta,$$

$$a_1+2a_2+\ldots+na_n = a_{n-1}+2a_{n-2}+\ldots+na_0 = w$$

must be satisfied.

The greater part of the section deals with the linear relations between these invariants; as an example it is proved that there can be no *gauche* invariant of the quintic of degree less than 18.

The discussion in this paper is limited entirely to the question of linear independence: to make the method applicable to the question of reducibility it would be necessary to investigate the product of two concomitants of the forms considered.

There is one case, however, in which such an investigation is not required—that of a single quadratic in any number of variables. This is the subject of Section VI.: it is there shown that for any given degree and orders in the different kinds of variables there is only one linearly independent concomitant; hence, this is reducible or otherwise according as there is or is not a product of concomitants having the same degree and orders.

20. Let F be any rational integral homogeneous function of the coefficients of certain q-ary quantics, and $f = HGf_1$ the function equivalent to it, which is obtained in the manner explained above, and is linear in each of n sets of q-ary variables. These sets of variables will be denoted by the letters $a_1, a_2, \dots, a_n$. When it is necessary to speak of the variables of the set a_r, they will be written $a_{r,1}, a_{r,2}, \dots, a_{r,q}$.

The substitutions employed will be those of the symmetric group of the letters $a_1, a_2, \dots, a_n$; *i.e.*, they will interchange entire sets.

The expression G denotes the fact that certain of these sets correspond to the same quantic: thus, if the set $a_1, a_2, \dots, a_r$ correspond to a single quantic of order r, then $\{a_1 a_2 \dots a_r\}$ is one of the factors of G. The expression H was defined as the substitutional equivalent to the expression H_1, which occurs in the result $H_1\phi$ of introducing the symbolical notation in F. Let $\mathbf{a}_1, \mathbf{a}_2, \dots, \mathbf{a}_m$ represent the sets of variables in $H_1\phi$; then, if F is of degree s in the coefficients of a certain quantic order r, s of these sets belong to this quantic. Suppose that these are the sets $\mathbf{a}_1, \mathbf{a}_2, \dots, \mathbf{a}_s$; then H_1 has the factor $\{\mathbf{a}_1 \mathbf{a}_2 \dots \mathbf{a}_s\}$. To each of these sets in $H_1\phi$, r sets of f correspond; for the sake of argument the sets corresponding to $\mathbf{a}_h$ will be written

$$a_{(h-1)r+1},\ a_{(h-1)r+2},\ \dots,\ a_{hr}.$$

Then the factor of H corresponding to the factor $\{\mathbf{a}_1 \mathbf{a}_2 \dots \mathbf{a}_s\}$ of H_1 may be written $\{\mathbf{a}_1 \mathbf{a}_2 \dots \mathbf{a}_s\}$, on the understanding that each transposition $(\mathbf{a}_h \mathbf{a}_k)$ is to be replaced by the product

$$(a_{(h-1)r+1}\, a_{(k-1)r+1})(a_{(h-1)r+2}\, a_{(k-1)r+2}) \dots (a_{hr}\, a_{kr}).$$

The expression HG depends in no way on the form of f; so that, if ψ is any function homogeneous and linear in each of the sets of variables $a_1, a_2, \dots, a_n$, then $HG\psi$ represents a function of the coefficients of the original quantics of the same degree in each as F.

Consider $\{a_1 a_2 \dots a_n\} f_1$: this is a sum of terms each of which is a term of the n-ic

$$a_{1_x} a_{2_x} \dots a_{n_x} = a_x.$$

Consider next the expression

$$\{a_3 \dots a_n\}\{a_1 a_2\}' f_1:$$

this is a sum of terms of $\quad {}_1\mathbf{a}_{1x}^{n-2}\, {}_2\mathbf{a}_{2x},$

where $$ {}_1x = x, \quad {}_2x = (xy) = \left\| \begin{matrix} x_1 x_2 \ldots x_q \\ y_1 y_2 \ldots y_q \end{matrix} \right\|, $$

$$ {}_1\mathbf{a}_{1x}^{n-2} = a_{3_x} a_{4_x} \ldots a_{n_x}, $$

$$ {}_2a_{2x} = (a_1 a_2)_{2x} = \begin{vmatrix} a_{1,1} a_{1,2} \\ a_{2,1} a_{2,2} \end{vmatrix} \begin{vmatrix} x_1 x_2 \\ y_1 y_2 \end{vmatrix} + \ldots . $$

For, if $a_{1,r} a_{2,s} B$ is a term of f_1, then $\begin{vmatrix} a_{1,r} a_{1,s} \\ a_{2,r} a_{2,s} \end{vmatrix} B$ is the corresponding term of $\{a_1 a_2\}' f_1$. More generally, suppose that N consists of n_1 groups degree 1, n_2 groups degree 2, ..., n_k groups degree k; $k \not> q$. Then PNf_1 is a sum of multiples of terms of

$$ (P)\ {}_1\mathbf{a}_{1x}^{n_1}\, {}_2\mathbf{a}_{2x}^{n_2} \ldots {}_k a_{kx}^{n_k} $$

where $$ {}_1\mathbf{a}_{1x}^{n} = a_{1_x} a_{2_x} \ldots a_{n_{1x}}, $$

$a_1, a_2, \ldots, a_{n_1}$ being the letters which occur in the groups of degree 1 in N; where

$$ {}_2a_{2x}^{n_2} = \Pi\, (ab)_{2x}, $$

$\Pi\,\{ab\}'$ being the product of groups degree 2 in N, and so on. For, if we replace (ab) by C—a new kind of variable—we obtain n_2 letters $c_1, c_2, \ldots, c_{n_2}$, and from the form of P we obtain

$$ P\phi\,(c_1, c_2, \ldots, c_n) = \frac{1}{n_2!} P\,\{c_1 c_2 \ldots c_{n_2}\}\,\phi\,(c_1, c_2, \ldots, c_{n_2}). $$

And hence, if ϕ is linear in each set of variables c, $P\phi$ may in general be obtained by polarization from a function degree n_2 in a single set of variables c.

The groups of N degrees 3, 4, ... may be dealt with in the same way. If N contain a group degree $>q$, we know by § 8 of my former paper that $Nf = 0$.

The expressions ${}_1\mathbf{a}_{1x}, {}_2a_{2x}, \ldots, {}_k\mathbf{a}_{kx}$, if the sets of variables $a_1, a_2, \ldots$ are all cogredient with each other, and contragredient to the sets $x, y, z, \ldots$, are unaltered by a linear transformation.

The expression $$ {}_1a_{1x}^{n_1}\, {}_2a_{2x}^{n_2} \ldots {}_k a_{kx}^{n_k} = E $$

is then a concomitant of the forms

$$ a_{1_x}, a_{2_x}, \ldots, a_{n_x}. $$

If we operate on E with HG, it will become the symbolical representation of a concomitant of the q-ary quantics. Hence $HGPNf_1$, and therefore $HGTf_1$, represents a sum of numerical multiples of the coefficients of certain concomitants of the quantics from which we started. But

$$f = HGf_1 = \Sigma AHGTf_1.$$

Hence any rational integral algebraic function of the coefficients of certain q-ary quantics may be linearly expressed in terms of coefficients of concomitants of these quantics.

21. If f be itself a concomitant, then $HGPNf$ is invariantive. But $HGPNf$ has just been shown to be a linear function of the coefficients of a certain concomitant. Let us write

$$HGPNf = \lambda_0 A_0 + \lambda_1 A_1 + \dots,$$

where A_0, A_1, ... are the coefficients referred to, and the λ's certain quantities independent of $a_1, a_2, \dots, a_n$. Now make any linear transformation: the coefficients A_0, A_1, ... are not left unchanged, but the whole expression is left unchanged; hence the quantities $\lambda_0, \lambda_1, \dots$ must be contragredient to A_0, A_1, ... ; no relation can exist between different coefficients A, since the weight of each is different. The quantities λ are independent of the quantities $a_1, a_2, \dots, a_n$, and are functions of the remaining variables $x, y, z, \dots$, defined as being contragredient to the coefficients of A_0, A_1, ... in the concomitant

$${}_1a^{n_1}_{1x}\, {}_2a^{n_2}_{2x} \dots {}_ka^{n_k}_{kx}.$$

Consider any one of them, and polarize it so as to make it linear in each of the sets of variables it contains. Let us call the function in this form Λ, and use the letters $y_1, y_2, \dots, y_m$ to denote the various sets of variables which it contains.

Operate on Λ with the series

$$1 = \Sigma AT.$$

Then we may consider separately expressions

$$P_1 N_1 \Lambda.$$

Now, if s be any substitution affecting the letters $y_1, y_2, \dots$, the expression $s\Lambda$ is cogredient with Λ.

Hence $P_1 N_1 \Lambda$ is cogredient with Λ or zero. But $P_1 N_1 \Lambda$ is obtained by polarization from terms

$${}_1x^{m_1}_{r_1}\, {}_2x^{m_2}_{r_1} \dots {}_kx^{m_k}_{r_k},$$

where m_1 is the number of groups of N degree 1, and so on. Hence Λ is cogredient with

$$_1x^{m_1}\,_2x^{m_2}\,\ldots\,_kx^{m_k}.$$

Therefore this term is cogredient with

$$_1x^{n_1}\,_2x^{n_2}\,\ldots\,_kx^{n_k}.$$

But it is obvious that two such variable expressions cannot be cogredient unless $m_1 = n_1$, $m_2 = n_2$, ..., for the numbers of terms of $_1x$, $_2x$, ... are different; so that in any other case it would not be possible to make the terms of the two expressions correspond.

Hence, if P_1N_1 belong to any other expression T except that one which is such that each term contains n_1 negative symmetric groups degree 1, n_2 of degree 2, and so on, then

$$P_1N_1\Lambda = 0.$$

Hence Λ itself may be obtained by polarization from

$$_1x^{n_1}\,_2x^{n_2}\,\ldots\,_kx^{n_k}.$$

Therefore the concomitant $HGPNf$ is obtained by polarization from

$$_1a^{n_1}_{_1x}\,_2a^{n_2}_{_2x}\,\ldots\,_ka^{n_k}_{_kx}.$$

But
$$f = \frac{1}{\rho}HGf = \frac{1}{\rho}\Sigma AHGTf,$$

where $(HG)^2 = \rho HG$—a relation due to the form of HG.

Hence every concomitant of a system of q-ary quantics can be expressed symbolically as a sum of symbolical products, the factors being of the forms

$$a_{1,x} = \Sigma a_{1,r}x_r,\quad (a_1a_2)_{2x} = \Sigma\begin{vmatrix} a_{1,r} & a_{1,s} \\ a_{2,r} & a_{2,s}\end{vmatrix}\begin{vmatrix} x_r & x_s \\ y_r & y_s\end{vmatrix},\quad \&c.,$$

and, finally,
$$(a_1a_2\,\ldots\,a_q) = \begin{vmatrix} a_{1,1} & a_{1,2} & \ldots & a_{1,q} \\ a_{2,1} & a_{2,2} & \ldots & a_{2,q} \\ \ldots & \ldots & & \ldots \\ a_{q,1} & a_{q,2} & \ldots & a_{q,q}\end{vmatrix}.$$

The variables $_1x$, $_2x$, ... may be regarded as entirely new variables defined as being cogredient with x, (xy), (xyz), ...

22. In every case except the last a linear transformation leaves the symbolical factor unaltered. In the last case the transformed factor

is equal to the original factor multiplied by the modulus of transformation. This constitutes a proof of the fact that, if an algebraic function of the coefficients and variables of certain q-ary quantics is unaltered by any linear transformation, except for a factor which contains only the coefficients of transformation, then this factor must be a power of the modulus of transformation.

V. *Particular Application to Invariants of Binary Forms.*

23. Let I be any invariant of certain binary forms; we may suppose it to be represented symbolically in the ordinary way. By polarization this may be represented as a sum of symbolical products containing each symbolical letter linearly.

Let one such term be*

$$i = (\mathrm{a}_1\mathrm{b}_1)(\mathrm{a}_2\mathrm{b}_2)\dots(\mathrm{a}_m\mathrm{b}_m)$$
$$= \frac{1}{2^m}\{a_1b_1\}'\{a_2b_2\}'\dots\{a_mb_m\}'(\mathrm{a}_1\mathrm{b}_1)(\mathrm{a}_2\mathrm{b}_2)\dots(\mathrm{a}_m\mathrm{b}_m).$$

Operate on this with the series

$$1 = \Sigma AT;$$

then every term Ti before $T_{2,2,\dots,2}\,i$ vanishes, owing to the existence of the factor of i,
$$\{a_1b_1\}'\dots\{a_mb_m\}'.$$

Every term Ti after this factor vanishes, because the sets of variables a_1, b_1, ... have only two variables in each set; and therefore a negative symmetric group of degree three must annihilate i. ($T'_3 i = 0$ owing to the fundamental identity between symbolical factors—which is the same thing as the above statement.) Hence

$$i = A_{2,2,\dots,2}\,T_{2,2,\dots,2}\,i$$
$$= \frac{1}{2^m}\sqrt{A_{2,2,\dots,2}}\{a_1b_1\}'\dots\{a_mb_m\}'\begin{Bmatrix} a_1a_2\dots a_m \\ b_1b_2\dots b_m\end{Bmatrix}(\mathrm{a}_1\mathrm{b}_1)(\mathrm{a}_2\mathrm{b}_2)\dots(\mathrm{a}_m\mathrm{b}_m).$$

24. Let us write

$$\begin{Bmatrix} a_1a_2\dots a_m \\ b_1b_2\dots b_m\end{Bmatrix}(\mathrm{a}_1\mathrm{b}_1)\dots(\mathrm{a}_m\mathrm{b}_m) = \begin{Bmatrix} a_1a_2\dots a_m \\ b_1b_2\dots b_m\end{Bmatrix}_1$$
$$= (-1)^m\begin{Bmatrix} b_1b_2\dots b_m \\ a_1a_2\dots a_m\end{Bmatrix}_1.$$

* Roman letters are here used to distinguish algebraic symbolical factors from substitutions.

This expression may be looked upon as an algebraic function; so that, if we multiply it by any substitution, the result may be obtained by an interchange of the letters. Otherwise a substitution which multiplies it may be regarded as multiplying the substitutional part PN of the expression. Hence

$$\begin{Bmatrix} a_2 a_1 a_3 \ldots a_m \\ b_1 b_2 b_3 \ldots b_m \end{Bmatrix}_1 = (a_1 a_2) \begin{Bmatrix} a_1 a_2 a_3 \ldots a_m \\ b_1 b_2 b_3 \ldots b_m \end{Bmatrix}_1$$

$$= \begin{Bmatrix} a_1 a_2 a_3 \ldots a_m \\ b_1 b_2 b_3 \ldots b_m \end{Bmatrix}_1;$$

so that the letters in either row may be interchanged without altering the value of the function. The fundamental identity has been used completely and no longer need be remembered, for the form of the function is such that it puts in evidence the fact that it is annihilated by any negative symmetric group of degree greater than three. The expression, however, satisfies an identity, for, by § 10,

$$\{a_1 a_2 \ldots a_m b_r\} \begin{Bmatrix} a_1 a_2 \ldots a_m \\ b_1 b_2 \ldots b_m \end{Bmatrix}_1 = 0$$

and

$$\{b_1 b_2 \ldots b_m a_r\} \begin{Bmatrix} a_1 a_2 \ldots a_m \\ b_1 b_2 \ldots b_m \end{Bmatrix}_1 = 0;$$

and, by § 12, this expression satisfies no other relation.

25. Expanding the product of § 23,

$$\{a_1 b_1\}' \{a_2 b_2\}' \ldots \{a_m b_m\}',$$

we may obtain $i = (a_1 b_1)(a_2 b_2) \ldots (a_m b_m)$

as a sum of expressions such as

$$\begin{Bmatrix} a_1 a_2 \ldots a_m \\ b_1 b_2 \ldots b_m \end{Bmatrix}_1.$$

Since I is an invariant of certain binary forms,

$$I = HGi,$$

where G is the product of certain positive symmetric groups and H is an expression which interchanges some of these groups. Hence every invariant of a system of binary quantics may be expressed in terms of the forms

$$HG \begin{Bmatrix} a_1 a_2 \ldots a_m \\ b_1 b_2 \ldots b_m \end{Bmatrix}_1.$$

26. Consider in particular the invariants of a single binary n-ic. We must take the expression

$$HG\begin{Bmatrix} a_1 a_2 \ldots a_m \\ b_1 b_2 \ldots b_m \end{Bmatrix}_1,$$

where the $2m$ letters are arranged in sets of n, and G consists of the product of the positive symmetric groups of these sets.

H is a substitutional expression which interchanges these sets in all possible ways.

Let δ be the degree of the invariant; then

$$2m = \delta n,$$

and m is the weight.

Let us suppose that of these δ sets there are α_0 such that all the letters of the set are in the lower row of the above expression; α_1 such that 1 letter is in the upper row and $n-1$ in the lower row; and so on. In general, let there be α_r sets having r letters in the upper and $n-r$ in the lower row. These numbers completely define the expression, for the letters of either row may be interchanged in all possible ways, the letters of any one set possess no individuality, and no set as a whole possesses individuality.

We may conveniently use the notation

$$f(\alpha_0, \alpha_1, \ldots, \alpha_n)$$

to denote this expression.

Every invariant of the n-ic may be expressed linearly in terms of invariants

$$f(\alpha_0, \alpha_1, \alpha_2, \ldots, \alpha_n).$$

The numbers $\alpha_0, \alpha_1, \ldots, \alpha_n$ must satisfy the relations

$$\alpha_0+\alpha_1+\alpha_2+\ldots+\alpha_n = \delta,$$

$$\alpha_1+2\alpha_2+\ldots+n\alpha_n = n\alpha_0+(n-1)\,\alpha_1+\ldots+\alpha_{n-1} = m,$$

the number of letters in either row.

27. These invariants satisfy certain identical relations. We have seen in § 24 that

$$[1+(a_1 b_r)+(a_2 b_r)+\ldots+(a_m b_r)]\begin{Bmatrix} a_1 a_2 \ldots a_m \\ b_1 b_2 \ldots b_m \end{Bmatrix}_1 = 0.$$

Suppose that b_r is a letter of a set which has r letters in the upper row; then b_r is moved up to the upper row, and the letters of the upper row take its place in all possible ways.

If a letter of one of those sets which have s letters above is moved down, then the number a_s is diminished by unity, and the number a_{s-1} increased by unity. At the same time the number a_r is diminished by unity, and the number a_{r+1} increased by unity. But a letter of those sets which have s letters above may be chosen in sa_s ways; hence the identity will be of the form

$$\sum_{s \neq r, r+1} sa_s f(a_0, a_1, \ldots, a_{s-1}+1, a_s-1, \ldots, a_r-1, a_{r+1}+1, \ldots, a_n)$$
$$+r(a_r-1) f(a_0, a_1, \ldots, a_{r-1}+1, a_r-2, a_{r+1}+1, \ldots, a_n)$$
$$+(r+1)(a_{r+1}+1) f(a_0, a_1, \ldots, a_{r-1}, a_r, a_{r+1}, \ldots, a_n) = 0.$$

If in this we write a_r+1 for a_r and $a_{r+1}-1$ for a_{r+1}, we obtain the identity $\sum_s sa_s f(a_0, a_1, a_2, \ldots, a_{s-1}+1, a_s-1, \ldots, a_n) = 0.$

In the same way, from the identity

$$[1+(a_r b_1)+(a_r b_2)+\ldots+(a_r b_m)] \begin{Bmatrix} a_1 a_2 \ldots a_m \\ b_1 b_2 \ldots b_m \end{Bmatrix}_1 = 0,$$

we obtain

$$\sum_s (n-s)\, a_s f(a_0, a_1, a_2, \ldots, a_{s-1}, a_s-1, a_{s+1}+1, \ldots, a_n) = 0.$$

These two are the only linear relations between the invariants of a particular degree, unless we include the following

$$\begin{Bmatrix} a_1 a_2 \ldots a_m \\ b_1 b_2 \ldots b_m \end{Bmatrix}_1 = (-1)^m \begin{Bmatrix} b_1 b_2 \ldots b_m \\ a_1 a_2 \ldots a_m \end{Bmatrix},$$

and hence $f(a_0, a_1, \ldots, a_n) = (-1)^m f(a_n, a_{n-1}, \ldots, a_0).$

The second of the above identities may be written

$$na_0 f(a_0-1, a_1+1, a_2, \ldots, a_n)$$
$$= -(n-1)\, a_1 f(a_0, a_1-1, a_2+1, \ldots, a_n)$$
$$-(n-2)\, a_2 f(a_0, a_1, a_2-1, a_3+1, \ldots, a_n)$$
$$-\ldots;$$

so that, if $a_1 \neq 0$, the invariant $f(a_0, a_1, \ldots, a_n)$ may be expressed in terms of invariants in which the value of a_0 is increased, of a_1 is decreased, while that of a_n is not decreased.

Similarly, if $a_{n-1} \neq 0$, we may express this invariant in terms of invariants in which a_n is increased, a_{n-1} decreased, and a_0 is not decreased.

Proceeding step by step, we may decrease a_1 and a_{n-1} to zero, at the same time increasing a_0 and a_n. Thus every invariant may be expressed in terms of such as have a_1 and a_{n-1} zero.

Hence, if $n<4$, there is not more than one invariant for each degree.

28. Although the above identities contain all the linear relations between the invariants f, yet there are other relations—not independent of these—which it is useful to have.

Consider the result of multiplying

$$\left\{\begin{matrix} a_1 a_2 \dots a_m \\ b_1 b_2 \dots b_m \end{matrix}\right\}_1 = I$$

by any positive symmetric group Γ. We may regard the product as the sum of the results of operating on I with each of the substitutions of Γ; or else we may multiply the substitutional part of I by Γ as explained (Section III.), and so express the product as a sum of such terms as I, but each of which have all the letters of Γ in the same row; the two expressions for the product must be equal.

In particular, if the degree of Γ is greater than m, then the second expression is zero.

Consider now the invariant

$$f(a_0, a_1, \dots, a_n) = HGI,$$

and let us find the result of writing ΓI for I. Let the degree of Γ be ϖ, and let us suppose that the letters of Γ occupy ρ places in the upper row of I, and $\varpi-\rho$ in the lower row. Then ΓI is the sum of the results of arranging the ϖ letters of Γ in all possible ways in the ϖ places assigned to them in I. Of these $\binom{\varpi}{\rho}$ are different arrangements.

The letters of I will be supposed to be distributed in the manner defined by certain numbers,

$$a_{0,0}, a_{0,1}, \dots, a_{0,n};\ a_{1,0}, \dots, a_{1,n-1};\ \dots;\ a_{n,0},$$

where the number $a_{\lambda,\mu}$ signifies that there are $a_{\lambda,\mu}$ sets of letters such that μ letters from each set appear in Γ, λ appear in the upper row of I but not in Γ, and the rest, $n-\lambda-\mu$ in number, appear in the lower row of I but not in Γ.

We must consider some one term of the sum $HG\Gamma I$: this will be defined completely when the positions of the letters of Γ are given. These will be given by the numbers $a_{\lambda,\mu,\nu}$, the meaning of which symbol is that of the $a_{\lambda,\mu}$ sets of letters defined as above there are

$a_{\lambda,\mu,\nu}$ in the particular expression under consideration which have ν of the μ letters belonging to Γ in the upper row and the remaining $\mu-\nu$ in the lower row.

Now the $a_{\lambda,\mu}$ sets of letters can be divided up into $a_{\lambda,\mu,0}, a_{\lambda,\mu,1}, \ldots, a_{\lambda,\mu,\mu}$ sets (of which the sum must be $a_{\lambda,\mu}$), in $\dfrac{a_{\lambda,\mu}!}{a_{\lambda,\mu,0}!\, a_{\lambda,\mu,1}! \ldots a_{\lambda,\mu,\mu}!}$ ways.

Further, in one of the $a_{\lambda,\mu,\nu}$ sets the ν letters which are to appear in the upper row of the invariant may be chosen in $\binom{\mu}{\nu}$ ways. Hence the number of different arrangements of the members of these $a_{\lambda,\mu}$ sets which will fulfil the requirements of the term considered is

$$\frac{a_{\lambda,\mu}!}{a_{\lambda,\mu,0}!\, a_{\lambda,\mu,1}! \ldots a_{\lambda,\mu,\mu}!}\binom{\mu}{0}^{a_{\lambda,\mu,0}}\binom{\mu}{1}^{a_{\lambda,\mu,1}}\ldots\binom{\mu}{\mu}^{a_{\lambda,\mu,\mu}}.$$

Hence the total number of such terms is

$$\prod_{\lambda=0}^{\lambda=n-1}\prod_{\mu=1}^{\mu=n-\lambda}\left\{a_{\lambda,\mu}!\prod_{\nu=0}^{\nu=\mu}\frac{\binom{\mu}{\nu}^{a_{\lambda,\mu,\nu}}}{a_{\lambda,\mu,\nu}!}\right\}.$$

Consider then the coefficient of the term $f(\gamma_0, \gamma_1, \ldots, \gamma_n)$: it is obtained as the sum of the coefficients of the terms for which the quantities $a_{\lambda,\mu,\nu}$ satisfy the relations

$$\sum_{\lambda=0}^{\lambda=p}\sum_{\mu=p-\lambda}^{\mu=n-\lambda} a_{\lambda,\mu,p-\lambda} = \gamma_p.$$

This sum may be seen at once to be the coefficient of $x_0^{\gamma_0}x_1^{\gamma_1}\ldots x_n^{\gamma_n}$ in the expansion of

$$x_0^{a_{0,0}}\, x_1^{a_{1,0}} \ldots x_n^{a_{n,0}},$$

$$(x_0+x_1)^{a_{0,1}}(x_1+x_2)^{a_{1,1}} \ldots (x_{n-1}+x_n)^{a_{n-1,1}},$$

$$\left(x_0+\binom{2}{1}x_1+x_2\right)^{a_{0,2}}\left(x_1+\binom{2}{1}x_2+x_3\right)^{a_{1,2}} \ldots \left(x_{n-2}+\binom{2}{1}x_{n-1}+x_n\right)^{a_{n-2,2}},$$

$$\ldots \quad \ldots \quad \ldots \quad \ldots \quad \ldots \quad \ldots \quad \ldots \quad \ldots \quad \ldots \quad \ldots$$

$$\left(x_0+\binom{n}{1}x_1+\binom{n}{2}x_2+\ldots+x_n\right)^{a_{0,n}}.$$

This coefficient has only taken into account the different terms of ΓI but each term is repeated $\rho!(\varpi-\rho)!$ times; the coefficient just found must then be multiplied by this.

It is interesting to notice that, if we write

$$x_0 = 1,\ x_1 = x,\ x_2 = x^2,\ \ldots,\ x_n,$$

then the coefficient of x^w, where w is the weight of the invariant considered, in the above expansion is $\binom{\varpi}{\rho}$. This is as it should be, since this coefficient represents the sum of the coefficients which have to be considered.

29. To find when $\varpi \not> m$ the second form in which the expression $HG\Gamma I$ may be written.

Let us write $$I = \begin{Bmatrix} a_1 a_2 \dots a_m \\ b_1 b_2 \dots b_m \end{Bmatrix}_1,$$

$$\Gamma = \{a_1 a_2 \dots a_\rho b_1 b_2 \dots b_{\varpi-\rho}\}.$$

Then $$\Gamma I = \rho!\,[1+(b_{\varpi-\rho}a_1)+(b_{\varpi-\rho}a_2)+\dots+(b_{\varpi-\rho}b_{\varpi-\rho-1})]$$
$$\times[1+(b_{\varpi-\rho-1}a_1)+\dots+(b_{\varpi-\rho-1}b_{\varpi-\rho-2})]$$
$$\dots \quad \dots \quad \dots \quad \dots \quad \dots \quad \dots$$
$$\times[1+(b_1a_1)+(b_1a_2)+\dots+(b_1a_\rho)]$$
$$\times\begin{Bmatrix} a_1 a_2 \dots a_m \\ b_1 b_2 \dots b_m \end{Bmatrix}_1.$$

Now $$[1+(b_1a_1)+\dots+(b_1a_\rho)]\begin{Bmatrix} a_1 a_2 \dots a_m \\ b_1 b_2 \dots b_m \end{Bmatrix}_1$$
$$= -[(b_1a_{\rho+1})+(b_1a_{\rho+2})+\dots+(b_1a_m)]\begin{Bmatrix} a_1 a_2 \dots a_m \\ b_1 b_2 \dots b_m \end{Bmatrix}_1.$$

Taking each term we obtain

$$\Gamma I = (-)^{\varpi-\rho}\rho!\,\Sigma I', \tag{IV.}$$

where I' has all the letters of Γ in the upper row, and where the places which $b_1 b_2 \dots b_{\varpi-\rho}$ occupy in the lower row of I are filled up with $\varpi-\rho$ of the letters $a_{\rho+1} \dots a_m$, in all possible ways. The number of *different* ways in which this can be done is $\binom{m-\rho}{\varpi-\rho}$, but the number of terms in the $\Sigma I'$ of equation (IV.) is

$$(m-\rho)(m-\rho-1)\dots(m-\varpi+1) = \binom{m-\rho}{\varpi-\rho}(\varpi-\rho)!.$$

Hence each different term is affected by the coefficient

$$(-)^{\varpi-\rho}(\varpi-\rho)!\,\rho!.$$

Hence $$I = (-1)^{\varpi-\rho}(\varpi-\rho)!\,\rho!\,\Sigma\lambda f(\beta_0, \beta_1, \dots, \beta_n),$$

where the coefficient λ is obtained in the same way as the corresponding coefficient of the last paragraph; it is equal to the coefficient of $x_0^{\beta_0} x_1^{\beta_1} \dots x_n^{\beta_n}$ in the expansion of

$$x_0^{a_{0,0}} x_1^{a_{0,1}} x_2^{a_{0,2}} \dots x_n^{a_{0,n}},$$

$$(x_0+x_1)^{a_{1,0}} (x_1+x_2)^{a_{1,1}} \dots (x_{n-1}+x_n)^{a_{1,n-1}},$$

$$\left(x_0+\binom{2}{1} x_1+x_2\right)^{a_{2,0}} \left(x_1+\binom{2}{1} x_2+x_3\right)^{a_{2,1}} \dots \left(x_{n-2}+\binom{2}{1} x_{n-1}+x_n\right)^{a_{2,n-2}},$$

$$\dots \quad \dots \quad \dots \quad \dots \quad \dots \quad \dots \quad \dots \quad \dots \quad \dots \quad \dots$$

$$\left(x_0+\binom{n}{1} x_1+\binom{n}{2} x_2+\dots+x_n\right)^{a_{n,0}}.$$

Let us call this coefficient $\phi_2(\beta_0, \beta_1, \dots, \beta_n)$, and the corresponding coefficient of the last paragraph $\phi_1(\gamma_0, \gamma_1, \dots, \gamma_n)$; then we obtain the identity

$$\Sigma\phi_1(\gamma_0, \gamma_1, \dots, \gamma_n) f(\gamma_0, \gamma_1, \dots, \gamma_n)$$
$$= (-1)^{\varpi-\rho} \Sigma\phi_2(\beta_0, \beta_1, \dots, \beta_n) f(\beta_0, \beta_1, \dots, \beta_n).$$

30. Let us consider the case

$$a_{\lambda,\mu} = 0, \quad \text{unless} \quad \mu = 0 \text{ or } \lambda = 0,$$

and $$a_{0,\mu} = 0, \quad \text{unless} \quad \mu = 0 \text{ or } n.$$

Then $$\varpi = n a_{0,n},$$

$$m-\rho = \Sigma \lambda a_{\lambda,0}.$$

The two sides of the identity will be given by the expansions

$$x_0^{a_{0,0}} x_1^{a_{1,0}} \dots x_n^{a_{n,0}} \left(x_0+nx_1+\binom{n}{2} x_2+\dots+x_n\right)^{a_{0,n}}$$

and $(-1)^{\varpi-\rho} x_0^{a_{0,0}} x_n^{a_{0,n}} (x_0+x_1)^{a_{1,0}} (x_0+2x_1+x_2)^{a_{2,0}} \dots$

$$\dots (x_0+nx_1+\dots+x_n)^{a_{n,0}}.$$

Any term of either expansion

$$\lambda x_0^{\gamma_0} x_1^{\gamma_1} \dots x_n^{\gamma_n},$$

where $$\gamma_1+2\gamma_2+\dots+n\gamma_n = m,$$

represents a term of the identity, viz.,

$$\lambda f(\gamma_0, \gamma_1, \dots, \gamma_n).$$

Now it will be regarded as a reduction when $f(a_0, a_1, \dots, a_n)$ is expressed in terms of invariants in which one or both of a_0, a_n is increased; also if this invariant is expressed in terms of invariants $f(\gamma_0, \gamma_1, \ldots, \gamma_n)$ such that the first argument γ_2, which differs from the corresponding argument a_r, exceeds it, whether we start comparing them at one end or at the other. In such a case we shall say that $f(\gamma_0, \gamma_1, \ldots, \gamma_n)$ is more simple than $f(a_0, a_1, \ldots, a_n)$.

Let us suppose that $a_{0,n} > a_{n,0}$.

Now $$a_{1,0}+2a_{2,0}+\ldots+na_{n,0} = m-p.$$

Hence $x_0^{\beta_0+a_{0,0}} x_1^{\beta_1+a_{1,0}} \ldots x_n^{\beta_n+a_{n,0}}$, a term of the expansion of

$$x_0^{a_{0,0}} x_1^{a_{1,0}} \ldots x_n^{a_{n,0}} (x_0+nx_1+\ldots+x_n)^{a_{0,n}},$$

will give an invariant

$$\beta_1+2\beta_2+\ldots+n\beta_n = p$$

and $$\beta_0+\beta_1+\ldots+\beta_n = a_{0,n}.$$

Let us suppose that $p = ra_{0,n}$;

then $$\binom{n}{r}^{a_{0,n}} f(a_{0,0}, a_{1,0}, \ldots, a_{r,0}+a_{0,n}, a_{r+1,0}, \ldots, a_{n,0})$$

is a term of the left-hand side of the identity. Every other term on this side is more simple; if not, we must be able to choose positive numbers $\beta_0, \beta_1, \ldots, \beta_r$, such that

$$\beta_0+\beta_1+\ldots+\beta_r = a_{0,n}$$

and $$\beta_1+2\beta_2+\ldots+r\beta_r = ra_{0,n};$$

or else positive numbers $\beta_r, \beta_{r+1}, \ldots, \beta_n$

such that $$\beta_r+\beta_{r+1}+\ldots+\beta_n = a_{0,n},$$

$$r\beta_r+(r+1)\beta_{r+1}+\ldots+n\beta_n = ra_{0,n}.$$

The only solution in either case is $\beta_r = a_{0,n}$. By hypothesis the number $a_{0,n} > a_{n,0}$, and so all the terms on the other side of the identity are more simple than that under consideration.

Hence, if $a_r > a_n$, we may express $f(a_0, a_1, \ldots, a_n)$ in terms of simpler invariants.

In the same way we may deduce from this identity that, if $a_r+a_{r+1} > a_n$, the invariant $f(a_0, a_1, \ldots, a_n)$ may be expressed in terms of simpler invariants.

These reductions will evidently be true sometimes when

$$a_r+a_{r+1} = a_n.$$

Now $$f(a_0, a_1, \ldots, a_n) = (-1)^m f(a_n, a_{n-1}, \ldots, a_0).$$
Hence we may perform a similar reduction, if $a_r + a_{r+1} > a_0$.

31. Let us consider some particular cases.

For the quartic we may have $a_1 = 0$, $a_3 = 0$, and $a_2 \not> a_0$ or a_4. The invariants are evidently

$$f(1, 0, 0, 0, 1),\quad f(1, 0, 1, 0, 1),\quad f(2, 0, 0, 0, 2),$$

and so on.

For the quintic we may take $a_1 = 0$, $a_4 = 0$, $a_2 + a_3 \not> a_0$ or a_4. Let us find the degree of the lowest *gauche* invariant. For a *gauche* invariant the weight m is odd; hence

$$f(a_0, a_1, \ldots, a_n) = -f(a_n, a_{n-1}, \ldots, a_0).$$

Hence, if $$a_0 = a_n,\ a_1 = a_{n-1}, \ldots,$$
$$f(a_0, a_1, \ldots, a_n) = 0.$$

Now $$m = a_1 + 2a_2 + \ldots + na_n = na_0 + (n-1)\, a_1 + \ldots + a_{n-1};$$

and therefore $$n(a_0 - a_n) + (n-2)(a_1 - a_{n-1}) + \ldots = 0.$$

For the quintic, if $$a_1 = 0 = a_4,$$
$$5(a_0 - a_5) + (a_2 - a_3) = 0.$$

Hence for the lowest *gauche* invariant

$$a_3 - a_2 = 5(a_0 - a_5) \neq 0.$$

Also we may take $a_2 + a_3 \not> a_0$ or a_5; hence, if $a_3 > a_2$, $a_0 > 5$.

Consider $$a_2 = 0,\quad a_3 = 5,\quad a_0 = 6,\quad a_5 = 5.$$

This is the first case in which the equations

$$a_0 = a_5,\quad a_2 = a_3$$

need not be satisfied; but it does not give a *gauche* invariant. The earliest case is of degree 18, there being only one invariant of this degree, viz.,

$$f(7, 0, 0, 5, 0, 6).$$

VI. *The Concomitants of a single q-ary Quadratic.*

32. We have already proved that any function of the coefficients of a q-ary quantic may be expressed as a linear function of the coefficients of concomitants of the form $PN\phi$.

When this function is itself a concomitant then the function of the coefficients becomes $PN\phi$ itself. The manner in which the variables appear is known from the form of PN.

Hence the expression

$$\left\{\begin{matrix} a_{1,1}\, a_{1,2} \cdots\cdots\cdots a_{1,\beta_1} \\ a_{2,1}\, a_{2,2} \cdots a_{2,\beta_2} \\ \cdots \quad \cdots \quad \cdots \\ a_{k,1} \cdots a_{k,\beta_k} \end{matrix}\right\}$$

defines a concomitant; and all concomitants may be expressed in terms of these.

33. For the case of a quadratic, these concomitants are operated upon by an operator G which connects the letters together two and two. For a single quadratic we have further an operator H which interchanges the sets of two in all possible ways. The concomitant will be completely defined when we know the numbers $a_{1,1}$ of sets having both letters in the first row, $a_{2,2}$ of sets having both letters in the second row, $a_{1,2}$ of sets having one letter in the first and one in the second row, and so on.

We will write the concomitant then in the form

$$f(a_{1,1},\ a_{2,2},\ \ldots,\ a_{k,k};\ \ a_{1,2},\ \ldots,\ a_{k-1,k}),$$

where $a_{r,s}$ is the number of sets of letters having one letter in the r-th and one in the s-th row.

Consider the identity corresponding to

$$\{a_{1,1}\, a_{1,2} \ldots a_{1,\beta_1} b\}\, I = 0,$$

where b is a letter of $a_{1,2}$. We obtain

$$\begin{aligned} &2(a_{1,1}+1)\, f(a_{1,1},\ a_{2,2},\ \ldots;\ a_{1,2},\ a_{1,3},\ \ldots) \\ &\quad + (a_{1,2}-1)\, f(a_{1,1}+1,\ a_{2,2}+1,\ \ldots;\ a_{1,2}-2,\ a_{1,3},\ \ldots) \\ &\quad + \sum_{r=3}^{k} a_{1,r} f(a_{1,1}+1,\ a_{2,2},\ \ldots;\ a_{1,2}-1,\ a_{1,3},\ \ldots,\ a_{1,r}-1,\ \ldots,\ a_{2,r}+1, \\ &\qquad\qquad\qquad\qquad\qquad\qquad\qquad\qquad\qquad\qquad \ldots,\ a_{k-1,k}) = 0. \end{aligned}$$

Hence, unless $\qquad a_{1,2} = 0 = a_{1,3} \ldots = a_{1,k},$

we can increase $a_{1,1}$. If β_1 is odd, the letters of the top row cannot all be made to belong to the sets $a_{1,1}$; and, by the above identity, the concomitant vanishes.

Similarly, we find by taking the letters of the second row with one out of a lower row, that $\alpha_{2,r}$ may be taken to be zero: and that the concomitant vanishes unless β_2 is even.

Hence, in order to give a non-zero concomitant, the number of letters in each row of PN must be even. Also we may suppose that there are no sets $\alpha_{r,s}$, $r \neq s$.

There is then only one independent form, viz.,

$$f\left(\frac{\beta_1}{2}, \frac{\beta_2}{2}, \ldots, \frac{\beta_k}{2}; 0, 0, \ldots, 0\right).$$

34. If there is a product of concomitants of the quadratic which is of the same degree, and of the same order in each kind of variable, as the above, then the above is reducible.

Hence the only irreducible forms for a single q-ary quadratic are given by

$$f(1, 1, \ldots, 1; 0, 0, \ldots, 0),$$

where the number of rows of the PN from which the expression is obtained is $\not> q$.

Thus the irreducible concomitants of a q-ary quadratic are q in number; they may be written symbolically

$$a_{1^x}^2, \ (ab)_{2^x}^2, \ \ldots, \ (a_1 a_2 \ldots a_{q-1})_{q-1^x}^2, \ (a_1 a_2 \ldots a_q)^2.$$

ON QUADRATIC INVARIANT TYPES

By *A. Young, M.A.*, Clare College, Cambridge

THE invariants under consideration here are those which are linear in the coefficients of each of the various quadratics concerned. It is well known that if the quadratics are

$$a_x^2,\ b_x^2,\ c_x^2,\ \ldots,$$

the only irreducible invariants are of the two types

$$(ab)^2,\ (ab)(bc)(ca).$$

Now any symbolical product of factors (ab), in which each letter appears twice, represents a quadratic invariant type. But any such invariant is merely a product of invariants of the type

$$(ab)(bc)(cd)\ldots(ka).$$

The object of the present note is to find the actual expression for this invariant when the degree is even, in terms of invariants of the irreducible types.

The symbolical factors satisfy the fundamental identity

$$(ab)(cd) + (ac)(db) + (ad)(bc) = 0,$$

and it is by this alone that the reduction must be made.

The notation

$$\{abcd...k\} \equiv (ab)(bc)(cd)...(ka)$$

will be used.

Using the fundamental identity for the pair of factors $(ab)(cd)$, we obtain

$$\{abcd...k\} + \{acbd...k\} = -(bc)^2\{ad...k\} \quad ......(1).$$

Taking the pair of factors $(ab)(de)$, we obtain

$$\{abcde...k\} - \{adcbe...k\} = \{bcd\}\{ae...k\} \quad ......(2),$$

and so on.

Now it will be convenient to write the quadratics in the form

$$a^2_{1_x},\ a^2_{2_x},\ ...,$$

where $\qquad a_{1_x} = a_{1,1}x_1 + a_{1,2}x_2.$

Consider the invariant type of degree $2n$,

$$\{a_1a_2...a_{2n}\}$$

by equation (1) we obtain

$$\{a_1a_2a_3...a_{2n}\} + \{a_2a_1a_3...a_{2n}\} = -(a_1a_2)^2\{a_3a_4...a_{2n}\}$$

$$-\{a_1a_3a_4...a_{2n}a_2\} - \{a_3a_1a_4...a_{2n}a_2\} = +(a_1a_3)^2\{a_2a_4...a_{2n}\}$$

$$\{a_1a_4a_5...a_3\} + \{a_4a_1a_5...a_3\} = -(a_1a_4)^2\{a_2a_3a_5...a_{2n}\}$$

....

$$\{a_1a_{2n}a_2...a_{2n-1}\} + \{a_{2n}a_1a_2...a_{2n-1}\} = -(a_1a_{2n})^2\{a_2a_3...a_{2n-1}\},$$

the signs being taken alternately positive and negative. Adding this system of equations, we obtain

$$2\{a_1a_2a_3...a_{2n}\} = -(a_1a_2)^2\{a_3a_4...a_{2n}\}$$
$$+(a_1a_3)^2\{a_2a_4...a_{2n}\} - ... - (a_1a_{2n})^2\{a_2a_3...a_{2n-1}\} \quad ...(3).$$

It may now be deduced from this equation that

$$2^{n-1}\{a_1a_2a_3\ldots a_{2n}\}$$

is a square root of the skew symmetric determinant

$$\left|\begin{array}{ccccc} 0, & (a_1a_2)^2, & (a_1a_3)^2, & \ldots, & (a_1a_{2n})^2 \\ -(a_1a_2)^2, & 0, & (a_2a_3)^2, & \ldots, & (a_2a_{2n})^2 \\ -(a_1a_3)^2, & -(a_2a_3)^2, & 0, & \ldots, & (a_3a_{2n})^2 \\ \ldots & \ldots & \ldots & \ldots & \ldots \\ -(a_1a_{2n})^2, & -(a_2a_{2n})^2, & -(a_3a_{2n})^2, & \ldots, & 0 \end{array}\right|.$$

For let us assume that the theorem is true when the degree of the type is even and less than $2n$; then by a well known property of Pfaffians,* a square root of the above determinant is given by

$$-(a_1a_2)^2\,2^{n-1}\{a_3a_4\ldots a_{2n}\}$$
$$+(a_1a_3)^2\,2^{n-1}\{a_2a_4\ldots a_{2n}\}-\ldots-(a_1a_{2n})^2\,2^{n-1}\{a_2a_3\ldots a_{2n-1}\}.$$

Hence by (3) the theorem is true for degree $2n$.

It is easy to verify that it is true for degree 4, in fact (3) gives

$$2\{a_1a_2a_3a_4\}=(a_1a_2)^2(a_3a_4)^2-(a_1a_3)^2(a_2a_4)^2+(a_1a_4)^2(a_2a_3)^2.$$

Therefore the theorem is true for all even degrees.

The relation proved is an algebraic identity, it will still be true if we write

$$a_{r,2}=1, \quad r=1,2,\ldots,2n;$$

hence, by squaring we obtain the identity

$$2^{2n-2}(a_1-a_2)^2(a_2-a_3)^2\ldots(a_{2n}-a_1)^2$$

$$=\left|\begin{array}{ccccc} 0, & (a_1-a_2)^2, & (a_1-a_3)^2, & \ldots, & (a_1-a_{2n})^2 \\ -(a_1-a_2)^2, & 0, & (a_2-a_3)^2, & \ldots, & (a_2-a_{2n})^2 \\ \ldots & \ldots & \ldots & \ldots & \ldots \\ -(a_{2n}-a_1)^2, & -(a_{2n}-a_2)^2, & -(a_{2n}-a_3)^2, & \ldots, & 0 \end{array}\right|.$$

---

* Scott, *Theory of Determinants.*

THE EXPANSION OF THE *n*th POWER OF A DETERMINANT

By *A. Young, M.A.*, Clare College, Cambridge

THE expansion to be proved is

$$\begin{vmatrix} a_1, & a_2, & a_3 \\ b_1, & b_2, & b_3 \\ c_1, & c_2, & c_3 \end{vmatrix}^n = \Sigma (-)^{\epsilon+\iota} \frac{n! \begin{vmatrix} \epsilon, & \zeta \\ \theta, & \iota \end{vmatrix}_\alpha}{\alpha!\,\epsilon!\,\zeta!\,\theta!\,\iota!} a_1^\alpha a_2^\beta a_3^\gamma b_1^\delta b_2^\epsilon b_3^\zeta c_1^\eta c_2^\theta c_3^\iota,$$

where

$$\begin{vmatrix} \epsilon, & \zeta \\ \theta, & \iota \end{vmatrix}_\alpha = \epsilon_\alpha \iota_\alpha - \binom{\alpha}{1} \epsilon_{\alpha-1} \iota_{\alpha-1} \zeta\theta + \binom{\alpha}{2} \epsilon_{\alpha-2} \iota_{\alpha-2} \zeta_2 \theta_2 - \ldots,$$

in formal analogy to the binomial theorem, the meaning of the suffixes being defined by

$$x_m = x(x-1)(x-2)\ldots(x-m+1);$$

and where the above summation extends to all terms for which

$$\begin{aligned} \alpha+\beta+\gamma &= n, & \alpha+\delta+\zeta &= n, \\ \delta+\epsilon+\zeta &= n, & \beta+\epsilon+\eta &= n, \\ \zeta+\eta+\theta &= n, & \gamma+\zeta+\theta &= n. \end{aligned}$$

To establish the theorem it will be useful to use the notation

$$\begin{pmatrix} & n & \\ \alpha, & \beta, & \gamma \\ \delta, & \epsilon, & \zeta \\ \eta, & \theta, & \iota \end{pmatrix}$$

to represent the coefficient of

$$a_1^\alpha a_2^\beta a_3^\gamma b_1^\delta b_2^\epsilon b_3^\zeta c_1^\eta c_2^\theta c_3^\iota$$

in the expansion of Δ^n, Δ being the determinant considered.

Using the method of induction, we assume the truth of the theorem for the expansion of Δ^{n-1}, and then multiply each side of the expression obtained by Δ. We obtain, by equating coefficients,

$$\begin{pmatrix} & n & \\ \alpha, & \beta, & \gamma \\ \delta, & \epsilon, & \zeta \\ \eta, & \theta, & \iota \end{pmatrix} = \begin{pmatrix} & n-1 & \\ \alpha-1, & \beta\ , & \gamma \\ \delta\ , & \epsilon-1, & \zeta \\ \eta\ , & \theta\ , & \iota-1 \end{pmatrix} - \begin{pmatrix} & n-1 & \\ \alpha-1, & \beta\ , & \gamma \\ \delta\ , & \epsilon\ , & \zeta-1 \\ \eta\ , & \theta-1, & \iota \end{pmatrix}$$

$$+ \begin{pmatrix} & n-1, & \\ \alpha\ , & \beta-1, & \gamma \\ \delta\ , & \epsilon\ , & \zeta-1 \\ \eta-1, & \theta\ , & \iota \end{pmatrix} - \begin{pmatrix} & n-1, & \\ \alpha\ , & \beta-1, & \gamma \\ \delta-1, & \epsilon\ , & \zeta \\ \eta\ , & \theta\ , & \iota-1 \end{pmatrix}$$

$$+ \begin{pmatrix} & n-1 & \\ \alpha\ , & \beta\ , & \gamma-1 \\ \delta-1, & \epsilon\ , & \zeta \\ \eta\ , & \theta-1, & \iota \end{pmatrix} - \begin{pmatrix} & n-1 & \\ \alpha\ , & \beta\ , & \gamma-1 \\ \delta\ , & \epsilon-1, & \zeta \\ \eta-1, & \theta\ , & \iota \end{pmatrix} \text{...(I.)}$$

$$= \frac{(-)^{\epsilon+\iota}(n-1)!}{\alpha!\,\epsilon!\,\zeta!\,\theta!\,\iota!}\left[\alpha\epsilon\iota \begin{vmatrix} \epsilon-1, & \zeta \\ \theta\ , & \iota-1 \end{vmatrix}_{\alpha-1} - \alpha\zeta\theta \begin{vmatrix} \epsilon\ , & \zeta-1 \\ \theta-1, & \iota \end{vmatrix}_{\alpha-1}\right.$$

$$\left. + \zeta \begin{vmatrix} \epsilon, & \zeta-1 \\ \theta, & \iota \end{vmatrix}_{\alpha} + \iota \begin{vmatrix} \epsilon, & \zeta \\ \theta, & \iota-1 \end{vmatrix}_{\alpha} + \theta \begin{vmatrix} \epsilon\ , & \zeta \\ \theta-1, & \iota \end{vmatrix}_{\alpha} + \epsilon \begin{vmatrix} \epsilon-1, & \zeta \\ \theta\ , & \iota \end{vmatrix}\right]$$

$$= \frac{(-)^{\epsilon+\iota}(n-1)!}{\alpha!\,\epsilon!\,\zeta!\,\theta!\,\iota!}\left[\alpha \sum_{r=0}^{\alpha-1} \binom{\alpha-1}{r} (-)^r \epsilon_{\alpha-r}\iota_{\alpha-r}\zeta_r\theta_r\right.$$

$$- \alpha \sum_{r=0}^{\alpha-1} \binom{\alpha-1}{r} (-)^r \epsilon_{\alpha-1-r}\iota_{\alpha-1-r}\zeta_{r+1}\theta_{r+1}$$

$$\left. + \sum_{r=0}^{\alpha} \binom{\alpha}{r} (-)^r \left\{\begin{matrix} \epsilon_{\alpha-r}\iota_{\alpha-r}\zeta_{r+1}\theta_r + \epsilon_{\alpha-r}\iota_{\alpha-r+1}\zeta_r\theta_r \\ + \epsilon_{\alpha-r}\iota_{\alpha-r}\zeta_r\theta_{r+1} + \epsilon_{\alpha-r+1}\iota_{\alpha-r}\zeta_r\theta_r \end{matrix}\right\}\right]$$

$$= \frac{(-)^{\epsilon+\iota}(n-1)!}{\alpha!\,\epsilon!\,\zeta!\,\theta!\,\iota!}\left[\sum_{r=0}^{\alpha} (-)^r \binom{\alpha}{r} \epsilon_{\alpha-r}\iota_{\alpha-r}\zeta_r\theta_r \left\{\begin{matrix} \alpha-r+r+\zeta-r+\iota-\alpha+r \\ +\theta-r+\epsilon-\alpha+r \end{matrix}\right\}\right]$$

.........(II.).

Now the expression in the inner bracket

$$= -\alpha+\zeta+\iota+\theta+\epsilon = -\alpha+n-\gamma+n-\beta = n.$$

Hence

$$\begin{pmatrix} & n & \\ \alpha, & \beta, & \gamma \\ \delta, & \epsilon, & \zeta \\ \eta, & \theta, & \iota \end{pmatrix} = \frac{(-)^{\epsilon+\iota}\, n!\begin{vmatrix} \epsilon, & \zeta \\ \theta, & \iota \end{vmatrix}_a}{\alpha!\,\epsilon!\,\zeta!\,\theta!\,\iota!}.$$

The theorem is thus true *in general* for the expansion of Δ^n, if it is true for the expansion of Δ^{n-1}. We have assumed in the proof that no one of the letters $\alpha, \beta, \gamma, \ldots$, is zero. It is necessary to consider these special cases.

(i) $\alpha=0$. The first two coefficients in (I) are missing. Hence the coefficient becomes

$$\frac{(-)^{\epsilon+\iota}(n-1)!}{\epsilon!\,\zeta!\,\theta!\,\iota!}(\zeta+\iota+\theta+\epsilon)=\frac{(-)^{\epsilon+\iota}\, n!}{\epsilon!\,\zeta!\,\theta!\,\iota!},$$

which is the coefficient required :—taking

$$\begin{vmatrix} \epsilon, & \zeta \\ \theta, & \iota \end{vmatrix}_0 = 1.$$

(ii) $\beta=0$. The third and fourth coefficients are missing in (I). The expression in the inner bracket in (II) is then

$$\theta+\epsilon=n.$$

Hence the coefficient is as before. The cases

$$\gamma=0,\ \delta=0,\ \eta=0$$

are identical with this.

(iii) $\epsilon=0$. The first and last coefficients in (I) are absent. The expression in the inner bracket of (II) becomes

$$\zeta+\iota+\theta-\alpha=n.$$

The other special cases in which two or more letters vanish may be discussed in the same way. Hence the theorem is true for all coefficients of Δ^n if it is true for all coefficients of Δ^{n-1}.

It is obviously true when $n=1$, hence it is true in general.

Now if two rows in Δ are interchanged, Δ^n is changed to $(-)^n\Delta^n$, hence

$$\begin{pmatrix} & n & \\ \delta, & \epsilon, & \zeta \\ \alpha, & \beta, & \gamma \\ \eta, & \theta, & \iota \end{pmatrix} = (-)^n \begin{pmatrix} & n & \\ \alpha, & \beta, & \gamma \\ \delta, & \epsilon, & \zeta \\ \eta, & \theta, & \iota \end{pmatrix}.$$

Similarly

$$\left(\begin{matrix} & n & \\ \beta, & \alpha, & \gamma \\ \epsilon, & \delta, & \zeta \\ \theta, & \eta, & \iota \end{matrix}\right) = (-)^n \left(\begin{matrix} & n & \\ \alpha, & \beta, & \gamma \\ \delta, & \epsilon, & \zeta \\ \eta, & \theta, & \iota \end{matrix}\right).$$

Hence

$$\frac{(-)^{\epsilon+\iota}\, n! \begin{vmatrix} \epsilon, & \zeta \\ \theta, & \iota \end{vmatrix}_\alpha}{\alpha!\,\epsilon!\,\zeta!\,\theta!\,\iota!} = \frac{(-)^{\zeta+\theta}\, n! \begin{vmatrix} \zeta, & \delta \\ \iota, & \eta \end{vmatrix}_\beta}{\beta!\,\zeta!\,\delta!\,\iota!\,\eta!} = \dots.$$

In calculating the value of the coefficient it will then be simplest to choose that one of the indices α, β, ... which is least, and calculate by means of that index and its minor.

The process of calculating the coefficients of

$$(a_1b_2c_3 + a_1b_3c_2 + a_2b_3c_1 + a_2b_1c_3 + a_3b_1c_2 + a_3b_2c_1)^n$$

is almost identical: using the symbol

$$\left\{\begin{matrix} & n & \\ \alpha, & \beta, & \gamma \\ \delta, & \epsilon, & \zeta \\ \eta, & \theta, & \iota \end{matrix}\right\}$$

to denote the coefficient of the same term as before, we find that

$$\left\{\begin{matrix} & n & \\ \alpha, & \beta, & \gamma \\ \delta, & \epsilon, & \zeta \\ \eta, & \theta, & \iota \end{matrix}\right\} = \frac{n!\,[\epsilon\iota + \zeta\theta]_\alpha}{\alpha!\,\epsilon!\,\iota!\,\zeta!\,\theta!},$$

where

$$[\epsilon\iota + \zeta\theta]_\alpha = \epsilon_\alpha\iota_\alpha + \binom{\alpha}{1}\epsilon_{\alpha-1}\iota_{\alpha-1}\zeta_1\theta_1 + \dots.$$

"The Maximum Order of an Irreducible Covariant of a System of Binary Forms." By A. YOUNG, M.A., Clare College, Cambridge. Communicated by Major P. A. MACMAHON, D.Sc., F.R.S. Received September 26, 1903.

It has been suggested to me that an incidental result of a paper I have recently communicated to the London Mathematical Society may be of interest. In the paper in question it is proved that all covariants of a system of binary n-ics are linearly expressible in terms of—

(i) Covariants of the form

$$(a_1a_2)^{\lambda_1}(a_2a_3)^{\lambda_2}\ \ldots\ (a_{\delta-1}a_\delta)^{\lambda_{\delta-1}}a_{1_x}^{n-\lambda_1}a_{2_x}^{n-\lambda_1-\lambda_2}\ \ldots\ a_{\delta_x}^{n-\lambda_{\delta-1}},$$

where

$$\lambda_1 \nless 2^{\delta-2},\quad \lambda_2 \nless 2^{\delta-3},\quad \ldots\quad \lambda_{\delta-1} \nless 1$$

(ii) Covariants having a symbolical factor

$$(ab)^{\lambda}(bc)^{n-\lambda}(ca)^{\rho}$$

(iii) Products of covariants.

Mr. J. H. Grace, in a note appended to the paper, has deduced from this result a means of calculating the maximum order, in the variables, of an irreducible covariant of a system of quantics. If no quantic of the system is of order exceeding n, the maximum order of an irreducible covariant is the greatest of the numbers

$$n\delta - 2^{\delta} + 2,$$

where δ is an integer,—in fact, the degree in the coefficients of the covariant in question.

If $n = 2^i + k$ where $k \ngtr 2^i$, it will be seen that the maximum is

$$(i+1)(2^i + k) - 2^{i+1} + 2.$$

The covariant of maximum order is then

$$(a_1a_2)^{2^{i-1}}(a_2a_3)^{2^{i-2}}\ \ldots\ (a_{i-1}a_i)^2(a_ia_{i+1})a_{1_x}^{n-2^{i-1}}\ \ldots\ a_{i+1_x}^{n-1}.$$

There are strong reasons for believing this covariant to be actually irreducible; in the contrary case a reduction is obtained for certain forms classed as perpetuants.*

* MacMahon, 'Camb. Phil. Trans.,' vol. 19, p. 234; Grace, 'Lond. Math. Soc. Proc.,' vol. 35, p. 107.

The following is a table giving the maximum order for all values of n from 1 to 100, together with the degree, δ, of the covariant of maximum order:—

n.	Max. order.	δ.	n.	Max. order.	δ.	n.	Max. order.	δ.	n.	Max. order.	δ.	n.	Max. order.	δ.
1	1	1	21	75	5	41	184	6	61	304	6	81	441	7
2	2	1, 2	22	80	5	42	190	6	62	310	6	82	448	7
3	4	2	23	85	5	43	196	6	63	316	6	83	455	7
4	6	2, 3	24	90	5	44	202	6	64	322	6, 7	84	462	7
5	9	3	25	95	5	45	208	6	65	329	7	85	469	7
6	12	3	26	100	5	46	214	6	66	336	7	86	476	7
7	15	3	27	105	5	47	220	6	67	343	7	87	483	7
8	18	3, 4	28	110	5	48	226	6	68	350	7	88	490	7
9	22	4	29	115	5	49	232	6	69	357	7	89	497	7
10	26	4	30	120	5	50	238	6	70	364	7	90	504	7
11	30	4	31	125	5	51	244	6	71	371	7	91	511	7
12	34	4	32	130	5, 6	52	250	6	72	378	7	92	518	7
13	38	4	33	136	6	53	256	6	73	385	7	93	525	7
14	42	4	34	142	6	54	262	6	74	392	7	94	532	7
15	46	4	35	148	6	55	268	6	75	399	7	95	539	7
16	50	4, 5	36	154	6	56	274	6	76	406	7	96	546	7
17	55	5	37	160	6	57	280	6	77	413	7	97	553	7
18	60	5	38	166	6	58	286	6	78	420	7	98	560	7
19	65	5	39	172	6	59	292	6	79	427	7	99	567	7
20	70	5	40	178	6	60	298	6	80	434	7	100	574	7

ON COVARIANT TYPES OF BINARY n-ICS*

By A. Young

[Communicated April 16th, 1903.—Received April 22nd, 1903.—Received in revised form August 3rd, 1903.]

1. The "finiteness" of the complete system of concomitants of a single binary form was first proved by Gordan† in the year 1868. Before that time it was believed that quantics of order higher than 4 did not possess a finite system of concomitants. Since then many different proofs of this central fact in the theory of forms have been given by various writers. Its truth has been established for ternary forms and forms having any greater number of variables; for forms possessing two or more different *sets* of variables, whether cogredient or otherwise; for any simultaneous system of forms; and for types of concomitants when the actual number of base forms becomes infinite, but the order of each is finite. Information, with references, as to the various proofs of the finiteness is to be found in any of the various editions of Meyer's "Bericht über den gegenwärtigen Stand der Invariantentheorie."‡

The majority of the proofs, especially of the later ones, keep the attention fixed on the collective idea of finiteness, and give little or no information as to the actual composition, formation, or extent of the complete system. There are two exceptions, which stand out pre-eminently from the above general rule: these are the proof given by Gordan and that due to Jordan.

Gordan proved the theorem by actually giving a process by means of which the complete system might be established, and by showing that this process yielded only a finite number of irreducible concomitants. In the later editions of his proof the process was considerably modified, but the principle remained the same.

---

* The paper as originally written consisted of §§ 2–8. The introduction (§ 1) and the table (§ 9) have been written at the suggestion of the referees.

† *Jour. für Math.* (*Crelle*), Vol. LXIX., pp. 323–354.

‡ The original edition appeared in 1892 in the *Jahresbericht der Deutschen Math.-Vereinigung*, Bd. I.; French revised edition by Fehr, 1897 (Gauthier-Villars, Paris); Italian revised edition by Vivanti, 1899 (Pellerano, Naples); and, finally, the article "Invariantentheorie" by Meyer in the *Encycl. Math. Wiss.*, Bd. I., p. 320 (I. B. 2), 1899.

Jordan's proof was published in 1876.* It is remarkable that it has not attracted more attention. The first half of the paper referred to is devoted to an introduction to the theory of symbolic algebra. The author then proceeded to examine the covariants of degree 3. His discussion was exhaustive and established the following theorem :—

If a_x^p, b_x^q, c_x^r be any three binary forms, the order of each being $\geqslant n$, then the various covariants furnished by the expression

$$(ab)^\lambda (bc)^\mu (ca)^\nu a_x^{p-\nu-\lambda} b_x^{q-\lambda-\mu} c_x^{r-\mu-\nu} \quad (\lambda+\mu+\nu=n)$$

may be linearly expressed in terms of the $n+1$ covariants C^ρ_{abc} $(\rho=0,1,\ldots,n)$,

where $$C^\rho_{abc}=(ab)^{n-\rho}(bc)^\rho a_x^{p-n+\rho} b_x^{q-n} c_x^{r-\rho}.$$

They may also be linearly expressed as a function of the covariants C^ρ_{abc}, C^ψ_{bca}, C^ψ_{cab}, where $\rho \not> \frac{1}{3}n$. If $n=3k+2$, these last covariants will be all independent. If $n=3k+1$, they will be connected by a relation which will permit us to express the sum $C^k_{abc}+C^k_{bca}+C^k_{cab}$ as a linear function of those C's for which $\rho<k$. If $n=3k$, there will be two relations permitting us to express two of C^k_{abc}, C^k_{bca}, C^k_{cab} as a function of the third and of those covariants C for which ρ is less than k.†

In the next paragraph covariants of a system of quantics, the order of each being at least $2l$, were considered. It was proved that these belonged to three classes : (i.) covariants which have a factor $(ab)^{l+1}$; (ii.) covariants which have a factor $(ab)^l (bc)^l (ca)^\rho$ $(\rho>0)$; (iii.) covariants and products of covariants of the form $(ab)^\mu (bc)^\nu (cd)^{\mu'} (de)^{\nu'} \ldots a_x^{p-\mu} \ldots$; the symbols $a, b, c, \ldots$ corresponding to any quantics of the system, and the exponents $\mu, \nu, \mu', \nu', \ldots$ being different from zero and satisfying the inequalities $\nu \leqslant \frac{1}{2}\mu$, $\nu' \leqslant \frac{1}{2}\mu'$, $\nu'' \leqslant \frac{1}{2}\mu''$, $\ldots$; $\mu \leqslant l$, $\mu' \leqslant \mu-\nu-\epsilon$, $\mu'' \leqslant \mu'-\nu'-\epsilon'$, $\ldots$, where $\epsilon^{(i)}$ is equal to 0 or 1, according as $\mu^{(i)}$ is odd or even.

In the next paragraph an upper limit to the degree in the coefficients of irreducible covariants of the third class was obtained.

In the last paragraph the finiteness was proved for any simultaneous system, the proof being equally applicable to the finiteness of types of concomitants, for the letters of a symbolical product were regarded as not interchangeable. In addition, an upper limit to the order (in the variables) of an irreducible concomitant of a system of binary forms, the order of

* *Liouville's Jour. de Math.*, 1876.

† This theorem is fundamental to the method used in the present paper. It is extended in Jordan's second paper, *Liouville's Jour. de Math.*, 1879. It was rediscovered by Stroh. *Math. Ann.*, Bd. XXXI., pp. 444–454, who proved it in a more general form than that given above, using an entirely different method.

each being equal to or less than n, was obtained. This limit, which for any value of n could be calculated by a definite arithmetical process, is the same as that given here for the first eleven values of n, but for $n = 12$, and for higher values of n, Jordan's upper limit is here shown to be too high.

In *Liouville's Journal de Mathématiques*, 1879, Jordan wrote a second paper to determine an upper limit for the degree of an irreducible concomitant.

Upper limits for the degree and order of an irreducible concomitant have also been given by Sylvester.* He assigned the limit

$$\Sigma\left(\frac{n^2+1}{2}\right)+\Sigma\left(\frac{\nu^2}{2}\right)-2$$

as an upper limit to the order of a concomitant of a simultaneous system of binary forms of odd orders $n, n', n'', \ldots$, and even orders $\nu, \nu', \nu'', \ldots$; but he gave no proof. The limit is higher than that given by Jordan.

Our knowledge of complete systems of concomitants of binary forms is still limited to a few cases. Complete systems have been calculated for single forms of orders from 1 to 8 inclusive. For two quantics (h, k) the systems are known for the cases (2, 1), (2, 2), (2, 3), (3, 3), (3, 4), (2, 5), (2, 6), (4, 4). Systems of types of concomitants are known for the first four orders separately. The system for any number of quadratics and linear forms has been found. Also rules have been given by which the additional concomitants may be at once obtained when a linear or quadratic form is added to an already known complete system. References in nearly all the cases quoted are to be found in Meyer's *Bericht*.

2. The main object of the present paper is to obtain the complete irreducible system of types of covariants of binary forms of order n whose grade does not exceed $\frac{1}{2}n$.

The theorem to be proved is in reality an extension of the theorem proved for perpetuant types by Grace,† and follows his work closely both in statement and in demonstration.

The determinant factors of the symbolical products used here are alone written down, for, the orders of the various quantics considered being known, the remaining factors can at once be supplied.

* *Proc. Roy. Soc.*, Vol. XXVII., pp. 11, 12 (1878); *Comptes Rendus*, Vol. LXXXVI., pp. 1437–1441, 1491–1492, 1519–1522 (1878).

† *Proc. London Math. Soc.*, Vol. XXXV., pp. 107–111.

3. Any covariant which is of unit degree in the coefficients of each of the quantics $a_{1_x}^{n_1}, a_{2_x}^{n_2}, \ldots, a_{\delta_x}^{n_\delta}$ can be expressed linearly in terms of (i.) covariants of the form $(a_1 a_2)^{\lambda_1}(a_2 a_3)^{\lambda_2} \ldots (a_{\delta-1} a_\delta)^{\lambda_{\delta-1}}$, where $\lambda_1 \nless 2^{\delta-2}$, $\lambda_2 \nless 2^{\delta-3}, \ldots, \lambda_{\delta-1} \nless 1$, and the arrangement of the letters $a_1, a_2, \ldots, a_\delta$ is fixed; (ii.) covariants which have a factor of the form $(a_h a_k)^\lambda (a_k a_l)^{n_k-\lambda}$; (iii.) products of covariants of lower total degree.

Assuming that this theorem is true when the number of quantics concerned is less than δ, we shall first prove its truth when the number is equal to δ.

Any covariant of the kind under discussion may be expressed in terms of transvectants of the form $(a_{1_x}^{n_1}, C_{\delta-1})^\mu$, where $C_{\delta-1}$ is a covariant of unit degree in the coefficients of each of the quantics $a_{2_x}^{n_2}, a_{3_x}^{n_3}, \ldots, a_{\delta_x}^{n_\delta}$. Hence, on the above assumption, the transvectant written down can be expressed linearly in terms of transvectants

$$[a_{1_x}^{n_1}, (a_2 a_3)^{\lambda_2}(a_3 a_4)^{\lambda_3} \ldots (a_{\delta-1} a_\delta)^{\lambda_{\delta-1}}]^\mu,$$

and of covariants belonging to the second and third classes.*

Now all possible transvectants of this form are under consideration; hence each may be replaced by one of its terms.

Unless $\lambda_2+\mu < n_2$, the transvectant contains a term belonging to the second class. Hence all the covariants considered can be expressed in terms of covariants of the form $(a_1 a_2)^\mu (a_2 a_3)^{\lambda_2} \ldots (a_{\delta-1} a_\delta)^{\lambda_{\delta-1}}$, and of covariants belonging to the second and third classes.

4. Let

$$(a_4 a_5)^{\lambda_4} \ldots (a_{\delta-1} a_\delta)^{\lambda_{\delta-1}} a_{x}^{n_4-\lambda_4} a_{5_y}^{n_5-\lambda_4-\lambda_5} a_{6_y}^{n_6-\lambda_5-\lambda_6} \ldots a_{\delta_y}^{n_\delta-\lambda_{\delta-1}} \equiv a_4^{n_4-\lambda_4} q_y^\rho.$$

Then $(a_1 a_2)^\mu (a_2 a_3)^{\lambda_2} \ldots (a_{\delta-1} a_\delta)^{\lambda_{\delta-1}}$ differs from any other term of

$$[(a_1 a_2)^\mu (a_2 a_3)^{\lambda_2},\ a_{4_x}^{n_4-\lambda_4} q_y^\rho]_{y=x}^{\lambda_3} \qquad \text{(I.)}$$

by covariants in which the number of factors involving a_1, a_2, a_3 only is greater than $\lambda_2+\mu$.

Consider the transvectant $[(ab)^\lambda (bc)^\kappa,\ a_{4_x}^{n_4-\lambda_4} q_y^\rho]_{y=x}^{\lambda_3}$, where $\kappa < 2^{\delta-3}$ and a, b, c are the letters a_1, a_2, a_3 in some order.

The systems of covariants $(ab)^\lambda (bc)^\kappa$ and $[(ab)^\lambda, c_x^{m_3}]^\kappa$, where $\lambda+\kappa$ has a constant value and κ remains less than $2^{\delta-3}$, are equivalent—in the

* It is easy to verify that, if $C_{\delta-1}$ is reducible, the transvectant $(a_{1_x}^{n_1}, C_{\delta-1})^\mu$ belongs, on the assumption made, to the second or third class.

sense that any member of one system can be expressed linearly in terms of members of the other.

Let us introduce a new symbol α where $\alpha_x^{m_1+m_2-2\lambda}$ is the covariant $(ab)^\lambda$; m_1, m_2 being the orders of a and b. Then it will be sufficient to consider the transvectants $\left[(\alpha c)^\kappa,\ a_{4_x}^{n_4-\lambda_4} q_y^\rho\right]_{y=x}^{\lambda_3}$, where $\kappa < 2^{\delta-3}$.

But on the assumption made, since this is a covariant of the $\delta-1$ quantics $\alpha_x^{m_1+m_2-2\lambda}$, $c_x^{m_3}$, $a_{4_x}^{n_4}$, ..., $a_{\delta_x}^{n_\delta}$, it can be expressed in terms of (i.) covariants $(\alpha c)^{\mu_2}(ca_4)^{\mu_3}(a_4 a_5)^{\mu_4} \ldots (a_{\delta-1}a_\delta)^{\mu_{\delta-1}}$, where $\mu_2 \nless 2^{\delta-3}$, $\mu_3 \nless 2^{\delta-4}$, ..., $\mu_{\delta-1} \nless 1$; (ii.) covariants which have a factor of the form $(a_h a_k)^\nu (a_k a_l)^{n_k-\nu}$; (iii.) products of covariants of lower total degree.

Amongst the covariants here in the second class, there are included those which have a factor $(a_h \alpha)^\nu (\alpha a_k)^{m_1+m_2-2\lambda-\nu}$; but we may suppose that $m_1 \ngtr m_2$ and $\nu \nless m_1+m_2-2\lambda-\nu$, and hence that $\nu \nless m_1-\lambda$. Then every such covariant can be expressed in terms of such as have a factor of the form $(a_h a)^{m_1-\lambda}(ab)^\lambda$, and is therefore properly to be included in the second class.

5. It follows from what we have just proved that, if either μ or λ_2 is less than $2^{\delta-3}$, the transvectant (I.) can be expressed in terms of covariants which contain a greater number of factors involving a_1, a_2, a_3 only. We may then suppose that $\lambda_2+\mu \nless 2.2^{\delta-3}$. Again, the covariant $(a_1a_2)^\mu(a_2a_3)^{\lambda_2}$ belongs to the second class unless n_1, n_2, n_3 are each greater than $\lambda_2+\mu$. For, if $n_1 \leqslant \lambda_2+\mu$,

$$(a_1a_2)^\mu(a_2a_3)^{\lambda_2} = (a_1a_2)^\mu\left[(a_1a_3)-(a_1a_2)\right]^{n_1-\mu}(a_2a_3)^{\lambda_2+\mu-n_1}$$
$$= \Sigma A_r(a_1a_3)^r(a_1a_2)^{n_1-r}(a_2a_3)^{\lambda_2+\mu-n_1}.$$

In this case the transvectant (I.), and every one of its terms, belongs to the second class.

When n_1, n_2, n_3 are all greater than $\lambda_2+\mu$, and $\lambda_2+\mu$ is not less than $2^{\delta-2}$, we may express the covariant $(a_1a_2)^\mu(a_2a_3)^{\lambda_2}$ linearly in terms of the covariants*

$$(a_2a_3)^{\lambda_2+\mu},\quad (a_2a_3)^{\lambda_2+\mu-1}(a_3a_1),\quad \ldots,\quad (a_2a_3)^{\lambda_2+\mu-2^{\delta-3}+1}(a_3a_1)^{2^{\delta-3}-1};$$
$$(a_3a_1)^{\lambda_2+\mu},\quad (a_3a_1)^{\lambda_2+\mu-1}(a_1a_2),\quad \ldots,\quad (a_3a_1)^{\lambda_2+\mu-2^{\delta-3}+1}(a_1a_2)^{2^{\delta-3}-1};$$
$$(a_1a_2)^{\lambda_2+\mu},\quad (a_1a_2)^{\lambda_2+\mu-1}(a_2a_3),\quad \ldots,\quad (a_1a_2)^{2^{\delta-2}}(a_2a_3)^{\lambda_2+\mu-2^{\delta-2}}.$$

Now, by § 3, if C be any one of the covariants in the first two rows just written down, the number of factors involving a_1, a_2, a_3 only in the transvectant $(C,\ a_{4_x}^{n_4-\lambda_4} q_y^\rho)_{y=x}^{\lambda_3}$ can be increased.

* Stroh, *Math. Ann.*, Bd. XXXI., pp. 444–454; Jordan, *Liouville's Jour. de Math.*, 1876, 1879.

Hence these covariants can ultimately be expressed in terms of covariants of the second and third classes, and of covariants which contain the factor $(a_1 a_2)^{\lambda_1}$, where $\lambda_1 \not< 2^{\delta-2}$.

If, now, we write $a_x^{n_1+n_2-2\lambda_1} \equiv (a_1 a_2)^{\lambda_1}$, and use the assumption for covariants of total degree $\delta-1$, we see, as in § 3, that all covariants can be expressed linearly in terms of covariants of the second and third classes, and of covariants of the form $(a_1 a_2)^{\lambda_1} (a_2 a_3)^{\lambda_2} \dots (a_{\delta-1} a_\delta)^{\lambda_{\delta-1}}$, where

$$\lambda_1 \not< 2^{\delta-2}, \quad \lambda_2 \not< 2^{\delta-3}, \quad \dots, \quad \lambda_\delta \not< 1.$$

Thus the theorem is true when there are δ quantics involved, provided that it is true when any $\delta-1$ quantics are involved.

6. When only three quantics are involved we have to consider covariants of the form $(a_1 a_2)^\lambda (a_2 a_3)^\mu (a_3 a_1)^\nu$.

If $\lambda+\mu = n_2$, this covariant belongs to the second class.

If $\lambda+\mu < n_2$, the index of $(a_3 a_1)$ can be diminished by means of the identity

$$(a_3 a_1) = -(a_1 a_2) - (a_2 a_3).$$

Hence we have only to consider covariants $(a_1 a_2)^\lambda (a_2 a_3)^\mu$, where $\lambda+\mu < n_2$.

If $\mu = n_3$, this covariant belongs to the second class.

If $\mu < n_3$ and $\lambda = 1$, we can, by means of the relation

$$(a_1 a_3)^{\mu+1} = \{(a_1 a_2)+(a_2 a_3)\}^{\mu+1}$$
$$= (a_2 a_3)^{\mu+1} + \binom{\mu+1}{1}(a_1 a_2)(a_2 a_3)^\mu + \binom{\mu+1}{2}(a_1 a_2)^2 (a_2 a_3)^{\mu-2} + \dots,$$

express the covariant in terms of members of the first and third classes.

Hence the theorem is true when three quantics are involved; it is therefore true in general.

7. Let us suppose that all the quantics considered in the theorem just proved are of the same order n. Then the covariants of the second class all contain a factor of the form $(ab)^\lambda (bc)^{n-\lambda}$. These can be expressed in terms of transvectants, such as $[(ab)^\lambda (bc)^{n-\lambda} (ca)^r, C]^\varpi$, where r is positive or zero.

Now Jordan* has proved that the covariant $(ab)^\lambda (bc)^{n-\lambda} (ca)^r$ can be expressed in terms of covariants $(ab)^\mu (bc)^{n-\mu} (ca)^r$, where $\mu \geqslant \frac{1}{2}n$, and $\mu - r \geqslant 2(n-\mu-r)$, *i.e.*, $r \not< 2n-3\mu$.

Hence all covariant types of binary forms of order n can be expressed in terms of (i.) covariant types of the form $(a_1 a_2)^{\lambda_1} (a_2 a_3)^{\lambda_2} \dots (a_{\delta-1} a_\delta)^{\lambda_{\delta-1}}$,

* *Loc. cit.*

where $\lambda_1 \nless 2^{\delta-2}$, $\lambda_2 \nless 2^{\delta-3}$, ..., $\lambda_{\delta-1} \nless 1$; (ii.) covariant types which have a factor $(ab)^{\lambda}(bc)^{n-\lambda}(ca)^{r}$, where $\lambda \nless \frac{1}{2}n$, $r \nless 2n-3\lambda$; (3) products of covariants of lower degree. This theorem at once gives a system of types in terms of which all irreducible types of grade not exceeding $\frac{1}{2}n$ may be expressed.

8. It appears practically certain that the theorem for perpetuant types is exact, *i.e.*, that the covariant types for binary forms of infinite order $(a_1 a_2)^{\lambda_1}(a_2 a_3)^{\lambda_2} \ldots (a_{\delta-1} a_\delta)^{\lambda_\delta}$, where $\lambda_1 \nless 2^{\delta-2}$, $\lambda_2 \nless 2^{\delta-3}$, ..., $\lambda_\delta \nless 1$, and the order of the letters is fixed beforehand, are both independent and irreducible (this system being equivalent to that used by Grace). If this is the case, covariants of this form must be independent and irreducible for quantics of finite order. It does not follow, however, that they cannot be expressed in terms of covariants of higher grade, or else in terms of covariants belonging to the second class. In fact, as it is easy to verify, if $\lambda_1 + \lambda_r \geqslant n$, such a covariant type of a system of binary n-ics can be expressed in terms of members of the second class, and of covariants $(a_1 a_2)^{\lambda_1'}(a_2 a_3)^{\lambda_2'} \ldots (a_{\delta-1} a_\delta)^{\lambda_{\delta-1}'}$, where one or more of the differences $\lambda_1' - \lambda_1$, $\lambda_2' - \lambda_2$, ..., $\lambda_{r-1}' - \lambda_{r-1}$ is positive, and the rest are zero.

PERPETUANT SYZYGIES

By A. Young *and* P. W. Wood

[Received and Read May 12th, 1904.]

The methods used in the following discussion are entirely based on the symbolical notation, perpetuants linear in the coefficients of each quantic concerned being alone considered (perpetuant types). The symbolical forms of perpetuant types (which had previously all been identified by other methods) were calculated by Grace.* His result—" Any perpetuant linear in the coefficients of each of δ quantics, denoted by the letters $a_1, a_2, \ldots, a_\delta$, can be expressed linearly in terms of products and of perpetuants $(a_1 a_2)^{\lambda_1} (a_1 a_3)^{\lambda_2} \ldots (a_1 a_\delta)^{\lambda_{\delta-1}}$, where $\lambda_1 \geqslant 2^{\delta-2}$, $\lambda_2 \geqslant 2^{\delta-3}$, $\ldots$, $\lambda_{\delta-1} \geqslant 1$, the order of the letters being fixed beforehand"—is fundamental to our present purpose and will for convenience be quoted as " The Perpetuant Type Theorem."

The difficulty of dealing with the actual syzygies is avoided as follows: all possible products of irreducible forms, for a given degree and weight, are arranged in a predetermined sequence, so that each individual product has a definite place in that sequence. Any syzygy may then be regarded as expressing that one of its products, which comes first in the predetermined sequence, in terms of products which come after it. The first product will be called reducible, by virtue of its being linearly expressible in terms of later products.

Thus, instead of actually finding syzygies, we seek to discover what products are reducible and what products are irreducible. Now all products of perpetuant types of degree δ can be expressed in terms of the forms

$$(a_1 a_2)^{\lambda_1} (a_1 a_3)^{\lambda_2} \ldots (a_1 a_\delta)^{\lambda_{\delta-1}}, \ \ldots;$$

where the indices λ are only restricted to be positive integers or zeros, the order of the letters being fixed; such forms are all linearly independent, for, if there is a relation between them, there must be a purely algebraic relation between the symbolical products written down, and this is

* *Proc. London Math. Soc.*, Vol. xxxiv.

easily* shown to be impossible. The generating function for all linearly independent perpetuant types and products of perpetuant types of degree δ is therefore $\frac{1}{(1-x)^{\delta-1}}$; it is known† that the generating function for actually irreducible forms of degree δ is $\frac{x^{2^{\delta-1}-1}}{(1-x)^{\delta-1}}$. So the generating function for irreducible products must be

$$\frac{1-x^{2^{\delta-1}-1}}{(1-x)^{\delta-1}}.$$

It is our object to identify these irreducible products: when this has been done for any degree, all the independent syzygies of this degree will have been identified, there being one such syzygy for each reducible product of irreducible forms. Now there cannot be fewer irreducible forms (types) and products of degree δ than the number enumerated by the generating function $\frac{1}{(1-x)^{\delta-1}}$; hence, when we have reduced all forms except this number, we shall have a proof of the irreducibility of the remaining products, as well as a new proof of the irreducibility of the forms already classified as perpetuant types.

Our results are only complete as far as degree 8: a large class of products has been discussed in general, and, though it cannot be asserted that the generating functions obtained for these products are exact, yet the fact that the particular cases of them are exact as far as degree 8 establishes a strong presumption that such is always the case.

The syzygies are all derived from those of degree 4 due to Stroh,‡ and the Jacobian identity

$$(bc)+(ca)+(ab) \equiv 0.$$

The syzygies of degree 4 have been completely discussed already§ by a different method, but we have discussed them here fully in illustration of the general methods employed in the present paper.

---

* For, if there is a relation, let $(a_1a_2)^{\lambda_1}\Sigma N(a_1a_3)^{\lambda_2}\ldots(a_1a_\delta)^{\lambda_{\delta-1}}$ be those terms for which the index of (a_1a_2) is lowest; then we may divide out by $(a_1a_2)^{\lambda_1}$ and put $a_1 = a_2$ after division; we get $\Sigma N(a_1a_3)^{\lambda_2}\ldots(a_1a_\delta)^{\lambda_{\delta-1}} = 0$. Proceeding in this way, we ultimately see that every coefficient N is zero.

† The irreducibility of these forms has been finally demonstrated by Wood, "On the Irreducibility of Perpetuant Types," *Proc. London Math. Soc.*, Vol. 1, Ser. 2.

‡ *Math. Ann.*, Bd. XXXIII., S. 61–107, § 18.

§ Wood, *Proc. London Math. Soc.*, Vol. 2, Ser. 2.

The above methods are apparently insufficient to obtain all the syzygies for the reduction of products of three forms each of degree 3, and here arises the difficulty in the treatment of products of degree 9.

It seems probable that the generating function for all irreducible products and types of degree δ, where the products have no factor of degree less than $(n+1)$, is

$$\frac{x^{\binom{\delta}{1}+\binom{\delta}{2}+\dots+\binom{\delta}{\kappa}}}{(1-x)^{\delta-1}}, \qquad \kappa \leqslant \frac{\delta}{2};$$

and this result has been proved true for the cases $\kappa = 1$, 2, and $\kappa \geqslant \frac{\delta}{3}$. The question will be discussed at the end of Section VI.

The paper has been divided into sections as follows:—

CONTENTS.

Previous Literature on Perpetuant Syzygies.

Previously published papers, of which none date since 1887, differ fundamentally from the present in three respects: (1) by the use of the literal notation and the theory of partitions; (2) by making the discovery of the syzygies a basis for the enumeration of the irreducible perpetuants; (3) by actual expression, apart from mere enumeration, of the syzygies.

The present paper starts from a knowledge of the symbolical form of the perpetuants and deduces a means of enumerating the syzygies; their actual expressions can be obtained from Section III. All the papers quoted are to be found in the *American Journal of Mathematics.*

Sylvester, Vol. v., "On Subinvariants."

This paper was the starting point of all investigations on perpetuants and treated of certain syzygies of degrees 5, 6, and 7: an error in degree 7 was subsequently corrected by Hammond.

Cayley, Vol. VII., "A Memoir on Seminvariants."

This deals with a method of suitably expressing certain syzygies of degrees 5 and 6.

Hammond, Vol. v., "On the Solution of the Differential Equation of Sources."
This paper obtains the results of Sylvester afresh, and corrects the error in the treatment of degree 7.
——— Vol. VIII., "On Perpetuants, with Applications to the Theory of Finite Quantics."
This paper sketches a general method of classifying and expressing certain syzygies, but gives no definite result for any degree greater than 8.
MacMahon, Vol. x., "Expression of Syzygies among Perpetuants by means of Partitions."
The paper contains, *inter alia*, the actual expression of a number of sextic syzygies enumerated by the generating function

$$\frac{x^6+x^9+x^{11}+x^{12}+x^{13}+x^{14}+x^{15}+x^{16}+x^{17}}{(1-x^2)(1-x^4)(1-x^6)}.$$

Several sextic syzygies, however, remain unexpressed.

I. EXTENSION OF THE PERPETUANT TYPE THEOREM.

1. Consider any perpetuant

$$(a_1 a_2)^{\lambda_1} (a_1 a_3)^{\lambda_2} \dots (a_1 a_\kappa)^{\lambda_{\kappa-1}} \equiv C_\kappa .$$

We assume throughout that the letters are taken in a definite order fixed beforehand. If C_κ appears as a factor of a product P, we shall find certain conditions, ultimately affecting the indices λ, for the "irreducibility" of P; here we use the word "irreducibility" in a perfectly general sense, as implying that P cannot be expressed in terms of other forms obeying certain definite laws (*v.* Note, p. 225).

Notation.—The symbol $[a_{r_1} a_{r_2} \dots a_{r_l}]$ denotes any covariant, reducible or otherwise, involving the symbolical letters $a_{r_1}, a_{r_2}, \dots, a_{r_l}$; and the symbol $[a_{r_1} a_{r_2} \dots a_{r_l}]'$ denotes any covariant, reducible or otherwise, involving all the symbolical letters $a_1, a_2, \dots, a_\kappa$ concerned except $a_{r_1}, a_{r_2}, \dots, a_{r_l}$.

If the transvectant $([a_{r_1} \dots a_{r_l}], [a_{r_1} \dots a_{r_l}]')^\lambda$ replace C_κ in P, we may assume from our previous knowledge that, unless $\lambda \geqslant \mu_{r_1 r_2 \dots r_l}$, P is, according to some definition, reducible; here $\mu_{r_1 r_2 \dots r_l}$ is a quantity supposed known from other investigations: its value will depend on our definition of "reducibility."

We shall obtain all such transvectants of covariants involving some of the letters of C_κ with covariants involving the remaining letters, if we suppose that $a_{r_1}, a_{r_2}, \dots, a_{r_l}$ are some or all of $a_2, a_3, \dots, a_\kappa$, so that a_1 appears always in $[a_{r_1} a_{r_2} \dots a_{r_l}]'$.

To every choice of the letters $a_{r_1}, a_{r_2}, \dots, a_{r_l}$ corresponds a quantity $\mu_{r_1 r_2 \dots r_l}$, so that the number of such quantities $\mu_{r_1 r_2 \dots r_l}$ is $2^{\kappa-1}-1$; for there are $\binom{\kappa-1}{l}$ quantics $\mu_{r_1 r_2 \dots r_l}$ with l suffixes, and l may have any value from 1 to $\kappa-1$.

The order of the suffixes may be conveniently fixed by the rule

$$r_1 < r_2 < r_3 \dots < r_l.$$

We shall use σ_r to denote the sum of all the μ's whose first suffix is r, $\sigma_{r,s}$ to denote the sum of all the μ's whose first two suffixes are r, s, and so on.

Thus $$\sigma_\kappa = \mu_\kappa, \qquad \sigma_{\kappa-1} = \mu_{\kappa-1} + \mu_{\kappa-1,\kappa},$$

and so on. We shall establish the following

THEOREM.—*The conditions of irreducibility* of the product P are, as far as the factor $C_\kappa \equiv (a_1 a_2)^{\lambda_1} (a_1 a_3)^{\lambda_2} \dots (a_1 a_\kappa)^{\lambda_{\kappa-1}}$ is concerned,*

$$\lambda_{\kappa-1} \geqslant \sigma_\kappa, \quad \lambda_{\kappa-2} \geqslant \sigma_{\kappa-1}, \quad \dots, \quad \lambda_1 \geqslant \sigma_2.$$

We have

$$C_\kappa \equiv ([a_\kappa]', [a_\kappa])^{\lambda_{\kappa-1}} + \Sigma \nu ([a_\kappa]', [a_\kappa])^{\lambda_{\kappa-1}-\epsilon},$$

and therefore by hypothesis C_κ is reducible unless $\lambda_{\kappa-1} \geqslant \mu_\kappa$.

Put $$(a_1 a_2)^{\lambda_1} (a_1 a_3)^{\lambda_2} \dots (a_1 a_{\kappa-2})^{\lambda_{\kappa-3}} \equiv a \equiv [a_{\kappa-1} a_\kappa]'.$$

Then C_κ differs from $(a a_{\kappa-1})^{\lambda_{\kappa-2}} (a a_\kappa)^{\lambda_{\kappa-1}}$ by terms which we have just defined as reducible.

If $\lambda_{\kappa-2} + \lambda_{\kappa-1}$ is less than either of $\mu_{\kappa-1}$ or $\mu_{\kappa-1,\kappa}$, then obviously C_κ is reducible: if $\lambda_{\kappa-2} + \lambda_{\kappa-1}$ is greater than each of $\mu_{\kappa-1}$ and $\mu_{\kappa-1,\kappa}$, then, by Stroh's series, the covariant

$$(a a_{\kappa-1})^{\lambda_{\kappa-2}} (a a_\kappa)^{\lambda_{\kappa-1}}$$

is linearly expressible in terms of

$$(a_{\kappa-1} a_\kappa)^{\lambda_{\kappa-2}+\lambda_{\kappa-1}}, \ \left((a_{\kappa-1} a_\kappa)^{\lambda_{\kappa-2}+\lambda_{\kappa-1}-1}, a\right), \ \dots, \ \left((a_{\kappa-1} a_\kappa)^{\lambda_{\kappa-2}+\lambda_{\kappa-1}-\mu_{\kappa-1,\kappa}+1}, a\right)^{\mu_{\kappa-1,\kappa}-1};$$

$$(a a_\kappa)^{\lambda_{\kappa-2}+\lambda_{\kappa-1}}, \ \left((a a_\kappa)^{\lambda_{\kappa-2}+\lambda_{\kappa-1}-1}, [a_{\kappa-1}]\right), \ \dots, \ \left((a a_\kappa)^{\lambda_{\kappa-2}+\lambda_{\kappa-1}-\mu_{\kappa-1}+1}, [a_{\kappa-1}]\right)^{\mu_{\kappa-1}-1};$$

$$(a a_{\kappa-1})^{\lambda_{\kappa-2}+\lambda_{\kappa-1}}, \ (a a_{\kappa-1})^{\lambda_{\kappa-2}+\lambda_{\kappa-1}-1} (a a_\kappa), \ \dots, \ (a a_{\kappa-1})^{\mu_{\kappa-1}+\mu_{\kappa-1,\kappa}} (a a_\kappa)^{\lambda_{\kappa-2}+\lambda_{\kappa-1}-\sigma_{\kappa-1}}.$$

Now by hypothesis each form in the first two rows is reducible, and therefore $(a a_{\kappa-1})^{\lambda_{\kappa-2}} (a a_\kappa)^{\lambda_{\kappa-1}}$ is reducible unless $\lambda_{\kappa-2} \geqslant \mu_{\kappa-1} + \mu_{\kappa-1,\kappa}$.

---

* We always consider $(a_1 a_2)^{\lambda_1} (a_1 a_3)^{\lambda_2} \dots (a_1 a_\kappa)^{\lambda_{\kappa-1}}$ reducible if we can express it in terms of forms $(a_1 a_2)^{\mu_1} (a_1 a_3)^{\mu_2} \dots (a_1 a_\kappa)^{\mu_{\kappa-1}}$ such that the first of the quantities $\mu_1 - \lambda_1$, $\mu_2 - \lambda_2$, ..., $\mu_{\kappa-1} - \lambda_{\kappa-1}$ which is not zero is positive.

Hence a necessary condition for irreducibility is

$$\lambda_{\kappa-2} \geqslant \sigma_{\kappa-1}.$$

Assume the theorem true for the indices λ_{r-1} where $r > s$; then, as before, if we put $a \equiv [a_1 a_2 \ldots a_{s-1}]$, we have to consider the form

$$(aa_s)^{\lambda_{s-1}} (aa_{s+1})^{\lambda_s} \ldots (aa_\kappa)^{\lambda_{\kappa-1}}.$$

By Stroh's series this can be expressed in terms of

1. $(a_s a_{s+1})^{\lambda_s+\lambda_{s-1}} [a_s a_{s+1}]', ([a_s a_{s+1}], [a_s a_{s+1}]'), \ldots, ([a_s a_{s+1}], [a_s a_{s+1}]')^{\sigma_{s,\, s+1}-1}$;
2. $(aa_{s+1})^{\lambda_s+\lambda_{s-1}} \ldots, \;([a_s]', [a_s]), \;\ldots, \;([a_s]', [a_s])^{\sigma_s-\sigma_{s,\, s+1}-1}$;
3. $(aa_s)^{\lambda_s+\lambda_{s-1}} \ldots, (aa_s)^{\lambda_s+\lambda_{s-1}-1}(aa_{s+1}) \ldots, \ldots, (aa_s)^{\sigma_s}(aa_{s+1})^{\lambda_s+\lambda_{s-1}-\sigma_s}(aa_{s+2})^{\lambda_{s+1}} \ldots.$

We proceed to show that each form in the first two rows is reducible.

(1) Any member of the first row is

$$([a_s a_{s+1}], [a_s a_{s+1}]')^\nu,$$

where $$[a_s a_{s+1}] \equiv (a_s a_{s+1})^{\lambda_{s-1}+\lambda_s-\nu} \equiv \beta, \text{ say,}$$

and $$[a_s a_{s+1}]' \equiv (aa_{s+2})^{\lambda_{s+1}} \ldots (aa_\kappa)^{\lambda_{\kappa-1}},$$

so that we need only consider the form

$$(a\beta)^\nu (aa_{s+2})^{\lambda_{s+1}} \ldots (aa_\kappa)^{\lambda_{\kappa-1}}.$$

In addition to the actual reductions implied by hypothesis, we must also consider those reductions which consist in increasing the earlier indices at the expense of the later ones. (This is a common feature of all inductive proofs, since these reduced indices may be raised to their proper limits by virtue of the inductive proof; *v.* Note, p. 225.)

So, since we have assumed the theorem true for the index λ_s (of a factor followed by not more than $(\kappa-s-1)$ other factors), we must have $\nu \geqslant \sigma_{s,\, s+1}$, if the form last written down is irreducible. Hence every form in the first row is reducible.

(2) If we put $$\beta \equiv (aa_{s+1})^{\lambda_s+\lambda_{s-1}-\nu},$$

$$[a_s]' \equiv (\beta a_{s+2})^{\lambda_{s+1}} (\beta a_{s+3})^{\lambda_{s+2}} \ldots (\beta a_\kappa)^{\lambda_{\kappa-1}};$$

and we have to consider forms

$$(\beta a_s)^\nu (\beta a_{s+2})^{\lambda_{s+1}} \ldots (\beta a_\kappa)^{\lambda_{\kappa-1}}.$$

The condition for the irreducibility of this form is by hypothesis

$\nu \geqslant$ sum of μ's whose first suffix is s, and which do not contain a suffix $s+1$,

i.e., $\nu \geqslant \sigma_s - \sigma_{s,\, s+1}$.

Hence every form in the second row is reducible: and, since the index of (aa_s) in the last row is in every term $\geqslant \sigma_s$, we have as the condition, affecting λ_{s-1}, of irreducibility,

$$\lambda_{s-1} \geqslant \sigma_s.$$

We have demonstrated the truth of the conditions for the indices $\lambda_{\kappa-1}$ and $\lambda_{\kappa-2}$, and therefore, by induction, the theorem is universally true.

2. Cor. 1.—The perpetuant type theorem is immediately deducible from the preceding result: for, if C_κ is the only factor of the product P, every quantity μ is unity, since in this case the transvectants are reducible only if they are of zero order.

Hence

σ_s = number of ways of choosing all, none, or any of the suffixes $s+1, \ldots, \kappa$,

$= 2^{\kappa-s}$,

and therefore the conditions for the irreducibility of C_κ are

$$\lambda_{s-1} \geqslant 2^{\kappa-s}, \qquad s = 2, \ldots, \kappa.$$

Cor. 2.—An especially important case arises when the product P is (for any reason) reducible if C_κ is a Jacobian: in this case every quantity μ is 2, and

$$\sigma_s = 2^{\kappa-s+1},$$

and therefore $\qquad \lambda_{s-1} \geqslant 2^{\kappa-s+1}, \qquad s = 2, \ldots, \kappa.$

In this case the minimum weight of C_κ for irreducibility is $2^\kappa - 2$.

(Reference may be made to a paper by Grace, "Extension of Two Theorems on Covariants": *Proc. London Math. Soc.*, Ser. 2, Vol. 1.)

II. Syzygies of Degrees 3 and 4 respectively.

3. The syzygies of degrees 3 and 4 offer no difficulties, and may be actually written down apart from the consideration of their reducing special forms: those of degree 4 have been fully discussed already.* A discussion for degree 4 is here given from a different point of view in order to exhibit in a simple case the methods by which syzygies of higher degree are treated.

Degree 3.—There are no syzygies except for weight unity, for which there is a single syzygy:—

$$(bc)+(ca)+(ab) \equiv 0.$$

* Wood, "On Perpetuant Syzygies of Degree 4," *Proc. London Math. Soc.*, Ser. 2, Vol. 2.

The generating function for irreducible forms is $\frac{x^3}{(1-x)^2}$; and, since there are three linearly independent product forms $(bc)^\omega$, $(ca)^\omega$, $(ab)^\omega$ for all values of ω, except $\omega = 1, 0$, for which there are only two and one respectively, the generating function for irreducible product forms is

$$3x^2/(1-x)+2x+1.$$

Hence the generating function for all forms is

$$\frac{x^3}{(1-x)^2}+\frac{3x^2}{1-x}+2x+1 = \frac{1}{(1-x)^2}.$$

4. *Degree* 4.—There are five different kinds of type forms and products of forms of degree 4: they may be conveniently written

$$C_4,\ C_2^2,\ C_1 C_3,\ C_1^2 C_2,\ C_1^4;$$

where C_r denotes any perpetuant of degree r, C_r^2 denotes the product of two perpetuants of degree r, and so on. This notation will be preserved throughout.

A product is in the first place considered reducible if it is expressible in terms of products of other kinds which follow it in the above sequence. For weight ω the syzygies are nine in number and consist of—

(1) The three syzygies, due to Stroh, of the form

$$\{a_1 a_2 a_3 a_4\}_\omega \equiv \{(a_1 a_2)+(a_3 a_4)\}^\omega - \{(a_1 a_4)+(a_3 a_2)\}^\omega = 0.$$

(2) The six syzygies, arising from the Jacobian identity, of the form

$$(a_1 a_2)^{\omega-1}(a_3 a_4) = (a_1 a_2)^{\omega-1}(a_2 a_4) - (a_1 a_2)^{\omega-1}(a_2 a_3).$$

These last syzygies reduce all products C_2^2, when one of the factors C_2 is a Jacobian, by expressing each such product in terms of products $C_1 C_3$.

In order to fix what products are reduced by Stroh's syzygies, we arrange the products C_2^2 in the following order:—

$$(a_1 a_4)^\lambda (a_2 a_3)^\mu,\quad (a_1 a_3)^\lambda (a_2 a_4)^\mu,\quad (a_1 a_2)^\lambda (a_3 a_4)^\mu,$$

and define—

(i.) $(a_1 a_4)^\lambda (a_2 a_3)^\mu$ as reducible if it is expressible in terms of products $(a_1 a_3)^\lambda (a_2 a_4)^\mu$, $(a_1 a_2)^\lambda (a_3 a_4)^\mu$ and products $C_1 C_3$, &c.

(ii.) $(a_1 a_3)^\lambda (a_2 a_4)^\mu$ as reducible if it is expressible in terms of products $(a_1 a_2)^\lambda (a_3 a_4)^\mu$ and products $C_1 C_3$, &c.

Finally, we shall arrange the products $(a_1 a_r)^\lambda (a_s a_t)^\mu$ in ascending values of λ, and define any such form as reducible if it is expressible in terms of similar forms with greater values of λ.

The syzygy $\{a_1 a_2 a_3 a_4\}_\omega \equiv 0$ will reduce the form $(a_1 a_4)^2 (a_2 a_3)^{\omega-2}$, since $(a_1 a_4)(a_2 a_3)^{\omega-1} = \Sigma C_1 C_3$, by the Jacobian identity; the syzygy $\{a_1 a_2 a_4 a_3\}_\omega \equiv 0$ reduces in the same way the form $(a_1 a_3)^2 (a_2 a_4)^{\omega-2}$, and the syzygy $\{a_1 a_3 a_2 a_4\}_\omega \equiv 0$ reduces

$$\binom{\omega}{2}(a_1 a_4)^2 (a_2 a_3)^{\omega-2} + \binom{\omega}{3}(a_1 a_4)^3 (a_2 a_3)^{\omega-3}.$$

Moreover, from $\{a_1 a_2 a_3 a_4\}_\omega \equiv 0$, we have reduced

$$\binom{\omega}{2}(a_1 a_4)^2 (a_3 a_2)^{\omega-2} + \binom{\omega}{3}(a_1 a_4)^3 (a_3 a_2)^{\omega-3},$$

i.e.,
$$\binom{\omega}{2}(a_1 a_4)^2 (a_2 a_3)^{\omega-2} - \binom{\omega}{3}(a_1 a_4)^3 (a_2 a_3)^{\omega-3}.$$

Hence both $(a_1 a_4)^2 (a_2 a_3)^{\omega-2}$ and $(a_1 a_4)^3 (a_2 a_3)^{\omega-3}$ are, in accordance with our definition, reducible products.

Hence the irreducible products C_2^2 are

$$(a_1 a_2)^\lambda (a_3 a_4)^\mu, \qquad \lambda \geqslant 2, \quad \mu \geqslant 2;$$

$$(a_1 a_3)^\lambda (a_2 a_4)^\mu, \qquad \lambda \geqslant 3, \quad \mu \geqslant 2;$$

$$(a_1 a_4)^\lambda (a_2 a_3)^\mu, \qquad \lambda \geqslant 4, \quad \mu \geqslant 2;$$

and so the generating function for all products C_2^2 is

$$\frac{x^4}{(1-x)^2} + \frac{x^5}{(1-x)^2} + \frac{x^6}{(1-x)^2} = \frac{x^4 - x^7}{(1-x)^3}.$$

Finally, when the weight is unity we have only forms $C_1^2 C_2$; by the Jacobian transformation, we can compel any one definite letter to appear in the factor C_2; so for weight unity there are only three such forms, while for all other weights there are six: the generating function for the products $C_1^2 C_2$ is therefore

$$6x^2 / (1-x) + 3x.$$

The results for degree 4 may be summarized thus:—

Forms.	Generating Functions.
C_4	$\dfrac{x^7}{(1-x)^3}$
C_2^2	$\dfrac{x^4-x^7}{(1-x)^3}$
$C_1 C_3$	$\dfrac{4x^3}{(1-x)^2}$
$C_1^2 C_2$	$\dfrac{6x^2}{1-x}+3x$
C_1^4	1

Hence the generating function for the total number of forms of degree 4 is

$$\frac{x^7+x^4-x^7}{(1-x)^3}+\frac{4x^3}{(1-x)^2}+\frac{6x^2}{1-x}+3x+1=\frac{1}{(1-x)^3}.$$

III. Definition of Reducibility.—General Methods of Reduction by Differential Operators.

5. The product forms will be arranged in a definite sequence to be particularized immediately, and a product will be defined as reducible if it is linearly expressible in terms of product forms which follow it in this definite sequence. The sequence of the *letters* involved in a symbolical expression will be that defined by their suffixes, viz., $a_1, a_2, \ldots, a_\delta$.

The criteria for determining the relative positions of any two products A and B in the sequence will be made to depend successively on:—

(i.) The partial degrees of the factors of each product.

(ii.) The arrangement of the letters among the factors of each product.

(iii.) The weights of the factors of each product.

(iv.) The indices of the symbolical determinants of the factors of each product.

(i.) If $A \equiv C_{m_1} C_{m_2} \ldots C_{m_\kappa}, \qquad m_1 \leqslant m_2 \leqslant m_3 \ldots \leqslant m_\kappa;$

$B \equiv C_{n_1} C_{n_2} \ldots C_{n_l}, \qquad n_1 \leqslant n_2 \leqslant n_3 \ldots \leqslant n_l;$

then A precedes B, if the first of the quantities

$$m_1 - n_1,\ m_2 - n_2,\ \ldots,\ m_r - n_r,\ \ldots,$$

which is not zero, is positive.

(ii.) If
$$\left.\begin{aligned} A &\equiv C_{m_1} C_{m_2} \ldots C_{m_\kappa} \\ B &\equiv C'_{m_1} C'_{m_2} \ldots C'_{m_\kappa} \end{aligned}\right\},\quad m_1 \leqslant m_2 \leqslant m_3 \ldots \leqslant m_\kappa,$$

so that the factors of A and B are of the same partial degrees, then the arrangement of the letters is taken into consideration.

Case I.—$m_1 < m_2$.

Suppose C_{m_1} contains the letters $a_{r_1}, a_{r_2}, \ldots, a_{r_{m_1}}$,

and C'_{m_2} contains the letters $a_{s_1}, a_{s_2}, \ldots, a_{s_{m_1}}$;

where $r_1, r_2, \ldots, r_{m_1}$, or $s_1, s_2, \ldots, s_{m_1}$ are any m_1 of the suffixes 1, 2, 3, ..., δ, such that

$$r_1 < r_2 < r_3 \ldots < r_{m_1},$$

and
$$s_1 < s_2 < s_3 \ldots < s_{m_1}.$$

Then A precedes B, if the first of the quantities

$$r_1 - s_1,\ r_2 - s_2,\ \ldots,\ r_{m_1} - s_{m_1}$$

which is not zero is negative.

This assumes that C_{m_1} and C'_{m_1} do not contain the same set of m_1 letters.

Case II.—$m = m_1 = m_2 = \ldots = m_\theta < m_{\theta+1} \leqslant m_{\theta+2} \ldots \leqslant m_\kappa$.

In the first place the sequence is to be determined as in Case I. by all the letters occurring in the factors $C_{m_1} C_{m_2} \ldots C_{m_\theta}$.

When the letters in $C_{m_1} C_{m_2} \ldots C_{m_\theta}$ are the same as those in $C'_{m_1} C'_{m_2} \ldots C'_{m_\theta}$, it is necessary to distinguish between the factors $C_{m_1}, C_{m_2}, \ldots, C_{m_\theta}$, in order to determine whether A precedes B or not. Let a_ϕ be the first of the set of letters which occurs in $C_{m_1} C_{m_2} \ldots C_{m_\theta}$; then we define C_{m_1} to be that factor which contains a_ϕ; similarly C'_{m_1} is that factor of the product $C'_{m_1} C'_{m_2} \ldots C'_{m_\theta}$ which contains a_ϕ. Let the letters of C_{m_1}, C'_{m_1} be respectively $a_\phi, a_{r_2}, a_{r_3}, \ldots, a_{r_m}$ and $a_\phi, a_{s_2}, a_{s_3}, \ldots, a_{s_m}$, where

$$\phi < r_2 < r_3 \ldots < r_m,$$

$$\phi < s_2 < s_3 \ldots < s_m.$$

Then A precedes B if the first of the differences

$$r_2-s_2,\ r_3-s_3,\ \dots,\ r_m-s_m$$

which is not zero is *positive*.

(As an example, see the arrangement of the sequence of products C_2^2 at the foot of p. 228.)

If the letters of C_{m_1} and C'_{m_1} are the same, we apply the same test to the products $C_{m_2}C_{m_3}\dots C_{m_\theta}$, $C'_{m_2}C'_{m_3}\dots C'_{m_\theta}$. These two products now contain the same set of letters; we choose C_{m_2}, C'_{m_2} to be the two factors which contain the first letter of this set. Then, unless the letters of C_{m_2}, C'_{m_2} are all the same, the sequence is determined by these letters, in the same way as before. Otherwise we must apply the same test to the products $C_{m_3}\dots C_{m_\theta}$, $C'_{m_3}\dots C'_{m_\theta}$.

Finally, when the letters of the various corresponding factors of $C_{m_1}C_{m_2}\dots C_{m_\theta}$ and $C'_{m_1}C'_{m_2}\dots C_{m_\theta}$ are in every case the same, the tests which have just been laid down must be applied to the products $C_{m_{\theta+1}}\dots C_{m_\kappa}$, $C'_{m_{\theta+1}}\dots C'_{m_\kappa}$ in order to determine whether A precedes B or not.

(iii.) If A and B each have their factors of the same partial degrees in the same sets of letters, let

$$A\equiv C_{m_1}C_{m_2}\dots C_{m_\kappa},\qquad B\equiv C'_{m_1}C'_{m_2}\dots C'_{m_\kappa},\qquad m_1\leqslant m_2\dots\leqslant m_\kappa,$$

where, if certain of the m's are equal, the sequence of the factors $C_{m_1}, C_{m_2}, \dots, C_{m_\kappa}$ is determined as in (ii.), and further let

$C_{m_1}, C_{m_2}, \dots, C_{m_\kappa}$ be of total weights $\omega_1, \omega_2, \dots, \omega_\kappa$ respectively,

and $C'_{m_1}, C'_{m_2}, \dots, C'_{m_\kappa}$ be of total weights $\omega'_1, \omega'_2, \dots, \omega'_\kappa$ respectively.

Then A precedes B if the first of the quantities

$$\omega_1-\omega'_1,\ \omega_2-\omega'_2,\ \dots,\ \omega_\kappa-\omega'_\kappa$$

which is not zero is negative.

(iv.) Finally, we have to consider the case of products whose factors are of the same weights in the same sets of letters. Suppose

$$\left.\begin{aligned} C_{m_r} &\equiv (a_{(r,1)}a_{(r,2)})^{\lambda_{(r,1)}}(a_{(r,1)}a_{(r,3)})^{\lambda_{(r,2)}}\dots(a_{(r,1)}a_{(r,m_r)})^{\lambda_{(r,m_r-1)}} \\ C'_{m_r} &\equiv (a_{(r,1)}a_{(r,2)})^{\lambda'_{(r,1)}}(a_{(r,1)}a_{(r,3)})^{\lambda'_{(r,2)}}\dots(a_{(r,1)}a_{(r,m_r)})^{\lambda'_{(r,m_r-1)}} \end{aligned}\right\},$$

$$r=1,2,\dots,\kappa$$

where $$(r, 1) < (r, 2) \ldots < (r, m_r-1) < (r, m_r).$$

Then A precedes B, if the first of the quantities

$$\begin{array}{llllll}
(1) & \lambda_{(1,1)}-\lambda'_{(1,1)}, & \lambda_{(1,2)}-\lambda'_{(1,2)}, & \ldots, & \lambda_{(1,m_1)}-\lambda'_{(1,m_1)}, \\
(2) & \lambda_{(2,1)}-\lambda'_{(2,1)}, & \lambda_{(2,2)}-\lambda'_{(2,2)}, & \ldots, & \lambda_{(2,m_2)}-\lambda'_{(2,m_2)}, \\
\ldots & \ldots & \ldots & \ldots & \ldots \\
(r) & \lambda_{(r,1)}-\lambda'_{(r,1)}, & \lambda_{(r,2)}-\lambda'_{(r,2)}, & \ldots, & \lambda_{(r,m_r)}-\lambda'_{(r,m_r)}, \\
\ldots & \ldots & \ldots & \ldots & \ldots \\
(\kappa) & \lambda_{(\kappa,1)}-\lambda'_{(\kappa,1)}, & \lambda_{(\kappa,2)}-\lambda'_{(\kappa,2)}, & \ldots, & \lambda_{(\kappa,m_\kappa)}-\lambda'_{(\kappa,m_\kappa)}
\end{array}$$

which is not zero is negative.

A reference to the reductions in the case of degree 5 (in the succeeding section) should go some way towards making the above definition of reducibility quite clear.

General Methods of Reduction by the use of Differential Operators.

6. All the reductions made use of depend ultimately on Stroh's syzygies of degree 4: consider the syzygy

$$\{(ab)+(cd)\}^\lambda - \{(ad)+(cb)\}^\lambda = 0,$$

or $$\sum_{r=0}^{r=\lambda} \binom{\lambda}{r} (ab)^r (cd)^{\lambda-r} - \sum_{r=0}^{r=\lambda} \binom{\lambda}{r} (ad)^r (cb)^{\lambda-r} = 0.$$

Here the letters a, b, c, d denote any of the letters a_1, a_2, ..., a_8 used elsewhere in the present paper. This syzygy may be written

$$e^{(ab)D}(cd)^\lambda - e^{(ad)B}(cb)^\lambda = 0,$$

where $$B \equiv x_1 \frac{\partial}{\partial b_2} - x_2 \frac{\partial}{\partial b_1}, \qquad D \equiv x_1 \frac{\partial}{\partial d_2} - x_2 \frac{\partial}{\partial d_1};$$

and therefore $$B . b_x^m = 0, \quad D(cd) = c_x, \quad \ldots .$$

Let Γ_1 be any covariant containing the letters a, b, but not the letters c, d, and let Γ_2 be any covariant containing the letters c, d, but none of the letters of Γ_1. Then we may write

$$\Gamma_1 = \Sigma P_1 (ab)^{\mu_1}, \qquad \Gamma_2 = \Sigma P_2 (cd)^{\mu_2},$$

where P_1 does not contain the letter b in any determinantal factor, and P_2 does not contain the letter d in any determinantal factor.

Then we have a syzygy

$$P_1 . e^{(ab)D} \{P_2(cd)^\lambda\} = P_1 . e^{(ad)B} \{P_2(cb)^\lambda\} ; \qquad \text{(I.)}$$

and this syzygy is a relation between products of covariants of the form $\Gamma_1\Gamma_2$ and the same products having the letters b and d interchanged.

Now we have arranged our products in a fixed sequence, we shall suppose that the products on the right-hand side of the relation (I.) come after those on the left in this fixed sequence. The syzygy may then be conveniently written

$$P_1 e^{(ab)D} \{P_2(cd)^\lambda\} = R,$$

where R denotes any products which come in the fixed sequence after the products on the left-hand side: the sum of the products on the left-hand side is, in fact, according to our definition, reducible.

7. In the first place we shall consider the reduction of products $C_2 C_\kappa$: let $C_2 C_\kappa = (ab)^\mu (d_\kappa d_1)^{\lambda_1} (d_\kappa d_2)^{\lambda_2} \dots (d_\kappa d_{\kappa-1})^{\lambda_{\kappa-1}}$: the sequence of such products is primarily determined by the letters contained in the factor C_2: we suppose that the product in which the letters are ab precedes any of the products in which the letters of C_2 are

$$ad_1,\ ad_2,\ \dots,\ ad_r \quad \text{respectively.}$$

This is in accordance with our defined sequence of products when all the letters $d_1, d_2, \dots, d_r$ come after the letter b. In this case we have a series of relations of the form

$$e^{(ab)D_1} C_\kappa = R, \quad e^{(ab)D_2} C_\kappa = R, \quad \dots, \quad e^{(ab)D_r} C_\kappa = R,$$

where

$$D_r = x_1 \frac{\partial}{\partial d_{r,2}} - x_2 \frac{\partial}{\partial d_{r,1}}.$$

We proceed to show that *these r relations are all linearly independent*: this result is of fundamental importance.

The operators $D_1, D_2, \dots, D_r$ are obviously all commutative. The first relation may be written

$$C_\kappa + (ab) . D_1 C_\kappa + \frac{(ab)^2}{2!} D_1^2 C_\kappa + \dots = R,$$

or, since the first two terms are obviously reducible, the second by the Jacobian identity, it may be written

$$\frac{(ab)^2}{2!} D_1^2 C_\kappa + \frac{(ab)^3}{3!} D_1^3 C_\kappa + \dots = R, \qquad \text{(II.)}$$

and this relation is true, whatever covariant C_κ may be involving the letters $d_1, d_2, \dots, d_\kappa$.

8. Consider the equation $D_1 C_\kappa = C'_\kappa$

as an equation to determine C_κ.

If

$$C'_\kappa = (d_\kappa d_1)^{\mu_1} (d_\kappa d_2)^{\mu_2} \dots (d_\kappa d_{\kappa-1})^{\mu_{\kappa-1}},$$

then one solution is obviously

$$C_\kappa = \frac{1}{\mu_1+1} (d_\kappa d_1)^{\mu_1+1} (d_\kappa d_2)^{\mu_2} \dots (d_\kappa d_{\kappa-1})^{\mu_{\kappa-1}}.$$

Let $C_\kappa^{(1)}$ be any other solution; then we can always write

$$C_\kappa^{(1)} = \sum_\rho N_\rho (d_\kappa d_1)^{\mu_{\rho,1}} (d_\kappa d_2)^{\mu_{\rho,2}} \dots (d_\kappa d_{\kappa-1})^{\mu_{\rho,\kappa-1}},$$

and hence

$$\sum_\rho N_\rho \mu_{\rho,1} (d_\kappa d_1)^{\mu_{\rho,1}-1} (d_\kappa d_2)^{\mu_{\rho,2}} \dots (d_\kappa d_{\kappa-1})^{\mu_{\rho,\kappa-1}} \equiv (d_\kappa d_1)^{\mu_1} \dots (d_\kappa d_{\kappa-1})^{\mu_{\kappa-1}}.$$

The covariants in this identity are all independent: it follows, therefore, that C_κ can differ from $\frac{1}{\mu_1+1} (d_\kappa d_1)^{\mu_1+1} (d_\kappa d_2)^{\mu_2} \dots (d_\kappa d_{\kappa-1})^{\mu_{\kappa-1}}$ only by covariants which are products of the quantic d_1 with covariants of degree $\kappa-1$. And so, neglecting reducible terms, $D_1^{-1} C_\kappa$ has a unique meaning: in the same way $D_1^{-2} C_\kappa$, $D_1^{-3} C_\kappa$, ... have each a unique meaning. So in (II.) we may replace C_κ by $D_1^{-2} C_\kappa$: the resulting relation is

$$\frac{(ab)^2}{2!} C_\kappa + \frac{(ab)^3}{3!} D_1 C_\kappa + \dots = R,$$

and, in the same way, from the relation $e^{(ab) D_2} C_\kappa = R$ we obtain

$$\frac{(ab)^2}{2!} C_\kappa + \frac{(ab)^3}{3!} D_2 C_\kappa + \dots = R.$$

Therefore, on subtraction,

$$\frac{(ab)^3}{3!} (D_1 - D_2) C_\kappa + \frac{(ab)^4}{4!} (D_1^2 - D_2^2) C_\kappa + \dots = R,$$

or $$\left[\frac{(ab)^3}{3!} + \frac{(ab)^4}{4!} (D_1 + D_2) + \frac{(ab)^5}{5!} (D_1^2 + D_1 D_2 + D_2^2) \dots\right] (D_1 - D_2) C_\kappa = R.$$

Writing this $$\chi (D_1 - D_2) C_\kappa = R,$$

and replacing C_κ by $D_1^{-1} C_\kappa$, we have

$$\chi (1 - D_2 D_1^{-1}) C_\kappa = R.$$

In this replace C_κ by $D_2 D_1^{-1} C_\kappa$: we have

$$\chi (D_2 D_1^{-1} - D_2^2 D_1^{-2}) C_\kappa = R.$$

So, adding, we get $$\chi(1-D_2^2 D_1^{-2})\, C_\kappa = R.$$

Proceeding in this way, we obtain

$$\chi(1-D_2^\rho D_1^{-\rho})\, C_\kappa = R.$$

Since $D_1^{-\rho}$ has no effect on the index of $(d_\kappa d_2)$ in the covariant C_κ, ρ may be taken so large that $D_2^\rho D_1^{-\rho} C_\kappa = 0$: so the last relation is

$$\chi C_\kappa = R,$$

or
$$\left[\frac{(ab)^3}{3!}+\frac{(ab)^4}{4!}(D_1+D_2)+\dots\right] C_\kappa = R.$$

9. Now let us assume that, in general, we can deduce from the σ relations
$$e^{(ab)D_1} C_\kappa = R, \quad e^{(ab)D_2} C_\kappa = R, \quad \dots, \quad e^{(ab)D_\sigma} C_\kappa = R$$
the relation :—

$$\left\{\frac{(ab)^{\sigma+1}}{(\sigma+1)!}+\frac{(ab)^{\sigma+2}}{(\sigma+2)!}(D_1+D_2+\dots+D_\sigma)+\dots\right.$$
$$\left.+\frac{(ab)^{\sigma+\tau}}{(\sigma+\tau)!}\Big(\sum_{\varpi_1+\dots+\varpi_\sigma=\tau-1} D_1^{\varpi_1} D_2^{\varpi_2}\dots D_\sigma^{\varpi_\sigma}\Big)+\dots\right\} C_\kappa = R. \qquad \text{(III.)}$$

We shall show that, if we take another relation $e^{(ab)D_{\sigma+1}} C_\kappa = R$, then the resulting relation will be of the same form as (III.) with σ changed to $\sigma+1$.

On our assumption we can deduce from the σ relations

$$e^{(ab)D_1} C_\kappa = R, \quad \dots, \quad e^{(ab)D_{\sigma-1}} C_\kappa = R, \quad e^{(ab)D_{\sigma+1}} C_\kappa = R$$

the relation :—

$$\left\{\frac{(ab)^{\sigma+1}}{(\sigma+1)!}+\frac{(ab)^{\sigma+2}}{(\sigma+2)!}(D_1+D_2+\dots+D_{\sigma-1}+D_{\sigma+1})+\dots\right.$$
$$\left.+\frac{(ab)^{\sigma+\tau}}{(\sigma+\tau)!}\Big(\sum_{\varpi_1+\dots+\varpi_{\sigma-1}+\varpi_{\sigma+1}=\tau-1} D_1^{\varpi_1}\dots D_{\sigma-1}^{\varpi_{\sigma-1}} D_{\sigma+1}^{\varpi_{\sigma+1}}\Big)+\dots\right\} C_\kappa = R.$$

Subtracting this last relation from (III.), we obtain

$$\left\{\frac{(ab)^{\sigma+2}}{(\sigma+2)!}+\frac{(ab)^{\sigma+3}}{(\sigma+3)!}\left(\sum_{\rho=1}^{\rho=\sigma+1} D_\rho\right)+\dots\right.$$
$$\left.+\frac{(ab)^{\sigma+\tau}}{(\sigma+\tau)!}\left(\sum_{\varpi_1+\dots+\varpi_\sigma+\varpi_{\sigma+1}=\tau-2} D_1^{\varpi_1}\dots D_{\sigma+1}^{\varpi_{\sigma+1}}\right)+\dots\right\}(D_{\sigma-1}-D_{\sigma+1})\, C_\kappa = R,$$

and by the same argument as before we can show that $(D_{\sigma-1}-D_{\sigma+1})\, C_\kappa$

may be replaced by C_κ. Now we have seen that (III.) is true when $\sigma = 1, 2$: it is therefore true for all values of σ.

An immediate consequence of this is the

THEOREM.—The r relations

$$e^{(ab)D_1} C_\kappa = R, \qquad e^{(ab)D_2} C_\kappa = R, \qquad \ldots, \qquad e^{(ab)D_r} C_\kappa = R$$

are in general linearly independent, and together reduce the product $(ab)^\lambda C_\kappa$, when $\lambda \leqslant r+1$.

10. Certain special cases remain for consideration.

(i.) The relation $e^{(ab)D_\kappa} C_\kappa = R$ is true if the product $C_2 C_\kappa$, in which the letters of C_2 are (ab), precedes the product $C_2 C_\kappa$, in which the letters of C_2 are (ad_κ). We have previously written C_κ in the form

$$(d_\kappa d_1)^{\lambda_1} (d_\kappa d_2)^{\lambda_2} \ldots (d_\kappa d_{\kappa-1})^{\lambda_{\kappa-1}},$$

but it may be equally well written in the form

$$\Sigma N (d_1 d_\kappa)^{\mu_1} (d_1 d_2)^{\mu_2} \ldots (d_1 d_{\kappa-1})^{\mu_{\kappa-1}}.$$

Now the result of operating on each of the terms of this sum is R, and therefore the result of operating on the whole sum, or C_κ, is also R: in other words, our preceding results are not affected by the form of C_κ.

(ii.) If the product $(ab)^\lambda C_\kappa$ comes before the product $(d_\sigma b)^\lambda C_\kappa$, we have in the same way another syzygy

$$e^{(ba)D_\sigma} C_\kappa = R$$

or

$$e^{(ab)(-D_\sigma)} C_\kappa = R;$$

all such syzygies are independent of those already considered: we may, in fact, write $D_{\kappa+1} \equiv -D_\sigma$, and then the extension of the result embodied in (III.) to the case where the operators are $D_1, D_2, \ldots, D_\sigma, D_{\kappa+1}$ is almost identical with the previous investigation. Hence we deduce that, if r_1 letters of C_κ come after a, and r_2 letters of C_κ come after b, then

$$(ab)^\lambda C_\kappa = R, \qquad \text{if } \lambda \leqslant r_1 + r_2 + 1.$$

(iii.) We may suppose that the letters a and b refer to covariants of the original quantics and not to the quantics themselves, and that one of the letters—say, d_1—refers to a covariant involving the letters, say, $e_1, e_2, \ldots, e_l$. Then, if the product $(ab)^\lambda C_\kappa$ precedes the product $(ad)^\lambda C'_\kappa$, we have a syzygy

$$e^{(ab)D_1} C_\kappa = R,$$

or, from the identity $(D_1+D_2+\ldots+D_\kappa)\,C_\kappa = 0$, which is easily verified,

$$c^{(ab)(-D_2-D_3-\ldots-D_\kappa)}\,C_\kappa = R.$$

Let P be the covariant obtained from C_κ by replacing d_1 by the letters $e_1, e_2, \ldots, e_l$; then

$$(D_2+D_3+\ldots+D_\kappa+E_1+E_2+\ldots+E_l)\,P = 0,$$

and hence the syzygy may be written

$$c^{(ab)(E_1+E_2+\ldots+E_l)}\,C_\kappa = R.$$

Now C_κ may be represented as a sum of transvectants of covariants of a definite set of r of its letters with covariants of the remainder, and so we may also obtain syzygies of the form

$$c^{(ab)(D_1+D_2+\ldots+D_r)}\,C_\kappa = R,$$

where $d_1, d_2, \ldots, d_r$ are any of the letters of C_κ (the existence of any such relation will, of course, depend on the sequence of the products). We have to show that all such syzygies are linearly independent. We write $D_{\kappa+1}$ for any such composite operator as $(D_1+D_2+\ldots+D_r)$, and observe that all operators, simple or composite, are commutative. The proof of the relation (III.) then proceeds on the same lines as before for both simple and composite operators: only one point arises:—Has the operation

$$D_{\kappa+1}^{-1}\,C_\kappa \equiv (D_1+D_2+\ldots+D_r)^{-1}\,C_\kappa$$

a perfectly definite meaning?

To prove that it has, we write C_κ in the form $\Sigma\, N(\alpha, \beta)^\lambda$, where β is a covariant of the letters $d_1, d_2, \ldots, d_r$, and α is a covariant of the remaining letters: then, in the same way as before, we see that

$$(D_1+D_2+\ldots+D_r)^{-1}\,(\alpha\beta)^\lambda$$

differs from $(\lambda+1)^{-1}(\alpha\beta)^{\lambda+1}$ only by products of covariants of $d_1, d_2, \ldots, d_r$ with covariants of the remaining letters, and so the meaning is for our purpose definite and unique.

IV. SYZYGIES OF DEGREE 5.

11. We shall treat the syzygies of degree 5 at some length in explanation of the general principles set forth in Section III.

The product forms to be considered are of the following classes:—

$$C_2C_3,\; C_1C_4,\; C_1C_2^2,\; C_1^2C_3,\; C_1^3C_2,\; C_1^5.$$

The generating functions for all the irreducible products, except C_2C_3 and $C_1^3C_2$, are known from the results for degrees 3 and 4.

(i.) *Generating Function for Products* $C_2 C_3$.

The $C_2 C_3$ products are to be arranged in a sequence to be determined thus:

(*a*) by virtue of the letters involved in the factor C_2: the following is the sequence:—

(i.)	(ii.)	(iii.)	(iv.)	(v.)
$(a_1 a_2)^\nu C_3$,	$(a_1 a_3)^\nu C_3$,	$(a_1 a_4)^\nu C_3$,	$(a_1 a_5)^\nu C_3$,	$(a_2 a_3)^\nu C_3$,
(vi.)	(vii.)	(viii.)	(ix.)	(x.)
$(a_2 a_4)^\nu C_3$,	$(a_2 a_5)^\nu C_3$,	$(a_3 a_4)^\nu C_3$,	$(a_3 a_5)^\nu C_3$,	$(a_4 a_5)^\nu C_3$.

(*b*) If two products $C_2 C_3$ have the same letters in the C_2 factor, that product whose C_2 factor is of smaller weight precedes the other.

Any $C_2 C_3$ product is then defined as reducible, if it is expressible linearly in terms of products of other classes and of $C_2 C_3$ products which follow it in the sequence just determined.

First consider the products $(a_1 a_2)^\nu C_3$: using the notation of the preceding section, we have six linearly independent relations (§ 9)

$$e^{\pm(a_1 a_2) D_3} C_3 = R, \qquad e^{\pm(a_1 a_2) D_4} C_3 = R, \qquad e^{\pm(a_1 a_2) D_5} C_3 = R,$$

where, for instance, in the relation $e^{(a_1 a_2) D_3} C_3 = R$, R represents $C_2 C_3$ product forms such as $(a_1 a_3)^\nu C_3$ together with product forms $C_1 C_2^2$, $C_1^2 C_3$, and, in the relation $e^{-(a_1 a_2) D_3} C_3 = R$, R represents $C_2 C_3$ product forms such as $(a_2 a_3)^\nu C_3$ with product forms $C_1 C_2^2$, $C_1^2 C_3$. So, for irreducibility, we must have $\nu > 6+1$, that is, $\nu \geqslant 8$.

Moreover, C_3 must not be a Jacobian of the form

$$\big((a_r a_s)^\lambda,\ a_t\big),\ \text{where } r,\ s,\ t \text{ are } 3,\ 4,\ 5 \text{ in any order;}$$

for
$$(a_1 a_2)^\nu \big((a_r a_s)^\lambda,\ a^t\big) = (a_r a_s)^\lambda \big((a_1 a_2)^\nu,\ a_t\big) - a_t\big((a_1 a_2)^\nu,\ (a_r a_s)^\lambda\big)$$
$$= R, \quad \text{whatever } r \text{ and } s \text{ may be.}$$

Hence, by the second Corollary of Section I. (§ 2), the weight of C_3 is at least 6, and therefore the minimum weight of an irreducible product $(a_1 a_2)^\nu C_3$ is 14.

The treatment of the products $(a_1 a_3)^\nu C_3$ is similar: here (by § 9) there are reductions corresponding to the interchange of a_3 with a_4, a_5, and the interchange of a_1 with a_2, a_4, a_5. The five linearly independent relations are

$$e^{(a_3 a_1) D_2} C_3 = R, \qquad e^{\pm(a_1 a_3) D_4} C = R, \qquad e^{\pm(a_1 a_3) D_5} C_3 = R.$$

Hence $\nu \geqslant 7$, and, as before, C_3 must not be a Jacobian of the form $\big((a_r a_s)^\lambda,\ a_t\big)$, r, s, t being 2, 4, 5 in any order: therefore the minimum weight of C_3 is 6, and the minimum weight of an irreducible product $(a_1 a_3)^\nu C_3$ is 13.

Similarly for an irreducible product $(a_1 a_4)^\nu C_3$, $\nu \geqslant 6$, and the minimum weight of C_3 is 6; while, for an irreducible product $(a_1 a_5)^\nu C_3$, $\nu \geqslant 5$, and the minimum weight of C_3 is 6.

Next consider the products $(a_2 a_3)^\nu C_3$: here there are reductions corresponding to the interchanges of a_2 with a_4, a_5 and the interchanges of a_3 with a_4, a_5. The four linearly independent relations are

$$e^{\pm(a_2 a_3) D_4} C_3 = R, \qquad e^{\pm(a_2 a_3) D_5} C_3 = R; \qquad \text{so that } \nu \geqslant 6.$$

Moreover, C_3 must not be a Jacobian of the form $\left((a_4 a_5)^\lambda, a_1\right)$. Hence, by the second Corollary of Section I. (§ 2), the weight of C_3 is at least 4: therefore the minimum weight of an irreducible product $(a_2 a_3)^\nu C_3$ is 10. In the same way, for an irreducible product $(a_2 a_4)^\nu C_3$, $\nu \geqslant 5$, and the weight of $C_3 \geqslant 4$, while, for an irreducible product $(a_2 a_5)^\nu C_3$, $\nu \geqslant 4$, and the weight of $C_3 \geqslant 4$.

Now consider the products $(a_3 a_4)^\nu C_3$: we have for the reduction of such forms two linearly independent relations

$$e^{\pm(a_3 a_4) D_5} C_3 = R;$$

so that $\nu \geqslant 4$, while, by the perpetuant type theorem, C_3 is of weight 3 at least. So, for $(a_3 a_5)^\nu C_3$, $\nu \geqslant 3$, and C_3 is of weight 3 at least.

Finally, for the products $(a_4 a_5)^\nu C_3$, we have $\nu \geqslant 2$, for, if $\nu = 1$, there is a Jacobian identity; also the weight of C_3 is 3 at least.

12. So the irreducible products $C_2 C_3$ may be tabulated thus:—

$(a_1 a_2)^\nu (a_3 a_4)^\lambda (a_3 a_5)^\mu$,	$\nu \geqslant 8, \ \lambda \geqslant 4, \ \mu \geqslant 2$;
$(a_1 a_3)^\nu (a_2 a_4)^\lambda (a_2 a_5)^\mu$,	$\nu \geqslant 7, \ \lambda \geqslant 4, \ \mu \geqslant 2$;
$(a_1 a_4)^\nu (a_2 a_3)^\lambda (a_2 a_5)^\mu$,	$\nu \geqslant 6, \ \lambda \geqslant 4, \ \mu \geqslant 2$;
$(a_1 a_5)^\nu (a_2 a_3)^\lambda (a_2 a_4)^\mu$,	$\nu \geqslant 5, \ \lambda \geqslant 4, \ \mu \geqslant 2$;
$(a_2 a_3)^\nu (a_1 a_4)^\lambda (a_1 a_5)^\mu$,	$\nu \geqslant 6, \ \lambda \geqslant 3, \ \mu \geqslant 1$;
$(a_2 a_4)^\nu (a_1 a_3)^\lambda (a_1 a_5)^\mu$,	$\nu \geqslant 5, \ \lambda \geqslant 3, \ \mu \geqslant 1$;
$(a_2 a_5)^\nu (a_1 a_3)^\lambda (a_1 a_4)^\mu$,	$\nu \geqslant 4, \ \lambda \geqslant 3, \ \mu \geqslant 1$;
$(a_3 a_4)^\nu (a_1 a_2)^\lambda (a_1 a_5)^\mu$,	$\nu \geqslant 4, \ \lambda \geqslant 2, \ \mu \geqslant 1$;
$(a_3 a_5)^\nu (a_1 a_2)^\lambda (a_1 a_4)^\mu$,	$\nu \geqslant 3, \ \lambda \geqslant 2, \ \mu \geqslant 1$;
$(a_4 a_5)^\nu (a_1 a_2)^\lambda (a_1 a_3)^\mu$,	$\nu \geqslant 2, \ \lambda \geqslant 2, \ \mu \geqslant 1$.

Here the minimum value of ν is determined by the reductions of

Section III., while the minimum values of the indices λ, μ are determined by the theorem of Section I. (§§ 1, 2).

The generating functions for the products $C_2 C_3$ are, therefore, as follows :—

Product.	G. F.	Product.	G. F.
$(a_1 a_2)^\nu C_3$	$\dfrac{x^{14}}{(1-x)^3}$	$(a_2 a_4)^\nu C_3$	$\dfrac{x^9}{(1-x)^3}$
$(a_1 a_3)^\nu C_3$	$\dfrac{x^{13}}{(1-x)^3}$	$(a_2 a_5)^\nu C_3$	$\dfrac{x^8}{(1-x)^3}$
$(a_1 a_4)^\nu C_3$	$\dfrac{x^{12}}{(1-x)^3}$	$(a_3 a_4)^\nu C_3$	$\dfrac{x^7}{(1-x)^3}$
$(a_1 a_5)^\nu C_3$	$\dfrac{x^{11}}{(1-x)^3}$	$(a_3 a_5)^\nu C_3$	$\dfrac{x^6}{(1-x)^3}$
$(a_2 a_3)^\nu C_3$	$\dfrac{x^{10}}{(1-x)^3}$	$(a_4 a_5)^\nu C_3$	$\dfrac{x^5}{(1-x)^3}$

Hence the generating function for all products $C_2 C_3$ is

$$\frac{x^5+x^6+\ldots+x^{13}+x^{14}}{(1-x)^3}=\frac{x^5-x^{15}}{(1-x)^4}.$$

(ii.) *Generating Function for Products $C_1^3 C_2$.*

Here the only possible reductions arise from the Jacobian identity, when the weight is unity. In this case we can insist on a_1 appearing in the symbolical determinant factor; so that there are only four linearly independent products of unit weight, while there are ten of any other weight.

Hence the generating function is $10x^2/(1-x)+4x$.

The generating functions for types and products of degree 5 may be tabulated thus :—

Forms.	G. F.	Forms.	G. F.
C_5	$\dfrac{x^{15}}{(1-x)^4}$	$C_1^2 C_3$	$\dfrac{10x^3}{(1-x)^2}$
$C_2 C_3$	$\dfrac{x^5-x^{15}}{(1-x)^4}$	$C_1^3 C_2$	$\dfrac{10x^2}{1-x}+4x$
$C_1 C_4$	$\dfrac{5x^7}{(1-x)^3}$	C_1^5	1
$C_1 C_2^2$	$\dfrac{5(x^4-x^7)}{(1-x)^3}$		

The sum of all these generating functions is $\frac{1}{(1-x)^4}$, which is known to be the generating function for the enumeration of the linearly independent covariants of degree 5: from this it follows that the perpetuant type of theorem is exact, and that we have found all the syzygies of degree 5.

V. (i.) PRODUCTS HAVING NO FACTOR OF DEGREE LESS THAN 2.—(ii.) PRODUCTS HAVING NO FACTOR OF DEGREE LESS THAN 3.

13. In the present section we shall show that—

(i.) The generating function for types and irreducible products of degree δ having no factor C_1 is

$$\frac{x^\delta}{(1-x)^{\delta-1}}.$$

(ii.) The generating function for types and irreducible products of degree δ having no factor C_1 or C_2 is

$$\frac{x^{\binom{\delta}{1}+\binom{\delta}{2}}}{(1-x)^{\delta-1}}.$$

It seems probable that, in general, the generating function for types and irreducible products of degree δ having no factor of degree κ or less is

$$\frac{x^{\binom{\delta}{1}+\binom{\delta}{2}+\ldots+\binom{\delta}{\kappa}}}{(1-x)^{\delta-1}}.$$

This general result is verified in Section VI. in the case of

$$3\kappa+2 \geqslant \delta \geqslant 2\kappa.$$

(i.) *Products having no Factor of Degree less than* 2.

14. Those products which do contain a factor C_1 are all of the form

$$C_1^r . P_{\delta-r}, \quad r = 1, 2, \ldots, \delta-2, \delta.$$

Let V_δ be the generating function for all types and products not having a factor C_1; then the expression

$$V_\delta+\binom{\delta}{1}V_{\delta-1}+\binom{\delta}{2}V_{\delta-2}+\ldots+\binom{\delta}{r}V_{\delta-r}+\ldots+\binom{\delta}{\delta-2}V_2+V_0$$

is clearly the generating function for all types and irreducible products of degree δ, and so we must have

$$\frac{1}{(1-x)^{\delta-1}} = V_\delta + \binom{\delta}{1} V_{\delta-1} + \ldots + \binom{\delta}{r} V_{\delta-r} + \ldots + \binom{\delta}{\delta-2} V_2 + V_0.$$

In the first place $V_0 = 1$: consider $\binom{\delta}{\delta-2} V_2$, the generating function for the products $C_1^{\delta-2} C_2$; if this product is of unit weight, so that C_2 is a Jacobian, we can insist on a definite letter being present in the symbolical determinant; and therefore the coefficient of x in $\binom{\delta}{\delta-2} V_2$ is $(\delta-1)$; for any other weight the products $C_2 C_1^{\delta-2}$ are irreducible, and so

$$\binom{\delta}{\delta-2} V_2 = (\delta-1)\,x + \binom{\delta}{\delta-2} \frac{x^2}{1-x}.$$

By reference to our previous results, we find that

$$V_3 = \frac{x^3}{(1-x)^2}, \qquad V_4 = \frac{x^4}{(1-x)^3}, \qquad V_5 = \frac{x^5}{(1-x)^4}.$$

Let us assume that

$$V_{\delta'} = \frac{x^{\delta'}}{(1-x)^{\delta'-1}} \quad \text{for the values } 3,\, 4,\, 5,\, \ldots,\, \delta-1 \text{ of } \delta';$$

then, from the above relation, we have

$$V_\delta = \frac{\{x+(1-x)\}^\delta}{(1-x)^{\delta-1}} - \left\{ \binom{\delta}{1} \frac{x^{\delta-1}}{(1-x)^{\delta-2}} + \ldots + \binom{\delta}{r} \frac{x^{\delta-r}}{(1-x)^{\delta-r-1}} + \ldots + (\delta-1)\,x + 1 \right\}$$

$$= \frac{x^\delta}{(1-x)^{\delta-1}}, \text{ on expanding the first term.}$$

Hence by induction the result is true in general.

Therefore :—

The generating function for all types and irreducible products of degree δ which contain no factor of unit degree is $\dfrac{x^\delta}{(1-x)^{\delta-1}}$.

(ii.) *Products having no Factor of Degree less than* 3.

15. We proceed to find the generating functions for the following products :—

(1) C_2^m, (2) $C_2^m P_{\delta-2m}$, (3) $C_2^m C_3$, (4) $C_2^m C_4$;

where $P_{\delta-2m}$ is any product of degree $\delta-2m$ which contains no factor of degree < 3 and is neither C_3 nor C_4.

It will be seen that the peculiarity in the treatment of the products $C_2^m C_3$ and $C_2^m C_4$ arises from the Jacobian identity; it may be easily verified that this special identity yields no exceptional reductions of other products $C_2^m P_r$, even if P_r is one of the products C_3^2, C_4^2, $C_3 C_4$, $C_3 C_\kappa$, $C_4 C_\kappa$ (see § 17, Note).

(1) *Generating Function for C_2^m.*

We know that the generating function for C_2^2 is

$$\frac{x^4-x^7}{(1-x)^3} = \frac{x^4(1-x)(1-x^3)}{(1-x)^4}.$$

Consider first the product C_2^3: we have the following products:—

$$(a_1 a_6)^\lambda C_2^2, \quad (a_1 a_5)^\lambda C_2^2, \quad (a_1 a_4)^\lambda C_2^2, \quad (a_1 a_3)^\lambda C_2^2, \quad (a_1 a_2)^\lambda C_2^2.$$

Now any of these products is reducible if it is expressible in terms of products following it: thus for $(a_1 a_2)^\lambda C_2^2$ there is no such reduction, so that $\lambda \geqslant 2$, and the generating function for such products is $\frac{x^2 V_2}{1-x}$, where V_2 is the generating function of C_2^2; for $(a_1 a_3)^\lambda C_2^2$ there is a reduction corresponding to the interchange of a_2 and a_3, so that $\lambda \geqslant 3$, and so on; finally, for $(a_1 a_6)^\lambda C_2^2$ there are reductions corresponding to the interchanges of a_6 with a_2, a_3, a_4, a_5 respectively, and therefore $\lambda \geqslant 6$; hence the generating function for all these products is

$$\frac{x^2+x^3+x^4+x^5+x^6}{1-x} V_2 = \frac{x^2-x^7}{(1-x)^2} V_2 = \frac{x^6(1-x)(1-x^3)(1-x^5)}{(1-x)^6}.$$

Assume that the generating function for C_2^{m-1} is

$$V_{m-1} = x^{2m-2}\frac{(1-x)(1-x^3)\dots(1-x^{2m-3})}{(1-x)^{2m-2}}.$$

Then the products C_2^m are

$$(a_1 a_{2m})^\lambda C_2^{m-1}, \quad (a_1 a_{2m-1})^\lambda C_2^{m-1}, \quad \dots, \quad (a_1 a_r)^\lambda C_2^{m-1}, \quad \dots, \quad (a_1 a_2)^\lambda C_2^{m-1}.$$

Here, as before, any product is reducible if it is expressible in terms of products following it: thus for $(a_1 a_r)^\lambda C_2^{m-1}$ there are $(r-2)$ reductions corresponding to the interchanges of a_r with $a_2, a_3, \dots, a_{r-1}$ respectively, and so $\lambda \geqslant r$; for, by Section III., these reductions are all independent; the generating function for this set is $\frac{x^r}{1-x}V_{m-1}$, and therefore the generating function for all products C_2^m is

$$V_m = \frac{x^2+x^3+\ldots+x^{2m}}{1-x}V_{m-1} = \frac{(x^2-x^{2m+1})}{(1-x)^2}V_{m-1}$$

$$= \frac{x^{2m}(1-x)(1-x^3)\ldots(1-x^{2m-1})}{(1-x)^{2m}}.$$

Therefore, by induction, the generating function for products C_2^m is

$$\frac{x^{2m}(1-x)(1-x^3)\ldots(1-x^{2m-1})}{(1-x)^{2m}}.$$

(2) *Generating Function for $C_2^m P_{\delta-2m}$ (where $P_{\delta-2m}$ is any form or product, other than C_3 or C_4, which contains neither C_1 nor C_2).*

16. *Lemma.*—If the roots of $y^n-p_1y^{n-1}+p_2y^{n-2}-\ldots+(-)^np_n=0$ are $1, x, x^2, \ldots, x^{n-1}$, then

(1) $$p_\kappa = x^{\binom{\kappa}{2}}\frac{(1-x^n)(1-x^{n-1})\ldots(1-x^{n-\kappa+1})}{(1-x)(1-x^2)\ldots(1-x^\kappa)}.$$

(2) $$p_n+(1-x)p_{n-2}+(1-x)(1-x^3)p_{n-4}+\ldots +(1-x)(1-x^3)\ldots(1-x^{n-1}) = 1,\ n \text{ even};$$

$$p_n+(1-x)p_{n-2}+(1-x)(1-x^3)p_{n-4}+\ldots +(1-x)(1-x^3)\ldots(1-x^{n-2})p_1 = 1,\ n \text{ odd}.$$

(1) Assuming the theorem true for all values up to and including n, we shall show that it is also true for the value $(n+1)$; it is obviously true when $n=1, 2$.

If $y^{n+1}-q_1y^n+q_2y^{n-1}+\ldots+(-)^{n-1}q_{n+1}$

$$\equiv \left(y^n-p_1y^{n-1}+p_2y^{n-2}+\ldots+(-)^np_n\right)(y-x^n),$$

$$q_\kappa = p_\kappa+x^np_{\kappa-1},\quad \kappa = 1, 2, \ldots, n,$$

$$q_{n+1} = p_nx^n;$$

and therefore, by hypothesis, if $\kappa = 1, 2, \ldots, n$,

$$q_\kappa = x^{\binom{\kappa}{2}} \frac{(1-x^n)\ldots(1-x^{n-\kappa+1})}{(1-x)\ldots(1-x^\kappa)} + x^{n+\binom{\kappa-1}{2}} \frac{(1-x^n)\ldots(1-x^{n-\kappa+2})}{(1-x)\ldots(1-x^{\kappa-1})}$$

$$= x^{\binom{\kappa}{2}} \frac{(1-x^n)\ldots(1-x^{n-\kappa+2})}{(1-x)\ldots(1-x^\kappa)} \{1-x^{n-\kappa+1}+x^{n-\kappa+1}(1-x^\kappa)\}$$

$$= x^{\binom{\kappa}{2}} \frac{(1-x^{n+1})(1-x^n)\ldots(1-x^{n-\kappa+2})}{(1-x)(1-x^2)\ldots(1-x^\kappa)},$$

and $$q_{n+1} = x^n . x^{\binom{n}{2}} = x^{\binom{n+1}{2}}.$$

Therefore the result is true universally.

(2) Let the roots of

$$y^{n+1}-q_1 y^n+\ldots+(-)^{n+1} q_{n+1} = 0 \quad \text{be} \quad 1, x, x^2, \ldots, x^n,$$

and those of

$$y^{n-1}-r_1 y^{n-2}+\ldots+(-)^{n-1} r_{n-1} = 0 \quad \text{be} \quad 1, x, x^2, \ldots, x^{n-2}.$$

Assuming the result for the p's and the r's, it will be sufficient to prove it for the q's; for brevity we consider only the case of n being even. Then

$$q_{n+1}+(1-x)q_{n-1}+(1-x)(1-x^3)q_{n-3}+\ldots+(1-x)(1-x^3)\ldots(1-x^{n-1})q_1$$
$$= x^n p_n+(1-x)\{p_{n-1}+x^n p_{n-2}\}+\ldots+(1-x)(1-x^3)\ldots(1-x^{n-1})(p_1+x^n),$$

and $$p_\kappa = x^{\binom{\kappa}{2}} \frac{(1-x^n)\ldots(1-x^{n-\kappa+1})}{(1-x)\ldots(1-x^\kappa)} = \frac{1-x^n}{1-x^{n-\kappa}} r_\kappa$$

by the first part of the Lemma. Therefore

$$q_{n+1}+(1-x)q_{n-1}+\ldots+(1-x)(1-x^3)\ldots(1-x^{n-1})q_1$$
$$= x^n\{p_n+(1-x)p_{n-2}+\ldots+(1-x)(1-x^3)\ldots(1-x^{n-1})\}$$
$$+(1-x)p_{n-1}+(1-x)(1-x^3)p_{n-3}+\ldots+(1-x)(1-x^3)\ldots(1-x^{n-1})p_1$$
$$= x^n+(1-x^n)\{r_{n-1}+(1-x)r_{n-3}+\ldots+(1-x)(1-x^3)\ldots(1-x^{n-3})r_1\}$$
$$= x^n+(1-x^n) \text{ (by hypothesis)}$$
$$= 1.$$

17. Let V be the generating function of $P_{\delta-2m}$, $\delta = 2m+r$ being the total degree, and let the letters in C_2^m be

$$a_{\kappa_1}, a_{\kappa_2}, \ldots, a_{\kappa_{2m}}, \quad \text{where} \quad \kappa_1 < \kappa_2 < \kappa_3 \ldots < \kappa_{2m}.$$

Then, by definition, the interchange of the letter $a_{\kappa_{2m}}$ with any one of the letters $a_{\kappa_{2m}+1}, a_{\kappa_{2m}+2}, \ldots, a_{\kappa_\delta}$ implies a reduction, and, by Section III., corresponding to each of these interchanges the weight of C_2^m is increased by unity. So in general, by interchanging $a_{\kappa_{2m-\epsilon}}$ with succeeding letters occurring in $P_{\delta-2m}$, the minimum weight for irreducibility is increased by $(\delta-\kappa_{2m-\epsilon}-\epsilon)$, since this is the possible number of interchanges. Here ϵ may be any one of 0, 1, 2, ..., $2m-1$, and therefore, by interchanging all the letters of C_2^m with the letters of $P_{\delta-2m}$ succeeding, the weight is increased by

$$(\delta-\kappa_1)+(\delta-\kappa_2)+\ldots+(\delta-\kappa_{2m})-\binom{2m}{2}.$$

If the $2m$ letters of C_2^m are fixed, the generating function is

$$\frac{x^{2m}(1-x)(1-x^3)\ldots(1-x^{2m-1})}{(1-x)^{2m-1}}V;$$

and therefore, if these $2m$ letters are chosen in all possible ways and interchanged with the letters of $P_{\delta-2m}$, the generating function required is

$$V.x^{-\binom{2m}{2}}\frac{x^{2m}(1-x)(1-x^3)\ldots(1-x^{2m-1})}{(1-x)^{2m-1}}\Sigma x^{(\delta-\kappa_1)+(\delta-\kappa_2)+\ldots+(\delta-\kappa_{2m})},$$

where the Σ applies to all possible choices of the suffixes $\kappa_1, \kappa_2, \ldots, \kappa_{2m}$ from the suffixes 1, 2, 3, ..., δ.

The numbers $\delta-\kappa_1$, $\delta-\kappa_2$, ..., $\delta-\kappa_{2m}$ are all different and each is one of the numbers 0, 1, 2, ..., $\delta-1$; so, by the Lemma,

$$\Sigma x^{\delta-\kappa_1+\delta-\kappa_2+\ldots+\delta-\kappa_{2m}} = p_{2m} = x^{\binom{2m}{2}}\frac{(1-x^\delta)\ldots(1-x^{\delta-2m+1})}{(1-x)\ldots(1-x^{2m})}.$$

Therefore the generating function for $C_2^m P_{\delta-2m}$ is

$$V\frac{x^{2m}(1-x)(1-x^3)\ldots(1-x^{2m-1})}{(1-x)^{2m}}\frac{(1-x^\delta)(1-x^{\delta-1})\ldots(1-x^{\delta-2m+1})}{(1-x)(1-x^2)\ldots(1-x^{2m})},$$

where V is the generating function for $P_{\delta-2m}$ and $\delta = 2m+r$ is the total degree.

Note.—If $P_{\delta-2m}$ contains a factor C_3 or C_4 which is a Jacobian, any extra reductions due to a Jacobian transformation will be included in V, the generating function for $P_{\delta-2m}$.

(3) *Generating Function for* $C_2^m C_3$.

18. A different kind of reduction is introduced only in the case where C_3 is a Jacobian: three cases must be separately considered—

(*a*) C_3 is a Jacobian and does not contain a_1: here there is a new reduction.

(β) C_3 is a Jacobian and contains a_1, but not a_2: here there is a new reduction if, and only if, C_3 is of the form (C_2, a_1).

(γ) C_3 contains both a_1 and a_2: here there is never any new reduction.

(α) If the product is irreducible, C_3 can be written in the form $(a_{r_1} a_{r_2})^\lambda (a_{r_1} a_{r_3})^\mu$, $\lambda \geqslant 4$, $\mu \geqslant 2$, by Section I., Cor. 2.

The number of reductions due to replacing the letters of C_3 by earlier letters contained in C_2^m is

$$(r_1-1)+(r_2-2)+(r_3-3) = (r_1-2)+(r_3-2)+(r_3-2),$$

and therefore the generating function for such products is

$$\frac{x^6}{(1-x)^2} \frac{x^{2m}(1-x)\dots(1-x^{2m-1})}{(1-x)^{2m}} \sum_{r_1, r_2, r_3=2}^{2m+3} x^{(r_1-2)+(r_2-2)+(r_3-2)}$$

$$= x^{2m+6} \frac{(1-x)(1-x^3)\dots(1-x^{2m-1})}{(1-x)^{2m+2}} \frac{x^3(1-x^{2m+2})(1-x^{2m+1})(1-x^{2m})}{(1-x)(1-x^2)(1-x^3)},$$

by the Lemma (§ 16).

(β) If the product is irreducible, C_3 can be written in the form

$$(a_1 a_{r_1})^\lambda (a_1 a_{r_2})^\mu, \qquad \lambda \geqslant 3, \quad \mu \geqslant 1,$$

by Section I., § 1, and the number of reductions due to replacing a_{r_1}, a_{r_2} by earlier letters contained in C_2^m is

$$(r_1-2)+(r_2-3) = (r_1-3)+(r_2-3)+1;$$

therefore the generating function for such products is

$$\frac{x^5}{(1-x)^2} \frac{x^{2m}(1-x)(1-x^3)\dots(1-x^{2m-1})}{(1-x)^{2m}} \sum_{r_1, r_2=3}^{2m+3} x^{(r_1-3)+(r_2-3)}$$

$$= x^{2m+5} \frac{(1-x)(1-x^3)\dots(1-x^{2m-1})}{(1-x)^{2m+2}} \frac{x(1-x^{2m+1})(1-x^{2m})}{(1-x)(1-x^2)}.$$

(γ) Here it makes no difference to the irreducibility of the product whether C_3 is a Jacobian or not; so we may write

$$C_3 = (a_1 a_2)^\lambda (a_1 a_r)^\mu, \qquad \lambda \geqslant 2, \quad \mu \geqslant 1;$$

and the number of reductions due to replacing a_r by earlier letters contained in C_2^m is $(r-3)$; and therefore the generating function for such products is

$$\frac{x^3}{(1-x)^2} \frac{x^{2m}(1-x)(1-x^3)\dots(1-x^{2m-1})}{(1-x)^{2m}} \sum_{r=3}^{2m+3} x^{r-3}$$

$$= x^{2m+3} \frac{(1-x)(1-x^3)\dots(1-x^{2m+1})}{(1-x)^{2m+3}}.$$

So, finally, the generating function for all products $C_2^m C_8$ is

$$\frac{x^{2m+3}(1-x)(1-x^3)\dots(1-x^{2m+1})}{(1-x)^{2m+3}}\left\{1+\frac{x^3(1-x^{2m})}{1-x^2}+x^6\frac{(1-x^{2m})(1-x^{2m+2})}{(1-x^2)(1-x^3)}\right\}$$

$$=\frac{x^{\delta}}{(1-x)^{\delta-1}}\left\{(1-x)(1-x^3)\dots(1-x^{\delta-4})p_3+(1-x)(1-x^3)\dots(1-x^{\delta-2})p_1\right\},$$

where $\delta = 2m+3$ is the total degree, and (as in the Lemma of § 16)

$$p_\kappa = x^{\binom{\kappa}{2}}\frac{(1-x^{\delta})(1-x^{\delta-1})\dots(1-x^{\delta-\kappa+1})}{(1-x)(1-x^2)\dots(1-x^{\kappa})}.$$

(4) *Generating Function for $C_2^m C_4$.*

19. A different kind of reduction from those considered in § 17 is introduced when C_4 is a Jacobian of the form (C_2, C_2), and does not contain a_1, and in no other case.

If C_4 does not contain a_1, then, by Section I., if the product is irreducible, it can be written in the form

$$C_4 \equiv (a_{r_1} a_{r_2})^{\lambda_1}(a_{r_1} a_{r_3})^{\lambda_2}(a_{r_1} a_{r_4})^{\lambda_3}, \quad \text{where} \quad \lambda_1 \geqslant 6, \ \lambda_2 \geqslant 3, \ \lambda_3 \geqslant 1,$$

and, in the same way as before, we find that the generating function for such products is

$$\frac{x^{10}}{(1-x)^3}\ \frac{x^{2m}(1-x)(1-x^3)\dots(1-x^{2m-1})}{(1-x)^{2m}}\Sigma x^{(r_1-1)+(r_2-2)+(r_3-3)+(r_4-4)}$$

$$= x^{2m+10}\frac{(1-x)(1-x^3)\dots(1-x^{2m-1})}{(1-x)^{2m+3}}\ \frac{x^4(1-x^{2m+3})\dots(1-x^{2m})}{(1-x)(1-x^2)(1-x^3)(1-x^4)}.$$

If C_4 does contain the letter a_1, we write

$$C_4 \equiv (a_1 a_{r_1})^{\lambda_1}(a_1 a_{r_2})^{\lambda_2}(a_1 a_{r_3})^{\lambda_3}, \quad \text{where} \quad \lambda_1 \geqslant 4, \ \lambda_2 \geqslant 2, \ \lambda_3 \geqslant 1;$$

and the generating function for such products is

$$\frac{x^7}{(1-x)^3}\ \frac{x^{2m}(1-x)(1-x^3)\dots(1-x^{2m-1})}{(1-x)^{2m}}\Sigma x^{(r_1-2)+(r_2-3)+(r_3-4)}$$

$$= x^{2m+7}\frac{(1-x)(1-x^3)\dots(1-x^{2m-1})}{(1-x)^{2m+3}}\ \frac{(1-x^{2m+3})(1-x^{2m+2})(1-x^{2m+1})}{(1-x)(1-x^2)(1-x^3)}.$$

Hence, adding and simplifying these two results, we find for the complete generating function of products $C_2^m C_4$,

$$\frac{x^\delta}{(1-x)^{\delta-1}}\Big\{(1-x)(1-x^3)\dots(1-x^{\delta-5})p_4+(1-x)(1-x^3)\dots(1-x^{\delta-3})p_2$$
$$+(1-x)(1-x^3)\dots(1-x^{\delta-1})\Big(1-\frac{1}{1-x}\Big)\Big\},$$

where $\delta = 2m+4$ is the total degree, and p_2, p_4 are defined above (§ 16).

20. We are now in a position to find V_δ, the generating function for the types and those irreducible products which contain no factor of degree less than 3; for we know that $\frac{x^\delta}{(1-x)^{\delta-1}}$ is the generating function for the irreducible forms with no factor C_1, and every such product is of the form $C_2^m P_{\delta-2m}$. We must consider two cases according as δ is even or odd.

δ even.—

$$\frac{x^\delta}{(1-x)^{\delta-1}} = V_\delta+\frac{x^2(1-x)}{(1-x)^2}\,\frac{(1-x^\delta)(1-x^{\delta-1})}{(1-x)(1-x^2)}V_{\delta-2}+\dots$$
$$+\frac{x^{2m}(1-x)(1-x^3)\dots(1-x^{2m-1})}{(1-x)^{2m}}\,\frac{(1-x^\delta)(1-x^{\delta-1})\dots(1-x^{\delta-2m+1})}{(1-x)(1-x^2)\dots(1-x^{2m})}V_{\delta-2m}+\dots$$
$$+\frac{x^\delta}{(1-x)^{\delta-1}}\Big\{(1-x)(1-x^3)\dots(1-x^{\delta-5})p_4+(1-x)(1-x^3)\dots(1-x^{\delta-3})p_2$$
$$+(1-x)(1-x^3)\dots(1-x^{\delta-1})\Big(1-\frac{1}{1-x}\Big)\Big\}$$
$$+x^\delta\,\frac{(1-x)(1-x^3)\dots(1-x^{\delta-1})}{(1-x)^\delta};$$

the last two terms being the generating functions for $C_2^n C_4$ and C_2^{n+2} respectively, where $\delta = 2n+4$.

δ odd.—

$$\frac{x^\delta}{(1-x)^{\delta-1}} = V_\delta+\frac{x^2(1-x)}{(1-x)^2}\,\frac{(1-x^\delta)(1-x^{\delta-1})}{(1-x)(1-x^2)}V_{\delta-2}+\dots$$
$$+\frac{x^{2m}(1-x)(1-x^3)\dots(1-x^{2m-1})}{(1-x)^{2m}}\,\frac{(1-x^\delta)(1-x^{\delta-1})\dots(1-x^{\delta-2m+1})}{(1-x)(1-x^2)\dots(1-x^{2m})}V_{\delta-2m}+\dots$$
$$+\frac{x^\delta}{(1-x)^{\delta-1}}\{(1-x)(1-x^3)\dots(1-x^{\delta-4})p_3+(1-x)(1-x^3)\dots(1-x^{\delta-2})p_1\};$$

the last term being the generating function for $C_2^n C_3$, where $\delta = 2n+3$.

Now assume that, for $r < \delta$ and > 4,

$$V_r = \frac{x^{\binom{r}{1}+\binom{r}{2}}}{(1-x)^{r-1}},$$

and it is easily verified, from the two results of the preceding Lemma (§ 16), that, whether δ is even or odd,

$$V_\delta = \frac{x^{\binom{\delta}{1}+\binom{\delta}{2}}}{(1-x)^{\delta-1}}.$$

Hence the result is established in general.

VI. PRODUCTS OF THE KINDS (1) C_κ^2; (2) $C_\kappa C_m$, $\kappa < m < 2\kappa$; (3) $C_\kappa C_{2\kappa}$.

21. The general principle involved by the present application of Stroh's syzygies of degree 4 may be explained thus: Suppose we are seeking for the syzygies in which a product $C_\kappa C_m$ occurs. Let us divide up the letters of C_κ into two sets,

$$a_1, a_2, \ldots, a_{r_1}, \qquad b_1, b_2, \ldots, b_{r_2};$$

and those of C_m into two sets,

$$c_1, c_2, \ldots, c_{s_1}, \qquad d_1, d_2, \ldots, d_{s_2}.$$

Let α, β, γ, δ be covariants, whose symbolical characters are comprised in the sets $a_1, a_2, \ldots, a_{r_1}$: $b_1, b_2, \ldots, b_{r_2}$: $c_1, c_2, \ldots, c_{s_1}$: $d_1, d_2, \ldots, d_{s_2}$ respectively. Then C_κ can be expressed in terms of transvectants of the form $(\alpha, \beta)^\lambda$, and C_m can be expressed in terms of transvectants of the form $(\gamma, \delta)^\mu$.

Consider the Stroh syzygy

$$\{(\alpha\beta)+(\gamma\delta)\}^\omega - \{(\alpha\delta)+(\gamma\beta)\}^\omega = 0,$$

which may be written $\quad e^{(\alpha\beta) D_\delta}(\gamma\delta)^\omega = e^{(\alpha\delta) D_\beta}(\gamma\beta)^\omega,$

where
$$D_\beta = x_1\frac{\partial}{\partial\beta_2} - x_2\frac{\partial}{\partial\beta_1},$$

$$D_\delta = x_1\frac{\partial}{\partial\delta_2} - x_2\frac{\partial}{\partial\delta_1}.$$

Now let us suppose $(\alpha\delta)^\rho(\gamma\beta)^\sigma$ is a product which, by the definition of Section III., succeeds the product $(\alpha\beta)^{\rho'}(\gamma\delta)^{\sigma'}$ in the sequence of products: then we have the relation

$$e^{(\alpha\beta) D_\delta}(\gamma\delta)^\omega = \text{reducible terms}$$

or
$$e^{(\alpha\beta) D_\delta}(c_1, c_2, \ldots, c_{s_1}, \delta) = \text{reducible terms},$$

where $(c_1, c_2, \ldots, c_{s_1}, \delta)$ is the symbolical product obtained by replacing γ by the letters of which it is composed. Moreover, since $c_1, c_2, \ldots, c_{s_1}, \delta$ are the only letters involved in this symbolical product, we know that

$$(D_{c_1}+D_{c_2}+\ldots+D_{c_{s_1}}+D_\delta)(c_1, c_2, \ldots, c_{s_1}, \delta) = 0.$$

Therefore we may write the relation

$$\left.\begin{aligned} e^{-(\alpha\beta)\left\{D_{c_1}+D_{c_2}+\ldots+D_{c_{s_1}}\right\}} C_m &= \text{reducible terms} \\ \text{or} \qquad e^{(\alpha\beta)\left\{D_{d_1}+D_{d_2}+\ldots+D_{d_{s_2}}\right\}} C_m &= \text{reducible terms} \end{aligned}\right\}. \tag{1}$$

This relation gives us a reduction for the product $C_\kappa C_m$ by virtue of the fact that the product

$$(a_1, a_2, \ldots, a_{r_1},\ b_1, b_2, \ldots, b_{r_2})(c_1, c_2, \ldots, c_{s_1},\ d_1, d_2, \ldots, d_{s_2})$$

precedes, in our fixed arrangement, the product

$$(a_1, a_2, \ldots, a_{r_1},\ d_1, d_2, \ldots, d_{s_2})(b_1, b_2, \ldots, b_{r_1},\ c_1, c_2, \ldots, c_{s_1}).$$

In this way we obtain a series of relations such as (1): such relations have been considered in Section III. [§ 10, (iii.)], and have there been shewn to be all independent. It follows, therefore, that we obtain one reduction for $C_\kappa C_m$ corresponding to every product $C_{\kappa'} C_{m'}$ following $C_\kappa C_m$ in the sequence of products already determined, provided that each of $C_{\kappa'}$, $C_{m'}$ contains letters from each of the factors C_κ, C_m; and for every such reduction the minimum weight of C_κ for irreducibility is increased by unity.

(i.) *Generating Function for C_κ^2.*

22. Let

$$\begin{aligned} C_\kappa^2 &\equiv (a_1 a_{r_1})^{\lambda_1}(a_1 a_{r_2})^{\lambda_2} \ldots (a_1 a_{r_{\kappa-1}})^{\lambda_{\kappa-1}}(b_1 b_2)^{\mu_1}(b_1 b_3)^{\mu_2} \ldots (b_1 b_\kappa)^{\mu_{\kappa-1}}, \\ &\equiv C_\kappa^{(1)} C_\kappa^{(2)}, \end{aligned}$$

where $r_1 < r_2 < \ldots < r_{\kappa-1}$, and the letters involved are $a_1, a_2, \ldots, a_{2\kappa}$ in some order. If either $C_\kappa^{(1)}$ or $C_\kappa^{(2)}$ is a Jacobian, the product is reducible, since it is expressible in terms of products $C_{\kappa-\epsilon} C_{\kappa+\epsilon}$: hence, by the second Corollary of Section I., each of $C_\kappa^{(1)}$ or $C_\kappa^{(2)}$ is of weight $(2^\kappa-2)$ at least.

In the first place we suppose the letters of $C_\kappa^{(1)}$ and of $C_\kappa^{(2)}$ to be fixed and we disregard their arrangement: we then find how many reductions are possible by virtue of expressing $C_\kappa^{(1)} C_\kappa^{(2)}$ in the form $\Sigma C_{\kappa-\epsilon} C_{\kappa+\epsilon}$. To this end we consider all possible products $C_r C_{2\kappa-r}$, such that r takes any value from 2 to $\kappa-1$, where each of C_r, $C_{2\kappa-r}$ contains letters from each of the factors $C_\kappa^{(1)}$, $C_\kappa^{(2)}$: then we know that, corresponding to every such product, there is a reduction for $C_\kappa^{(1)} C_\kappa^{(2)}$, and that these reductions are all independent; so that, corresponding to each such product, the minimum weight of $C_\kappa^{(1)}$ for the irreducibility of $C_\kappa^{(1)} C_\kappa^{(2)}$ is increased by unity.

The letters of C_r may be chosen in $\binom{2\kappa}{r}$ ways, but we must exclude every case where these r letters are all included among the letters of $C_\kappa^{(1)}$ or among those of $C_\kappa^{(2)}$, for in such cases there is no corresponding reduction of the product $C_\kappa^{(1)} C_\kappa^{(2)}$; hence the number of choices for $C_r C_{2\kappa-r}$ is

$$\binom{2\kappa}{r} - \binom{\kappa}{r} - \binom{\kappa}{r}.$$

So the total number of reductions of this kind for $C_\kappa^{(1)} C_\kappa^{(2)}$ is

$$\sum_{r=2}^{r=\kappa-1} \left\{ \binom{2\kappa}{r} - 2\binom{\kappa}{r} \right\} = \tfrac{1}{2}\left[2^{2\kappa} - 4\kappa - 2 - \binom{2\kappa}{\kappa}\right] - 2[2^\kappa - \kappa - 2]$$

$$= 2^{2\kappa-1} - 2^{\kappa+1} - \tfrac{1}{2}\binom{2\kappa}{\kappa} + 3.$$

Further, by the Jacobian identities, which are independent of these reductions, the minimum weight of $C_\kappa^{(1)}$ is $2^\kappa - 2$. So, whatever are the letters of $C_\kappa^{(1)}$, the minimum weight of $C_\kappa^{(1)}$ to ensure the irreducibility of $C_\kappa^{(1)} C_\kappa^{(2)}$ is

$$2^{2\kappa-1} - 2^\kappa - \tfrac{1}{2}\binom{2\kappa}{\kappa} + 1.$$

It remains to discover how many products $P_1 P_2$ follow $C_\kappa^{(1)} C_\kappa^{(2)}$ in the sequence of products already determined, where P_1 and P_2 are each of degree κ, and P_1 is taken as that factor which contains a_1.

If $r_1 > 2$, there is a reduction corresponding to every product $P_1 P_2$, where P_1 contains a_2, and the number of such products is $\binom{2\kappa-2}{\kappa-2}$.

If $r_1 > 3$, there is a reduction corresponding to every product $P_1 P_2$, where P_1 contains a_3 and not a_2, and the number of such products is $\binom{2\kappa-3}{\kappa-2}$.

Continuing in this way, we obtain

$$\binom{2\kappa-2}{\kappa-2} + \binom{2\kappa-3}{\kappa-2} + \ldots + \binom{2\kappa-r_1+1}{\kappa-2} = \binom{2\kappa-1}{\kappa-1} - \binom{2\kappa-r_1+1}{\kappa-1}$$

reductions.

If we have any further reductions, P_1 must contain both a_1 and a_{r_1}, and we get a reduction if P_1 contains also a_{r_1+1} where $r_1+1 < r_2$; in this way we have

$$\binom{2\kappa-r_1-1}{\kappa-3}+\binom{2\kappa-r_1-2}{\kappa-3}+\ldots+\binom{2\kappa-r_2+1}{\kappa-3}=\binom{2\kappa-r_1}{\kappa-2}-\binom{2\kappa-r_2+1}{\kappa-2}$$

reductions.

The same argument is pursued for each of the letters of $C_\kappa^{(1)}$ in turn: hence the number of reductions, corresponding to the products $C_\kappa^2 = P_1P_2$, which follow $C_\kappa^{(1)}C_\kappa^{(2)}$ in the determined sequence is

$$\binom{2\kappa-1}{\kappa-1}-\binom{2\kappa-r_1+1}{\kappa-1}+\binom{2\kappa-r_1}{\kappa-2}-\binom{2\kappa-r_2+1}{\kappa-2}+\ldots-\binom{2\kappa-r_{\kappa-1}+1}{1}$$

$$=\binom{2\kappa-1}{\kappa-1}-\binom{2\kappa-r_1}{\kappa-1}-\binom{2\kappa-r_2}{\kappa-2}-\binom{2\kappa-r_3}{\kappa-3}-\ldots-\binom{2\kappa-r_{\kappa-1}}{1}-1.$$

Hence, if the letters of $C_\kappa^{(1)}$ are $a_1, a_{r_1}, \ldots, a_{r_{\kappa-1}}$, its minimum weight is

$$2^{2\kappa-1}-2^\kappa-\left[\binom{2\kappa-r_1}{\kappa-1}+\binom{2\kappa-r_2}{\kappa-2}+\ldots+\binom{2\kappa-r_{\kappa-1}}{1}\right],$$

and, since the minimum weight of $C_\kappa^{(2)}$ is $(2^\kappa-2)$, the generating function for the irreducible products C_κ^2 is

$$\frac{x^{2^{2\kappa-1}-2}}{(1-x)^{2\kappa-2}}\sum_{r_1, r_2, \ldots, r_{\kappa-1}} x^{-\binom{2\kappa-r_1}{\kappa-1}-\binom{2\kappa-r_2}{\kappa-2}-\ldots-\binom{2\kappa-r_{\kappa-1}}{1}},$$

where $r_1, r_2, \ldots, r_{\kappa-1}$ take all values satisfying

$$1 < r_1 < r_2 < \ldots < r_{\kappa-1} < 2\kappa+1.$$

We proceed to show that

$$S \equiv \sum_{r_1, r_2, \ldots, r_{\kappa-1}} x^{-\binom{2\kappa-r_1}{\kappa-1}-\binom{2\kappa-r_2}{\kappa-2}-\ldots-\binom{2\kappa-r_{\kappa-1}}{1}} = \frac{1-x^{-\binom{2\kappa-1}{\kappa-1}}}{1-x^{-1}}.$$

First find the sum for the possible values $r_{\kappa-2}+1, r_{\kappa-2}+2, \ldots, 2\kappa$ of $r_{\kappa-1}$; we have

$$\sum_{r_{\kappa-1}} x^{-\binom{2\kappa-r_{\kappa-1}}{1}} = \frac{1-x^{-\binom{2\kappa-r_{\kappa-2}}{1}}}{1-x^{-1}};$$

and therefore

$$S \equiv \frac{1}{1-x^{-1}}\left\{\sum_{r_1, r_2, \ldots, r_{\kappa-2}}\left(x^{-\binom{2\kappa-r_1}{\kappa-1}-\ldots-\binom{2\kappa-r_{\kappa-2}}{2}} - x^{-\binom{2\kappa-r_1}{\kappa-1}-\ldots-\binom{2\kappa-r_{\kappa-3}}{3}-\binom{2\kappa-r_{\kappa-2}+1}{2}}\right)\right\};$$

also $$\sum_{r_{\kappa-2}=r_{\kappa-3}+1}^{2\kappa-1}\left\{x^{-\binom{2\kappa-r_{\kappa-2}}{2}}-x^{-\binom{2\kappa-r_{\kappa-2}+1}{2}}\right\} = 1-x^{-\binom{2\kappa-r_{\kappa-3}}{2}};$$

and therefore

$$S \equiv \frac{1}{1-x^{-1}} \left\{ \sum_{r_1, r_2, \ldots, r_{\kappa-3}} \left(x^{-\binom{2\kappa-r_1}{\kappa-1} - \ldots - \binom{2\kappa-r_{\kappa-3}}{3}} - x^{-\binom{2\kappa-r_1}{\kappa-1} - \ldots - \binom{2\kappa-r_{\kappa-3}+1}{3}} \right) \right\}$$

$$\equiv \frac{1}{1-x^{-1}} \sum_{r_1, r_2, \ldots, r_{\kappa-\epsilon}} \left\{ x^{-\binom{2\kappa-r_1}{\kappa-1} - \ldots - \binom{2\kappa-r_{\kappa-\epsilon}}{\epsilon}} - x^{-\binom{2\kappa-r_1}{\kappa-1} - \ldots - \binom{2\kappa-r_{\kappa-\epsilon}+1}{\epsilon}} \right\},$$

by repeated application of the same method.

Hence, finally,

$$S \equiv \frac{1}{1-x^{-1}} \sum_{r_1=2}^{r_1=\kappa+2} \left\{ x^{-\binom{2\kappa-r_1}{\kappa-1}} - x^{-\binom{2\kappa-r_1+1}{\kappa-1}} \right\} = \frac{1}{1-x^{-1}} \left\{ 1 - x^{-\binom{2\kappa-1}{\kappa-1}} \right\}.$$

So the generating function for C_κ^2 is

$$\frac{x^{2^{2\kappa-1}-2}}{(1-x)^{2\kappa-2}} \, \frac{1-x^{-\binom{2\kappa-1}{\kappa-1}}}{1-x^{-1}} = \frac{x^{2^{2\kappa-1}-1-\frac{1}{2}\binom{2\kappa}{\kappa}} \left\{ 1 - x^{\frac{1}{2}\binom{2\kappa}{\kappa}} \right\}}{(1-x)^{2\kappa-1}},$$

since $$\binom{2\kappa-1}{\kappa-1} = \binom{2\kappa-1}{\kappa} = \tfrac{1}{2}\binom{2\kappa}{\kappa}.$$

(ii.) *Generating Function for $C_\kappa C_m$, where $\kappa < m < 2\kappa$.*

23. In the first place we suppose the letters of C_κ and C_m to be fixed, and we disregard their arrangement; we seek for products $C_r C_s$ $(r+s = m+\kappa)$, such that r takes all values from 2 to $\kappa-1$, where each of $C_r C_s$ contains letters from each of $C_\kappa C_m$; then, as before, for every such product the minimum weight of C_κ, for the irreducibility of the product $C_\kappa C_m$, is increased by unity.

The letters of C_r may be chosen in $\binom{\kappa+m}{r}$ ways; but we must exclude the cases where these r letters are all included among the letters of C_κ or among those of C_m; hence the total number of such products and therefore of reductions is

$$\sum_{r=2}^{r=\kappa-1} \left\{ \binom{\kappa+m}{r} - \binom{\kappa}{r} - \binom{m}{r} \right\} = \sum_{r=1}^{r=\kappa-1} \left\{ \binom{m+\kappa}{r} - \binom{m}{r} \right\} - (2^\kappa - 2).$$

Independently of these reductions the minimum weight of C_κ must be $2^\kappa - 2$, for, if C_κ is a Jacobian, the product $C_\kappa C_m$ is certainly reducible; hence the minimum weight of C_κ for irreducibility is

$$\sum_{r=1}^{r=\kappa-1} \left\{ \binom{m+\kappa}{r} - \binom{m}{r} \right\}.$$

Again, if C_m is a Jacobian of the form (C_l, C_{m-l}), then

$$C_\kappa C_m = (C_l, C_\kappa)\, C_{m-l} - (C_{m-l}, C_\kappa)\, C_l,$$

and this is a reduction if both l and $m-l$ are less than κ $(m < 2\kappa)$. So l may have the values $\kappa-1$, $\kappa-2$, ..., $m-\kappa+1$ (it is to be remarked that the values θ, $m-\theta$ of l lead to the same result); now the letters of C_m can be chosen for C_l in $\binom{m}{l}$ ways, so that the number of ways in which C_m can be chosen to give rise to a Jacobian reduction, whatever be the letters of C_κ, is

$$\tfrac{1}{2}\left\{\binom{m}{\kappa-1}+\binom{m}{\kappa-2}+\dots+\binom{m}{m-\kappa+1}\right\}$$

$$= \tfrac{1}{2}\{2^m-2\} - \left\{\binom{m}{\kappa}+\binom{m}{\kappa+1}+\dots+\binom{m}{m-1}\right\}.$$

Moreover, by the perpetuant type theorem, the minimum weight of C_m is $2^{m-1}-1$, and all these reductions are independent. Hence, without regarding the arrangement of the letters, we see that the minimum weight of the product $C_\kappa C_m$ for irreducibility is

$$\sum_{r=1}^{r=\kappa-1}\left\{\binom{m+\kappa}{r}-\binom{m}{r}\right\}+2^m-2-\sum_{r=\kappa}^{r=m-1}\binom{m}{r}$$

$$= \sum_{r=1}^{r=\kappa-1}\binom{m+\kappa}{r}+2^m-2-(2^m-2)$$

$$= \sum_{r=1}^{r=\kappa-1}\binom{m+\kappa}{r}.$$

In considering the arrangement of the letters, there is one reduction corresponding to each product $C'_\kappa C'_m$ which follows in the sequence of products the product $C_\kappa C_m$ which we are considering: this is obvious if C'_κ contain letters from C_κ and C_m. If, however, C'_κ contain only letters from C_m, there is still one reduction; for, if the product $C_\kappa C_m$ is irreducible, C_m must not be expressible in terms of a Jacobian of which C'_κ is one of the forms.

Now the total number of forms $C_\kappa C_m$ is $\binom{m+\kappa}{\kappa}$, and the extra reductions due to the arrangements of the letters number in the various cases

$$0,\ 1,\ 2,\ \dots,\ \binom{m+\kappa}{\kappa}-1, \text{ respectively};$$

hence the generating function for $C_\kappa C_m$ is

$$\frac{x^{\binom{m+\kappa}{\kappa-1}+\binom{m+\kappa}{\kappa-2}+\ldots+\binom{m+\kappa}{1}}\left\{1+x+x^2+\ldots+x^{\binom{m+\kappa}{\kappa}-1}\right\}}{(1-x)^{m+\kappa-2}}$$

$$=\frac{x^{\binom{m+\kappa}{\kappa-1}+\binom{m+\kappa}{\kappa-2}+\ldots+\binom{m+\kappa}{1}}\left\{1-x^{\binom{m+\kappa}{\kappa}}\right\}}{(1-x)^{m+\kappa-1}}.$$

This result proves that the generating function for products of degree δ having no factor of degree less than $(\kappa+1)$, provided $\kappa \geqslant \frac{1}{3}\delta$ (and of course $\leqslant \frac{1}{2}\delta$), is

$$\frac{x^{\binom{\delta}{1}+\binom{\delta}{2}+\ldots+\binom{\delta}{\kappa}}}{(1-x)^{\delta-1}}$$

(see also Section V., § 13).

The case of $\kappa=\frac{1}{2}\delta$ requires a word of explanation: the irreducible forms considered in this case are the perpetuant types of degree δ; their generating function is known to be

$$\frac{x^{2^{\delta-1}-1}}{(1-x)^{\delta-1}},$$

and this has been proved to be exact;* when δ is odd and $=2n+1$, the generating function as given by the preceding result is

$$\frac{x^{\binom{2n+1}{1}+\ldots+\binom{2n+1}{n-1}+\binom{2n+1}{n}}}{(1-x)^{2n}}=\frac{x^{2^{\delta-1}-1}}{(1-x)^{\delta-1}}.$$

A slight exception occurs in the case where δ is even and $=2n$; here the generating function is

$$\frac{x^{2^{\delta-1}-1}}{(1-x)^{\delta-1}}=\frac{x^{\binom{2n}{1}+\binom{2n}{2}+\ldots+\binom{2n}{n-1}+\frac{1}{2}\binom{2n}{n}}}{(1-x)^{2n-1}}.$$

(iii.) *Generating Function for $C_\kappa C_{2\kappa}$.*

24. Similar methods are applied in the investigation of products $C_\kappa C_{2\kappa}$, but the resulting generating function does not in general admit of ex-

* Wood, "On the Irreducibility of Perpetuant Types," *Proc. London Math. Soc.*, Ser. 2, Vol. 2.

pression in a simple form; in the special case of products C_3C_6 we shall evaluate the generating function for the purpose of the subsequent discussion of syzygies of degree 9.

The number of reductions possible, where the letters of $C_\kappa C_{2\kappa}$ are fixed, is

$$\sum_{r=2}^{r=\kappa-1}\left\{\binom{3\kappa}{r}-\binom{2\kappa}{r}-\binom{\kappa}{r}\right\}+2^\kappa-2+2^{2\kappa-1}-1;$$

for C_κ must not be a Jacobian, and the minimum weight of $C_{2\kappa}$ is $2^{2\kappa-1}-1$. Hence the minimum weight is, for a fixed choice of letters,

$$\sum_{r=1}^{r=\kappa-1}\binom{3\kappa}{r}+\tfrac{1}{2}\binom{2\kappa}{\kappa}-\tfrac{1}{2}(2^{2\kappa}-2)-(2^\kappa-2)+2^\kappa-2+2^{2\kappa-1}-1$$

$$=\sum_{r=1}^{r=\kappa-1}\binom{3\kappa}{r}+\tfrac{1}{2}\binom{2\kappa}{\kappa}.$$

Now suppose that the letters of C_κ are $a_{r_1}, a_{r_2}, \ldots, a_{r_\kappa}$, where

$$1<r_1<r_2\ldots<r_\kappa;$$

then the number of products $C'_\kappa C'_{2\kappa}$, such that C'_κ contains letters of C_κ and of $C_{2\kappa}$, which follow the product $C_\kappa C_{2\kappa}$ in the sequence already determined is easily seen to be

$$N\equiv\binom{3\kappa-r_1}{\kappa}-\binom{2\kappa-r_1+1}{\kappa}+\binom{3\kappa-r_2}{\kappa-1}+\ldots+\binom{3\kappa-r_\kappa}{1};$$

and, as before, we can show that

$$\sum_{r_1, r_2, \ldots, r_\kappa} x^N=\sum_{r_1} x^{\binom{3\kappa-r_1}{\kappa}-\binom{2\kappa-r_1+1}{\kappa}}\,\frac{1-x^{\binom{3\kappa-r_1}{\kappa-1}}}{1-x};$$

here r_1 may have any value from 2 to $2\kappa+1$.

If C_κ contains a_1 (corresponding to $r_1=1$), we get a term

$$x^{\binom{3\kappa-1}{\kappa}-\frac{1}{2}\binom{2\kappa}{\kappa}}\,\frac{1-x^{\binom{3\kappa-1}{\kappa-1}}}{1-x};$$

where the extra index $\frac{1}{2}\binom{2\kappa}{\kappa}$ arises from the fact that, if $C_{2\kappa}$ does not contain a_1 and can be expressed as a Jacobian (C_κ, C'_κ), there is a reduction.

Hence the generating function for irreducible products $C_\kappa C_{2\kappa}$ is

$$\frac{x^{\binom{3\kappa}{1}+\binom{3\kappa}{2}+\dots+\binom{3\kappa}{\kappa-1}+\frac{1}{2}\binom{2\kappa}{\kappa}}}{(1-x)^{3\kappa-1}}\left\{x^{\binom{3\kappa-1}{\kappa}-\frac{1}{2}\binom{2\kappa}{\kappa}}\left(1-x^{\binom{3\kappa-1}{\kappa-1}}\right)\right.$$

$$\left.+\sum_{r=2}^{r=2\kappa+1} x^{\binom{3\kappa-r}{\kappa}-\binom{2\kappa-r+1}{\kappa}}\left(1-x^{\binom{3\kappa-r}{\kappa-1}}\right)\right\}.$$

Putting $\kappa=3$, we find for the irreducible products C_3C_6 of degree 9 the generating function

$$\frac{x^{55}(1-x^{10}+x^9-x^{19}+x^{16}-x^{31}+x^{25}-x^{74})}{(1-x)^8}.$$

Summary of Results giving Generating Functions for the Enumeration of certain Classes of Irreducible Products of total Degree δ.

Products.	Generating Functions.
All products containing no factor C_1 (§ 14)	$\dfrac{x^\delta}{(1-x)^{\delta-1}}$
C_2^n (where $\delta=2n$) (§ 15)	$\dfrac{x^\delta(1-x^3)(1-x^5)\dots(1-x^{\delta-1})}{(1-x)^{\delta-1}}$
$C_2^m C_3$ (where $\delta=2m+3$) (§ 18)	$\dfrac{x^\delta(1-x^3)(1-x^5)\dots(1-x^{\delta-2})}{(1-x)^{\delta-1}}\left\{1+\dfrac{x^3(1-x^{\delta-3})}{1-x^2}+x^6\dfrac{(1-x^{\delta-3})(1-x^{\delta-1})}{(1-x^2)(1-x^3)}\right\}$
$C_2^m C_4$ (where $\delta=2m+4$) (§ 19)	$\dfrac{x^{\delta+3}(1-x^3)(1-x^5)\dots(1-x^{\delta-5})(1-x^{\delta-3})(1-x^{\delta-2})(1-x^{\delta-1})}{(1-x)^{\delta-1}(1-x^2)(1-x^3)} \times\left\{1+\dfrac{x^7(1-x^{\delta-4})}{(1-x^4)}\right\}$
$C_2^m P_{\delta-2m}$ ($P_{\delta-2m}$ contains no factor of degree less than 3 and is neither C_3 nor C_4) (§ 17)	$\dfrac{x^{2m}(1-x^3)(1-x^5)\dots(1-x^{2m-1})}{(1-x)^{2m-1}}\,\dfrac{(1-x^\delta)(1-x^{\delta-1})\dots(1-x^{\delta-2m+1})}{(1-x)(1-x^2)\dots(1-x^{2m})}V$ (where V is the generating function of $P_{\delta-2m}$)
All products containing no factor of degree less than 3 (§ 20)	$\dfrac{x^{\binom{\delta}{1}+\binom{\delta}{2}}}{(1-x)^{\delta-1}}$
C_κ^2 (where $\delta=2\kappa$) (§ 22)	$\dfrac{x^{2^{\delta-1}-1-\frac{1}{2}\binom{\delta}{\frac{1}{2}\delta}}\left(1-x^{\frac{1}{2}\binom{\delta}{\frac{1}{2}\delta}}\right)}{(1-x)^{\delta-1}}$
$C_\kappa C_m$ (where $\delta=\kappa+m$, $\kappa<m<2\kappa$) (§ 23)	$\dfrac{x^{\binom{\delta}{1}+\binom{\delta}{2}+\dots+\binom{\delta}{\kappa-1}}\left(1-x^{\binom{\delta}{\kappa}}\right)}{(1-x)^{\delta-1}}$

VII. SYZYGIES OF DEGREES 6, 7, AND 8.—NOTE ON SYZYGIES OF DEGREE 9.

25. We are now in a position to write down from the preceding general results the generating functions for the enumeration of the irreducible products of degrees 6, 7, and 8 respectively.

Degree 6.

	Products.	Generating Functions.	How obtained.
Type forms :—	C_6	$\dfrac{x^{31}}{(1-x)^5}$	Perpetuant type theorem
Products :—	C_3^2	$\dfrac{x^{21}-x^{31}}{(1-x)^5}$	C_κ^2, $\kappa=3$ (Section VI., § 22)
	$C_2 C_4$	$\dfrac{x^9+x^{11}-x^{14}-x^{21}}{(1-x)^5}$	$C_2^m C_4$, $m=1$ (Section V., § 19)
	C_2^3	$\dfrac{x^6-x^9-x^{11}+x^{14}}{(1-x)^5}$	C_2^m, $m=3$ (Section V., § 15)
	$C_1 C_5$	$\dfrac{6x^{15}}{(1-x)^4}$	Perpetuant type theorem
	$C_1 C_2 C_3$	$\dfrac{6(x^5-x^{15})}{(1-x)^4}$	$C_2 C_3$, $m=1$ (Section V., § 18)
	$C_1^2 C_4$	$\dfrac{15x^7}{(1-x)^3}$	Perpetuant type theorem
	$C_1^2 C_2^2$	$\dfrac{15(x^4-x^7)}{(1-x)^3}$	C_2^m, $m=2$ (Section V., § 15)
	$C_1^3 C_3$	$\dfrac{20x^3}{(1-x)^2}$	Perpetuant type theorem
	$C_1^4 C_2$	$\dfrac{15x}{1-x}-10x$	$C_1^{n-2} C_2$, $n=6$ (Section V., § 14)
	C_1^6	1	

Also—

(i.) Sum of generating functions for forms C_6, C_3^2 is $x^{21}/(1-x)^5$ (see Section V., § 13).

(ii.) Sum of generating functions for forms C_6, C_3^2, $C_2 C_4$, C_2^3 is $x^6/(1-x)^5$ (see Section V., § 13).

(iii.) Sum of all generating functions is $1/(1-x)^5$, so that all the enumeration of the irreducible forms is exact and all the syzygies for degree 6 have been identified.

One point calls for comment: in reducing the products $C_2 C_4$, if $C_2 = (a_5 a_6)^\nu$, the only restriction on ν is (Section V., § 19) $\nu \geqslant 2$. Now consider the form

$$C_2 C_4 = (a_5 a_6)^2 \{(a_1 a_2)^3, (a_3 a_4)^3\},$$

which arises in products of weight 9. By the ordinary Jacobian identity we have

$$(a_5 a_6)^2 \{(a_1 a_2)^3, (a_3 a_4)^3\} \equiv (a_1 a_2)^3 \{(a_5 a_6)^2, (a_3 a_4)^3\} - (a_3 a_4)^3 \{(a_5 a_6)^2, (a_1 a_2)^3\}.$$

Now each of the Jacobians $\{(a_5 a_6)^2, (a_3 a_4)^3\}$, $\{(a_5 a_6)^2, (a_1 a_2)^3\}$ is of degree 4 and weight 6, and therefore is expressible as a sum of product forms. Hence the product $(a_5 a_6)^2 \{(a_1 a_2)^3, (a_3 a_4)^3\}$ is reducible: this is not in opposition to the general result, as we can shew that the Jacobian $\{(a_1 a_2)^3, (a_3 a_4)^3\}$, although of degree 4 and weight 7, is itself reducible. For, by the perpetuant type theorem,

$$\begin{aligned}\{(a_1 a_2)^3, (a_3 a_4)^3\} &\equiv \kappa (a_1 a_2)^4 (a_2 a_3)^2 (a_3 a_4) + \text{product forms} \\ &\equiv \kappa \{(a_1 a_2)^4, (a_3 a_4)\}^2 + \lambda \{(a_1 a_2)^5, (a_3 a_4)\} + \mu \{(a_1 a_2)^4, (a_3 a_4)^2\} \\ &\quad + \text{product forms},\end{aligned}$$

and every term of $\{(a_1 a_2)^5, (a_3 a_4)\}$ is a product form. Hence

$$\{(a_1 a_2)^3, (a_3 a_4)^3\} \equiv \kappa \{(a_1 a_2)^4, (a_3 a_4)\}^2 + \mu \{(a_1 a_2)^4, (a_3 a_4)^2\} + \text{product forms}.$$

In this result interchange a_1 and a_2 and subtract; then

$$2 \{(a_1 a_2)^3, (a_3 a_4)^3\} = \text{product forms};$$

and therefore the Jacobian is reducible.

In the treatment of product forms of degree 3κ ($\kappa \geqslant 3$) no such difficulty can arise, for, although we have the analogous result

$$C'_\kappa (C_\kappa, \bar{C}_\kappa) = \Sigma \text{ (product of three factors)},$$

where C'_κ is of weight $2^{2\kappa-2}-2$, and each of C_κ, $\bar{C}_\kappa$ is of weight $2^{2\kappa-2}-1$, it will be seen that the minimum weight, as determined in Section V., § 24, of the factor $C'_\kappa C_{2\kappa}$ is in all cases greater than $2^{2\kappa-2}-2$; so that there is no occasion to investigate the reducibility of the Jacobian form $(C_\kappa, \bar{C}_\kappa)$ of weight $2^{2\kappa-1}-1$ and degree 2κ (see § 24).

Degree 7.

26.

	Products.	Generating Functions.	How obtained.
Type forms:—	C_7	$\dfrac{x^{63}}{(1-x)^6}$	Perpetuant type theorem
Products:—	$C_3 C_4$	$\dfrac{x^{28}-x^{63}}{(1-x)^6}$	$C_\kappa C_m$, $\kappa = 3$, $m = 4$ (Section VI., § 23)
	$C_2 C_5$	$\dfrac{x^{17}(1-x^7)(1+x^2+x^4)}{(1-x)^6}$	$C_2^m C_r$, $m = 1$, $r = 5$ (Section V., § 17)
	$C_2^2 C_3$	$\dfrac{x^7(1-x^5)(1-x^7)(1+x^5+x^7)}{(1-x)^6}$	$C_2^m C_3$, $m = 2$ (Section V., § 18)
	$C_1 C_6$	$\dfrac{7x^{31}}{(1-x)^5}$	Results for degree 6
	$C_1 C_3^2$	$\dfrac{7(x^{21}-x^{31})}{(1-x)^5}$	
	$C_1 C_2 C_4$	$\dfrac{7(x^9+x^{11}-x^{14}-x^{21})}{(1-x)^5}$	
	$C_1 C_2^3$	$\dfrac{7(x^6-x^9-x^{11}+x^{14})}{(1-x)^5}$	
	$C_1^2 C_5$	$\dfrac{21x^{15}}{(1-x)^4}$	
	$C_1^2 C_2 C_3$	$\dfrac{21(x^5-x^{15})}{(1-x)^4}$	
	$C_1^3 C_4$	$\dfrac{35x^7}{(1-x)^3}$	
	$C_1^3 C_2^2$	$\dfrac{35(x^4-x^7)}{(1-x)^3}$	
	$C_1^4 C_3$	$\dfrac{35x^3}{(1-x)^2}$	
	$C_1^5 C_2$	$\dfrac{21x}{1-x}-15x$	$C_1^{n-2}C_2$, $n = 7$ (Section V., § 14)
	C_1^7	1	

Also—

(i.) Sum of generating functions for forms C_7, $C_3 C_4$ is $x^{28}/(1-x)^6$ (see Section V., § 13).

(ii.) Sum of generating functions for forms C_7, C_3C_4, C_2C_5, $C_2^2C_3$ is $x^7/(1-x)^6$ (see Section V., § 13).

(iii.) Sum of all generating functions is $1/(1-x)^6$; so that all the syzygies are enumerated above.

Degree 8.

It is clear that we need only determine the generating functions for the products C_8, C_4^2, C_3C_5, C_2C_6, $C_2C_3^2$, $C_2^2C_4$, C_2^4; since all other products have a factor C_1, and the corresponding generating functions may be determined from the results for degree 7 (see Section V., § 14).

	Products	Generating Functions	How obtained
Type forms:—	C_8	$\frac{x^{127}}{(1-x)^7}$	Perpetuant type theorem
Products :—	C_4^2	$\frac{x^{92}-x^{127}}{(1-x)^7}$	C_κ^2, $\kappa=4$ (Section VI., § 22)
	C_3C_5	$\frac{x^{36}-x^{92}}{(1-x)^7}$	$C_\kappa C_m$, $\kappa=3$, $m=5$ (Section VI., § 23)
	C_2C_6	$\frac{x^2(1-x^7)(1-x^8)}{(1-x^2)(1-x)^2}\ \frac{x^{31}}{(1-x)^5}$	$C_2^mC_r$, $m=1$, $r=6$ (Section V., § 17)
	$C_2C_3^2$	$\frac{x^2(1-x^7)(1-x^8)}{(1-x^2)(1-x)^2}\ \frac{x^{21}-x^{31}}{(1-x)^5}$	$C_2^mC_r$, $m=1$, $r=6$ (Section V., § 17)
	$C_2^2C_4$	$\frac{x^{11}(1-x^5)(1-x^6)(1-x^{14})}{(1-x^2)(1-x)^7}$	$C_2^mC_4$, $m=2$ (Section V., § 19)
	C_2^4	$\frac{x^8(1-x^3)(1-x^5)(1-x^7)}{(1-x)^7}$	C_2^m, $m=4$ (Section V., § 15)

Also—

(i.) Sum of generating functions for forms C_8, C_4^2 (containing no factor of degree 3 or less) is

$$\frac{x^{92}}{(1-x)^7}=\frac{x^{\binom{8}{1}+\binom{8}{2}+\binom{8}{3}}}{(1-x)^7}.$$

(ii.) Sum of generating functions for forms C_8, C_4^2, C_3C_5 is $x^{36}/(1-x)^7$ (see Section V., § 13).

(iii.) Sum of all generating functions above is $x^8/(1-x)^7$ (see Section V., § 13).

Note on Syzygies of Degree 9.

27. The only types and products of degree 9 which have no factor of degree less than 3 are C_9, C_4C_5, C_3C_6, C_3^3; and therefore the generating function for all such irreducible forms is $x^{45}/(1-x)^8$. Also the generating function for C_9 is $x^{255}/(1-x)^8$: the generating function for C_4C_5 is $\dfrac{x^{129}(1-x^{126})}{(1-x)^8}$, while the generating function for C_3C_6 is (§ 24)

$$\frac{x^{55}(1-x^{10}+x^{9}-x^{19}+x^{16}-x^{31}+x^{25}-x^{74})}{(1-x)^8}.$$

Hence the generating function for the products C_3^3 should be

$$\frac{x^{45}-x^{255}-x^{129}(1-x^{126})-x^{55}(1-x^{10}+x^{9}-x^{19}+x^{16}-x^{31}+x^{25}-x^{74})}{(1-x)^8}$$

$$=\frac{x^{45}(1-x^{10}+x^{20}-x^{19}+x^{29}-x^{26}+x^{41}-x^{35})}{(1-x)^8}.$$

This generating function may be written in the form

$$\frac{\begin{array}{l}x^{45}(1+2x+3x^2+4x^3+5x^4+6x^5+7x^6+8x^7+9x^8+10x^9+10x^{10}+10x^{11}\\ \quad+10x^{12}+10x^{13}+10x^{14}+10x^{15}+10x^{16}+10x^{17}+10x^{18}+9x^{19}+9x^{20}\\ \quad+9x^{21}+9x^{22}+9x^{23}+9x^{24}+9x^{25}+8x^{26}+7x^{27}+6x^{28}+6x^{29}+6x^{30}+6x^{31}\\ \quad+6x^{32}+6x^{33}+6x^{34}+5x^{35}+4x^{36}+3x^{37}+2x^{38}+x^{39})\end{array}}{(1-x)^6}$$

The sum of the coefficients in the numerator is 280, the same as the number of possible different arrangements of the letters in the factors of C_3^3.

The first few terms may be accounted for by the methods already used, thus

$(a_1a_2a_3)(a_4a_5a_6)(a_7a_8a_9)$ is of minimum weight 45,

$$\left.\begin{array}{l}(a_1a_2a_4)(a_3a_5a_6)(a_7a_8a_9)\\(a_1a_2a_3)(a_4a_5a_7)(a_6a_8a_9)\end{array}\right\}\quad ,, \quad ,, \quad 46,$$

$$\left.\begin{array}{l}(a_1a_2a_5)(a_3a_4a_6)(a_7a_8a_9)\\(a_1a_2a_4)(a_3a_5a_7)(a_6a_8a_9)\\(a_1a_2a_3)(a_4a_5a_8)(a_6a_7a_9)\end{array}\right\}\quad ,, \quad ,, \quad 47,$$

$$\left.\begin{array}{l}(a_1a_2a_6)(a_3a_4a_5)(a_7a_8a_9)\\(a_1a_2a_5)(a_3a_4a_7)(a_6a_8a_9)\\(a_1a_2a_4)(a_3a_5a_8)(a_6a_7a_9)\\(a_1a_2a_3)(a_4a_5a_9)(a_6a_7a_8\end{array}\right\}\quad ,, \quad ,, \quad 48.$$

But our known reductions give six products of minimum weight 49 (instead of five products): they are

$$(a_1 a_2 a_7)(a_3 a_4 a_5)(a_6 a_8 a_9),$$

$$(a_1 a_2 a_6)(a_3 a_4 a_7)(a_5 a_8 a_9),$$

$$(a_1 a_2 a_5)(a_3 a_4 a_8)(a_6 a_7 a_9),$$

$$(a_1 a_2 a_4)(a_3 a_5 a_9)(a_6 a_7 a_8),$$

$$(a_1 a_2 a_3)(a_4 a_6 a_7)(a_5 a_8 a_9),$$

$$(a_1 a_3 a_4)(a_2 a_5 a_6)(a_7 a_8 a_9),$$

and one of these forms should be of minimum weight 50 at least.

This would seem to imply that Stroh's syzygies of degree 4 as used hitherto are insufficient for the investigation of the products C_3^3: it seems probable that similar difficulties will arise successively in the treatment of products C_4^4, C_5^5, ..., C_m^m, ..., that is, whenever we arrive at a degree m^2.

[*Note.*—The symbol $(a_1 a_2 a_3)$ is used simply to denote any covariant linear in the coefficients of each of the quantics represented by the letters a_1, a_2, a_3.]

ON CERTAIN CLASSES OF SYZYGIES

By A. Young

[Communicated November 10th, 1904.—Received January 9th, 1905.]

The object of this paper is to extend so far as possible the results obtained for perpetuant syzygies in a recent paper* by Mr. Wood and myself to syzygies between concomitants of binary forms of finite order. In using the symbolical notation for binary forms it is unnecessary to write down any symbolical factors involving the variables, for these may be supplied when the orders of the forms are known. In this case the symbolical expressions used in the paper referred to for covariants of forms of infinite order may also be regarded as covariants of forms

$$a_{1_x}^{n_1},\quad a_{2_x}^{n_2},\quad \ldots,$$

all of finite order; provided that in any such expressions no letter a_r occurs to a degree greater than n_r, the order of the corresponding quantic.

The result to be proved is that the syzygies obtained for perpetuants are true when the orders of some or all of the quantics concerned are finite, if those forms in which any letter a_r appears to a degree greater than the order n_r of the corresponding quantic are ignored, except for forms which have a factor $(a_r a_s)^{\lambda}(a_s a_t)^{n_s-\lambda}$.

Thus, if $\Sigma P+\Sigma P' = 0$ be one of the perpetuant syzygies obtained in the former paper, where ΣP is the sum of those products in which the letter a_1 does not appear to a degree greater than n_1, the letter a_2 does not appear to a degree greater than n_2, and so on, and $\Sigma P'$ is the sum of those products in which some letter a_r appears to a degree greater than n_r, then for the quantics $a_{1_x}^{n_1}$, $a_{2_x}^{n_2}$, ... we have a syzygy

$$\Sigma P = \Sigma Q$$

where each of the terms Q has a factor of the form $(a_r a_s)^{\lambda}(a_s a_t)^{n_s-\lambda}$.

---

* *Proc. London Math. Soc.*, Ser. 2, Vol. 2, p. 221.

The result is exactly analogous to the extension of reducibility results for perpetuants to forms of finite order.* It will be convenient here to call covariants such as Q forms of the second class, as was done in the paper referred to.

To obtain the result it is necessary to prove it true for Stroh's syzygies, and this constitutes practically the only difficulty.

1. Stroh's fundamental syzygy

$$\{(ab)+(cd)\}^{w}-\{(ad)+(cb)\}^{w}=0$$

is true for forms of finite order only when w is less than the order of each of the forms concerned. We proceed to obtain a corresponding syzygy, true for all values of w; in fact, it is to be proved that

$$\Sigma\binom{w}{r}(ab)^{r}(cd)^{w-r}-\Sigma\binom{w}{r}(ad)^{r}(cb)^{w-r}=\Sigma Q$$

where each of the terms Q on the right has a factor of the form

$$(a_r a_s)^{\lambda}(a_s a_t)^{n_s-\lambda}$$

and the sums on the left extend only to such terms as can possibly represent covariants of the quantics in question. Thus, if $a_x^{n_1}$, $b_x^{n_2}$, $c_x^{n_3}$, $d_x^{n_4}$ are the quantics, then $r \not> n_1$ or n_2 in the first sum, while it is $\not< w-n_3$ or $w-n_4$; and similarly for the second sum.

2. We shall first prove the identity ($n_4 < w$)

$$\sum_{r=0}^{n_4}\binom{w}{r}\left[(ab)^{w-r}(cd)^{r}-(cb)^{w-r}(ad)^{r}\right]$$

$$=\sum_{r=0}^{w-n_4-1}\binom{w}{r}(ac)^{w-n_4-r}(cb)^{r}\sum_{s=0}^{n_4}\binom{w-n_4-1-r+s}{s}(ad)^{n_4-s}(cd)^{s}$$

where it is supposed that none of n_1, n_2, n_3 are less than w.

Let us assume this true when $n_4 > a$, and prove it for the case when $n_4=a-1$.

---

* A. Young, *Proc. London Math. Soc.*, Ser. 2, Vol. 1.

Since it is true when $n_4 = a$, we have

$$\sum_{r=0}^{a-1} \binom{w}{r} [(ab)^{w-r}(cd)^r - (cb)^{w-r}(ad)^r]$$

$$= -\binom{w}{a}[(ab)^{w-a}(cd)^a - (cb)^{w-a}(ad)^a]$$

$$+ \sum_{r=0}^{w-a-1} \binom{w}{r} (ac)^{w-a-r}(cb)^r \sum_{s=0}^{a} \binom{w-a-1-r+s}{s} (ad)^{a-s}(cd)^s$$

$$= -\sum_{r=0}^{w-a-1} \binom{w}{a}\binom{w-a}{r} (ac)^{w-a-r}(cb)^r(cd)^a$$

$$+ \binom{w}{a}(cb)^{w-a}(ac)[(ad)^{a-1}+(ad)^{a-2}(cd)+\ldots+(cd)^{a-1}]$$

$$+ \sum_{r=0}^{w-a-1} \binom{w}{r} (ac)^{w-a-r}(cb)^r \sum_{s=0}^{a} \binom{w-a-1-r+s}{s} (ad)^{a-s}(cd)^s$$

$$= \binom{w}{a}(cb)^{w-a}(ac) \sum_{s=0}^{a-1} (ad)^{a-1-s}(cd)^s$$

$$+ \sum_{r=0}^{w-a-1} \binom{w}{r} (ac)^{w-a-r}(cb)^r \sum_{s=0}^{a} \binom{w-a-1-r+s}{w-a-1-r} [(ad)^{a-s}(cd)^s - (cd)^a]^*$$

$$= \binom{w}{a}(cb)^{w-a}(ac) \sum_{s=0}^{a-1} (ad)^{a-1-s}(cd)^s$$

$$+ \sum_{r=0}^{w-a-1} \binom{w}{r} (ac)^{w-a-r+1}(cb)^r \sum_{s=0}^{a-1} (ad)^{a-s-1}(cd)^s \sum_{t=0}^{s} \binom{w-a-1-r+t}{w-a-1-r}$$

$$= \sum_{r=0}^{w-a} \binom{w}{r} (ac)^{w-a+1-r}(cb)^r \sum_{s=0}^{a-1} \binom{w-a-r+s}{w-a-r} (ad)^{a-s-1}(cd)^s.$$

Dividing this result by a_x, we see that the theorem is true for $n_4 = a-1$.

Hence, if the relation is true when $n_4 = a$, it is true when $n_4 = a-1$.

It is easily verified when $n_4 = w-1$; it is true for $n_4 = w$; hence it is always true.

3. In exactly the same way it may be shown that, if n_2 and n_4 are each less than w, but neither n_1 nor n_4 is less than w, then

$$\sum_{r=w-n_2}^{n_4} \binom{w}{r} [(ab)^{w-r}(cd)^r - (cb)^{w-r}(ad)^r]$$

$$= \sum_{r=0}^{w-n_4-1} \sum_{s=0}^{n_4} \binom{w}{r}\binom{w-n_4-1-r+s}{s} (ac)^{w-n_4-r}(cb)^r(ad)^{n_4-s}(cd)^s$$

$$- \sum_{r=0}^{w-n_2-1} \sum_{s=0}^{n_2} \binom{w}{r}\binom{w-n_2-1-r+s}{s} (ac)^{w-n_2-r}(cd)^r(ab)^{n_2-s}(cb)^s.$$

---

* For $\sum_{s=0}^{s} \binom{w-a-1-r+s}{w-a-1-r} = \binom{w-r}{w-a-r} = \dfrac{\binom{w}{a}\binom{w-a}{r}}{\binom{w}{r}}$.

4. To extend the result of § 3 to the case in which n_1, n_2, n_4 are all less than w, but $n_3 \geqslant w$, we start with the result of § 3, where n_1 may be supposed equal to w, and then transform the relation to one which has a factor a_x; this factor being removed, we have a relation for the case $n_1 = w-1$. Proceeding thus step by step, a relation is established for all cases when $n_1 < w$.

LEMMA.—The following relation is required:—

$$(ac)^{w-n_4-r}(cb)^r(ad)^{n_4-s}(cd)^s$$

$$= \sum_{t=0}^{\lambda}\binom{n_4-s}{t}(ac)^{w-r-s-t}(cb)^r(cd)^{s+t}$$

$$+(-)^{\lambda+1}\sum_{t=\lambda+1}^{n_4-s}(-)^t\binom{t-1}{\lambda}\binom{n_4-s}{t}(ac)^{w-n_4-r}(cb)^r(cd)^{s+t}(ad)^{n_4-s-t}.$$

We assume that this is true for a particular value of λ, and then shew that it is true for $\lambda+1$.

Thus, on this assumption,

$$(ac)^{w-n_4-r}(cb)^r(ad)^{n_4-s}(cd)^s$$

$$= \sum_{t=0}^{\lambda}\binom{n_4-s}{t}(ac)^{w-r-s-t}(cb)^r(cd)^{s+t}$$

$$+\binom{n_4-s}{\lambda+1}(ac)^{w-n_4-r}(cb)^r(cd)^{s+\lambda+1}\{(ac)^{n_4-s-\lambda-1}-[(ad)-(cd)]^{n_4-s-\lambda-1}$$
$$+(ad)^{n_4-s-\lambda-1}\}$$

$$+(-)^{\lambda+1}\sum_{t=\lambda+2}^{n_4-s}(-)^t\binom{t-1}{\lambda}\binom{n_4-s}{t}(ac)^{w-n_4-r}(cb)^r(cd)^{s+t}(ad)^{n_4-s-t}$$

$$= \sum_{t=0}^{\lambda+1}\binom{n_4-s}{t}(ac)^{w-r-s-t}(cb)^r(cd)^{s+t}$$

$$+(-)^{\lambda+2}\sum_{t=\lambda+2}^{n_4-s}(-)^t\left\{\binom{n_4-s}{\lambda+1}\binom{n_4-s-\lambda-1}{t-\lambda-1}-\binom{t-1}{\lambda}\binom{n_4-s}{t}\right\}$$
$$\times(ac)^{w-n_4-r}(cb)^r(cd)^{s+t}(ad)^{n_4-s-t}$$

$$= \sum_{t=0}^{\lambda+1}\binom{n_4-s}{t}(ac)^{w-r-s-t}(cb)^r(cd)^{s+t}$$

$$+(-)^{\lambda+2}\sum_{t=\lambda+2}^{n_4-s}(-)^t\binom{t-1}{\lambda+1}\binom{n_4-s}{t}(ac)^{w-n_4-r}(cb)^r(cd)^{s+t}(ad)^{n_4-s-t}.$$

Hence the lemma is true for $\lambda+1$ if it is true for any particular value of λ. But it is evidently true when $\lambda = 0$; hence it is always true.

It follows in exactly the same way that

$$(ac)^{w-n_2-r}(cd)^r(ab)^{n_2-s}(cb)^s$$

$$= \sum_{t=0}^{\lambda}\binom{n_2-s}{t}(ac)^{w-r-s-t}(cd)^r(cb)^{s+t}$$

$$+(-)^{\lambda+1}\sum_{t=\lambda+1}^{n_2-s}(-)^t\binom{t-1}{\lambda}\binom{n_2-s}{t}(ac)^{w-n_2-r}(cd)^r(cb)^{s+t}(ab)^{n_2-s-t}.$$

5. Now apply these results to modify the relation of § 3, so that it involves not more than n_1 a's in any term on the right. Then

$$\sum_{r=w-n_2}^{n_4}\binom{w}{r}\left[(ab)^{w-r}(cd)^r-(cb)^{w-r}(ad)^r\right]$$

$$= \sum_{r=0}^{w-n_4-1}\sum_{s=0}^{w-r-n_1-1}\sum_{t=w-r-s-n_1}^{n_4-s}(-)^{t+w-r-s-n_1}\binom{w}{r}\binom{w-n_4-1-r+s}{s}$$

$$\times\binom{t-1}{w-r-s-n_1-1}\binom{n_4-s}{t}(ac)^{w-n_4-r}(cb)^r(ad)^{n_4-s-t}(cd)^{s+t}$$

$$+\sum_{r=0}^{w-n_4-1}\sum_{s=w-r-n_1 \text{ or } 0}^{n_4}\binom{w}{r}\binom{w-n_4-1-r+s}{s}(ac)^{w-n_4-r}(cb)^r(ad)^{n_4-s}(cd)^s$$

$$-\sum_{r=0}^{w-n_2-1}\sum_{s=0}^{w-r-n_1-1}\sum_{t=w-r-s-n_1}^{n_2-s}(-)^{t+w-r-s-n_1}\binom{w}{r}\binom{w-n_2-1-r+s}{s}$$

$$\times\binom{t-1}{w-r-s-n_1-1}\binom{n_2-s}{t}(ac)^{w-n_2-r}(cd)^r(ab)^{n_2-s-t}(cb)^{s+t}$$

$$-\sum_{r=0}^{w-n_2-1}\sum_{s=w-r-n_1 \text{ or } 0}^{n_2}\binom{w}{r}\binom{w-n_2-1-r+s}{s}(ac)^{w-n_2-r}(cd)^r(ab)^{n_2-s}(cb)^s$$

$$+\sum_{r=0}^{w-n_4-1}\sum_{s=0}^{w-r-n_1-1}\sum_{t=0}^{w-r-s-n_1-1}\binom{w}{r}\binom{w-n_4-1-r+s}{s}\binom{n_4-s}{t}$$

$$\times(ac)^{w-r-s-t}(cb)^r(cd)^{s+t}$$

$$-\sum_{r=0}^{w-n_2-1}\sum_{s=0}^{w-r-n_1-1}\sum_{t=0}^{w-r-s-n_1-1}\binom{w}{r}\binom{w-n_2-1-r+s}{s}\binom{n_2-s}{t}$$

$$\times(ac)^{w-r-s-t}(cd)^r(cb)^{s+t}.$$

Let us first suppose that $n_1 \geqslant$ both n_2 and n_4. Then the left-hand side of the relation will require no alteration. Now the relation is an identity; therefore, if a c is introduced into every determinant factor by means of relations of the form $(ab) = (ac)+(cb)$, then the sum of the coefficients of each of the resulting terms $(ac)^\lambda(cb)^\mu(cd)^\nu$ must be separately zero.

In particular, all those terms which have a factor $(ac)^{n_1+1}$ must vanish separately; now no such terms can arise from any part of either side of the relation except in its last two lines, and in these lines every term contains a factor $(ac)^{n_1+1}$. Moreover these terms are unaffected by the operation of introducing c into every determinant factor; their sum is therefore zero.

Hence, when $n_1 > n_2$ and $n_1 > n_4$,

$$\sum_{r=w-n_2}^{n_4} \binom{w}{r} [(ab)^{w-r}(cd)^r - (cb)^{w-r}(ad)^r]$$

$$= \sum_{r=0}^{w-n_4-1} \sum_{s=0}^{w-r-n_1-1} \sum_{t=w-r-s-n_1}^{n_4-s} (-)^{t+w-r-s-n_1} \binom{w}{r} \binom{w-n_4-1-r+s}{s}$$

$$\times \binom{t-1}{w-r-s-n_1-1} \binom{n_4-s}{t} (ac)^{w-n_4-r}(cb)^r(ad)^{n_4-s-t}(cd)^{s+t}$$

$$+ \sum_{r=0}^{w-n_4-1} \sum_{s=w-r-n_1 \text{ or } 0}^{n_4} \binom{w}{r} \binom{w-n_4-1-r+s}{s} (ac)^{w-n_4-r}(cb)^r(ad)^{n_4-s}(cd)^s$$

$$- \sum_{r=0}^{w-n_2-1} \sum_{s=0}^{w-r-n_1-1} \sum_{t=w-r-s-n_1}^{n_2-s} (-)^{t+w-r-s-n_1} \binom{w}{r} \binom{w-n_2-1-r+s}{s}$$

$$\times \binom{t-1}{w-r-s-n_1-1} \binom{n_2-s}{t} (ac)^{w-n_2-r}(cd)^r(ab)^{n_2-s-t}(cb)^{s+t}$$

$$- \sum_{r=0}^{w-n_2-1} \sum_{s=w-r-n_1 \text{ or } 0}^{n_2} \binom{w}{r} \binom{w-n_2-1-r+s}{s} (ac)^{w-n_2-r}(cd)^r(ab)^{n_2-s}(cb)^s.$$

If $n_1 < n_2$, there will be terms on the right which when c is introduced into every determinant factor will yield terms having a factor $(ac)^{n_1+1}$.

We may bring over these terms to the right-hand side, and, using the results of § 4, we may replace them by others by means of the relation

$$\sum_{r=w-n_2}^{w-n_1-1} \binom{w}{r} (ab)^{w-r}(cd)^r$$

$$= \sum_{r=w-n_2}^{w-n_1-1} \sum_{t=0}^{w-r-n_1-1} \binom{w}{r} \binom{w-r}{t} (ac)^{w-r-t}(cd)^r(cb)^t$$

$$+ \sum_{r=w-n_2}^{w-n_1-1} \sum_{t=w-r-n_1}^{w-r} (-)^{t+w-r-n_1} \binom{w}{r} \binom{t-1}{w-r-n_1-1} \binom{w-r}{t}$$

$$\times (cd)^r(cb)^t(ab)^{w-r-t}.$$

The first of these two sums will disappear with the last two lines of the former sums, and so the relation holds good with an alteration of limits on the left and with the subtraction of the second of the above sums from the right.

Each term of this second sum can be written in the form

$$\Sigma N(cd)^r (ac)^s (ab)^{n_1-s} (cb)^{w-r-n_1}.$$

If $n_1 < n_4$, another alteration in the limits must be made on the left, and terms of the form

$$\Sigma N(cb)^{w-r} (ac)^s (ad)^{n_1-s} (cd)^{r-n_1}$$

introduced on the right.

6. We have next to consider the case when w is greater than the order of each of the quantics. The method of proof here does not require any assumption as to the relative magnitudes of w, n_1, n_2, n_3, n_4; so that the theorem is proved again for the three cases already discussed; on the other hand, the proof given here affords no information as to what are the terms of the second class on the right-hand side of Stroh's syzygy.

We shall assume that the theorem has been proved when the total weight of the syzygy is w and the orders of the quantics are n_1, n_2, n_3+1, n_4; and then shew that it is true when the weight of the syzygy is w and the orders of the quantics are n_1, n_2, n_3, n_4.

We have then a syzygy

$$\text{(I.)} \qquad \Sigma \binom{w}{r} (ab)^{w-r} (cd)^r - \Sigma \binom{w}{r} (cb)^{w-r} (ad)^r = \Sigma T$$

where T is a covariant of $a_x^{n_1}$, $b_x^{n_2}$, $c_x^{n_3+1}$, $d_x^{n_4}$ of the second class; and the sums on the left extend only to such symbolical products as may be interpreted.

In will in general happen that when the order of c becomes n_3 either one or two terms on the left of (I.) cease to be interpretable. Let T' be the sum of these terms; they must be removed to the right-hand side of the relation.

Now (I.) is an algebraical identity between the symbols a, b, c, d, x; hence, if we can express the forms $\Sigma T - T'$ in terms of a sum of symbolical products which represent covariants of $a_x^{n_1}$, $b_x^{n_2}$, $c_x^{n_3}$, $d_x^{n_4}$ of the second class multiplied by a factor c_x, we shall have a new relation, from which a factor c_x may be removed; the result is then the syzygy required.

We proceed to prove that this can be done. Those factors of a symbolical product which make the covariant belong to the second class are here referred to as the second class factors. No covariant of $\Sigma T - T'$ need be considered which has already a factor c_x.

If a covariant of $\Sigma T - T'$ has a factor $(\epsilon\zeta)$, which does not appear amongst its second class factors, and also a factor b_x, we will use the identity for $(\epsilon\xi)\,b_x$, viz., $(\epsilon\xi)b_x = (\epsilon b)\xi_x + (b\xi)\epsilon_x$.

Thus all the covariants of $\Sigma T - T'$ can be expressed in terms of—

(1) Covariants of $a_x^{n_1}$, $b_x^{n_2}$, $c_x^{n_3}$, $d_x^{n_4}$ of the second class multiplied by c_x.

(2) Covariants of $a_x^{n_1}$, $b_x^{n_2}$, $c_x^{n_3+1}$, $d_x^{n_4}$ of the second class which have no factor b_x or c_x.

(3) Covariants of $a_x^{n_1}$, $b_x^{n_2}$, $c_x^{n_3+1}$, $d_x^{n_4}$ of the second class which contain the letter b in all factors other than the second class factors and have no factor c_x.

7. The covariants can be further restricted. Covariants of set (3) contain a factor $(bc)^{n_3+1}$ unless the letter c appears in the original second class factors.

If the second class factors were originally $(c\epsilon)^\lambda(\epsilon\xi)^{n-\lambda}$ (where n is the order of ϵ), we have now a factor $(bc)^{n_3+1-\lambda}(c\epsilon)^\lambda$. By means of repeated use of the identity for $(c\epsilon)b_x$ such a covariant can be expressed in terms of covariants (1), (2), and

(4) Covariants which have a factor $(bc)^{n_3+1}$.

It remains to consider covariants which have a factor $(ac)^\lambda(cd)^{n_3+1-\lambda}$. Here we begin by introducing d into as many determinant factors as possible; the covariant is thus expressed in terms of covariants (1), of covariants which have no factor c_x or d_x, and of covariants which have a factor $(cd)^{n_3+1}$.

By means of the identity for $(ac)(bd)$ we may express those covariants which have no factor c_x or d_x in terms of covariants which have one or other of the factors $(bc)^\lambda(cd)^{n_3+1-\lambda}$, $(cd)^\lambda(da)^{n_4-\lambda}$.

Covariants having either one of these factors or else the factor $(cd)^{n_3-1}$ have already been discussed.

We are thus left with covariants (1), (2), and (4).

Covariants of set (2) may be written

$$(bc)^\lambda(cd)^\mu(da)^\nu(ab)^\rho(bd)^\sigma(ac)^\tau;$$

where $\lambda+\mu+\tau = n_3+1, \qquad \lambda+\rho+\sigma = n_2,$

and one of the indices is zero.

By means of the identity for $(ab)(cd)$ we may express this in terms of similar covariants in which either $\rho = 0$ or $\mu = 0$. When $\rho = 0$ there is a factor $(bc)^{\lambda_1}(bd)^{n_2-\lambda_1}$, and by means of the identity for $(ac)\,d_x$ we may express the covariant in terms of covariants (1) and of

(5) Covariants which have no factor c_x, b_x, or d_x;

(6) Covariants which have no factor c_x or b_x and which contain the letter a only in factors (ad), a_x.

When $\mu = 0$ there is a factor $(bc)^{\lambda}(ca)^{n_3+1-\lambda}$, and by means of repeated use of the identity

$$(ac)(ab)\,d_x = (ac)(db)\,a_x + (bc)(ad)\,a_x + (ab)(ad)\,c_x$$

the covariant can be expressed in terms of covariants (1), (4), (5) and of covariants of the same kind for which $\rho = 0$.

Thus we can express all covariants of the sum $\Sigma T - T'$ in terms of covariants (1), (4), (5), (6).

Now the relation from which we started may be regarded as an identity expressing $\Sigma T - T'$ as a sum of terms having a factor c_x. Hence the sum of the covariants of sets (4), (5), and (6) which now appear on the right-hand side of our relation must have a factor c_x. It is shewn in the next paragraph that this means that the sum is zero; hence $\Sigma T - T'$ is equal to a sum of covariants (1).

The theorem is then true when the order of c is n_3, provided it is true when the order of c is n_3+1. Hence, if it is true when the order of c is w, it is always true. But it is true when the order of no one of the quantics is less than w: hence by repeated use of the induction it is true always.

8. *A sum of covariants, linear in the coefficients of each of the four quantics*

$$a_x^{n_1},\quad b_x^{n_2},\quad c_x^{n_3+1},\quad d_x^{n_4},$$

each of which is of one of the following kinds:—

(1) *Covariants which have no factor c_x, b_x, or d_x;*

(2) *Covariants which have no factor c_x, b_x and in which the letter a appears only in factors (ad), a_x;*

(3) *Covariants which have a factor $(bc)^{n_3+1}$;—*

cannot be expressed as a sum of symbolical products having a factor c_x, unless this sum is zero.

The covariants (1) can be expressed as transvectants

$$\left[(cb)^{\lambda_1}(cd)^{\lambda_2}(bd)^{\lambda_3},\ a_x^{n_1}\right]^{n_3+n_2+n_4-2\lambda_1-2\lambda_2-2\lambda_3},$$

each of which only contains a single term. These transvectants may be expressed in terms of the transvectants

$$\left[(cb)^{\lambda_1}(cd)^{n_3-\lambda_1}(bd)^{\lambda_2},\ a_x^{n_1}\right]^{n_2+n_4-n_3-2\lambda_2},$$

and of transvectants $\left[(cb)^{\lambda_1}(cd)^{\lambda_2},\ a_x^{n_1}\right]^{n_3+n_2+n_4-2\lambda_1-2\lambda_2}.$

The covariants (2) can be expressed as transvectants

$$[(cb)^{\lambda_1}(cd)^{n_3-\lambda_1}(bd)^{n_2-\lambda_1},\ a_x^{n_1}]^{\rho},$$

each of which contains only a single term.

The covariants (3) can be expressed in terms of transvectants

$$[(cb)^{n_3+1}(bd)^{\lambda_1},\ a_x^{n_1}]^{\rho}.$$

Now transvectants of $a_x^{n_1}$ which have not the same index are independent; hence, if the sum of covariants has a factor c_x, the sums of the equivalent transvectants of $a_x^{n_1}$ for each index must have a factor c_x. Hence in all cases in which the index of the transvectant is equal to the order of the covariant on the right the sum of the transvectants must be zero. For other cases, if w is the weight of the covariant, and ρ the index of the transvectant, then the index of (bd) is known to be $w-\rho-n_3-1$, and the index of (cb) is the least of the numbers $n_2-(w-\rho-n_3-1)$ and n_3+1; so that there is only one such term. This term does not contain a factor c_x; so that, if the sum of covariants considered has a factor c_x, the coefficient of each of these transvectants must be zero; *i.e.*, the sum must be zero.

COROLLARY.—*The same is true if one or more of the letters a, b, c, d refers not to a single fundamental quantic, but to either of two quantics: thus we may suppose that a is written for* α *or* α' [*and in this case it may be supposed that the covariant has a factor* $(\alpha\alpha')^{\mu_1}$], *provided further that there is no factor* α_x *or else that the letter* α' *appears only in factors* $(\alpha\alpha')$, α'_x, *and that this restriction is to be applied in every case where a letter does not refer to a single fundamental quantic.*

The method of proof is nearly identical with that which precedes. The covariants (1) when

$$\nu_1-\mu_1 \geqslant n_3+n_2+n_4-2\lambda_1-2\lambda_2-2\lambda_3 \geqslant \rho, \text{ say},$$

can be expresssed as transvectants

$$[(cb)^{\lambda_1}(cd)^{\lambda_2}(bd)^{\lambda_3},\ (\alpha\alpha')^{\mu_1}\alpha_x^{\nu_1-\mu_1}\alpha_y'^{\nu_1-\mu_1}]^{\rho}_{y=x};$$

and when $\nu_1-\mu_1 < \rho$ these covariants can be expressed as transvectants

$$[(cb)^{\lambda_1}(cd)^{\lambda_2}(bd)^{\lambda_3},\ (\alpha\alpha')^{\mu_1}\alpha_x^{\nu_1-\mu_1}\alpha_x'^{\rho+\mu_1-\nu_1}\alpha_y'^{\nu'_1+\nu_1-2\mu_1-\rho}]^{\rho}_{y=x}.$$

The letters on the left-hand side of the transvectants may here be simple or composite: it does not affect matters.

Similarly covariants (2) can be expressed as transvectants

$$(P,\ Q)^{\rho}_{y=x},$$

where

$$P = (cb)^{\lambda_1} (cd)^{n_3-\lambda_1} (bd)^{n_2-\lambda_1} \Big[d_x^{n_4+2\lambda_1-n_2-n_3}, \quad (\delta\delta')^{\mu_4} \delta_x^{\nu_4+2\lambda_1-n_2-n_3-\mu_4} \delta_y'^{\nu'-\mu_4},$$

$$(\delta\delta')^{\mu_4} \delta_x^{\nu_4+2\lambda_1-n_2-n_3-n_4} \delta_x'^{\rho+n_2+n_3+n_4-\nu_4-2\lambda_1} \delta_y'^{\cdots},$$

$$\text{or} \quad (\delta\delta')^{\mu_4} \delta_x'^{\rho} \delta_y'^{\nu'_4-\rho-\mu_4} \Big],$$

according to the magnitude of ν_4, and where

$$Q = (aa')^{\mu_1} (a_x^{\nu_1-\mu_1} a_y'^{\nu'_1-\mu_1} \text{ or } a_x^{\nu_1-\mu_1} a_x'^{\rho+\mu_1-\nu_1} a_y'^{\nu'_1+\nu_1-\rho-2\mu_1}).$$

The covariants (3) can be expressed in terms of transvectants in an exactly similar way. When this has been done the argument is the same as in the preceding theorem.

If a represents a fundamental quantic, the theorem follows exactly in the same way.

9. When one of the letters a, b, c, d refers to a covariant of the original quantics Stroh's syzygies for perpetuants are true as before. Unfortunately forms which are of the second class when c is one of the original quantics are not necessarily of the second class when c is a covariant. For, although a form having a factor $(ac)^{\lambda}(cb)^{n_3-\lambda}$ can be proved to be really of the second class when c is a covariant of order n_3, yet forms having a factor $(ab)^{\lambda}(bc)^{n_2-\lambda}$ are by no means necessarily of the second class. As we shall require Stroh's syzygy in the cases when some or all of a, b, c, d refer to covariants, it is necessary to establish it in these cases. The method of proof is essentially the same as that used in §§ 6, 7.

We shall first confine our attention to the cases when a, b, c, d represent either covariants of the second degree or else fundamental quantics.

If $c = (\gamma\gamma')^{\mu_3} \gamma_x^{\nu_3+1-\mu_3} \gamma_x'^{\nu'_3-\mu_3}$, Stroh's syzygy will be written

$$\Sigma \binom{\lambda}{r} (ab)^r (\gamma\gamma')^{\mu_3} (\gamma d)^{\lambda-r} - \Sigma \binom{\lambda}{r} (ad)^r (\gamma\gamma')^{\mu_3} (\gamma b)^{\lambda-r} = \Sigma T.$$

In the same way, if $a = (aa')^{\mu_1}$, the letter a' will appear in the factors (aa'), a'_x only on the left-hand side.

It will be assumed that the syzygy is true when the order of γ is ν_3+1, and then it will be shewn to be true when the order of γ is ν_3. It is, however, necessary to further limit the forms ΣT which are of the second class. The precise limitations imposed depend on the number of the quantics a, b, c, d which are fundamental, and are stated later in §§ 12–15.

It is sufficient here to notice that in the beginning of the induction, when a, b, d are fundamental quantics and $\lambda = \nu_3+1$, then the terms $\Sigma T - T'$ only contain γ' in their second class factors when it appears in the factor $(\gamma\gamma')$.

10. LEMMA.—All covariants linear in the coefficients of each of the quantics $a_x^{n_1}$, $b_x^{n_2}$, $c_x^{n_3+1}$, $d_x^{n_4}$ [where $c_x^{n_3+1} \equiv (\gamma\gamma')^{\mu_3}\gamma_x^{\nu_3+1-\mu_3}\gamma'^{\nu'_3-\mu_3}$ and a, b, d are fundamental quantics or else covariants of degree 2], which belong to the second class owing to factors of the form $(\epsilon\zeta)^\lambda(\zeta\eta)^{n-\lambda}$, or of the form $(\epsilon\zeta)^\lambda(\zeta\zeta')^{n-\lambda}$, where n is the order of the quantic ζ, and ϵ, ζ, η refer to any of the quantics $a_x^{n_1}$, $b_x^{n_2}$, $\gamma_x^{\nu_3+1}$, $d_x^{n_4}$ if fundamental (or if one or more of these are covariants to $\alpha_x^{\nu_1}$, $\beta_x^{\nu_2}$, or $\delta_x^{\nu_4}$ as the case may be), can be expressed in terms of—

(1) Covariants of the same character when $\gamma_x^{\nu_3}$ is written for $\gamma_x^{\nu_3+1}$, which have a factor γ_x.

(2) Covariants which have no factor c_x, b_x, or d_x; where, for instance, c_x stands for either γ_x or γ'_x.

(3) Covariants which have no factor c_x, b_x and in which the letters a only appear in factors (ad), a_x.

(4) Covariants which have a factor $(\beta\gamma')^{\nu'_2-\lambda}(\gamma'\gamma)^\lambda$ or a factor $(\beta\gamma')^{\nu_2-\lambda}(\beta\beta')^\lambda$, or, if b is a fundamental quantic, a factor $(b\gamma')^{\nu'_3-\lambda}(\gamma'\gamma)^\lambda$. Provided further that, if $b \equiv (\beta\beta')^{\mu_2}\beta_x^{\nu_2-\mu_2}\beta'^{\nu'_2-\mu_2}_x$, then $\nu'_2 \geqslant \nu_2$, and that in the similar cases the orders of α and δ are not greater than those of α', δ' respectively.

In the course of the proof additional factors $(\alpha\alpha')$, $(\beta\beta')$, $(\gamma\gamma')$, $(\delta\delta')$ will from time to time make their appearance; it will be regarded as a reduction whenever such an additional factor appears without destroying the restrictions imposed on the covariants under consideration: none of the operations employed involve the loss of such a factor.

We shall first show that any covariant of the kind considered which has a factor γ'_x may be expressed in terms of covariants (1) and of covariants of the same kind which have an additional factor $(\gamma\gamma')$. All such covariants may then be neglected.

For consider such a covariant. Either there is a factor $(\gamma\gamma')^{\mu_3}(\gamma\epsilon)^{\nu_3+1-\mu_3}$, or else there is a factor $(\gamma\eta)$ which does not belong to the second class factors.

In the first case a reduction is obtained by means of the identity for $(\gamma\epsilon)\gamma'_x$; in the second case by the identity for $(\gamma\eta)\gamma'_x$. In either case the statement is true.

The next step is to remove all factors b_x. If there is a factor b_x, any factor $(\epsilon\zeta)$ which does not contain b and does not belong to the second class factors may be removed by the identity for $(\epsilon\zeta)b_x$; and every time a factor c_x appears there is a reduction. Thus we are eventually left with covariants (1), with covariants which have no factor b_x or c_x, and with covariants which have the letter b (β or β') in every determinant factor except the second class factors. If b is a fundamental quantic, all such covariants have a factor $(b\gamma')^{\nu'_3-\mu_3}(\gamma\gamma')^{\mu_3}$, and are therefore included in (4). When $b \equiv (\beta\beta')^{\mu_2}$, the covariants can be expressed in terms of such as have a factor $(\beta\gamma')^{\nu'_3-\mu_3}(\gamma\gamma')^{\mu'}$ or a factor $(\beta\gamma')^{\nu_2-\mu_2}(\beta\beta')^{\mu_2}$, and of covariants of exactly the same nature as to the second class factors which have additional factors $(\beta\beta')$. Thus, in addition to covariants (1) and (4), we have only to consider covariants which have no factor b_x or c_x.

Let us suppose that $d_x \equiv (\delta\delta')^{\mu_4}$, that there is a factor $(\epsilon\delta')$ where $\epsilon \neq \delta$, and that there is a factor δ_x. By means of the identity for $(\epsilon\delta')\delta_x$, all factors δ_x may be removed, or else all determinant factors except $(\delta\delta')$ in which δ' appeared; it will be assumed that this operation has been carried out, as also the corresponding operation affecting the letters a.

We have now to consider covariants

$$\text{(I.)} \qquad (bc)^{\lambda}(cd)^{\mu}(da)^{\nu}(ab)^{\rho}(bd)^{\sigma}(ac)^{\tau},$$

where $\lambda+\mu+\tau = n_3+1, \qquad \lambda+\rho+\sigma = n_2,$

and where the letter c stands for either of the letters γ, γ'; the letter b stands for either of the letters β, β'; and so on.

By means of the identity for $(bd)(ac)$, these covariants can be expressed in terms of similar covariants, in which either $\sigma = 0$ or $\tau = 0$.

When $\sigma = 0$, $\qquad \rho+\lambda = n_2$.

If $a = (\alpha\alpha')^{\mu_1}$, either there is no factor α_x or every letter α in the factors (ab), (ad), (ac) may be taken to be α. In the first case we may use the identities

$$(\alpha'b)(\alpha\epsilon)-(\alpha b)(\alpha'\epsilon) = -(\alpha\alpha')(b\epsilon),$$

and hence these covariants can be expressed in terms of covariants which have a factor $(b\alpha)^{\nu_1-\mu_1}(\alpha\alpha')^{\mu_1}$, of covariants of the above form in which all the factors (ab) are (αb), and of covariants of exactly the same nature which have additional factors $(\alpha\alpha')$.

The first set of covariants can be expressed in terms of covariants having a factor $(\beta\alpha)^{\nu_1-\mu_1}(\alpha\alpha')^{\mu_1}$, or of covariants having a factor $(\beta\alpha)^{\nu_2-\mu_2}(\beta\beta')^{\mu_2}$—there being no factor b_x or c_x in either case—and of covariants having no factor b_x or c_x but having additional factors $(\beta\beta')$. When there is a factor $(\beta\alpha)^{\nu_1-\mu_1}(\alpha\alpha')^{\mu_1}$ or a factor $(\beta\alpha)^{\nu_2-\mu_2}(\beta\beta')^{\mu_2}$ we may use the identity for $(ac)d_x$ to express the covariant in terms of covariants (2), of covariants (I.) in which $\tau = 0$, and of covariants (1).

We need only then consider the case in which the factor $(ab)^\rho(bc)^\lambda$ is $(ab)^\rho(bc)^\lambda$.

Reasoning as before, it may be shown that we may further suppose this factor to be $(ab)^\rho(b\gamma)^\lambda$.

If b represents a fundamental quantic, we may at once express such a covariant in terms of covariants (2), of covariants (I.) for which $\tau = 0$, and of covariants (1), by using the identity for $(ac)d_x$. The same is true when $b = (\beta\beta')^{\mu_2}$, for it is evident (since $\lambda + \rho = n_2$) that we may express such covariants in terms of covariants (I.) which have either a factor $(a\beta)^{\nu_2 - \mu'_2}(\beta\beta')^{\mu'_2}$ or a factor $(\gamma\beta)^{\nu_2-\mu'_2}(\beta\beta')^{\mu'_2}$, where $\mu'_2 \geqq \mu_2$, the order of β having been assumed to be equal to or less than that of β'.

When $\tau = 0$, $\lambda + \mu = n_3 + 1$.

Proceeding as before, we can express all such covariants in terms of covariants (I.) which have one of the following factors

$$(\beta\beta')^{\mu_2}(\beta\gamma)^{\nu_2-\mu_2},\quad (\gamma\gamma')^{\mu_3}(\gamma\beta)^{\nu_3+1-\mu_3},\quad (\gamma\gamma')^{\mu_3}(\gamma\delta)^{\nu_3+1-\mu_3},\quad (\delta\delta')^{\mu_4}(\delta\gamma)^{\nu_4-\mu_4},$$

and of covariants (I.) for which $\tau = 0$, but which have additional factors (aa'), ...

In the first two cases we may remove factors $(ab)d_x$ one pair at a time; one of the resulting forms has a factor b_x, and this may be removed by the identity for $(cd)b_x$ unless there is a factor $(bc)^{n_3+1}$. Thus we are eventually left with covariants (1), (2), (3) and covariants which have a factor $(bc)^{n_3+1}$.

When there is a factor $(\gamma\gamma')^{\mu_3}(\gamma\delta)^{\nu_3+1-\mu_3}$ or a factor $(\delta\delta')^{\mu_4}(\gamma\delta)^{\nu_4-\mu_4}$ we may proceed just as in the first two cases, if there is a factor $(\gamma'\delta)$, $(\gamma\delta')$, or $(\gamma'\delta')$. Otherwise there is a factor $(\gamma\gamma')^{\mu_3}(\beta\gamma')^{\nu_3-\mu_3}$, and the covariant is one of the forms required. This completes the proof of the lemma, for covariants which have a factor $(bc)^{n_3+1}$ can be expressed in terms of covariants (4) and covariants (I.) which have additional factors $(\beta\beta')$.

11. *Covariants which have a factor* $(\beta\gamma')^{\nu'_3-\mu_3}(\gamma\gamma')^{\mu_3}$ *or a factor* $(\beta\gamma')^{\nu_2-\mu_2}(\beta\beta')^{\mu_2}$, *or, if* b *represents a fundamental quantic, a factor* $(b\gamma')^{\nu'_3-\mu_3}(\gamma'\gamma)^{\mu_3}$, *can be expressed in terms of*

(1) *Covariants of the same character which possess also a factor* γ_x.

(2) *Covariants which have no factor* c_x, b_x, *or* d_x.

(3) *Covariants which have no factor* c_x *or* b_x *and in which the letters* a *only appear in factor* (ad), a_x.

(4) *Covariants which have a factor* $(cb)^{n_3+1}$.

To prove this, take a covariant which has one of the three given factors. Any factor $(\epsilon\zeta)\beta_x$, $[\epsilon \neq \beta,\ \zeta \neq \beta]$ may be removed by means of the identity. We then obtain covariants (1), (4) and covariants of the kind considered which have no factor β_x. Let us suppose that one of these latter covariants has a factor $(\epsilon\zeta)$ which contains neither β nor β'; then, by means of the identity for $(\epsilon\zeta)\beta'_x$, this may be expressed in terms of covariants (1), (4) and of covariants which still have the original factor, but have no factor b_x. The identity for $(ab)(cd)$ must then be used to reduce the index of (ab) or that of (cd) to zero in the last case. If the index of (ab) is zero, we may express the covariants in terms of covariants (1), (2), (3) by means of the identity for $(ac)d_x$. If the index of (cd) is zero, the same identity may be used to reduce the index of (ac) or of d_x to zero; this leaves us with covariants (1), (2) and covariants which have no factor (ac). In these latter the identity for $(ab)(cd)$ must be used again: the process is one which may be repeated, for the index of (ab) is continually being decreased. Thus we are eventually left with covariants (1), (2), (3) and covariants which have no factors (ac) or (cd), and these are covariants (4).

12. Consider now the case when a, b, d refer to fundamental quantics and $c = (\gamma\gamma')^{\mu_3}$. We shall assume that when the order of γ is ν_3+1 the terms ΣT on the right-hand side of the syzygy (see § 9) are of the second class and only contain the letter γ' in second class factors, when it appears in the factor $(\gamma\gamma')$; or else contain a factor $(b\gamma')^{\nu_3'-\mu_3}(\gamma\gamma')^{\mu_3}$. But, by §§ 10, 11, these terms can be expressed in terms of

(1) Covariants (1) of § 10.

(2) Covariants which have no factor c_x, b_x, or d_x.

(3) Covariants which have no factor c_x or b_x and in which the letters a only appear in the factors (ad), a_x.

(4) Covariants which have a factor $(cb)^{n_3+1}$.

Now the assumed syzygy tells us that the sum of the covariants (2), (3), (4) has a factor c_x; hence by § 8 this sum is zero.

We may then divide each side of the resulting relation by γ_x, and thus obtain a relation of exactly the same form as that from which we started; but in which the order of γ is ν_3.

The assumption made is true when the induction begins, and hence is always true.

13. Let a, d represent fundamental quantics, $b = (\beta\beta')^{\mu_2}$, $c = (\gamma\gamma')^{\mu_3}$. The induction proceeds on the same lines as before; at each step the

order of γ is decreased by unity. At the commencement of the induction the order of γ is equal to the weight of the syzygy. Now when this is the case the letter γ' may be ignored: so that practically c may be regarded as a fundamental quantic. In order to see that in this case the syzygy is true, and also to find the character of the terms of the second class on the right, we must interchange b and c in the result of § 12. Thus the terms on the right only contain the letter β' in their second class factors, when it appears in the factor $(\beta\beta')$, or else contain a factor

$$(\gamma\beta')^{\nu'_2-\mu_2}(\beta\beta')^{\mu_2}.$$

Covariants which have a factor

$$(\gamma\beta')^{\nu'_2-\mu_2}(\beta\beta')^{\mu_2}$$

can be expressed in terms of

(7) Covariants which have a factor

$$(\gamma\beta')^{\nu'_2-\mu_2}(\beta\beta')^{\mu_2}\gamma_x,$$

of covariants (1), (5), (6) of §§ 10, 11, and of covariants which have no factors b_x or c_x.

These latter covariants have been discussed in § 10. Hence, if at any stage of the induction the covariants on the right are of the kind discussed in the lemma of § 10 or else have one or other of the factors

$$(\beta\gamma')^{\nu'_3-\mu_3}(\gamma\gamma')^{\mu_3},\quad (\beta\gamma')^{\nu_2-\mu_2}(\beta\beta')^{\mu_2},\quad (\gamma\beta')^{\nu'_2-\mu_2}(\beta\beta')^{\mu_2},$$

then these covariants can be expressed in terms of covariants belonging to the sets (1), (2), (3), (5), (6), (7) of §§ 10, 11, 13.

This sum can be expressed by § 8 as a sum of covariants of the sets (1), (5), and (7), and hence the covariants on the right are of the same nature when we diminish the order of γ by unity.

The theorem is then true in this case.

14. Let a represent a fundamental quantic, and

$$b = (\beta\beta')^{\mu_2},\qquad c = (\gamma\gamma')^{\mu_3},\qquad d = (\delta\delta')^{\mu_4}.$$

The induction proceeds as before, commencing when the order of γ is so large that c may be regarded as a fundamental quantic. The terms now on the right-hand side of the syzygy must be obtained by interchanging c and d in § 13. They must then consist of covariants of the kind discussed in the lemma of § 10 and covariants which have one or other of the factors

$$(\beta\delta')^{\nu'_4-\mu_4}(\delta\delta')^{\mu_4},\qquad (\beta\delta')^{\nu_2-\mu_2}(\beta\beta')^{\mu_2},\qquad (\delta\beta')^{\nu'_2-\mu_2}(\beta\beta')^{\mu_2}.$$

It will then be assumed that always in this case the covariants on the right are of the kind discussed in § 10 or contain one of the following factors

$$(\beta\delta')^{\nu'_4-\mu_4}(\delta\delta')^{\mu_4}, \quad (\beta\delta')^{\nu_2-\mu_2}(\beta\beta')^{\mu_2}, \quad (\delta\beta')^{\nu'_2-\mu_2}(\beta\beta')^{\mu_2},$$

$$(\beta\gamma')^{\nu'_3-\mu_3}(\gamma\gamma')^{\mu_3}, \quad (\beta\gamma')^{\nu_2-\mu_2}(\beta\beta')^{\mu_2}, \quad (\gamma\beta')^{\nu'_2-\mu_2}(\beta\beta')^{\mu_2}.$$

The induction may now be established as before.

15. Let each of the letters a, b, c, d represent a covariant of degree 2. Then the process of proof is the same as before, the covariants on the right being of the same kind as the covariants on the right in the case of § 14, or else containing one of the factors

$$(\beta\alpha')^{\nu'_1-\mu_1}(\alpha\alpha')^{\mu_1}, \quad (\beta\alpha')^{\nu_2-\mu_2}(\beta\beta')^{\mu_2}, \quad (\alpha\beta')^{\nu'_2-\mu_2}(\alpha\alpha')^{\mu_2}.$$

16. Finally it may be shown in the same way that, whatever covariants a, b, c, d may represent, the theorem is true.

For, if c is a covariant, it may always be written as a sum of terms of the second class and of covariants of the form

$$(\gamma\gamma_1)^{\lambda_1}(\gamma_1\gamma_2)^{\lambda_2}\dots(\gamma_k\gamma')^{\lambda_{k+1}},$$

the sequence of the letters being fixed, and possibly one or more of the indices zero.

The other covariants a, b, d being similarly expressed, the theorem is proved in exactly the same manner as the more elementary cases proved in §§ 8–15; the alterations required are practically only verbal.

17. It is necessary to extend the work in the paper on "Perpetuant Syzygies," step by step, to the case when the orders of the quantics concerned are finite. The generating functions found for perpetuant products cannot be true for products of covariants of forms of finite order, for the corresponding generating functions when expanded cannot go to infinity; all reference to the generating functions is then omitted.

Essentially the discussion of perpetuant products proceeded as ollows:—

(i.) All products were arranged in a fixed sequence, defined in Section III., § 5; and a product was defined as reducible if it could be expressed in terms of products which came after it in the fixed sequence.

(ii.) The discussion of products $(ab)^\lambda C_\kappa$ in Section III., §§ 6–10.

(iii.) The extension of the perpetuant type theorem, Section I.

(iv.) The application of (ii.) and (iii.) to any product $C_\kappa C_\lambda$ in Section VI.

The rest of the paper consisted of applications of these results to particular forms of products.

When the orders of the fundamental quantics are finite, there is nothing to prevent our taking the products in the sequence defined in (i.). A product will then be called reducible if it can be expressed in terms of products which come after it and of covariants of the second class.

(ii.) The introduction of differential operators in § 6, when writing Stroh's syzygy, may be made in the same way when the orders are finite; it being understood that symbolical products which have not a proper meaning (owing to a letter occurring to too great a degree) must be omitted; and that forms of the second class are neglected. It is useful to remark that, if P be a symbolical product of degree n_4 in a_4, we have a syzygy

$$\sigma^{(a_1 a_2) A_4} A_4^{-\rho} P = R,$$

in which the first ρ terms must be omitted.

The argument of §§ 7, 8, 9 simply concerns a set of linear equations, and holds good when the orders are finite, all non-interpretable forms arising from Stroh's syzygies being kept until the argument is finished. In the same way § 10 holds good.

18. (iii.) The method of proof of the extension of the perpetuant type theorem was identical with Grace's proof of the original theorem.* To prove the corresponding theorem for forms of finite order, a method identical with that used to extend the original theorem to forms of finite order† may be employed. But, as no new point arises in the course of proof, it is thought unnecessary to reproduce it.

(iv.) Having established the results (ii.) and (iii.) for forms of finite order, the argument of Section VI. for products $C_\kappa C_\lambda$ follows as well.

Thus we arrive at the conclusion that all products which would be reducible as products of perpetuants are reducible as products of covariants of forms of finite order.

19. It is possible, at least in a few of the simplest cases, to calculate generating functions for irreducible products, when the orders are finite.

Thus the generating function for forms $C_2 = (a_1 a_2)^\lambda$ $(n_1 \not> n_2)$ is $\dfrac{x - x^{n_1}}{1-x}$, the form $(a_1 a_2)^{n_1}$ being of the second class.

* *Proc. London Math. Soc.*, Vol. xxxv.

† A. Young, *Proc. London Math. Soc.*, Ser. 2, Vol. 1.

Consider next the forms C_2^2:—When $\lambda > n_2-2$ and $n_2 \not> n_3$ the first term of the syzygy

$$\Sigma \binom{\lambda}{r} (a_1 a_4)^r (a_2 a_3)^{\lambda-r} = R$$

which is interpretable is

$$\binom{\lambda}{\lambda-n_2} (a_1 a_4)^{\lambda-n_2} (a_2 a_3)^{n_2},$$

and this is of the second class. Thus we have a reduction for $(a_1 a_4)^{\lambda-n_2+1} (a_2 a_3)^{n_2-1}$. And from the syzygy

$$\Sigma \binom{\lambda}{r} (a_1 a_4)^r (a_3 a_2)^{\lambda-r} = R$$

we have also a reduction for $(a_1 a_4)^{\lambda-n_2+2} (a_2 a_3)^{n_2-2}$.

Hence, for all values of λ,

$$(a_1 a_4)^\lambda (a_2 a_3)^\mu = R$$

when $\mu > n_2-3$.

Similarly $(a_1 a_3)^\lambda (a_2 a_4)^\mu = R$, $n_2 \not> n_4$ when $\mu > n_2-2$.

Consider the generating function for $(a_1 a_2)^\lambda (a_3 a_4)^\mu$, $n_1 \not> n_2$, $n_3 \not> n_4$. We first have to include all terms having a factor $(a_1 a_2)^2 (a_3 a_4)^2$, then to exclude those which have a factor $(a_1 a_2)^2 (a_3 a_4)^{n_3}$, then to exclude those which have a factor $(a_1 a_2)^{n_1} (a_3 a_4)^2$, and finally to include those which have a factor $(a_1 a_2)^{n_1} (a_3 a_4)^{n_3}$ and have been excluded twice.

Thus the generating function is

$$\frac{x^4}{(1-x)^2} - \frac{x^{n_3+2}}{(1-x)^2} - \frac{x^{n_1+2}}{(1-x)^2} + \frac{x^{n_1+n_3}}{(1-x)^2} = \frac{x^4(1-x^{n_1-2})(1-x^{n_3-2})}{(1-x)^2}.$$

Assuming that n_1, n_2, n_3, n_4 are in ascending order of magnitude, we see that the generating function for $(a_1 a_3)^\lambda (a_2 a_4)^\mu$ is

$$\frac{x^4(1-x^{n_1-3})(1-x^{n_2-2})}{(1-x)^2}$$

and that that for $(a_1 a_4)^\lambda (a_2 a_3)^\mu$ is

$$\frac{x^5(1-x^{n_1-4})(1-x^{n_2-3})}{(1-x)^2}.$$

Hence the generating function for all products C_2^2 is

$$\frac{x^4(1-x^{n_1-2})(1-x^{n_3-2})}{(1-x)^2} + \frac{x^5(1-x^{n_1-3})(1-x^{n_2-3})}{(1-x)^2} + \frac{x^6(1-x^{n_1-4})(1-x^{n_2-4})}{(1-x)^2}.$$

The generating functions for products C_2^m may be calculated in the same way, *e.g.*, that for $(a_1a_3)^\lambda (a_2a_6)^\mu (a_4a_5)^\nu$ is

$$\frac{x^9(1-x^{n_1-3})(1-x^{n_2-5})(1-x^{n_4-4})}{(1-x)^3},$$

for this product is always reducible if $\lambda \geqslant n_1$, if $\mu \geqslant n_2-1$, or if $\nu \geqslant n_4-2$.

20. A covariant of degree 3 can be expressed in terms of members of the second class if its weight is greater than the order of any one of the quantics concerned. The generating function of covariants C_3, the quantics concerned being $a_{1_x}^{n_1}$, $a_{2_x}^{n_2}$, $a_{3_x}^{n_3}$, $n_1 \not> n_2 \not> n_3$, is

$$\frac{x^3-(n_1-2)x^{n_1}+(n_1-3)x^{n_1+1}}{(1-x)^2}.$$

Consider products C_2C_3, in particular the set

$$(a_1a_2)^\nu (a_3a_4)^\lambda (a_3a_5)^\mu,$$

where n_1, n_2, n_3, n_4, n_5 are in ascending order of magnitude. The six syzygies

$$e^{\pm(a_1a_2)A_3} A_3^{-\rho} (a_3a_4)^\lambda (a_3a_5)^{n_3-\lambda} = R,$$

$$e^{\pm(a_1a_2)A_4} A_4^{-\rho} (a_3a_4)^\lambda (a_3a_5)^{n_3-\lambda} = R,$$

$$e^{\pm(a_1a_2)A_5} A_5^{-\rho} (a_3a_4)^\lambda (a_3a_5)^{n_3-\lambda} = R$$

reduce all forms $$(a_1a_2)^{\rho+\sigma} (a_3a_4)^{\lambda-\sigma} (a_3a_5)^{n_3-\lambda}$$

where $\sigma = 1, 2, 3, 4, 5, 6$. The argument for the linear independence of these syzygies is identical with that for the independence of perpetuant syzygies. We thus see that all products

$$(a_1a_2)^\nu (a_3a_4)^\lambda (a_3a_5)^\mu$$

are reducible for which $\lambda+\mu > n_3-7$.

The products when the letters are arranged in any one of the other possible manners may be treated in the same way. In § 12 (p. 240) of the paper on "Perpetuant Syzygies" a table is given of the limits of the indices ν, λ, μ for irreducibility in the various cases. The general result here is that, if the product is reducible when $\nu \leqslant \nu_1$, then it is also reducible when $\lambda+\mu > n-\nu_1$, where n is the order of the quantic of lowest order that occurs in C_3.

To prove this fact, it is merely necessary to remark that the syzygies which give the reductions in the first case are the same (except for their weight) as those which give the reductions in the latter case.

21. The main theorem of this paper applied to a single quantic shews that its covariants may be treated as perpetuants so far as the known results for perpetuants as regards reducibility or syzygies are concerned, provided that forms having a factor $(ab)^{\frac{1}{2}n}$ are neglected. The method by which this result has been arrived at is of such a general nature that it would appear almost certain that when the, as yet unknown, syzygies of degree 9 and of greater degree are discovered these also may be extended to covariants of forms of finite order by the same process.

III. *On Relations among Perpetuants.* By A. Young

[*Received* August 2, 1904.]

Any covariant type of a system of quantics of infinite order can be expressed in terms of covariants of the form

$$(a_1a_2)^{\lambda_1}(a_1a_3)^{\lambda_2}\dots(a_1a_\delta)^{\lambda_{\delta-1}}*, \qquad \text{(i)}$$

or of the form

$$(a_1a_2)^{\lambda_1}(a_2a_3)^{\lambda_2}\dots(a_{\delta-1}a_\delta)^{\lambda_{\delta-1}}, \qquad \text{(ii)}$$

the sequence of the letters being fixed beforehand.

All such forms are linearly independent, for it is evident that no linear algebraical relation can exist between symbolical products of the form (i) when the sequence of letters is fixed.

The conditions for irreducibility of either (i) or (ii) are

$$\lambda_1 \nless 2^{\delta-2},\ \lambda_2 \nless 2^{\delta-3},\ \dots,\ \lambda_{\delta-1} \nless 1\dagger.$$

Again if all the letters are interchangeable, the conditions of irreducibility become

$$\left.\begin{aligned}
\lambda_{\delta-1} &= 1+\xi_{\delta-1}\\
\lambda_{\delta-2} &= 2+\xi_{\delta-2}+\xi_{\delta-1}\\
&\dots\dots\dots\dots\\
\lambda_r &= 2^{\delta-r-1}+\xi_r+\xi_{r+1}+\dots+\xi_{\delta-1}\\
&\dots\dots\dots\dots\\
\lambda_2 &= 2^{\delta-3}+\xi_2+\xi_3+\dots+\xi_{\delta-1}\\
\lambda_1 &= 2^{\delta-2}+2(\xi_2+\xi_3+\dots+\xi_{\delta-1})+\xi_1
\end{aligned}\right\}, \qquad \text{(iii)}$$

where the ξ's are positive integers or zeros‡.

It has been pointed out that this result, which was proved originally for perpetuants belonging to a single quantic (in which case ξ_1 must be even), also gives the conditions for perpetuant types when these are expressed in terms of products of either of the forms (i) and (ii), the sequence of the letters not being fixed§.

* In writing down symbolical products we shall omit factors of the form $a_{1x}, a_{2x}, \dots$.

† Grace, *Proc. Lond. Math. Soc.*, Vol. xxxv., p. 107.

‡ *Ibid.* p. 319.

§ Grace and Young, *Algebra of Invariants*, p. 379.

The exact number of perpetuant types of degree δ and weight w is known to be

$$\binom{w - 2^{\delta-1} + 1 + \delta - 2}{\delta - 2}.$$

But when the sequence of the letters is not fixed it will be found that the conditions (iii) give too many perpetuant types; it is the first object of this paper to determine what are the relations among these forms.

It will be convenient to make use of the notation of the theory of substitutions; to avoid confusion symbols which refer to substitutions are printed in Roman type.

The symbol $\{\mathrm{a\,b} \ldots \mathrm{k}\}$ is used to denote the sum of the substitutions of the symmetric group of the letters $a, b, \ldots k$.

The symbol $\{\mathrm{a\,b} \ldots \mathrm{k}\}'$ denotes the sum of the substitutions of the alternating group of the letters $a, b, \ldots k$ minus the sum of the substitutions of these letters which do not belong to the alternating group.

In the last part of the paper the reducibility of certain transvectants is deduced from the results obtained.

1. Consider a perpetuant of degree δ

$$P \equiv (a_1 a_2)^{\lambda_1} (a_2 a_3)^{\lambda_2} \ldots (a_{\delta-1} a_\delta)^{\lambda_{\delta-1}}.$$

All perpetuants such as P will be supposed arranged in order according to the indices of the different factors, the sequence of the letters not being fixed. Thus if

$$Q \equiv (b_1 b_2)^{\mu_1} (b_2 b_3)^{\mu_2} \ldots (b_{\delta-1} b_\delta)^{\mu_{\delta-1}}$$

where $b_1, b_2, \ldots b_\delta$ are the letters $a_1, a_2, \ldots a_\delta$ arranged in some order; then Q will precede P provided that the first of the differences

$$\mu_1 - \lambda_1,\ \mu_2 - \lambda_2,\ \ldots \mu_{\delta-1} - \lambda_{\delta-1}$$

which does not vanish is positive.

If all these differences are zero, P and Q belong to the same set, and take the same position in our arrangement.

To express that the sum of certain forms like P is equal to a linear function of perpetuants which precede them in the chosen arrangement and of products of forms of lower degree, it will be convenient to write

$$\Sigma P = R.$$

The symbol R is throughout used in this sense. Thus unless the indices λ satisfy the conditions (iii) we have

$$P = R.$$

2. *Covariants of degree three.*

All perpetuants of degree three can be expressed in terms of those of the form

$$(a_1 a_2)^\lambda (a_2 a_3)^\mu, \quad \lambda \not< 2\mu.$$

If $\lambda = 2\mu$,

$$(a_1 a_2)^{2\mu} (a_2 a_3)^{\mu} - (a_1 a_3)^{2\mu} (a_3 a_2)^{\mu} = R.$$

If $\lambda = 2\mu + 1$,

$$(a_1 a_2)^{2\mu+1} (a_2 a_3)^{\mu} + (a_2 a_3)^{2\mu+1} (a_3 a_1)^{\mu} + (a_3 a_1)^{2\mu+1} (a_1 a_2)^{\mu} = R.$$

These facts are well known. They may be deduced at once from Stroh's series*.

The relations may be written

$$(a_1 a_2)^{2\mu} (a_2 a_3)^{\mu} = \tfrac{1}{6} \{\mathrm{a}_1 \mathrm{a}_2 \mathrm{a}_3\} (a_1 a_2)^{2\mu} (a_2 a_3)^{\mu} + R,$$

$$\{\mathrm{a}_1 \mathrm{a}_2 \mathrm{a}_3\}' (a_1 a_2)^{2\mu+1} (a_2 a_3)^{\mu} = R.$$

When $\lambda > 2\mu + 1$, we have three independent forms $(a_1 a_2)^{\lambda} (a_2 a_3)^{\mu}$.

Hence the number of perpetuants of degree three and weight $w = 3k + 2$ is $3k$; for we may take $\mu = 1, 2, \ldots k$.

The number of perpetuants of weight $w = 3k + 1$ is $3(k-1) + 2 = 3k - 1$. And the number when $w = 3k$ is $3(k-1) + 1 = 3k - 2$.

In every case the number is $w - 2$; and this is known otherwise to be the exact number of perpetuants of degree three and weight w. Hence there can be no relations between these perpetuants other than those just enumerated.

3. Let $$P \equiv (a_1 a_2)^{\lambda_1} (a_2 a_3)^{\lambda_2} \ldots (a_{\delta-1} a_\delta)^{\lambda_{\delta-1}}$$

be a perpetuant of degree δ, whose indices λ satisfy the conditions (iii): we proceed to prove that if $\xi_{r-1} = 0\ (r > 2)$, then

$$\{\mathrm{a}_r \mathrm{a}_{r+1}\}' P = R.$$

Let the symbol α refer to the perpetuant

$$(a_1 a_2)^{\lambda_1} (a_2 a_3)^{\lambda_2} \ldots (a_{r-2} a_{r-1})^{\lambda_{r-2}}$$

when considered as a single binary form of infinite order. Then

$$(\alpha a_r)^{\lambda_{r-1}} (a_r a_{r+1})^{\lambda_r} \ldots (a_{\delta-1} a_\delta)^{\lambda_{\delta-1}} - P = R.$$

Now $\lambda_{r-1} = 2^{\delta-r-1} + \lambda_r$, since $\xi_{r-1} = 0$, hence by Stroh's series

$$\left\{ - \sum_{=0}^{\lambda_r} \binom{\lambda_{r-1} + \lambda_r}{i} \binom{2\lambda_r - i}{\lambda_r} (\alpha a_r)^{\lambda_{r-1} + \lambda_r - i} (a_r a_{r+1})^i \right.$$

$$+ (-)^{\lambda_r} \sum_{i=0}^{2^{\delta-r-1}-1} \binom{\lambda_{r-1} + \lambda_r}{i} \binom{\lambda_r + 2^{\delta-r-1} - 1 - i}{\lambda_r} (a_r a_{r+1})^{\lambda_{r-1} + \lambda_r - i} (a_{r+1} \alpha)^i$$

$$\left. + (-)^{\lambda_r} \sum_{i=0}^{\lambda_r} \binom{\lambda_{r-1} + \lambda_r}{i} \binom{\lambda_r + 2^{\delta-r-1} - 1 - i}{2^{\delta-r-1} - 1} (a_{r+1} \alpha)^{\lambda_r + \lambda_{r-1} - i} (\alpha a_r)^i \right\}$$

$$\times (a_{r+1} a_{r+2})^{\lambda_{r+1}} (a_{r+2} a_{r+3})^{\lambda_{r+2}} \ldots (a_{\delta-1} a_\delta)^{\lambda_{\delta-1}} = 0. \quad \ldots\ldots\ldots\ldots\ldots\ldots\ldots\ldots\text{(iv)}$$

But $$(a_r a_{r+1})^{\mu_1} (a_{r+1} \alpha)^{\mu_2} (a_{r+1} a_{r+2})^{\lambda_{r+1}} \ldots (a_{\delta-1} a_\delta)^{\lambda_{\delta-1}}$$

$$= (a_r a_{r+1})^{\mu_1} (a_{r+1} \alpha)^{\mu_2} (\alpha a_{r+2})^{\lambda_{r+1}} \ldots (a_{\delta-1} a_\delta)^{\lambda_{\delta-1}} + R;$$

and when $\mu_2 < 2^{\delta-r-1}$ the perpetuant on the right-hand side can be expressed in terms

* *Math. Ann.*, Bd. 31; *Algebra of Invariants*, p. 64.

of forms which contain a greater number of factors involving a_r, a_{r+1}, α only. Thus we see that all the terms of the second sum in (iv) may be included in the symbol R; in fact this relation becomes

$$\{-(\alpha a_r)^{\lambda_{r-1}}(a_r a_{r+1})^{\lambda_r} + (-)^{\lambda_r}(a_{r+1}\alpha)^{\lambda_{r-1}}(\alpha a_r)^{\lambda_r}\}(a_{r+1}a_{r+2})^{\lambda_{r+1}} \dots (a_{\delta-1}a_\delta)^{\lambda_{\delta-1}} = R.$$

Whence $$\{\mathrm{a}_r\mathrm{a}_{r+1}\}'(\alpha a_r)^{\lambda_{r-1}}(a_r a_{r+1})^{\lambda_r}(a_{r+1}a_{r+2})^{\lambda_{r+1}} \dots (a_{\delta-1}a_\delta)^{\lambda_{\delta-1}} = R;$$

and therefore $$\{\mathrm{a}_r\mathrm{a}_{r+1}\}' P = R.$$

4. When $\xi_1 = 0$, we have $\lambda_1 = 2\lambda_2$.

In § 2 we saw that $$\{\alpha\beta\}'(a_1 a_2)^{2\lambda_2}(a_2 a_3)^{\lambda_2} = R$$

where α, β are any two of the letters a_1, a_2, a_3.

Hence also $$\{\alpha\beta\}'(a_1 a_2)^{2\lambda_2}(a_2 a_3)^{\lambda_2}(a_3 a_4)^{\lambda_3} \dots (a_{\delta-1}a_\delta)^{\lambda_{\delta-1}} = R;$$

for $$(a_1 a_3)^{2\lambda_2}(a_3 a_2)^{\lambda_2}(a_3 a_4)^{\lambda_3} \dots (a_{\delta-1}a_\delta)^{\lambda_{\delta-1}} - (a_1 a_3)^{2\lambda_2}(a_3 a_2)^{\lambda_2}(a_2 a_4)^{\lambda_3} \dots (a_{\delta-1}a_\delta)^{\lambda_{\delta-1}} = R.$$

Therefore when $\xi_1 = 0$

$$\{\alpha\beta\}' P = R, \text{ where } \alpha, \beta \text{ are any two of } a_1, a_2, a_3.$$

Similarly from the fact that

$$\{\mathrm{a}_1\mathrm{a}_2\mathrm{a}_3\}'(a_1 a_2)^{2\lambda_2+1}(a_2 a_3)^{\lambda_2} = R,$$

we deduce that if $\xi_1 = 1$

$$\{\mathrm{a}_1\mathrm{a}_2\mathrm{a}_3\}' P = R.$$

The following relations have been obtained:

(a) $\xi_{r-1} = 0$, $(r > 2)$, $\{\mathrm{a}_r\mathrm{a}_{r+1}\}' P = R$.

(b) $\xi_1 = 0$, $\{\mathrm{a}_1\mathrm{a}_2\}' P = R$, $\{\mathrm{a}_2\mathrm{a}_3\}' P = R$.

(c) $\xi_1 = 1$, $\{\mathrm{a}_1\mathrm{a}_2\mathrm{a}_3\}' P = R$.

(d) ξ_1 even, $\{\mathrm{a}_1\mathrm{a}_2\}' P = R$,
ξ_1 odd, $\{\mathrm{a}_1\mathrm{a}_2\} P = R$.

It remains to shew that there are no more relations.

5. Assuming that the relations just enumerated are all that exist between the forms which satisfy the conditions (iii), we proceed to prove that the number of these forms which are linearly independent is $\binom{w - 2^{\delta-1} + 1 + \delta - 2}{\delta - 2}$; w being the weight and δ the degree.

Let $$\lambda_1 = 2^{\delta-2} + \mu_1, \quad \lambda_2 = 2^{\delta-3} + \mu_2, \dots \lambda_{\delta-1} = 1 + \mu_{\delta-1}.$$

The conditions (iii) become

$$\mu_2 \not< \mu_3 \not< \mu_4 \dots \not< \mu_{\delta-1}, \quad \mu_1 \not< 2\mu_2$$

together with the fact that the μ's are positive integers.

Consider first those forms for which

$$\mu_{r+1} = 0 = \mu_{r+2} = \ldots = \mu_{\delta-1},$$

and $\mu_1, \mu_2, \ldots \mu_r$ are all different from zero.

Let $$P = (a_1 a_2)^{\lambda_1} (a_2 a_3)^{\lambda_2} \ldots (a_{\delta-1} a_\delta)^{\lambda_{\delta-1}}$$

be one of these forms; then by § 3

$$\{ab\}' P = R,$$

when a, b are any two of the letters $a_{r+2}, a_{r+3}, \ldots a_\delta$.

The letters $a_1, a_2, \ldots a_{r+1}$ can be chosen in $\binom{\delta}{r+1}$ ways. When this set of letters has been selected, the number of forms corresponding to given values of $\mu_1, \mu_2, \ldots \mu_r$ depends (i) on what consecutive pairs of μ's are equal, and (ii) on whether $\mu_1 - 2\mu_2 = 0$, 1 or > 1. This number is then quite independent of δ, provided that $\delta \geqslant r+1$. Also the set of values which can be given to $(\mu_1, \mu_2, \ldots \mu_r)$ is independent of δ.

Hence if $\phi(r, \varpi)$ is the number of independent forms P of degree δ for which $\Sigma\mu = \varpi$, $\mu_r > 0$, and

$$\mu_{r+1} = \mu_{r+2} = \ldots = \mu_{\delta-1} = 0,$$

then $\phi(r, \varpi)$ is the number of independent forms of degree $r+1$, for which no μ is zero, and $\Sigma\mu = \varpi$.

Now if μ_r is not zero, the number of forms of degree $r+1$, corresponding to a given set of values $(\mu_1, \mu_2, \ldots \mu_r)$ of the μ's, is the same as the number of forms corresponding to the set of values $(\mu_1 - 2, \mu_2 - 1, \ldots \mu_r - 1)$. Hence $\phi(r, \varpi)$ is the total number of forms of degree $r+1$, for which $\Sigma\mu = \varpi - r - 1$.

Again, the total number of forms of degree δ is equal to the number of forms for which μ_1 is the last non-zero μ, together with the number of forms for which μ_2 is the last non-zero μ, and so on. Hence this number

$$= \phi(1, \varpi)\binom{\delta}{2} + \phi(2, \varpi)\binom{\delta}{3} + \ldots + \phi(\delta-2, \varpi)\binom{\delta}{\delta-1} + \phi(\delta-1, \varpi)\binom{\delta}{\delta}.$$

Now we shall assume that the total number of forms of degree $r < \delta$ and weight w is

$$\binom{w - 2^{r-1} + 1 + r - 2}{r-2} = \binom{\varpi' + r - 2}{r-2},$$

where $\varpi' = \Sigma\mu$: also that the number of forms of degree δ, for which $\Sigma\mu = \varpi'$, $\varpi' < \varpi$, is

$$\binom{\varpi' + \delta - 2}{\delta - 2}.$$

Then by hypothesis

$$\phi(r, \varpi') = \binom{\varpi' - 2}{r-1}$$

when $r+1 < \delta$, and when $r+1 = \delta$, but $\varpi' < \varpi + \delta$.

Hence the number of forms of degree δ, for which $\Sigma\mu = \varpi$, is

$$\binom{\delta}{2} + \binom{\varpi - 2}{1}\binom{\delta}{3} + \binom{\varpi - 2}{2}\binom{\delta}{4} + \dots + \binom{\varpi - 2}{\delta - 3}\binom{\delta}{\delta - 1} + \binom{\varpi - 2}{\delta - 2},$$

and this is the coefficient of x^{ϖ} in the expansion of $(1+x)^{\delta} \times (x+1)^{\varpi - 2}$.

Therefore the number required

$$= \binom{\varpi + \delta - 2}{\varpi} = \binom{\varpi + \delta - 2}{\delta - 2} = \binom{w - 2^{\delta - 1} + 1 + \delta - 2}{\delta - 2}.$$

Hence the number of independent forms of weight w and degree δ is

$$\binom{w - 2^{\delta-1} + 1 + \delta - 2}{\delta - 2},$$

provided that this is true for weight $< w$ and degree δ, and also for degree $< \delta$ and any weight.

Now if $w < 2^{\delta-1} + 1$, ϖ is negative, and there are no forms of weight w and degree δ; the formula is then true for weight $w < 2^{\delta-1} - 1$ and degree δ; hence it is true for degree δ and any weight if it is true for degree $< \delta$ and any weight. But it is evidently true for degree 2, hence it is always true.

Thus on the assumption that there are no relations other than those of § 4, we find that the number of independent irreducible forms is the same as the exact number of perpetuants. It follows that there can be no other relations between the forms considered.

6. In consequence of the relations of § 4

$$(a_1 a_2)^{2^{\delta-2}} (a_2 a_3)^{2^{\delta-3}} \dots (a_{\delta-1} a_\delta)$$

can be written in the form

$$\{\mathrm{a}_1 \mathrm{a}_2 \dots \mathrm{a}_\delta\} P + R,$$

where P is a numerical multiple of the above form, and R is a sum of products of perpetuants.

Again, if

$$Q = (a_1 a_2)^{2^{\delta-2}+1} (a_2 a_3)^{2^{\delta-3}} \dots (a_{\delta-2} a_{\delta-1})^2 (a_{\delta-1} a_\delta),$$

we have

$$\{\mathrm{a}_1 \mathrm{a}_2\} Q = R, \qquad \{\alpha\beta\}' Q = R,$$

$$\{\mathrm{a}_1 \mathrm{a}_2 \mathrm{a}_3\}' Q = R,$$

where α, β are any two of the letters $a_3, a_4, \dots a_\delta$.

Now using the substitutional series

$$1 = \Sigma A_{\alpha_1, \alpha_2, \dots \alpha_h} T_{\alpha_1, \alpha_2, \dots \alpha_h}\text{*},$$

in which the letters affected are $a_1, a_2, \dots a_\delta$; we have

$$Q = \Sigma A_{\alpha_1, \alpha_2, \dots \alpha_h} T_{\alpha_1, \alpha_2, \dots \alpha_h} Q.$$

* Young, *Proc. Lond. Math. Soc.*, Vol. XXXIII., p. 133, *et seq.*

But from the above equations

$$Q = \{a_1 a_2\}' \{a_3 a_4 \dots a_\delta\} \frac{1}{2 \cdot (\delta - 2)!} Q + R.$$

Hence every $T_{a_1, a_2, \dots a_h} Q$ is equal to R, except $T_{\delta-1, 1}$ and $T_{\delta-2, 1, 1}$.

Now $$T_{\delta-1, 1} \{a_1 a_2\}' \{a_3 a_4 \dots a_\delta\} = \{a_1 a_2\}' \{a_3 a_4 \dots a_\delta\} T_{\delta-1, 1}$$
$$= \lambda \{a_1 a_2\}' [\{a_1 a_3 a_4 \dots a_\delta\} + \{a_2 a_3 a_4 \dots a_\delta\}].$$

Therefore $$T_{\delta-1, 1} Q = \mu \{a_1 a_2\}' [\{a_1 a_3 a_4 \dots a_\delta\} + \{a_2 a_3 a_4 \dots a_\delta\}] Q + R$$
$$= \mu \{a_1 a_2\}' \{a_2 a_3 a_4 \dots a_\delta\} \{a_1 a_2\}' Q + R.$$

Again $$T_{\delta-2, 1, 1} Q = \frac{1}{2(\delta - 2)!} \{a_1 a_2\}' \{a_3 a_4 \dots a_\delta\} T_{\delta-2, 1, 1} Q + R$$
$$= \nu \{a_3 a_4 \dots a_\delta\} \{a_1 a_2 a_3\}' \{a_3 a_4 \dots a_\delta\} Q + R.$$

But $$\{a_1 a_2 a_3\}' Q = R,$$

therefore $$\{a_1 a_2 a_3\}' . \{a_3 a_4 \dots a_\delta\} Q = (\delta - 2)! \{a_1 a_2 a_3\}' Q + R = R;$$

and therefore $$T_{\delta-2, 1, 1} Q = R.$$

Hence $$Q = A \{a_1 a_2\}' \{a_2 a_3 a_4 \dots a_\delta\} Q + R,$$

where A is numerical; and R is a sum of products of perpetuants.

7. We proceed now to consider certain transvectants. The order of each of the quantics involved is supposed to be greater than the weight of the covariant in which it occurs; in this case theorems proved for perpetuants will be true.

(i) Consider first the transvectants

$$C \equiv ((a_1 a_2), a_{3_x}^{n_3})^\lambda.$$

If $\lambda = 1$, the covariant is of degree three and weight two, and hence must be reducible,—for the minimum weight of an irreducible perpetuant of degree three is three.

If $\lambda = 2$, and the orders of the quantics represented by a_1, a_2 are the same,

$$C = (a_1 a_2)^2 (a_2 a_3) + R.$$

Also $$\{a_1 a_2\} C = 0.$$

But by §§ 4, 6 $$\{a_1 a_2\} (a_1 a_2)^2 (a_2 a_3) = 2 (a_1 a_2)^2 (a_2 a_3) + R;$$

therefore in this case C is reducible.

(ii) Transvectants $$C \equiv ((a_1 a_2), (a_3 a_4))^\lambda$$

are reducible when $\lambda = 1, 2, 3, 4$, owing to the fact that the minimum weight of an irreducible form of degree four is seven.

Let us suppose that the quantics concerned are

$$a_{1_x}^{n_1}, \quad a_{2_x}^{n_2}, \quad a_{3_x}^{n_3}, \quad a_{4_x}^{n_4}.$$

Then if $n_1 = n_2$ $$\{a_1 a_2\} C = 0.$$

Hence, if $\lambda = 5$, we have by § 6

$$C = \tfrac{1}{2} \{a_1 a_2\} C + R = R.$$

Similarly, if $n_1 = n_2$ and $n_3 = n_4$,

$$\{a_1 a_2\} C = 0, \qquad \{a_3 a_4\} C = 0;$$

then if $\lambda = 6$

$$C = \Sigma A \{ab\}' \{bcd\} C + R = R$$

(where a, b, c, d are the letters a_1, a_2, a_3, a_4 in some order).

Again, if $n_1 = n_2 = n_3 = n_4$ and $\lambda = 7$,

$$\{a_1 a_2\} C = 0, \quad \{a_3 a_4\} C = 0, \quad [1 + (a_1 a_3)(a_2 a_4)] C = 0.$$

Also C can be expressed in terms of forms

$$(ab)^6 (bc)^2 (cd),$$

and of products of forms.

Hence
$$C = A_4 T_4 C + A_{3,1} T_{3,1} C + \Sigma A \{ab\}' \{cd\}' \{ac\} \{bd\} C + R.$$

But from the above equations

$$T_4 C = 0, \qquad T_{3,1} C = 0;$$

and we have only to consider expressions like

$$\{a_1 a_2\}' \{a_3 a_4\}' \{a_1 a_3\} \{a_2 a_4\} C = \tfrac{1}{2} \{a_1 a_2\}' \{a_3 a_4\}' \{a_1 a_3\} \{a_2 a_4\} [1 + (a_1 a_3)(a_2 a_4)] C = 0.$$

And hence C is reducible in this case.

(iii) Transvectants $\qquad C \equiv ((a_1 a_2), \quad (a_3 a_4)^2)^\lambda$.

If $\lambda = 1, 2, 3$, C is reducible owing to the fact that the weight is less than seven.

If $\lambda = 4$, and $n_1 = n_2$, C is reducible since

$$\{a_1 a_2\} C = 0.$$

(iv) Transvectants $\qquad C \equiv ((a_1 a_2)^2, \quad (a_3 a_4)^2)^\lambda$.

If $\lambda = 1, 2$, C is reducible owing to the fact that the weight is less than seven.

If $\lambda = 3$, and $n_1 = n_2 = n_3 = n_4$, C is reducible since

$$[1 + (a_1 a_3)(a_2 a_4)] C = 0.$$

(v) Transvectants $\qquad C \equiv ((a_1 a_2), \quad (a_3 a_4)^3)^\lambda$.

C is reducible if $\lambda = 1, 2, 3, 4$, and $n_1 = n_2$, $n_3 = n_4$.

(vi) Transvectants $\qquad C \equiv ((a_1 a_2)^2, \quad (a_3 a_4)^3)^\lambda$.

C is reducible if $\lambda = 1, 2$ and $n_3 = n_4$.

(vii) Transvectants $\qquad C \equiv ((a_1 a_2)^3, \quad (a_3 a_4)^3)^\lambda$.

C is reducible if $\lambda = 1, 2, 3$, and $n_1 = n_2 = n_3 = n_4$.

ON BINARY FORMS

By A. Young

[Read January 22nd, 1914.]

The object of this paper is to develop a method of attacking some of the problems in the theory of binary forms. Problems connected with the enumeration of complete systems are particularly in view.

Every method introduced requires some justification for its existence; its utility needs to be judged by results. In this case the method is at once applied to covariant types of degree four of the binary form of order n, and the complete irreducible set of these is obtained.

The preliminary analysis is concerned with the theory of perpetuants, and incidentally the complete system of perpetuant syzygies for every degree and weight is obtained. It appears that all perpetuant syzygies of the first kind can be obtained symbolically from those due to Stroh, and that consequently the extension to any degree of the work * of Mr. Wood and myself, for the first eight degrees, depends solely on accurate enumeration, and does not require the introduction of any new principle or the discovery of a different type of syzygy.

I. *Explanation of Method.*

1. We are concerned here entirely with the symbolical notation. Its introduction by Aronhold at once gave a method by which all covariants could be mathematically expressed. At the same time in the calculus it provides every form considered has the covariant property. But it has the drawback that a great many unnecessary forms appear in any discussion. Various methods have or can be suggested by which the forms considered may be limited to a linearly independent set. But such methods cannot avail much in most problems unless it is possible to express the product of two forms so expressed in terms of the corresponding forms.

---

* *Proc. London Math. Soc.*, Ser. 2, Vol. 2.

Grace,* in applying the symmetrical notation to MacMahon's theory of perpetuants, has succeeded in doing this for the case when the order of every quantic considered is infinite. In this case he selected one quantic $a_{1_x}^{\infty}$ for particular attention, introducing the symbol a_1 into every determinant factor, by means of the equation

$$(a_2 a_3)\, a_{1_x} = (a_1 a_3)\, a_{2_x} - (a_1 a_2)\, a_{3_x}.$$

Thus the only symbolical products he had to consider were of the form (omitting factors a_x)

$$(a_1 a_2)^{\lambda_2} (a_1 a_3)^{\lambda_3} \ldots (a_1 a_\delta)^{\lambda_\delta}.$$

These, when perpetuant *types* are under consideration, are all linearly independent. There are no superfluous forms.

Now, when we come to forms of finite order, we cannot, as a rule, apply this method as it stands, for the reason that there are not a sufficient number of factors a_{1_x} in order to be able to introduce the letter a_1 into every determinant factor. In fact, if we can do so, n_1, the order of the corresponding quantic, must be equal to or greater than the weight of the covariant considered.

Let w be the weight of the covariant C, then if we multiply C symbolically by $a_{1_x}^{w-n_1}$, we can express $a_{1_x}^{w-n_1} C$ in the form

$$\Sigma N (a_1 a_2)^{\lambda_2} (a_1 a_3)^{\lambda_3} \ldots (a_1 a_\delta)^{\lambda_\delta} a_{2_x}^{n_2-\lambda_2} a_{3_x}^{n_3-\lambda_3} \ldots a_{\delta_x}^{n_\delta-\lambda_\delta},$$

where N is numerical.

We have thus, as in the case of perpetuants, a linearly independent set of symbolical products

$$(a_1 a_2)^{\lambda_2} (a_1 a_3)^{\lambda_3} \ldots (a_1 a_\delta)^{\lambda_\delta}$$

to consider. But there is this difference: separate products do not represent actual covariants, but only certain linear functions of such products. We shall proceed to shew how every such product may be made to represent a covariant or else a form which we shall call a *fundamental form*.

After that we shall proceed to shew how products of covariants may be dealt with, as in the case of perpetuants.

2. Let us consider covariant types of degree δ; that is, covariants

* *Proc. London Math. Soc.*, Vol. xxxv, p. 107.

linear in the coefficients of each of the quantics

$$a_{1_x}^{n_1},\ a_{2_x}^{n_2},\ \ldots,\ a_{\delta_x}^{n_\delta}.$$

It is supposed, to start with, that these quantics are arranged in a fixed sequence.

Let us fix our attention on some covariant type expressed in the ordinary manner as a single symbolical product. We say that this covariant is a term of the continued transvectant

$$((\ldots((a_1 a_2)^{\lambda_2},\ a_3)^{\lambda_3},\ a_4)^{\lambda_4},\ \ldots,\ a_\delta)^{\lambda_\delta}$$

(using the single symbolical letter to denote the corresponding quantic). This statement is nearly obvious. An immediate proof is obtained by induction. Assume it true for degree δ; then, if C be a symbolical product representing a covariant of degree $\delta+1$, C is a term of a transvectant

$$(P,\ a_{\delta+1})^{\lambda_{\delta+1}},$$

and, since P is a symbolical product representing a covariant of degree δ, the theorem in question is true for P, and therefore it is also true for C.

Now the fact that every term of a transvectant differs from the whole transvectant, by a linear function of transvectants of lower index, leads us at once to the fact that any term of the continued transvectant

$$((\ldots((a_1 a_2)^{\lambda_2},\ a_3)^{\lambda_3},\ a_4)^{\lambda_4},\ \ldots,\ a_\delta)^{\lambda_\delta}$$

differs from the whole transvectant by a linear function of forms

$$((\ldots((a_1 a_2)^{\mu_2},\ a_3)^{\mu_3},\ a_4)^{\mu_4},\ \ldots,\ a_\delta)^{\mu_\delta},$$

which are such that the first of the differences

$$\lambda_\delta-\mu_\delta,\ \lambda_{\delta-1}-\mu_{\delta-1},\ \ldots,\ \lambda_2-\mu_2,$$

which does not vanish is positive.

We are then at liberty to express every covariant type of degree δ in terms of continued transvectants of the above form.

3. Let us now return to the consideration of a single symbolical product which represents a covariant type C of degree δ. Let the weight of C be w.

The symbolical product $a_{1_x}^{w-n}C$ can be expressed in the form

$$\Sigma N(a_1 a_2)^{\lambda_2}(a_1 a_3)^{\lambda_3}\ldots(a_1 a_\delta)^{\lambda_\delta},$$

where N is numerical: by repeated use of the equation

$$(a_r a_s)\, a_{1_x} = (a_1 a_s)\, a_{r_x} - (a_1 a_r)\, a_{s_x}.$$

We shall arrange the products in a definite sequence by saying that

$$(a_1 a_2)^{\lambda_2} (a_1 a_3)^{\lambda_3} \dots (a_1 a_\delta)^{\lambda_\delta}$$

precedes $$(a_1 a_2)^{\mu_2} (a_1 a_3)^{\mu_3} \dots (a_1 a_\delta)^{\mu_\delta},$$

provided that the first of the differences

$$\lambda_\delta - \mu_\delta,\ \lambda_{\delta-1} - \mu_{\delta-1},\ \dots,\ \lambda_2 - \mu_2,$$

which does not vanish is positive.

The continued transvectants will be supposed arranged in sequence according to the same law.

Now it is to be observed that a continued transvectant is defined by the same set of numbers $\lambda_2, \lambda_3, \dots, \lambda_\delta$, as a product

$$(a_1 a_2)^{\lambda_2} (a_1 a_3)^{\lambda_3} \dots (a_1 a_\delta)^{\lambda_\delta}.$$

If the continued transvectant be expressed as a sum of the products considered (by multiplying it by $a_{1_x}^{w-n_1}$), the first of the products in our sequence to appear will be that which is defined by the same numbers.

Now every continued transvectant represents a covariant type; but only certain linear functions of the products (viz., such as are divisible by $a_{1_x}^{w-n_1}$) represent actual covariants. The difference between the two cases being accounted for by the fact that there are certain limitations to be imposed on the indices of the transvectant; whilst the only limitations to the indices of the product are those expressed by the inequalities

$$\lambda_2 \ngtr n_2,\ \lambda_3 \ngtr n_3,\ \dots,\ \lambda_\delta \ngtr n_\delta.$$

These limitations are also necessary for the transvectant, but in addition we must have

(i) $\lambda_2 \ngtr n_1,\ 2\lambda_2 + \lambda_3 \ngtr n_1 + n_2,\ 2\lambda_2 + 2\lambda_3 + \lambda_4 \ngtr n_1 + n_2 + n_3,\ \dots,$

$$2\lambda_2 + 2\lambda_3 + \dots + 2\lambda_{\delta-1} + \lambda_\delta \ngtr n_1 + n_2 + n_3 + \dots + n_{\delta-1}.$$

In the case of products we shall use the term *fundamental forms* to denote products for which the set of inequalities (i) is not satisfied.

4. We proceed to shew that corresponding to every other product, that is to every product for which the inequalities (i) are satisfied, there is

a unique covariant which can be represented as a linear function of that product and of fundamental forms. We have seen that the transvectant

$$((\ldots((a_1 a_2)^{\lambda_2}, a_3)^{\lambda_3}, a_4)^{\lambda_4}, \ldots, a_\delta)^{\lambda_\delta}$$

can be expressed as a linear function of our products of which the first term is

$$(a_1 a_2)^{\lambda_2}(a_1 a_3)^{\lambda_3} \ldots (a_1 a_\delta)^{\lambda_\delta}.$$

Let
$$N(a_1 a_2)^{\mu_2}(a_1 a_3)^{\mu_3} \ldots (a_1 a_\delta)^{\mu_\delta}$$

be the next term in the order of our sequence to appear; if it is not a fundamental form we may subtract the covariant

$$N((\ldots((a_1 a_2)^{\mu_2}, a_3)^{\mu_3}, a_4)^{\mu_4}, \ldots, a_\delta)^{\mu_\delta}$$

from both sides of our equation.

Proceeding thus step by step, we arrive at the truth of the above statement. That the covariant is unique is evident from the fact that every covariant can be expressed in terms of the transvectants considered, and that these transvectants can be expressed in terms of the covariants found, and *vice versa*.

5. Let us use the notation

$$(\lambda_2, \lambda_3, \ldots, \lambda_\delta)$$

to denote the covariant corresponding to

$$(a_1 a_2)^{\lambda_2}(a_1 a_3)^{\lambda_3} \ldots (a_1 a_\delta)^{\lambda_\delta},$$

i.e. the covariant obtained from this product by the addition of a linear function of fundamental forms.

Then we have a set of linearly independent covariant types of degree δ in terms of which every such covariant type may be linearly expressed. And this set is composed of the forms

$$(\lambda_2, \lambda_3, \ldots, \lambda_\delta),$$

where
$$\lambda_2 \not> n_2, \; \lambda_3 \not> n_3, \; \ldots, \; \lambda_\delta \not> n_\delta,$$

and the λ's further satisfy conditions (i).

It will be convenient to have a notation for the covariant

$$(\lambda_2, \lambda_3, \ldots, \lambda_\delta),$$

in which the letters corresponding to the different quantics appear; we shall for this purpose use the notation

$$\left(\frac{a_2^{\lambda_2} a_3^{\lambda_3} \dots a_\delta^{\lambda_\delta}}{a_1}\right) \equiv (\lambda_2, \lambda_3, \dots, \lambda_\delta).$$

In order to discover what forms are reducible, or to find relations between products of forms, it is necessary to be able to express the product of any two of our forms as a linear function of the forms of a higher degree.

Thus, for example, the product

$$\left(\frac{a_2^{\lambda_2} a_3^{\lambda_3} \dots a_\delta^{\lambda_\delta}}{a_1}\right) (a_{\delta+1} a_{\delta+2})^{\lambda} = \Sigma (-)^i \binom{\lambda}{i} \left(\frac{a_2^{\lambda_2} a_3^{\lambda_3} \dots a_\delta^{\lambda_\delta} a_{\delta+1}^{i} a_{\delta+2}^{\lambda-i}}{a_1}\right).$$

The case of perpetuants is much simpler than that of forms of finite order, and the analysis in this case is a necessary preliminary to that of the more difficult case.

II. *Perpetuants.*

6. Grace proved that the perpetuants

$$(a_1 a_2)^{\lambda_2} (a_1 a_3)^{\lambda_3} \dots (a_1 a_\delta)^{\lambda_\delta}$$

can be expressed in terms of products of perpetuants and of forms of this kind for which

$$\lambda_2 \geqslant 2^{\delta-2}, \ \lambda_3 \geqslant 2^{\delta-3}, \ \dots, \ \lambda_\delta \geqslant 2^0.$$

This is the result. The method by which the result was obtained (by means of certain relations due to Stroh) is not the method we require here. We shall therefore proceed to establish the same result by a slightly different method for the sake of the analysis. The analysis will be capable of application to forms of finite order.

7. It is our aim at the outset to express every possible product of two forms as a linear function of forms

$$(a_1 a_2)^{\lambda_2} (a_1 a_3)^{\lambda_3} \dots (a_1 a_\delta)^{\lambda_\delta}.$$

In order to do this we must separate the letters $a_1, a_2, \dots, a_\delta$ into two sets. We may write them

$$a_1, \ a_{r_2}, \ \dots, \ a_{r_e},$$

$$a_{s_1}, \ a_{s_2}, \ \dots, \ a_{s_\eta}.$$

Then we consider the product of any covariant type of the one set by any covariant type of the other set.

The product to be considered is of the form

$$(a_1 a_{r_2})^{\lambda_{r_2}} (a_1 a_{r_3})^{\lambda_{r_3}} \dots (a_1 a_{r_\epsilon})^{\lambda_{r_\epsilon}} (a_{s_1} a_{s_2})^{\lambda_{s_2}} (a_{s_1} a_{s_3})^{\lambda_{s_3}} \dots (a_{s_1} a_{s_\eta})^{\lambda_{s_\eta}}$$

$$= \Sigma(-)^{i_2+i_3+\dots+i_\eta} \binom{\lambda_{s_2}}{i_2} \dots \binom{\lambda_{s_\eta}}{i_\eta} (a_1 a_{r_2})^{\lambda_{r_2}} \dots (a_1 a_{r_\epsilon})^{\lambda_{r_\epsilon}}$$

$$\times (a_1 a_{s_1})^{i_2+\dots+i_\eta} (a_1 a_{s_2})^{\lambda_{s_2}-i_2} \dots (a_1 a_{s_\eta})^{\lambda_{s_\eta}-i_{s_\eta}}$$

$$= e^{-(a_1 a_{s_1})\partial/[\partial(a_1 a_{s_2})] - (a_1 a_{s_1})\partial/[\partial(a_1 a_{s_3})] - \dots - (a_1 a_{s_1})\partial/[\partial(a_1 a_{s_\eta})]}$$

$$\times (a_1 a_{r_2})^{\lambda_{r_2}} \dots (a_1 a_{r_\epsilon})^{\lambda_{r_\epsilon}} (a_1 a_{s_2})^{\lambda_{s_2}} \dots (a_1 a_{s_\eta})^{\lambda_{s_\eta}}.$$

Let us suppose that $s_1 = 2$, and let us use the notation

$$D_s \equiv \frac{\partial}{\partial(a_1 a_s)}.$$

Then without fear of ambiguity we may write our result [replacing $(a_1 a_2)$ by a_2 in the exponential index]

$$e^{-a_2 D_{s_2} - a_2 D_{s_3} - \dots - a_2 D_{s_\eta}} (0, \lambda_3, \lambda_4, \dots, \lambda_\delta)$$

$$= \left(\frac{a_{r_2}^{\lambda_{r_2}} a_{r_3}^{\lambda_{r_3}} \dots a_{r_\epsilon}^{\lambda_{r_\epsilon}}}{a_1}\right) \left(\frac{a_{s_2}^{\lambda_{s_2}} a_{s_3}^{\lambda_{s_3}} \dots a_{s_\eta}^{\lambda_{s_\eta}}}{a_2}\right);$$

since $$(\lambda_2, \lambda_3, \dots, \lambda_\delta) \equiv (a_1 a_2)^{\lambda_2} (a_1 a_3)^{\lambda_3} \dots (a_1 a_\delta)^{\lambda_\delta}$$

for perpetuants.

8. We thus have a set of equations

$$e^{-a_2 D_{s_2} - a_2 D_{s_3} - \dots - a_2 D_{s_\eta}} (0, \lambda_3, \lambda_4, \dots, \lambda_\delta) = R$$

to consider, where $$s_2, s_3, \dots, s_\eta$$

are any, all or none of the numbers

$$3, 4, \dots, \delta.$$

Since each of the $\delta - 2$ numbers may be taken or left we obtain $2^{\delta-2}$ equations. We shall shew that the $2^{\delta-2}$ equations are, in general, independent and are just sufficient to express every form

$$(\lambda_2, \lambda_3, \dots, \lambda_\delta),$$

for which $\lambda_2 < 2^{\delta-2}$ in terms of similar forms for which $\lambda_2 \geqslant 2^{\delta-2}$ and of products of forms of lower order.

In order to prove this we must arrange our equations in a particular manner. We begin with the equation

$$(0,\ \lambda_3,\ \ldots,\ \lambda_\delta) = R,$$

representing the fact that this form has the quantic $a_{2_x}^{\infty}$ for a factor.

The next equation will be

$$e^{-a_2 D_\delta}(0,\ \lambda_3,\ \ldots,\ \lambda_\delta) = R,$$

or $(0,\ \lambda_3,\ \ldots,\ \lambda_\delta) - \lambda_\delta(1,\ \lambda_3,\ \ldots,\ \lambda_\delta - 1) + \binom{\lambda_\delta}{2}(2,\ \lambda_3,\ \ldots,\ \lambda_\delta - 2) - \ldots = R.$

This equation with the help of that already used reduces $(1,\ \lambda_3,\ \ldots,\ \lambda_\delta - 1)$; *i.e.*, it expresses this form in terms of earlier forms in the sequence and of products of forms.

We next consider

$$e^{-a_2 D_{\delta-1}}(0,\ \lambda_3,\ \ldots,\ \lambda_{\delta-1},\ \lambda_\delta) = R,$$

and it is easy to see that this reduces the form

$$(2,\ \lambda_3,\ \ldots,\ \lambda_{\delta-1} - 2,\ \lambda_\delta).$$

When we come to our next equation

$$e^{-a_2 D_{\delta-1} - a_2 D_\delta}(0,\ \lambda_3,\ \ldots,\ \lambda_{\delta-1},\ \lambda_\delta) = R,$$

it is necessary to take it in conjunction with the last. We have, on subtracting,

$$\begin{aligned}
&\left[e^{-a_2 D_{\delta-1} - a_2 D_\delta} - e^{-a_2 D_{\delta-1}}\right](0,\ \lambda_3,\ \ldots,\ \lambda_{\delta-1},\ \lambda_\delta) \\
&\quad = \lambda_\delta(1,\ \lambda_3,\ \ldots,\ \lambda_{\delta-1},\ \lambda_\delta - 1) - \lambda_\delta\lambda_{\delta-1}(2,\ \lambda_3,\ \ldots,\ \lambda_{\delta-1} - 1,\ \lambda_\delta - 1) \\
&\qquad + \lambda_\delta\binom{\lambda_{\delta-1}}{2}(3,\ \lambda_3,\ \ldots,\ \lambda_{\delta-1} - 2,\ \lambda_\delta - 1) - \ldots \\
&\qquad + \text{terms in which the last argument is less than } \lambda_\delta - 1 \\
&\quad = R.
\end{aligned}$$

Also

$$\begin{aligned}
&\left[e^{-a_2 D_{\delta-1}} - 1\right](0,\ \lambda_3,\ \ldots,\ \lambda_{\delta-1} + 1,\ \lambda_\delta - 1) \\
&\quad = -(\lambda_{\delta-1} + 1)(1,\ \lambda_3,\ \ldots,\ \lambda_{\delta-1},\ \lambda_\delta - 1) + \binom{\lambda_{\delta-1} + 1}{2}(2,\ \lambda_3,\ \ldots,\ \lambda_{\delta-1} - 1,\ \lambda_\delta - 1) \\
&\qquad - \binom{\lambda_{\delta-1} + 1}{3}(3,\ \lambda_3,\ \ldots,\ \lambda_{\delta-1} - 2,\ \lambda_\delta - 1) + \ldots \\
&\quad = R.
\end{aligned}$$

Using the results of our first two equations we may write these two equations

$$\lambda_{\delta-1}(2, \lambda_3, \ldots, \lambda_{\delta-1}-1, \lambda_\delta-1)-\binom{\lambda_{\delta-1}}{2}(3, \lambda_3, \ldots, \lambda_{\delta-1}-2, \lambda_\delta-1) = R,$$

$$\binom{\lambda_{\delta-1}+1}{2}(2, \lambda_3, \ldots, \lambda_{\delta-1}-1, \lambda_\delta-1)$$
$$-\binom{\lambda_{\delta-1}+1}{3}(3, \lambda_3, \ldots, \lambda_{\delta-1}-2, \lambda_\delta-1) = R.$$

These two equations are proved to be independent by calculating the determinant formed by the coefficients—its value is $\frac{1}{2}\lambda_{\delta-1}\binom{\lambda_{\delta-1}+1}{3}$.

Thus we can express

$$(2, \lambda_3, \ldots, \lambda_{\delta-1}, \lambda_\delta) \quad \text{and} \quad (3, \lambda_3, \ldots, \lambda_{\delta-1}, \lambda_\delta)$$

in terms of forms $\qquad (\mu_2, \lambda_3, \ldots, \lambda_{\delta-2}, \mu_{\delta-1}, \mu_\delta)$,

and of products of forms; where $\mu_2 \nless 4$ and the first of the differences

$$\mu_\delta-\lambda_\delta, \quad \mu_{\delta-1}-\lambda_{\delta-1}$$

which does not vanish is negative.

In general we shall consider the equation

$$e^{-a_2D_{r_1}-a_2D_{r_2}-\ldots-a_2D_{r_\epsilon}}(0, \lambda_3, \ldots, \lambda_\delta) = R \quad (r_1 < r_2 < \ldots < r_\epsilon)$$

before the equation

$$e^{-a_2D_{s_1}-a_2D_{s_2}-\ldots-a_2D_{s_\eta}}(0, \lambda_3, \ldots, \lambda_\delta) = R \quad (s_1 < s_2 < \ldots < s_\eta),$$

if $r_1 > s_1$.

If $r_1 = s_1$ we consider the two equations simultaneously. In fact, we have a set of $2^{\delta-r_1}$ simultaneous equations in which the first operator in the exponential index is D_{r_1}.

9. THEOREM.—*The* $2^{\delta-r}$ *equations*

$$e^{-a_2D_{s_1}-a_2D_{s_2}-\ldots-a_2D_{s_\eta}}(0, \lambda_3, \ldots, \lambda_\delta) = R,$$

where $s_1, s_2, \ldots, s_\eta$ *are all, any or none of the numbers* $r+1, r+2, \ldots, \delta$ *are just sufficient to express all forms*

$$(\lambda_2, \lambda_3, \ldots, \lambda_r, \lambda_{r+1}, \ldots, \lambda_\delta),$$

for which $\lambda_2 < 2^{\delta-r}$ *in terms of products of forms, and of forms*

$$(\mu_2, \lambda_3, \ldots, \lambda_r, \mu_{r+1}, \ldots, \mu_\delta),$$

where $\mu_2 \geqslant 2^{\delta-r}$, *and the first of the differences*

$$\lambda_\delta - \mu_\delta,\ \lambda_{\delta-1} - \mu_{\delta-1},\ \ldots,\ \lambda_{r+1} - \mu_{r+1},$$

which does not vanish is positive.

Let us assume the theorem to be true as it stands for a particular value of r. We proceed to show then that it is true when r is changed to $r-1$.

Consider the equations

$$e^{-a_2 D_r - a_2 D_{s_1} - a_2 D_{s_2} - \ldots - a_2 D_{s_\eta}} \quad (0, \lambda_3, \ldots, \lambda_\delta) = R,$$

for which $s_1, s_2, \ldots, s_\eta$ are all, any or none of the numbers

$$r+1,\ r+2,\ \ldots,\ \delta.$$

The equations may be written

$$e^{-a_2 D_{s_1} - a_2 D_{s_2} - \ldots - a_2 D_{s_\eta}} [e^{-a_2 D_r} (0, \lambda_3, \ldots, \lambda_\delta)] = R,$$

and when they are written in this way they are identical in form with the set of equations for which we have just assumed our theorem true. Hence, on making use of the assumption, we find that

$$e^{-a_2 D_r} (\lambda_2, \lambda_3, \ldots, \lambda_r, \lambda_{r+1}, \ldots, \lambda_\delta) = R,$$

if $\lambda_2 < 2^{\delta-r}$; and that the symbol R here stands for products of forms and numerical multiples of forms

$$(\mu_2, \lambda_3, \ldots, \lambda_r, \mu_{r+1}, \ldots, \mu_\delta),$$

where $\mu_2 \geqslant 2^{\delta-r}$ and the first of the differences

$$\lambda_\delta - \mu_\delta,\ \lambda_{\delta-1} - \mu_{\delta-1}, \ldots,\ \lambda_{r+1} - \mu_{r+1}$$

which does not vanish is positive.

We thus have $2^{\delta-r}$ equations to consider of a simplified form, in which the covariants we consider differ only in the arguments λ_2 and λ_r, the general equation of the set being

$$\Sigma(-)^r \binom{\lambda_r}{\xi} (\lambda_2 + \xi, \lambda_3, \ldots, \lambda_{r-1}, \lambda_r - \xi, \lambda_{r+1}, \ldots, \lambda_\delta) = R.$$

Using our assumption again we see that we have a reduction for all those terms for which $\lambda_2 + \xi < 2^{\delta-r}$, and, in fact, we may suppose that these reductions are inserted, taken over to the other side of the equation, and included in the general symbol R. Taking then the first

$2^{\delta-r}$ terms of each of our equations, we have a set of $2^{\delta-r}$ linear equations to solve for the $2^{\delta-r}$ variables

$$(2^{\delta-r}+\xi,\ \lambda_3,\ \ldots,\ \lambda_{r-1},\ \lambda_r-\xi-2^{\delta-r},\ \lambda_{r+1},\ \ldots,\ \lambda_\delta)\quad (\xi=0,\ 1,\ \ldots,\ 2^{\delta-r}-1).$$

If the determinant formed by the coefficients of these $2^{\delta-r}$ variables in the several equations is not zero, then the equations give a reduction for every one of these covariants.

The determinant in question is

$$\begin{vmatrix} \binom{\lambda_r}{2^{\delta-r}} & \binom{\lambda_r}{2^{\delta-r}+1} & \cdots & \binom{\lambda_r}{2^{\delta-r+1}-1} \\ \binom{\lambda_r-1}{2^{\delta-r}-1} & \binom{\lambda_r-1}{2^{\delta-r}} & \cdots & \binom{\lambda_r-1}{2^{\delta-r+1}-2} \\ \cdots & \cdots & \cdots & \cdots \\ \binom{\lambda_r-m}{2^{\delta-r}-m} & \binom{\lambda_r-m}{2^{\delta-r}+1-m} & \cdots & \binom{\lambda_r-m}{2^{\delta-r+1}-1-m} \\ \cdots & \cdots & \cdots & \cdots \\ \binom{\lambda_r-2^{\delta-r}+1}{1} & \binom{\lambda_r-2^{\delta-r}+1}{2} & \cdots & \binom{\lambda_r-2^{\delta-r}+1}{2^{\delta-r}} \end{vmatrix}$$

$$=\frac{\lambda_r!\,(\lambda_r-1)!\,\ldots\,(\lambda_r-2^{\delta-r}+1)!}{(\lambda_r-2^{\delta-r})!\,(\lambda_r-2^{\delta-r}-1)!\,\ldots\,(\lambda_r-2^{\delta-r+1}+1)!}$$

$$\times\frac{1!\,2!\,\ldots\,(2^{\delta-r}-1)!}{2^{\delta-r}!\,(2^{\delta-r}+1)!\,\ldots\,(2^{\delta-r+1}-1)!}$$

$$\times\begin{vmatrix} 1 & 1 & \cdots & 1 \\ \binom{2^{\delta-r}}{1} & \binom{2^{\delta-r}+1}{1} & \cdots & \binom{2^{\delta-r+1}-1}{1} \\ \cdots & \cdots & \cdots & \cdots \\ \binom{2^{\delta-r}}{m} & \binom{2^{\delta-r}+1}{m} & \cdots & \binom{2^{\delta-r+1}-1}{m} \\ \cdots & \cdots & \cdots & \cdots \\ \binom{2^{\delta-r}}{2^{\delta-r}-1} & \binom{2^{\delta-r}+1}{2^{\delta-r}-1} & \cdots & \binom{2^{\delta-r+1}-1}{2^{\delta-r}-1} \end{vmatrix}$$

$$=\frac{\binom{\lambda_r}{2^{\delta-r}}\binom{\lambda_r}{2^{\delta-r}+1}\cdots\binom{\lambda_r}{2^{\delta-r+1}-1}}{\binom{\lambda_r}{1}\binom{\lambda_r}{2}\cdots\binom{\lambda_r}{2^{\delta-r}-1}}.$$

This is not zero unless $\lambda_r < 2^{\delta-r+1}-1$; but in this case our equations only involve $\lambda_r - 2^{\delta-r}+1$ *variables* of the form

$$(2^{\delta-r}+\xi,\ \lambda_3,\ \ldots,\ \lambda_{r-1},\ \lambda_r-\xi-2^{\delta-r},\ \lambda_{r+1},\ \ldots,\ \lambda_\delta),$$

i.e., those for which ξ has the values 0, 1, 2, ..., $\lambda_r-2^{\delta-r}$. (If $\lambda_r < 2^{\delta-r}$ none of these forms occur.)

To solve our equations for these, we take the first $\lambda_r-2^{\delta-r}+1$ equations and calculate the determinant formed by the coefficients. Its value, obtained as above, is

$$\frac{\binom{\lambda_r}{2^{\delta-r}}\binom{\lambda_r}{2^{\delta-r}+1}\cdots\binom{\lambda_r}{\lambda_r}}{\binom{\lambda_r}{1}\binom{\lambda_r}{2}\cdots\binom{\lambda_r}{\lambda_r-2^{\delta-r}}}.$$

Thus in any case the solution of our equations gives

$$(2^{\delta-r}+\xi,\ \lambda_3,\ \ldots,\ \lambda_{r-1},\ \lambda_r-\xi-2^{\delta-r},\ \lambda_{r+1},\ \ldots,\ \lambda_\delta) = R,$$

when $$\xi < 2^{\delta-r} \quad\text{and}\quad \lambda_r \nless \xi+2^{\delta-r}.$$

The terms included in the symbol R are either products or forms

$$(\mu_2,\ \lambda_3,\ \ldots,\ \lambda_{r-1},\ \mu_r,\ \mu_{r+1},\ \ldots,\ \mu_\delta),$$

which occur later in our sequence than the term on the left, and for which $\mu_2 \nless 2^{\delta-r}$. By repeated application of this result to all terms on the right for which $\mu_2 < 2^{\delta-r+1}$ we find that we may restrict μ_2 to be equal to or greater than $2^{\delta-r+1}$.

Thus, if the theorem is true for any particular value of r, it is true for $r-1$; but we have seen that it is true when $r=\delta$ or $r=\delta-1$. Hence it is true in general.

In particular we deduce that the form $(\lambda_2, \lambda_3, \ldots, \lambda_\delta)$ can be expressed in terms of products of later forms in the sequence when $\lambda_2 < 2^{\delta-2}$.

10. The equations

$$e^{-a_2D_{s_1}-a_2D_{s_2}-\ldots-a_2D_{s_\eta}}(0,\ \lambda_3,\ \ldots,\ \lambda_\delta) = R$$

result in establishing reductions which depend solely on the value of λ_2.

We have another set of equations

$$e^{-a_rD_{t_1}-a_rD_{t_2}-\ldots-a_rD_{t_\zeta}}(\lambda_2,\ \lambda_3,\ \ldots,\ \lambda_{r-1},\ 0,\ \lambda_{r+1},\ \ldots,\ \lambda_\delta) = R,$$

where $$t_\zeta > t_{\zeta-1} > \ldots t_2 > t_1 > r;$$

which establish reductions dependent on the value of λ_r. They give a reduction when $\lambda_r < 2^{\delta-r}$.

We shall consider all our equations in regular sequence, and those equations which affect the value of λ_r will be considered *before* those which affect the value of λ_s when $r > s$.

Thus, when we examine any form

$$(\lambda_2, \lambda_3, \ldots, \lambda_\delta),$$

we may find that it is reducible because $\lambda_r < 2^{\delta-r}$ and also because $\lambda_s < 2^{\delta-s}$. Then, if $r > s$, we shall suppose that the form is reduced by the λ_r equations; it is then necessary for a complete discussion of these forms to discover what the λ_s equations may mean. In the case of perpetuants we know from the well known facts of the subject that these λ_s equations cannot introduce any new reductions, for all reducible forms have been reduced, and that therefore they must lead to syzygies. But, so far as the present investigation has gone, it might happen that they lead to new reductions. Indeed, in the case of forms of finite order the discussion may be carried on on precisely similar lines, and then it will frequently be found that these λ_s equations lead to new reductions and not to syzygies. We have shewn (§ 7) that every possible product of perpetuants of total degree s can be expressed in the form

$$e^{-a_r D_{s_1} - a_r D_{s_2} - \ldots - a_r D_{s_\eta}} (\lambda_2, \lambda_3, \ldots, \lambda_{r-1}, 0, \lambda_{r+1}, \ldots, \lambda_\delta),$$

where $$r < s_1 < s_2 < \ldots < s_\eta.$$

Hence a complete discussion of our equations involves not only a complete discussion of the question of reducibility, but also of that of syzygies as well.

We shall proceed to prove the following theorem:

The equation

$$e^{-a_r D_{s_1} - a_r D_{s_2} - \ldots - a_r D_{s_\eta}} (\lambda_2, \ldots, \lambda_{r-1}, 0, \lambda_{r+1}, \ldots, \lambda_\delta) = R,$$

where $$r < s_1 < s_2 < \ldots < s_\eta,$$

reduces to a syzygy when $\lambda_\sigma < 2^{\delta-\sigma+1}$, *or when* $\lambda_\tau < 2^{\delta-\tau}$, *where* σ *is any one of the numbers* $s_1, s_2, \ldots, s_\eta$, *and* τ *is one of the numbers* $r+1, r+2, \ldots, \delta$, *which is not included in the set* $s_1, s_2, \ldots, s_\eta$.

11. Let us first consider the equation ($r < s$)

$$\begin{aligned} e^{-a_r D_s} (\lambda_2, \lambda_3, \ldots, \lambda_{r-1}, 0, \lambda_{r+1}, \ldots, \lambda_\delta) &= R \\ &= (a_r a_s)^{\lambda_s} (a_1 a_2)^{\lambda_2} (a_1 a_3)^{\lambda_3} \ldots (a_1 a_{r-1})^{\lambda_{r-1}} (a_1 a_{r+1})^{\lambda_{r+1}} \ldots \\ &\qquad (a_1 a_{s-1})^{\lambda_{s-1}} (a_1 a_{s+1})^{\lambda_{s+1}} \ldots (a_1 a_\delta)^{\lambda} \\ &\equiv [\lambda_2, \lambda_3, \ldots, \lambda_{r-1}, 0, \lambda_{r+1}, \ldots, \lambda_{s-1}, \underline{\lambda_s}, \lambda_{s+1}, \ldots, \lambda_\delta], \text{ say.} \end{aligned}$$

Consider the identity

$$(a_1 a_2)^{\lambda_2} (a_1 a_3)^{\lambda_3} \ldots (a_1 a_{r-1})^{\lambda_{r-1}} \left\{ (a_1 a_{r+2}) - (a_1 a_{r+1}) \right\}^{\lambda_{r+2}} (a_1 a_{r+3})^{\lambda_{r+3}} \ldots$$
$$(a_r a_s)^{\lambda_s} (a_1 a_{s+1})^{\lambda_{s+1}} \ldots (a_1 a_\delta)^{\lambda_\delta}$$
$$= (a_1 a_2)^{\lambda_2} \ldots (a_1 a_{r-1})^{\lambda_{r-1}} (a_{r+1} a_{r+2})^{\lambda_{r+2}} (a_1 a_{r+3})^{\lambda_{r+3}} \ldots (a_1 a_{s-1})^{\lambda_{s-1}}$$
$$\times \left\{ (a_1 a_s) - (a_1 a_r) \right\}^{\lambda_s} (a_1 a_{s+1})^{\lambda_{s+1}} \ldots (a_1 a_\delta)^{\lambda_\delta}.$$

Expanding the braces on each side by the binomial theorem, we obtain a syzygy.

The syzygy at once gives us the relation between the equations

$$\Sigma (-)^i \binom{\lambda_{r+2}}{i} e^{-a_r D_s} (\lambda_2, \ldots, \lambda_{r-1}, 0, i, \lambda_{r+2}-i, \lambda_{r+3}, \ldots, \lambda_\delta)$$

$$= \Sigma (-)^j \binom{\lambda_s}{j} e^{-a_{r+1} D_{r+2}} (\lambda_2, \ldots, \lambda_{r-1}, j, 0, \lambda_{r+2}, \ldots, \lambda_{s-1}, \lambda_s - j, \lambda_{s+1}, \ldots, \lambda_\delta).$$

Now every equation on the right-hand side is discussed before any of those on the left since $r+1 > r$. Hence this syzygy yields the relation

$$e^{-a_{r+1} D_{r+2}} [\lambda_2, \ldots, \lambda_{r-1}, 0, 0, \lambda_{r+2}, \ldots, \lambda_{s-1}, \underline{\lambda_s}, \lambda_{s+1}, \ldots, \lambda_\delta] = R.$$

And in general when $\sigma_2 > \sigma_1 > r$, and neither σ_1 or σ_2 is equal to s, we obtain just such another syzygy which yields the relation

$$e^{-a_{\sigma_1} D_{\sigma_2}} ([\lambda_2, \ldots, \lambda_{r-1}, 0, \lambda_{r+1}, \ldots, \lambda_{s-1}, \underline{\lambda_s}, \lambda_{s+1}, \ldots, \lambda_\delta]_{\lambda_{\sigma_1}=0}) = R.$$

The result may be at once extended to a slightly more general syzygy to which the relation

$$e^{-a_\sigma D_{\sigma_1} - a_\sigma D_{\sigma_2} - \ldots - a_\sigma D_{\sigma_\kappa}} ([\lambda_2, \ldots, \lambda_{r-1}, 0, \lambda_{r+1}, \ldots, \underline{\lambda_s}, \ldots, \lambda_\delta]_{\lambda_\sigma=0}) = R$$

(where $r < \sigma < \sigma_1 < \ldots < \sigma_\kappa$, and none of the σ's which here appear is equal to s) corresponds.

Let us call these syzygies the perpetuant syzygies of the type A.

12. Consider the identity ($r < s$)

$$(a_1 a_2)^{\lambda_2} \ldots (a_1 a_{r-1})^{\lambda_{r-1}} \left\{ (a_r a_s) - (a_1 a_{r+1}) \right\}^{\lambda_s} (a_1 a_{r+2})^{\lambda_{r+2}} \ldots$$
$$(a_1 a_{s-1})^{\lambda_{s-1}} (a_1 a_{s+1})^{\lambda_{s+1}} \ldots (a_1 a_\delta)^{\lambda_\delta}$$
$$= (a_1 a_2)^{\lambda_2} \ldots (a_1 a_{r-1})^{\lambda_{r-1}} \left\{ (a_{r+1} a_s) - (a_1 a_r) \right\}^{\lambda_s} (a_1 a_{r+2})^{\lambda_{r+2}} \ldots$$
$$(a_1 a_{s-1})^{\lambda_{s-1}} (a_1 a_{s+1})^{\lambda_{s+1}} \ldots (a_1 a_\delta)^{\lambda_\delta}.$$

Expanding the braces on both sides we obtain a Stroh syzygy, and this at once gives the relation between our equations

$$\Sigma(-)^i \binom{\lambda_s}{i} e^{-a_r D_s}(\lambda_2, \ldots, \lambda_{r-1}, 0, i, \lambda_{r+2}, \ldots, \lambda_{s-1}, \lambda_s - i, \lambda_{s+1}, \ldots, \lambda_\delta)$$

$$= \Sigma(-)^i \binom{\lambda_s}{i} e^{-a_{r+1} D_s}(\lambda_2, \ldots, \lambda_{r-1}, i, 0, \lambda_{r+2}, \ldots, \lambda_{s-1}, \lambda_s - i, \lambda_{s+1}, \ldots, \lambda_\delta).$$

Every equation represented on the right is considered before any of those on the left of this relation : hence we may write it

$$\Sigma(-)^i \binom{\lambda_s}{i} [\lambda_2, \ldots, \lambda_{r-1}, 0, i, \lambda_{r+2}, \ldots, \lambda_{s-1}, \underline{\lambda_s - i}, \lambda_{s+1}, \ldots, \lambda_\delta] = R.$$

And although a slightly different meaning must be attached to the operator, we may, without fear of ambiguity, write this equation

$$e^{-a_{r+1} D_s}[\lambda_2, \ldots, \lambda_{r-1}, 0, 0, \lambda_{r+2}, \ldots, \lambda_{s-1}, \underline{\lambda_s}, \lambda_{s+1}, \ldots, \lambda_\delta] = R.$$

In the same way we obtain, whenever $s > \sigma$,

$$e^{-a_\sigma D_s}([\lambda_2, \ldots, \lambda_{r-1}, 0, \lambda_{r+1}, \ldots, \lambda_{s-1}, \underline{\lambda_s}, \lambda_{s+1}, \ldots, \lambda_\delta]_{\lambda_\sigma = 0}) = R,$$

and whenever $\sigma > s$,

$$e^{-a_s D_\sigma}[\lambda_2, \ldots, \lambda_{r-1}, 0, \lambda_{r+1}, \ldots, \lambda_{s-1}, \underline{0}, \lambda_{s+1}, \ldots, \lambda_\delta] = R.$$

That is, we obtain syzygies which yield these relations.

Now combining one of these syzygies with one of those of the last paragraph, we have a syzygy expressed by

$$(a_1 a_2)^{\lambda_2} \ldots (a_1 a_{r-1})^{\lambda_{r-1}} \{(a_1 a_{r+2}) - (a_1 a_{r+1})\}^{\lambda_{r+2}} \{(a_r a_s) - (a_1 a_{r+1})\}^{\lambda_s}$$
$$(a_1 a_{r+3})^{\lambda_{r+3}} \ldots (a_1 a_{s-1})^{\lambda_{s-1}} (a_1 a_{s+1})^{\lambda_{s+1}} \ldots (a_1 a_\delta)^{\lambda_\delta}$$
$$= (a_1 a_2)^{\lambda_2} \ldots (a_1 a_{r-1})^{\lambda_{r-1}} (a_{r+1} a_{r+2})^{\lambda_{r+2}} \{(a_{r+1} a_s) - (a_1 a_r)\}^{\lambda_s} (a_1 a_{r+3})^{\lambda_{r+3}} \ldots$$
$$(a_1 a_{s-1})^{\lambda_{s-1}} (a_1 a_{s+1})^{\lambda_{s+1}} \ldots (a_1 a_\delta)^{\lambda_\delta},$$

which yields a relation

$$e^{-a_{r+1} D_{r+2} - a_{r+1} D_s}[\lambda_2, \ldots, \lambda_{r-1}, 0, \lambda_{r+2}, \ldots, \lambda_{s-1}, \underline{\lambda_s}, \lambda_{s+1}, \ldots, \lambda_\delta] = R.$$

In this way we obtain syzygies to give each of the relations

$$e^{-a_\sigma D_{\sigma_1} - a_\sigma D_{\sigma_2} - \ldots - a_\sigma D_{\sigma_\kappa}}([\lambda_2, \ldots, \lambda_{r-1}, 0, \lambda_{r+1}, \ldots, \lambda_{s-1}, \underline{\lambda_s}, \lambda_{s+1}, \ldots, \lambda_\delta]_{\lambda_\sigma = 0}) = R,$$

when $r < \sigma < \sigma_1 < \sigma_2 < \ldots < \sigma_\kappa$ and $\sigma \neq s$.

These relations have already been fully discussed in § 9, when discussing the question of reducibility: we obtain from them at once the result

$$[\lambda_2, \ldots, \lambda_{r-1}, 0, \lambda_{r+1}, \ldots, \lambda_{s-1}, \underline{\lambda_s}, \lambda_{s+1}, \ldots, \lambda_\delta] = R,$$

when $\lambda_\sigma < 2^{\delta-\sigma}$, where $r < \sigma \neq s$.

We will call the syzygies of this paragraph the perpetuant syzygies of the type B.

13. In obtaining the limitations to the value of λ_s, and the corresponding syzygies, for the equation

$$e^{-a_r D_s}(\lambda_2, \ldots, \lambda_{r-1}, 0, \lambda_{r+1}, \ldots, \lambda_\delta) = R.$$

We shall simplify the work and not lose anything in generality if we suppose $r = 2$ and $s = 3$. Thus we consider

$$e^{-a_2 D_3}(0, \lambda_3, \lambda_4, \ldots, \lambda_\delta) \equiv [0, \underline{\lambda_3}, \lambda_4, \ldots, \lambda_\delta] = R.$$

If $\lambda_3 = 0$, our equation becomes

$$(0, 0, \lambda_4, \ldots, \lambda_\delta) = a_{2_x}^{\infty}\left(\frac{a_3^0 a_4^{\lambda_4} \ldots a_\delta^{\lambda_\delta}}{a_1}\right);$$

but we already know from a previous equation that

$$(0, 0, \lambda_4, \ldots, \lambda_\delta) = a_{3_x}^{\infty}\left(\frac{a_2^0 a_4^{\lambda_4} \ldots a_\delta^{\lambda_\delta}}{a_1}\right).$$

Thus the equation simply gives the obvious syzygy

$$a_{2_x}^{\infty}\left(\frac{a_3^0 a_4^{\lambda_4} \ldots a_\delta^{\lambda_\delta}}{a_1}\right) = a_{3_x}^{\infty}\left(\frac{a_3^0 a_4^{\lambda_4} \ldots a_\delta^{\lambda_\delta}}{a_1}\right).$$

If $\lambda_3 = 1$, our equation becomes

$$(0, 1, \lambda_4, \ldots, \lambda_\delta) - (1, 0, \lambda_4, \ldots, \lambda_\delta)$$

$$= (a_2 a_3)\left(\frac{a_4^{\lambda_4} \ldots a_\delta^{\lambda_\delta}}{a_1}\right) = a_{2_x}^{\infty}\left(\frac{a_3^1 a_4^{\lambda_4} \ldots a_\delta^{\lambda_\delta}}{a_1}\right) - a_{3_x}^{\infty}\left(\frac{a_2^1 a_4^{\lambda_4} \ldots a_\delta^{\lambda_\delta}}{a_1}\right);$$

giving again a syzygy. This syzygy is the Jacobian syzygy.

Consider the two identities

$$\{(a_1 a_4) - (a_2 a_3)\}^{\lambda_4}(a_1 a_5)^{\lambda_5} \ldots (a_1 a_\delta)^{\lambda_\delta} = \{(a_1 a_2) + (a_3 a_4)\}^{\lambda_4}(a_1 a_5)^{\lambda_5} \ldots (a_1 a_\delta)^{\lambda_\delta},$$

and

$$\{(a_1 a_4) + (a_2 a_3)\}^{\lambda_4}(a_1 a_5)^{\lambda_5} \ldots (a_1 a_\delta)^{\lambda_\delta} = \{(a_1 a_3) + (a_2 a_4)\}^{\lambda_4}(a_1 a_5)^{\lambda_5} \ldots (a_1 a_\delta)^{\lambda_\delta},$$

when these are expanded they yield Stroh syzygies. These syzygies give us the relations

$$e^{-a_3D_4}[0, \underline{0}, \lambda_4, \ldots, \lambda_\delta] = R,$$

and
$$e^{+a_3D_4}[0, \underline{0}, \lambda_4, \ldots, \lambda_\delta] = R.$$

And in general we find in this way syzygies which give the relations

$$e^{-a_3D_\sigma}[0, \underline{0}, \lambda_4, \ldots, \lambda_\delta] = R,$$

and
$$e^{+a_3D_\sigma}[0, \underline{0}, \lambda_4, \ldots, \lambda_\delta] = R,$$

for
$$\sigma = 4, 5, \ldots, \delta.$$

Further, from the syzygies

$$\{(a_1a_4)-(a_2a_3)\}^{\lambda_4}\{(a_1a_5)-(a_2a_3)\}^{\lambda_5}(a_1a_6)^{\lambda_6}\ldots(a_1a_\delta)^{\lambda_\delta}$$
$$= \{(a_1a_2)+(a_3a_4)\}^{\lambda_4}\{(a_1a_2)+(a_3a_5)\}^{\lambda_5}(a_1a_6)^{\lambda_6}\ldots(a_1a_\delta)^{\lambda_\delta},$$

and $\{(a_1a_4)+(a_2a_3)\}^{\lambda_4}\{(a_1a_5)+(a_2a_3)\}^{\lambda_5}(a_1a_6)^{\lambda_6}\ldots(a_1a_\delta)^{\lambda_\delta}$

$$= \{(a_1a_3)+(a_2a_4)\}^{\lambda_4}\{(a_1a_3)+(a_2a_5)\}^{\lambda_5}(a_1a_6)^{\lambda_6}\ldots(a_1a_\delta)^{\lambda_\delta},$$

we obtain the relations

$$e^{\pm(a_3D_4+a_3D_5)}[0, \underline{0}, \lambda_4, \ldots, \lambda_\delta] = R.$$

Proceeding thus we can write down a set of syzygies which give us the relations
$$e^{\pm(a_3D_{s_1}+a_3D_{s_2}+\ldots+a_3D_{s_\eta})}[0, \underline{0}, \lambda_4, \ldots, \lambda_\delta] = R,$$

where $s_1, s_2, \ldots, s_\eta$ are all, any or none of $4, 5, \ldots, \delta$.

These syzygies we shall refer to as the perpetuant syzygies of the type C.

14. It is necessary to discuss the equations just found.

We shall arrange them in a sequence as we have done the other equations:

Thus the equations

$$e^{\pm(a_3D_{r_1}+a_3D_{r_2}+\ldots+a_3D_{r_\epsilon})}[0, \underline{0}, \lambda_4, \ldots, \lambda_\delta] = R \quad (r_1 < r_2 < \ldots < r_\epsilon),$$

will be discussed before the equations

$$e^{\pm(a_3D_{s_1}+a_3D_{s_2}+\ldots+a_3D_{s_\eta})}[0, \underline{0}, \lambda_4, \ldots, \lambda_\delta] = R \quad (s_1 < s_2 < \ldots < s_\eta),$$

when
$$r_1 > s_1.$$

But, if $r_1 = s_1$, the equations are discussed simultaneously.

Thus the first pair of equations to be discussed is

$$e^{\pm a_3 D_\delta}[0, \underline{0}, \lambda_4, \ldots, \lambda_\delta] = R.$$

Whence $\quad [0, \underline{0}, \lambda_4, \ldots, \lambda_{\delta-1}, \lambda_\delta] \pm \lambda_\delta [0, \underline{1}, \lambda_4, \ldots, \lambda_{\delta-1}, \lambda_\delta - 1]$

$$+\binom{\lambda_\delta}{2}[0, \underline{2}, \lambda_4, \ldots, \lambda_{\delta-1}, \lambda_\delta - 2] \pm \binom{\lambda_\delta}{3}[0, \underline{3}, \lambda_4, \ldots, \lambda_{\delta-1}, \lambda_\delta - 3] + \ldots = R,$$

giving immediate reductions for

$$[0, \underline{2}, \lambda_4, \ldots, \lambda_{\delta-1}, \lambda_\delta - 2],$$

and

$$[0, \underline{3}, \lambda_4, \ldots, \lambda_{\delta-1}, \lambda_\delta - 3].$$

The forms $[0, \underline{\lambda_3}, \lambda_4, \ldots, \lambda_\delta]$ being arranged in sequence according to the same rules as the forms $(\lambda_2, \lambda_3, \ldots, \lambda_\delta)$.

15. LEMMA.—*The* $2^{\delta-r+1}$ *equations*

$$e^{\pm(a_3 D_{s_1} + a_3 D_{s_2} + \ldots + a_3 D_{s_\eta})}[0, \underline{0}, \lambda_4, \ldots, \lambda_\delta] = 0,$$

where $s_1, s_2, \ldots, s_\eta$ *are all, any or none of the numbers* $r+1, r+2, \ldots, \delta$ *are just sufficient to express all forms*

$$[0, \underline{\lambda_3}, \lambda_4, \ldots, \lambda_r, \lambda_{r+1}, \ldots, \lambda_\delta],$$

for which $\lambda_3 < 2^{\delta-r+1}$, *in terms of forms*

$$[0, \underline{\mu_3}, \lambda_4, \ldots, \lambda_r, \mu_{r+1}, \ldots, \mu_\delta],$$

where $\mu_3 \nless 2^{\delta-r+1}$ *and the first of the differences*

$$\lambda_\delta - \mu_\delta, \ \lambda_{\delta-1} - \mu_{\delta-1}, \ \ldots, \ \lambda_{r+1} - \mu_{r+1},$$

which does not vanish is positive.

Let us assume the truth of this proposition for a particular value of r, and then consider the $2^{\delta-r+1}$ equations

$$e^{\pm(a_3 D_r + a_3 D_{s_1} + \ldots + a_3 D_{s_\eta})}[0, \underline{0}, \lambda_4, \ldots, \lambda_\delta] = 0,$$

where $s_1, s_2, \ldots, s_\eta$ are all, any or none of the numbers $r+1, r+2, \ldots, \delta$.

Let us write

$$e^{-a_3 D_r}[0, \underline{0}, \lambda_4, \ldots, \lambda_{r-1}, \lambda_r, \lambda_{r+1}, \ldots, \lambda_\delta] \equiv [0, \underline{0}, \lambda_4, \ldots, \lambda_{r-1}, \underline{\lambda_r}, \lambda_{r+1}, \ldots, \lambda_\delta],$$

and

$$e^{+a_3D_r}[0, \underline{0}, \lambda_4, \ldots, \lambda_{r-1}, \lambda_r, \lambda_{r+1}, \ldots, \lambda_\delta] \equiv [0, \underline{0}, \lambda_4, \ldots, \lambda_{r-1}, \underline{\lambda'_r}, \lambda_{r+1}, \ldots, \lambda_\delta].$$

Then we have two sets of equations

$$e^{-a_3D_{s_1}-a_3D_{s_2}-\ldots-a_3D_{s_\eta}}[0, \underline{0}, \lambda_4, \ldots, \lambda_{r-1}, \underline{\lambda_r}, \lambda_{r+1}, \ldots, \lambda_\delta] = 0,$$

and $$e^{+a_3D_{s_1}+a_3D_{s_2}+\ldots+a_3D_{s_\eta}}[0, \underline{0}, \lambda_4, \ldots, \lambda_{r-1}, \underline{\lambda'_r}, \lambda_{r+1}, \ldots, \lambda_\delta] = 0.$$

From the theorem of § 9 we know that the solution of these equations expresses all forms ($\lambda_3 < 2^{\delta-r}$)

$$[0, \underline{\lambda_3}, \lambda_4, \ldots, \lambda_{r-1}, \underline{\lambda_r}, \lambda_{r+1}, \ldots, \lambda_\delta],$$

in terms of forms $[0, \underline{\mu_3}, \lambda_4, \ldots, \lambda_{r-1}, \underline{\lambda_r}, \mu_{r+1}, \ldots, \mu_\delta]$;

and all forms ($\lambda_3 < 2^{\delta-r}$)

$$[0, \underline{\lambda_3}, \lambda_4, \ldots, \lambda_{r-1}, \underline{\lambda'_r}, \lambda_{r+1}, \ldots, \lambda_\delta],$$

in terms of forms $[0, \underline{\mu_3}, \lambda_4, \ldots, \lambda_{r-1}, \underline{\lambda'_r}, \mu_{r+1}, \ldots, \mu_\delta]$;

where in both cases $\mu_3 \nless 2^{\delta-r}$, and the first of the differences

$$\lambda_\delta - \mu_\delta, \lambda_{\delta-1} - \mu_{\delta-1}, \ldots, \lambda_{r+1} - \mu_{r+1},$$

which does not vanish is positive.

We thus obtain two sets of equations

$$e^{\pm a_3D_r}[0, \underline{\lambda_3}, \lambda_4, \ldots, \lambda_{r-1}, \lambda_r, \ldots, \lambda_\delta] = R,$$

where $$\lambda_3 = 0, 1, \ldots, 2^{\delta-r}-1.$$

Expanding them out, we have

$$\Sigma(-)^i \binom{\lambda_r}{i}[0, \underline{\lambda_3+i}, \lambda_4, \ldots, \lambda_{r-1}, \lambda_r - i, \lambda_{r+1}, \ldots, \lambda_\delta] = R,$$

and $$\Sigma \binom{\lambda_r}{i}[0, \underline{\lambda_3+i}, \lambda_4, \ldots, \lambda_{r-1}, \lambda_r - i, \lambda_{r+1}, \ldots, \lambda_\delta] = R.$$

Now using the assumption made we see that these equations may be regarded as equations to give the values of

$$[0, \underline{2^{\delta-r+1}+\xi}, \lambda_4, \ldots, \lambda_{r-1}, \lambda_r - 2^{\delta-r+1} - \xi, \lambda_{r+1}, \ldots, \lambda_\delta] = R,$$

$$\xi = 0, 1, 2, \ldots, 2^{\delta-r+1}-1.$$

Adding and subtracting our equations in pairs, we obtain two new sets;

one of which connects those forms for which ξ is even, and the other those forms for which ξ is odd.

They may be written

$$\cosh a_3 D_r [0, \underline{\lambda_3}, \lambda_4, \ldots, \lambda_\delta] = R,$$

and

$$\sinh a_3 D_r [0, \underline{\lambda_3}, \lambda_4, \ldots, \lambda_\delta] = R.$$

We desire to prove the linear independence of each set.

For this purpose we must calculate the determinants formed by the coefficients. In the first case the determinant is

$$\begin{vmatrix} \binom{\lambda_r}{2^{\delta-r+1}} & \binom{\lambda_r}{2^{\delta-r+1}+2} & \cdots & \binom{\lambda_r}{2^{\delta-r+1}+2\sigma} & \cdots & \binom{\lambda_r}{2^{\delta-r+2}-2} \\ \binom{\lambda_r-1}{2^{\delta-r+1}-1} & \binom{\lambda_r-1}{2^{\delta-r+1}+1} & \cdots & \binom{\lambda_r-1}{2^{\delta-r+1}+2\sigma-1} & \cdots & \binom{\lambda_r-1}{2^{\delta-r+2}-3} \\ \cdots & \cdots & \cdots & \cdots & \cdots & \cdots \\ \binom{\lambda_r-\tau}{2^{\delta-r+1}-\tau} & \binom{\lambda_r-\tau}{2^{\delta-r+1}+2-\tau} & \cdots & \binom{\lambda_r-\tau}{2^{\delta-r+1}+2\sigma-\tau} & \cdots & \binom{\lambda_r-1}{2^{\delta-r+2}-2-\tau} \\ \cdots & \cdots & \cdots & \cdots & \cdots & \cdots \\ \binom{\lambda_r-2^{\delta-r}+1}{2^{\delta-r}+1} & \binom{\lambda_r-2^{\delta-r}+1}{2^{\delta-r}+3} & \cdots & \binom{\lambda_r-2^{\delta-r}+1}{2^{\delta-r}+1+2\sigma} & \cdots & \binom{\lambda_r-2^{\delta-r}+1}{2^{\delta-r+2}-1-2^{\delta-r}} \end{vmatrix}$$

$$= \frac{\binom{\lambda_r}{2^{\delta-r+1}}\binom{\lambda_r}{2^{\delta-r+1}+2}\cdots\binom{\lambda_r}{2^{\delta-r+2}-2}}{\binom{\lambda_r}{1}\binom{\lambda_r}{2}\cdots\binom{\lambda_r}{2^{\delta-r}-1}}\,\Delta.$$

Where, on changing columns into rows and rows into columns,

$$\Delta = \begin{vmatrix} 1 & \binom{2^{\delta-r+1}}{1} & \binom{2^{\delta-r+1}}{2} & \cdots & \binom{2^{\delta-r+1}}{2^{\delta-r}-1} \\ 1 & \binom{2^{\delta-r+1}+2}{1} & \binom{2^{\delta-r+1}+2}{2} & \cdots & \binom{2^{\delta-r+1}+2}{2^{\delta-r}-1} \\ \cdots & \cdots & \cdots & \cdots & \cdots \\ 1 & \binom{2^{\delta-r+1}+2\sigma}{1} & \binom{2^{\delta-r+1}+2\sigma}{2} & \cdots & \binom{2^{\delta-r+1}+2\sigma}{2^{\delta-r}-1} \\ \cdots & \cdots & \cdots & \cdots & \cdots \\ 1 & \binom{2^{\delta-r+2}-2}{1} & \binom{2^{\delta-r+2}-2}{2} & \cdots & \binom{2^{\delta-r+2}-2}{2^{\delta-r}-1} \end{vmatrix}$$

We shall now consider the more general determinant

$$\Delta_k = \begin{vmatrix} 1 & \binom{n}{1} & \binom{n}{2} & \cdots & \binom{n}{k-1} \\ 1 & \binom{n+2}{1} & \binom{n+2}{2} & \cdots & \binom{n+2}{k-1} \\ \cdots & \cdots & \cdots & \cdots & \cdots \\ 1 & \binom{n+2k-2}{1} & \binom{n+2k-2}{2} & \cdots & \binom{n+2k-2}{k-1} \end{vmatrix}$$

Subtract each row from that immediately below it, then the $(\sigma+1)$-th row becomes

$$0,\ \binom{n+2\sigma-1}{0}+\binom{n+2\sigma-2}{0},\ \ldots,\ \binom{n+2\sigma-1}{\tau-1}+\binom{n+2\sigma-2}{\tau-1},\ \ldots,$$
$$\binom{n+2\sigma-1}{k-2}+\binom{n+2\sigma-2}{k-2};$$

since
$$\binom{n+2\sigma}{\tau} = \binom{n+2\sigma-1}{\tau}+\binom{n+2\sigma-1}{\tau-1}$$
$$= \binom{n+2\sigma-2}{\tau}+\binom{n+2\sigma-2}{\tau-1}+\binom{n+2\sigma-1}{\tau-1}.$$

Next we repeat the process of subtracting each row from the next below, leaving the first two rows unaltered. The $(\tau+1)$-th element of the $(\sigma+1)$-th row becomes now

$$\binom{n+2\sigma-2}{\tau-2}+2\binom{n+2\sigma-3}{\tau-2}+\binom{n+2\sigma-4}{\tau-2}.$$

We keep on repeating the process, each time leaving one more row unchanged. After t subtractions the $(\tau+1)$-th element of the $(\sigma+1)$-row becomes

$$\binom{n+2\sigma-t}{\tau-t}+\binom{t}{1}\binom{n+2\sigma-t-1}{\tau-t}+\binom{t}{2}\binom{n+2\sigma-t-2}{\tau-t}+\ldots$$
$$+\binom{t}{t}\binom{n+2\sigma-2t}{\tau-t}.$$

This $(\sigma+1)$-th row is not left unchanged until $t=\sigma$, and so its final form will be obtained by giving t the value σ. The $(\tau+1)$-th element is then zero when $\tau < \sigma$, and its value when $\tau = \sigma$ is

$$1+\binom{\sigma}{1}+\binom{\sigma}{2}+\ldots+\binom{\sigma}{\sigma} = 2^{\sigma}.$$

Thus we eventually transform Δ_k into a determinant in which every element below the leading diagonal is zero, and where the elements of this diagonal are

$$1,\ 2,\ 2^2,\ \ldots,\ 2^{k-1}.$$

Hence $$\Delta_k = 2^{0+1+2+\ldots+k-1} = 2^{\binom{k}{2}}.$$

Hence the determinant formed by the coefficients of our equations which we wished to calculate

$$= \frac{\binom{\lambda_r}{2^{\delta-r+1}}\binom{\lambda_r}{2^{\delta-r+1}+2}\cdots\binom{\lambda_r}{2^{\delta-r+2}-2}}{\binom{\lambda_r}{1}\binom{\lambda_r}{2}\cdots\binom{\lambda_r}{2^{\delta-r}-1}}\, 2^{\binom{2^{\delta-r}}{2}}.$$

The determinant of the coefficients of the other set of equations can, in a similar manner, be shewn to be

$$\frac{\binom{\lambda_r}{2^{\delta-r+1}+1}\binom{\lambda_r}{2^{\delta-r+1}+3}\cdots\binom{\lambda_r}{2^{\delta-r+2}-1}}{\binom{\lambda_r}{1}\binom{\lambda_r}{2}\cdots\binom{\lambda_r}{2^{\delta-r}-1}}\, 2^{\binom{2^{\delta-r}}{2}}.$$

Thus we obtain

$$[0,\ \underline{2^{\delta-r+1}+\xi},\ \lambda_4,\ \ldots,\ \lambda_{r-1},\ \lambda_r-2^{\delta-r+1}-\xi,\ \lambda_{r+1},\ \ldots,\ \lambda_\delta] = R,$$

for all values $\xi = 0, 1, 2, \ldots, 2^{\delta-r+1}-1$; provided $\lambda_r \nless 2^{\delta-r+2}-1$. If λ_r is less than this value, we can remove some of our equations, for there are fewer forms to solve for. The determinants, when we take the same number of equations (starting from the beginning), as there are forms, can easily be calculated, and are found not to be zero.

Hence the equations give

$$[0,\ \underline{\lambda_3},\ \lambda_4,\ \ldots,\ \lambda_{r-1},\ \lambda_r,\ \ldots,\ \lambda_\delta] = R,$$

provided $$\lambda_3 < 2^{\delta-r+2}.$$

Where R consists of forms

$$[0,\ \underline{\mu_3},\ \lambda_4,\ \ldots\ \lambda_{r-1},\ \mu_r,\ \ldots,\ \mu_\delta],$$

for which $\mu_3 \nless 2^{\delta-r+1}$, and where the first of the differences

$$\lambda_\delta - \mu_\delta,\ \lambda_{\delta-1} - \mu_{\delta-1},\ \ldots,\ \lambda_r - \mu_r,$$

which does not vanish is positive.

We may apply this result again to all forms on the right-hand side for which $\mu_3 < 2^{\delta-r+2}$, and thus ultimately we obtain the condition $\mu_3 < 2^{\delta-r+2}$. Thus, if the lemma is true for a particular value of r, it is true when we replace r by $r-1$. Now it is true when $r = \delta - 1$; hence it is always true. Thus the truth of the lemma is established.

16. We may now apply the lemma to the equations of § 14. We find at once that the syzygies obtained in § 13 are sufficient to express the equation $(\lambda_3 < 2^{\delta-2})$,

$$e^{-a_2 D_3} (0, \lambda_3, \lambda_4, \ldots, \lambda_\delta) = R$$

in terms of equations already considered and of equations

$$e^{-a_2 D_3} (0, \mu_3, \mu_4, \ldots, \mu_\delta) = R,$$

where $\mu_3 \nless 2^{\delta-2}$, and the first of the differences

$$\lambda_\delta - \mu_\delta,\ \lambda_{\delta-1} - \mu_{\delta-1},\ \ldots,\ \lambda_4 - \mu_4,$$

which does not vanish is positive.

The equation $\qquad e^{-a_2 D_3} (0, \lambda_3, \lambda_4, \ldots, \lambda_\delta) = R,$

then may be said to yield a syzygy when

$$\lambda_3 < 2^{\delta-2}, \text{ or } \lambda_4 < 2^{\delta-4}, \text{ or } \lambda_5 < 2^{\delta-5}, \ \ldots, \text{ or } \lambda_\delta < 1.$$

Thus the theorem enunciated in § 10 is true for the equation

$$e^{-a_2 D_3} (0, \lambda_3, \lambda_4, \ldots, \lambda_\delta) = R.$$

And in just the same way it can be established for

$$e^{-a_r D_s} (\lambda_2, \lambda_3, \ldots, \lambda_{r-1}, 0, \lambda_{r+1}, \ldots, \lambda_\delta) = R \quad (s > r).$$

17. Let us now consider the equation

$$e^{-a_2 D_{s_1} - a_2 D_{s_2} - \ldots - a_2 D_{s_\eta}} (0, \lambda_3, \ldots, \lambda_\delta) \equiv [0, \lambda_3, \ldots, \lambda_\delta] = R$$
$$(s_1 < s_2 < \ldots < s_\eta).$$

By means of the perpetuant syzygies of the types A and B, discussed in §§ 11, 12, we obtained relations by which we can reduce our equation when $\lambda_\sigma < 2^{\delta-\sigma}$, where σ is any one of the numbers 3, 4, ..., δ which is not included among the numbers $s_1, s_2, \ldots, s_\eta$.

We may then confine our attention to those syzygies which will give limitations to the value of λ_σ when σ is one of the numbers $s_1, s_2, \ldots, s_\eta$.

To fix our ideas let us put $s_1 = 3$, and consider the syzygies which will affect λ_3.

We obtain first certain syzygies of the type C,

$$\{(a_1 a_{r_1}) - (a_2 a_3)\}^{\lambda_{r_1}} (a_2 a_{s_2})^{\lambda_{s_2}} (a_2 a_{s_3})^{\lambda_{s_3}} \ldots (a_2 a_{s_\eta})^{\lambda_{s_\eta}} (a_1 a_{r_2})^{\lambda_{r_2}} \ldots$$

$$= \{(a_3 a_{r_1}) + (a_1 a_2)\}^{\lambda_{r_1}} \{(a_1 a_{s_2}) - (a_1 a_2)\}^{\lambda_{s_2}} \ldots \{(a_1 a_{s_\eta}) - (a_1 a_2)\}^{\lambda_{s_\eta}} (a_1 a_{r_2})^{\lambda_{r_2}} \ldots,$$

where $r_1, r_2, \ldots$ are just those of the numbers 4, 5, ..., δ, which are not included in the set $s_2, s_3, \ldots, s_\eta$.

Similarly we have

$$\{(a_1 a_{r_1}) - (a_2 a_3)\}^{\lambda_{r_1}} \{(a_1 a_{r_2}) - (a_2 a_3)\}^{\lambda_{r_2}} (a_2 a_{s_2})^{\lambda_{s_2}} \ldots (a_2 a_{s_\eta})^{\lambda_{s_\eta}} (a_1 a_{r_3})^{\lambda_{r_3}} \ldots$$

$$= \{(a_3 a_{r_1}) + (a_1 a_2)\}^{\lambda_{r_1}} \{(a_3 a_{r_2}) + (a_1 a_2)\}^{\lambda_{r_2}} \{(a_1 a_{s_2}) - (a_1 a_2)\}^{\lambda_{s_2}} \ldots$$
$$\{(a_1 a_{s_\eta}) - (a_1 a_2)\}^{\lambda_{s_\eta}} (a_1 a_{r_3})^{\lambda_{r_3}} \ldots .$$

Then we have a set of syzygies we will call syzygies of the type D: such are

$$\{(a_2 a_{s_2}) - (a_2 a_3)\}^{\lambda_{s_2}} (a_2 a_{s_3})^{\lambda_{s_3}} \ldots (a_2 a_{s_\eta})^{\lambda_{s_\eta}} (a_1 a_{r_1})^{\lambda_{r_1}} (a_1 a_{r_2})^{\lambda_{r_2}} \ldots$$

$$= (a_3 a_{s_2})^{\lambda_{s_2}} \{(a_1 a_{s_3}) - (a_1 a_2)\}^{\lambda_{s_3}} \ldots \{(a_1 a_{s_\eta}) - (a_1 a_2)\}^{\lambda_{s_\eta}} (a_1 a_{r_1})^{\lambda_{r_1}} (a_1 a_{r_2})^{\lambda_{r_2}} \ldots .$$

We obtain fresh syzygies by replacing any term $(a_2 a_s)$ on the left by $\{(a_2 a_s) - (a_2 a_3)\}$, and making the corresponding change on the right of $\{(a_1 a_s) - (a_1 a_2)\}$ into $(a_3 a_s)$. Or we may change on the left $(a_1 a_r)$ into $\{(a_1 a_r) - (a_2 a_3)\}$, and at the same time on the right $(a_1 a_r)$ into $\{(a_3 a_r) + (a_1 a_2)\}$.

In this way we obtain a set of syzygies which will give us the $2^{\delta-3}$ relations between our equations

$$e^{-a_3 D_{\sigma_1} - a_3 D_{\sigma_2} - \ldots - a_3 D_{\sigma_k}} [0, 0, \lambda_4, \ldots, \lambda_\delta] = R,$$

where $\sigma_1, \sigma_2, \ldots, \sigma_k$ are any, all or none of the numbers 4, 5, ..., δ.

Again, we have the syzygies of the type C,

$$\{(a_1 a_{r_1}) + (a_2 a_3)\}^{\lambda_{r_1}} \{(a_1 a_{r_2}) + (a_2 a_3)\}^{\lambda_{r_2}} (a_2 a_{s_2})^{\lambda_{s_2}} \ldots (a_2 a_{s_\eta})^{\lambda_{s_\eta}} (a_1 a_{r_3})^{\lambda_{r_3}} \ldots$$

$$= \{(a_2 a_{r_1}) + (a_1 a_3)\}^{\lambda_{r_1}} \{(a_2 a_{r_2}) + (a_1 a_3)\}^{\lambda_{r_2}} (a_2 a_{s_2})^{\lambda_{s_2}} \ldots (a_2 a_{s_\eta})^{\lambda_{s_\eta}} (a_1 a_{r_3})^{\lambda_{r_3}} \ldots,$$

for example. This will give the relation

$$e^{+a_3 D_{r_1} + a_3 D_{r_2}} [0, 0, \lambda_4, \ldots, \lambda_\delta] = R.$$

And so we obtain syzygies which yield

$$e^{a_3 D_{\sigma_1} + a_3 D_{\sigma_2} + \ldots + a_3 D_{\sigma k}} [0, 0, \lambda_4, \ldots, \lambda_\delta] = R,$$

where $\sigma_1, \sigma_2, \ldots, \sigma_k$ are all or any of the numbers $r_1, r_2, r_3, \ldots$.

We have then certain syzygies which we shall include in the type D; an example of these is

$$\{(a_1 a_{r_1}) + (a_2 a_3)\}^{\lambda_{r_1}} \{(a_2 a_{s_2}) - (a_2 a_3)\}^{\lambda_{s_2}} (a_2 a_{s_3})^{\lambda_{s_3}} \ldots$$
$$(a_2 a_{s_\eta})^{\lambda_{s_\eta}} (a_1 a_{r_2})^{\lambda_{r_2}} (a_1 a_{r_3})^{\lambda_{r_3}} \ldots$$
$$= \{(a_2 a_{r_1}) + (a_1 a_3)\}^{\lambda_{r_1}} \{(a_1 a_{s_2}) - (a_1 a_3)\}^{\lambda_{s_2}} (a_2 a_{s_3})^{\lambda_{s_3}} \ldots$$
$$(a_2 a_{s_\eta})^{\lambda_{s_\eta}} (a_1 a_{r_2})^{\lambda_{r_2}} (a_1 a_{r_3})^{\lambda_{r_3}} \ldots.$$

This particular syzygy yields

$$e^{a_3 D_{r_1} - a_3 D_{s_1}} [0, 0, \lambda_4, \ldots, \lambda_\delta] = R.$$

The syzygies of which this is an example yield the set of relations

$$e^{a_3 D_{\rho_1} + a_3 D_{\rho_2} + \ldots + a_3 D_{\rho k} - a_3 D_{\sigma_1} - a_3 D_{\sigma_2} - \ldots - a_3 D_{\sigma k}} [0, 0, \lambda_4, \ldots, \lambda_\delta] = R,$$

where $\rho_1, \rho_2, \ldots, \rho_k$ are any of $r_1, r_2, r_3, \ldots$, and $\sigma_1, \sigma_2, \ldots, \sigma_k$ are any of $s_1, s_2, \ldots, s_\eta$.

Lastly, we have a set of syzygies we shall call syzygies of the type E. They are really forms of the Jacobian syzygy, an example of these is

$$(a_2 a_3) \{(a_2 a_{s_2}) - (a_2 a_3)\}^{\lambda_{s_2}} \{(a_2 a_{s_3}) - (a_2 a_3)\}^{\lambda_{s_3}} (a_2 a_{s_4})^{\lambda_{s_4}} \ldots$$
$$(a_2 a_{s_\eta})^{\lambda_{s_\eta}} (a_1 a_{r_1})^{\lambda_{r_1}} (a_1 a_{r_2})^{\lambda_{r_2}} \ldots$$
$$= (a_1 a_3) \{(a_1 a_{s_2}) - (a_1 a_3)\}^{\lambda_{s_2}} \{(a_1 a_{s_3}) - (a_1 a_3)\}^{\lambda_{s_3}} (a_1 a_{r_1})^{\lambda_{r_1}} (a_1 a_{r_2})^{\lambda_{r_2}} \ldots$$
$$(a_2 a_{s_4})^{\lambda_{s_4}} \ldots (a_2 a_{s_\eta})^{\lambda_{s_\eta}}$$
$$-(a_1 a_2)(a_3 a_{s_2})^{\lambda_{s_2}} (a_3 a_{s_3})^{\lambda_{s_3}} \{(a_1 a_{s_4}) - (a_1 a_2)\}^{\lambda_{s_4}} \ldots$$
$$\{(a_1 a_{s_\eta}) - (a_1 a_2)\}^{\lambda_{s_\eta}} (a_1 a_{r_1})^{\lambda_{r_1}} (a_1 a_{r_2})^{\lambda_{r_2}} \ldots,$$

whence $$e^{-a_3 D_{s_2} - a_3 D_{s_3}} [0, 1, \lambda_4, \ldots, \lambda_\delta] = R.$$

and so, in general, we have syzygies which yield

$$e^{-a_3 D_{\sigma_1} - a_3 D_{\sigma_2} - \ldots - a_3 D_{\sigma k}} [0, 1, \lambda_4, \ldots, \lambda_\delta] = R,$$

where $\sigma_1, \sigma_2, \ldots, \sigma_k$ are any, all or none of $s_2, s_3, \ldots, s_\eta$.

18. We have to prove that the $2^{\delta - t + 1}$ equations

$$\text{(i)} \quad e^{-a_3 D_{\sigma_1} - a_3 D_{\sigma_2} - \ldots - a_3 D_{\sigma k}} [0, 0, \lambda_4, \ldots, \lambda_\delta] = 0,$$

where $\sigma_1, \sigma_2, \ldots, \sigma_k$ are all, any or none of $t+1, t+2, \ldots, \delta$;

$$\text{(ii)} \quad e^{a_3 D_{\rho_1} + a_3 D_{\rho_2} + \ldots + a_3 D_{\rho_k} - a_3 D_{\sigma_1} - a_3 D_{\sigma_2} - \ldots - a_3 D_{\sigma_k}} [0, 0, \lambda_4, \ldots, \lambda_\delta] = 0,$$

where $\rho_1, \rho_2, \ldots, \rho_k$ are all or any of the numbers $r_1, r_2, r_3, \ldots$ which are contained in $t+1, t+2, \ldots, \delta$; and $\sigma_1, \sigma_2, \ldots, \sigma_k$ are all, any or none of the numbers $s_1, s_2, \ldots, s_\eta$ which are contained in $t+1, t+2, \ldots, \delta$;

$$\text{(iii)} \quad e^{-a_3 D_{\sigma_1} - a_3 D_{\sigma_2} - \ldots - a_3 D_{\sigma_k}} [0, 1, \lambda_4, \ldots, \lambda_\delta] = 0,$$

where $\sigma_1, \sigma_2, \ldots, \sigma_k$ are all, any or none of the numbers $s_1, s_2, \ldots, s_\eta$ which are contained in $t+1, t+2, \ldots, \delta$:

are just sufficient to express all forms

$$[0, \lambda_3, \lambda_4, \ldots, \lambda_t, \lambda_{t+1}, \ldots, \lambda_\delta],$$

for which $\lambda_3 < 2^{\delta-t+1}$, in terms of forms

$$[0, \mu_3, \lambda_4, \ldots, \lambda_t, \mu_{t+1}, \ldots, \mu_\delta],$$

where $\mu_3 \nless 2^{\delta-t+1}$, and the first of the differences

$$\lambda_\delta - \mu_\delta, \lambda_{\delta-1} - \mu_{\delta-1}, \ldots, \lambda_{t+1} - \mu_{t+1}$$

which does not vanish is positive.

The proof follows the lines of the proof of the Lemma of § 15, and we need not give it in full.

We assume that the theorem is true for a particular value of t, and then proceed to prove the next step. We have two cases here.

(i) $t = r$; then applying the theorem of § 9, we show that

$$e^{-a_3 D_r} [0, \lambda_3, \lambda_4, \ldots, \lambda_\delta] = R \quad \text{for} \quad \lambda_3 = 0, 1, \ldots, 2^{\delta-r}-1.$$

We obtain, in the same way, for the same values of λ_3,

$$e^{+a_3 D_r} [0, \lambda_3, \lambda_4, \ldots, \lambda_\delta] = R,$$

for the proof of the theorem of § 9 is not altered if the sign of certain of the operators is changed throughout. From these two equations we obtain the result by the reasoning of § 15.

(ii) $t = s$; our assumption gives at once

$$e^{a_3 D_s} [0, \lambda_3, \lambda_4, \ldots, \lambda_\delta] = R \quad \text{for} \quad \lambda_3 = 0, 1, \ldots, 2^{\delta-s+1}.$$

Then, applying the theorem of § 9, we find the truth of the statement of this paragraph.

Thus, in either case, the induction proceeds step by step, and, as the theorem is true for the simplest case of $t = \delta - 1$, it is always true.

19. We apply this result to the relations of § 17, and we at once obtain the truth of the theorem of § 10 for the equation

$$e^{-a_2 D_3 - a_2 D_{s_2} - \dots - a_2 D_{s_\eta}} (0, \lambda_3, \dots, \lambda_\delta) = R$$

so far as the argument λ_3 is concerned.

The proof follows the same lines for the arguments $\lambda_{s_2}, \dots, \lambda_{s_\eta}$. But it is necessary now in order to complete the proof to add a fresh convention.

We have so far regarded the equations

$$e^{-a_r D_{\sigma_1} - a_r D_{\sigma_2} - \dots - a_r D_{\sigma_h}} (\lambda_2, \lambda_3, \dots, \lambda_{r-1}, 0, \lambda_{r+1}, \dots, \lambda_\delta) = R$$

$$(r < \sigma_1 < \sigma_2 < \dots < \sigma_h),$$

$$e^{-a_r D_{\tau_1} - a_r D_{\tau_2} - \dots - a_r D_{\tau_k}} (\lambda_2, \lambda_3, \dots, \lambda_{r-1}, 0, \lambda_{r+1}, \dots, \lambda_\delta) = R$$

$$(r < \tau_1 < \tau_2 < \dots < \tau_k),$$

as simultaneous when $\sigma_1 = \tau_1$.

We must now arrange all our equations in sequence according to the law that the first of the above equations precedes the second if the first of the numbers

$$\sigma_1 - \tau_1, \quad \sigma_2 - \tau_2, \quad \dots,$$

which does not vanish is positive, and this rule will be made complete if we introduce the symbols σ_{h+1}, τ_{k+1}, each of which is supposed to be numerically greater than any given number.

Thus, when $h = 0$, we have the equation

$$(\lambda_2, \lambda_3 \dots, \lambda_{r-1}, 0, \lambda_{r+1}, \dots, \lambda_\delta) = R,$$

for which σ_1 exceeds any given number, and which therefore precedes all the other equations at the moment under consideration.

We deal with our equations in regular order, beginning with the earliest in the sequence. Each equation will reduce a fresh form or else with the previous equations in the sequence it must give rise to a syzygy.

The truth of the theorem of § 10 is established now for every possible case, exactly as we have established it for those cases we have discussed.

20. Having arrived at the truth of the theorem of § 10, let us consider the equation

$$e^{-a_r D_{s_1} - a_r D_{s_2} - \dots - a_r D_{s_\eta}} (\lambda_2, \lambda_3, \dots, \lambda_{r-1}, 0, \lambda_{r+1}, \dots, \lambda_\delta) = R$$

$$(r < s_1 < s_2 < \dots < s_\eta).$$

In general it has been taken as one of $2^{\delta-s_1}$ equations which will reduce forms

$$(\lambda_2, \lambda_3, \ldots, \lambda_{r-1}, \lambda_r, \lambda_{r+1}, \ldots, \lambda_\delta) = R,$$

when
$$\lambda_r = 2^{\delta-s_1},\ 2^{\delta-s_1}+1,\ \ldots,\ 2^{\delta-s_1+1}-1.$$

We have in the last paragraph introduced a convention by which these $2^{\delta-s_1}$ equations are arranged in a definite sequence. We may then associate each equation with a definite form which it reduces. We shall suppose that the earliest equation will reduce the form with the lowest value of λ_r, and so on. This supposition gives consistent results, for the determinants of the coefficients involved are easily seen to be different from zero—in general. By this arrangement the equation

$$e^{-a_r D_{s_1} - a_r D_{s_2} - \ldots - a_r D_{s_\eta}} (\lambda_2, \lambda_3, \ldots, \lambda_{r-1}, 0, \lambda_{r+1}, \ldots, \lambda_\delta) = R$$

reduces the form

$$(\lambda_2, \lambda_3, \ldots, \lambda_{r-1}, 2^{\delta-s_1}+2^{\delta-s_2}+\ldots+2^{\delta-s_\eta}, \lambda_{r+1}, \ldots, \lambda_{s_1}-2^{\delta-s_1}, \ldots,$$
$$\lambda_{s_2}-2^{\delta-s_2}, \ldots);$$

i.e. it expresses this form in terms of later members of our sequence of forms and of products of forms of lower degree.

If $\lambda_s < 2^{\delta-s+1}$ or if $\lambda_\tau < 2^{\delta-\tau}$, where $\tau > r$ and is not one of $s_1, s_2, \ldots, s_\eta$, then this form has been reduced by a previous equation. But, in either of these cases, there is a syzygy by means of which this equation can be expressed in terms of previous equations, as we have shewn in our theorem of § 10.

Thus, to every equation we have a definite reduction or a syzygy.

21. Now let us review the perpetuant types of degree δ.

Firstly, they can all, reducible or irreducible, be expressed linearly in terms of the forms

$$(\lambda_2, \lambda_3, \ldots, \lambda_\delta),$$

and these forms are all linearly independent. *Secondly*, any product of perpetuant types of total degree δ can be expressed as a product of *two* perpetuants, neither of which is necessarily irreducible; and, when this product is expressed in terms of our standard forms of degree δ, it can be written, without ambiguity,

$$e^{-a_r D_{s_1} - a_r D_{s_2} - \ldots - a_r D_{s_\eta}} (\lambda_2, \ldots, \lambda_{r-1}, 0, \lambda_{r+1}, \ldots, \lambda_\delta)$$
$$(r < s_1 < s_2 < \ldots < s_\eta < \delta+1).$$

Thirdly, the complete discussion of the equations

$$e^{-a_r D_{s_1} - a_r D_{s_2} - \dots - a_r D_{s_r}} (\lambda_2, \dots, \lambda_{r-1}, 0, \lambda_{r+1}, \dots, \lambda_\delta) = R,$$

involves, firstly, the discovery of the laws of reducibility and irreducibility, and, secondly, the discovery of all the syzygies of the first kind.

The laws of reducibility established by Grace follow from this. And we have now shewn that all syzygies of the first kind can very simply be deduced from those of Stroh and the Jacobian form of syzygy.

III. *Forms of Finite Order.*

22. The discussion for forms of finite order follows identically the same lines as that for perpetuants. We express all covariants of degree δ in terms of the forms

$$(\lambda_2, \lambda_3, \dots, \lambda_\delta)$$

defined as in § 5. We then consider every possible product of *two* covariants of total degree δ, and we express it in terms of our standard forms. The equations which we get in this way will give us the laws of reducibility of our standard forms, and also will yield every syzygy for this degree.

The discussion is rendered more complicated by the fact that

$$(\lambda_2, \lambda_3, \dots, \lambda_\delta)$$

is no longer equal to the simple product

$$(a_1 a_2)^{\lambda_2} (a_1 a_3)^{\lambda_3} \dots (a_1 a_\delta)^{\lambda_\delta},$$

but is equal to this plus a linear function of the fundamental forms.

If the set of inequalities

$$\lambda_2 \not> n_1, \; 2\lambda_2 + \lambda_3 \not> n_1 + n_2, \; 2\lambda_2 + 2\lambda_3 + \lambda_4 \not> n_1 + n_2 + n_3, \; \dots,$$
$$2\lambda_2 + 2\lambda_3 + 2\lambda_4 + \dots + 2\lambda_{\delta-1} + \lambda_\delta \not> n_1 + n_2 + n_3 + \dots + n_{\delta-1},$$

is not satisfied,

$$(a_1 a_2)^{\lambda_2} (a_1 a_3)^{\lambda_3} \dots (a_1 a_\delta)^{\lambda_\delta}$$

is itself a fundamental form; and we must write

$$(\lambda_2, \lambda_3, \dots, \lambda_\delta) = 0.$$

The analysis for perpetuants must then be modified in two ways.

Firstly, the product ($s < s_1 < s_2 < \ldots < s_\eta < \delta+1$)

$$\left(\frac{a_{r_1}^{\lambda_{r_1}} a_{r_2}^{\lambda_{r_2}} \ldots a_{r_\epsilon}^{\lambda_{r_\epsilon}}}{a_1}\right)\left(\frac{a_{s_1}^{\lambda_{s_1}} a_{s_2}^{\lambda_{s_2}} \ldots a_{s_\eta}^{\lambda_{s_\eta}}}{a_s}\right)$$

is equal to a sum of forms, of which the *earliest* are

$$e^{-a_s D_{s_1} - a_s D_{s_2} - \ldots - a_s D_{s_\eta}} (\lambda_2, \ldots, \lambda_{s-1}, 0, \lambda_{s+1}, \ldots, \lambda_\delta),$$

in general; but which contains other terms too.

Secondly, if the numbers $\lambda_{s_1}, \lambda_{s_2}, \ldots, \lambda_{s_\eta}$ do not satisfy the set of inequalities

$$\lambda_{s_1} \not> n_s,\quad 2\lambda_{s_1}+\lambda_{s_2} \not> n_s+n_{s_1},\quad 2\lambda_{s_1}+2\lambda_{s_2}+\lambda_{s_3} \not> n_s+n_{s_1}+n_{s_2},\quad \ldots,$$

$$2\lambda_{s_1}+2\lambda_{s_2}+\ldots+2\lambda_{s_{\eta-1}}+\lambda_{s_\eta} \not> n_s+n_{s_1}+\ldots+n_{s_{\eta-1}},$$

then
$$\left(\frac{a_{s_1}^{\lambda_{s_1}} a_{s_2}^{\lambda_{s_2}} \ldots a_{s_\eta}^{\lambda_{s_\eta}}}{a_s}\right) = 0,$$

in this case there is no equation.

Thus many of the equations obtained for the case of perpetuants do not exist for forms of finite order; the corresponding reductions either do not exist or else they are brought about by other equations. Thus, equations which for perpetuants yielded syzygies may now yield reductions. It will frequently be found that the reduction which corresponds to such an equation is most simply found by a consideration of what the corresponding perpetuant syzygy becomes when the orders of the quantics take the finite values of the case in hand.

The forms
$$(\lambda_2, \lambda_3, \ldots, \lambda_\delta)$$

are arranged in sequence according to the same law as for perpetuants. Also the law of sequence of equations is still adhered to. It is useful to remember that no form can be reducible for quantics of finite order, which is not so for perpetuants, and also that an equation which produces a reduction for perpetuants must reduce the same or an earlier form (if it exists at all) for quantics of finite order.

23. At the outset the question rises: Can we find an explicit expression for
$$(\lambda_2, \lambda_3, \ldots, \lambda_\delta)$$

in terms of $$(a_1 a_2)^{\lambda_2} (a_1 a_3)^{\lambda_3} \ldots (a_1 a_\delta)^{\lambda_\delta},$$

and the fundamental forms?

We proceed to find such an expression for the case when n_1 alone is finite and the orders of all the other quantics are infinite. In this case we observe that a fundamental form is simply a form

$$(a_1 a_2)^{\lambda_2} (a_1 a_3)^{\lambda_3} \ldots (a_1 a_\delta)^{\lambda_\delta},$$

for which $\lambda_2 > n_1$.

We proceed to prove the following theorem :—

When the orders $n_2, n_3, \ldots, n_\delta$ of the quantics concerned are greater than the weight of the covariants under consideration, while the order n_1 is less than this quantity, the covariant

$$(\lambda_2, \lambda_3, \ldots, \lambda_\delta)$$

may be represented by the sum

$$(a_1 a_2)^{\lambda_2} (a_1 a_3)^{\lambda_3} \ldots (a_1 a_\delta)^{\lambda_\delta}$$

$$+\Sigma(-)^i \binom{n_1-\lambda_2+i-1}{i-1} \binom{\lambda_3}{j_3} \binom{\lambda_4}{j_4} \cdots \binom{\lambda_\delta}{j_\delta} (a_1 a_2)^{n_1+i} (a_1 a_3)^{j_3} (a_1 a_4)^{j_4} \ldots (a_1 a_\delta)^{j_\delta},$$

where $$i = \Sigma\lambda - \Sigma j - n_1.$$

For simplicity we will take $\delta = 4$. And for this case we will prove the symbolical identity

(I) $$(a_1 a_2)^{\lambda_2} (a_1 a_3)^{\lambda_3} (a_1 a_4)^{\lambda_4}$$

$$+\Sigma(-)^i \binom{n_1-\lambda_2+i-1}{i-1} \binom{\lambda_3}{j} \binom{\lambda_4}{\rho-i-j} (a_1 a_2)^{n_1+i} (a_1 a_3)^j (a_1 a_4)^{\rho-i-j}$$

$$= \sum_{\xi=0}^{\lambda_3-\rho} \binom{\rho-1+\xi}{\rho-1} (a_2 a_3)^\rho (a_1 a_2)^{\lambda_2+\xi} (a_1 a_3)^{\lambda_3-\rho-\xi} (a_1 a_4)^{\lambda_4}$$

$$+\sum_{\xi=1}^{\lambda_4} \binom{\lambda_3}{j} \binom{\xi-1}{\rho-1-j} (a_2 a_3)^j (a_2 a_4)^{\rho-j} (a_1 a_2)^{\lambda_2+\lambda_3+\xi-\rho} (a_1 a_4)^{\lambda_4-\xi},$$

where $$\rho = \lambda_2+\lambda_3+\lambda_4-n_1.$$

The forms on the right are ordinary symbolical products which represent as they stand covariants of the quantics with which we are concerned. Let us assume the truth of this identity as it stands and then deduce that it is true when λ_4 is changed into λ_4+1 and n_1 into

n_1+1.   It is to be noticed that this change leaves ρ unchanged. To do this multiply the supposed identity by $(a_1 a_4)$. Then when the order of the a_1 quantic is n_1+1, those terms under the sign of summation on the left for which $i = 1$ are no longer fundamental, and those terms only.

From the identity

$$(a_2 a_3)^j (a_2 a_4)^{\rho-j} = \{(a_1 a_3)-(a_1 a_2)\}^j \{(a_1 a_4)-(a_1 a_2)\}^{\rho-j},$$

we obtain

$$(a_1 a_3)^j (a_1 a_4)^{\rho-j}$$
$$= (a_2 a_3)^j (a_2 a_4)^{\rho-j} - \sum_{i_1+i_2 \neq 0} (-)^{i_1+i_2} \binom{j}{i_1}\binom{\rho-j}{i_2}(a_1 a_2)^{i_1+i_2}(a_1 a_3)^{j-i_1}(a_1 a_4)^{\rho-j-i_2}.$$

We make use of this result and the identity becomes

$$(a_1 a_2)^{\lambda_2}(a_1 a_3)^{\lambda_3}(a_1 a_4)^{\lambda_4+1}$$

$$+ \sum_{i=2} (-)^i \binom{n_1-\lambda_2+i-1}{i-1}\binom{\lambda_3}{j}\binom{\lambda_4}{\rho-i-j}(a_1 a_2)^{n_1+i}(a_1 a_3)^j (a_1 a_4)^{\rho+1-i-j}$$

$$+ \sum_{i_1+i_2 \neq 0} \binom{\lambda_3}{j}\binom{\lambda_4}{\rho-1-j}(-)^{i_1+i_2}\binom{j}{i_1}\binom{\rho-j}{i_2}(a_1 a_2)^{n_1+1+i_1+i_2}$$
$$\times (a_1 a_3)^{j-i_1}(a_1 a_4)^{\rho-j-i_2}$$

$$= \sum_{\xi=0}^{\lambda_3-\rho} \binom{\rho-1+\xi}{\rho-1}(a_2 a_3)^\rho (a_1 a_2)^{\lambda_2+\xi}(a_1 a_3)^{\lambda_3-\rho-i}(a_1 a_4)^{\lambda_4+1}$$

$$+ \sum_{\xi=1}^{\lambda_4} \binom{\lambda_3}{j}\binom{\xi-1}{\rho-1-j}(a_2 a_3)^j (a_2 a_4)^{\rho-j}(a_1 a_2)^{\lambda_2+\lambda_3+\xi-\rho}(a_1 a_4)^{\lambda_4+1-\xi}$$

$$+ \sum \binom{\lambda_3}{j}\binom{\lambda_4}{\rho-1-j}(a_2 a_3)^j (a_2 a_4)^{\rho-j}(a_1 a_2)^{n_1+1}.$$

The right-hand side of our identity is already the same that we should get by writing λ_4+1 for λ_4, and n_1+1 for n_1 in the identity we want to prove. The coefficient of $(a_1 a_2)^{n_1+1+i}(a_1 a_3)^j (a_1 a_4)^{\rho-i-j}$ on the left is

$$(-)^{i+1}\binom{n_1-\lambda_2+i}{i}\binom{\lambda_3}{j}\binom{\lambda_4}{\rho-i-1-j}$$

$$+(-)^i \sum_k \binom{\lambda_3}{k}\binom{\lambda_4}{\rho-1-k}\binom{k}{j}\binom{\rho-k}{\rho-i-j}.$$

Now $$\sum_k \binom{\lambda_3}{k}\binom{\lambda_4}{\rho-1-k}\binom{k}{j}\binom{\rho-k}{\rho-i-j}$$

= the coefficient of $x^j y^{\rho-1} z^{\rho-i-j}$ in the expansion of

$$\{1+y(1+x)\}^{\lambda_3}\{1+y(1+z)\}^{\lambda_4}(1+z)$$

= the coefficient of $x^j y^{\rho-1} z^{\rho-i-j}$ in the expansion of

$$\left\{1+\frac{xy}{1+y}\right\}^{\lambda_3}\left\{1+\frac{zy}{1+y}\right\}^{\lambda_4}(1+y)^{\lambda_3+\lambda_4}(1+z)$$

$$=\binom{\lambda_3}{j}\left\{\binom{\lambda_4}{\rho-i-j}\binom{\lambda_3+\lambda_4-\rho+i}{i-1}+\binom{\lambda_4}{\rho-i-j-1}\binom{\lambda_3+\lambda_4-\rho+i+1}{i}\right\}.$$

Hence the coefficient of

$$(a_1a_2)^{n_1+1+i}(a_1a_3)^j(a_1a_4)^{\rho-i-j}$$

is $$(-)^i\binom{\lambda_3}{j}\binom{\lambda_4}{\rho-i-j}\binom{n_1-\lambda_2+i}{i-1}$$

$$+(-)^i\binom{\lambda_3}{j}\binom{\lambda_4}{\rho-i-j-1}\left\{\binom{n_1-\lambda_2+i+1}{i}-\binom{n_1-\lambda_2+i}{i}\right\}$$

$$=\quad(-)^i\binom{\lambda_3}{j}\binom{n_1-\lambda_2+i}{i-1}\left\{\binom{\lambda_4}{\rho-i-j}+\binom{\lambda_4}{\rho-i-j-1}\right\}$$

$$=\quad(-)^i\binom{n_1-\lambda_2+i}{i-1}\binom{\lambda_3}{j}\binom{\lambda_4+1}{\rho-i-j}.$$

The identity is true, then, when we replace λ_4 and n_1 by λ_4+1 and n_1+1.

If, then, it is true for certain values of λ_4 and n_1, it is still true if these values are both increased by unity, and therefore if they are both increased by any the same number.

(i) Let n_1 be greater than λ_4. Then, if the identity is true when $n_1-\lambda_4$ and 0 are written for n_1 and λ_4, it is true as it stands. It will be sufficient simply to discuss the case $\lambda_4=0$ and leave n_1 unaltered. The identity then becomes

(II) $$(a_1a_2)^{\lambda_2}(a_1a_3)^{\lambda_3}$$

$$+\Sigma(-)^i\binom{n_1-\lambda_2+i-1}{i-1}\binom{\lambda_3}{\rho-1}(a_1a_2)^{n_1+i}(a_1a_3)^{\rho-1}$$

$$=\sum_{\xi=0}^{\lambda_3-\rho}\binom{\rho-1+\xi}{\rho-1}(a_2a_3)^{\rho}(a_1a_2)^{\lambda_2+\xi}(a_1a_3)^{\lambda_3-\rho-\xi}.$$

To prove this we write the right-hand side in the form

$$\sum_{\xi=0}^{\lambda_3-\rho} \binom{\rho-1+\xi}{\rho-1} \{(a_1 a_3)-(a_1 a_2)\}^{\rho} (a_1 a_2)^{\lambda_2+\xi} (a_1 a_3)^{\lambda_3-\rho-\xi}$$

$$= \sum_{\xi=0}^{\lambda_3-\rho} \binom{\rho-1+\xi}{\rho-1} \Sigma (-)^{\zeta} \binom{\rho}{\zeta} (a_1 a_2)^{\lambda_2+\xi+\zeta} (a_1 a_3)^{\lambda_3-\xi-\zeta}.$$

The coefficient of $(a_1 a_2)^{\lambda_2+\eta} (a_1 a_3)^{\lambda_3-\eta}$

is $(-)^{\eta}\left[\binom{\rho}{\eta}-\binom{\rho}{\rho-1}\binom{\rho}{\eta-1}+\ldots+(-)^{\lambda_3-\rho}\binom{\rho-1+\lambda_3-\rho}{\rho-1}\binom{\rho}{\eta-\lambda_3+\rho}\right].$

To find the value of this we shall prove the identity

$$\binom{\rho}{i}-\binom{\rho}{1}\binom{\rho}{i-1}+\binom{\rho+1}{2}\binom{\rho}{i-2}-\ldots+(-)^{j}\binom{\rho+j-1}{j}\binom{\rho}{i-j}$$
$$= (-)^{j}\binom{i-1}{j}\binom{\rho+j}{i}.$$

Assume that it is true as it stands and add one more term $(-)^{j+1}\binom{\rho+j}{j+1}\binom{\rho}{i-j-1}$ to each side.

The right-hand side becomes

$$(-)^{j+1}\frac{(\rho+j)!\{\rho i-(j+1)(\rho+j+1-i)\}}{(j+1)!(\rho+j+1-i)!(i-j-1)!\,i} = (-)^{j+1}\binom{i-1}{j+1}\binom{\rho+j+1}{i},$$

and so the induction proceeds step by step: for the identity is obvious for $j=0$.

Making use of this result we find that the coefficient of $(a_1 a_2)^{\lambda_2+\eta} (a_1 a_3)^{\lambda_3-\eta}$ is

$$(-)^{\eta+\lambda_3-\rho}\binom{\eta-1}{\lambda_3-\rho}\binom{\lambda_3}{\eta},$$

which is the same as the coefficient of the corresponding term on the left-hand side of the identity, for $\lambda_2+\lambda_3 = n_1+\rho$. This coefficient is unity when η is zero, it is zero for $\eta = 1, 2, \ldots, n_1-\lambda_2$, and its value is

$$(-)^{i}\binom{n_1-\lambda_2+i-1}{i-1}\binom{\lambda_3}{\rho-i}$$

for $$\eta = n_1-\lambda_2+i.$$

The identity (I) is then true if $\lambda_4 = 0$, and therefore whenever $n_1 > \lambda_4$.

(ii) Let n_1 be equal to or less than λ_4. Then, if the identity is true when 0 and λ_4-n_1 are written for n_1 and λ_4, it is true as it stands. It will be sufficient to discuss the case $n_1=0$. It is just as easy to take the case $n_1<\lambda_2$. Here the left-hand side of (I) becomes

$$(a_1a_2)^{\lambda_2}(a_1a_3)^{\lambda_3}(a_1a_4)^{\lambda_4}+(-)^{\lambda_2-n_1}\binom{-1}{\lambda_2-n_1-1}(a_1a_2)^{\lambda_2}(a_1a_3)^{\lambda_3}(a_1a_4)^{\lambda_4},$$

every other term under the sign of summation vanishes. The left-hand side is therefore zero. On the right there are no terms in the first sum, for $\lambda_3-\rho$ is negative, and in the second sum every coefficient is zero for $\xi-1<\rho-1-j$, since j must be less than λ_3. Thus (I) is true when $n_1<\lambda_2$. [In the same way we see that the general expression in the enunciation of our theorem

$$(a_1a_2)^{\lambda_2}\dots(a_1a_\delta)^{\lambda_\delta}$$
$$+\Sigma(-)^i\binom{n_1-\lambda_2+i-1}{i-1}\binom{\lambda_3}{j_3}\binom{\lambda_4}{j_4}\dots\binom{\lambda_\delta}{j_\delta}(a_1a_2)^{n_1+i}(a_1a_3)^{j_3}\dots(a_1a_\delta)^{j_\delta},$$

vanishes when $n_1<\lambda_2$.]

The identity (I) is then true when $n_1 \not> \lambda_4$; it is therefore true for all values of n_1 and λ_4.

Now the identity (I) expresses the sum of

$$(a_1a_2)^{\lambda_2}(a_1a_3)^{\lambda_3}(a_1a_4)^{\lambda_4},$$

and certain fundamental forms as a sum of symbolical products which represent actual covariants of the quantics under discussion. This sum of covariants is then the covariant we have named

$$(\lambda_2,\ \lambda_3,\ \lambda_4).$$

The theorem is then true for degree 4. Assuming that it has been proved for degree $\delta-1$, it can be proved for degree δ in just the same way that it has been proved for degree 4. The actual form of the covariants on the right of the identity is not given, and it is not required. It is sufficient that the right-hand side of the identity should contain only symbolical products which represent actual covariants of the quantics concerned. There is no difficulty in obtaining the expression, but it is troublesome to write out, and no advantage is gained by doing so.

24. When the orders of all the quantics are finite the case is not so simple. For the discussion of the covariants of degree 4 we require

the linear function of fundamental forms that must be added to

$$(a_1 a_2)^{\lambda_2}(a_1 a_3)^{\lambda_3}$$

in order that the sum may really be a covariant of $a_{1_x}^{n_1}, a_{2_x}^{n_2}, a_{3_x}^{n_3}$. We shall prove that:—

The covariant

$$(\lambda_2, \lambda_3) = (a_1 a_2)^{\lambda_2}(a_1 a_3)^{\lambda_3}$$
$$+\Sigma(-)^i \binom{\lambda_3-\rho_1-\rho_2+i-1}{i-1}\binom{\lambda_3-\rho_2}{\rho_1-i}(a_1 a_2)^{n_1-\rho_2+i}(a_1 a_3)^{\rho_1+\rho_2-i},$$

where $\rho_i = \lambda_2+\lambda_3-n_i$, *or* 0, *according as* $\lambda_2+\lambda_3 >$ *or* $< n_i$.

In the first place the terms under the sign of summation are all fundamental forms, for

$$2(n_1-\rho_2+i)+\rho_1+\rho_2-i = 2n_1+\rho_1-\rho_2+i = n_1+n_2+i > n_1+n_2,$$

since the coefficient is zero unless $i > 0$.

Moreover the index of $(a_1 a_2)$ never exceeds $n_1-\rho_2+\rho_1 = n_2$, for $i \not> \rho_1$.

From the identity (II) of the last paragraph, we have for the case $\rho_2 = 0$,

$$(a_1 a_2)^{\lambda_2}(a_1 a_3)^{\lambda_3}+\Sigma(-)^i\binom{\lambda_3-\rho_1+i-1}{i-1}\binom{\lambda_3}{\rho_1-i}(a_1 a_2)^{n_1+i}(a_1 a_3)^{\rho_1-i}$$
$$=\sum_{\xi=0}^{\lambda_3-\rho_1}\binom{\rho_1-1+\xi}{\rho_1-1}(a_2 a_3)^{\rho_1}(a_1 a_2)^{\lambda_2+\xi}(a_1 a_3)^{\lambda_3-\rho_1-\xi},$$

an identity which establishes our theorem in this case. We shall take this as it stands and suppose that n_2 has its least possible value $\lambda_2+\lambda_3$.

Now in this replace λ_3 by $\lambda_3-\rho_2$, keeping λ_2 and ρ_1 unaltered; then n_1 must be replaced by $n_1-\rho_2$, since $n_1 = \lambda_2+\lambda_3-\rho_1$, and n_2 must be replaced by $n_2-\rho_2$; we have

$$(a_1 a_2)^{\lambda_2}(a_1 a_3)^{\lambda_3-\rho_2}+\Sigma(-)^i\binom{\lambda_3-\rho_1-\rho_2+i-1}{i-1}\binom{\lambda_3-\rho_2}{\rho_1-i}(a_1 a_2)^{n_1+i-\rho_2}(a_1 a_3)^{\rho_1-i}$$
$$=\sum_{\xi=0}^{\lambda_3-\rho_1-\rho_2}\binom{\rho_1-1+\xi}{\rho_1-1}(a_2 a_3)^{\rho_1}(a_1 a_2)^{\lambda_2+\xi}(a_1 a_3)^{\lambda_3-\rho_1-\rho_2-\xi}.$$

Now multiply this result through by $(a_1 a_3)^{\rho_2}$ and we have

$$(a_1 a_2)^{\lambda_2}(a_1 a_3)^{\lambda_3}+\Sigma(-)^i\binom{\lambda_3-\rho_1-\rho_2+i-1}{i-1}\binom{\lambda_3-\rho_2}{\rho_1-i}(a_1 a_2)^{n_1+i-\rho_2}(a_1 a_3)^{\rho_1+\rho_2-i}$$
$$=\sum_{\xi=0}^{\lambda_3-\rho_1-\rho_2}\binom{\rho_1-1+\xi}{\rho_1-1}(a_2 a_3)^{\rho_1}(a_1 a_2)^{\lambda_2+\xi}(a_1 a_3)^{\lambda_3-\rho_1-\xi}.$$

Since the right-hand side of this represents a covariant of the quantics concerned, it is $= (\lambda_2, \lambda_3)$. Q. E. D.

25. It will be sometimes useful to use the notation

$$(a_1 a_2)^{\lambda_2} (a_1 a_3)^{\lambda_3} + \Sigma(-)^i \binom{\lambda_3 - \rho_1 - \rho_2 + i - 1}{i-1} \binom{\lambda_3 - \rho_2}{\rho_1 - i} (a_1 a_2)^{n_1 + i - \rho_2} (a_1 a_3)^{\rho_1 + \rho_2 - i}$$

$$\equiv (a_1 a_2)_{\lambda_2} (a_1 a_3)_{\lambda_3}.$$

Also we shall copy the index notation of ordinary algebra further by writing

$$\{(a_1 a_3) - (a_1 a_2)\}_\lambda \equiv \Sigma(-)_i \binom{\lambda}{i} (a_1 a_2)_i (a_1 a_3)_{\lambda - i},$$

and also by writing

$$(a_1 a_2)_\mu (a_1 a_3)_\nu \{(a_1 a_3) - (a_1 a_2)\}_\lambda \equiv \Sigma(-)^i \binom{\lambda}{i} (a_1 a_2)_{\mu + i} (a_1 a_3)_{\nu + \lambda - i}.$$

When we confine ourselves to the operations of symbolical algebra this notation will not involve any assumptions.

We will now prove that with this notation

$$\text{(III)} \qquad \{(a_1 a_3) - (a_1 a_2)\}_\lambda = \{(a_1 a_3) - (a_1 a_2)\}^\lambda.$$

In other words we shall prove that

$$\Sigma(-)^i \binom{\lambda}{i} (a_1 a_2)_i (a_1 a_3)_{\lambda - i} = \Sigma(-)^i \binom{\lambda}{i} (a_1 a_2)^i (a_1 a_3)^{\lambda - i}.$$

In fact

$$\Sigma(-)^i \binom{\lambda}{i} (a_1 a_2)_i (a_1 a_3)_{\lambda - i} - \Sigma(-)^i \binom{\lambda}{i} (a_1 a_2)^i (a_1 a_3)^{\lambda - i}$$

$$= \Sigma(-)^{i+} \binom{\lambda}{i} \binom{n_2 - \rho_1 - i + j - 1}{j - 1} \binom{n_2 - i}{\rho_1 - j} (a_1 a_2)^{n_1 + j - \rho_2} (a_1 a_3)^{\rho_1 + \rho_2 - j}.$$

The coefficient of

$$(a_1 a_2)^{n_1 + j - \rho_2} (a_1 a_3)^{\rho_1 + \rho_2 - j}$$

$=$ the coefficient of $x^{j-1} y^{\rho_1 - j}$ in the expansion of

$$\{(1+x)(1+y) - 1\}^\lambda (1+x)^{j - \rho_1 - \rho_2 - 1} (1+y)^{-\rho_2} (-)^j$$

$=$ the coefficient of $x^{j-1} y^{\rho_1 - j}$ in the expansion of

$$\{x + y + xy\}^\lambda (1+x)^{j - \rho_1 - \rho_2 - 1} (1+y)^{-\rho_2} (-)^j$$

$= 0$, unless $j - 1 + \rho_1 - j \geqslant \lambda$,

i.e., unless $\qquad \lambda - n_1 > \lambda,$

all the coefficients on the right are then zero, and hence (III) is identically true.

Consider now the difference

$$(a_1 a_2)_\mu (a_1 a_3)_\nu \{(a_1 a_3)-(a_1 a_2)\}_{\lambda-\mu-\nu}-(a_1 a_2)^\mu (a_1 a_3)^\nu \{(a_1 a_3)-(a_1 a_2)\}^{\lambda-\mu-\nu}$$

$$= \Sigma(-)^i \binom{\lambda-\mu-\nu}{i} (-)^j \binom{\lambda-\mu-i-\rho_1-\rho_2+j-1}{j-1} \binom{\lambda-\mu-i-\rho_2}{\rho_1-j}$$

$$\times (a_1 a_2)^{n_1+j-\rho_2} (a_1 a_3)^{\rho_1+\rho_2-j}.$$

The coefficient of $\qquad (a_1 a_2)^{n_1+j-\rho_2} (a_1 a_3)^{\rho_1+\rho_2-j} (-)^j$

$=$ the coefficient of $x^{j-1} y^{\rho_1-j}$ in the expansion of

$$\{(1+x)(1+y)-1\}^{\lambda-\mu-\nu} (1+x)^{\nu-\rho_1-\rho_2+j-1} (1+y)^{\nu-\rho_2}$$

$= 0$, unless $\rho_1 > \lambda-\mu-\nu$,

i.e., unless $\qquad \mu+\nu > n_1.$

Hence

(IV) $\quad (a_1 a_2)_\mu (a_1 a_3)_\nu \{(a_1 a_3)-(a_1 a_2)\}_{\lambda-\mu-\nu}$

$$= (a_1 a_2)^\mu (a_1 a_3)^\nu \{(a_1 a_3)-(a_1 a_2)\}^{\lambda-\mu-\nu},$$

unless $\qquad \mu+\nu > n_1.$

IV. *Covariants of Degree 4.*

26. These may all be represented as linear functions of the covariants defined by

$$(\lambda_2,\ \lambda_3,\ \lambda_4).$$

We shall suppose the quantics of which these are covariants are

$$a_{1_x}^{n_1},\ a_{2_x}^{n_2},\ a_{3_x}^{n_3},\ a_{4_x}^{n_4}.$$

Then we obviously must have

$$\lambda_2 \ngtr n_2,\quad \lambda_3 \ngtr n_3,\quad \lambda_4 \ngtr n_4.$$

Also, if the set of inequalities

$$\lambda_2 \ngtr n_1,\quad 2\lambda_2+\lambda_3 \ngtr n_1+n_2,\quad 2\lambda_2+2\lambda_3+\lambda_4 \ngtr n_1+n_2+n_3$$

is not satisfied, the form $\qquad (\lambda_2,\ \lambda_3,\ \lambda_4) = 0.$

Otherwise these forms are linearly independent.

The first step towards discussing the problem of reducibility is to express all products of lower forms of total degree 4 in terms of these forms, just as we have done for perpetuants. We obtain thus a set of equations which we have to discuss.

As a basis of discussion the forms are arranged in sequence as in perpetuants; thus

$$(\lambda_2, \lambda_3, \lambda_4)$$

precedes

$$(\mu_2\ \mu_3,\ \mu_4),$$

if the first of the differences

$$\lambda_4-\mu_4,\ \lambda_3-\mu_3,\ \lambda_2-\mu_2$$

which does not vanish is positive. We seek to express earlier members of the sequence in terms of later members and of products of forms.

27. Let us first suppose that the factor containing a_1 is of degree 3.

Then, by the theorem of § 24,

$$\left(\frac{a_2^{\lambda_2} a_3^{\lambda_3}}{a_1}\right) = (a_1 a_2)^{\lambda_2} (a_1 a_3)^{\lambda_3}$$

$$+\Sigma(-)^i \binom{\lambda_3-\rho_1-\rho_2+i-1}{i-1} \binom{\lambda_3-\rho_2}{\rho_1-i} (a_1 a_2)^{n_1-\rho_2+i} (a_1 a_3)^{\rho_1+\rho_2-i}.$$

Hence

$$\left(\frac{a_2^{\lambda_2} a_3^{\lambda_3}}{a_1}\right) a_{4_x}^{n_4} = (\lambda_2, \lambda_3, 0),$$

for it can differ from this by fundamental forms only.

Again, if $\rho_1 = \lambda_2+\lambda_4-n_1$ and $\rho_2 = \lambda_2+\lambda_4-n_2$,

(V) $$\left(\frac{a_2^{\lambda_2} a_4^{\lambda_4}}{a_1}\right) a_{3_x}^{n_3} = (\lambda_2, 0, \lambda_4)$$

$$+\Sigma(-)^i \binom{\lambda_4-\rho_1-\rho_2+i-1}{i-1} \binom{\lambda_4-\rho_2}{\rho_1-i} (n_1-\rho_2+i,\ 0,\ \rho_1+\rho_2-i).$$

This gives a reduction for $(\lambda_2, 0, \lambda_4)$, provided

$$2\lambda_2+\lambda_4 \not> n_1+n_2.$$

Again,

(VI) $$\left(\frac{a_3^{\lambda_3} a_4^{\lambda_4}}{a_1}\right) a_{2_x}^{n_2} = (0, \lambda_3, \lambda_4)$$

$$+\Sigma(-)^i \binom{\lambda_4-\rho_1-\rho_3+i-1}{i-1} \binom{\lambda_4-\rho_3}{\rho_1-i} (0,\ n_1-\rho_3+i,\ \rho_1+\rho_3-i),$$

which gives a reduction for $(0, \lambda_3, \lambda_4)$, provided

$$2\lambda_3+\lambda_4 \not> n_1+n_3 \quad \text{and} \quad \lambda_3 \not> n_1.$$

28. If the factor containing a_1 is of degree 2, we have simply

(VII) $$(a_1 a_2)^{\lambda_2}(a_3 a_4)^{\lambda_4} = \Sigma(-)^i \binom{\lambda_4}{i}(a_1 a_2)^{\lambda_2}(a_1 a_3)^i (a_1 a_4)^{\lambda_4-i}$$

$$= e^{-a_3 D_4}(\lambda_2, 0, \lambda_4),$$

using the same notation as for perpetuants.

Next we have

(VIII) $$(a_1 a_3)^{\lambda_3}(a_2 a_4)^{\lambda_4} = e^{-a_2 D_4}(0, \lambda_3, \lambda_4),$$

and then

(IX) $$(a_1 a_4)^{\lambda_4}(a_2 a_3)^{\lambda_3} = e^{-a_2 D_3}(0, \lambda_3, \lambda_4).$$

Finally, the factor containing a_1 may be of degree 1 only, then we have

(X) $$\left(\frac{a_3^{\lambda_3} a_4^{\lambda_4}}{a_2}\right) a_{1_x}^{n_1} = e^{-a_2 D_3 - a_2 D_4}\Big[(0, \lambda_3, \lambda_4)$$

$$+\Sigma(-)^i \binom{\lambda_4-\rho_2-\rho_3+i-1}{i-1}\binom{\lambda_4-\rho_3}{\rho_2-i}(0, n_2-\rho_3+i, \rho_2+\rho_3-i)\Big],$$

provided $$2\lambda_3+\lambda_4 \not> n_2+n_3, \quad \lambda_3 \not> n_2.$$

Thus, we have obtained every possible reduction equation for degree 4. The equations are either the same as or modifications of the corresponding perpetuant equations.

In discussing the equations we shall confine ourselves to the case of most importance, viz., when

$$n_2 = n_3 = n_4 = n;$$

but the order n_1 may be any independent number.

29. As concerns λ_4, the only limit is the same as for perpetuants:

$$(\lambda_2, \lambda_3, \lambda_4)$$

is reducible if $\lambda_4 = 0$; otherwise we must go to λ_2 or λ_3.

For the limit of λ_3 for reducibility we have two equations, (V) and (VII).

From (V) we learn that

$$(\lambda_2, 0, \lambda_4) = R,$$

provided $$2\lambda_2+\lambda_4 \not> n+n_1.$$

Putting the value of $(\lambda_2, 0, \lambda_4)$ obtained from (V) in (VII), we find

$$(\lambda_2, 1, \lambda_4-1) = R,$$

provided $$2\lambda_2+\lambda_4 \not> n+n_1.$$

If $2\lambda_2+\lambda_4 > n+n_1$, we have, from (VII),

$$(\lambda_2, 0, \lambda_4) = R$$

(for λ_4 cannot exceed $n_4 = n$ in any case).

Thus always $$(\lambda_2, 0, \lambda_4) = R,$$

and $$(\lambda_2, 1, \lambda_4) = R,$$

provided $$\lambda_4 < n, \quad \text{and} \quad 2\lambda_2+\lambda_4 \not> n+n_1-1.$$

30. Let us now discuss the reducibility limits of λ_2.

We have the following equations

(VI) $$(0, \lambda_3, \lambda_4) = R,$$

when $2\lambda_3+\lambda_4 \not> n+n_1$, $\lambda_3 \not> n_1$,

(VIII) $$(0, \lambda_3, \lambda_4)-\lambda_4(1, \lambda_3, \lambda_4-1)+\ldots = R,$$

when $\lambda_3 \not> n_1$,

(IX) $$(0, \lambda_3, \lambda_4)-\lambda_3(1, \lambda_3-1, \lambda_4)+\binom{\lambda_3}{2}(2, \lambda_3-2, \lambda_4)-\binom{\lambda_3}{3}(3, \lambda_3-3, \lambda_4) +\ldots = R.$$

when $\lambda_4 \not> n_1$,

(X) $$(0, \lambda_3, \lambda_4)-\lambda_3(1, \lambda_3-1, \lambda_4)+\binom{\lambda_3}{2}(2, \lambda_3-2, \lambda_4)-\binom{\lambda_3}{3}(3, \lambda_3-3, \lambda_4)+\ldots$$

$$-\lambda_4(1, \lambda_3, \lambda_4-1)+\lambda_3\lambda_4(2, \lambda_3-1, \lambda_4-1)-\binom{\lambda_3}{2}\lambda_4(3, \lambda_3-2, \lambda_4-1)+\ldots = R,$$

when $2\lambda_3+\lambda_4 \not> 2n$.

Taking these last two equations together, we see that (IX) is true when *either* $\lambda_4 \not> n_1$, *or* $2\lambda_3+\lambda_4 \not> 2n$. And that when we replace these conditions by the original condition of (IX) we may replace (X) by

(XI) $$(1, \lambda_3, \lambda_4-1)-\lambda_3(2, \lambda_3-1, \lambda_4-1)+\binom{\lambda_3}{2}(3, \lambda_3-2, \lambda_4-1)-\ldots = R,$$

when $\lambda_4 \not> n_1$, and $2\lambda_3+\lambda_4 \not> 2n$.

Let us first see what the equations give us just as they stand.

$(0, \lambda_3, \lambda_4)$ is reducible if any one of our equations exists. Hence we see that it is reducible unless $\lambda_3 > n_1$, $\lambda_4 > n_1$, and $2\lambda_3+\lambda_4 > 2n$.

The reduction of $(3, \lambda_3, \lambda_4)$ requires the coexistence of equations of each of the four types, and there is only one way in which it can be reduced. It is easy to see that it is not reducible unless

$$\lambda_3 \ngtr n-3,\quad \lambda_3 \ngtr n_1-3,\quad \lambda_4 \ngtr n-1,\quad \lambda_4 \ngtr n_1-1,\quad 2\lambda_3+\lambda_4 \ngtr 2n-5,$$

$$2\lambda_3+\lambda_4 \ngtr n+n_1-5.$$

The conditions of reducibility are more complicated when $\lambda_2 = 1$ or 2; it will be convenient to separate the discussion into two cases.

(α) $n_1 \geqslant n$.—The equations (VIII) and (IX) always exist; together they reduce $(1, \lambda_3-1, \lambda_4)$. Then $(1, \lambda_3, \lambda_4)$ is reducible if $\lambda_3 < n$.

From (VI) and (VIII) we have a reduction for $(1, \lambda_3, \lambda_4)$, provided $\lambda_4 < n$ and $2\lambda_3+\lambda_4 \ngtr n+n_1-1$.

Thus $(1, \lambda_3, \lambda_4) = R$, when $\lambda_3 < n$, or when

$$\lambda_4 < n \quad\text{and}\quad 2\lambda_3+\lambda_4 \ngtr n+n_1-1.$$

From the first two equations with (XI) we find that $(2, \lambda_3, \lambda_4) = R$, when $\lambda_3 \ngtr n-1$, $\lambda_4 \ngtr n-1$, $2\lambda_3+\lambda_4 \ngtr n+n_1-3$.

In this case we observe that $(0, \lambda_3, \lambda_4)$ is always reducible.

(β) $n_1 < n$.—Here $(1, \lambda_3, \lambda_4)$ may be reduced by (VI) and (VIII), in which case we have the conditions

$$\text{(i)}\quad \lambda_3 \ngtr n_1,\quad \lambda_4 \ngtr n-1,\quad 2\lambda_3+\lambda_4 \ngtr n+n_1-1;$$

or by (VIII) and (IX) in which case the conditions are

$$\text{(ii)}\quad \lambda_3 \ngtr n_1-1,\quad \lambda_4 \ngtr n_1;$$

or

$$\text{(iii)}\quad \lambda_3 \ngtr n_1-1,\quad 2\lambda_3+\lambda_4 \ngtr 2n-2;$$

or else by (XI) when

$$\text{(iv)}\quad \lambda_4 \ngtr n_1-1,\quad 2\lambda_3+\lambda_4 \ngtr 2n-1.$$

Also $(2, \lambda_3, \lambda_4)$ may be reduced by (VI), (VIII) and (IX) when the conditions are

$$\text{(i)}\quad \lambda_3 \ngtr n_1-2,\quad \lambda_4 \ngtr n-1,\quad 2\lambda_3+\lambda_4 \ngtr n+n_1-3;$$

or, by (XI), (VIII) and (IX), when

$$\text{(ii)}\quad \lambda_3 \ngtr n_1-2,\quad \lambda_4 \ngtr n_1-1,\quad 2\lambda_3+\lambda_4 \ngtr 2n-3;$$

or else using (VI) and (VIII) to reduce the first term of (XI), we obtain the conditions

$$\text{(iii)}\quad \lambda_3 \not> n_1-1, \quad \lambda_4 \not> n_1-1, \quad 2\lambda_3+\lambda_4 \not> n+n_1-3.$$

31. It is necessary to examine equation (VI) a little more closely. The two conditions for its existence may be replaced by the single condition $\lambda_3 \not> n_1-\rho$.

When $\lambda_3 = n_1-\rho$, the equation takes the form

$$(0, \lambda_3, \lambda_4)-\rho_1(0, \lambda_3+1, \lambda_4-1) = R;$$

and when $\lambda_3 < n_1-\rho$, it takes the form

$$(0, \lambda_3, \lambda_4) = R;$$

where in each case R represents a linear function of products of forms and of forms $(0, \mu_3, \mu_4)$ for which $\mu_4 < \lambda_4-1$.

A difficulty apparently arises when we use (VI) and (VIII) in conjunction in the case $\lambda_3 = n_1-\rho$; for eliminating $(0, \lambda_3, \lambda_4)$, we have

$$\rho_1(0, \lambda_3+1, \lambda_4-1)-\lambda_4(1, \lambda_3, \lambda_4-1)+\ldots = R,$$

giving a reduction for $(0, \lambda_3+1, \lambda_4-1)$ instead of for $(1, \lambda_3, \lambda_4-1)$.

But in this case $(0, \lambda_3+1, \lambda_4-1)$ is reduced by another equation of the type (VIII), unless $\rho = 0$, and the reduction of $(1, \lambda_3, \lambda_4-1)$ then follows.

When $\rho = 0$, $\quad 2(\lambda_3+1)+(\lambda_4-1) \not> 2n-(\lambda_4-1)$,

and hence, from (IX), we have

$$(0, \lambda_3+1, \lambda_4-1)-(\lambda_3+1)(1, \lambda_3, \lambda_4-1)+\ldots = R.$$

Then, taking these equations in conjunction, we obtain the reductions exactly as stated in the last paragraph.

32. We have so far discussed our equations without any reference to the reductions already obtained when $\lambda_3 < 2$ or $\lambda_4 < 1$. Thus some of our forms will be reduced twice over. In the case of perpetuants the result of equating the different reductions was shewn to lead to a syzygy in every case. Now we shall find that it may lead to a syzygy or else it may lead to the reduction of a form not previously reduced.

Let us turn to equation (VI). Put $\lambda_3 = 0$ and use (V), thus

$$\text{(XII)}\quad R = (0, 0, \lambda_4) + \Sigma(-)^i \binom{\lambda_4 - \rho_1 + i - 1}{i-1}\binom{\lambda_4}{\rho_1 - i}(0, n_1+i, \rho_1 - i)$$

$$-(0, 0, \lambda_4) - \Sigma(-)^i \binom{\lambda_4 - \rho_1 + i - 1}{i-1}\binom{\lambda_4}{\rho_1 - i}(n_1+i, 0, \rho_1 - i),$$

giving a reduction for $(0, n_1+1, \lambda_4 - n_1 - 1)$ instead of a syzygy when $\lambda_4 > n_1+1$; it should be noted here that $\lambda_4 \ngtr n$.

Now this is already reduced by (IX) since $2(n_1+1)+\lambda_4 - n_1 - 1 \ngtr 2n$. Also we have a reduction for the form $(1, n_1, \lambda_4 - n_1 - 1)$ which occurs in this equation from (VI) and (VIII). Thus we obtain a reduction for $(2, n_1 - 1, \lambda_4 - n_1 - 1)$. This is the final reduction when $\lambda_4 > 2n_1$, but if $\lambda_4 \ngtr 2n_1$, we can use an equation of the type (XI), and so reduce the form $(3, n_1 - 2, \lambda_4 - n_1 - 1)$. These forms were not reduced in § 30.

The reduction when $\lambda_3 = 1$ is given by (VII). To find what (VI) gives us in this case, put $\lambda_2 = 0$ in (VII) and use (VI) for each term, thus (assuming $\rho = 0$)

$$a_{1_x}^{n_1} a_{2_x}^{n_2} (a_3 a_4)^{\lambda_4}$$

$$= \Sigma(-)^i \binom{\lambda_4}{i}(0, i, \lambda_4 - i)$$

$$= \Sigma(-)^i \binom{\lambda_4}{i}\left[\left(\frac{a_3^i a_4^{\lambda_4 - i}}{a_1}\right) a_{2_x}^{n_2} - \Sigma(-)^j \binom{\lambda_4 - i - \rho_1 + j - 1}{j-1}\binom{\lambda_4 - i}{\rho_1 - j}(0, n_1 + j, \rho_1 - j)\right]$$

$$= \Sigma(-)^i \binom{\lambda_4}{i}\left(\frac{a_3^i a_4^{\lambda_4 - i}}{a_1}\right) a_{2_x}^{n_2},$$

since the coefficient of $(0, n_1+j, \rho_1 - j)$ is zero. Thus in this case we only get a syzygy of a very obvious nature.

When ρ is not zero, we have only the case $\lambda_4 = n$, and then (VI) gives the reduction of $(0, 1, n)$ which has not been reduced by (VII).

When $\lambda_4 = 0$, (VI) only gives an obvious syzygy.

33. The equation (VIII) gives syzygies just as in the case of perpetuants when $\lambda_4 = 0$ or 1, or $\lambda_3 = 0$.

When $\lambda_3 = 1$, we reduced the equation in the perpetuant theory by means of the syzygy

$$\{(a_2a_4)-(a_1a_3)\}^{\lambda_4+1}-\{(a_3a_4)-(a_1a_2)\}^{\lambda_4+1}=0.$$

This holds good as it stands when $\lambda_4 \ngtr n-1$, and $\lambda_4 \ngtr n_1-1$. But it still furnishes an identity when $\lambda_4 > n_1-1$ and $\lambda_4 \ngtr n-1$.

We write this identity

$$\sum_{i=0}^{n_1}(-)^i\binom{\lambda_4+1}{i}\left[(a_2a_4)^{\lambda_4+1-i}(a_1a_3)^i-(a_3a_4)^{\lambda_4+1-i}(a_1a_2)^i\right]$$

$$=(-)^{n_1}\left[\binom{\lambda_4+1}{n_1+1}\{(a_2a_4)^{\lambda_4-n_1}(a_1a_3)^{n_1+1}-(a_3a_4)^{\lambda_4-n_1}(a_1a_2)^{n_1+1}\}\right.$$

$$\left.-\binom{\lambda_4+1}{n_1+2}\{(a_2a_4)^{\lambda_4-n_1-1}(a_1a_3)^{n_1+2}-(a_3a_4)^{\lambda_4-n_1-1}(a_1a_2)^{n_1+2}\}+\ldots\right].$$

Now, from (XII), we have (changing λ_4 into λ_4+1)

$$\binom{\lambda_4+1}{n_1+1}\{(0,\ n_1+1,\ \lambda_4-n_1)-(n_1+1,\ 0,\ \lambda_4-n_1)\}$$

$$-(n_1+1)\binom{\lambda_4+1}{n_1+2}\{(0,\ n_1+2,\ \lambda_4-n_1-1)-(n_1+2,\ 0,\ \lambda_4-n_1-1)\}+\ldots=R.$$

Hence on subtraction we obtain a syzygy if $\lambda_4 \ngtr n_1+1$; and a reduction for

$$(0,\ n_1+2,\ \lambda_4-n_1-1),$$

when $$\lambda_4 \ngtr n-1.$$

The reduction equation is

$$\text{(XIII)}\quad \binom{\lambda_4+1}{n_1+1}\{[e^{-a_2D_4}-1](0,\ n_1+1,\ \lambda_4-n_1)$$
$$-[e^{-a_3D_4}-1](n_1+1,\ 0,\ \lambda_4-n_1)\}$$
$$-\binom{\lambda_4+1}{n_1+2}\{[e^{-a_2D_4}-(n_1+1)](0,\ n_1+2,\ \lambda_4-n_1-1)$$
$$-[e^{-a_3D_4}-(n_1+1)](n_1+2,\ 0,\ \lambda_4-n_1-1)\}+\ldots=R.$$

With the help of (IX), this in general will reduce the form

$$(1,\ n_1+1,\ \lambda_4-n_1-1)$$

when $\lambda_4 > 2n_1$; but if otherwise we can use (XI) also and so reduce $(2,\ n_1,\ \lambda_4-n_1-1)$.

We must examine (XIII) further, owing to the presence of an exception. Expanding, we obtain

$$\text{(XIV)} \quad \sum_{i=1} (-)^i \binom{\lambda_4+1}{n_1+i} \left[\binom{n_1+i-1}{i-1} - 1 \right] [(0,\, n_1+i,\, \lambda_4-n_1-i+1) \\ -(n_1+i,\, 0,\, \lambda_4-n_1-i+1)] \\ - \sum_{j=1}^{n_1} \sum_{i=0}^{\lambda_4-n_1-j} (-)^{i+j} \frac{(\lambda_4+1)!}{j!\,(n_1+1+i)!\,(\lambda_4-n_1-j-i)!} \\ [(j,\, n_1+1+i,\, \lambda_4-n_1-i-j) - (n_1+1+i,\, j,\, \lambda_4-n_1-i-j)] = R.$$

When $n_1 = 1$, the left-hand side of (XIV) becomes

$$\sum_{i=1}^{\lambda_4} (-)^i \binom{\lambda_4+1}{i+1} (i-1)\, e^{-a_2D_3-a_2D_4} (0,\, 1+i,\, \lambda_4-i).$$

And since $2(1+i)+\lambda_4-i \not< 2n$ (for $\lambda_4 \not< n-1$) we can use (X), and thus obtain a syzygy. This furnishes then no extra reduction when $n_1 = 1$.

We have yet to consider the case $\lambda_4 = n$, that is the equation

$$e^{-a_2D_4}\,(0,\, 1,\, n) = R.$$

34. The equation (IX) gives syzygies which are quite obvious when $\lambda_4 = 0$ or $\lambda_3 < 2$.

For $\lambda_3 = 2$, we use the syzygy

$$\{(a_1a_4)-(a_2a_3)\}^{\lambda_4+2} + \{(a_1a_4)+(a_2a_3)\}^{\lambda_4+2} \\ = \{(a_3a_4)+(a_1a_2)\}^{\lambda_4+2} + \{(a_1a_3)+(a_2a_4)\}^{\lambda_4+2},$$

which reduces the equation when

$$\lambda_4 \not< n-2 \quad \text{and} \quad \lambda_4 \not< n_1-2.$$

The equation exists only when $\lambda_4 \not< n_1$. We can shew then that this furnishes a syzygy whenever our equation exists and $\lambda_4 \not< n-2$. For

$$0 = \{(a_1a_4)-(a_2a_3)\}^{\lambda_4+2} + \{(a_1a_4)+(a_2a_3)\}^{\lambda_4+2} \\ - \{(a_3a_4)+(a_1a_2)\}^{\lambda_4+2} - \{(a_2a_4)+(a_1a_3)\}^{\lambda_4+2} \\ = P + 2(a_1a_4)^{\lambda_4+2} - (a_1a_2)^{\lambda_4+2} - (a_1a_3)^{\lambda_4+2} \\ - (\lambda_4+2)(a_1a_2)^{\lambda_4+1}(a_3a_4) - (\lambda_4+2)(a_1a_3)^{\lambda_4+2}(a_2a_4)$$

(where P is used here and elsewhere to denote products of covariants)

$$= P + \{(a_1a_2)+(a_2a_4)\}^{\lambda_4+2} + \{(a_1a_3)+(a_3a_4)\}^{\lambda_4+2} - (a_1a_2)^{\lambda_4+2} \\ - (a_1a_3)^{\lambda_4+2} - (\lambda_4+2)(a_1a_2)^{\lambda_4+1}(a_3a_4) - (\lambda_4+2)(a_1a_3)^{\lambda_4+1}(a_2a_4) \\ = P + (\lambda_4+2)(a_2a_3)\{(a_1a_2)^{\lambda_4+1} - (a_1a_3)^{\lambda_4+1}\} \\ = P,$$

giving us a syzygy for all cases $\lambda_3 = 2$, $\lambda_4 \not< n_1$.

For $\lambda_3 = 3$, we use the syzygy

$$\begin{aligned}0 = \; & \{(a_1a_4)+(a_2a_3)\}^{\lambda_4+3}-\{(a_1a_4)-(a_2a_3)\}^{\lambda_4+3} \\ & -\{(a_1a_3)+(a_2a_4)\}^{\lambda_4+3}+\{(a_1a_2)+(a_3a_4)\}^{\lambda_4+3} \\ & +\{(a_1a_3)-(a_2a_4)\}^{\lambda_4+3}-\{(a_1a_2)-(a_3a_4)\}^{\lambda_4+3} \\ = \; & P+2(\lambda_4+3)\{(a_1a_4)^{\lambda_4+2}(a_2a_3)+(a_1a_3)^{\lambda_4+2}(a_4a_2)+(a_1a_2)^{\lambda_4+2}(a_3a_4)\} \\ = \; & P+2(\lambda_4+3)\left[\{(a_1a_2)+(a_2a_4)\}^{\lambda_4+2}(a_2a_4)-\{(a_1a_3)+(a_3a_4)\}^{\lambda_4+2}(a_3a_4)\right. \\ & \left. -(a_1a_3)^{\lambda_4-2}(a_2a_4)+(a_1a_2)^{\lambda_4+2}(a_3a_4)\right] \\ = \; & P+4\binom{\lambda_4+3}{2}\left[(a_1a_2)^{\lambda_4+1}(a_2a_4)^2-(a_1a_3)^{\lambda_4+1}(a_3a_4)^2\right] \\ & +2\binom{\lambda_4+3}{1}\left[(a_1a_2)^{\lambda_4+2}\{(a_2a_4)+(a_3a_4)\}-(a_1a_3)^{\lambda_4+2}\{(a_2a_4)+(a_3a_4)\}\right].\end{aligned}$$

If $\lambda_4 = n_1$, we obtain

$$\begin{aligned}&(n_1+2)\{(n_1+1,\ 0,\ 2)-(0,\ n_1+1,\ 2)\} \\ &-2(n_1+1)\{(n_1+2,\ 0,\ 1)-(0,\ n_1+2,\ 1)\} = P.\end{aligned}$$

And using (XII) we find we have a relation between products of covariants only, *i.e.*, a syzygy.

If $\lambda_4 = n_1-1$, we obtain

$$(n_1+1,\ 0,\ 1)-(0,\ n_1+1,\ 1) = P,$$

and again using (XII) we have a syzygy.

If $\lambda_4 < n-1$, there is a syzygy without the help of (XII). Thus we obtain a syzygy from (IX) in every case when $\lambda_3 < 4$, or $\lambda_4 < 1$, provided $\lambda_4 \not> n-2$.

We have still to consider the cases $\lambda_4 = n-1$ or n. In fact we have to consider the four equations

$$e^{-a_2D_3}(0,\ 2,\ n-1) = R, \quad e^{-a_2D_3}(0,\ 3,\ n-1) = R,$$

$$e^{-a_2D_3}(0,\ 2,\ n) = R, \quad e^{-a_2D_3}(0,\ 3,\ n) = R,$$

where it must be remembered in each case that $\lambda_4 \not> n_1$.

35. The equation (X) gives obvious syzygies when $\lambda_3 < 2$ or $\lambda_4 < 2$. For the other cases the syzygies

$$\{(a_2a_4)-(a_2a_3)\}^w = (a_3a_4)^w,$$

and $\{(a_2a_4)-(a_2a_3)\}^{w-1}(a_2a_3) = -(a_3a_4)^{w-1}(a_1a_2)-(a_3a_4)^w+(a_3a_4)^{w-1}(a_1a_4)$,

may be used, as for perpetuants; provided the weight w is not greater than n. When the weight is greater than n we find ourselves with five equations to deal with of just the same type as those of equation (IX).

We are thus left with ten equations to consider, four of weight $n+1$, four of weight $n+2$, and two of weight $n+3$.

36. For weight $n+1$ the equations, in the case of perpetuants were reduced by means of syzygies obtained from the symbolical identity

(XV) $$A_1(n+1)\left[(a_2a_3)\{(a_2a_4)-(a_2a_3)\}^n-(a_1a_3)\{(a_1a_4)-(a_1a_3)\}^n + (a_1a_2)(a_3a_4)^n\right]$$
$$+A_2\left[\{(a_2a_4)-(a_2a_3)\}^{n+1}-(a_3a_4)^{n+1}\right]$$
$$+A_3\left[\{(a_3a_4)-(a_1a_2)\}^{n+1}-\{(a_2a_4)-(a_1a_3)\}^{n+1}\right]$$
$$+A_4\left[\{(a_3a_4)+(a_1a_2)\}^{n+1}-\{(a_1a_4)-(a_2a_3)\}^{n+1}\right]$$
$$+A_5\left[\{(a_2a_4)+(a_1a_3)\}^{n+1}-\{(a_1a_4)+(a_2a_3)\}^{n+1}\right]$$
$$+(A_2-A_3-A_4)\left[(a_3a_4)^{n+1}-\{(a_1a_4)-(a_1a_3)\}^{n+1}\right]$$
$$+(-A_2+A_3-A_5)\left[(a_2a_4)^{n+1}-\{(a_1a_4)-(a_1a_2)\}^{n+1}\right]$$
$$+(-)^n\{-A_1(n+1)+A_2-A_4+(-)^nA_5\}\left[(a_2a_3)^{n+1} - \{(a_1a_3)-(a_1a_2)\}^{n+1}\right]=0.$$

From § 25 we see that

$$\{(a_2a_4)-(a_2a_3)\}^{n+1}$$
$$=\sum_{i=1}^{n}(-)^i\binom{n+1}{i}(a_2a_3)_i(a_2a_4)_{n+1-i}+(a_2a_4)^{n+1}+(a_3a_2)^{n+1} - \{1-(-)^n\}(a_2a_3)^n(a_2a_4);$$

and that

$$(a_2a_3)\{(a_2a_4)-(a_2a_3)\}^n$$
$$=\sum_{i=0}^{n-1}(-)^i\binom{n}{i}(a_2a_3)_{i+1}(a_2a_4)_{n-i}-(a_3a_2)^{n+1}-(-)^n(a_2a_3)^n(a_2a_4),$$

where $(a_2a_3)_i(a_2a_4)_{n+1-i}$ is an actual covariant of the three quantics concerned: in these we replace

$$(a_2a_3)^n(a_2a_4) \quad \text{by} \quad (a_2a_3)^n(a_1a_4)-\{(a_1a_3)-(a_1a_2)\}^n(a_1a_2),$$

and then substitute in (XV).

In the result the coefficient of each of $(a_3a_4)^{n+1}$, $(a_2a_4)^{n+1}$, $(a_1a_4)^{n+1}$, $(a_2a_3)^{n+1}$, $(a_1a_3)^{n+1}$ is zero. And, in fact, the identity is a syzygy as it stands for all values of the five constants when $n_1 > n$.

If $n_1 \not> n$, we need the following results from § 25,

$$(a_1a_3)\{(a_1a_4)-(a_1a_3)\}^n - \Sigma(-)^i \binom{n}{i}(a_1a_3)_{i+1}(a_1a_4)_{n-i} + (a_3a_1)^{n+1}$$

$$= (-)^n \sum_{i=1}^{n-n_1+1} (-)^i \binom{-n-3+n_1+i}{i-1}\binom{-1}{n+1-n_1-i}(a_1a_3)^{n_1+i-1}(a_1a_4)^{n+2-n_1-i}$$

$$= (-)^{n_1} \sum_{i=1}^{n-n_1+1} (-)^i \binom{n+1-n_1}{i-1}(a_1a_3)^{n_1+i-1}(a_1a_4)^{n+2-n_1-i};$$

and

$$\{(a_1a_4)-(a_1a_3)\}^{n+1} - \Sigma(-)^i\binom{n+1}{i}(a_1a_3)_i(a_1a_4)_{n+1-i} - (a_1a_4)^{n+1} - (a_3a_1)^{n+1}$$

$$= \sum_{i=1}^{n-n_1+1}(-)^i\left[\binom{n_1+i-2}{i-1}\binom{n}{n_1+i-1} - (-)^{n_1}\binom{n+1-n_1}{i-1}\right]$$

$$\times (a_1a_3)^{n_1+i-1}(a_1a_4)^{n+2-n_1-i}.$$

Making use of these results in (XV), and of the corresponding result for $\{(a_1a_4)-(a_1a_2)\}^{n+1}$, we obtain from (XV) in the notation of this paper,

(XVI)

$$-A_1(n+1)(-)^{n_1}\sum_{i=1}^{n-n_1+1}(-)^i\binom{n+1-n_1}{i-1}(0,\ n_1+i-1,\ n+2-n_1-i)$$

$$-(A_2-A_3-A_4)\sum_{i=1}^{n-n_1+1}(-)^i\left[\binom{n_1+i-2}{i-1}\binom{n}{n_1+i-1} - (-)^{n_1}\binom{n+1-n_1}{i-1}\right]$$

$$\times (0,\ n_1+i-1,\ n+2-n_1-i)$$

$$-(A_2-A_3+A_5)\left\{\binom{n}{n_1} - (-)^{n_1}\right\}(n_1,\ 0,\ n+1-n_1)$$

$$+\sum_{i=1}^{n-n_1}\binom{n+1}{n_1+i}\{A_5-(-)^{n_1+i}A_3\}\, e^{-a_2D_4}(0,\ n_1+i,\ n+1-n_1-i)$$

$$-\sum_{i=1}^{n-n_1}\binom{n+1}{n_1+i}\{A_5+(-)^{n_1+i}A_4\}\, e^{-a_2D_3}(0,\ n+1-n_1-i,\ n_1+i) = P.$$

It is evident that we only get syzygies (with the help of the regular equations) when $n_1 = n$.

In general when all the A's except A_1 are zero, we find

$$(0,\ n_1,\ n+1-n_1) - (n+1-n_1)(0,\ n_1+1,\ n-n_1) + \ldots = R.$$

From $A_1 = A_4 = A_5 = 0$, $A_2 = A_3$, we obtain

$$\binom{n+1}{n_1+1} e^{-a_2 D_4}(0, n_1+1, n-n_1)$$

$$-\binom{n+1}{n_1+2} e^{-a_2 D_4}(0, n_1+2, n-n_1-1)+\ldots = R.$$

From

$$A_3 = A_4 = A_5 = 0 \text{ and } A_1(n+1)(-)^{n_1}+A_2\left\{\binom{n}{n_1}-(-)^{n_1}\right\} = 0,$$

we obtain
$$\left\{\binom{n}{n_1}-(-)^{n_1}\right\}(n_1, 0, n+1-n_1)$$

$$+\left[n_1\binom{n}{n_1+1}-(n+1-n_1)\binom{n}{n_1}\right](0, n_1+1, n-n_1)+\ldots = R.$$

This is all we get in general, for (XVI) does not help us, as a rule, unless A_4 and A_5 are zero, as it is easy to see. Further the three equations are all we require; since when $n_1 < n-2$, the equations

$$e^{-a_2 D_3}(0, 2, n-1) = R, \quad e^{-a_2 D_3}(0, 3, n-2) = R$$

no longer exist.

Thus, when $n_1 < n-2$, we find three new reductions which easily may be shewn to be those for $(n_1, 1, n-n_1)$, $(2, n_1-1, n-n_1)$, the third being $(1, n_1, n-n_1)$, if $n > 2n_1-1$; but, if $n \leqslant 2n_1-1$, this is already reduced and our third reduction is $(3, n_1-2, n-n_1)$.

When $n_1 = n-1$, we have to look for five reductions or syzygies; the three equations obtained for the general case enable us to express each of $(0, n, 1)$, $(0, n-1, 2)$, and $(n-1, 0, 2)$ as a sum of products. Substitute their values in (XVI); and it reduces to

$$-\binom{n+1}{n}\{A_5+(-)^n A_4\} e^{-a_2 D_3}(0, 1, n) = P.$$

But using (VI) we find that

$$\begin{aligned} e^{-a_2 D_3}(0, 1, n) &= (0, 1, n)-(1, 0, n) \\ &= (n-1)(0, n-1, 2)-(n-2)(0, n, 1)-(n-1)(n-1, 0, 2)+P \\ &= P. \end{aligned}$$

And hence the extra equations give two syzygies here.

When $n_1 = n-2$, we have one extra equation to obtain.

It is plain that we must have $A_5-(-)^n A_4 = 0$ in (XVI). We find our equation by putting $A_1 = A_3 = 0$, $A_2 = A_4 = (-)^n$, $A_5 = 1$. And by means of this we can obtain the reduction of the extra form $(2, n-2, 1)$.

37. For weight $n+2$, we find that the equation

$$e^{-a_2D_3-a_2D_4}(0, 2, n) = R$$

is required for the ordinary reductions, unless $n_1 \geqslant n$. The equation

$$e^{-a_2D_3}(0, 2, n) = R$$

exists only when $n_1 \geqslant n$, and is then required for the reduction of $(1, 1, n)$. The equation

$$e^{-a_2D_3}(0, 3, n-1) = R$$

exists only when $n_1 \geqslant n-1$; and

$$e^{-a_2D_3-a_2D_4}(0, 3, n-1) = R,$$

which exists only when $n \geqslant 5$ is required for ordinary reductions when $n_1 < 4$.

Thus we require three reductions or syzygies when $n_1 \geqslant n$, two when $n_1 = n-1$, one only when $n-1 > n_1 > 3$, and none when $n_1 \leqslant 3$.

We replace n by $n+1$ in (XV); and observing that by § 25 we have

$$\begin{aligned}
&\{(a_2a_4)-(a_2a_3)\}^{n+2} \\
&= \sum_{i=2}^{n} (-)^i \binom{n+2}{i} (a_2a_3)_i (a_2a_4)_{n+2-i} + (a_2a_4)^{n+2} - (n+2)(a_2a_4)^{n+1}(a_2a_3) \\
&\qquad + (a_3a_2)^{n+2} + (n+2)(a_3a_2)^{n+1}(a_2a_4) \\
&\qquad + \{n^2-1-(-)^n(2n+3)\}(a_2a_3)^{n-1}(a_2a_4)^3 \\
&\qquad - \{n^2-n-3-(-)^n(3n+4)\}(a_2a_3)^n(a_2a_4)^2,
\end{aligned}$$

and

$$\begin{aligned}
&(a_2a_3)\{(a_2a_4)-(a_2a_3)\}^{n+1} \\
&= \sum_{i=1}^{n-1} (-)^i \binom{n+1}{i} (a_2a_3)_{i+1} (a_2a_4)_{n+1-i} + (a_2a_3)(a_2a_4)^{n+1} \\
&\qquad - (a_3a_2)^{n+2} - (n+1)(a_3a_2)^{n+1}(a_2a_4) \\
&\qquad - \{n-1-(-)^n(2n+1)\}(a_2a_3)^{n-1}(a_2a_4)^3 \\
&\qquad + \{n-2-(-)^n(3n+1)\}(a_2a_3)^n(a_2a_4)^2.
\end{aligned}$$

In these we replace

$$(a_2 a_3)^{n-1} (a_2 a_4)^3$$

by $(a_2 a_3)^{n-1}(a_1 a_4)^3$

$$-\{(a_1 a_3)-(a_1 a_2)\}^{n-1} \{3(a_1 a_2)(a_1 a_4)^2-3(a_1 a_2)^2(a_1 a_4)+(a_1 a_2)^3\},$$

and

$$(a_2 a_3)^n (a_2 a_4)^2$$

by $\qquad (a_2 a_3)^n (a_1 a_4)^2-\{(a_1 a_3)-(a_1 a_2)\}^n \{2(a_1 a_2)(a_1 a_4)-(a_1 a_2)^2\},$

and then substitute in our new identity.

In order that the identity may yield a relation between actual covariants, the constants must satisfy the conditions

(XVII) $\qquad A_1-A_2+A_3+A_5=0, \quad A_1-A_3+A_4=0.$

When $n_1 \geqslant n+2$, we find if n is even no syzygies, but reductions for the forms $(3, n-3, 2)$, $(2, n-1, 1)$, $(3, n-2, 1)$; and if n is odd there is a syzygy and the forms $(3, n-3, 2)$, $(2, n-1, 1)$ only are reducible.

When $n_1 = n+1$, and n is even, our identity furnishes reductions for $(3, n-3, 2)$, $(1, n, 1)$ and $(2, n-1, 1)$; but when n is odd there is a syzygy and reductions only for $(3, n-3, 2)$ and $(1, n, 1)$.

When $n_1 = n$, there are no syzygies, the reductions are $(n-1, 1, 2)$, $(2, n-2, 2)$, $(3, n-3, 2)$, when n is even, and $(2, n-2, 2)$, $(3, n-3, 2)$, $(1, n, 1)$ when n is odd.

When $n_1 = n-1$, we expect only two results from our identity, and we find that the constants must satisfy the additional condition $A_4+A_5=0$. And whether n is odd or even we find the new reductions to be $(3, n-4, 3)$ and $(1, n-1, 2)$.

When $n_1 < n-1$, we have one reduction only to look for, and we must have $A_4 = 0 = A_5$; and therefore $2A_1 = A_2 = 2A_3$. We find then a reduction for $(3, n_1-3, n-n_1+2)$, when $n_1 \nless \frac{n+3}{2}$, but for $(2, n_1-2, n-n_1+2)$, when $n_1 < \frac{n+3}{2}$; and no new reduction at all when $n_1 < 4$.

38. Lastly, when the weight is $n+3$, we find that the equation

$$e^{-a_2 D_3} (0, 3, n) = R,$$

which only exists when $n_1 \geqslant n$, is always required for the reduction of $(1, 2, n)$. The equation

$$e^{-a_2 D_3 - a_2 D_4} (0, 3, n) = R$$

is also required for the ordinary reductions unless $n_1 \geqslant n \geqslant 6$. To obtain the reduction or syzygy corresponding to this last case, we replace n by $n+2$ in (XV) and proceed as before; then we find that the constants must satisfy the two conditions (XVII), and also the further conditions $A_3+A_4=0$ and $A_3-A_5=0$; whence

$$\frac{A_1}{2}=\frac{A_2}{4}=\frac{A_3}{1}=\frac{A_4}{-1}=\frac{A_5}{1}.$$

When $n_1 > n$ *and* n *is odd*, the form $(3, n-4, 4)$ is reduced.

When $n_1 > n+1$ *and* n *is even*, the form $(3, n-3, 3)$ is reduced.

When $n_1 = n+1$ *and* n *is even*, the form $(2, n-2, 3)$ is reduced.

When $n_1 = n$, the form $(2, n-3, 4)$ is reduced, whether n be even or odd.

39. We can now sum up our results. As before stated, $(0, \lambda_3, \lambda_4)$ is reducible unless

$$\lambda_3 > n_1, \quad \lambda_4 > n_1, \quad \text{and} \quad 2\lambda_3+\lambda_4 > 2n;$$

it is therefore always reducible when $n_1 \geqslant n$.

The reducibility limits of $(1, \lambda_3, \lambda_4)$ are illustrated in Fig. 1: where

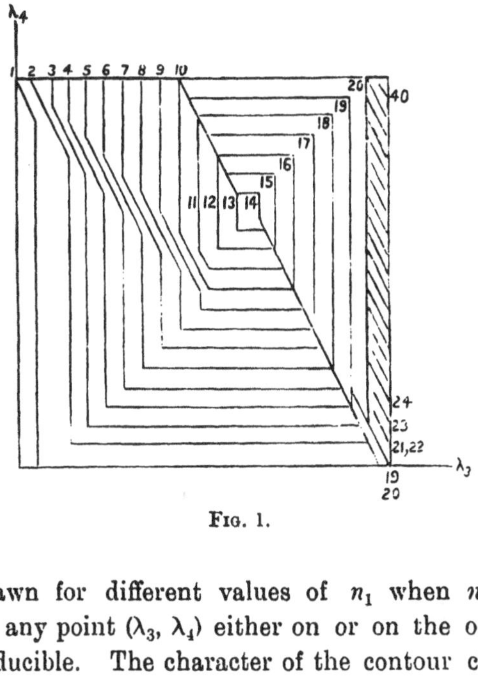

FIG. 1.

contours are drawn for different values of n_1 when $n=20$, the form corresponding to any point (λ_3, λ_4) either on or on the origin side of the contour being reducible. The character of the contour changes according to the value of n_1; thus the reducibility limits are, when

(i) $n_1 \leqslant \frac{n}{2}$, $\lambda_3 \not> n_1-1$ or $\lambda_3 = n_1$ or n_1+1 and $2\lambda_3+\lambda_4 \not> n+n_1$

or $\lambda_4 \not> n_1-1$, $2\lambda_3+\lambda_4 \not> 2n-1$.

(ii) $\frac{2n}{3} \geqslant n_1 > \frac{n}{2}$, $\lambda_3 \not> n_1-1$, $2\lambda_3+\lambda_4 \not> 2n-2$,

or $\lambda_4 \not> n_1-1$, $2\lambda_3+\lambda_4 \not> 2n-1$.

(iii) $n-2 \geqslant n_1 > \frac{2n}{3}$, $2\lambda_3+\lambda_4 \not> 2n-2$ or $\lambda_3 \not> n_1-1$, $\lambda_4 \not> n_1$

or $\lambda_4 \not> n_1-1$, $2\lambda_3+\lambda_4 \not> 2n-1$.

(iv) $n_1 = n-1$, a modification is introduced owing to the reducibility of $(1, n-1, 2)$; we have then

$n_1 = n-1$ or n, $2\lambda_3+\lambda_4 \not> 2n-2$ or $\lambda_3 \not> n_1-1$, $\lambda_4 \not> n_1$

or $\lambda_4 \not> n_1-1$, $2\lambda_3+\lambda_4 \not> 2n$.

(v) $n_1 = n+1$, $\lambda_3 \not> n-1$ or $2\lambda_3+\lambda_4 \not> 2n+1$.

(vi) $n_1 > n+1$, $\lambda_3 \not> n-1$ or $2\lambda_3+\lambda_4 \not> n+n_1-1$.

(vii) $n_1 > 2n$, every form is reducible.

The reducibility limits of $(2, \lambda_3, \lambda_4)$ and $(3, \lambda_3, \lambda_4)$ are traced in Figs. 2 and 3. It will be seen that in both these cases there is part

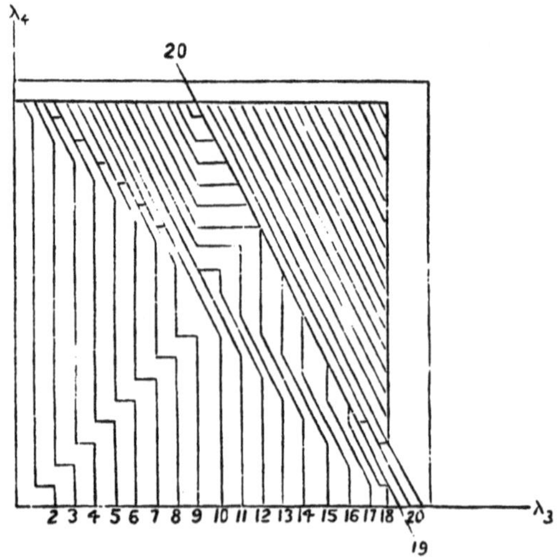

FIG. 2.

of the figure which corresponds to forms irreducible for all values of n_1.

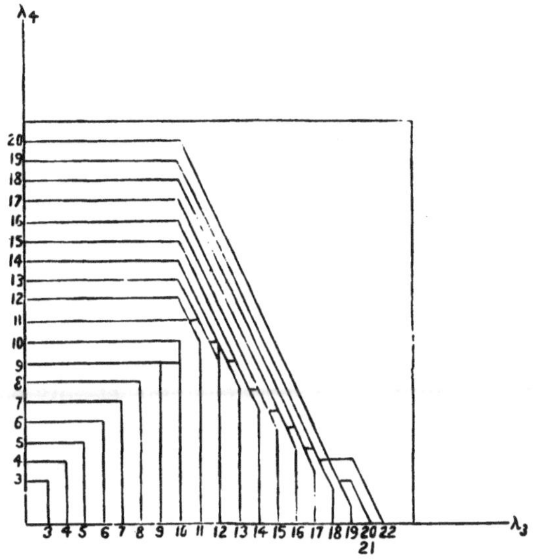

FIG. 3.

40. It is noteworthy that our special cases introduce the reductions of $(n_1, 1, n-n_1)$ when $n_1 < n$; and of $(n-1, 1, 2)$ when $n_1 = n$, and is even, which must be added to the reductions given in § 29.

THE ELECTROMAGNETIC PROPERTIES OF COILS OF CERTAIN FORMS

By A. Young

[Read May 9th, 1918.—Received September 20th, 1918.]

The magnetic potential of a closed circuit at any point outside the circuit is measured by ωi, where ω is the solid angle subtended by the circuit at the point, and i is the strength of the current. The magnetic potential of a coil can be obtained by summing the expressions ωi for each turn.

The induction parallel to the axis of a coil at any point is, of course, the ratio of the difference between the potentials at two neighbouring points—such that the straight line joining them is parallel to the axis—to the distance between the points. The effect would be the same if we took the difference between the potentials of two coils at the same point, the one coil being obtained from the other by a small displacement parallel to the axis. Thus, for a single layer coil of n turns to the centimetre, $n\Omega i$ is the magnetic induction parallel to the axis at a point, where Ω is the solid angle subtended by the coil itself at the point.

In what follows coils are considered in which each turn is in the form of a rectangle. The solid angle subtended by a rectangle at any point may be represented as the inverse tangent of a certain ratio. Hence is deduced by integration the magnetic potential of such a coil. The coil is looked upon as an infinitely thin current sheet.

A further double integration enables either the self induction of the coil to be written down, or else the mutual induction of two such coils which are such that the three principal planes are parallel; *i.e.* the planes at right angles to the axis and those through the axis at right angles to the sides of the rectangular turns.

In § 3 the potential is obtained for a coil of the same form but of finite thickness, and thence by double integration the mutual induction of two such coils whose axes and sides are parallel.

In § 4 the self induction of this coil is given.

In § 5 the question of the looseness of the windings is taken into account, and in the remaining paragraphs the discussion of § 5 is applied to the cases considered.

1. Firstly, consider a coil in which each turn is of the shape of a rectangle whose sides are $2a$, $2b$; the turns being closely wound, so that we may consider the coil as equivalent to an infinitely thin current sheet.

Take the axis of the coil for axis z, its middle point for origin, and the axes of x and y parallel to the sides $2a$, $2b$ respectively of the coil. The axes are so chosen that they form a right handed system. Now the solid angle subtended by a rectangle $ABCD$ at any point O on the straight line through A at right angles to its plane is

$$\tan^{-1}\frac{AB\,.\,AD}{OA\,.\,OC}.$$

The potential of the turn in the plane $z=\zeta$ (supposed to carry unit current in the positive direction from x to y round z) at the point (x, y, z) is

$$\tan^{-1}\frac{(a+x)(b+y)}{(z-\zeta)\sqrt{\{(a+x)^2+(b+y)^2+(z-\zeta)^2\}}}$$

$$+\tan^{-1}\frac{(a-x)(b+y)}{(z-\zeta)\sqrt{\{(a-x)^2+(b+y)^2+(z-\zeta)^2\}}}$$

$$+\tan^{-1}\frac{(a+x)(b-y)}{(z-\zeta)\sqrt{\{(a+x)^2+(b-y)^2+(z-\zeta)^2\}}}$$

$$+\tan^{-1}\frac{(a-x)(b-y)}{(z-\zeta)\sqrt{\{(a-x)^2+(b-y)^2+(z-\zeta)^2\}}}.$$

Now, if $$r^2=x^2+y^2+z^2,$$

$$\int\tan^{-1}\frac{xy}{zr}\,dz=z\tan^{-1}\frac{xy}{zr}-x\tanh^{-1}\frac{y}{r}-y\tanh^{-1}\frac{x}{r}.$$

Then the potential of the coil at (x, y, z) is

$$Vi=ni\left[\left[\left[\zeta\tan^{-1}\frac{\xi\eta}{\zeta r}-\xi\tanh^{-1}\frac{\eta}{r}-\eta\tanh^{-1}\frac{\zeta}{r}\right]_{\xi=x-a}^{x+a}\right]_{\eta=y-b}^{y+b}\right]_{\zeta=z-h}^{z+h},$$

where $r^2=\xi^2+\eta^2+\zeta^2$, n is the number of terms per unit length, $2h$ is the length of the coil, i is strength of the current: and the values of the variables inserted above and below the end brackets imply that we have to take the difference of the values of the functions when the variables have these values.

2. The magnetic induction parallel to the axis of the coil is $-\partial V/\partial z$. The mutual inductance of this coil and another—whose axis is parallel to its axis, whose turns are rectangles having sides parallel to the axes of x and y and of lengths $2a'$, $2b'$, whose length is $2h'$, whose centre is (α, β, γ) and which has n' turns per unit length—is

$$n' \int_{\alpha-a'}^{\alpha+a'} \int_{\beta-b'}^{\beta+b'} \int_{\gamma-h'}^{\gamma+h'} -\frac{\partial V}{\partial z}\, dx\, dy\, dz.$$

We require then the indefinite integral

$$\iint \left[-\zeta \tan^{-1}\frac{\xi\eta}{\zeta r} + \xi \tanh^{-1}\frac{\eta}{r} + \eta \tanh^{-1}\frac{\xi}{r} \right] d\xi\, d\eta$$

$$= -\xi\eta\zeta \tan^{-1}(\xi\eta/\zeta r) + \tfrac{1}{2}\eta(\xi^2-\zeta^2)\tanh^{-1}(\eta/r) + \tfrac{1}{2}\xi(\eta^2-\zeta^2)\tanh^{-1}(\xi/r)$$
$$+ (2\zeta^2-\xi^2-\eta^2)\, r/6$$
$$= \phi(\xi, \eta, \zeta).$$

Thus the mutual inductance of the two coils is

$$nn' \Big[\Big[\Big[\Big[\Big[\Big[\phi(\xi, \eta, \zeta)\Big]_{\xi=x-a}^{x+a}\Big]_{\eta=y-b}^{y+b}\Big]_{\zeta=z-h}^{z+h}\Big]_{x=\alpha-a'}^{\alpha+a'}\Big]_{y=\beta-b'}^{\beta+b'}\Big]_{z=\gamma-h'}^{\gamma+h'}.$$

Noticing that ϕ is unaltered by the change of sign of either of its three arguments, we can deduce the self inductance of the original coil in the form

$$8n^2 \Big[\Big[\Big[\phi(\xi, \eta, \zeta)\Big]_{\xi=0}^{2a}\Big]_{\eta=0}^{2b}\Big]_{\zeta=0}^{2h}.$$

If the second coil have its axis parallel to the axis y, and the sides of its rectangular turns parallel to the axes of x and z respectively, the mutual inductance becomes

$$nn' \Big[\Big[\Big[\Big[\Big[\Big[-\tfrac{1}{2}\xi\zeta^2 \tan^{-1}\frac{\xi\eta}{\zeta r} - \tfrac{1}{2}\xi\eta^2 \tan^{-1}\frac{\xi\zeta}{\eta r} - \tfrac{1}{6}\xi^3 \tan^{-1}\frac{\eta\zeta}{\xi r} + \xi\eta\zeta \tanh^{-1}\frac{\xi}{r}$$
$$+ \frac{3\xi^2\zeta-\zeta^3}{6}\tanh^{-1}\frac{\eta}{r} + \frac{3\xi^2\eta-\eta^3}{6}\tanh^{-1}\frac{\zeta}{r}$$
$$- \tfrac{1}{3}\eta\zeta r\Big]_{\xi=x-a}^{x+a}\Big]_{\eta=y-b}^{y+b}\Big]_{\zeta=z-h}^{z+h}\Big]_{x=\alpha-a'}^{\alpha+a'}\Big]_{y=\beta-h'}^{\beta+h'}\Big]_{z=\gamma-c'}^{\gamma+c'}.$$

3. Let us now suppose that the coil is equivalent to a current sheet not infinitely thin, but such that the section of it by a plane perpendicular to the axis is bounded by two rectangles with parallel sides of lengths $2a$, $2b$, and $2a+2t$, $2b+2t$ respectively.

To find the potential at (x, y, z) we require the integral

$$\int\left[\zeta\tan^{-1}\frac{(\xi+t)(\eta+t)}{\zeta r}-(\xi+t)\tanh^{-1}\frac{\eta+t}{r}-(\eta+t)\tanh^{-1}\frac{\xi+t}{r}\right]dt$$

$$=\tfrac{1}{2}\pi\zeta t+\zeta(\xi+t)\tan^{-1}[\{(2t+\xi+\eta)(t+\eta)-r^2\}/\zeta r]$$
$$+\zeta(\eta+t)\tan^{-1}[\{(2t+\xi+\eta)(t+\xi)-r^2\}/\zeta r]$$
$$+\tfrac{1}{2}[\zeta^2-(t+\xi)^2]\tanh^{-1}\{(t+\eta)/r\}+\tfrac{1}{2}[\zeta^2-(t+\eta)^2]\tanh^{-1}\{(t+\xi)/r\}$$
$$-(1/2\sqrt{2})\{2\zeta^2-(\xi-\eta)^2\}\tanh^{-1}\{(2t+\xi+\eta)/r\sqrt{2}\},$$

where $$r^2=(\xi+t)^2+(\eta+t)^2+\zeta^2.$$

Let us call this function V_t; then the potential of the coil at (x, y, z) is the sum of the four values of

$$ni[V_t-V_0]_{\zeta=z-h}^{z+h},$$

obtained by giving to ξ the values $a\pm x$, and to η the values $b\pm y$ simultaneously. Here n is the number of turns per unit area of the cross section through the axis at right angles to one of the sides of the coil. In using the function V_t it will be observed that the term $\frac{1}{2}\pi\zeta t$ only adds to the potential a term independent of (x, y, z), and therefore can be neglected.

In order to find inductances in this case, we want to integrate with respect to X and Y the function

$$V'=ZX\tan^{-1}[\{(X+Y)Y-R^2\}/ZR]+ZY\tan^{-1}[\{(X+Y)X-R^2\}/ZR]$$
$$-(1/2\sqrt{2})\{2Z^2-(X-Y)^2\}\tanh^{-1}\{(X+Y)/R\sqrt{2}\}$$
$$+\tfrac{1}{2}(Z^2-X^2)\tanh^{-1}(Y/R)+\tfrac{1}{2}(Z^2-Y^2)\tanh^{-1}(X/R),$$

where $$R^2=X^2+Y^2+Z^2.$$

We find

$$6\int V'dX=(3X^2-Z^2)Z\tan^{-1}[\{(X+Y)Y-R^2\}/ZR]$$
$$-(3Y^2-6XY-Z^2)Z\tan^{-1}[\{(X+Y)X-R^2\}/ZR]$$
$$-(3/\sqrt{2})\{6Z^2(X-Y)-(X-Y)^3\}\tanh^{-1}\{(X+Y)/R\sqrt{2}\}+(Y^2-Z^2)R$$
$$+(3Z^2-X^2)X\tanh^{-1}(Y/R)+(Y^3-3XY^2+3XZ^2-3YZ^2)\tanh^{-1}(X/R),$$

and $$24 \iint V' dX dY$$

$$= 4Z\{(3Y-X)X^2+(X-Y)Z^2\}\tan^{-1}[\{(X+Y)Y-R^2\}/ZR]$$
$$+4Z\{(3X-Y)Y^2+(Y-X)Z^2\}\tan^{-1}[\{(X+Y)X-R^2\}/ZR]$$
$$-(1/\sqrt{2})\{4Z^4-12Z^2(X-Y)^2+(X-Y)^4\}\tanh^{-1}\{(X+Y)/R\sqrt{2}\}$$
$$+\{X^3+Y^3-2(X+Y)Z^2\}R$$
$$+\{2Z^4+12Z^2XY-6Z^2Y^2+Y^4-4XY^3\}\tanh^{-1}(X/R)$$
$$+\{2Z^4+12Z^2XY-6Z^2X^2+X^4-4YX^3\}\tanh^{-1}(Y/R)$$
$$= \psi(X, Y, Z).$$

For the mutual inductance of this coil and a second which is equivalent to a thin rectangular current sheet, the axes of the coils being parallel and also their plane boundaries, we have

$$nn'\Big[\Big[\Big[\Big[\Big[\Big[\Big[\psi(X, Y, Z)\Big]_{X=t+a-\xi}^{t+a+\xi}\Big]_{Y=t+b-\eta}^{t+b+\eta}\Big]_{Z=\zeta-h}^{\zeta+h}\Big]_{\xi=\alpha-a'}^{\alpha+a'}\Big]_{\eta=\beta-b'}^{\beta+b'}\Big]_{\zeta=\gamma-h'}^{\gamma+h'}\Big]_{t=0}^{t}:$$

where (α, β, γ) is the centre; $2a'$, $2b'$ the lengths of the sides of a turn; $2h'$ the axial length; and n' the number of turns per unit axial length of the second coil.

4. The self inductance of the coil considered is a sum of two parts of which the first is

$$L_1 = 2n^2t\Big[\Big[\Big[\Big[\Big[\Big[\psi(X, Y, Z)\Big]_{X=t+a-\xi}^{t+a+\xi}\Big]_{Y=t+b-\eta}^{t+b+\eta}\Big]_{\xi=-a}^{a}\Big]_{\eta=-b}^{b}\Big]_{t=0}^{t}\Big]_{Z=0}^{h}.$$

The second part L_2 is that due to the lines of induction which cut the conducting material.

This is a sum of terms of which the typical one is obtained from the linkage of the flux due to the current in that part of the coil which is bounded by the rectangles whose sides are $2a+2t$, $2b+2t$, and $2a+2\tau$, $2b+2\tau$, with the lines of induction which pass through the area bounded by the rectangles whose sides are $2a+2\tau$, $2b+2\tau$, and $2a+2\tau-2\delta\tau$, $2b+2\tau-2\delta\tau$, the rectangles being symmetrically situated with respect to the coil, and in the same section perpendicular to the axis; and τ varying from 0 to t.

The result after integration is

$$L_2 = 8n^2 \Big[\Big[-F_1(\vartheta, -2t-2a, -2t-2b, z) + F_1(\vartheta, -t-2a, -t-2b, z)$$

$$-F_1(\vartheta, 0, 0, z) + F_1(\vartheta, -t, -t, z) + F_2(\vartheta, -2t-2b, z)$$

$$+F_2(\vartheta, -2t-2a, z) + F_2(\vartheta, 2b, z) + F_2(\vartheta, 2a, z)$$

$$-tF_3(\vartheta, 2b, z) - tF_3(\vartheta, 2a, z) \Big]_{\vartheta=0}^{t} \Big]_{z=0}^{2h},$$

where $$F_1(\vartheta, \alpha, \beta, z)$$

$$= -z\{\tfrac{1}{4}\vartheta^4 + \tfrac{1}{3}(\alpha+\beta)\vartheta^3 + \tfrac{1}{2}\alpha\beta\vartheta^2\} \tan^{-1}\frac{zr}{(\vartheta+\alpha)(\vartheta+\beta)}$$

$$+(1/60)\{5\alpha^3(\alpha-2\beta)z - 10\alpha(\alpha-\beta)z^3 + z^5\} \tan^{-1}[\{(2\vartheta+\alpha+\beta)(\vartheta+\beta) - r^2\}/zr]$$

$$+(1/60)\{5\beta^3(\beta-2\alpha)z - 10\beta(\beta-\alpha)z^3 + z^5\} \tan^{-1}[\{(2\vartheta+\alpha+\beta)(\vartheta+\alpha) - r^2\}/zr]$$

$$-(1/120)[12(\vartheta+\alpha)^5 + 15(\beta-2\alpha)(\vartheta+\alpha)^4 + 20\alpha(\alpha-\beta)(\vartheta+\alpha)^3$$

$$-z^2\{20(\vartheta+\alpha)^3 + 30(\beta-2\alpha)(\vartheta+\alpha)^2 + 60\alpha(\alpha-\beta)(\vartheta+\alpha)\}$$

$$-5z^4(\beta-2\alpha)] \tanh^{-1}\{(\vartheta+\beta)/r\}$$

$$-(1/120)[12(\vartheta+\beta)^5 + 15(\alpha-2\beta)(\vartheta+\beta)^4 + 20\beta(\beta-\alpha)(\vartheta+\beta)^3$$

$$-z^2\{20(\vartheta+\beta)^3 + 30(\alpha-2\beta)(\vartheta+\beta)^2 + 60\beta(\beta-\alpha)(\vartheta+\beta)\}$$

$$-5z^4(\alpha-2\beta)] \tanh^{-1}\{(\vartheta+\alpha)/r\}$$

$$+(1/240\sqrt{2})(\alpha+\beta)\{5(\alpha-\beta)^4 - 60z^2(\alpha-\beta)^2 + 16z^4\} \tanh^{-1}\{(2\vartheta+\alpha+\beta)/r\sqrt{2}\}$$

$$+r^5/60 - (\alpha+\beta)r^3\vartheta/48 - (\alpha^2+\beta^2+3\alpha\beta+13z^2)r^3/120$$

$$-(\alpha+\beta)\{(\alpha-\beta)^2 - 7z^2\} r\vartheta/48$$

$$-[5(\alpha^2+\beta^2)(\alpha-\beta)^2 - z^2\{44(\alpha^2+\beta^2) - 18\alpha\beta\} - 14z^4] r/240,$$

where $$r^2 \equiv (\vartheta+\alpha)^2 + (\vartheta+\beta)^2 + z^2.$$

$$F_2(\vartheta,\ \beta,\ z)$$

$$\begin{aligned}
&= (1/6)\,\{\vartheta^3(3\vartheta+2\beta)z-\vartheta^2z^3\}\tan^{-1}[\{\beta(\vartheta+\beta)-r^2\}/zr] \\
&\quad +(1/6)\,\{3\vartheta^2(\vartheta+\beta)^2z-\vartheta^2z^3\}\tan^{-1}\{(\beta\vartheta+r^2)/zr\} \\
&\quad +(1/240\sqrt{2})\,\{-3(8\vartheta-\beta)(2\vartheta+\beta)^4+60(4\vartheta-\beta)(2\vartheta+\beta)^2z^2 \\
&\qquad\qquad -260\beta z^4\}\tanh^{-1}(\beta/r\sqrt{2}) \\
&\quad +(1/24)\,\{(4\vartheta+3\beta)\,\vartheta^4-6(2\vartheta+\beta)\,\vartheta^2z^2-\beta z^4\}\tanh^{-1}\{(\vartheta+\beta)/r\} \\
&\quad -(1/6)(\vartheta+\beta)\vartheta^2\,\{(\vartheta+\beta)^2-3z^2\}\tanh^{-1}(\vartheta/r) \\
&\quad +(1/240)\,\{-5\beta^4z+20\beta^2z^3+188z^5\}\tan^{-1}\{\beta(2\vartheta+\beta)/2zr\} \\
&\quad -(1/120\sqrt{2})\,\{\beta^5+40\beta^3z^2+140\beta z^4\}\tanh^{-1}\{(2\vartheta+\beta)/r\sqrt{2}\} \\
&\quad +(1/720)\,\{18\vartheta^3+22\beta\vartheta^2+28\beta^2\vartheta-13\beta^3+306\vartheta z^2-262\beta z^2\}\,\beta r,
\end{aligned}$$

where $$r^2 \equiv \vartheta^2+(\vartheta+\beta)^2+z^2.$$

$$F_3(\vartheta,\ \beta,\ z)$$

$$\begin{aligned}
&= (1/6)\,\{4\vartheta^3+3\beta\vartheta^2-(2\vartheta+\beta)z^2\}\,z\tan^{-1}[\{\beta(\vartheta+\beta)-r^2\}/zr] \\
&\quad +(1/6)\,\{(\vartheta+\beta)^2(4\vartheta+\beta)-(2\vartheta+\beta)z^2\}\,z\tan^{-1}\{(\beta\vartheta+r^2)/zr\} \\
&\quad -(1/24\sqrt{2})\,\{-3(2\vartheta+\beta)^4+36(2\vartheta+\beta)^2z^2+52z^4\}\tanh^{-1}(\beta/r\sqrt{2}) \\
&\quad +(1/24)\,\{5\vartheta^4+4\beta\vartheta^3-6\vartheta(3\vartheta+2\beta)z^2+z^4\}\tanh^{-1}\{(\vartheta+\beta)/r\} \\
&\quad +(1/24)\,\{-(\vartheta+\beta)^3(5\vartheta+\beta)+6(\vartheta+\beta)(3\vartheta+\beta)z^2-z^4\}\tanh^{-1}(\vartheta/r) \\
&\quad +(1/72)\,\{\vartheta^2+\beta\vartheta+5\beta^2+83z^2\}\,\beta r,
\end{aligned}$$

where $$r^2 \equiv \vartheta^2+(\vartheta+\beta)^2+z^2.$$

5. So far the current has been supposed to be uniformly distributed in the space occupied by the coil. In practice this space is partly occupied by insulation. A closer approximation to the truth, and one which will exhibit the degree of error introduced by the assumption made, may be obtained in the following manner.

Let $\phi(z)$ be a function of z which with all its derivatives is finite and continuous throughout the region considered.

Consider the sum

$$\begin{aligned}
S = {} & \phi(z+h)-\phi(z+h-k)+\phi(z+h-k-k')-\phi(z+h-2k-k') \\
& +\phi(z+h-2k-2k')-\phi(z+h-3k-2k')+\ldots-\phi(z-h),
\end{aligned}$$

where $$2h = \nu k+(\nu-1)k'.$$

Using Taylor's theorem we may write

$$\phi(z+h)-\phi(z+h-k)=k\phi'(z+h)-(k^2/2!)\phi''(z+h)+(k^3/3!)\phi'''(z+h)-\ldots;$$

and therefore

$$S = k\{\phi'(z+h)+\phi'(z+h-k-k')+\phi'(z+h-2k-2k')+\ldots +\phi'(z-h+k)\}$$

$$-(k^2/2!)\{\phi''(z+h)+\phi''(z+h-k-k')+\ldots\}+\ldots \qquad \text{(I)}$$

$$= \frac{k}{k+k'}\{\phi(z+h)-\phi(z-h)\}$$

$$+\frac{kk'}{2(k+k')}\{\phi'(z+h)-\phi'(z-h)\}$$

$$-\frac{kk'(k-k')}{12(k+k')}\{\phi''(z+h)-\phi''(z-h)\}$$

$$-\frac{k^2k'^2}{24(k+k')}\{\phi'''(z+h)-\phi'''(z-h)\}+\ldots. \qquad \text{(II)}$$

Let the coefficient of $\phi^{(m)}(z+h)-\phi^{(m)}(z-h)$ in the series (II) be

$$(-)^m(k+k')^m F_m/m!.$$

To calculate F_m we expand each term of (II) thus

$$\phi^{(m)}(z+h)-\phi^{(m)}(z-h)$$

$$=\sum_{r=1}^{\infty}\frac{(k+k')^r}{r!}(-)^{r+1}\left[\phi^{(m+r)}(z+h)+\phi^{(m+r)}(z+h-k-k')+\ldots\right],$$

and then compare the result with (I), equating the coefficients of

$$\left[\phi^{(n)}(z+h)+\phi^{(n)}(z+h-k-k')+\ldots\right].$$

Hence

$$\left(\frac{k}{k+k'}\right)^n = \sum_{m=0}^{n-1}\binom{n}{m}F_m.$$

The solution of this set of difference equations is

$$F_m=\frac{1}{m+1}\left\{\left(\frac{k}{k+k'}\right)^{m+1}-\frac{m+1}{2}\left(\frac{k}{k+k'}\right)^m+B_1\binom{m+1}{2}\left(\frac{k}{k+k'}\right)^{m-1}\right.$$

$$\left.-B_3\binom{m+1}{4}\left(\frac{k}{k+k'}\right)^{m-3}+\ldots\right\},$$

where B_1, B_3, &c., are Bernoulli's numbers.

6. Let us now apply this to the potential of the coil considered in § 1. Instead of looking on the coil as a single thin current sheet, we will look upon it as made up of a series of thin current sheets distributed in bands of breadth k, separated by bands of insulation of breadth k', thus approximating to the case of a coil wound with a single layer of very thin conducting tape.

The potential function

$$\phi(z) = ni\left[\left[z\tan^{-1}\frac{\xi\eta}{zr} - \xi\tanh^{-1}\frac{\eta}{r} - \tanh^{-1}\frac{\xi}{r}\right]_{\xi=x-a}^{x+a}\right]_{\eta=y-b}^{y+b},$$

considered as a function of z and all its derivatives, is finite and continuous except for points where $z = 0$, and either $x = \pm a$ or $y = \pm b$.

Observing that

$$\phi''(z) = -ni\left[\left[\frac{\xi\eta}{r}\left(\frac{1}{z^2+\xi^2}+\frac{1}{z^2+\eta^2}\right)\right]_{\xi=x-a}^{x+a}\right]_{\eta=y-b}^{y+b},$$

we see that $\phi(z+k)$ can be expanded in ascending powers of k in a convergent series whenever the four inequalities

$$z^2+(x\pm a)^2 > k^2, \quad z^2+(y\pm b)^2 > k^2$$

are all satisfied.

In obtaining the expression for S given in the last paragraph, Taylor's expansions for

$$\phi\{z+h-k-r(k+k')+k\} \text{ in ascending powers of } k,$$

and for

$$\phi\{z+h-k-k'-r(k+k')+k+k'\} \text{ in ascending powers of } k+k',$$

are used; where r is a positive integer less than $2h/(k+k')$.

We see then that we may use the expression found for S, provided the distance of the point (x, y, z) from the current sheet is greater than $k+k'$. Thus if this condition is satisfied

$$\left[\phi(\zeta)+\tfrac{1}{2}k'\phi'(\zeta)-\tfrac{1}{12}k'(k-k')\,\phi''(\zeta)-\tfrac{1}{24}kk'^2\phi'''(\zeta)+\dots\right]_{\zeta=z-h}^{z+h}$$

will be a closer approximation to the true potential.

When this condition is not satisfied we may proceed thus: consider a point (x, y, z) near the top boundary of the coil, but such that its distance from the second layer is everywhere greater than $k+k'$

Then we may take the expression

$$\left[\phi(\zeta)+\tfrac{1}{2}k'\phi'(\zeta)\dots\right]_{\zeta=z-h}^{z+h-k-k'}$$

as the potential of the coil when the top layer is removed. The potential of the top layer by itself is

$$\phi(z+h)-\phi(z+h-k).$$

The potential of the whole coil at this point is the sum of these two expressions.

The case where the point (x, y, z) is near any other part of the coil may be treated in a similar manner.

7. The mutual inductance of two coils whose axes and plane sides are parallel may be treated in the same way. The difficulty arising from divergent series and infinities may be avoided too, for the case when one coil partly encloses the other, by supposing the magnetic force due to the *outer* coil linked with the current in the *inner* coil, and taking the summations in the order due to this arrangement.

The result is given by the expression found in § 2, with the function

$$\left[1+\tfrac{1}{2}k'\frac{\partial}{\partial\zeta}-\tfrac{1}{12}k'(k-k')\frac{\partial^2}{\partial\zeta^2}+\ldots\right]$$
$$\times\left[1+\tfrac{1}{2}k_1'\frac{\partial}{\partial\zeta}-\tfrac{1}{12}k_1'(k_1-k_1')\frac{\partial^2}{\partial\zeta^2}+\ldots\right]\phi(\xi,\eta,\zeta),$$

substituted for $\phi(\xi, \eta, \zeta)$: where k_1, k_1' are the breadths of the bands of conducting tape and insulation respectively of the second coil.

To find the self inductance of a coil, we must take the magnetic force due to the whole coil and sum the linkages between the limits for ξ and η: $a-l$, $l-a$, and $b-l$, $l-b$ respectively. This will give us an expression

$$8n^2\left[\left[\left[\left\{1+\tfrac{1}{2}k'\frac{\partial}{\partial\zeta}-\tfrac{1}{12}k'(k-k')\frac{\partial^2}{\partial\zeta^2}+\ldots\right\}^2\phi(\xi,\eta,\zeta)\right]_{\xi=l}^{2a-l}\right]_{\eta=l}^{2b-l}\right]_{\zeta=0}^{2h}, \qquad \text{(III)}$$

where l must be greater than $k+k'$.

We may consider the linkages due to that part of the flux which is near the conductor, and has been excluded from the above, separately.

The magnetic force at a point near one side of the coil consists of two parts, that due to the side to which it is near and that due to the rest of the coil. The latter part will be accounted for if the upper limits in (III) for ξ and η be changed to $2a$ and $2b$, this introduces no infinities. The former part varies from zero—when the point lies actually in an insulating strip—up to $2\pi ni\{(k+k')/k\}$, when the point lies just outside a conducting strip; in the former case the magnetic force is not required for the linkage. Thus this former part will add to the inductance an amount less than

$$16\pi n^2 hl(a+b).$$

8. The same considerations may be applied to the case of a thick coil discussed in §§ 3 and 4. By replacing V_t by

$$\left\{1+\tfrac{1}{2}k'\frac{\partial}{\partial\zeta}-\tfrac{1}{12}k'(k-k')\frac{\partial^2}{\partial\zeta^2}+\ldots\right\}V_t,$$

in § 3, we obtain the potential of a coil formed of a single layer of wire of rectangular cross section $t\times k$ with a thickness k' of insulation between, at any point not too near the conductor.

To find the potential when there are several layers each of thickness t_1 separated by insulation thickness t_2, we apply to V_t the additional operator

$$\left\{1+\tfrac{1}{2}t_1\frac{\partial}{\partial t}-\tfrac{1}{12}t_2(t_1-t_2)\frac{\partial^2}{\partial t^2}+\ldots\right\}.$$

The case when (x, y, z) is near the coil, and also the inductance, may be considered as before.

TERNARY PERPETUANTS

By ALFRED YOUNG

[Received December 7th, 1922.—Read December 14th, 1922.]

1. In what follows, functions of the coefficients of the quantics and of the variables are alone considered, and line coordinates are not introduced at all. So that pure covariants alone are treated of and not contravariants or mixed concomitants. Further, the covariants are supposed to be linear in the coefficients of each quantic, that is they are covariant types.

In § 2 a set of symbolical products is obtained in terms of which all perpetuant types of given degree can be linearly expressed.

In § 3 it is proved that the members of this set are linearly independent.

In § 4 a generating function is found which gives the number of members of this set of given weight and degree.

In § 5 various useful expressions for this generating function are given.

In § 6 a reduced generating function is obtained, giving the number of perpetuant types of given weight and degree, which are linearly independent of those which have one of the quantics as a factor.

In § 7 the generating functions as far as degree 8 are given with the numerical coefficients calculated.

In § 8 perpetuants of degree 4 are discussed.

In §§ 9, 10, 11, 12 forms of degrees 5, 6, 7, 8, and of the lowest weights are considered. In the cases of degrees 7 and 8 it is shown that the generating functions indicate the existence of syzygies which are obtained.

In § 13 a lower limit is obtained for the weight of an irreducible perpetuant of degree r. It is shown that if r is even, no perpetuant of weight less than $r-3$ can be irreducible, and if r is odd, the lower limit is $r-2$. It is not by any means established that there are always irreducible perpetuants of these weights.

In § 14 it is proved that there is no family of syzygies for ternary forms analogous to the Stroh syzygies for binary forms.

2. We use the symbolical notation. As a rule the factors a_x are not written, the covariants being completely defined by the determinant factors of the symbolical product.

The quantics dealt with will be written

$$a_{1x}^n, a_{2x}^n, \ldots, a_{rx}^n$$

where n is indefinitely large—or any way at least as large as the weight of the covariants discussed.

We arrange the quantics in a fixed sequence, and for convenience take the sequence to be that of the numerical order of the suffixes.

Then the letter a_1 may be introduced into every determinant factor by means of the fundamental identity; this is supposed done.

Next a product $(a_1 a_p a_q)(a_1 a_r a_s)$ satisfies the identity

$$(a_1 a_p a_q)(a_1 a_r a_s) + (a_1 a_q a_r)(a_1 a_p a_s) + (a_1 a_r a_p)(a_1 a_q a_s) = 0.$$

Let $$s > r > q > p.$$

Then we see that a product of two factors, such that the two letters other than a_1 in one lie between the two corresponding letters of the other factor, can be expressed in terms of pairs of factors for which this is not the case. Hence we can express all our forms in terms of symbolical products, each determinant factor of which contains a_1, and which are such that no pair of factors have the second and third letters of one lying between the second and third letters of the other. It is evident that the fundamental identity is not capable of giving us anything further. But a formal proof of the linear independence of the members of the set is given in the next paragraph.

In general this set of symbolical products does not consist of a single row of determinant factors raised to different powers—as is the case with binary perpetuants; but it consists of a certain number of such rows.

Thus for degree 5 the perpetuants are expressed in terms of the forms

$$(a_1a_2a_3)^{\lambda_1}(a_1a_2a_4)^{\lambda_2}(a_1a_2a_5)^{\lambda_3}(a_1a_3a_5)^{\lambda_4}(a_1a_4a_5)^{\lambda_5}$$

and $$(a_1a_2a_3)^{\lambda_1}(a_1a_2a_4)^{\lambda_2}(a_1a_3a_4)^{\lambda_3}(a_1a_3a_5)^{\lambda_4}(a_1a_4a_5)^{\lambda_5};$$

there being two rows of factors.

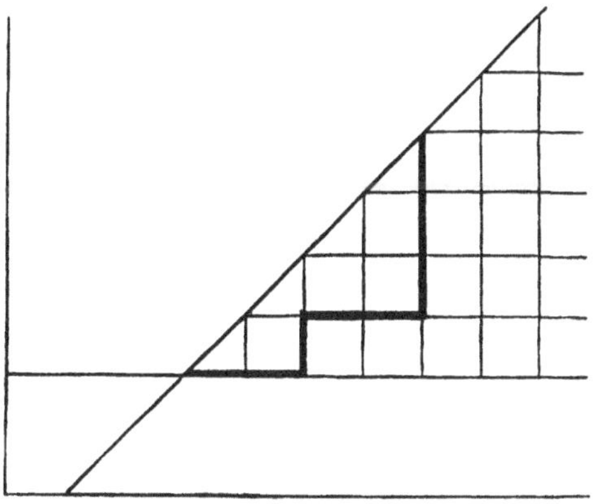

We may represent the different factors of a product by points on a diagram. Thus, arrange the letters of a factor in ascending order of magnitude of the suffixes, and let the point (q, p) stand for the factor $(a_1a_pa_q)$. Then all the points will lie between the straight lines $y = 2$ and $y = x-1$, or on the boundary of the enclosed space. Then in any product we arrange the factors in the sequence of the second letters: when the second letters are equal we arrange according to the sequence of the last letters. Then, if straight lines be drawn joining points representing consecutive factors starting from the first [which is usually (3, 2)] as we pass from point to point the values of both x and y are never on the decrease, owing to the "between-ness" rule.

We may pass from a factor $(a_1a_pa_q)$ to the next $(a_1a_ra_s)$ by a series of steps, increasing the second or third suffix by unity at each step and leaving the other unchanged. In other words we can pass from one point on the diagram of the product to the next by a series of steps, each parallel to one axis and increasing one coordinate by unity.

Thus, if r is the degree of the covariant, we can represent it as a product of powers of $(2r-5)$ factors beginning with $(a_1a_2a_3)^{\lambda}$ and ending with

$(a_1 a_{r-1} a_r)^{\mu}$; the passage from one factor to the next, corresponding to a unit step parallel to one axis—(some of the factors, of course, may be omitted, the corresponding power being zero).

In general, the journey from (2, 3) to $(r-1, r)$ may be performed in this manner in

$$\binom{2r-8}{r-4} - \binom{2r-8}{r-6}$$

ways. And there is this number of different rows of factors to be considered for degree r. As this particular fact is not required for our purpose, it is thought unnecessary to prove it.

The figure gives the diagram for

$$(a_1 a_2 a_3)^{\lambda_1} (a_1 a_2 a_4)^{\lambda_2} (a_1 a_2 a_5)^{\lambda_3} (a_1 a_3 a_5)^{\lambda_4} (a_1 a_3 a_6)^{\lambda_5}$$

$$\times (a_1 a_3 a_7)^{\lambda_6} (a_1 a_4 a_7)^{\lambda_7} (a_1 a_5 a_7)^{\lambda_8} (a_1 a_6 a_7)^{\lambda_9}.$$

3. To prove the linear independence of the set of forms just obtained, let us suppose that there is a linear relation between our set of forms for degree r, but none of a lower degree. If every term in the relation has the factor $(a_1 a_{r-1} a_r)$, we can divide by this factor and still have an identical relation; hence we may assume that there is at least one term without this factor. The linear relation will remain true if we replace every a_r by a_{r-1}, and the "between-ness" rule will still hold for the resulting terms. But as there is no such linear relation of degree $r-1$, it must be now a formal identity, such that the coefficient of each *different* product is zero. The relation of degree r may be written

$$\Sigma AP (a_1 a_{p_1} a_{r-1})^{\lambda_1} (a_1 a_{p_2} a_{r-1})^{\lambda_2} \ldots (a_1 a_{q_1} a_r)^{\mu_1} (a_1 a_{q_2} a_r)^{\mu_2} = 0$$

where P represents factors containing neither a_r nor a_{r-1}, and A is numerical.

By the "between-ness" rule no q is less than any p. In order that this may become a formal identity when a_r is replaced by a_{r-1}, this will have to be the case with each set of factors which have the part P in common, separately. Confining our attention to one such set, we see that in the original relation this requires that corresponding to one such term written above, there must be another *different* one in which one or more pairs of factors $(a_1 a_p a_{r-1})(a_1 a_q a_r)$ is replaced by the pair $(a_1 a_q a_{r-1})(a_1 a_p a_r)$ if $p = q$, this makes no change and if $p \neq q$ this is not possible owing to the "between-ness" rule. Hence the supposed identity for degree r cannot exist, unless there is one for degree $r-1$. But there is none for degree 3 or degree 4 and therefore our set of forms is linearly independent.

4. Let us call the generating function for linearly independent perpetuants of degree r, G_r. The number of different forms belonging to a single row of factors is $(1-x)^{-2r+5}$; when, however, one or more factors of a row are absent, the rows will be found to overlap.

Let $G_{r,s}$ be the generating function for those products of degree r which contain the factor $(a_1 a_2 a_s)$, where $s \geqslant 5$, but which contain no factor $(a_1 a_2 a_t)$ where $t > s$. And let $G_{r,4}$ be the generating function for those products of degree r which contain no factor $(a_1 a_2 a_t)$, $t > 4$: in $G_{r,4}$ there are included products which contain no factor $(a_1 a_2 a_4)$.

Looking at the diagram of § 2, $G_{r,s}$ is the generating function of those products for which the line on the diagram contains the point $(s, 2)$ but leaves the straight line $y = 2$ at this point, when $s > 4$; but, if $s = 4$, it contains no point on $y = 2$ beyond $(s, 2)$, but need not contain this point.

We proceed to obtain formulæ expressing the generating functions for perpetuants of the r forms represented by $a_1, a_2, \ldots, a_r$ in terms of the generating functions for perpetuants of the $r-1$ forms left when that represented by a_2 is removed.

Since every perpetuant belongs to one of the classes $G_{r,s}$ $(s = 4, 5, \ldots, r)$, and these classes do not overlap,

$$G_r = G_{r,4} + G_{r,5} + \ldots + G_{r,r}.$$

Now the "between-ness" rule does not exclude any factor $(a_1 a_s a_t)$, $t > s > 2$, on account of the presence of the two factors $(a_1 a_2 a_3)(a_1 a_2 a_4)$, so every product giving a perpetuant of the forms represented by

$$a_1, a_3, \ldots, a_r$$

multiplied by $(a_1 a_2 a_3)^\lambda (a_1 a_2 a_4)^\mu$ will give a perpetuant of the class $G_{r,4}$.

Hence $$G_{r,4} = (1-x)^{-2} G_{r-1}.$$

The only factor excluded by the "between-ness" rule when $(a_1 a_2 a_5)$ is present is $(a_1 a_3 a_4)$, thus any perpetuant of class $G_{r,5}$ may be obtained by taking a perpetuant of $a_1, a_3, \ldots, a_r$, replacing the factor $(a_1 a_3 a_4)^\lambda$ by

$(a_1 a_2 a_4)^\lambda$, and multiplying by the factor $(a_1 a_2 a_3)^\mu (a_1 a_2 a_5)^{\nu+1}$, thus

$$G_{r,5} = x(1-x)^{-2} G_{r-1} = x(1-x)^{-2}(G_{r-1,4}+G_{r-1,5}+\ldots+G_{r-1,r-1}).$$

Consider any perpetuant of class $G_{r,6}$. First drop the factor

$$(a_1 a_2 a_3)^\lambda (a_1 a_2 a_4)^\mu,$$

then replace $(a_1 a_2 a_5)^\nu$ by $(a_1 a_3 a_4)^\nu$; now there is a factor $(a_1 a_2 a_6)^{\rho+1}$, we will replace this by $(a_1 a_3 a_5)^\rho (a_1 a_3 a_6)$—originally there were no factors $(a_1 a_3 a_4)$, $(a_1 a_3 a_5)$ in the perpetuant owing to the "between-ness" rule. We are left with a perpetuant of the forms $a_1, a_3, a_4, \ldots, a_r$ which contains one factor $(a_1 a_3 a_6)$; that is it belongs to one of the classes

$$G_{r-1,5},\quad G_{r-1,6},\quad \ldots,\quad G_{r-1,r-1}.$$

If it contains no factor $(a_1 a_3 a_t)$, $t > 6$, it belongs to $G_{r-1,5}$. Moreover, the steps by which we obtained it from the original perpetuant are strictly and uniquely reversible, the number of such perpetuants is then given by $G_{r-1,5}$. If it contains a factor $(a_1 a_3 a_h)$, but no factor $(a_1 a_3 a_t)$, $t > h$, it belongs to $G_{r-1,h-1}$, but here we must remember there is a factor $(a_1 a_3 a_6)$ which is not contained in every perpetuant of $G_{r-1,h-1}$, the number of perpetuants leading to this class is then $xG_{r-1,h-1}$. Hence

$$G_{r,6} = (1-x)^{-2}(G_{r-1,5}+xG_{r-1,6}+xG_{r-1,7}+\ldots+xG_{r-1,r-1})$$

and in the same way we find in general

$$G_{r,s} = (1-x)^{-2}(G_{r-1,s-1}+xG_{r-1,s}+xG_{r-1,s+1}+\ldots+xG_{r-1,r-1})$$

$$G_{r,r} = x(1-x)^{-2r+5}.$$

Now
$$G_3 = \frac{1}{1-x},\qquad G_4 = \frac{1}{(1-x)^3},$$

$$G_5 = G_{5,4}+G_{5,5} = (1+x)(1-x)^{-5}.$$

We proceed to prove that*

$$G_r = \left((z^{r-4})\right)(1+z)^{r-4}(x+z)^{r-4}(1-z^{-2}x)(1-x)^{-2r+5}$$

and that

$$G_{r,s} = \left((z^{r-5})\right) x(1+z)^{r-s}(x+z)^{r-5}(1-z^{-2}x)(1-x)^{-2r+5}.$$

* The notation $((z^n))\,F(z)$ is used to denote the coefficient of z^n in the expansion of $F(z)$.

Assume the truth of these formulæ for a given value of r. Then

$$G_{r+1,5} = x(1-x)^{-2}G_r$$

$$= \left((z^{r-4})\right) x(1+z)^{r-4}(x+z)^{r-4}(1-z^{-2}x)(1-x)^{-2r+3}.$$

$$G_{r+1,s} = (1-x)^{-2}\{G_{r,s-1}+xG_{r,s}+\ldots+xG_{r,r}\}$$

$$= \left((z^{r-5})\right) x(x+z)^{r-5}(1-z^{-2}x)$$

$$\times\left\{(1+z)^{r-s+1}+x\frac{(1+z)^{r-s+1}-1}{z}\right\}(1-x)^{-2r+3}$$

$$= \left((z^{r-4})\right) x(x+z)^{r-4}(1+z)^{r-s+1}(1-z^{-2}x)(1-x)^{-2r+3}.$$

For the next step we require

$$\left((z^{r-1})\right)(1+z)^r(x+z)^r(1-z^{-2}x)$$

$$= \binom{r}{1}x+\binom{r}{1}\cdot\binom{r}{2}x^2+\binom{r}{2}\cdot\binom{r}{3}x^3+\ldots$$

$$-x\left\{\binom{r}{1}+\binom{r}{2}\cdot\binom{r}{1}x+\binom{r}{3}\cdot\binom{r}{2}x^2+\ldots\right\}=0,$$

and hence

$$\left((z^{r-3})\right)(x+z)^{r-4}(1-z^{-2}x)(1-x)^{-2r+3}(1+z)^{r-4}z^2$$

$$= \left((z^{r-5})\right)(x+z)^{r-4}(1+z)^{r-4}(1-z^{-2}x)(1-x)^{-2r+3} = 0.$$

Then

$$G_{r+1} = G_{r+1,4}+G_{r+1,5}+\ldots+G_{r+1,r+1}$$

$$= \left((z^{r-4})\right)(x+z)^{r-4}(1-z^{-2}x)(1-x)^{-2r+3}$$

$$\times\{(1+z)^{r-4}+x(1+z)^{r-4}+\ldots+x\}$$

$$= \left((z^{r-3})\right)(x+z)^{r-4}(1-z^{-2}x)(1-x)^{-2r+3}\{z(1+z)^{r-4}+x(1+z)^{r-3}-x\}$$

$$= \left((z^{r-3})\right)(x+z)^{r-4}(1-z^{-2}x)(1-x)^{-2r+3}(1+z)^{r-4}\{z+x+xz\};$$

by the result just proved above we may add the term z^2 in the last bracket without affecting the result. Hence

$$G_{r+1} = \left((z^{r-3})\right)(x+z)^{r-4}(1-z^{-2}x)(1-x)^{-2r+3}(1+z)^{r-4}\{z+x+xz+z^2\}$$

$$= \left((z^{r-3})\right)(1+z)^{r-3}(x+z)^{r-3}(1-z^{-2}x)(1-x)^{-2r+3}.$$

Thus, if our formulæ are true for degree r, they are also true for degree $r+1$. But they are true for degree 5, and therefore they are true in general.

5. We may put the generating function in slightly different forms which will be useful later. Thus

$$G_r(1-x)^{2r-5} = \left((z^{r-4})\right)(1+z)^{r-4}(x+z)^{r-4}(1-z^{-2}x)$$

$$= \sum_{a=0}^{r-4}\left\{\binom{r-4}{a}\binom{r-4}{a} - \binom{r-4}{a-1}\binom{r-4}{a+1}\right\}x^a$$

$$= \Sigma\,\frac{1}{a}\binom{r-3}{a+1}\binom{r-4}{a-1}x^a = \frac{1}{r-3}\,\Sigma\binom{r-3}{a+1}\binom{r-3}{a}x^a$$

$$= \Sigma\,\frac{1}{a}x^a\sum_{s=0}^{a-1}\binom{a}{s}\binom{2a-s}{a+1}\binom{r-3}{2a-s}\ {}^{*}$$

$$= \sum_a \frac{1}{a}x^a\sum_{s=0}^{a-1}\binom{a}{s}\binom{2a-s}{a+1}\sum_{\sigma=0}^{2a-s}(-)^{\sigma}\binom{\sigma+2}{2}\binom{r}{2a-\sigma-s}\ {}^{\dagger}$$

$$= \sum_a\sum_\lambda B_{a,\lambda}\binom{r}{\lambda}x^a,$$

where $$B_{a,\lambda} = \frac{1}{a}\sum_s(-)^{s+\lambda}\binom{a}{s}\binom{2a-s}{a+1}\binom{2a-s-\lambda+2}{2}.$$

Thus $B_{a,\lambda}$ is entirely independent of r and a function of a and λ alone. The lower limit of s is zero, the upper limit is either $a-1$ or $2a-\lambda+2$, whichever is the smaller.

When $2a-\lambda+2 \geqslant a-1$, a simple form can be found for $B_{a,\lambda}$,

---

* For $\binom{r-4}{a-1} = \left((x^{a-1})\right)(1+x)^{r-4} = \left((x^{a-1})\right)(1+x)^{r-4-a}(1+x)^a = \Sigma\binom{a}{s}\binom{r-4-a}{a-1-s}$.

† For $\binom{r-3}{2a-s} = \left((x^{2a-s})\right)(1+x)^{-3}(1+x)^r$.

i.e. when $\lambda \not> a+3$. For then

$$B_{a,\lambda} = \frac{1}{a}\sum_{s=0}^{a-1}(-)^{s+\lambda}\binom{a}{s}\binom{2a-s}{a+1}\left\{\binom{2a-s+2}{2}-\lambda\binom{2a-s+1}{1}+\binom{\lambda}{2}\right\}$$

$$= (-)^{\lambda}\frac{1}{a}\left\{\binom{a+3}{2}\binom{a+2}{3}-\lambda\binom{a+2}{1}\binom{a+1}{2}+\binom{\lambda}{2}a\right\}$$

$$= (-)^{\lambda}\left\{\binom{a+4}{4}+\binom{a+3}{4}-\lambda\binom{a+2}{2}+\binom{\lambda}{2}\right\}$$

provided $a > 0$: this is also true if $a = 0$ and $\lambda \not> 2$.

The case $a = 0$ should be treated separately. The constant term in $G_r(1-x)^{-2r+5}$ is 1. Hence $B_{0,\lambda} = 0$ unless $\lambda = 0$ and then it is unity. Again it is to be observed that $B_{a,\lambda} = 0$ when $\lambda > 2a$, and

$$B_{a,2a} = \frac{1}{a}\binom{2a}{a+1}.$$

6. Let us now exclude all forms which have one of the quantics as a factor, and seek for the generating function of the perpetuants that are left. We shall call this the reduced generating function G'_r.

Its value is

$$G'_r = G_r - rG_{r-1} + \binom{r}{2}G_{r-2} - \ldots + (-)^t\binom{r}{t}G_{r-t} + \ldots$$
$$+ (-)^{r-1}G_1 + (-)^r G_0 + (-)^r x.$$

We obtain it by considering a definite product of λ quantics and a form of degree $r-\lambda$ having no separate quantic as a factor. The first term in the above sum includes this product once, the second excludes it λ times, the third includes it $\binom{\lambda}{2}$ times, and so on; finally it is found to be included

$$1-\lambda+\binom{\lambda}{2}-\ldots = (1-1)^{\lambda} = 0$$

times.

The last term of all has to be added owing to the Jacobian syzygy of weight unity,

$$(abc)d_x - (bcd)a_x + (cda)b_x - (dab)c_x = 0.$$

Were it not for this syzygy G_r would include $\binom{r}{3}$ forms of weight one, whereas it only includes $\binom{r-1}{2}$ such forms. In G'_r we have to calculate the term to be added so that there is no term of weight 1. Taking the expression written down, we find the coefficient of x to be

$$\binom{r-1}{2}-r\binom{r-2}{2}+\binom{r}{2}\binom{r-3}{2}-\ldots+(-)^{r-3}\binom{r}{r-3}\binom{2}{2}+(-1)^r$$

$$=((x^2))\Big\{(1+x)^{r-1}-r(1+x)^{r-2}+\binom{r}{2}(1+x)^{r-3}-\ldots$$

$$+(-)^{r-3}\binom{r}{r-3}(1+x)^2+(-)^r x^2\Big\}$$

$$=((x^2))\left[\frac{\{(1+x)-1\}^r}{1+x}-\frac{(-1)^r}{1+x}+(-)^r x^2\right]=0.$$

It should be observed that the exactness of the value given for the reduced generating function is upset by a syzygy, but that the syzygy, having the effect of reducing the number of independent products, will make it appear that there are fewer irreducible forms than there really are. To find the value of G'_r, let us consider the sum

$$\sum_{t=0}^{r-3}(-)^t\binom{r}{t}G_{r-t}(1-x)^{2r-5}$$

$$=\Sigma(-)^t\binom{r}{t}\left[G_{r-t}(1-x)^{2r-2t-5}\right](1-x)^{2t}$$

$$=\sum_{t,\,w,\,s,\,\lambda}(-)^{s+t}\binom{2t}{s}\binom{r}{t}\binom{r-t}{\lambda}B_{w-s,\,\lambda}\,x^w$$

$$=\sum_{t,\,w,\,s,\,\lambda,\,\sigma}(-)^{s+t}\binom{s-\sigma}{\sigma}2^{s-2\sigma}\binom{t}{s-\sigma}\binom{r}{\lambda}\binom{r-\lambda}{t}B_{w-s,\,\lambda}\,x^{w}*$$

$$=\Sigma(-)^{s+t}2^{s-2\sigma}\binom{s-\sigma}{\sigma}\binom{r}{\lambda}\binom{r-\lambda}{s-\sigma}\binom{r-\lambda-s+\sigma}{t-s+\sigma}B_{w-s,\,\lambda}\,x^w. \quad \text{(A)}$$

* For $\binom{2t}{s}=((x^s))(1+2x+x^2)^t$.

Now when $r-\lambda-s+\sigma$ is positive and $\lambda \geqslant 3$ (noticing that $s \geqslant \sigma$)

$$\sum_{t=0}^{r-3} \binom{r-\lambda-s+\sigma}{t-s+\sigma}(-)^t = (-)^{s-\sigma}(1-1)^{r-\lambda-s+\sigma} = 0\,;$$

when $\lambda = 2$, this sum is $\qquad (-)^{r+1}$;

when $\lambda = 1$, the sum is $\qquad (-)^{r+1}(r-2-s+\sigma)$;

when $\lambda = 0$, the sum is

$$(-)^{r+1}\binom{r-1-s+\sigma}{2}.$$

Also $r-\lambda-s+\sigma$ can never be negative; it can, however, be zero. Let Γ represent the sum of those terms of the series we are discussing, for which $r-\lambda-s+\sigma = 0$.

Then

$$\sum_{t=0}^{r-3} (-)^t \binom{r}{t} G_{r-t}(1-x)^{2r-5} - \Gamma$$

$$= \sum_{w,\, s,\, \sigma} (-)^{r+s+1} x^w 2^{s-2\sigma} \binom{s-\sigma}{\sigma} \left[\binom{r}{2}\binom{r-2}{s-\sigma} B_{w-s,\,2} \right.$$

$$\left. + r\binom{r-1}{s-\sigma}(r-2-s+\sigma) B_{w-s,\,1} + \binom{r}{s-\sigma}\binom{r-1-s+\sigma}{2} B_{w-s,\,0}\right]$$

$$= \Sigma (-)^{r+s+1} x^w 2^{s-2\sigma} \binom{s-\sigma}{\sigma} \left[\binom{r}{2}\binom{r-2}{s-\sigma} \{B_{w-s,\,2} + 2B_{w-s,\,1} + B_{w-s,\,0}\} \right.$$

$$\left. - r\binom{r-1}{s-\sigma}\{B_{w-s,\,1} + B_{w-s,\,0}\} + \binom{r}{s-\sigma} B_{w-s,\,0}\right]$$

$$= \Sigma (-)^{r+s+1} x^w 2^{s-2\sigma} \binom{s-\sigma}{\sigma} \left[\binom{r}{2}\binom{r-2}{s-\sigma} - \binom{r}{1}\binom{r-1}{s-\sigma}\binom{w-s+2}{2} \right.$$

$$\left. + \binom{r}{s-\sigma}\binom{w-s+3}{3}\frac{w-s+2}{2}\right]$$

$$= \Sigma (-)^{r+s+1} x^w$$

$$\left[\binom{r}{2}\binom{2r-4}{s} - r\binom{2r-2}{s}\binom{w-s+2}{2} + \binom{2r}{s}\binom{w-s+3}{3}\frac{w-s+2}{2}\right]^*$$

---

* For $\qquad \Sigma \binom{s-\sigma}{\sigma}\binom{r-2}{s-\sigma} 2^{s-2\sigma} = ((x^s))\,(1+2x+x^2)^{r-2}.$

Now the limits of summation for s are 0 and w. Then we have

$$\sum_{s=0}^{w} (-)^s \binom{2r-4}{s} = (-)^w \binom{2r-5}{w},$$

$$\Sigma (-)^s \binom{2r-2}{s}\binom{w-s+2}{2} = \left((x^w)\right)(1-x)^{2r-2}(1-x)^{-3} = (-)^w \binom{2r-5}{w},$$

$$\Sigma (-)^s \binom{2r}{s}\binom{w-s+3}{3}\frac{w-s+2}{2}$$

$$= \left((x^w)\right)(1-x)^{2r}(1-x)^{-5} + \left((x^{w-1})\right)(1-x)^{2r}(1-x)^{-5}$$

$$= (-)^w \left\{\binom{2r-5}{w} - \binom{2r-5}{w-1}\right\}.$$

Hence

$$\sum_{t=0}^{r-3} (-)^t \binom{r}{t} G_{r-t}(1-x)^{2r-5} - \Gamma$$

$$= \Sigma (-)^{r+w+1} x^w \left[\binom{2r-5}{w}\left\{\binom{r}{2} - \binom{r}{1} + 1\right\} - \binom{2r-5}{w-1}\right]$$

$$= -(-)^r (1-x)^{2r-5}\left[\binom{r-1}{2} + x\right].$$

Hence we find $$G'_r(1-x)^{2r-5} = \Gamma$$

where Γ is the sum of terms from (A) for which $r-\lambda-s+\sigma = 0$. Since

$$r-\lambda-s+\sigma = (t-s+\sigma)+(r-\lambda-t)$$

and neither of these latter quantities, indicated by the brackets, can be negative, they must separately vanish. Thus

$$\Gamma = \Sigma (-)^\sigma 2^{r-\lambda-\sigma}\binom{r-\lambda}{\sigma}\binom{r}{\lambda} B_{w-r+\lambda-\sigma,\,\lambda}\, x^w.$$

Now if $$\lambda > 2a, \quad B_{a,\lambda} = 0,$$

thus we may limit the series to

$$\lambda \not> 2(w-r+\lambda-\sigma)$$

or $$\lambda \not< 2(r+\sigma-w).$$

Also from the way the series was formed

$$\lambda \not> r-\sigma,$$

therefore $$2(r+\sigma-w) \not> r-\sigma,$$

and $$w \not< \frac{r+3\sigma}{2}.$$

Hence the series Γ cannot begin before the term $x^{r/2}$.

Suppose that r is even and equal to $2w$, then to find the coefficient of x^w, we see that $\sigma=0$, $\lambda=2w$, the first term of Γ is

$$B_{w,\,2w}\,x^w = \frac{1}{w}\binom{2w}{w+1}x^w.$$

Next let r be odd and equal to $2w-1$, for the first term we have $\sigma=0$, $\lambda=r$ or $r-1$, and hence this term is

$$\{2rB_{w-1,\,r-1}+B_{w,\,r}\}\,x^w = \binom{2w}{w+1}\frac{3w-5}{2w}x^w.$$

I have not succeeded in calculating the general term of Γ in a concise form. The last six terms are

$$(-)^r\Bigg[\left\{\binom{2r-6}{3}\binom{r}{3}-\binom{2r-6}{1}\binom{r}{4}\right.$$
$$\left.-\binom{2r-5}{4}\binom{r-1}{2}-\binom{2r-5}{5}+\binom{r}{5}\right\}x^{2r-9}$$
$$+\left\{\binom{r-1}{4}(3r-35)-\binom{2r-5}{4}\right\}x^{2r-8}$$
$$+\left\{\binom{2r-5}{3}-\binom{r-1}{3}(4r-15)\right\}x^{2r-7}$$
$$+\left\{5\binom{r-1}{3}-\binom{2r-5}{2}\right\}x^{2r-6}-\binom{r-3}{2}x^{2r-5}-x^{2r-4}\Bigg].$$

Also, if r is even and equal to 2ρ, the coefficient of $x^{\rho+1}$ is

$$3\binom{2\rho-1}{\rho+2}\frac{(\rho+1)(3\rho-4)}{\rho-1};$$

if r is odd and equal to $2\rho-1$, the coefficient of $x^{\rho+1}$ is

$$-\binom{2\rho-1}{\rho+2}\frac{9\rho^3-36\rho^2-11\rho+74}{2(\rho-1)}.$$

The last terms of Γ are easier to calculate because they only contain expressions $B_{a,\lambda}$, $\lambda \not> a+3$ in the last two cases, for which a simple expression has been found; and in the other cases the values of $B_{a,\lambda}$ are easy to obtain because $\lambda-a$ is small.

7. The generating functions up to degree 8 are

$$G_3 = \frac{1}{1-x}, \quad G_4 = \frac{1}{(1-x)^3}, \quad G_5 = \frac{1+x}{(1-x)^5}, \quad G_6 = \frac{1+3x+x^2}{(1-x)^7},$$

$$G_7 = \frac{1+6x+6x^2+x^3}{(1-x)^9}, \quad G_8 = \frac{1+10x+20x^2+10x^3+x^4}{(1-x)^{11}}.$$

The reduced generating functions up to degree 8 are

$$G'_3 = \frac{x}{1-x}, \quad G'_4 = \frac{2x^2-x^4}{(1-x)^3}, \quad G'_5 = \frac{10x^3-10x^4+x^5+x^6}{(1-x)^5}$$

$$G'_6 = \frac{5x^3+30x^4-55x^5+29x^6-3x^7-x^8}{(1-x)^7}$$

$$G'_7 = \frac{49x^4+35x^5-189x^6+176x^7-64x^8+6x^9+x^{10}}{(1-x)^9}$$

$$G'_8 = \frac{14x^4+280x^5-252x^6-364x^7+685x^8-430x^9+120x^{10}-10x^{11}-x^{12}}{(1-x)^{11}}.$$

The reduced generating functions for degrees 3, 4, 5, represent irreducible perpetuants, that for degree 6 may be further reduced by subtracting the generating function for products $(abc)^r(def)^{w-r}$; now

$$(abc)(def)^\lambda = (def)^\lambda[(abd)-(acd)+(bcd)],$$

hence we must only take account of these products for which $r \geqslant 2$ and $w-r \geqslant 2$, that is we must subtract

$$\tfrac{1}{2}\binom{6}{3}\frac{x^4}{(1-x)^2}$$

from G'_6. Thus we find a final generating function for irreducible perpetuants degree 6

$$G''_6 = \frac{5x^3+20x^4-5x^5-71x^6+97x^7-51x^8+10x^9}{(1-x)^7}.$$

8. *Forms of degree* 4.—There are only four possible determinant

factors in the symbolical product, viz. (bcd), (adc), (abd), (acb); let us call them for short u_1, u_2, u_3, u_4. Then

$$u_1+u_2+u_3+u_4 = 0.$$

It will now be shown that products of the u's of total degree w may be expressed in terms of such products as have a factor $u^{w/2}$ at least. In other words perpetuants of degree 4 and weight w are expressible in terms of those which are of grade $w/2$ at least.* By means of the identity all products of degree w can be expressed in terms of the products

$$u_1^{w-\alpha-\beta}\, u_2^{\alpha}\, u_3^{\beta},$$

a linearly independent set of number $\binom{w+2}{2}$.

Let w be even and equal to 2ρ. The number of independent products is

$$\binom{2\rho+2}{2} = 3\binom{\rho+1}{2}+\binom{\rho+2}{2}.$$

The number of independent products having a factor u_4^{α} is

$$\binom{2\rho-\alpha+2}{2}.$$

Consider those products which have a factor u_4^{ρ}, those which have a factor $u_1^{\rho+1}$ or a factor $u_2^{\rho+1}$, and those which have a factor $u_3^{\rho+1}$. We have thus a set of $\binom{2\rho+2}{2}$ products, if they are linearly independent, all products of degree $w = 2\rho$ can be expressed in terms of them. We will write those products which have a factor u_4^{ρ} in the form

$$u_4^{2\rho},\quad u_4^{2\rho-1}u_2,\quad u_4^{2\rho-1}u_3,\quad \dots,\quad u_4^{\rho}u_2^{\rho},\quad u_4^{\rho}u_2^{\rho-1}u_3,\quad \dots,\quad u_4^{\rho}u_3^{\rho},$$

and then express them in terms of products $u_1^{\alpha}u_2^{\beta}u_3^{\gamma}$, thus

$$u_4^{2\rho-\beta'-\gamma'}u_2^{\beta'}u_3^{\gamma'} = \binom{2\rho-\beta'-\gamma'}{\beta-\beta',\ \gamma-\gamma'}\, u_1^{2\rho-\beta-\gamma}u_2^{\beta}u_3^{\gamma}.$$

Regard these identities as a set of equations for expressing the products $u_1^{2\rho-\beta-\gamma}u_2^{\beta}u_3^{\gamma}$ for which neither of the three indices is greater than ρ in terms of members of our chosen set of products.

---

* An extension of Jordan's Lemma for Binary Forms. Another proof has been given by Wood, *Proc. London Math. Soc.*, Ser. 2, Vol. 1, p. 345.

Let Δ be the determinant of $\binom{\rho+2}{2}$ rows and columns, whose general element is $\binom{2\rho-\beta'-\gamma'}{\beta-\beta',\ \gamma-\gamma'}$, the values of β, γ being the same throughout each column, and the values of β', γ' being the same throughout each row; then our set of equations is independent if $\Delta \neq 0$. This then is the condition of independence of the members of our set.

Divide each row by $(2\rho-\beta'-\gamma')!\,\beta'!\,\gamma'!$, multiply each column by $(2\rho-\beta-\gamma)!\,\beta!\,\gamma!$, and call the resulting determinant Δ'. Then the general element of Δ' is $\binom{\beta}{\beta'}\binom{\gamma}{\gamma'}$ where

$$\beta'+\gamma' \leqslant \rho, \qquad \rho \leqslant \beta+\gamma \leqslant 2\rho,$$

and neither β nor γ exceeds ρ.

It is now convenient to fix the sequence of rows and columns in Δ' thus:—

The first $\rho+1$ rows have $\gamma' = 0$ and $\beta' = 0, 1, \ldots, \rho$, in order;

the next ρ rows have $\gamma' = 1$ and $\beta' = 0, 1, \ldots, \rho-1$, in order;

... ...

the last row has $\gamma' = \rho$, $\beta' = 0$.

The first column has $\beta = 0$, $\gamma = \rho$;

the second and third columns have $\beta = 1$, $\gamma = \rho-1, \rho$, in this order;

... ...

the last $\rho+1$ columns have $\beta = \rho$, $\gamma = 0, 1, \ldots, \rho$, in order.

Now expand Δ' as a sum of products of determinants thus:—Take first a minor of $\rho+1$ rows and columns from the first $\rho+1$ rows as the first determinant in the product; take next a minor of ρ rows and columns from the next ρ rows as the second determinant and so on, the last factor in each product being a single element out of the last row.

Consider then the minor determinants formed from the first $\rho+1$ rows, every column in any one of these has a common factor $\binom{\gamma}{0}$ (the fact that it happens to be unity should be disregarded for the moment); after removing these factors any column consists of the elements

$$\binom{\beta}{\beta'},\ \beta' = 0, 1, \ldots, \rho;$$

and as the values of β are restricted to 0, 1, ..., ρ, the resulting determinants are all

$$\begin{vmatrix} \binom{0}{0} & \binom{1}{0} & \cdots & \binom{\rho}{0} \\ \binom{0}{1} & \binom{1}{1} & \cdots & \binom{\rho}{1} \\ \cdots & \cdots & \cdots & \\ \binom{0}{\rho} & \binom{1}{\rho} & \cdots & \binom{\rho}{\rho} \end{vmatrix} = D_1$$

or zero. The first column of Δ' must supply the first column of D_1, for it is the only one for which $\beta = 0$. Hence also $\binom{\rho}{0}$ is a factor of Δ', for $\gamma = \rho$ for the first column.

Take now the next ρ rows, we may remove the factors $\binom{\gamma}{1}$ from the columns of the minor determinants, and we find ourselves always left with the determinant

$$\begin{vmatrix} \binom{1}{0} & \binom{2}{0} & \cdots & \binom{\rho}{0} \\ \binom{1}{1} & \binom{2}{1} & \cdots & \binom{\rho}{1} \\ \cdots & \cdots & \cdots & \cdots \\ \binom{1}{\rho-1} & \binom{2}{\rho-1} & \cdots & \binom{\rho}{\rho-1} \end{vmatrix} = D_2$$

or zero. For we may leave out of account the first column of Δ', that being required for the first determinant if the resulting product is not to be zero. There are only two columns for which $\beta = 1$, one of these must supply the second column of D_1, the other the first column of D_2. In one case the column factors give $\binom{\rho-1}{0}\binom{\rho}{1}$ and in the other $-\binom{\rho}{0}\binom{\rho-1}{1}$. Together the second and third columns furnish the factor

$$\begin{vmatrix} \binom{\rho-1}{0} & \binom{\rho}{0} \\ \binom{\rho-1}{1} & \binom{\rho}{1} \end{vmatrix}.$$

Proceeding thus we find $\quad \Delta' = D_1^2 D_2^2 \ldots D_{\rho+1}^2$.

None of these determinants is zero,* and therefore Δ' is not zero, and

* Cf. Grace and Young, *Algebra of Invariants*, pp. 217–218.

therefore also Δ is not zero.

Hence the members of the chosen set are linearly independent, and all perpetuants of degree 4 and even weight w can be expressed in terms of those of grade $w/2$ at least.

When w is odd, let $$w = 2\rho+1.$$

Then $$\binom{2\rho+3}{2} = 3\binom{\rho+2}{2}+\binom{\rho+1}{2}$$

and we can prove in just the same way that all products of the u's of degree $2\rho+1$ can be expressed linearly in terms of those products which have one of the factors $u_1^{\rho+1}$, $u_2^{\rho+1}$, $u_3^{\rho+1}$, $u_4^{\rho+2}$.

The perpetuant type of weight 2 is of importance for our purposes. It may be written $$(abc)(abd) \equiv \{ab, cd\}.$$

Then. $$\{(abc)-(abd)\}^2 = \{(cdb)-(cda)\}^2$$

whence $$2\{ab, cd\}-2\{cd, ab\} = (abc)^2+(abd)^2-(cdb)^2-(cda)^2. \tag{I}$$

We will use the notation (ab) to denote a transposition of two letters and {abc ...} to denote the sum of the substitutions of the symmetric group of the letters a, b, c, ... and {abc ...}′ to denote the sum of the substitutions of the alternating group of these letters minus the sum of those substitutions which do not belong to the alternating group. To avoid confusion substitutional symbols will be printed in Roman type. Then

$$\{\mathrm{abc}\}\{ab, cd\} = 2(abc)\{(abd)+(bcd)+(cad)\} = 2(abc)^2. \tag{II}$$

Thus there are two independent perpetuants of this type—as the generating function indicates. They may be taken to be $\{ab, cd\}$, $\{ac, bd\}$.

9. *Forms of degree* 5.—There are 10 irreducible forms of weight 3 to be looked for, but none of weight 2.

To reduce $(abc)(ade)$:*

$$\begin{aligned}&(bdc)(bde)-(bec)(bed)\\ &\quad = \{-(abc)+(abd)-(acd)\}\{(abd)-(abe)+(ade)\}\\ &\qquad\qquad -\{-(abc)+(abe)-(ace)\}\{-(abd)+(abe)-(ade)\}.\end{aligned}$$

* The reducibility of this form was proved by Turnbull, *Proc. London Math. Soc.*, Ser. 2, Vol. 9, p. 115. The reduction of this form and also the corresponding form in any number of variables was proved by Gilham, *Proc. London Math. Soc.*, Sér. 2, Vol. 20, p. 326. Gilham deduces the fact that the Jacobian of a Jacobian is reducible for any number of variables; also his theorem that the product of two Jacobians can be expressed as a sum of three-term products gives a syzygy of degree 6 and weight 2.

Hence on multiplying out

$3(abc)(ade)$

$$= 2(abc)(ade)-(abe)(acd)+(abd)(ace)$$

$$= -\{bd, ce\}+\{be, cd\}+2\{ab, ce\}+\{ae, bc\}-2\{ab, cd\}-\{ad, bc\}$$

$$+\{ae, bd\}-\{ad, be\}-\{ae, cd\}+\{ad, ce\}+(abd)^2-(abe)^2. \qquad \text{(III)}$$

It follows from this that any perpetuant of degree r which contains a pair of factors $(abc)(ade)$, such that the letters b, c, d, e occur in no other determinant factor, must be reducible.

Now a perpetuant of degree r and weight w has w determinant factors, each may be supposed to contain the letter a; there are then $2w$ places in these factors to be filled by the remaining $r-1$ letters. Let ρ letters appear more than once, then $r-1-\rho$ letters appear once each. There will then be a pair of factors $(abc)(ade)$, such that the letters b, c, d, e do not occur again unless

$$r-\rho \leqslant w+2.$$

But
$$r-1-\rho+2\rho \leqslant 2w.$$

Hence the perpetuant is reducible unless

$$2r \leqslant 3(w+1)$$

or a minimum weight for irreducibility is $\frac{2}{3}r-1$.

Consider the forms of weight 3. We notice that there are just 10 forms

$$(abc)^2(ade),$$

for the letters d, e can be selected in 10 ways. We proceed to show that all perpetuants of this degree and weight can be expressed in terms of these and of products of perpetuants. The possible kinds of products besides that written down are

$$(abc)(abd)(abe) \quad \text{and} \quad (abc)(abd)(ace).$$

Let us write
$$(abc)(abd)(ace) = \{a, bc, de\}.$$

Then
$$\{a, bc, de\} = -\{a, cb, ed\}.$$

Also
$$(abc)[(abd)(ace)-(abc)(ade)-(abe)(acd)] = 0,$$

or
$$\{de\}'\{a, bc, de\} = (abc)^2(ade).$$

Hence $\{\mathrm{bc}\}\{a, bc, de\} = (abc)^2(ade).$ (IV)

Also from $\{(abc)-(abd)\}^2 = \{(cdb)-(cda)\}^2$

we find, on multiplying each side by (ace),

$$2\{a, bc, de\}+2\{c, da, be\}$$
$$= (abc)^2(ace)-(acd)^2(ace)+(abd)^2(ace)-(bcd)^2(ace);$$

hence, using (IV),

$$\{\mathrm{ad}\}\, 2\{a, bc, de\}$$
$$= \{\mathrm{ad}\}[(abc)^2(ace)-(acd)^2(ace)+(abd)^2(ace)-(bcd)^2(ace)-2(cda)^2(cbe)]$$

and therefore also

$$\{\mathrm{ae}\}\, 2\{a, bc, de\}$$
$$= \{\mathrm{ae}\}[-(abc)^2(abd)+(abe)^2(abd)-(ace)^2(abd)+(bce)^2(abd)+2(bea)^2(bcd)].$$

Now

$$\{\mathrm{ade}\}-(\mathrm{ad})\{\mathrm{ae}\}-(\mathrm{ae})\{\mathrm{ad}\}$$
$$= 1+(\mathrm{ad})+(\mathrm{ae})+(\mathrm{de})+(\mathrm{ade})+(\mathrm{aed})-(\mathrm{ad})-(\mathrm{ad})(\mathrm{ae})-(\mathrm{ae})-(\mathrm{ae})(\mathrm{ad})$$
$$= \{\mathrm{de}\};$$

hence

$$4\{a, bc, de\}$$
$$= 2[\{\mathrm{de}\}+\{\mathrm{de}\}']\{a, bc, de\}$$
$$= 2(abc)^2(ade)-2(acd)^2(cbe)-2(ace)^2(cbd)-2(bde)^2(bca)$$
$$+2(abd)^2(bce)+2(abe)^2(bcd)+2(cde)^2(cba)$$
$$+(abc)^2\{(ace)+(acd)\}-(acd)^2\{(ace)+(dce)\}-(ace)^2\{(acd)+(ecd)\}$$
$$-(bcd)^2\{(acd)+(bde)\}+(bce)^2\{(bde)-(ace)\}-(bde)^2\{(dba)+(eba)\}$$
$$+(abd)^2\{(abe)+(dbe)-(abc)-(bcd)\}+(abe)^2\{(abd)+(ebd)-(abc)-(ebc)\}$$
$$+(cde)^2\{(eca)+(dca)-(dcb)-(ecb)\} \quad \text{(V)}$$

which is obtained after a few trifling reductions. Also

$$(abc)(abd)(abe) = (abc)(abd)\{(cbe)+(ace)+(abc)\}$$
$$= -\{b, ac, de\}+\{a, bc, de\}+(abc)^2(abd). \quad \text{(VI)}$$

Thus all perpetuants of degree 5 and weight 3 can be expressed in terms of perpetuants of the type $(abc)^2(ade)$ and of products.

10. *Forms of degree* 6.—The generating function shows that five irreducible forms are to be expected of weight 3.

The only possible kind of symbolical product to be considered is

$$(abc)(abd)(aef) \equiv X.$$

It is easy to see that

$$\{ab\}'X = (abc)(abd)(abf) - (abc)(abd)(abe),$$

$$2[1-(ac)(bd)]X = [(abc)^2+(abd)^2-(bcd)^2-(acd)^2](aef)$$
$$+2(cda)(cdb)[(cfa)-(cea)],$$

$$\{cef\}'X = 0.$$

Hence we see that, if α, β, γ be any three of the six letters,

$$\{\alpha\beta\gamma\}'X = R$$

where R stands for a sum of products of perpetuants. Also, if α, β, γ, δ be any four of the letters

$$\{\alpha\beta\gamma\delta\} X = R,$$

for, if three of these letters be taken from a, b, c, d, the reduction is made by (I), otherwise two of the letters are e and f, and the transposition (ef) changes the sign of X. Thus, if we operate on X with the substitutional series*

$$1 = \Sigma A_{\alpha_1, \alpha_2}, \ldots, T_{\alpha_1, \alpha_2}, \ldots,$$

we find

$$X = A_{3,3} T_{3,3} X + R,$$

since every other T contains a factor $\{\alpha\beta\gamma\}'$ or a factor $\{\alpha\beta\gamma\delta\}$.

Now $$T_{3,3} = \Sigma \{abc\}\{def\}\{ad\}'\{be\}'\{cf\}'.$$

Let $$N = \Sigma \{ad\}'\{be\}'\{cf\}',$$

then $$T_{3,3} = \Sigma \{abc\}\{def\} N = N\Sigma \{abc\}\{def\}$$

and every substitution is permutable with N.

---

* Grace and Young, *Algebra of Invariants*, p. 356; *Proc. London Math. Soc.*, Vol. 33, p. 133; Vol. 34, p. 361.

There are only six terms in this sum which, operating on X, give a result other than zero. Of these

$$\{ace\}\{bdf\}NX = \tfrac{1}{2}\{ace\}\{bdf\}N\{ab\}X+R$$

$$= \tfrac{1}{2}\{ace\}\{bdf\}N\{1+(ba)+(bc)+(be)-(bc)-(be)\}X+R$$

$$= \tfrac{1}{2}\{bdf\}[\tfrac{1}{2}\{abce\}-\{ace\}\{(bc)+(be)\}]NX+R$$

$$= -\tfrac{1}{2}(bc)\{abe\}\{cdf\}NX+R.$$

Thus we find

$$X = A[1-\tfrac{1}{2}(bc)-\tfrac{1}{2}(bd)]\{ef\}'\{abe\}\{cdf\}NX+R$$

where A is numerical.

We may therefore consider the form

$$\{abe\}\{cdf\}N(abc)(abd)(aef) = \begin{bmatrix} abe \\ cdf \end{bmatrix} \text{ say,}$$

as the representative perpetuant of this degree and order.

There are just five independent forms. First, it is immaterial which triad of letters is placed at the top: let that one which contains a be at the top. The interchange of letters in either triad leaves the form unaltered.

It is possible to express all forms in terms of those which have a in the top row and b in the lower row. Thus

$$[1+(bc)+(bd)+(bf)]\begin{bmatrix} abe \\ cdf \end{bmatrix} = R,$$

and hence

$$\begin{bmatrix} abe \\ cdf \end{bmatrix} = R-\begin{bmatrix} ace \\ bdf \end{bmatrix}-\begin{bmatrix} ade \\ cbf \end{bmatrix}-\begin{bmatrix} afe \\ cdb \end{bmatrix}.$$

Amongst these latter forms there is one relation,

$$R = \{cdef\}\begin{bmatrix} acd \\ bef \end{bmatrix}$$

$$= 4\left\{\begin{bmatrix} acd \\ bef \end{bmatrix}+\begin{bmatrix} ace \\ bdf \end{bmatrix}+\begin{bmatrix} acf \\ bde \end{bmatrix}+\begin{bmatrix} ade \\ bcf \end{bmatrix}+\begin{bmatrix} adf \\ bce \end{bmatrix}+\begin{bmatrix} aef \\ bcd \end{bmatrix}\right\}.$$

Thus there are just five linearly independent forms. The generating function proves that no reduction has been overlooked and the number is exact.

11. *Forms of degree* 7.—It has been shown that $\frac{2}{3}r-1$ is a minimum weight for irreducibility of forms of degree r; there are, then, no irreducible forms of degree less than 4.

Let a be placed in each determinant factor, then there are eight more places to be filled up with six letters. There are four possible kinds of product—

$$\text{(i)} \quad (abc)(abd)(abe)(afg),$$

$$\text{(ii)} \quad (abc)^2(ade)(afg),$$

$$\text{(iii)} \quad (abc)(abd)(ace)(afg),$$

$$\text{(iv)} \quad (abc)(abd)(aef)(aeg);$$

(ii) is reducible at once by (III), (iii) is reducible on multiplying (V) by (afg), in the same way (I) is reduced by (VI), and we are left with (iv).

Then $$\{\mathrm{df}\}'(abc)(abd)(aef)(aeg) = (abc)(abe)(aeg)(adf) = R.$$

Similarly the result of operating with $\{\mathrm{cf}\}'$, $\{\mathrm{cg}\}'$, and $\{\mathrm{dg}\}'$, is R in each case. Hence

$$(abc)(abd)(aef)(aeg) = \tfrac{1}{24}\{\mathrm{cdfg}\}(abc)(abd)(aef)(aeg)+R.$$

But by (II) we find

$$R = \{\mathrm{bcd}\}(abc)(abd)(aef)(aeg)$$

$$= \tfrac{1}{12}[1+(\mathrm{bc})+(\mathrm{bd})]\{\mathrm{cdfg}\}(abc)(abd)(aef)(aeg)+R.$$

Hence also

$$R = \tfrac{1}{3}\{\mathrm{bcdf}\}(abc)(abd)(aef)(aeg)$$

$$= \tfrac{1}{12}[1+(\mathrm{bc})+(\mathrm{bd})+(\mathrm{bf})]\{\mathrm{cdfg}\}(abc)(abd)(aef)(aeg)+R,$$

by subtracting the first of these results from the second we find

$$R = \tfrac{1}{12}(\mathrm{bf})\{\mathrm{cdfg}\}(abc)(abd)(aef)(aeg),$$

and therefore $(abc)(abd)(aef)(aeg)$ is reducible, and there are no irreducible perpetuants of weight 4 and degree 7.

Now G'_7 the reduced generating function enumerates products $P_3 . P_4$ as well as irreducible perpetuants P_7. The generating function for products $P_3 . P_4$ is

$$\binom{7}{3}\frac{x^2}{1-x}\cdot\frac{2x^2-x^4}{(1-x)^3}.$$

Subtracting this from G_7', we obtain

$$G_7'' = \frac{-21x^4+385x^5-854x^6+701x^7-64x^8-274x^9+176x^{10}-35x^{11}}{(1-x)^9}.$$

The first term being negative indicates a syzygy. To obtain this, we have by (III)

$$3(abc)^2(aef)(adg) = -(abc)^2[\{de, fg\} - \{df, eg\}]+R_1,$$

where R_1 stands for products of perpetuants of which one of the factors is of degree unity (*i.e.* one of the quantics).

Again, from (V)

$$2\{a, bc, de\}(afg)$$

$$= (abc)^2(ade)(afg)-(acd)^2(cbe)(afg)-(ace)^2(cbd)(afg)-(bde)^2(bca)(afg)$$

$$+(abd)^2(bce)(afg)+(abe)^2(bcd)(afg)+(cde)^2(cba)(afg)+R_1.*$$

Hence

$$-6\{a, bc, de\}(afg)$$

$$= \{\mathrm{fg}\}'[(abc)^2\{df, eg\}-(acd)^2\{bf, eg\}-(ace)^2\{bf, dg\}+(bde)^2\{af, cg\}$$

$$+(abd)^2\{cf, eg\}+(abe)^2\{cf, dg\}-(cde)^2\{af, bg\}]+R_1.$$

Operate on both sides with $\{\mathrm{efg}\}'$, the left-hand side vanishes identically owing to the fundamental identity, and we obtain

$$\{\mathrm{efg}\}'\{\mathrm{ad}\}\{\mathrm{bc}\}(ace)^2\{bf, dg\} = R,$$

the required syzygy.

12. *Forms of degree* 8.—The first term in G_8' is $14x^4$, but this includes products $P_4.P_4$, the generating function for which is

$$\frac{1}{2}\binom{8}{4}\left\{\frac{2x^2-x^4}{(1-x)^3}\right\}^2.$$

On subtraction the generating function is found to begin with the term $-136x^4$, and hence a syzygy must be sought. Since $\frac{2}{3}r-1$ is greater than 4, there can be no irreducible perpetuant of weight less than 5.

* A term such as $(bcd)^2(bde)(afg)$ may be included in R_1 owing to the identity

$$(afg) = (bfg)-(abf)+(abg).$$

To find the syzygy of weight 4, consider the identity

$$(abc)(efh)\{(abd)(efg)-(abg)(efd)-(abe)(fgd)-(abf)(edg)\} = 0.$$

After using (III) we obtain

$$3\{ab, cd\}\{ef, gh\}-3\{ab, cg\}\{ef, dh\}$$

$$= \{ab, ce\}[\{fd, gh\}-\{fg, dh\}]+\{ab, cf\}[\{ed, gh\}-\{eg, dh\}]+R_1.$$

Operate with $\{\mathrm{fgd}\}'$ and we obtain

$$\{\mathrm{fgd}\}'\{ab, cd\}\{ef, gh\} = R_1.$$

It is then clear that

$$\{\alpha\beta\gamma\}'\{ab, cd\}\{ef, gh\} = R_1$$

when α, β, γ are any three of the eight letters a, b, c, d, e, f, g, h. Also, if P be a positive symmetric group containing three letters out of either of the two factors, by (II)

$$P\{ab, cd\}\{ef, gh\} = R_1.$$

Operate then on $\{ab, cd\}\{ef, gh\} \equiv X$ with the series

$$1 = \Sigma A_{a_1, a_2}, \ldots, T_{a_1, a_2}, \ldots,$$

every term but $T_{4,4}$ gives R_1 as a result.

Now $$T_{4,4} = \Sigma\{\mathrm{abcd}\}\{\mathrm{efgh}\}\{\mathrm{ae}\}'\{\mathrm{bf}\}'\{\mathrm{cg}\}'\{\mathrm{dh}\}'.$$

Let $$N = \Sigma\{\mathrm{ae}\}'\{\mathrm{bf}\}'\{\mathrm{cg}\}'\{\mathrm{dh}'\}'.$$

We have only to consider substitutional expressions in which each positive symmetric group contains two letters out of each product, any other case gives R_1, by (II).

Also $$\{\mathrm{abeg}\}\{\mathrm{cdfh}\}N[1+(\mathrm{eg})+(\mathrm{fg})]X = R_1,$$

and hence $$\{\mathrm{abeg}\}\{\mathrm{cdfh}\}NX = -\tfrac{1}{2}(\mathrm{fg}).\{\mathrm{abef}\}\{\mathrm{cdgh}\}NX,$$

and remembering that by (I)

$$[1-(\mathrm{eg})(\mathrm{fh})]X = R_1,$$

we see that we have only one form to consider,

$$\begin{bmatrix} abef \\ cdgh \end{bmatrix} = \{\mathrm{abef}\}\{\mathrm{cdgh}\}N\{ab, cd\}\{ef, gh\};$$

and in terms of this and products R_1 every perpetuant product $P_4.P_4$ of weight 4 can be expressed.

The number of linearly independent forms

$$\begin{bmatrix} abef \\ cdgh \end{bmatrix}$$

can be obtained thus. The form is annihilated by any positive symmetric group of degree 5.

Operate with {bcdgh} ; we obtain a sum of five terms, four of which have the letter b in the lower row. Thus we need only consider those forms which have a in the upper and b in the lower row. Next those forms which have both c and d in the lower row can be expressed in terms of such as have at least one of c and d in the upper row. In the case of

$$\begin{Bmatrix} aefg \\ bcdh \end{Bmatrix}$$

this is effected by operating with {efgcd}.

Lastly, amongst forms having c above and d below there is one relation to express them in terms of those which have both c and d above; and the same is true of forms having d above and c below.

Thus there are 4 forms both c and d above, 5 forms c above and d below, and 5 forms d above and c below—14 independent forms altogether. Hence there are 14 products $P_4.P_4$ linearly independent of each other and of products R_1. This is the correct number as given by G_8'. There are thus 136 linearly independent relations given by the syzygy.

13. It is possible now to obtain a higher limit to the weight of an irreducible perpetuant of degree r. We shall set before ourselves the object of introducing as many squared factors as possible in the symbolical product. The letter a is introduced into each of the factors which only appear to the first power.

Let us suppose that there are four such factors. We may for the moment regard the product of these four factors as a perpetuant: it is of degree not greater than 9. If there are two factors $(abc)(ade)$ such that the letters b, c, d, e do not appear in the other two factors, then by using (III) we may express this product as a sum of products of four factors involving fewer letters. If three of the factors are $(abc)(abd)(ace)$ then, by (V), they can be expressed in terms of products having a squared factor. By (VI) the same is true if three of the factors are $(abc)(abd)(abe)$.

There cannot, then, be more than seven letters involved, and by referring to § 11 we see that in this case the only unconsidered possibility is

$$(abc)(abd)(aef)(aeg);$$

this, by § 11, is expressible in terms of products of four factors with fewer letters and of products with a squared factor. If six letters are involved we see by § 10 that we may suppose the factors to be

$$(abc)(abd)(aef)$$

and one other. If this contains $(ab\ )$, $(ac\ )$, or $(ad\ )$ a squared factor may be introduced as we have seen above, and the only other possibility is (aef), introducing $(aef)^2$.

Thus a symbolical product which has four factors of the first degree only can be expressed in terms of products having not more than two of the first degree. Let us suppose there are λ squared factors and three factors of the first degree {which must be of the form $(abc)(abd)(aef)$ by § 10}, the maximum degree is

$$r = 2\lambda+6,$$

the weight is $w = 2\lambda+3$. In this case the degree is even and the minimum weight is $r-3$.

If there are only two factors of the first degree they must be $(abc)(abd)$. The maximum degree is

$$r = 2\lambda+4$$

and the weight is $w = 2\lambda+2$; this gives a higher weight for perpetuants of even degree.

If there is only one factor of the first degree the maximum degree is $r = 2\lambda+3$, and the weight is $w = 2\lambda+1$. Thus we see that there can be no irreducible perpetuant of even degree r of weight less than $r-3$, and of odd degree r of weight less than $r-2$.

14. The question of syzygies naturally raises the point whether it is possible to find syzygies for ternary forms in the same manner as Stroh did for binary forms. Stroh* used the symbolical identity

$$(ab)c_x d_x+(cd)a_x b_x = (ad)c_x b_x+(cb)a_x d_x,$$

---

* *Math. Ann.*, Vol. 33, pp. 61–107 (§§ 18–22); Vol. 34, pp. 306–320, 354–370; Vol. 36, pp. 262–288. Grace and Young, *Algebra of Invariants*, p. 140.

simply raising both sides to the power w and expanding by the binomial theorem.

It was shown by Mr. Wood and myself* that all perpetuant syzygies of degree up to 8 at least, for binary forms, may be deduced from these Stroh syzygies and the Jacobian syzygy. We proceed to prove that no such family of syzygies exists for ternary forms.

The problem resolves itself into finding an identity of the form

$$[a_1 a_2 \dots a_r b_1 b_2 \dots b_s] + [c_1 c_2 \dots c_t d_1 d_2 \dots d_u]$$
$$= [a_1 a_2 \dots a_r d_1 d_2 \dots d_u] + [c_1 c_2 \dots c_t b_1 b_2 \dots b_s] \quad \text{(VII)}$$

where

$$[a_1 a_2 \dots a_r b_1 b_2 \dots b_s]$$
$$= \Sigma A_{1,2,3}(a_1 a_2 a_3) + \Sigma B_{1,2,3}(b_1 b_2 b_3) + \Sigma E_{1,2;1}(a_1 a_2 b_1) + \Sigma F_{1;1,2}(a_1 b_1 b_2), \quad \text{(VIII)}$$

the other brackets representing similar expressions, each term containing one determinant factor only, and the coefficients A, B, &c., being numerical.

Consider the effect of operating with the substitutional series

$$1 = \Sigma A_{a_1, a_2, \dots, a_h} T_{a_1, a_2, \dots, a_h}$$

for the $r+s+t+u = n$ letters involved. Any negative symmetric group of four letters annihilates every term of (VII). Consider the group $\{efgh\}'$; if two of the letters e, f are outside the determinant factor the term contains e and f symmetrically, and will therefore be annihilated.

Otherwise it is annihilated owing to the fundamental identity, which may be written

$$\{efgh\}'(efg)h_x = 0.$$

Hence, if S be any term of (VII),

$$T_{a_1, a_2, \dots, a_h} S = 0,$$

unless $h < 4$, for T is a sum of terms each having a negative symmetric group degree h for factor.

Again, if N_3 is a negative symmetric group degree 3,

$$T_{a_1, a_2, \dots, a_h} N_3 = 0$$

---

* *Proc. London Math. Soc.*, Ser. 2, Vol. 2, p. 221.

unless $h > 2$. And since

$$(efg) = \tfrac{1}{6}\{\mathrm{efg}\}'(efg),$$

$$T_{a_1, a_2, \ldots, a_h} S = 0,$$

unless $h > 2$. That is, we need only consider those terms of the substitutional series for which $h = 3$.

Let us suppose that $a_2 > 1$, then every term of $T_{a_1, a_2, \ldots, a_h}$ has a factor of the form $\{\mathrm{efg}\}'\{\mathrm{hi}\}'$.

But $\{\mathrm{efg}\}'\{\mathrm{hi}\}'S$ is obviously zero unless S is of the form $(efh)g_x i_x, \ldots$, but even here we have by the fundamental identity

$$\{\mathrm{efg}\}'\{\mathrm{hi}\}'(efh)g_x i_x = \{\mathrm{efg}\}'\{\mathrm{hi}\}'(efg)h_x i_x = 0.$$

Thus we see that the substitutional theorem gives

$$S = A_{n-2, 1, 1}\, T_{n-2, 1, 1}\, S,$$

for every other term vanishes.

Now, $T_{n-2, 1, 1}$ is a sum of terms each of which has as a factor a positive symmetric group containing $n-2$ letters—that is, all the letters but two. This group may then be defined by the missing letters; if these be e and f we may write it $[\mathrm{ef}]$. Then it is clear that

$$[\mathrm{ef}]S = 0,$$

unless both e and f appear in the determinant factor of S.

Now, if Q be a sum of terms such as S, the necessary and sufficient condition for the identity $Q = 0$ to be true is that $[\mathrm{ef}]Q = 0$ for every choice of the letters e and f. It is necessary, for if $Q = 0$, obviously $[\mathrm{ef}]Q = 0$; it is sufficient because

$$Q = \Sigma ATQ = A_{n-2, 1, 1}\, T_{n-2, 1, 1}\, Q = A_{n-2, 1, 1}\, \Sigma H[\mathrm{ef}]\, Q$$

where H is a substitutional expression.

Let us apply this to the proposed identity,

$$[\mathrm{ef}][a_1 a_2 \ldots a_r b_1 b_2 \ldots b_s] = 0,$$

unless both e and f are found amongst the letters $a_1 \ldots a_r b_1 \ldots b_s$.

Hence, operating on (VII) with $[\mathrm{a}_\lambda \mathrm{b}_\mu]$, we find

$$[\mathrm{a}_\lambda \mathrm{b}_\mu][a_1 a_2 \ldots a_r b_1 b_2 \ldots b_s] = 0,$$

for $\lambda = 1, 2, \ldots, r$, $\mu = 1, 2, \ldots, s$. But this set of equations is the condition that

$$[a_1 a_2 \ldots a_r b_1 b_2 \ldots b_s] = [a_1 a_2 \ldots a_r] + [b_1 b_2 \ldots b_s],$$

that is, a sum of two sets of terms, one of which involves the a's only in the determinant factor, the other the b's only, as we proceed to show.

In the determinant factors $(a_\lambda a_\mu b_\nu)$ and $(a_\lambda b_\mu b_\nu)$ we may introduce the letter a_1 by means of the fundamental identity. Let us suppose this to have been done in the first place. Then, using the notation of (VIII), we see that $E_{\lambda, \mu; \nu}$ is zero unless λ or μ is unity, and that $F_{\lambda; \mu, \nu}$ is zero unless $\lambda = 1$.

Operating with $[\mathrm{a}_\lambda \mathrm{b}_\mu]$, we find

$$\sum_x E_{\lambda, x; \mu} - \sum_x F_{\lambda; \mu, x} = 0.$$

But
$$F_{\lambda; \mu, x} = 0$$

unless $\lambda = 1$; hence
$$\sum_x E_{\lambda, x; \mu} = 0$$

unless $\lambda = 1$.

Again, if $\lambda \neq 1$,
$$\sum_x E_{\lambda, x; \mu} \equiv E_{\lambda, 1; \mu};$$

hence all the E's are zero.

Now use the identity

$$(a_1 b_\mu b_\nu) = (a_1 b_1 b_\nu) - (a_1 b_1 b_\mu) + (b_1 b_\mu b_\nu),$$

and we find as before that all terms $(a\, b_\mu b_\nu)$ disappear, and hence

$$[a_1 a_2 \ldots a_r b_1 b_2 \ldots b_s] = \Sigma A_{1, 2, 3} (a_1 a_2 a_3) + \Sigma B_{1, 2, 3} (b_1 b_2 b_3).$$

Thus any relation of the form (VII) that can be obtained is incapable of giving syzygies, as it only yields obvious identities between products of forms.

The argument applies in the same way to quaternary forms and to forms with any higher number of variables.

X. *The Linear Invariants of Ten Quaternary Quadrics*

By Mr H. W. TURNBULL, Trinity College, and Dr ALFRED YOUNG, Clare College

[*Received* 4 May, *Read* 26 July, 1926.]

INTRODUCTION.

In the year 1825 a problem was proposed by l'Académie de Bruxelles which has never been satisfactorily answered, although many attempts have been made. In the original form* the problem was to find the general property holding between ten points on a surface of the second degree. Soon afterwards* the question was modified, and required the three dimensional analogue of Pascal's theorem for conics. As Chasles remarks, it is far easier to answer the latter question than the former: in fact he gives in Note XXXII various results which meet the case. So also have several† other geometers.

From the outset Chasles perceived the fundamental importance of the first question for the future progress of the theory of quadric surfaces. He observed that for conics, Pascal's theorem has two essential aspects, giving

(1) The relation between six arbitrary points on a conic.

(2) The relation between an arbitrary triangle and a conic.

Hence, in space, there will presumably be

(1) The relation between ten arbitrary points on a quadric surface.

(2) The relation between an arbitrary tetrahedron and a quadric surface.

For the conic these two relations are practically the same but for the quadric surface they are essentially distinct. Thus relation (2) is readily found, whereas (1) presents great difficulties. In fact (2) can be quoted‡ as follows, from Note XXXII of Chasles' *Aperçu*:

Let six edges of an arbitrary tetrahedron meet a quadric surface in twelve points, lying by threes on four planes, such that each plane contains three of the points lying respectively on three edges issuing from a vertex of the tetrahedron. Let each plane be associated with the corresponding opposite plane of the tetrahedron, cutting it in a line.

Then the four lines so found are generators of the same system of a quadric.

A geometrical proof follows directly by applying Pascal's theorem to the six points on three coplanar edges of the tetrahedron. For its Pascal line cuts each of the four lines in question, so that the four lines have at least four transversals.

But relation (1) is something very different, and the comment of Chasles is noteworthy:

* Cf. Chasles, *Aperçu Historique des méthodes en Géométrie* (Paris, 1875) (reprinted from the memoir in the *Mém. couronnées par l'Acad. de Bruxelles* 1837), p. 245, §§ 49, 50 and Note XXXII.

† Serret, *Comptes Rendus*, 82 (1876), 162–165, 208–210: "Une classe particulière de décagones," and "Sur une nouvelle analogie aux théorèmes de Pascal et de Brianchon." Hesse, *Crelle*, 24 (1842), 36. Steiner, *Crelle*, 68 (1868), 191–192. Sautreaux Felix, *Bulletin de l'Acad. de Bruxelles*, 2, 45 (1878), 426–430. K. Rohn, *Leipziger Berichte*, 46 (1894), p. 160; J. Thomae, *ibid.*, 44 (1892), p. 543, 49 (1897), p. 315, *Wiener Berichte*, 110 (1901), p. 204.

‡ Cf. Baker, *Principles of Geometry*, Vol. III. p. 53.

"La recherche de cette relation (1) est bien digne d'occuper les géomètres. Sans doute nous n'avons point encore tous les éléments nécessaires pour cette recherche: c'est une raison pour étudier sous tous les rapports, sous toutes les faces, les propriétés des surfaces du second degré. Aucune théorie, aucune découverte, quelque minime qu'elle paraisse d'abord, n'est à négliger; car, à défaut d'une application immédiate, chaque vérité partielle a du moins l'avantage d'être un anneau de la chaîne continue qui lie entre elles toutes les vérités de cette vaste théorie."

In view of these remarks, and because certain light can be thrown on this theory by the use of algebra, we submit the following pages. Algebraically the problem is solved by finding a symbolic form, free from unnecessary factors, for the invariant linear in the coefficients of ten quadrics. This invariant is the determinant of the coefficients. Hitherto the only symbolic forms discovered have had redundant factors. Reiss* first gave such a result, in a work wherein the corresponding determinants for six conics and for ten ternary cubics, however, were successfully treated. Hunyady† and others have worked at this problem but without finding the desired result. Now these investigations have all started from a geometrical theorem which has been expressed algebraically by the vanishing of a certain determinant. Thus for conics, Pascal's theorem involving six points gives this starting point, and in fact led Reiss to the required algebraical formula, involving a six-rowed determinant. The failure for the case of ten points on a quadric arises from our ignorance of a true analogue of Pascal's theorem, *free from irrelevant elements*. We have therefore reversed this process and in so doing have succeeded in constructing synthetically the desired algebraic form, free from irrelevant factors. This result gives explicitly the relation between ten points on a quadric surface, by means of the vanishing of a 240-term series. It corresponds with the twenty-term series for the cubic and the two-term series for the conic, given by Reiss. Whether our series can be further simplified is uncertain; but there are algebraic reasons for thinking that the series cannot be expressed by a small number of terms comparable in simplicity to the ternary case.

Part I deals with the algebraic problem of establishing the 240-term series K for the determinant of ten quadrics, § 5 (11) and (12). This also gives the relation between ten points on one quadric. Incidentally, § 4 (7) and (9) give the relation when the last three, 7, 8, 9, of the ten points 0, 1, 2, ... 9 are collinear:

$$\begin{aligned} 0 = G = &(0123)(0456)(1589)(6289)(3489) \\ &+(0134)(0562)(1689)(2389)(4589) \\ &+(0145)(0623)(1289)(3489)(5689) \\ &+(0156)(0234)(1389)(4589)(6289) \\ &+(0162)(0345)(1489)(5689)(2389). \end{aligned}$$

This is generated by the circular substitutional operator

$$\{(\mathit{23456})\}$$

acting upon any one of its five terms.

Part II gives some geometrical consequences chiefly connected with this function G.

Part III gives the more general but closely allied relations between all possible invariants linear in ten quadrics. There are in all twelve distinct types as shewn in §16, found by the methods of quantitative substitutional analysis‡. This leads to an alternative series for K but it has 286×12 terms.

* *Math. Annalen*, 2 (1870), 383–426 (397). † *Crelle*, 89 (1880), 47–69.

‡ Grace and Young, *Algebra of Invariants* (1903), 339–359.

Analytically the key to the problem is a full knowledge of the relations among the subgroups of the substitution group for permuting ten symbols in 10! ways. We do not consider that this matter has been entirely cleared up.

I

THE DETERMINANT OF THE COEFFICIENTS OF TEN QUATERNARY QUADRICS.

§1. The quadric in n homogeneous variables,

$$A = \sum_{i,k} a_{ik} x_i x_k, \quad \text{...(1)}$$

$$i, k, = 1, 2, \ldots n,$$

has $N = \frac{n(n+1)}{2}$ independent terms. Consequently, the coefficients of N such quadrics may be arranged as a determinant of order N, whose vanishing expresses the fact that the N quadrics are linearly related. Now this determinant is an invariant for linear transformations of the variables. If then we write the coefficients in symbolic form as

$$a_{ik} = a_i a_k \quad \text{...(2)}$$

and use different letters $a, b, c, \ldots$ to denote the different quadrics, we may express the determinant of the coefficients as an aggregate of terms, all of whose factors are bracket factors.

Thus for three binary quadrics A, B, C,

$$\begin{vmatrix} a_{11} & a_{12} & a_{22} \\ b_{11} & b_{12} & b_{22} \\ c_{11} & c_{12} & c_{22} \end{vmatrix} = \begin{vmatrix} a_1^2 & a_1 a_2 & a_2^2 \\ b_1^2 & b_1 b_2 & b_2^2 \\ c_1^2 & c_1 c_2 & c_2^2 \end{vmatrix} = -(bc)(ca)(ab), \quad \text{.....................(3)}$$

where $(bc) = (b_1 c_2 - b_2 c_1)$.

Again, for six ternary quadrics A, B, C, D, E, F,

$$\Gamma = \begin{vmatrix} a_{11} & a_{12} & a_{13} & a_{22} & a_{23} & a_{33} \\ b_{11} & b_{12} & b_{13} & b_{22} & b_{23} & b_{33} \\ c_{11} & c_{12} & c_{13} & c_{22} & c_{23} & c_{33} \\ d_{11} & d_{12} & d_{13} & d_{22} & d_{23} & d_{33} \\ e_{11} & e_{12} & e_{13} & e_{22} & e_{23} & e_{33} \\ f_{11} & f_{12} & f_{13} & f_{22} & f_{23} & f_{33} \end{vmatrix} \quad \text{.....................................(4)}$$

$$= |a_1^2 \; b_1 b_2 \; c_1 c_3 \; d_2^2 \; e_2 e_3 \; f_3^2|$$

$$= -(abc)(aef)(bfd)(cde)$$

$$+ (def)(dbc)(eca)(fab),$$

where $(abc) = |a_1 b_2 c_3| = \Sigma \pm a_1 b_2 c_3$.

There is no difficulty in proving the binary formula. The ternary case, which can be verified in various ways, is due to Reiss*.

The question immediately arises, what is the corresponding symbolic expression for the determinant of the coefficients of ten quaternary quadrics

$$A, B, C, D, E, F, G, H, J, K\,?$$

* *Math. Annalen, loc. cit.* p. 401.

This determinant,

$$\begin{vmatrix} a_{11} & a_{12} & a_{13} & a_{14} & a_{22} & a_{23} & a_{24} & a_{33} & a_{34} & a_{44} \\ b_{11} & b_{12} & b_{13} & b_{14} & b_{22} & b_{23} & b_{24} & b_{33} & b_{34} & b_{44} \\ c_{11} & c_{12} & c_{13} & c_{14} & c_{22} & c_{23} & c_{24} & c_{33} & c_{34} & c_{44} \\ d_{11} & d_{12} & d_{13} & d_{14} & d_{22} & d_{23} & d_{24} & d_{33} & d_{34} & d_{44} \\ e_{11} & e_{12} & e_{13} & e_{14} & e_{22} & e_{23} & e_{24} & e_{33} & e_{34} & e_{44} \\ f_{11} & f_{12} & f_{13} & f_{14} & f_{22} & f_{23} & f_{24} & f_{33} & f_{34} & f_{44} \\ g_{11} & g_{12} & g_{13} & g_{14} & g_{22} & g_{23} & g_{24} & g_{33} & g_{34} & g_{44} \\ h_{11} & h_{12} & h_{13} & h_{14} & h_{22} & h_{23} & h_{24} & h_{33} & h_{34} & h_{44} \\ j_{11} & j_{12} & j_{13} & j_{14} & j_{22} & j_{23} & j_{24} & j_{33} & j_{34} & j_{44} \\ k_{11} & k_{12} & k_{13} & k_{14} & k_{22} & k_{23} & k_{24} & k_{33} & k_{34} & k_{44} \end{vmatrix} = \Delta, \text{ say}, \quad\ldots\ldots\ldots\ldots(5)$$

becomes, in the symbolic notation,

$$\begin{vmatrix} a_1^2 & a_1a_2 & a_1a_3 & a_1a_4 & a_2^2 & a_2a_3 & a_2a_4 & a_3^2 & a_3a_4 & a_4^2 \\ b_1^2 & b_1b_2 & b_1b_3 & b_1b_4 & b_2^2 & b_2b_3 & b_2b_4 & b_3^2 & b_3b_4 & b_4^2 \\ c_1^2 & c_1c_2 & c_1c_3 & c_1c_4 & c_2^2 & c_2c_3 & c_2c_4 & c_3^2 & c_3c_4 & c_4^2 \\ d_1^2 & d_1d_2 & d_1d_3 & d_1d_4 & d_2^2 & d_2d_3 & d_2d_4 & d_3^2 & d_3d_4 & d_4^2 \\ e_1^2 & e_1e_2 & e_1e_3 & e_1e_4 & e_2^2 & e_2e_3 & e_2e_4 & e_3^2 & e_3e_4 & e_4^2 \\ f_1^2 & f_1f_2 & f_1f_3 & f_1f_4 & f_2^2 & f_2f_3 & f_2f_4 & f_3^2 & f_3f_4 & f_4^2 \\ g_1^2 & g_1g_2 & g_1g_3 & g_1g_4 & g_2^2 & g_2g_3 & g_2g_4 & g_3^2 & g_3g_4 & g_4^2 \\ h_1^2 & h_1h_2 & h_1h_3 & h_1h_4 & h_2^2 & h_2h_3 & h_2h_4 & h_3^2 & h_3h_4 & h_4^2 \\ j_1^2 & j_1j_2 & j_1j_3 & j_1j_4 & j_2^2 & j_2j_3 & j_2j_4 & j_3^2 & j_3j_4 & j_4^2 \\ k_1^2 & k_1k_2 & k_1k_3 & k_1k_4 & k_2^2 & k_2k_3 & k_2k_4 & k_3^2 & k_3k_4 & k_4^2 \end{vmatrix} = \Delta', \text{ say}. \quad\ldots\ldots\ldots\ldots(6)$$

It appears at once that Δ' may be expressed as a sum of terms

$$\Sigma\lambda\,(abcd)\,(aefg)\,(bejk)\,(cfkh)\,(dghj) \quad\ldots\ldots\ldots\ldots(7)$$

wherein λ is a numerical coefficient, and no two symbols a, b are convolved twice in one term. For it follows directly by making a general linear transformation of the variables x that Δ itself is an invariant, and is therefore by the fundamental theorem expressible as an aggregate of terms whose factors are of the type

$$(abcd) = |\,a_1\,b_2\,c_3\,d_4\,|.$$

Since each term contains twenty symbols $a, a, b, b, \ldots k, k$ they must appear in five such factors. But Δ is unaltered by an alternation, that is any combination of transpositions such as interchanging a and b and changing the sign. Accordingly we write the supposed symbolic expression for Δ' as

$$\Delta' = \Sigma I_1 + \Sigma I',$$

where I_1 denotes a term such as in (7) and I' denotes a term with some letters convolved twice, and we then operate* on this identity with

$$\{\mathrm{a\,b\,c\,d\,e\,f\,g\,h\,j\,k}\}',$$

which denotes the negative substitution group of these ten letters, i.e. the sum of all possible 10!

* *Cf.* Grace and Young, *loc. cit.* p. 345.

alternations, those which arise from an odd number of transpositions being affected with a minus sign. In this way we obtain

$$10!\Delta' = \Sigma' I_1,$$

expressing Δ' entirely in terms of the type I_1. For this operator annihilates all terms such as I'.

Hence a series such as (7) exists for the symbolical expression of Δ in invariant form. The question now resolves itself into what is the least number of terms requisite for this series? Before proceeding with this, we note an alternative geometrical interpretation leading to Δ'.

If in (6) the symbols $\quad a_1,\ a_2,\ a_3,\ a_4$

are interpreted as the homogeneous coordinates of a point a, with like treatment for $b, c, \dots k$, then the equation

$$\Delta' = 0$$

directly expresses the condition that ten points should lie on a quadric surface.

If one set $\quad k_1,\ k_2,\ k_3,\ k_4,$

say, denotes current coordinates, the same equation is the equation of the unique quadric through the other given nine points.

Also if a linear transformation

$$a \rightarrow \bar{a},$$

say, is made, then $b, c, \dots k$, being cogredient with a, undergo the same transformation, of which Δ' is an invariant. But, as it stands, Δ' does not shew explicitly the invariant condition, until Δ' is thrown into the form (7), when the equation

$$\Delta' = 0$$

immediately gives the geometrical condition holding between ten points on a quadric. The consideration of the simpler ternary case makes this clear.

Thus, in (4), the determinant

$$\Gamma = | a_1^2 \quad b_1 b_2 \quad c_1 c_3 \quad d_2^2 \quad e_2 e_3 \quad f_3^2 |$$

vanishes if six points of a plane lie on a conic, where

$$a = \{a_1,\ a_2,\ a_3\},$$

b, c, d, e, f denote the six points.

But the determinant Γ has also been expressed as

$$(def)(dbc)(eca)(fab) - (abc)(aef)(bfd)(cde). \qquad \dots\dots(8)$$

Now this last expression vanishes if certain three points are in line, namely, the points α, β, γ given by

$$\left.\begin{aligned} \alpha_i &= (dcf)\,e_i - (dce)\,f_i \\ \beta_i &= (cae)\,b_i - (cab)\,e_i \\ \gamma_i &= (afb)\,d_i - (afd)\,b_i \end{aligned}\right\}, \qquad \dots\dots(9)$$

$$i = 1, 2, 3,$$

for the condition $(\alpha\beta\gamma) = 0$, expressed in terms of a, b, c, d, e, f, immediately leads to this result. But this is precisely Pascal's theorem for the hexagon

$$a, f, e, b, d, c.$$

Yet without having expression (8) as a link between them, it is not apparent how closely allied is the theorem to the six-rowed determinant Γ.

§ 2. *Construction of the symbolic invariant form for* Δ. We approach this problem by a kind of induction, from the simpler case to the general.

Let Δ' be written briefly as

$$[a\,b\,c\,d\,e\,f\,g\,h\,j\,k]^2:$$

then the equation of the quadric through all but the first of these ten points may be written

$$[x\,b\,c\,d\,e\,f\,g\,h\,j\,k]^2=0. \qquad\qquad (1)$$

Now suppose the last three points h, j, k are in line, so that we may write

$$j_i=\lambda h_i+\mu k_i \quad (i=1,\,2,\,3,\,4)$$

or briefly

$$j=\lambda h+\mu k.$$

Then if this value of j is substituted in the last but one row of Δ', we obtain a quadratic in $\lambda:\mu$, where the coefficients of λ^2 and μ^2 vanish, and there is left

$$\lambda\mu\,\Delta_0,$$

where the last three rows in Δ_0 are

$$\begin{pmatrix} h_1^2 & h_1h_2 & h_1h_3 & \dots & h_4^2 \\ 2h_1k_1 & h_1k_2+h_2k_1 & h_1k_3+h_3k_1 & \dots & 2h_4k_4 \\ k_1^2 & k_1k_2 & k_1k_3 & \dots & k_4^2 \end{pmatrix}, \qquad\qquad (2)$$

and the other seven rows in Δ' and Δ_0 are alike.

Now, after expansion, all the third order determinants of this array are cubic functions of h and k with three linear factors

$$(hk)_{pq}\,(hk)_{rs}\,(hk)_{tu},$$

where the suffixes are 1, 2, 3, 4 in various orders, while

$$p\neq q,\quad r\neq s,\quad t\neq u$$

and

$$(hk)_{pq}=h_pk_q-h_qk_p.$$

So, if Δ_0 is expanded by Laplace's method, as a sum of products of two minors of orders seven and three, Δ_0 appears as a cubic function of line coordinates

$$(hk).$$

We may therefore suppose that the symbolic invariant expression for Δ_0 involves only terms of type

$$(a'\,b'\,c'\,d')(a'\,e'\,f'\,g')(b'\,f'\,h\,k)(c'\,g'\,h\,k)(d'\,e'\,h\,k), \qquad\qquad (3)$$

where the accented letters denote a, b, c, d, e, f, g in some order or other.

The sequel shews that Δ_0 is expressible as a five-term series obtained from

$$-(abcd)(aefg)(bfhk)(gchk)(dehk)$$

by a *cyclic* interchange $\{(\mathrm{c\,d\,e\,f\,g})\}$:

namely,

the order in the first term is	$c\,d\,e\,f\,g$,	
" " " " second " "	$d\,e\,f\,g\,c$,	
" " " " third " "	$e\,f\,g\,c\,d$,	............(4)
" " " " fourth " "	$f\,g\,c\,d\,e$,	
" " " " fifth " "	$g\,c\,d\,e\,f$.	

We need the fundamental quaternary identity

$$(abcd)(e\theta)=(ebcd)(a\theta)+(aecd)(b\theta)+(abed)(c\theta)+(abce)(d\theta),$$

sometimes written

$$\{\mathrm{abcde}\}'(abcd)(e\theta)=0.$$

Also it is convenient to replace the letters $a, b, c, \dots$ by the digits 0, 1, 2, ... 9.

§ 3. *The modified problem. Construction of* Δ_0. We consider point 7 to be on the line 89 while points 0, 1, 2, 3, 4, 5, 6 are arbitrary, so that, after what has preceded, Δ_0 will be expressible as

$$\Sigma\,(ijkl)\,(i'j'k'l')\,(mn\,89)\,(m'n'\,89)\,(m''n''\,89), \qquad \text{(1)}$$

where the letters $i, \ldots n''$ denote fourteen symbols 0, 0, 1, 1, ... 6, 6, in any order. The number of terms is at present unspecified, but there is no loss of generality in taking $i = i' = 0$, for by the fundamental identity the symbol 0 may be introduced into both the first and second factors of a term.

There are now ten possible pairs of such factors:

$$\left.\begin{aligned}
(0123)(0456) &= r_1 \text{ say},\\
(0134)(0256) &= r_2,\\
(0142)(0356) &= r_3,\\
(0145)(0236) &= r_4,\\
(0125)(0346) &= r_5,\\
(0135)(0426) &= r_6,\\
(0162)(0345) &= r_7,\\
(0163)(0425) &= r_8,\\
(0164)(0235) &= r_9,\\
(0165)(0432) &= r_{10}.
\end{aligned}\right\} \qquad \text{(2)}$$

Of these five are linearly independent and the other five are expressible in terms of them. We take

$$r_1,\ r_2,\ r_4,\ r_7,\ r_{10} \qquad \text{(3)}$$

and express the other five as

$$\left.\begin{aligned}
r_3 &= r_{10} - r_2 - r_1\\
r_5 &= r_2 - r_7 - r_{10}\\
r_6 &= -r_2 - r_4 + r_7\\
r_8 &= -r_1 + r_4 - r_7\\
r_9 &= r_1 - r_{10} - r_4
\end{aligned}\right\}. \qquad \text{(4)}$$

These results are easily verified by the fundamental identity.

Now consider the expression of thirty terms

$$\Sigma\,(0123)(0456)\,[(1589)(2689)(3489) + (1489)(2589)(3689) + (1689)(2489)(3589)],$$

where the symbols appear according to the following arrangement of ten rows of three columns of terms:

$$\left.\begin{array}{ll}
\text{(i)} & 0123.0456.15.26.34 + 14.25.36 + 16.24.35\\
\text{(ii)} & 0134.0256.12.36.45 + 15.32.46 + 16.35.42\\
\text{(iii)} & 0142.0356.15.46.23 + 13.45.26 + 16.43.25\\
\text{(iv)} & 0145.0236.13.42.56 + 12.46.53 + 16.43.52\\
\text{(v)} & 0125.0346.14.26.53 + 13.24.56 + 16.23.54\\
\text{(vi)} & 0135.0426.14.36.52 + 12.34.56 + 16.32.54\\
\text{(vii)} & 0162.0345.14.65.23 + 13.64.25 + 15.63.24\\
\text{(viii)} & 0163.0425.14.65.32 + 12.64.35 + 15.62.34\\
\text{(ix)} & 0164.0235.13.62.45 + 12.65.43 + 15.63.42\\
\text{(x)} & 0165.0432.13.64.52 + 12.63.54 + 14.62.53
\end{array}\right\} \qquad \text{(5)}$$

In each row (for example the ith row) the symbols 1, 2, 3, 4, 5, 6 are grouped in pairs, each pair involving one symbol from each factor of the leading factors r_i. Also in row (i) 1, 2, 3 occur first in the three factors of a term while 4, 5, 6 are permuted cyclically. Either the positive cycle

$$5 \quad 6 \quad 4 \qquad 4 \quad 5 \quad 6 \qquad 6 \quad 4 \quad 5$$

or the negative cycle

$$5 \quad 4 \quad 6 \qquad 4 \quad 6 \quad 5 \qquad 6 \quad 5 \quad 4$$

may be taken, so giving two equivalent expressions for row (i) since

$$\left.\begin{aligned} &(15)(26)(34)+(14)(25)(36)+(16)(24)(35) \\ =\,&(15)(24)(36)+(14)(26)(35)+(16)(25)(34) \end{aligned}\right\} \qquad \text{......................(6)}$$

identically, where

$$(ij)$$

is short for

$$(ij89).$$

Similarly for each row.

This explains the general plan of the array (5) although it does not shew why this array is selected. Now Σ has these properties:

(I) *Σ does not vanish identically.*

(II) *Σ vanishes if any pair of the nine symbols 0, 1, 2, 3, 4, 5; 6, 8, 9 are equivalent.*

Accordingly Σ must be a numerical multiple of the required invariant. Before proving these two properties which justify the selection of the terms (5), we remark that the terms were chosen by trial. The question was, how many terms to write down in each row of (5)? Analogy from the ternary case made it probable that symbols convolved in the first two bracket factors should be separated in the others. Consequently, owing to relations like (6), three terms in a row seemed worth trying. A trial row (i) as in (5) was written down, and other rows were constructed as far as possible by symmetry. Thus row (ii) is suggested from row (i) by interchanging 2 and 4 throughout. This method worked well and it was found at all stages of the construction that property II was holding. Next, for a special case it was found that (5) did not vanish identically. Accordingly it was worth while simplifying (5) by means of (4). Thus the third row of (5) which is

$$r_3[(1589)(4689)(2389)+\text{etc.}]$$

is expressible by (4) in terms of the tenth, second and first rows, and so on for the rest.

This procedure actually reduces the thirty-term series Σ to five terms.

§ 4. *Simplification of Σ.* In fact, if the substitutions (4) be made in the series (5), so as to reduce the number of rows from ten to five, the result is

$$\left.\begin{aligned} \Sigma = \;\; &(0123)(0456)A \\ +&(0134)(0562)B \\ +&(0145)(0623)C \\ +&(0156)(0234)D \\ +&(0162)(0345)E \end{aligned}\right\} = \begin{aligned} &r_1\,A \\ +&r_3\,B \\ +&r_4\,C \\ +&r_{10}\,D \\ +&r_7\,E \end{aligned}$$

where, in the contracted notation of array (5),

A is six times 15 . 26 . 34
B is six times 16 . 23 . 54
C is six times 12 . 43 . 56
D is six times 13 . 62 . 54
E is six times 14 . 23 . 65

For example, after using (4) we obtain

$$\begin{aligned} A = {} & 15.26.34 + 14.25.36 + 16.24.35 - 13.45.26 \\ & - 15.46.23 \qquad - 16.43.25 \\ & - 15.62.34 - 14.65.32 \qquad - 12.64.35 \\ & + 15.63.42 \qquad + 13.62.45 + 12.65.43 \end{aligned}$$

$$= 62\{15.43 + 15.43 + 13.45 + 13.45\} + 63\{15.42 + 14.52\} + 64\{15.23 + 12.53\} + 65\{14.23 + 12.43\} + 61\{24.53 + 25.43\}$$

$$= (4)\,62.15.43 + (2)\,14.35.62 + 15.62.43 + 14.62.53 + 62.15.43 + 62.14.53$$

by grouping the 5th and 7th, the 6th and 9th, the 8th and 11th, the 10th and 12th terms together, and also rearranging the first four terms. Here (4) denotes the coefficient 4 for the term 62.15.43. This last expression is six times 62.15.43.

Similarly for the other expressions B, C, D, E.

This leads to the following series for Σ,

$$\left.\begin{aligned} -\tfrac{1}{6}\Sigma = {} & (0123)(0456)(1589)(6289)(3489) \\ & + (0134)(0562)(1689)(2389)(4589) \\ & + (0145)(0623)(1289)(3489)(5689) \\ & + (0156)(0234)(1389)(4589)(6289) \\ & + (0162)(0345)(1489)(5689)(2389) \end{aligned}\right\} \qquad (7)$$

$= G$, say.

It is now easy to verify that G does not vanish identically but vanishes if any two of its nine symbols are equivalent. For in the special case when

$$0 = \lambda 2 + \mu 3,$$

and

$$5 = \lambda 4 + \mu 6, \qquad (8)$$

$$G = -\lambda^2\mu^2\,[(1234)(6891) - (1236)(4891)]\,(2346)(4689)(8923).$$

This certainly does not vanish identically when the points 1, 2, 3, 4, 6, 8, 9 are independent, for it does not vanish when 1689 are coplanar.

But expression (7) vanishes if for symbol 1 we substitute any of the following:

$$0,\ 2,\ 3,\ 4\ \ 5,\ 6,\ 8,\ 9,\ \lambda 8 + \mu 9.$$

Regarding 1 as current coordinates, we infer that the equation

$$G = 0$$

is that of a quadric through the points 0, 2, 3, 4, 5, 6 and the line (89). Hence G and Δ_0, which are of the same degree in each symbol for point or line, can only differ by a numerical factor. In fact

$$G = -\Delta_0, \qquad (9)$$

as is seen by comparing coefficients of

$$a_1^2 b_1 b_2\, c_1 c_3\, d_1 d_4\, e_2^2 f_2 f_3\, g_2 g_4$$

on both sides of (8) written in full, namely

$$\Sigma\,(abcd)\,(aefg)\,(bfhk)\,(gchk)\,(dehk)$$

$$= \begin{vmatrix}
a_1^2, & a_1a_2, & a_1a_3, & a_1a_4, & a_2^2, & a_2a_3, & a_2a_4, & a_3^2, & a_3a_4, & a_4^2 \\
b_1^2, & b_1b_2, & b_1b_3, & b_1b_4, & b_2^2, & b_2b_3, & b_2b_4, & b_3^2, & b_3b_4, & b_4^2 \\
c_1^2, & c_1c_2, & c_1c_3, & c_1c_4, & c_2^2, & c_2c_3, & c_2c_4, & c_3^2, & c_3c_4, & c_4^2 \\
d_1^2, & d_1d_2, & d_1d_3, & d_1d_4, & d_2^2, & d_2d_3, & d_2d_4, & d_3^2, & d_3d_4, & d_4^2 \\
e_1^2, & e_1e_2, & e_1e_3, & e_1e_4, & e_2^2, & e_2e_3, & e_2e_4, & e_3^2, & e_3e_4, & e_4^2 \\
f_1^2, & f_1f_2, & f_1f_3, & f_1f_4, & f_2^2, & f_2f_3, & f_2f_4, & f_3^2, & f_3f_4, & f_4^2 \\
g_1^2, & g_1g_2, & g_1g_3, & g_1g_4, & g_2^2, & g_2g_3, & g_2g_4, & g_3^2, & g_3g_4, & g_4^2 \\
h_1^2, & h_1h_2, & h_1h_3, & h_1h_4, & h_2^2, & h_2h_3, & h_2h_4, & h_3^2, & h_3h_4, & h_4^2 \\
k_1^2, & k_1k_2, & k_1k_3, & k_1k_4, & k_2^2, & k_2k_3, & k_2k_4, & k_3^2, & k_3k_4, & k_4^2 \\
2h_1k_1, & h_1k_2+h_2k_1, & h_1k_3+h_3k_1, & h_1k_4+h_4k_1, & 2h_2k_2, & h_2k_3+h_3k_2, & h_2k_4+h_4k_2, & 2h_3k_3, & h_3k_4+h_4k_3, & 2h_4k_4
\end{vmatrix} \quad \text{............(10)}$$

where Σ denotes the five-term cyclic substitution group

$$\{(c\ d\ e\ f\ g)\}.$$

Thus the coefficient in question appears only in the first term on the left of (10) and yields

$$(h_4k_3 - h_3k_4)^3.$$

This is the same as the three-rowed determinant

$$\begin{vmatrix} h_3^2 & h_3h_4 & h_4^2 \\ k_3^2 & k_3k_4 & k_4^2 \\ 2h_3k_3 & h_3k_4 + h_4k_3 & 2h_4k_4 \end{vmatrix}$$

which occurs at the lower end of the leading diagonal on the right side of (10). Hence relation (8) is established.

Before dealing with certain geometrical consequences of this result we pass to the general case and deduce from (10) the general formula for Δ' with its ten independent rows.

§ 5. *Construction of the general symbolic form* Δ'. We start with the five-term function G, § 4 (7), which is a cubic in both 8 and 9. This may be expressed by writing it

$$G\,(89,\ 89,\ 89). \quad \text{............(1)}$$

If we replace this by

$$G\,(78,\ 89,\ 97) \quad \text{............(2)}$$

and operate with the negative symmetric substitution group

$$\{789\}', \quad \text{............(3)}$$

we obtain a series, H say, of thirty terms which is an alternating function of 7, 8, 9, and also of

$$0,\ 1,\ 2,\ 3,\ 4,\ 5,\ 6,$$

for this last is a property of G which is not affected by the new operation. But H does not however alternate in all ten symbols simultaneously.

This process replaces each term of G by six terms with the groupings

$$\left.\begin{matrix} 78.89.97 & 97.78.89 & 89.97.78 \\ 78.97.89 & 89.78.97 & 97.89.78 \end{matrix}\right\} \quad \text{............(4)}$$

It is equivalent to polarizing the function G with the operator

$$\left(x\frac{\partial}{\partial y}\right)\left(x\frac{\partial}{\partial z}\right),$$

where x, y, z denote the points 7, 8, 9 respectively.

If now we form the 240-term series K where

$$K = \{1 - (70) - (71) - (72) - (73) - (74) - (75) - (76)\}\, H, \quad\quad (5)$$

obtained from H by interchanging the symbol 7 in turn with each of the seven indicated symbols, the result is a non-zero function which alternates in the eight symbols

$$0, 1, 2, 3, 4, 5, 6, 7.$$

For the negative symmetric group of $n+1$ symbols

$$\{a_0 a_1 a_2 \ldots a_n\}'$$

may always be developed from that of n symbols

$$\{a_1 a_2 \ldots a_n\}'$$

in this manner, since

$$[1 - (a_0 a_1) - (a_0 a_2) - \ldots - (a_0 a_n)]\, \{a_1 a_2 \ldots a_n\}' = \{a_0 a_1 a_2 \ldots a_n\}'.$$

It may now be proved that *this 240-term series K is a numerical multiple of the determinant Δ'.*

To do this we consider the symbols 0, 1, ... 9 as belonging to ten quadrics, so that K is a symbolic expression for an invariant linear in the coefficients of ten quadrics. Hence by Peano's theorem* K is expressible as a linear function of Δ' and invariants which only arise when some of the symbols refer to the same quadric, that is to say are equivalent symbols.

Now there are only seven possible types of invariants of this degree, as appears at once when an attempt is made to write them down. These are

$$\left.\begin{aligned}
I_1 &= (0123)(0456)(1478)(2579)(3689) \\
I_2 &= (0123)(0145)(6728)(6749)(3589) = (\overline{01}, \overline{67}, 24, 3589) \\
I_3 &= (0123)(0167)(2389)(4568)(4579) = (\overline{01}, \overline{23}, \overline{45}, 67, 89) \\
I_4 &= (0123)(0124)(6735)(8945)(6789) = (\overline{012}, 34:5, \overline{67}, \overline{89}) \\
I_5 &= (0126)(0127)(3458)(3459)(6789) = (\overline{012}, \overline{345}, 6789) \\
I_6 &= (0123)(2345)(4567)(6789)(8901) = (\overline{01}, \overline{23}, \overline{45}, \overline{67}, \overline{89}) \\
I_7 &= (0123)^2(4567)(6789)(8945) \qquad = (0123)^2(\overline{45}, \overline{67}, \overline{89})
\end{aligned}\right\} \quad\quad (6)$$

Here I_1 has no pair of symbols convolved twice; I_2 has one pair and consequently two pairs so convolved, as indicated by the vincula for $\overline{01}$ and $\overline{67}$. I_4 has, besides, a trio $\overline{012}$ convolved twice. And so on.

Now if any two symbols in I_1 are equivalent, 0 and 1 say, we have by the fundamental identity

$$\{04561\}'(0456)(1478) = 0,$$

$$2I_1 \equiv 0 \bmod I_2.$$

Further we may prove that I_2 is expressible in terms of I_3 even when all the symbols are independent. Hence by Peano's theorem

$$I_1 = \lambda\Delta + \sum_{i=3}^{7} \lambda_i I_i, \quad\quad (7)$$

where each λ is numerical.

Thus to reduce I_2, the identity of three terms arising from the six terms of

$$\{234\}'(0123)(0145) = 0,$$

i.e. $\quad (0123)(0145) + (0134)(0125) + (0142)(0135) = 0$

gives $\quad I_2 \equiv -(\overline{01}, \overline{67}, 42, 3598) \bmod I_3.$

Rewritten $\quad I_2 = (\overline{01}, \overline{67}, 42, 5398)$ „ .

Whence $\quad I_2 \equiv -(\overline{01}, \overline{67}, 42, 5389)$ „ .

Also $\quad I_2 \equiv (\overline{01}, \overline{67}, 24, 5389)$ „ .

* Grace and Young, *loc. cit.* p. 358.

So, neglecting terms in I_3, the invariant I_2 alternates (with change of sign) in 3, 5 and also in 8, 9; and is therefore symmetrical (without change of sign) in 4, 2. Again since

$$\{0128\}'(0123)(3689)=0$$

then $$I_2 \equiv -(\overline{01}, \overline{67}, 84, 3529) \bmod I_3.$$

Writing for short $I_2=[24, 8, 9]$, these results give

$$I_2 \equiv [24, 8, 9] \equiv -[24, 9, 8] \equiv [42, 8, 9] \equiv -[42, 9, 8]$$

and $$I_2 \equiv -[84, 2, 9].$$

Thus $$I_2 \equiv [42, 8, 9] \equiv -[82, 4, 9] \equiv -[28, 4, 9] \equiv [28, 9, 4] \equiv -[98, 2, 4].$$

This last is symmetrical in 9, 8 whereas I_2 alternates. Hence

$$\{89\}' I_2 \equiv -[98, 2, 4]+[89, 2, 4],$$

or $$I_2 \equiv 0 \bmod I_3. \quad \text{.......................}(8)$$

Now operate on both sides of identity (7) with the substitution group

$$\{\mathrm{a\ b\ c\ d\ e\ f\ g\ h}\}'$$

of any eight symbols chosen from among 0, 1, ... 9. Since only two of the ten symbols are excluded from this operation, a reference to the list (6) shows that at least one of the sets

$$\overline{01} \qquad \overline{23} \qquad \overline{45}$$

of I_3 is included in the operation, $\overline{01}$ say. But

$$\{01\}' I_3$$

vanishes identically. Therefore the whole operation annihilates I_3. Likewise for I_4, I_5, I_6, I_7.

Also $$\{\mathrm{a\ b\ c\ d\ e\ f\ g\ h}\}' \lambda\Delta' = 8!\,\lambda\Delta';$$

accordingly $$\{\mathrm{a\ b\ c\ d\ e\ f\ g\ h}\}' I_1 = 8!\,\lambda\Delta'. \quad \text{.......................}(9)$$

But K consists entirely of 240 terms like I_1 and is such that

$$\{01234567\}' K = 8!\,K. \quad \text{.......................}(10)$$

It is therefore a multiple, μ say, of Δ' and we may write

$$\left.\begin{aligned} H &= \{(23456)\}\{789\}'(0123)(0456)(1578)(6289)(3497),\\ K &= \{1-(07)-(17)-(27)-(37)-(47)-(57)-(67)\}H\\ &= \mu\Delta'. \end{aligned}\right\} \quad \text{.......................}(11)$$

The following special case shows that $\mu=-20$, which finally gives

$$\Delta' = -\tfrac{1}{20}K. \quad \text{.......................}(12)$$

Calculation of the numerical coefficient μ.

Let the symbols of ten quadrics be

$$a, b, c, d, e, f, g, h, j, k,$$

equivalent respectively to the previous symbols

$$1, 2, 3, 4, 5, 6, 7, 8, 9, 0.$$

If $a_{11}, \ldots k_{44}$ denote the hundred actual coefficients, we take the special case when all but those in the leading diagonal of Δ properly arranged vanish, while the diagonal coefficients are each unity, so that

$$\Delta = |\,a\ b\ c\ d\ e\ f\ g\ h\ j\ k\,| = 1$$

and $$a_{11}=b_{22}=c_{33}=d_{44}=e_{12}=f_{13}=g_{14}=h_{23}=j_{24}=k_{34}=1.$$

Symbolically we may write

$$1_1 1_1 = 2_2 2_2 = 3_3 3_3 = 4_4 4_4 = 5_1 5_2 = 6_1 6_3 = 7_1 7_4 = 8_2 8_3 = 9_2 9_4 = 0_3 0_4, \quad \text{...........}(13)$$

while all other such products are zero. Hence the following bracket factors, when expanded, are zero for all values of m:

$$(034m), (714m), (823m), (924m), (512m), (613m). \quad \text{......................(14)}$$

Also such factors as $(134m)$ in this case contain, when expanded, only one non-vanishing term

$$1_1 . m_2 . 3_3 . 4_4.$$

Let $[1\, m\, 3\, 4]$ denote this term.

Consider now the terms of K broken up as in (11) into eight sets. Each set, such as H, may be arranged in an array of six columns and five rows, due to the five terms of G, §4 (7), and the six groupings, §5 (4).

In the first set, H, three rows, the second, fourth and fifth, have $(034m)$ and are therefore zero. The first and third rows produce zero and -3, as follows:

(1) *First row.* The only non-zero arrangement of 0, 1, 2, 3, 4, 5 in the first term

$$(0123)(0456)(1578)(6289)(3497)$$

is

$$(1230)(5\cdot 04)(15\cdot\cdot)(\cdot 2\cdot\cdot)(\cdot\cdot 34),$$

for, by (13), symbol 1 must occupy the first place twice, so 5 in the third factor must be in the second place, since $5_1 5_2$ alone is non-zero. But this arrangement forces 6 to the second place in the second factor, whereas $6_2 6_p = 0$. So the first row vanishes.

(2) *Third row.* This is

$$\{\overline{78}.\overline{89}.\overline{97}\}(0145)(0623)(1278)(3489)(5697)$$
$$= -[1504][6230][1289][7834][5967]$$
$$-[1504][6230][1289][7934][5867]$$
$$-[1504][6230][1287][7934][5869]$$
$$= -1 - 1 - 1 = -3.$$

Next, in the second set $\qquad -(07)H$

all rows but the fourth are zero, while the leading term of the fourth row is

$$-(7156)(7234)(1308)(4589)(6290)$$

which yields

$$-[1567][7234][1830][5984][6209]$$
$$= -1.$$

Four such terms of the row are found to be equal, and two others vanish. Consequently $-(07)H = -4$.

Similar analysis leads to the results

$$\begin{aligned} H &= -3, \\ -(07)H &= -4, \\ -(17)H &= -2, \\ -(27)H &= 0, \\ -(37)H &= -4, \\ -(47)H &= -4, \\ -(57)H &= 0, \\ -(67)H &= -3. \end{aligned}$$

Hence by addition $\qquad K = -20.$

But in general $\qquad K = \mu\Delta.$

Thus $\qquad \mu = -20.$

II

GEOMETRICAL RESULTS.

§ 6. We now possess the projective invariant conditions that various lines and points should lie on a quadric surface. In ascending degree of algebraic simplicity they are as follows:

(1) *The condition that ten points should lie on a quadric.*

(2) *The condition that seven points and one line should lie on a quadric.*

(3) *The condition that four points and two lines should lie on a quadric.*

(4) *The condition that one point and three lines should lie on a quadric.*

Each is a particular case of that which precedes it. Using the results of § 5 (11) and § 4 (7), (8), the conditions are

$$\left.\begin{array}{ll}(1) & K=0,\\ (2) & G=0,\\ (3) & \begin{vmatrix}(2314)(8954), & (2316)(8956)\\ (2354)(8914), & (2356)(8916)\end{vmatrix}=0,\\ (4) & (1234)(6891)-(1236)(4891)=0.\end{array}\right\} \qquad (1)$$

Here in (1) none of the points are collinear,

in (2) points 789 are collinear,

in (3) points 789 and also 023 are collinear,

in (4) points 789 and also 023 and also 456 are collinear.

But in each case the formula fails if the lines in question are coplanar. Thus condition (3) which may be written*

$$\{23, 89, 1456\}^2 \qquad (2)$$

follows from (2) by substituting $\lambda 2+\mu 3$ for 0 and then removing the factor (2389). Also (4) is given by § 4 (8), when three like factors are removed. These two conditions were already known: the two more general ones are believed to be new.

The equation $G=0$ shews that:

Every line which lies on a quadric surface through seven arbitrary points belongs to a cubic line complex uniquely determined by the points.

The same equation also gives *a ruler construction for a quadric surface through a given line and six arbitrary points,* or what is the same thing it gives a collinearity test whereby a seventh given point is brought on to the same quadric. The construction is naturally elaborate, but it is straightforward and runs as follows.

First substitute

$$x1+y2+z3 \qquad (3)$$

for the symbol 0 in the expression G, § 4 (7). This gives the equation of the conic section of G in the plane 123 in homogeneous coordinates x, y, z referred to the triangle 123. Thus

$$\begin{aligned} G = \;& y(2134)(0562)(1689)(2389)(4589)\\ &+x(0145)(1623)(1289)(3489)(5689)\\ &+x(0156)(1234)(1389)(4589)(6289)\\ &+z(3162)(0345)(1489)(5689)(2389),\end{aligned}$$

and on further substituting for 0 as before, this becomes

$$Ayz+Bzx+Cxy. \qquad (4)$$

* Cf. *Proc. Camb. Phil. Soc.* XXII. (1925), 481–487.

We find
$$A=-(2389)\begin{vmatrix}(2314)(8954), & (2316)(8956)\\ (2354)(8914), & (2356)(8916)\end{vmatrix}$$
$$=-(2389)\{23,\ 89,\ 1456\}^2,$$
while
$$B=-(3189)\{31,\ 89,\ 2456\}^2, \quad \text{.......................................(5)}$$
and
$$C=-(1289)\{12,\ 89,\ 3456\}^2.$$

Now these right-hand factors of A, B, C are the associated* functions of the doubly infinite system of quadrics through the eight points

1, 2, 3, 4, 5, 6, 8, 9.

These functions all vanish when all quadrics through any seven of the points go through the eighth. They satisfy* the identity
$$\{23,\ 89,\ 1456\}^2+\{31,\ 89,\ 2456\}^2+\{12,\ 89,\ 3456\}^2=0. \quad \text{.................(6)}$$
Hence
$$\frac{A}{(2389)}+\frac{B}{(3189)}+\frac{C}{(1289)}=0. \quad \text{...............................(7)}$$
Now let P, Q, R, P', Q', R' denote the following six points,
$$\left.\begin{matrix}P=(54,\ 236) & Q=(54,\ 316) & R=(54,\ 126)\\ P'=(14,\ 896) & Q'=(24,\ 896) & R'=(34,\ 896)\end{matrix}\right\}, \quad \text{.....................(8)}$$
and p, the line common to the planes 123, 589.

Then the following identity is true, as is easy to verify,
$$\{23,\ 89,\ 1456\}^2=(PP'p), \quad \text{..................................(9)}$$
which gives Hesse's theorem for eight associated points, that the line (123, 589) intersects the join of (14, 896) and (54, 236).

Further, since the point (123, 89) may be written
$$(2389)\,1+(3189)\,2+(1289)\,3,$$
its coordinates (x', y', z') are given by these bracket factors which, we notice, appear in A, B, C. The conic section circumscribing triangle 123 is therefore
$$(PP'p)\,x'yz+(QQ'p)\,y'zx+(RR'p)\,z'xy=0. \quad \text{.......................(10)}$$
This conic passes through (x', y', z') owing to relation (6) above, as is also geometrically obvious because the whole line 89 lies on the quadric.

Now let M denote the point (23, 189) and T the point where the tangent at 1 cuts the line 23. Then from (10) we deduce the value of the cross ratio of the four points 23 LM,
$$\{23,\ LM\}=-\frac{(RR'p)}{(QQ'p)}. \quad \text{..................................(11)}$$
This determines L uniquely if the right-hand member here is known. But this is so, for, by (9) and (5) above, it is
$$\frac{(8936)(1254)}{(8926)(1354)}\left\{\frac{\dfrac{(8954)(1256)}{(8956)(1254)}-\dfrac{(3126)(8934)}{(3124)(8936)}}{\dfrac{(8954)(3156)}{(8956)(3154)}-\dfrac{(3126)(8924)}{(3124)(8926)}}\right\} \quad \text{.....................(12)}$$
or, say,
$$\rho\,.\,\frac{\rho_1-\rho_2}{\rho_3-\rho_4}.$$

* Cf. *Proc. Camb. Phil. Soc. loc. cit.*

Here each symbol ρ denotes a known cross ratio. Thus, for example,

$$\rho = \frac{(8936)(1254)}{(8926)(1354)} = [896.154, 3.2], \text{ say,}$$

denoting the cross ratio, on line 23, determined by two planes and these two points. Also this function of five cross ratios ρ leads by a definite ruler construction* to a single ratio σ. So the point T is found.

Thence by Pascal's theorem any number of points on the conic 123 are found, for the equivalent of five points is given. In particular this gives the point X, not on line 89, where a plane through 89 cuts the conic. A similar construction gives the point X' where the same plane cuts another such conic 234, say. Hence the plane cuts the quadric G in the line 89 and a second line XX'.

In this way the quadric G is constructed by finding all positions of XX'.

Also *the condition that seven points* 0, 1, 2, 3, 4, 5, 6 *and one line* 89 *should lie on a quadric surface is that any one point* 0 *should be collinear with the points* XX' *derived from the remaining six points and line as above.*

Here we have an answer to the Brussels question mentioned at the outset, but for the case one degree removed from the general case. A complete answer could be given if, for the quadric $K=0$, we could construct a cross ratio function analogous to $\rho(\rho_1-\rho_2)/(\rho_3-\rho_4)$. This is theoretically quite possible but it is of enormous practical difficulty.

III

THE INVARIANTS LINEAR IN TEN QUADRICS.

§7. Let us consider the invariants of ten quadrics linear in each. We have first that already mentioned §1 (7)

$$(0123)(0456)(1478)(2579)(3689).$$

This may be conveniently written in the form of a star

0
7 1 4 8
2 6
9
3 5

where the digits which are collinear in the star appear in the same determinant factor when the invariant is written out in full. The invariant is unaltered by any of the substitutions of a group of order 5!, which is simply transitive. Looking at the star we see that the invariant will be unaltered by a rotation of the star as a whole, or by a reflexion. Also by a substitution of the form

$$-(14)(25)(36).$$

These are sufficient to generate the group. We call this class of invariant I_1.

* Cf. Schur, *Grundlagen der Geometrie* (Leipzig und Berlin, 1909), 53–59. Also Baker, *Principles of Geometry*, Vol. I.

The only other possible types of invariants are (as given in § 5):

$$I_2 = (0123)(0145)(6728)(6749)(3589) = (\overline{01}, \overline{67}, 24, 3589)$$
$$I_3 = (0123)(0167)(2389)(4568)(4579) = (\overline{01}, \overline{23}, \overline{45}, 67, 89)$$
$$I_4 = (0123)(0124)(6735)(8945)(6789) = (\overline{012}, 34:5, \overline{67}, \overline{89})$$
$$I_5 = (0126)(0127)(3458)(3459)(6789) = (\overline{012}, \overline{345}, 6789)$$
$$I_6 = (0123)(2345)(4567)(6789)(8901) = (\overline{01}, \overline{23}, \overline{45}, \overline{67}, \overline{89})$$
$$I_7 = (0123)^2(4567)(6789)(8945) \qquad = (0123)^2(45, 67, 89).$$

We consider it a reduction if we can express invariants in terms of actual products—i.e. here in terms of forms I_7. Also, if we can increase the number of triads of letters, such that the members of a triad are all associated in two determinant factors: that is, if we can express forms I_1, I_2, I_3, I_4, I_6 in terms of I_5 and I_7 or the other forms in terms of I_4, I_5, I_7. Further, we aim at increasing the number of pairs convolved twice. So that we take the invariants in the order I_1, I_2, I_3, I_6, I_4, I_5, I_7 and seek to express those of earlier forms in terms of those of later forms.

The group of the star I_1 just written down may be very conveniently exhibited as follows. We write down the digits of the star in the tableau

$$\begin{array}{cc|c} & \multicolumn{1}{c}{0} & \\ 1 & 4 & 9 \\ 2 & 5 & 8 \\ 3 & 6 & 7. \end{array}$$

Then the twelve operations which leave 0 unchanged are those which interchange the three rows of the tableau, and the first two columns, it being understood that to substitutions made up of an odd number of transpositions a minus sign is prefixed. Any operation of the group which interchanges 0 and one of the digits in the first two columns is compounded of these, and of that which makes this interchange and at the same time interchanges diagonally the elements of the minor of the particular digit chosen. The operations which interchange 0 and one of the digits in the last column are a little more complicated in statement—but can easily be derived. We have no occasion to go into that matter.

Consider first I_7.

By the fundamental identity

$$\{689\}'(4567)(8945) = 0.$$

Hence $$\{689\}(\overline{45}, \overline{67}, \overline{89}) = 0.$$

Hence every positive symmetric group of three letters operating on $(\overline{45}, \overline{67}, \overline{89})$ produces zero. Hence using the notation* of substitutional analysis,

$$(\overline{45}, \overline{67}, \overline{89}) = A_{2,2,2}T_{2,2,2}(\overline{45}, \overline{67}, \overline{89}).$$

In the same way for I_6 $$\{34567\}'(0123)(4567) = 0.$$

Whence $$\{345\}(\overline{01}, \overline{23}, \overline{45}, \overline{67}, \overline{89})$$
$$= 2\{67\}(\overline{345}, 27:6, \overline{01}, \overline{89}). \quad\text{.................................}(6_1)$$

It is easy to deduce that the operation of $\{45\alpha\}$, where α is any one of the other digits, produces I_4. For example, if $\alpha = 1$, we have from the above

$$(\overline{01}, \overline{23}, \overline{45}, \overline{67}, \overline{89}) + (\overline{02}, \overline{13}, \overline{45}, \overline{67}, \overline{89}) + (\overline{03}, \overline{12}, \overline{45}, \overline{67}, \overline{89})$$
$$= \{45\}(\overline{123}, 05:4, \overline{89}, \overline{67}),$$

and operation with $\{451\}$ gives the result.

* Grace and Young, *Algebra of Invariants*, 351–356.

Operation with $\{124\}$ on this also shews that

$$\{124\}\,(\overline{01},\ \overline{23},\ \overline{45},\ \overline{67},\ \overline{89}) = \Sigma\, I_4.$$

Thus if G_3 be any positive symmetric group of degree three,

$$G_3 I_6 = \Sigma\, I_4.$$

Hence $$I_6 = A_{2,2,2,2,2}\, T_{2,2,2,2,2}\, I_6 + \Sigma\, I_4. \qquad \ldots\ldots(6_2)$$

For forms I_5 we use

$$\{3458\}'\,(3459)\,(6789) = 0,$$

whence $$\{3458\}\,(\overline{012},\ \overline{345},\ 6789) = 0, \quad \text{also} \quad \{89\}\,(\overline{012},\ \overline{345},\ 6789) = 0. \qquad \ldots\ldots(5_1)$$

The digits here appear in two distinct groups of 5 digits each, viz. 01267 and 34589. If the digits of T are those of one group

$$I_5 = A_{3,1,1}\, T_{3,1,1}\, I_5.$$

There are other relations for forms I_5, but they are not immediately apparent.

Consider next I_1,

$$\{3456\}'\,(0123)\,(0456) = 0,$$

whence $$\{36\} \begin{array}{ccccccc} & & & 0 & & & \\ 7 & & 1 & & 4 & & 8 \\ & & 2 & & 6 & & \\ & & & 9 & & & \\ & 3 & & & & 5 & \end{array} = (\overline{14},\ \overline{36},\ 08,\ 2759) - (\overline{25},\ \overline{36},\ 09,\ 1748), \qquad \ldots\ldots(1_1)$$

and hence $$I_1 = \frac{1}{10!}\{0123456789\}'\, I_1 + \Sigma\, I_2.$$

Now $\{0123456789\}'\, I_1$ is the invariant Δ with which we are really concerned. Since every one of the other kinds of invariant contains at least one pair of letters convolved twice, the negative symmetric group annihilates them all, so that Δ is entirely independent of them, and we may write

$$I_1 = \lambda\Delta + \Sigma\, I_2.$$

We really want to invert this equation to the form

$$\Delta = \Sigma\, I_1,$$

where the Σ contains as few terms as possible.

The primary object of the study of forms I_2 etc. is to obtain the relations between different forms I_1.

Consider the forms I_4; we have the following equations:

From $\{01243\}'\,(0124)\,(6735) = 0$,

$$\{0123\}\,(\overline{012},\ 34:5,\ \overline{67},\ \overline{89}) = 6\,(0123)^2\,(\overline{45},\ \overline{67},\ \overline{89}). \qquad \ldots\ldots(4_1)$$

From $\{8673\}'\,(6735)\,(8945) = 0$,

$$\{678\}\,(\overline{012},\ 34:5,\ \overline{67},\ \overline{89}) = 2\,(\overline{012},\ \overline{678},\ 3459). \qquad \ldots\ldots(4_2)$$

From $\{567\}'\,(6789)\,(8945) = 0$,

$$\{567\}\,(\overline{012},\ 34:5,\ \overline{67},\ \overline{89}) = 0. \qquad \ldots\ldots(4_3)$$

From $\{36789\}'\,(0123)\,(6789) = 0$,

$$\{367\}\,(\overline{012},\ 34:5,\ \overline{67},\ \overline{89}) = -2\,\{89\}\,(\overline{012},\ \overline{367},\ 8495). \qquad \ldots\ldots(4_4)$$

From $\{3394\}'\,(6735)\,(8945) = 0$,

$$\{34\}'\,(\overline{012},\ 34:5,\ \overline{67},\ \overline{89}) = -\{89\}\,(\overline{012},\ \overline{678},\ 3495). \qquad \ldots\ldots(4_5)$$

From $\{34567\}'\,(0123)\,(4567) = 0$,

$$\{45\}\,\{67\}\,[(\overline{012},\ 34:5,\ \overline{67},\ \overline{89}) + (\overline{012},\ 36:7,\ \overline{45},\ \overline{89})] = 2\,(0123)^2\,(\overline{45},\ \overline{67},\ \overline{89}). \quad \ldots(4_6)$$

Now operate on (4_6) with $\{\mathit{345}\}$ and use (4_4). This gives

$$2\,\{\mathit{345}\}\,(\overline{012},\ 34:5,\ \overline{67},\ \overline{89}) = \{\mathit{345}\}\,(0123)^2(\overline{45},\ \overline{67},\ \overline{89}) + 2\,\{\mathit{67}\}\,\{\mathit{89}\}\,(\overline{012},\ \overline{345},\ 8697). \quad \ldots(4_7)$$

We will use the notation

$$(\overline{012},\ \overline{34}:5,\ \overline{67},\ \overline{89}) = \{\mathit{34}\}\,(\overline{012},\ 34:5,\ \overline{67},\ \overline{89}).$$

Then $\{\mathit{345}\}\,(\overline{012},\ 34:5,\ \overline{67},\ \overline{89}) = (\overline{012},\ \overline{45}:3,\ \overline{67},\ \overline{89})$

$$+\,\{\mathit{45}\}\,[2\,(\overline{012},\ 34:5,\ \overline{67},\ \overline{89}) + (\overline{012},\ 43:5,\ \overline{67},\ \overline{89}) - (\overline{012},\ \overline{34}:5,\ \overline{67},\ \overline{89})],$$

whence using (4_5)

$$4\,\{\mathit{45}\}\,(\overline{012},\ 34:5,\ \overline{67},\ \overline{89}) + 2\,(\overline{012},\ \overline{45}:3,\ \overline{67},\ \overline{89})$$

$$= \{\mathit{345}\}\,(0123)^2(\overline{45},\ \overline{67},\ \overline{89}) + \{\mathit{45}\}\,\{\mathit{67}\}\,\{\mathit{89}\}\,[(\overline{012},\ \overline{345},\ 8697) - (\overline{012},\ \overline{678},\ 3495)].$$

Hence (4_6) becomes

$$2\,(\overline{012},\ \overline{45}:3,\ \overline{67},\ \overline{89}) + 2\,(\overline{012},\ \overline{67}:3,\ \overline{45},\ \overline{89})$$

$$= [\{\mathit{345}\} + \{\mathit{367}\} - 4]\,(0123)^2(\overline{45},\ \overline{67},\ \overline{89})$$

$$+ \{\mathit{45}\}\,\{\mathit{67}\}\,\{\mathit{89}\}\,[(\overline{012},\ \overline{345},\ 8697) + (\overline{012},\ \overline{367},\ 8495) - (\overline{012},\ \overline{678},\ 3495) - (\overline{012},\ \overline{458},\ 3697)].$$

Then operate with $1 + (\mathit{468})\,(\mathit{579}) + (\mathit{486})\,(\mathit{597})$,

$$4\,(\overline{012},\ \overline{45}:3,\ \overline{67},\ \overline{89}) + 4\,(\overline{012},\ \overline{67}:3,\ \overline{89},\ \overline{45}) + 4\,(\overline{012},\ \overline{89}:3,\ \overline{45},\ \overline{67})$$

$$= [2\,\{\mathit{345}\} + 2\,\{\mathit{367}\} + 2\,\{\mathit{389}\} - 12]\,(0123)^2(\overline{45},\ \overline{67},\ \overline{89})$$

$$+ \{\mathit{45}\}\,\{\mathit{67}\}\,\{\mathit{89}\}\,[2\,(\overline{012},\ \overline{345},\ 8697) + 2\,(\overline{012},\ \overline{367},\ 4859) + 2\,(\overline{012},\ \overline{389},\ 6475)],$$

for the sum of the six terms obtained from the last two terms of our equation is zero by (5_1).

Hence subtracting the former equation from half of this

$$2\,(\overline{012},\ \overline{89}:3,\ \overline{45},\ \overline{67})$$

$$= 2\,\{\mathit{89}\}\,(0128)^2(\overline{39},\ \overline{45},\ \overline{67})$$

$$+ \{\mathit{45}\}\,\{\mathit{67}\}\,\{\mathit{89}\}\,[(\overline{012},\ \overline{389},\ 6475) + (\overline{012},\ \overline{678},\ 3495) + (\overline{012},\ \overline{458},\ 3697)].$$

Adding to an equation derived from (4_5) and using (5_1) again, we have

$$2\,(\overline{012},\ 89:3,\ \overline{45},\ \overline{67}) = \{\mathit{89}\}\,(0128)^2(\overline{45},\ \overline{67},\ \overline{39}) + \{\mathit{45}\}\,\{\mathit{67}\}\,[(\overline{012},\ \overline{389},\ 6475)$$

$$- (\overline{012},\ \overline{689},\ 3475) - (\overline{012},\ \overline{489},\ 3657) - \tfrac{1}{2}\,(\overline{012},\ \overline{456},\ 8973)]. \quad \ldots\ldots\ldots(4_8)$$

That is $I_4 = \Sigma\, I_6 + \Sigma\, I_7.$

Let us now substitute the value found for I_4 in the six original equations we obtained. The equations (4_2), (4_3) and (4_5) are identically satisfied; (4_4) gives an equation that can very easily be derived from that obtained from (4_6). Consider (4_1); we may regard 0, 1, 2, 3 as symbols belonging to the same quadric, also 6, 7 as symbols belonging to a second, and 8, 9 as referring to a third quadric. Thus it is convenient to write (4_1)

$$4\,(\overline{aaa},\ aa:b,\ \overline{\beta\beta},\ \overline{\gamma\gamma}) = (aaaa)^2(\overline{ab},\ \overline{\beta\beta},\ \overline{\gamma\gamma})$$

and the notation explains itself. Using (5_1) we have from (4_8) and (4_1):

$$\{\beta\gamma\}\,[8\,(\overline{aaa},\ \overline{aab},\ \beta\gamma\beta\gamma) - 6\,(\overline{aaa},\ \overline{\beta aa},\ b\gamma\beta\gamma) - 4\,(\overline{aaa},\ \overline{\beta\beta\gamma},\ aa\gamma b)$$

$$+ (aaaa)^2(\overline{ab},\ \overline{\beta\beta},\ \overline{\gamma\gamma}) + 2\,(aaaa)^2(\overline{ba},\ \overline{\beta\beta},\ \overline{\gamma\gamma})] = 0,$$

or $\{\beta\gamma\}\,[8\,(\overline{aaa},\ \overline{aab},\ \beta\gamma\beta\gamma) + 8\,(\overline{aaa},\ \overline{a\beta\beta},\ b\gamma a\gamma) + 8\,(\overline{aaa},\ \overline{a\beta\beta},\ b\gamma a\gamma)$

$$+ 2\,(aaaa)^2(\overline{ba},\ \overline{\beta\beta},\ \overline{\gamma\gamma}) + (aaaa)^2(\overline{ab},\ \overline{\beta\beta},\ \overline{\gamma\gamma})] = 0. \quad \ldots\ldots\ldots(5_2)$$

Similarly (4_6) yields

$$\{\beta\gamma\}\,[12\,(\overline{aaa},\ \overline{a\beta\beta},\ \gamma\delta\gamma\delta) + 4\,(\overline{aaa},\ \overline{a\delta\delta},\ \beta\gamma\beta\gamma) + 4\,(\overline{aaa},\ \overline{\beta\delta\delta},\ \beta\gamma a\gamma)$$

$$+ 4\,(\overline{aaa},\ \overline{\beta\gamma\gamma},\ \beta\delta a\delta) - 4\,(\overline{aaa},\ \overline{\gamma\gamma\delta},\ a\beta\delta\beta)$$

$$+ (aaaa)^2(\overline{\beta\beta},\ \overline{\gamma\gamma},\ \overline{\delta\delta}) + 2\,(aaa\beta)^2(\overline{a\beta},\ \overline{\gamma\gamma},\ \overline{\delta\delta})] = 0. \quad \ldots\ldots\ldots(5_3)$$

Operate with $\{\beta\gamma\delta\}$,

$$\{\beta\gamma\delta\}\,[16\,(\overline{\alpha\alpha\alpha},\ \overline{\alpha\beta\beta},\ \gamma\delta\gamma\delta)+8\,(\overline{\alpha\alpha\alpha},\ \overline{\beta\delta\delta},\ \beta\gamma\alpha\gamma)+(\alpha\alpha\alpha\alpha)^2\,(\overline{\beta\beta},\ \overline{\gamma\gamma},\ \overline{\delta\delta})+2\,(\alpha\alpha\alpha\beta)^2\,(\overline{\alpha\beta},\ \overline{\gamma\gamma},\ \overline{\delta\delta})]=0.$$

Hence subtracting twice (5_3)

$$\{\beta\gamma\}\,[8\,(\overline{\alpha\alpha\alpha},\ \overline{\alpha\beta\beta},\ \gamma\delta\gamma\delta)+8\,(\overline{\alpha\alpha\alpha},\ \overline{\alpha\delta\delta},\ \beta\gamma\beta\gamma)+8\,(\overline{\alpha\alpha\alpha},\ \overline{\delta\beta\beta},\ \delta\gamma\alpha\gamma)$$
$$+8\,(\overline{\alpha\alpha\alpha},\ \overline{\gamma\gamma\delta},\ \alpha\beta\delta\beta)+(\alpha\alpha\alpha\alpha)^2\,(\overline{\beta\beta},\ \overline{\gamma\gamma},\ \overline{\delta\delta})+2\,(\alpha\alpha\alpha\delta)^2\,(\overline{\alpha\delta},\ \overline{\beta\beta},\ \overline{\gamma\gamma})]=0. \quad\ldots\ldots\ldots(5_4)$$

Now from (6_1)

$$2\,(\overline{01},\ \overline{23},\ \overline{45},\ \overline{67},\ \overline{89})-2\,(\overline{01},\ \overline{45},\ \overline{23},\ \overline{67},\ \overline{89})$$
$$=\{345\}\,(\overline{01},\ \overline{23},\ \overline{45},\ \overline{67},\ \overline{89})-\{245\}\,(\overline{01},\ \overline{24},\ \overline{35},\ \overline{67},\ \overline{89})$$
$$=2\,\{67\}\,(\overline{345},\ 27:6,\ \overline{01},\ \overline{89})-2\,\{01\}\,(\overline{245},\ 30:1,\ \overline{67},\ \overline{89})$$
$$=2\,(2345)^2\,(\overline{01},\ \overline{67},\ \overline{89})-2\,\{01\}\,\{23\}\,(\overline{245},\ 30:1,\ \overline{67},\ \overline{89}) \text{ by } (4_6),$$

or

$$2\,\{\beta\gamma\}'\,(\overline{\alpha\alpha},\ \overline{\beta\beta},\ \overline{\gamma\gamma},\ \overline{\delta\delta},\ \overline{\epsilon\epsilon})$$
$$=2\,(\beta\beta\gamma\gamma)^2\,(\overline{\alpha\alpha},\ \overline{\delta\delta},\ \overline{\epsilon\epsilon})-8\,(\overline{\beta\gamma\gamma},\ \beta\alpha:\alpha,\ \overline{\delta\delta},\ \overline{\epsilon\epsilon})$$
$$=-2\,(\beta\beta\gamma\gamma)^2\,(\overline{\alpha\alpha},\ \overline{\delta\delta},\ \overline{\epsilon\epsilon})-4\,(\beta\gamma\gamma\alpha)^2\,(\overline{\alpha\beta},\ \overline{\delta\delta},\ \overline{\epsilon\epsilon})$$
$$-16\,(\overline{\beta\gamma\gamma},\ \overline{\alpha\alpha\beta},\ \delta\epsilon\delta\epsilon)+16\,(\overline{\beta\gamma\gamma},\ \overline{\delta\alpha\beta},\ \alpha\epsilon\delta\epsilon)+16\,(\overline{\beta\gamma\gamma},\ \overline{\epsilon\alpha\beta},\ \alpha\delta\epsilon\delta)+8\,(\overline{\beta\gamma\gamma},\ \overline{\delta\delta\epsilon},\ \beta\alpha\epsilon\alpha)$$
$$=8\,(\overline{\beta\gamma\gamma},\ \overline{\delta\delta\epsilon},\ \beta\alpha\epsilon\alpha)+8\,(\overline{\beta\gamma\gamma},\ \overline{\delta\delta\alpha},\ \beta\epsilon\alpha\epsilon)+8\,(\overline{\beta\gamma\gamma},\ \overline{\epsilon\epsilon\alpha},\ \beta\delta\alpha\delta). \quad\ldots\ldots\ldots(6_3)$$

This after using (5_4).

Consider then $I_2=(\overline{01},\ \overline{67},\ 24,\ 3589)$;

it is unaltered by the substitutions of a group order 16, viz.:

$$\{(01),\ (67),\ (24)\,(35)\,(89),\ (06)\,(17)\,(38)\,(59)\}.$$

From $\{245\}'\,(0123)\,(0145)=0$, we obtain

$$\{35\}\,(\overline{01},\ \overline{67},\ 24,\ 3589)=(\overline{01},\ \overline{35},\ \overline{67},\ 42,\ 98),$$

similarly—or operating on this with $(06)\,(17)\,(38)\,(59)$,

$$\{89\}\,(\overline{01},\ \overline{67},\ 24,\ 3589)=(\overline{67},\ \overline{89},\ \overline{01},\ 42,\ 53) \quad\ldots\ldots\ldots(2_1)$$

From $\{2589\}'\,(0123)\,(3589)=0$, we have

$$\{28\}\,(\overline{01},\ \overline{67},\ 24,\ 3589)=(\overline{28},\ \overline{67},\ \overline{01},\ 35,\ 94)-(\overline{015},\ 34:9,\ \overline{28},\ \overline{67}),$$

similarly—or operating with $(24)\,(35)\,(89)$,

$$\{49\}\,(\overline{01},\ \overline{67},\ 24,\ 3589)=(\overline{49},\ \overline{67},\ \overline{01},\ 53,\ 82)-(\overline{013},\ 52:8,\ \overline{49},\ \overline{67}) \quad\ldots\ldots\ldots(2_2)$$

Then we find

$$2\,(\overline{01},\ \overline{67},\ 24,\ 3589)=\{49\}\,(\overline{01},\ \overline{67},\ 24,\ 3589)-\{28\}\,(\overline{01},\ \overline{67},\ 29,\ 3584)+\{24\}\,(\overline{01},\ \overline{67},\ 89,\ 3524)$$
$$-\{48\}\,(\overline{01},\ \overline{67},\ 89,\ 3542)+\{29\}\,(\overline{01},\ \overline{67},\ 49,\ 3582)-\{35\}\,(\overline{01},\ \overline{67},\ 42,\ 3589)+\{89\}\,(\overline{01},\ \overline{67},\ 42,\ 5389)$$
$$=(\overline{49},\ \overline{67},\ \overline{01},\ 53,\ 82)-(\overline{28},\ \overline{67},\ \overline{01},\ 35,\ 49)+(\overline{67},\ \overline{24},\ \overline{01},\ 98,\ 53)-(\overline{48},\ \overline{67},\ \overline{01},\ 35,\ 29)$$
$$+(\overline{29},\ \overline{67},\ \overline{01},\ 53,\ 84)-(\overline{01},\ \overline{35},\ \overline{67},\ 24,\ 98)+(\overline{67},\ \overline{89},\ \overline{01},\ 24,\ 35)$$
$$-(\overline{013},\ 52:8,\ \overline{49},\ \overline{67})+(\overline{015},\ 39:4,\ \overline{28},\ \overline{67})+(\overline{015},\ 39:2,\ \overline{48},\ \overline{67})-(\overline{013},\ 54:8,\ \overline{29},\ \overline{67}). \ldots(2_3)$$

Or

$$I_2=\Sigma I_3+\Sigma I_4{}^{*}.$$

We shall also find useful:

From $\{03689\}'\,(1402)\,(3689)=0$,

$$\{360\}\,(\overline{14},\ \overline{36},\ 08,\ 2759)=-2\,(\overline{360},\ 85:7,\ \overline{14},\ \overline{29})-2\,(\overline{148},\ \overline{360},\ 2759). \quad\ldots\ldots\ldots(2_5)$$

From $\{5368\}'\,(3689)\,(2759)=0$,

$$\{365\}\,(\overline{14},\ \overline{36},\ 08,\ 2759)=-2\,(\overline{365},\ 09:2,\ \overline{14},\ \overline{78}). \quad\ldots\ldots\ldots(2_6)$$

* This is of course not inconsistent with (8), § 5.

The invariant $I_3 = (\overline{01}, \overline{23}, \overline{45}, 67, 89)$ is unaltered by the substitutions of the group order 32,

$$\{(\mathit{01}), (\mathit{23}), (\mathit{45}), (\mathit{67})(\mathit{89}), (\mathit{02})(\mathit{13})(\mathit{68})(\mathit{79})\}.$$

It is subject to the following equations:

From $\{\mathit{678}\}'(4568)(4579) = 0$,

$$\{\mathit{67}\}(\overline{01}, \overline{23}, \overline{45}, 67, 89) = (\overline{01}, \overline{23}, \overline{89}, \overline{45}, \overline{67}). \quad \ldots\ldots (3_1)$$

From $\{\mathit{9456}\}(2389)(4568) = 0$,

$$\{\mathit{459}\}(\overline{01}, \overline{23}, \overline{45}, 67, 89) = 2(\overline{459}, 78:6, \overline{01}, \overline{23}). \quad \ldots\ldots (3_2)$$

From $\{\mathit{60123}\}'(0123)(4568) = 0$,

$$\{\mathit{016}\}(\overline{01}, \overline{23}, \overline{45}, 67, 89) = -2\{\mathit{23}\}(\overline{016}, 27:9, \overline{38}, \overline{45}). \quad \ldots\ldots (3_3)$$

From $\{\mathit{01672}\}'(0167)(2389) = 0$,

$$\{\mathit{012}\}(\overline{01}, \overline{23}, \overline{45}, 67, 89) = -2(\overline{012}, 36:8, \overline{79}, \overline{45}) - 2(\overline{012}, 37:9, \overline{68}, \overline{45}). \quad \ldots\ldots (3_4)$$

From $\{\mathit{62389}\}'(0167)(2389) = 0$,

$$\{\mathit{68}\}(\overline{01}, \overline{23}, \overline{45}, 67, 89) = -\{\mathit{23}\}(\overline{012}, 37:9, \overline{68}, \overline{45}) + (\overline{01}, \overline{23}, \overline{68}, \overline{45}, \overline{79}). \quad \ldots\ldots (3_5)$$

Operate on this last equation with $\{\mathit{018}\}$, we obtain, by (3_3),

$$\{\mathit{018}\}(\overline{01}, \overline{23}, \overline{45}, 67, 89) = 2\{\mathit{23}\}(\overline{018}, 27:9, \overline{36}, \overline{45}) - \{\mathit{018}\}\{\mathit{23}\}(\overline{012}, 37:9, \overline{68}, \overline{45})$$
$$+ \{\mathit{018}\}(\overline{01}, \overline{23}, \overline{68}, \overline{45}, \overline{79}). \quad \ldots\ldots (3_6)$$

Now by (6_1),

$$\{\mathit{238}\}(\overline{01}, \overline{23}, \overline{68}, \overline{45}, \overline{79}) = 2\{\mathit{01}\}(\overline{238}, 60:1, \overline{45}, \overline{79}),$$

and operating on this with $\{\mathit{018}\}$ we have,

$$\{\mathit{018}\}(\overline{01}, \overline{23}, \overline{68}, \overline{45}, \overline{79}) = -2\{\mathit{23}\}\{\mathit{79}\}(\overline{018}, 27:9, \overline{36}, \overline{45}) + 2\{\mathit{018}\}(\overline{238}, 60:1, \overline{45}, \overline{79}).$$

Hence

$$\{\mathit{018}\}(\overline{01}, \overline{23}, \overline{45}, 67, 89) = \{\mathit{018}\}\{\mathit{23}\}[(\overline{238}, 60:1, \overline{45}, \overline{79}) - (\overline{012}, 37:9, \overline{68}, \overline{45})]$$
$$+ 2\{\mathit{23}\}(\overline{018}, 29:7, \overline{36}, \overline{45}).$$

And therefore equation (2_3) can be written

$$2(\overline{01}, \overline{67}, 24, 3589) = \tfrac{1}{2}\{\mathit{249}\}(\overline{49}, \overline{67}, \overline{01}, 53, 82) - \tfrac{1}{2}\{\mathit{248}\}(\overline{48}, \overline{67}, \overline{01}, 35, 29)$$
$$- (\overline{24}, \overline{67}, \overline{01}, 53, 89) + (\overline{24}, \overline{67}, \overline{01}, 35, 89) + (\overline{24}, \overline{67}, \overline{01}, 53, 98) - (\overline{01}, \overline{35}, \overline{67}, 24, 98)$$
$$+ (\overline{89}, \overline{67}, \overline{01}, 35, 24) - \{\mathit{24}\}[(\overline{013}, 52:8, \overline{49}, \overline{67}) - (\overline{015}, 39:4, \overline{28}, \overline{67})]$$
$$= 3(\overline{24}, \overline{67}, \overline{01}, 35, 89) + (\overline{01}, \overline{35}, \overline{67}, 24, 89) + (\overline{89}, \overline{67}, \overline{01}, 35, 24)$$
$$- (\overline{24}, \overline{67}, \overline{89}, \overline{01}, \overline{35}) - (\overline{35}, \overline{01}, \overline{24}, 67, 89)$$
$$+ \tfrac{1}{2}\{\mathit{249}\}\{\mathit{67}\}[(\overline{679}, 32:4, \overline{01}, \overline{58}) - (\overline{246}, 75:8, \overline{39}, \overline{01})] + \{\mathit{67}\}(\overline{249}, 68:5, \overline{73}, \overline{01})$$
$$- \tfrac{1}{2}\{\mathit{248}\}\{\mathit{67}\}[(\overline{678}, 32:4, \overline{01}, \overline{59}) - (\overline{246}, 75:9, \overline{38}, \overline{01})] - \{\mathit{67}\}(\overline{248}, 69:5, \overline{73}, \overline{01})$$
$$- \{\mathit{24}\}[(\overline{013}, 52:8, \overline{49}, \overline{67}) - (\overline{015}, 39:4, \overline{28}, \overline{67})]. \quad \ldots\ldots (2_4)$$

Now interchange $(\mathit{06})(\mathit{17})(\mathit{38})(\mathit{59})$ and subtract, therefore

$$0 = 3(\overline{24}, \overline{67}, \overline{01}, 35, 89) - 3(\overline{24}, \overline{01}, \overline{67}, 89, 35)$$
$$- (\overline{24}, \overline{67}, \overline{89}, \overline{01}, \overline{35}) + (\overline{24}, \overline{67}, \overline{89}, \overline{35}, \overline{01}) - (\overline{35}, \overline{01}, \overline{24}, \overline{67}, \overline{89}) + (\overline{35}, \overline{01}, \overline{24}, \overline{89}, \overline{67})$$
$$+ \tfrac{1}{2}\{\mathit{89}\}'\{\mathit{249}\}\{\mathit{67}\}[(\overline{679}, 32:4, \overline{01}, \overline{58}) - (\overline{246}, 75:8, \overline{39}, \overline{01}) + \tfrac{1}{3}(\overline{249}, 68:5, \overline{73}, \overline{01})]$$
$$- \tfrac{1}{2}\{\mathit{35}\}'\{\mathit{245}\}\{\mathit{01}\}[(\overline{015}, 82:4, \overline{67}, \overline{93}) - (\overline{240}, 19:3, \overline{85}, \overline{67}) + \tfrac{1}{3}(\overline{245}, 03:9, \overline{18}, \overline{67})]$$
$$- \{\mathit{24}\}[(\overline{013}, 52:8, \overline{49}, \overline{67}) - (\overline{015}, 39:4, \overline{28}, \overline{67}) - (\overline{678}, 92:3, \overline{45}, \overline{01}) + (\overline{679}, 85:4, \overline{23}, \overline{01})].$$

But from (3_5)

$$(\overline{24}, \overline{01}, \overline{67}, 89, 35) - (\overline{24}, \overline{01}, \overline{67}, 35, 89)$$
$$= -\{\mathit{01}\}[(\overline{240}, 19:5, \overline{83}, \overline{67}) - (\overline{240}, 13:8, \overline{95}, \overline{67})] + (\overline{24}, \overline{01}, \overline{83}, \overline{67}, \overline{95}) - (\overline{01}, \overline{24}, \overline{83}, \overline{67}, \overline{95}).$$

Hence using (6_3) we obtain

$$(\overline{24}, \overline{67}, \overline{01}, 35, 89) - (\overline{24}, \overline{01}, \overline{67}, 35, 89) = \Sigma I_4 = \Sigma I_5 + \Sigma I_7. \quad\ldots\ldots\ldots\ldots\ldots(3_7)$$

It is unnecessary for our purposes to print the full expression, which is somewhat cumbersome.

Hence by (3_4), (3_3) and (3_2) it follows that

$$\{01a\}\, I_3 = \Sigma I_5 + \Sigma I_7,$$

where a is any one of the digits other than 0 or 1.

Now consider $$T_{a_1, a_2, \ldots a_h} \{01a\} = \Sigma NP\, \{01a\},$$

when $\alpha_1 \geqslant 3$. Let the first group of P (that of degree α_1) be {bcd...}. If this does not contain the digit 0, by means of the identity

$$NP\,[1 + (0b) + (0c) + (0d) + \ldots] = 0$$

we obtain $$NP = -\Sigma sN'P',$$

where s is a substitution, and the first group of every P' contains the digit 0. If the first group of P' does not contain the digit 1, let it be $\{0b'c'\ldots\}$. Then from the identity

$$sN'P'\,[1 + (10) + (1b') + (1c') + \ldots]\,\{01\} = 0$$

we have $$sN'P'\,\{01\} = -\tfrac{1}{2}\Sigma s'N''P''\,\{01\}$$

and the first group of each P'' contains both 0 and 1.

Hence $T_{a_1, a_2, \ldots a_h}(\overline{01}, \overline{23}, \overline{45}, 67, 89)$

$$= \Sigma\lambda s'N''P''(\overline{01}, \overline{23}, \overline{45}, 67, 89) = \Sigma I_5 + \Sigma I_7,$$

provided $\alpha_1 \geqslant 3$ (where λ is numerical).

In just the same way it may be shewn that

$$T_{2,2,2,}\ldots(\overline{01}, \overline{23}, \overline{45}, 67, 89) = \Sigma\, \lambda sNP\, (\overline{01}, \overline{23}, \overline{45}, 67, 89),$$

where the first three groups of each P are $\{01\}$, $\{23\}$, $\{45\}$.

Hence by (3_1) and (3_5)

$$T_{2,2,2,2,}\ldots I_3 = \Sigma I_5 + \Sigma I_6 + \Sigma I_7,$$

and since $$T_{2,2,2,2,}\ldots I_5 = 0 = T_{2,2,2,2,}\ldots I_7,$$

we have $$T_{2,2,2,2,}\ldots I_3 = \Sigma T_{2,2,2,2,}\ldots I_6,$$

and in fact $$T_{2,2,2,2,1,1} I_3 = 0.$$

In the same way it becomes evident that

$$T_{2, a_2, a_3, \ldots} I_3 = 0,$$

when $\alpha_2 = 1$, or when $\alpha_2 = 2$, $\alpha_3 = 1$.

Thus operating on I_3 with the identity

$$1 = \Sigma A_{a_1, a_2, \ldots a_h} T_{a_1, a_2, \ldots a_h}$$

we find $$I_3 = A_{2,2,2,1,1,1,1} T_{2,2,2,1,1,1,1} I_3 + \Sigma I_5 + \Sigma I_6 + \Sigma I_7.$$

Thus we have expressed I_2 in terms of forms $I_r\,(r > 2)$; I_3 in terms of $T_{2,2,2,1,4,1,1} I_3$ and I_5, I_6, I_7; I_4 in terms of I_5 and I_7; and I_6 in terms of $T_{2,2,2,2,2} I_6$ and I_5 and I_7 [see (6_2)]. Hence all our invariants can be expressed in terms of $T_{2,2,2,1,1,1,1} I_3$, $T_{2,2,2,2,2} I_6$, I_5, I_7 and Δ. Now I_5 is unaltered by the operations of two symmetric groups each of degree 3, *hence*

$$T_{a_1, a_2, \ldots} I_5 = 0,$$

unless $\alpha_1 > 3$, and if $\alpha_1 = 3$ unless $\alpha_2 = 3$. In the same way we see that

$$T_{a_1, \ldots} I_7 = 0,$$

unless $\alpha_1 \geqslant 4$. Any T except those thus specified and $T_{1,1,1,1,1,1,1,1,1,1}$ must annihilate all invariant types of degree 10.

§ 8. To calculate I_1, we will write

$$I_1 = \Sigma\, AT I_1$$

and examine the separate terms.

Consider first $T_{\alpha_1, \alpha_2, \dots \alpha_h} \Sigma\,(\mathrm{ab})$, where the sign of summation extends to every transposition of the letters in question.

Let us consider $NP\Sigma\,(\mathrm{ab})$, and let the first group of P be $\{\mathrm{a}_1\, \mathrm{a}_2 \dots \mathrm{a}_{\alpha_1}\}$. If both a and b belong to this group,

$$P(ab) = P,$$

and this happens in the case of $\binom{\alpha_1}{2}$ transpositions.

When a belongs to this group and b does not, we will consider the set of transpositions

$$(\mathrm{a}_1\, \mathrm{b}), (\mathrm{a}_2\, \mathrm{b}), (\mathrm{a}_3\, \mathrm{b}), \dots (\mathrm{a}_{\alpha_1}\, \mathrm{b}).$$

Then we know $\qquad NP\,[(\mathrm{a}_1\, \mathrm{b}) + (\mathrm{a}_2\, \mathrm{b}) + \dots + (\mathrm{a}_{\alpha_1}\, \mathrm{b})] = -NP;$

we may choose b in $\alpha_2 + \alpha_3 + \dots + \alpha_h$ ways, and thus have this number of sets. The remaining transpositions have no letter in common with the first group of P; we may then proceed in the same way with those of them which contain a letter from the second group of P, and so on. Thus we find

$$NP\,\Sigma\,(\mathrm{ab}) = \left[\binom{\alpha_1}{2} + \binom{\alpha_2}{2} + \dots + \binom{\alpha_h}{2} - \alpha_2 - 2\alpha_3 - 3\alpha_4 - \dots - (h-1)\,\alpha_h\right] NP.$$

And hence

$$T_{\alpha_1, \alpha_2, \dots \alpha_h} \Sigma\,(\mathrm{ab}) = \left[\Sigma \binom{\alpha}{2} - \Sigma\,(r-1)\,\alpha_r\right] T_{\alpha_1, \alpha_2, \dots \alpha_h}.$$

Let n be the total number of letters, then

$$T_{\alpha_1, \alpha_2, \dots \alpha_h} \Sigma\,\{\mathrm{ab}\} = \left[\binom{n}{2} + \Sigma \binom{\alpha}{2} - \Sigma\,(r-1)\,\alpha_r\right] T_{\alpha_1, \alpha_2, \dots \alpha_h}$$

$$= [n(\alpha_1 - 1) + (n - \alpha_1)(\alpha_2 - 1) + (n - \alpha_1 - \alpha_2)(\alpha_3 - 1) + \dots + \alpha_h(\alpha_h - 1)]\, T_{\alpha_1, \alpha_2, \dots \alpha_h},$$

which is zero when all the α's are unity, and positive in every other case, as was to be expected.

Now $I_1 = \begin{array}{ccccccc} & & & 0 & & & \\ 7 & & 1 & & 4 & & 8 \\ & & 2 & & 6 & & \\ & & & 9 & & & \\ & 3 & & & & 5 & \end{array}$ is unaltered by any of the operations of a group of order 5! (§ 7), viz.:

$$G = \{(07358)\,(12964),\ (78)\,(14)\,(26)\,(35),\ -(14)\,(25)\,(36)\}.$$

From (1_1) we obtain the value of $\{\mathrm{ab}\}\, I_1$ when a and b are collinear in the star. In the case of the digit *3* (for instance) there are only three digits not collinear with it, viz. *5*, *7* and *4*.

Now $\qquad \{35\} = \{356\} - (56)\,\{36\} - (36)\,\{56\}.$

Also $(09)\,(17)\,(35)\,(48)$ is an operation of the group G; it transforms (1_1) into

$$\{56\}\, I_1 = \{56\} \begin{array}{ccccccc} & & & 9 & & & \\ 1 & & 7 & & 8 & & 4 \\ & & 2 & & 6 & & \\ & & & 0 & & & \\ & 5 & & & & 3 & \end{array} = (\overline{78}, \overline{56}, 94, 2130) - (\overline{23}, \overline{56}, 90, 7184).$$

Hence $\{35\} I_1 = \frac{1}{2}\{356\}[(\overline{14}, 36, 08, 2759) - (\overline{25}, \overline{36}, 09, 1748)]$

$$- (\overline{14}, \overline{35}, 08, 2769) + (\overline{26}, \overline{35}, 09, 1748)$$
$$- (\overline{78}, \overline{53}, 94, 2160) + (\overline{26}, \overline{53}, 90, 7184)$$
$$= - (\overline{356}, 09 : 2, \overline{14}, \overline{78}) - (\overline{25}, \overline{36}, 09, 1748) - (\overline{23}, \overline{56}, 09, 1748)$$
$$+ (\overline{26}, \overline{35}, 09, 1748) - (\overline{14}, \overline{35}, 08, 2769) - (\overline{78}, \overline{35}, 94, 2160)$$

on using (2_6).

And therefore

$$\Sigma \{ab\} I_1 = \tfrac{1}{120} G\Sigma \{ab\} I_1 = \tfrac{1}{8} G[\{35\} + 2\{36\}] I_1$$
$$= \tfrac{1}{8} G[-(\overline{356}, 09 : 2, \overline{14}, \overline{78})$$
$$+ 2(\overline{14}, \overline{36}, 08, 2759) - 3(\overline{25}, \overline{36}, 09, 1748) - (\overline{23}, \overline{56}, 09, 1748)$$
$$+ (\overline{26}, \overline{35}, 09, 1748) - (\overline{14}, \overline{35}, 08, 2769) - (\overline{78}, \overline{35}, 94, 2160)].$$

The operation *(17)* *(48)* *(09)* *(35)* of G interchanges the last pair of forms in each of the last two rows, and the operation $-(12)(45)(89)$ of G interchanges the first two forms of the middle row, hence

$$\Sigma \{ab\} I_1 = \tfrac{1}{8} G[-(\overline{356}, 09 : 2, \overline{14}, \overline{78})$$
$$+ 6(\overline{14}, \overline{36}, 08, 2759) - (\overline{16}, \overline{34}, 08, 2759) - 2(\overline{14}, \overline{35}, 08, 2769)]. \quad \ldots(1_2)$$

§ 9. Consider then $T_{2,2,2,1,1,1,1} I_1 = \frac{1}{24} T_{2,2,2,1,1,1,1} \Sigma \{ab\} I.$

As in this paragraph only one particular T is referred to, we may drop the suffixes without fear of confusion. We have here

$$TI_4 = 0 = TI_5 = TI_6 = TI_7 = TG_3,$$

where G_3 is any positive symmetric group of three out of the ten digits.

Then from (2_4)

$$384TI_1 = TG[18(\overline{08}, \overline{14}, \overline{36}, 27, 59) + 6(\overline{14}, \overline{27}, \overline{36}, 08, 59) + 6(\overline{59}, \overline{36}, \overline{14}, 27, 08)$$
$$- 3(\overline{08}, \overline{16}, \overline{34}, 27, 59) - (\overline{16}, \overline{27}, \overline{34}, 08, 59) - (\overline{59}, \overline{34}, \overline{16}, 27, 08)$$
$$- 6(\overline{08}, \overline{14}, \overline{35}, 27, 69) - 2(\overline{14}, \overline{27}, \overline{35}, 08, 69) - 2(\overline{69}, \overline{35}, \overline{14}, 27, 08)].$$

The equations (3_1), (3_5) shew that $T(\overline{ab}, \overline{cd}, \overline{ef}, gh, ij)$ changes sign, but is otherwise unaltered when operated on by any transposition of g, h, i, j. Also from (3_3) interchanges of the pairs ab, cd, ef leave the form unaltered.

Using the operation *(08)* *(27)* *(34)* *(59)* of G, we have

$$TG(\overline{08}, \overline{14}, \overline{36}, 27, 59) = TG(\overline{08}, \overline{13}, \overline{46}, 27, 59) = -\tfrac{1}{2} TG(\overline{08}, \overline{16}, \overline{34}, 27, 59)$$

and similar results: hence

$$384TI_1 = TG[-12(\overline{16}, \overline{34}, \overline{08}, 27, 59) - 4(\overline{16}, \overline{34}, \overline{27}, 08, 59) - 4(\overline{16}, \overline{34}, \overline{59}, 08, 27)$$
$$- 6(\overline{14}, \overline{35}, \overline{08}, 27, 69) - 2(\overline{14}, \overline{35}, \overline{27}, 08, 69) - 2(\overline{14}, \overline{35}, \overline{69}, 08, 27)]$$
$$= TG[-12(\overline{16}, \overline{34}, \overline{08}, 27, 59) - 8(\overline{16}, \overline{34}, \overline{27}, 08, 59)$$
$$- 6(\overline{14}, \overline{35}, \overline{08}, 27, 69) - 2(\overline{14}, \overline{35}, \overline{27}, 08, 69) - 2(\overline{14}, \overline{35}, \overline{69}, 08, 27)]$$

because *(25)* *(79)* *(34)* *(16)* is an operation of G.

§ 10. Let us next consider $T_{2,2,2,2,2} I_1,$

the only other case for which $T_{\alpha_1, \alpha_2, \ldots} I_1$, is not zero, when $\alpha_1 < 3$.

Here again $TI_5 = TI_7 = 0.$

Consider $T(\overline{ab}, \overline{cd}, \overline{ef}, gh, ij).$

We will fix our attention on the last four letters. Let us use $T'_{\alpha_1} \ldots$ for the substitutional expressions relating to these letters alone.

Then $$1 = A_4 T_4' + A_{3,1} T'_{3,1} + A_{2,2} T'_{2,2} + A_{2,1,1} T'_{2,1,1} + A_{1,1,1,1} T'_{1,1,1,1},$$
$$T T_4' = 0 \text{ and } TT'_{3,1} = 0.$$

As regards $$T'_{2,1,1} = \Sigma \{gh\} \{ij\}'$$
we notice that if $T = \Sigma NP$ $$NP . \{ab\} \{cd\} \{ef\} \{gh\} \{ij\}' = 0.$$

For if P has not a factor $\{ab\}$, it has one $\{\underline{a\beta}\}$ and
$$NP . \{ab\} = NP [-(b\underline{\beta})] = -N (b\underline{\beta}) P',$$
where P' has a factor $\{ab\}$.

Thus $$NP \{ab\} \{cd\} \{ef\} \{gh\} \{ij\}' = NSP'' \{ab\} \{cd\} \{ef\} \{gh\} \{ij\}',$$
where P'' contains the factors $\{ab\}$, $\{cd\}$, $\{ef\}$, $\{gh\}$ and S is a substitution with a coefficient $(-\frac{1}{2})^\lambda$, where λ is the number of transpositions it contains. But the remaining factor of P'' must be $\{ij\}$ and $\{ij\} \{ij\}' = 0$.

Hence $$TT'_{2,1,1} (\overline{ab}, \overline{cd}, \overline{ef}, gh, ij) = 0$$
and similarly $$TT'_{1,1,1,1} (\overline{ab}, \overline{cd}, \overline{ef}, gh, ij) = 0$$
and $$TI_3 = A_{2,2} TT'_{2,2} I_3.$$

Now $T'_{2,2} = N'P'$ where $$P' = \{gh\} \{ij\} + \{gi\} \{hj\} + \{gj\} \{hi\}$$
and $$N' = \{gh\}' \{ij\}' + \{gi\}' \{hj\}' + \{gj\}' \{hi\}';$$
in this case $A_{2,2} = \frac{1}{18}$.

From (3_1) $$\{gh\} (\overline{ab}, \overline{cd}, \overline{ef}, gh, ij) = (\overline{ab}, \overline{cd}, \overline{ij}, \overline{ef}, \overline{gh}).$$

From (3_5) $$T \{gi\} (\overline{ab}, \overline{cd}, \overline{ef}, gh, ij) = T (\overline{ab}, \overline{cd}, \overline{gi}, \overline{ef}, \overline{hj}).$$

Also from (6_3) $$T (\overline{ab}, \overline{cd}, \overline{ef}, \overline{gh}, \overline{ij}) = T (\overline{ab}, \overline{cd}, \overline{ef}, \overline{ij}, \overline{gh}).$$

Hence since $$\{gj\} = \{gj\} [-(ij) + \{ij\}],$$
$$\begin{aligned} T \{gj\} (\overline{ab}, \overline{cd}, \overline{ef}, gh, ij) &= T [-(\overline{ab}, \overline{cd}, \overline{ef}, \overline{gj}, \overline{hi}) + \{gj\} (\overline{ab}, \overline{cd}, \overline{ef}, \overline{gh}, \overline{ij})] \\ &= T [(\overline{ab}, \overline{cd}, \overline{ef}, \overline{gh}, \overline{ij}) + (\overline{ab}, \overline{cd}, \overline{ef}, \overline{jh}, \overline{ig}) - (\overline{ab}, \overline{cd}, \overline{ef}, \overline{gj}, \overline{hi})] \\ &= -2T (\overline{ab}, \overline{cd}, \overline{ef}, \overline{gj}, \overline{hi}) \end{aligned}$$
on using (5_1).

Hence $$TP' (\overline{ab}, \overline{cd}, \overline{ef}, gh, ij) = -6T (\overline{ab}, \overline{cd}, \overline{ef}, \overline{gj}, \overline{hi})$$
and $$TN'P' (\overline{ab}, \overline{cd}, \overline{ef}, gh, ij) = -36T (\overline{ab}, \overline{cd}, \overline{ef}, \overline{gj}, \overline{hi})$$
and thus finally $$T_{2,2,2,2,2} (\overline{ab}, \overline{cd}, \overline{ef}, gh, ij) = -T (\overline{ab}, \overline{cd}, \overline{ef}, \overline{gj}, \overline{hi}).$$

Then (2_4) gives us
$$\begin{aligned} &2T (\overline{ab}, \overline{cd}, ef, gh\ ij) \\ &= T [-3 (\overline{ab}, \overline{cd}, \overline{ef}, \overline{gj}, \overline{hi}) - (\overline{ab}, \overline{cd}, \overline{gh}, \overline{ej}, \overline{fi}) - (\overline{ab}, \overline{cd}, \overline{ij}, \overline{gf}, \overline{eh}) - 2 (\overline{ab}, \overline{cd}, \overline{ef}, \overline{gh}, \overline{ij})] \\ &= T [-3 (\overline{ab}, \overline{cd}, \overline{ef}, \overline{gj}, \overline{hi}) + (\overline{ab}, \overline{cd}, \overline{gh}, \overline{ei}, \overline{fj}) + (\overline{ab}, \overline{cd}, \overline{ij}, \overline{fh}, \overline{eg})]. \end{aligned}$$

Hence from (1_2)
$$\begin{aligned} 30T_{2,2,2,2,2} I_1 &= T\Sigma \{ab\} I_1 \\ &= \tfrac{1}{8} TG [-18 (\overline{14}, \overline{36}, \overline{08}, \overline{29}, \overline{75}) + 6 (\overline{14}, \overline{36}, \overline{27}, \overline{05}, \overline{89}) + 6 (\overline{14}, \overline{36}, \overline{59}, \overline{02}, \overline{87}) \\ &\quad + 3 (\overline{16}, \overline{34}, \overline{08}, \overline{29}, \overline{75}) - (\overline{16}, \overline{34}, \overline{27}, \overline{05}, \overline{89}) - (\overline{16}, \overline{34}, \overline{59}, \overline{02}, \overline{87}) \\ &\quad + 6 (\overline{14}, \overline{35}, \overline{08}, \overline{29}, \overline{76}) - 2 (\overline{14}, \overline{35}, \overline{27}, \overline{06}, \overline{89}) - 2 (\overline{14}, \overline{35}, \overline{69}, \overline{02}, \overline{87})] \end{aligned}$$
and using the same reductions as in the last case,
$$\begin{aligned} &= \tfrac{1}{8} TG [12 (\overline{16}, \overline{34}, \overline{08}, \overline{29}, \overline{75}) - 8 (\overline{16}, \overline{34}, \overline{27}, \overline{05}, \overline{89}) \\ &\quad + 6 (\overline{14}, \overline{35}, \overline{08}, \overline{29}, \overline{76}) - 2 (\overline{14}, \overline{35}, \overline{27}, \overline{06}, \overline{89}) - 2 (\overline{14}, \overline{35}, \overline{69}, \overline{02}, \overline{87})]. \end{aligned}$$

Also because *(12) (49) (58) (30)* belongs to G,
$$2TG (\overline{14}, \overline{35}, \overline{69}, \overline{02}, \overline{87}) = 2TG (\overline{13}, \overline{46}, \overline{08}, \overline{29}, \overline{75}) = -TG (\overline{16}, \overline{34}, \overline{08}, \overline{29}, \overline{75});$$
hence $$\begin{aligned} 240TI_1 = TG [&13 (\overline{16}, \overline{34}, \overline{08}, \overline{29}, \overline{75}) - 8 (\overline{16}, \overline{34}, \overline{27}, \overline{05}, \overline{89}) \\ &+ 6 (\overline{14}, \overline{35}, \overline{08}, \overline{29}, \overline{76}) - 2 (\overline{14}, \overline{35}, \overline{27}, \overline{06}, \overline{89})]. \end{aligned}$$

IV

THE GROUP S OF TWELVE STARS.

§ 11. It is of interest to consider a sum of 12 stars I_1 instead of a single one. They form a group.

We consider the group obtained by adding to G the operation $-(09)(76)(34)(51)(28)$; that is a kind of inversion of the star

$$\begin{array}{ccccc} & & 0 & & \\ 7 & 1 & & 4 & 8 \\ & 2 & & 6 & \\ & & 9 & & \\ & 3 & & 5 & \end{array}$$

The group thus obtained is of order $2 \times 6!$. It is triply transitive, and contains an intransitive subgroup of degree 6 and order 2. We will use Γ to denote the sum of its operations. Then ΓI_1 contains 12 distinct stars. The group can be illustrated by a tableau

$$\begin{array}{ccc} & 0 & \\ 1 & 4 & 9 \\ 2 & 5 & 8 \\ 3 & 6 & 7 \end{array}$$

like that representing G. But here we include interchanges of all columns as well as of all rows, and also the change of rows into columns. There is one more operation needed, viz. that which changes the above tableau into

$$\begin{array}{ccc} & 0 & \\ 1 & 5 & 7 \\ 8 & 3 & 4 \\ 6 & 9 & 2 \end{array}$$

It has of course a negative sign.

The operations of Γ are made up of 13 conjugate sets as follows:

(1) Identity.
(2) $-(14)(25)(36)$ number 30.
(3) $(78)(14)(26)(35)$ reflection type, number 45.
(4) $-(09)(76)(34)(51)(82)$ inversion type, number 36.
(5) $(741)(025)(639)$ number 80.
(6) $-(741)(065923)$ number 240.
(7) $(7148)(0296)$ number 90.
(8) $-(7148)(0692)(35)$ number 90.
(9) $-(76138245)$ number 180.
(10) $(07358)(12964)$ rotation type, number 144.
(11) $-(0652914378)$ number 144.
(12) $(09)(73128546)$ number 180.
(13) $(7583)(1246)$ number 180.

It is worth while to notice that there is a group of order 6! contained in Γ which contains G, and gives 6 stars. And that this group also contains a group of order $\frac{1}{2}.6!$ which contains G and gives a sum of 3 stars.

The group Γ is simply isomorphic with a group of degree 12, viz. the group formed by the interchanges of the 12 stars, produced by the operations of Γ; for no operation of Γ leaves all 12 stars unaltered. This group of degree 12 is simply transitive.

As Γ is triply transitive, any three digits can be made collinear in one of the stars. The fourth associated with them is unique. Thus if *714* are chosen as collinear, *8* is collinear with them. And the single operation of Γ which leaves *7148* unchanged is $-(09)(26)(35)$.

It is a great help to recognize at once the possible configurations of digits in the original star which become collinear in some other star. These are *0789, 0149, 2536, 0269, 0359, 7148* and those obtained by rotation. There are thus 30 sets of collinear letters, and each set appears as collinear in two different stars.

Let S denote the sum of the 12 stars considered. Then

$$192\, T_{2,2,2,1,1,1,1} S$$
$$= T\Gamma[-6(\overline{16}, \overline{34}, \overline{08}, 27, 59) - 4(\overline{16}, \overline{34}, \overline{27}, 08, 59)$$
$$-3(\overline{14}, \overline{35}, \overline{08}, 27, 69) - (\overline{14}, \overline{35}, \overline{27}, 08, 69) - (\overline{14}, \overline{35}, \overline{69}, 08, 27)].$$

Now Γ contains $-(34)(08)(16)$, hence

$$\Gamma(\overline{16}, \overline{34}, \overline{08}, 27, 59) = 0.$$

By the operation $(16)(57)(29)(08)$ of Γ

$$T\Gamma(\overline{16}, \overline{34}, \overline{27}, 08, 59) = T\Gamma(\overline{16}, \overline{34}, \overline{59}, 08, 27)$$
$$= T\Gamma[\tfrac{1}{2}\{016\}(\overline{16}, \overline{34}, \overline{59}, 08, 27) - (\overline{06}, \overline{34}, \overline{59}, 18, 27) - (\overline{01}, \overline{34}, \overline{59}, 68, 27)].$$

Here the first term is zero, the last term is also zero because Γ contains $-(09)(76)(34)(51)(28)$; and using the operation $-(16)(59)(27)$ of Γ we have

$$T\Gamma(\overline{16}, \overline{34}, \overline{59}, 08, 27) = -T\Gamma(\overline{06}, \overline{34}, \overline{59}, 18, 27)$$
$$= -T\Gamma(\overline{01}, \overline{34}, \overline{59}, 68, 27) = \tfrac{1}{2} T\Gamma(\overline{16}, \overline{34}, \overline{59}, 08, 27) = 0.$$

For the third term of TS we use $-(09)(26)(35)$, then

$$T\Gamma(\overline{14}, \overline{35}, \overline{08}, 27, 69) = T\Gamma(\overline{14}, \overline{35}, \overline{98}, 27, 60) = -\tfrac{1}{2} T\Gamma(\overline{14}, \overline{35}, \overline{09}, 27, 68) = 0$$

owing to $-(09)(76)(34)(51)(28)$. By $-(09)(26)(35)$ again,

$$T\Gamma(\overline{14}, \overline{35}, \overline{27}, 08, 69) = T\Gamma(\overline{14}, \overline{35}, \overline{67}, 08, 29) = -\tfrac{1}{2} T\Gamma(\overline{14}, \overline{35}, \overline{26}, 08, 79) = 0$$

by $(14)(35)(26)(78)$. And lastly by $-(09)(14)(78)$,

$$T\Gamma(\overline{14}, \overline{35}, \overline{69}, 08, 27) = T\Gamma(\overline{14}, \overline{35}, \overline{60}, 98, 27) = -\tfrac{1}{2} T\Gamma(\overline{14}, \overline{35}, \overline{09}, 68, 27) = 0$$

by $-(09)(76)(34)(51)(28)$.

Hence $T_{2,2,2,1,1,1,1} S = 0$.

Let us consider now $T_{2,2,2,2,2} S$. Looking at the expression obtained for $T_{2,2,2,2,2} I_1$, the second term will now be

$$T\Gamma(\overline{16}, \overline{34}, \overline{27}, \overline{05}, \overline{89}) = T\Gamma[-(\overline{16}, \overline{34}, \overline{20}, \overline{57}, \overline{89}) - (\overline{16}, \overline{34}, \overline{25}, \overline{07}, \overline{89})];$$

the last term written down disappears because of $-(25)(36)(14)$.

Also owing to $-(09)(76)(34)(51)(28)$

$$0 = T\Gamma(\overline{16}, \overline{34}, \overline{20}, \overline{57}, \overline{89}) = T\Gamma(\overline{16}, \overline{34}, \overline{08}, \overline{29}, \overline{75}) = T\Gamma(\overline{14}, \overline{35}, \overline{08}, \overline{29}, \overline{76}).$$

Finally $-(05)(81)(63)(49)(27)$ is an operation of Γ (as we may see from the fact that

0
7 3 9 1
8 2
5
4 6

is a star of S, and this operation inverts it):

hence $$T\Gamma(\overline{14}, \overline{35}, \overline{27}, \overline{06}, \overline{89}) = 0.$$

Thus $$T_{2, 2, 2, 2, 2}S = 0.$$

§12. Consider $T_{\alpha_1}, \ldots$ where $\alpha_1 = 3$.

Here TI_7 is zero, also TI_5 unless $\alpha_2 = 3$.

Take first $T_{3, 3, 3, 1}$. Then TI_6 vanishes.

Let $I_5 = (\overline{abc}, \overline{def}, ghij)$.

If P does not contain {abc}, we can shew that

$$NP\,\{abc\} = \Sigma NsP'\,\{abc\},$$

where s is a substitution with a numerical coefficient, and P' contains {abc}. And in fact

$$NP\,(\overline{abc}, \overline{def}, ghij) = \Sigma Ns'P''\,(\overline{abc}, \overline{def}, ghij),$$

where P'' contains {abc} {def}. In our case it contains a third positive symmetric group of degree 3, of the letters g h i j. Now I_5 changes sign with the interchange of g and h, or of i and j; hence

$$P''I_5 = 0, \text{ and therefore } T_{3, 3, 3, 1}I_5 = 0.$$

Next take $T_{3, 3, 1, 1, 1, 1}$. As before we need only discuss

$$NP\,(\overline{abc}, \overline{def}, ghij), \text{ where } P = \{abc\}\,\{def\}.$$

Here $$N = \{\alpha_1\,\beta_1\,\mathrm{ghij}\}'\,\{\alpha_2\,\beta_2\}'\,\{\alpha_3\,\beta_3\}',$$

where α_1, α_2, α_3 are the letters a, b, c in some order, and β_1, β_2, β_3 are the letters d, e, f in some order. Then

$$NP(\alpha_1\,\beta_1)(\alpha_2\,\beta_2)(\alpha_3\,\beta_3)(\mathrm{gi})(\mathrm{hj}) = N(\alpha_1\,\beta_1)(\alpha_2\,\beta_2)(\alpha_3\,\beta_3)(\mathrm{gi})(\mathrm{hj})\,P = -NP.$$

But $$(\alpha_1\,\beta_1)(\alpha_2\,\beta_2)(\alpha_3\,\beta_3)(\mathrm{gi})(\mathrm{hj})(\overline{abc}, \overline{def}, ghij) = (\overline{def}, \overline{abc}, ijgh) = (\overline{abc}, \overline{def}, ghij).$$

Hence $$NPI_5 = -NPI_5 = 0, \text{ and } T_{3, 3, 1, 1, 1, 1}I_5 = 0.$$

Similarly we can shew that

$$T_{3, 3, 2, 2}I_5 = 0.$$

We are left only with $T_{3, 3, 2, 1, 1}I_5$, which is not zero.

The equations for I_5, viz. (5_1), (5_2), (5_3)—which alone are independent—give us nothing fresh here. The first two obviously cannot, as each is of the form $\Sigma G_4 I_5 = \Sigma I_7$.

We may prove as above that

$$T_{3, 3, 2, 1, 1}[(\overline{abc}, \overline{def}, ghij) + (\overline{abc}, \overline{def}, ijgh)] = 0.$$

Consider then $T_{3, 3, 2, 1, 1}\Gamma I_5$.

Since Γ is triply transitive we may choose the position of three digits in I_5 at will. Thus we may suppose I_5 to be (714, ..., ...).

Now the digit *8* is uniquely related to *714* in Γ being the digit collinear with this triad in one star of the set.

Also $TG_4 = 0$, where G_4 is any positive symmetric group of four digits; thus we may express TI_5 in the form

$$T\Sigma(\overline{714}, \overline{8\ldots}, \ldots).$$

There is a subgroup of Γ which just permutes the digits *7, 1, 4* amongst themselves, leaves *8* unchanged, and permutes the other 6 digits. It is of order 12; and leaving out the operations affecting *714* which do not alter our I_5, it is

$$1, (025)(963), (052)(936), (06)(29)(53), (03)(26)(59), (09)(23)(56),$$
$$-(09)(26)(53), -(02)(96), -(25)(63), -(50)(39), -(065923), -(032956) \quad \ldots(g_1).$$

The group is best expressed with reference to the tableau $\begin{matrix}0 & 2 & 5\\ 9 & 6 & 3\end{matrix}$, the operations all keeping the horizontal triads and the vertical pairs intact. We have then to consider the following distinct possibilities for the second triad of I_5: 809, 806, 802, and these alone.

Owing to the operation *(06)(29)(53)* of the subgroup of Γ, and the fact that

$$T(\overline{abc}, \overline{def}, ghij) = -T(\overline{abc}, \overline{def}, ijgh),$$

we see that the only form of the second of these possibilities which is not at once zero is

$$(\overline{714}, \overline{806}, 2953).$$

Now *(0184)(7396)* is an operation of Γ. Hence

$$T\Gamma(\overline{714}, \overline{806}, 2953) = T\Gamma(\overline{380}, \overline{417}, 2659) = T\Gamma(\overline{714}, \overline{803}, 5926)$$
$$= T\Gamma[-(36)(25)](\overline{714}, \overline{803}, 5926) = -T\Gamma(\overline{714}, \overline{806}, 2953) = 0.$$

Thus all the forms of the second possibility are zero.

Also it is easy to see that $T\Gamma(\overline{714}, \overline{809}, 2653) = 0$, and

$$T\Gamma(\overline{714}, \overline{809}, 2563) = 0.$$

The first possibility leaves us with one form

$$T\Gamma(\overline{714}, \overline{809}, 2356).$$

The operations $-(78)(23)(56)$, $(18)(74)(26)(09)$, $(48)(71)(09)(53)$ belong to Γ. Then the equation

$$T\Gamma\{7148\}(\overline{714}, \overline{8\ldots}, \ldots) = 0$$

becomes $$T\Gamma[1-(09)-(26)-(53)](\overline{714}, \overline{8\ldots}, \ldots) = 0.$$

Taking I_6 to be here $$(\overline{714}, \overline{802}, 6953)$$

the second and third terms belong to the second of the three possibilities considered and are therefore zero: hence

$$T\Gamma(\overline{714}, \overline{802}, 6953) = 0$$

and $$T\Gamma(\overline{714}, \overline{802}, 6593) = T\Gamma(\overline{714}, \overline{802}, 6395).$$

This last equation is obtainable otherwise by means of the subgroup of Γ given above. Finally the equation

$$T\Gamma\{0925\}(\overline{714}, \overline{809}, 2356) = 0$$

gives us $$T\Gamma[(\overline{714}, \overline{809}, 2356) + 4(\overline{714}, \overline{802}, 9356)] = 0.$$

We can get no further. Thus we can express all these forms in terms of one which we may take to be

$$T\Gamma(\overline{714}, \overline{809}, 2356),$$

which is not zero.

And $T\Gamma I_1$ is a numerical multiple of this.

§ 13. We have next to consider $T_{\alpha_1}, \ldots$ where $\alpha_1 = 4$.

It is easy to see that $$T_{4,3}, \ldots I_7 = 0,$$

and in fact that every $T_4, \ldots I_7$ is zero except $T_{4,2,2,2} I_7$.

In dealing with the cases of $\alpha_1 = 4$, it is convenient to classify the forms ΓNPI_6 and ΓNPI_7 according to the digits in the group of degree 4 in P. Since Γ is triply transitive three of these digits may be chosen at will. We take them to be *7, 1, 4*. The remaining digit may be *8*, which occupies a unique position, in that it is collinear with *714*; or else it may be one of the remaining six digits, and since the group we have called (g_1) is transitive in these six digits it will be sufficient to select one of them; we shall then take the first group of P to be either $\{7148\}$ or $\{7140\}$.

Let us write $\bar{N} = \Sigma N$; this is symmetric and hence commutative with every substitution. Consider the group of operations σ, which do not affect the first group of P and for which

$$\Gamma\bar{N}\sigma PI = \Gamma\bar{N}PI.$$

When the first group of P is $\{7148\}$, this group will be called (g_2); it is of order 48, contains (g_1) as a subgroup, and also the operations $-(09)$, $-(26)$, $-(53)$. When the first group of P is $\{7140\}$, the group (g_3) of σ is of order 8, viz.:

$$1,\ (8295),\ (8592),\ (89)(25),\ -(63)(25),\ -(63)(89),\ -(63)(82)(95),\ -(63)(85)(92).$$

The two cases are not entirely distinct, indeed the first can be expressed in terms of the second by means of the equation

$$\Gamma N[1+(07)+(01)+(04)+(08)]\{7148\}\ldots I = 0.$$

Hence when we can shew that $\Gamma\bar{N}PI$ vanishes in the case of the first group of P being $\{7140\}$ we shall know that

$$\Gamma TI = 0.$$

Now evidently $$T_{4,1,1,1,1,1,1}I_5 = 0.$$

Consider $T_{4,2,1,1,1,1}$. Here TI_6 is not zero.

But $$T\Gamma I_5 = \Gamma TI_5 = \Sigma\Gamma\bar{N}PI_5.$$

Let $P = \{abcd\}\{ef\}$. Then

$$PI_6 = \Sigma P\lambda(\overline{abc}, \overline{da\beta}, \gamma\delta\epsilon\zeta) + \Sigma P\lambda(\overline{aba}, \overline{cd\beta}, \gamma\delta\epsilon\zeta),$$

where λ is numerical. In the latter case (because $NPG_6 = 0$)

$$NP\{abacd\}(\overline{aba}, \overline{cd\beta}, \gamma\delta\epsilon\zeta) = 0,$$

i.e. $$NP[1+(\alpha a)+(\alpha b)+(\alpha c)+(\alpha d)](\overline{ab\alpha}, \overline{cd\beta}, \gamma\delta\epsilon\zeta) = 0,$$

or $$NP[3+(\alpha c)+(\alpha d)](\overline{ab\alpha}, \overline{cd\beta}, \gamma\delta\epsilon\zeta) = 0.$$

This expresses the second set of terms in terms of the first.

Again if G_3 is a positive symmetric group of three letters from e, f, g, h, i, j,

$$NPG_3 = 0.$$

Hence we may deduce

$$NPI_6 = \Sigma NP\lambda(\overline{abc}, \overline{def}, ghij)$$

and the interchange of any two of g, h, i, j merely changes the sign.

Consider then $$\Gamma\bar{N}\{7140\}\{\alpha\beta\}(\overline{714}, \overline{0\alpha\beta}, \gamma\delta\epsilon\zeta),$$

where $\alpha, \beta, \gamma, \delta, \epsilon, \zeta$ are the digits 2, 3, 5, 6, 8, 9 in some order. These forms are unaltered by the operations of (g_3). It follows that the form is zero when $\alpha\beta$ is any one of the pairs

(63), (82), (85), (92), (95) (from the last two operations of g_3),

or of the pairs (89) or (52). Now

$$\Gamma\bar{N}\{\alpha\beta\gamma\}PI_6 = 0,$$

thus

$$\Gamma\bar{N}\{7140\}[\{83\}(\overline{714}, \overline{083}, 2569) + \{86\}(\overline{714}, \overline{086}, 2539) + \{63\}(\overline{714}, \overline{063}, 2589)] = 0.$$

The third term is known to be zero, and the first two are equal because (g_3) contains $-(63)(25)$, and hence they are zero. In this way we see that every form is zero for which the first group of P is $\{7140\}$.

Hence $$\Gamma T_{4,2,1,1,1,1} I_5 = 0.$$

As regards $T_{4,2,2,1,1} I_5$ we may proceed as before to shew that
$$TI_5 = \Sigma\lambda\, N\{abcd\}\{ef\}\{gh\}(\overline{abc}, \overline{def}, gihj).$$

But now $$N\{ef\}\{gh\}\{abcd\}[1+(ea)+(eb)+(ec)+(ed)] = 0.$$

Hence we may write
$$TI_5 = \Sigma\lambda N\{abcd\}\{ef\}\{gh\}(\overline{abe}, \overline{cdf}, gihj).$$

Here N must contain $\{ij\}'$ and hence
$$NP(ac)(bd)(ef)(gh)(ij) = -NP,$$
whence $$T_{4,2,2,1,1} I_5 = 0.$$

Consider now $$T_{4,2,2,2};$$
here neither TI_5 nor TI_7 is zero.

Let $$P = \{abcd\}\{ef\}\{gh\}\{ij\},$$
then in equation (5_2) let the α's refer to the first group of P, the β's to the second group, the γs to the third and $a = i$, $b = j$.

Then since $NP\{ief\} = 0$, we have
$$NP[16(\overline{abc}, \overline{dij}, egfh) + 2(abci)^2(\overline{dj}, \overline{ef}, \overline{gh}) + (abcd)^2(\overline{ij}, \overline{ef}, \overline{gh})] = 0,$$
for N permutes the three groups of P of degree 2 in all possible ways. Thus
$$T_{4,2,2,2} I_5 = \Sigma T_{4,2,2,2} I_7$$
or this form is reducible.

Now consider ΓTI_7. As before we may shew that it is equal to
$$\Sigma\Gamma\lambda\overline{N}\{abcd\}\{ef\}\{gh\}\{ij\}(abcd)^2(\overline{ef}, \overline{gh}, \overline{ij}).$$

Consider then the case when the first group of P is $\{7140\}$. Looking at (g_3) we see that the form is zero if P also contains any one of the groups $\{82\}$, $\{85\}$, $\{29\}$, $\{59\}$, $\{36\}$, $\{89\}$, $\{25\}$. Thus for a non-zero form, no pair of the digits 8, 2, 5, 9 can be associated in P. But they cannot all be separated. Hence
$$\Gamma T_{4,2,2,2} I_7 = 0 = \Gamma T_{4,2,2,2} I_5.$$

The next case is $T_{4,3,1,1,1}$. And here TI_5 is not zero. The equations (5_2), (5_3), (5_4) give us nothing beyond what we know otherwise.

We may associate a unique form of I_5 with a given P. For (since $NPG_5 = 0$)
$$N\{\alpha_1\alpha_2\alpha_3\alpha_4\}\{\beta_1\beta_2\beta_3\}[2(\overline{\alpha_1\alpha_2\alpha_3}, \overline{\alpha_4\beta_1\beta_2}, \beta_3\, abc) + 3(\overline{\alpha_1\alpha_2\beta_1}, \overline{\alpha_3\alpha_4\beta_2}, \beta_3\, abc)] = 0,$$
and $$N\{\alpha_1\alpha_2\alpha_3\alpha_4\}\{\beta_1\beta_2\beta_3\}[2(\overline{\alpha_1\beta_1\beta_2}, \overline{\alpha_2\alpha_3\alpha_4}, \beta_3\, abc) + 3(\overline{\alpha_1\alpha_2\beta_1}, \overline{\alpha_3\alpha_4\beta_2}, \beta_3\, abc)] = 0.$$

Hence $$N\{\alpha_1\alpha_2\alpha_3\alpha_4\}\{\beta_1\beta_2\beta_3\}(\overline{\alpha_1\alpha_2\alpha_3}, \overline{\alpha_4\beta_1\beta_2}, \beta_3\, abc)$$
$$= NP(\overline{\alpha_1\beta_1\beta_2}, \overline{\alpha_2\alpha_3\alpha_4}, \beta_3\, abc) = -\tfrac{1}{3}NP(\overline{\alpha_2\alpha_3\alpha_4}, \overline{\beta_1\beta_2\beta_3}, bc\alpha_1 a).$$

We may therefore write $$TI_5 = \Sigma\lambda NPI_5,$$
where in this last sum P and I_5 are related as in the last form written down. Consider then $\Gamma\overline{N}PI_5$, where the first group of P is $\{7140\}$. To fix ideas the form may be represented by $\{\alpha\beta\gamma\}\,\delta\epsilon\zeta$, where the letters are the digits *2*, *3*, *5*, *6*, *8*, *9* in some order. Transpositions of α, β, γ leave the form unaltered, and of δ, ϵ, ζ change the sign, and the operation $\{\alpha\beta\gamma\delta\}$ produces zero. If *3* and *6* are outside the G_3 bracket, the operation (8295) of (g_3) shews that all the other digits

may be permuted without changing the value, and then the operation $\{8295\}$ shews that the form is zero. If both $8, 9$ or both $2, 5$ are inside the bracket the operation $(89)(25)$ of (g_3) shews that the form is zero. Thus we have a form $\{823\}\,956$; its sign is changed by (89) or (25); it is unaltered by (36). It is equal to $\{836\}\,925 = \{236\}\,985 = \ldots$. We have therefore just one form which is not zero.

The equations derived from $NPG_5 = 0$ enable us to express any form where the first group of P is collinear in terms of that just found. Or vice versa—they enable us to use the collinear form as the standard case. But they do not give us a fresh reduction.

In the case of $T_{4,3,2,1}\, I_5$ we find as above

$$N\{\alpha_1\alpha_2\alpha_3\alpha_4\}\{\beta_1\beta_2\beta_3\}\{\gamma_1\gamma_2\}[(\overline{\alpha_1\alpha_2\alpha_3}, \overline{\alpha_4\beta_1\beta_2}, \beta_3\gamma_1 a\gamma_2) + \tfrac{1}{3}(\overline{\alpha_1\alpha_2\alpha_3}, \overline{\beta_1\beta_2\beta_3}, a\gamma_2\alpha_4\gamma_1)] = 0.$$

Let us write $P = \{\alpha_1\alpha_2\alpha_3\alpha_4\}\{\beta_1\beta_2\beta_3\}\{\gamma_1\gamma_2\}$; then putting $a = \beta$ in (5_2) we obtain, since

$$TI_7 = 0,$$

$$NP[2(\overline{\alpha_1\alpha_2\alpha_3}, \overline{\alpha_4\beta_1 b}, \beta_2\gamma_1\beta_3\gamma_2) + (\overline{\alpha_1\alpha_2\alpha_3}, \overline{\alpha_4\beta_1\beta_2}, b\gamma_1\beta_3\gamma_2)$$
$$+ (\overline{\alpha_1\alpha_2\alpha_3}, \overline{\beta_1\beta_2\beta_3}, b\gamma_1\alpha_4\gamma_2) + (\overline{\alpha_1\alpha_2\alpha_3}, \overline{\beta_1\gamma_1\gamma_2}, b\beta_2\alpha_4\beta_3)] = 0.$$

Now $[2(\overline{\alpha_1\alpha_2\alpha_3}, \overline{\alpha_4\beta_1 b}, \beta_2\gamma_1\beta_3\gamma_2) + (\overline{\alpha_1\alpha_2\alpha_3}, \overline{\alpha_4\beta_1\beta_3}, \beta_2\gamma_1 b\gamma_2) + (\overline{\alpha_1\alpha_2\alpha_3}, \overline{b\beta_1\beta_3}, \beta_2\gamma_1\alpha_4\gamma_2)] = 0.$

Also as we have just seen (in the case of $T_{4,3,1,1,1}$, and the argument applies here)

$$NP[(\overline{\alpha_1\alpha_2\alpha_3}, \overline{\alpha_4\beta_1\beta_3}, \beta_2\gamma_1 b\gamma_2) - (\overline{\alpha_1\alpha_2\alpha_3}, \overline{\alpha_4\beta_1\beta_3}, b\gamma_2\beta_2\gamma_1)] = 0,$$

and

$$NP(\overline{\alpha_1\alpha_2\alpha_3}, \overline{\beta_1\gamma_1\gamma_2}, b\beta_2\alpha_4\beta_3) = -NP(\overline{\alpha_1\alpha_2\alpha_3}, \beta_1\beta_2\gamma_1, b\gamma_2\alpha_4\beta_3)$$
$$= \tfrac{1}{3} NP(\overline{\alpha_1\alpha_2\alpha_3}, \overline{\beta_1\beta_2\beta_3}, b\gamma_1\alpha_4\gamma_2),$$

and

$$NP(\overline{\alpha_1\alpha_2\alpha_3}, \overline{b\beta_1\beta_3}, \beta_2\gamma_1\alpha_4\gamma_2) = -\tfrac{1}{3} NP(\overline{\alpha_1\alpha_2\alpha_3}, \overline{\beta_1\beta_2\beta_3}, b\gamma_1\alpha_4\gamma_2).$$

Thus $NP(\overline{\alpha_1\alpha_2\alpha_3}, \overline{\beta_1\beta_2\beta_3}, b\gamma_1\alpha_4\gamma_2) = 0$; and hence

$$T_{4,3,2,1}\, I_5 = 0.$$

That $T_{4,3,3}\, I_5 = 0$ is almost self-evident from (5_1). If $P = \{\alpha_1\alpha_2\alpha_3\alpha_4\}\{\beta_1\beta_2\beta_3\}\{\gamma_1\gamma_2\gamma_3\}$, we may at once assign three α's to the first triad of I_5, and either three β's or one α and two β's to the second triad. Then three out of the last four letters will be γ. But as I_5 changes sign by the interchange of the ultimate or of the penultimate pair of its letters the form is zero.

For $T_{4,4,1,1}$ we have only to consider

$$N\{\alpha_1\alpha_2\alpha_3\alpha_4\}\{\beta_1\beta_2\beta_3\beta_4\}(\overline{\alpha_1\alpha_2\alpha_3}, \overline{\beta_1\beta_2\beta_3}, \beta_4 a\alpha_4 b),$$

and to notice that in this case $NP(\alpha_1\beta_1)(\alpha_2\beta_2)(\alpha_3\beta_3)(\alpha_4\beta_4)(ab) = -NP$, in order to see that

$$T_{4,4,1,1}\, I_5 = 0.$$

As regards $T_{4,4,2}$, TI_5 is not zero. Nor do the equations (5_2), (5_3), (5_4) yield any fresh information. We consider then $\Gamma\bar{N}PI_5$ where the first group of P is $\{7140\}$. Then looking at (g_3) we see that the form is zero when the last group of P is $\{36\}$, $\{82\}$, $\{59\}$, $\{85\}$, $\{29\}$, $\{89\}$ or $\{25\}$.

If $P = \{7140\}\{2596\}\{83\}$, we notice that

$$\Gamma\bar{N}[1 + (82) + (85) + (89) + (86)]\{7140\}\{2596\}\{83\}\, I_5 = 0.$$

The last term of this equation is zero, and the first four terms are all equal, owing to the operation (8295) of (g_3).

Hence $$\Gamma T_{4,4,2}\, I_5 = 0.$$

(Obviously here

$$TI_5 = \Sigma\lambda N\{\alpha_1\alpha_2\alpha_3\alpha_4\}\{\beta_1\beta_2\beta_3\beta_4\}\{\gamma_1\gamma_2\}(\overline{\alpha_1\alpha_2\alpha_3}, \overline{\beta_1\beta_2\beta_3}, \beta_4\gamma_1\alpha_4\gamma_2).)$$

§14. We have now to consider $T\alpha_1, \ldots$ when $\alpha_1 = 5$.

Here $T_5, \ldots I_7 = 0$, except in the case of $T_{5,2,2,1}\, I_7$. As regards $T_{5,1,1,1,1,1}$ we have only to consider

$$N\{\alpha_1\alpha_2\alpha_3\alpha_4\alpha_5\}\,(\overline{\alpha_1\alpha_2\alpha_3},\ \overline{\alpha_4\alpha_5 a},\ bcde).$$

Transpositions of the letters a, b, c, d, e change the sign. Since $NPG_6 = 0$

$$\begin{aligned} NP(\overline{\alpha_1\alpha_2\alpha_3},\ \overline{\alpha_4\alpha_5 a},\ bcde) &= -NP(\overline{\alpha_1\alpha_2 a},\ \overline{\alpha_3\alpha_4\alpha_5},\ bcde) \\ &= -NP(\overline{\alpha_3\alpha_4\alpha_5},\ \overline{\alpha_1\alpha_2 a},\ debc) = 0. \end{aligned}$$

And $$T_{5,1,1,1,1,1}\, I_5 = 0.$$

For $T_{5,2,1,1,1}$ we have at first sight to consider the possibilities for $P = \{\overline{\alpha_1\alpha_2\alpha_3\alpha_4\alpha_5}\}\{\overline{\beta_1\beta_2}\}$,

$$NP(\overline{\alpha_1\alpha_2\alpha_3},\ \overline{\alpha_4\alpha_5\beta_1},\ \beta_2 abc),\ NP(\overline{\alpha_1\alpha_2\alpha_3},\ \overline{\alpha_4\alpha_5\beta_1},\ ab\beta_2 c),\ NP(\overline{\alpha_1\alpha_2\alpha_3},\ \overline{\alpha_4\alpha_5 a},\ \beta_1 b\beta_2 c).$$

But proceeding as in the last case, we have

$$\begin{aligned} NP(\overline{\alpha_1\alpha_2\alpha_3},\ \overline{\alpha_4\alpha_5\beta_1},\ \beta_2 abc) &= -NP(\overline{\alpha_1\alpha_2\alpha_3},\ \overline{\alpha_4\alpha_5\beta_1},\ bc\beta_2 a) \\ &= NP(\overline{\alpha_1\alpha_2\alpha_3},\ \overline{\alpha_4\alpha_5\beta_1},\ ba\beta_2 c). \end{aligned}$$

And from the equation $NP\{\beta_1\beta_2 b\}\, I_5 = 0$, we obtain

$$NP(\overline{\alpha_1\alpha_2\alpha_3},\ \overline{\alpha_4\alpha_5 b},\ \beta_1 a\beta_2 c) + 2\,NP(\overline{\alpha_1\alpha_2\alpha_3},\ \overline{\alpha_4\alpha_5\beta_1},\ ba\beta_2 c) = 0.$$

Now in (5_2) put one γ equal to a and the other to c $\left(\text{we may polarize } a\dfrac{\partial}{\partial\gamma}\right)$ and remembering that if two of the letters a, b, c are in the second triad of I_5 the form is zero, we have

$$NP[(\overline{\alpha_1\alpha_2\alpha_3},\ \overline{\alpha_4\beta_1\beta_2},\ bca\alpha_5) + 2\,(\overline{\alpha_1\alpha_2\alpha_3},\ \overline{\alpha_4 c\alpha_5},\ b\beta_1 a\beta_2)] = 0,$$

i.e. $$NP(\overline{\alpha_1\alpha_2\alpha_3},\ \overline{\alpha_4\alpha_5\beta_1},\ bca\beta_2) + 2\,(\overline{\alpha_1\alpha_2\alpha_3},\ \overline{\alpha_4\alpha_5 c},\ b\beta_1 a\beta_2) = 0,$$

which with the equation already obtained shews that each term separately is zero.

Hence $$T_{5,2,1,1,1}\, I_5 = 0.$$

In the case of $T_{5,2,2,1}$, as we have already remarked, TI_7 is not zero. When

$$P = \{\alpha_1\alpha_2\alpha_3\alpha_4\alpha_5\}\{\beta_1\beta_2\}\{\gamma_1\gamma_2\},$$

we have to consider the possibilities:

$$NP(\overline{\alpha_1\alpha_2\alpha_3},\ \overline{\alpha_4\alpha_5 a},\ \beta_1\gamma_1\beta_2\gamma_2),\ NP(\overline{\alpha_1\alpha_2\alpha_3},\ \overline{\alpha_4\alpha_5\beta_1},\ a\gamma_1\beta_2\gamma_2),\ NP(\overline{\alpha_1\alpha_2\alpha_3},\ \overline{\alpha_4\alpha_5\beta_1},\ \beta_2\gamma_1 a\gamma_2).$$

By the methods used in the last two cases, we find that the first of these is zero; and the sum of the other two is also zero. Put $a = \alpha$ in (5_2), then

$$\begin{aligned} NP[16\,(\overline{\alpha_1\alpha_2\alpha_3},\ \overline{\alpha_4\beta_1\beta_2},\ b\gamma_1\alpha_5\gamma_2) &+ 16\,(\overline{\alpha_1\alpha_2\alpha_3},\ \overline{\alpha_4\gamma_1\gamma_2},\ b\beta_1\alpha_5\beta_2) \\ &+ 6\,(\alpha_1\alpha_2\alpha_3\alpha_4)^2\,(\overline{\alpha_5 b},\ \overline{\beta_1\beta_2},\ \overline{\gamma_1\gamma_2})] = 0, \end{aligned}$$

and since $NP(\beta_1\gamma_1)(\beta_2\gamma_2) = NP$, this may be written

$$NP(\overline{\alpha_1\alpha_2\alpha_3},\ \overline{\alpha_4\alpha_5\beta_1},\ b\gamma_1\beta_2\gamma_2) = \tfrac{3}{16}\,NP(\alpha_1\alpha_2\alpha_3\alpha_4)^2\,(\overline{\alpha_5 b},\ \overline{\beta_1\beta_2},\ \overline{\gamma_1\gamma_2}).$$

Thus we can express TI_8 in terms of TI_7. We will now consider ΓTI_7, and for this purpose we must examine the nature of the first group of P in relation to Γ. Three digits *714* may be selected at will. If the group also contains *8*, then it has a collinear set. The remaining digit may then be chosen arbitrarily for there is a subgroup of Γ, which merely permutes *7148* and is transitive in the remaining digits. All collinear sets can be transformed into each other by Γ, and hence all sets of five digits, such that four are collinear, can be transformed into each other by Γ.

If the group does not contain *8*, we may suppose it to contain *0*. The group will still be of the collinear type if the last symbol is *2*, *9* or *5*. The remaining cases, viz. 71406, 71403, are digits which are placed at the angles of a pentagon in one of the stars. Thus we have two cases of five digits, which may be referred to as the collinear quintet and the pentagon quintet respectively, Γ permutes each set of quintets transitively. Moreover when the five digits are taken away, the remaining five form a quintet of the same kind.

Consider then ΓNPI_7. And let us first suppose that the digits of the first group of P form a collinear quintet.

We may take them to be *71480*. Then the form will be unchanged by the operations of a group (g_5) of order 8: i.e. by

$$1, (2365), (2563), (26)(35), -(26), -(35), -(25)(36), -(23)(56).$$

This leaves the digit *9* untouched. We see at once that the form is zero if *9* is not included in the groups of P of order 2. We may for the moment define the form by these groups and that of order 1. Then we find

$$\{93\}\{56\}2 = -\{95\}\{36\}2 = -\{93\}\{52\}6 = \{95\}\{32\}6$$
$$= -\{92\}\{56\}3 = \{96\}\{52\}3 = \{92\}\{36\}5 = -\{96\}\{32\}5.$$

And there is just one form with a collinear quintet.

The equation

$$\Gamma\overline{N}[1+(97)+(91)+(94)+(98)+(90)]\{71480\}\{\quad\}\{\quad\}I_7 = 0$$

involves only collinear quintets, but is found to yield nothing. The other similar equations introduce quintets of both kinds.

When the digits of the first group of P form a pentagon quintet, we may take this to be 07358. The form will be unchanged by the operations of a group (g_6) of order 20, viz. 1, *(12964)* and its powers, *(1246)*, *(1642)*, *(14)(26)*, and the operations obtained from these by rotation. This group is doubly transitive.

Expressing the forms as before we see that there is a set of five equal of the type $\{14\}\{26\}9$, and that the remaining 10 are equal. Also owing to the equation $NG_3P = 0$, when none of the digits of G_3 appear in the first group of P, a form of the second type is equal to $-\frac{1}{2}$ times a form of the first.

Thus again there is but one form.

It remains to consider the equation

$$\Gamma\overline{N}[1+(90)+(97)+(93)+(95)+(98)]\{07358\}\{\quad\}\{\quad\}I_7 = 0;$$

from symmetry it is at once apparent (by looking at the star) that we need only consider this one equation of the type.

The first two terms belong to the pentagon quintets, the other four to the collinear quintets.

Look at the first pair of terms. Γ contains the operation $-(09)(26)(35)$. Hence the sum of the first two terms is

$$\Gamma\overline{N}[1-(26)]\{07358\}\{\quad\}\{\quad\}I_7.$$

For the other terms the operations of Γ,

$$(01349758)(26), (08745)(13629), -(013)(745698), -(07)(19)(43)(26)(58)$$

transform them respectively into forms which have $\{71480\}$ for the first group of P. Thus the sum of the last four terms is

$$\Gamma\overline{N}[(495)(13)+(54)(13)(962)-(31)(54)(96)-(519)(43)(26)]\{07358\}\{\quad\}\{\quad\}I_7$$
$$= \Gamma\overline{N}\{71480\}[(495)(13)-(54)(13)(92)-(31)(54)(96)+(519)(43)]\{\quad\}\{\quad\}I_7$$

after using (g_5).

So far we have left the groups of order 2 of P arbitrary.

Let us choose them to be $\{14\}\{29\}6$. Then the sum of the first two terms is zero. The others (defining them as before by the smaller groups of P) are

$$\{39\}\{25\}6-\{35\}\{92\}6-\{35\}\{26\}9+\{93\}\{25\}6;$$

the second and third terms are zero, and the first and last are equal. Hence all forms are zero when the first group of P is of the collinear type. Hence the sum of the first two terms of our equation is always zero.

Let the first term be $\{12\}\{49\}6$, then

$$\{12\}\{49\}6-\{16\}\{49\}2=0,$$

but the second of these we know to be equal to $-\frac{1}{2}$ times the first. Hence all the forms are zero and

$$\Gamma T_{5,2,2,1}I_5=\Gamma T_{5,2,2,1}I_7=0.$$

The next case is $T_{5,3,1,1}$ and here TI_5 is not zero.

Consider first the case when the first group of P in $\Gamma\overline{N}PI_5$ is a pentagon quintet: $\{07358\}$. Then defining the forms again by the smaller groups of P and using (g_6) we see that

$$\{142\}69=\{621\}49=-\{142\}96=-\{142\}69=0$$

and all forms in this case are zero.

In the collinear case, taking the first group of P to be $\{71480\}$, we see that if both of 2, 6 or both of 3, 5 are either inside or outside the second group of P the form is zero. And for the rest

$$\{239\}5.6=-\{639\}5.2=-\{259\}3.6=\{659\}3.2.$$

We now turn to the equation used for $T_{5,2,2,1}$ obtained from $\overline{N}G_6P=0$. We already know that the first two terms are always zero. We start with $P=\{07358\}\{142\}6.9$ and find

$$\{392\}6.5-\{359\}6.2-\{352\}9.6+\{932\}6.5=0,$$

giving us $2\{392\}6.5=0$.

Hence $$\Gamma T_{5,3,1,1}I_5=0.$$

For $T_{5,3,2}$ we have only to consider

$$N\{\alpha_1\alpha_2\alpha_3\alpha_4\alpha_5\}\{\beta_1\beta_2\beta_3\}\{\gamma_1\gamma_2\}(\overline{\alpha_1\alpha_2\alpha_3},\overline{\alpha_4\alpha_5\beta_1},\beta_2\gamma_1\beta_3\gamma_2)=-NP(\overline{\alpha_1\alpha_2\beta_1},\overline{\alpha_3\alpha_4\alpha_5},\beta_2\gamma_1\beta_3\gamma_2)=0,$$

to see that $$T_{5,3,2}I_5=0.$$

It is obvious at once that $$T_{5,4,1}I_5=0=T_{5,5}I_5.$$

§ 15. Consider now $T_{\alpha_1},\ldots$ when $\alpha_1=6$.

There is here only one case when $T_6,\ldots I_7$ is not zero. That is $T_{6,2,2}I_7$.

As to $T_{6,1,1,1,1}$, it can be shewn that TI_5 is not zero.

For the discussion of ΓNPI_5 we may regard it as the reciprocal of the case when the first group of P is of degree 4. Thus we mark two cases:

(i) The four digits outside the group of degree 6 are collinear. Let these be 7148. Now Γ has an operation $-(09)(26)(35)$; and hence the form must be zero. This is true whenever P has a group of degree 6, whatever the other groups may be.

(ii) The four digits outside the first group of P are 7140. Then Γ has an operation $(0471)(8592)$ which changes the sign of the form, which is therefore zero. Hence

$$\Gamma T_{6,1,1,1,1}I_5=0.$$

Next $$T_{6,2,1,1}I_5=0,$$

for we have to consider

$$N\{\underline{\alpha_1\alpha_2\alpha_3\alpha_4\alpha_5\alpha_6}\}\{\underline{\beta_1\beta_2}\}(\overline{\alpha_1\alpha_2\alpha_3},\overline{\alpha_4\alpha_5\alpha_6},\beta_1 a\beta_2 b).$$

And we have $$NP(\mathrm{ab})(\underline{\beta_1\beta_2})=-NP.$$

Lastly $T_{6,2,2}I_7$ is not zero.

We have from (5_2) $\quad 8(\overline{\alpha\alpha\alpha},\ \overline{\alpha\alpha\alpha},\ \overline{\beta\gamma\beta\gamma}) = -3(\alpha\alpha\alpha\alpha)^2(\overline{\alpha\alpha},\ \overline{\beta\beta},\ \overline{\gamma\gamma})$.

And hence $\quad TI_5 = \lambda TI_7$.

As to $\Gamma\overline{N}PI_7$ we have already seen that this is zero when the letters outside the first group of P are collinear. By the operation $-(14)(25)(36)$ of Γ we see that $\{14\}\{70\}$ is zero and also $\{14\}\{76\}$. By the operation $(0174)(8295)$ we see that $\{17\}\{40\} = \{01\}\{74\}$. And by the equation $NP\{017\} = 0$ we see that these are zero, and therefore every form in the second case.

Hence $\quad \Gamma T_{6,2,2}I_5 = 0 = \Gamma T_{6,2,2}I_7$.

There is no difficulty in seeing that

$$T_{\alpha_1,\alpha_2,\ldots}I_5 = 0 = T_{\alpha_1,\alpha_2,\ldots}I_7$$

when $\alpha_1 > 6$ or when $\alpha_1 = 6$ and $\alpha_2 > 2$.

SUMMARY.

§ 16. Summing up we have found twelve types of invariant. These are the determinant Δ, the forms

$$T_{2,2,2,1,1,1,1}I_3,\ T_{2,2,2,2,2}I_6,\ T_{3,3,2,1,1}I_5,\ T_{4,2,1,1,1,1}I_5,\ T_{4,3,1,1,1}I_5,\ T_{4,4,2}I_5,\ T_{5,3,1,1}I_5,\ T_{6,1,1,1,1}I_5$$

and the reducible types $\quad T_{4,2,2,2}I_7,\ T_{5,2,2,1}I_7,\ T_{6,2,2}I_7$.

But for an invariant ΓI we have only three types, viz. Δ and $\Gamma T_{3,3,2,1,1}I_5$, $\Gamma T_{4,3,1,1,1}I_5$.

Thus $\quad \Gamma I_1 = \lambda\Delta + \mu\Gamma T_{3,3,2,1,1}I_5 + \nu\Gamma T_{4,3,1,1,1}I_5$.

§ 17. To discuss the values of TI, or TS when $T = T_{\alpha_1,\alpha_2,\ldots}$ and $\alpha_1 > 2$, it will be advisable to obtain first the value of $T\Sigma\{abc\}$, where the Σ includes every positive symmetric group of degree 3. This is equal to $T\left[\binom{n}{3} + (n-2)\Sigma(ab) + \Sigma(abc)\right]$.

To obtain $NP.\Sigma(abc)$, we use a double suffix notation for the letters, in order to identify the group of P to which they belong; thus the groups of P will be supposed to be given by the rows of the tableau

$$\begin{array}{c} a_{1,1}\ a_{1,2}\ \ldots\ a_{1,\alpha_1} \\ a_{2,1}\ a_{2,2}\ \ldots\ a_{2,\alpha_2} \\ \ldots\ldots\ldots\ldots\ldots\ldots \end{array}$$

Then we have three kinds of substitutions (abc) which we may write

$$\text{(i) } (a_{r,\rho}\,a_{r,\sigma}\,a_{r,\tau}),\quad \text{(ii) } (a_{r,\rho}\,a_{r,\sigma}\,a_{s,\tau}),\quad \text{(iii) } (a_{r,\rho}\,a_{s,\sigma}\,a_{t,\tau}).$$

(i) $NP\Sigma(a_{r,\rho}\,a_{r,\sigma}\,a_{r,\tau}) = NP\left[2\Sigma\binom{\alpha_r}{3}\right]$.

(ii)
$$\begin{aligned} NP\Sigma(a_{r,\rho}\,a_{r,\sigma}\,a_{s,\tau}) &= NP\Sigma(a_{r,\rho}\,a_{r,\sigma})(a_{r,\rho}\,a_{r,\sigma}\,a_{s,\tau}) \\ &= NP[\Sigma(\alpha_r - 1 + \alpha_s - 1)(a_{r,\sigma}\,a_{s,\tau})] \\ &= NP[-\alpha_2(\alpha_1 + \alpha_2 - 2) - \alpha_3(\alpha_1 - 1 + \alpha_2 - 1 + 2\overline{\alpha_3 - 1}) \\ &\qquad - \alpha_4(\overline{\alpha_1 - 1} + \overline{\alpha_2 - 1} + \overline{\alpha_3 - 1} + 3\overline{\alpha_4 - 1}) - \ldots] \\ &= NP\left[-\Sigma 2(r-1)\binom{\alpha_r}{2} - \Sigma\alpha_r\alpha_s + \alpha_2 + 2\alpha_3 + \ldots + (r-1)\alpha_r + \ldots\right]. \end{aligned}$$

(iii) Let $r < s < t$, then we want

$$\begin{aligned} &NP\Sigma[(a_{r,\rho}\,a_{s,\sigma}\,a_{t,\tau}) + (a_{s,\sigma}\,a_{r,\rho}\,a_{t,\tau})] \\ &= NP\Sigma[(a_{r,\rho}\,a_{s,\sigma})(a_{s,\sigma}\,a_{t,\tau}) + (a_{r,\rho}\,a_{t,\tau})(a_{s,\sigma}\,a_{t,\tau})] \\ &= NP\Sigma[-2(s-1)(a_{s,\sigma}\,a_{t,\tau})] = NP\Sigma\, 2\binom{t-1}{2}\alpha_t. \end{aligned}$$

Hence $$NP\Sigma\,(abc) = NP\left[2\Sigma\binom{\alpha_r}{3} - 2\Sigma\,(r-1)\binom{\alpha_r}{2} + \Sigma\,(r-1)^2\,\alpha_r - \Sigma\alpha_r\alpha_s\right],$$

and $$T\Sigma\,\{abc\} = \left[\binom{n}{3} + (n-2)\,\Sigma\binom{\alpha_r}{2} - (n-2)\,\Sigma\,(r-1)\,\alpha_r + 2\Sigma\binom{\alpha_r}{3}\right.$$
$$\left. - 2\Sigma\,(r-1)\binom{\alpha_r}{2} + \Sigma\,(r-1)^2\,\alpha_r - \Sigma\alpha_r\alpha_s\right]T$$
$$= \left[\binom{n}{3} + 2\Sigma\binom{\alpha_r}{3} + \Sigma\,(n-2r)\binom{\alpha_r}{2} - \Sigma\,(r-1)\,(n-r-1)\,\alpha_r - \Sigma\alpha_r\alpha_s\right]T.$$

This is zero when $\alpha_1 < 3$, and only then.

Now $$\Sigma\,(ab)\,T_{3,3,2,1,1} = -\,7\,T_{3,3,2,1,1},$$
$$\Sigma\,(ab)\,.\,T_{4,3,1,1,1} = -\,3\,T_{4,3,1,1,1},$$
$$\Sigma\,(ab)\,\Delta = -\,45\Delta.$$

Also $$\Sigma\,(abc)\,.\,T_{3,3,2,1,1} = -\,8\,T_{3,3,2,1,1},$$
$$\Sigma\,(abc)\,.\,T_{4,3,1,1,1} = 0,$$
$$\Sigma\,(abc)\,\Delta = 240\Delta.$$

Then $$S = \lambda\Delta + \mu\Gamma T_{3,3,2,1,1}\,I_5 + \nu\Gamma T_{4,3,1,1,1}\,I_5,$$
$$\Sigma\,(ab)\,S = -\,45\lambda\Delta - 7\mu\Gamma T_{3,3,2,1,1}\,I_5 - 3\nu\Gamma T_{4,3,1,1,1},$$
$$\Sigma\,(abc)\,S = 240\,\lambda\Delta - 8\mu\Gamma T_{3,3,2,1,1}\,I_5,$$
$$\therefore\ [6 + 2\Sigma\,(ab) - \Sigma\,(abc)]\,S = -\,324\lambda\Delta.$$

This involves 286 sets of stars. There are 12 stars in each set; so it involves more than 12 times as many forms as the former expression for Δ. Looking at this other expression, viz.

$$K = -\,[1 - (07) - (17) - (27) - (37) - (47) - (57) - (67)]\,\{(23456)\}\,\{789\}' \begin{array}{ccccc} & & 0 & & \\ 8 & 1 & & 5 & 7 \\ & 2 & & 4 & \\ & & 9 & & \\ & 3 & & 6 & \end{array}$$

it is not difficult to see that no two terms belonging to the same set of stars appear in it. Thus in the first star 0123, 0456 are collinear; neither of these sets appears in a possible position for collinear digits in any other star of K.

ON QUANTITATIVE SUBSTITUTIONAL ANALYSIS

(Third Paper)

By ALFRED YOUNG

[Received 31 May, 1927.—Read 9 June, 1927.]

Two former papers have appeared on this subject in these *Proceedings*, the first in Ser. 1, 33 (1901), 97, and the second in Ser. 1, 34 (1902), 361. The object of these papers is the discussion and solution of substitutional equations of the form

$$\Sigma \lambda_r s_r X = R,$$

where s_r represents a permutation of the variables, X is an unknown function, and R a known function, λ_r being numerical. In the former papers expressions PN were introduced, which were derived from certain tableaux, here styled F, which each contained all the letters arranged in rows and columns, every row starting from the same left-hand column, and there being no more letters in any row than in a row above. For example, we may have F of the form

$$\begin{array}{l} a_{1,1}\, a_{1,2} \ldots \quad \ldots a_{1,\alpha_1} \\ a_{2,1}\, a_{2,2} \ldots a_{2,\alpha_2} \\ a_{3,1} \ldots \\ \ldots \qquad \ldots \\ a_{h,1} \ldots a_{h,\alpha_h}. \end{array}$$

In this case

$$P = \{\mathfrak{a}_{1,1}\, \mathfrak{a}_{1,2} \ldots \mathfrak{a}_{1,\alpha_1}\}\, \{\mathfrak{a}_{2,1} \ldots \mathfrak{a}_{2,\alpha_2}\} \ldots \{\mathfrak{a}_{h,1} \ldots \mathfrak{a}_{h,\alpha_h}\},$$

$$N = \{\mathfrak{a}_{1,1}\, \mathfrak{a}_{2,1} \ldots \mathfrak{a}_{h,1}\}'\, \{\mathfrak{a}_{1,2}\, \mathfrak{a}_{2,2} \ldots\}' \ldots,$$

where $\{\mathrm{abc} \ldots\}$ stands for the sum of the substitutions of the symmetric group of $a, b, c, \ldots$, which I call the positive symmetric group of $a, b, c, \ldots$. And $\{\mathrm{abc}\ldots\}'$ is the sum of the even substitutions of the symmetric group of $a, b, c, \ldots$, minus the sum of the odd substitutions of this group; I call this the negative symmetric group. Then P is the product of the positive symmetric groups of the rows of F, and N is the product of the negative symmetric groups of the columns. Further, $T_{a_1, a_2, \ldots, a_h}$ was defined as ΣPN, the sum extending to all expressions PN formed from F's which have a_1 letters in the first row, a_2 in the second row, and so on—where $a_1 \geqslant a_2 \geqslant a_3 \ldots \geqslant a_h$. It was shown that

$$1 = \Sigma A_{a_1, a_2, \ldots, a_h} T_{a_1, a_2, \ldots, a_h},$$

where $A_{a_1, a_2, \ldots, a_h}$ is a certain number; and that

$$PN \,.\, P'N' = 0,$$

when PN and $P'N'$ belong to different T's.

Thus a substitutional equation is equivalent to a number of entirely separate and independent equations

$$T \,.\, \Sigma \lambda_r s_r X = T \,.\, R,$$

there being no connexion between two such with different T's.

Unfortunately, belonging to any T a large number of expressions SPN made their appearance, which were connected by an intricate system of equations. The primary object of the present paper is to establish the fact that from among the possible tableaux F belonging to a given T a certain definite set (easily defined) $F_1, F_2, \ldots, F_f$ can be selected, from which we may derive forms $P_r \sigma_{r,s} N'_s$ $(r, s = 1, \ldots, f)$, where N'_s is obtained by a slight alteration from N_s, and $\sigma_{r,s}$ is that substitution which transforms the tableau F_s into F_r; these forms are all linearly independent; moreover,

$$P_r \sigma_{r,s} N'_s P_t \sigma_{t,u} N'_u = P_r \sigma_{r,u} N'_u \text{ or } 0,$$

according as t is or is not equal to s.

It follows at once that the solution of the original equation may be reduced to a certain system of matrix equations.

From this is deduced a fact as to the number of different values a given function may assume by interchange of variables, § 6.

It is shown, further, that the substitutional equation gives a set of f equations for a given T, each identical in form, but no two having a common member. Then in §§ 7, 11 it is shown how the equations may be expressed as equations between forms defined by the tableaux F.

Since, in practice, it is necessary to use forms defined by tableaux other than those of the standard set $F_1, \ldots, F_f$, convenient forms of the relations between these are worked out, §§ 8, 9, 10.

The results are illustrated, first by seeking to find what modifications may be introduced when it is known that the functions considered are unaltered by a cyclical interchange of all the variables. It is shown that a much smaller set of tableaux may be used, giving what I call reduced standard forms. And a theorem is proved establishing a relation between the number of reduced standard forms corresponding to any transitive group and the number of standard forms for one degree less. A further illustration is supplied by quartic invariant types, § 16. And, again, the theory is applied in some degree to a certain class of invariant types of the binary sextic, § 17.

When writing the two former papers I suffered from the disadvantage of being unacquainted with the closely related researches of the late Prof. Frobenius, published in the Berlin *Sitzungsberichte*, and beginning with "Über Gruppencharaktere", 1896; a lucid and more elementary exposition of the main features of Frobenius's theory of group characters was given by Schur*. But in a paper published in 1903 in the same place, "Über die charakteristischen Einheiten der symmetrischen Gruppe", Herr Frobenius referred to my two papers on substitutional analysis, and expounded very clearly the connexion between their contents and his own work. In view of his exposition there is no occasion for me to go over the same ground. But I feel that it is necessary to say a word in justification of my not adopting his notation throughout my work. The justification lies in my purpose of attempting to solve substitutional equations; and for this purpose it seems to me that it is necessary to use a notation which exhibits the relative positions of the letters. For the purpose of the general theory of groups it is different. Indeed, so far as that is concerned the general result of reduction of the group to matrices was accomplished by Frobenius in his researches, notably in his two papers "Über die Darstellung der endlichen Gruppen durch lineare Substitutionen", I (1897), II (1899), and all I can claim is to have expressed the matrices in a convenient form for calculation and solution of actual equations. Following his exposition in "Über die char. Einheiten", it will be seen that my expressions $P_r \sigma_{r,s} N'_s$ will appear as the elements of one of his irreducible group matrices. As I have occasion several times to refer to his papers I have done so in shortened form by the letters F.B.S., followed by the year of the Berlin *Sitzungsberichte* and the page referred to. Equations $\Sigma \lambda s X = R$ might be considered in the same way where s represents an operation of any

* "Neue Begründung der Theorie der Gruppen charaktere", *Berliner Sitzungsberichte* (1905), 406.

given group H, other than the symmetric group. The study of Frobenius's work shows how these can be solved, in much the same way as those here considered, with the help of group characters.

On the other hand, it will be seen how group characters would naturally turn up if we started with the object of solving such equations—instead of appearing, as they did, in the course of the solution of the problem to express any group as a linear substitution group, § 18.

My own two papers are quoted as Q.S.A. I and II.

It is perhaps as well to emphasize the fact that the theory is primarily meant to apply to the behaviour of functions when some of the variables are permuted among themselves. In the application, then, a function is supposed to stand on the right-hand side of the substitutional expression, and it is supposed to be operated on first by those operations which stand next it, and then the result is operated on by those further away. Thus

$$PNX = P(NX).$$

1. The form PN is derived from a tableau F*

$$\begin{array}{l} a_{1,1}\, a_{1,2} \ldots \quad \ldots \quad \ldots a_{1,\alpha_1} \\ a_{2,1}\, a_{2,2} \ldots \quad \ldots a_{2,\alpha_2} \\ \ldots \quad \ldots \quad \ldots \\ a_{h,1}\, a_{h,2} \ldots a_{h,\alpha_h} \end{array}$$

which contains α_1 letters in the first row, α_2 in the second, and so on; $\alpha_1 \geqslant \alpha_2 \geqslant \alpha_3 \ldots \geqslant \alpha_h$; every row commences from the same left-hand column, and the letters are the letters $a_1, a_2, \ldots, a_n$ in some arrangement. Then P is the product of the positive symmetric groups of the rows, and N is the product of the negative symmetric groups of the columns. Given the numbers $\alpha_1, \alpha_2, \ldots$ there are, of course $n!$ possible tableaux F. Amongst these we define certain particular tableaux as *standard*.

For this purpose we arrange the letters $a_1, a_2, \ldots, a_n$ in an agreed sequence; naturally the most convenient is the sequence of the suffixes. Then *a standard tableau is one in which in every row the letters appear in the order of the agreed sequence, and also in every column.* Thus when $n = 5$, $\alpha_1 = 3$, $\alpha_2 = 2$, there are 5 standard tableaux, viz.:

$$\begin{array}{lllll} a_1 a_2 a_3 & a_1 a_2 a_4 & a_1 a_2 a_5 & a_1 a_3 a_4 & a_1 a_3 a_5 \\ a_4 a_5 & a_3 a_5 & a_3 a_4 & a_2 a_5 & a_2 a_4 \end{array}.$$

Evidently in a standard tableau a_1 must appear in the top left-hand

* Q.S.A. II, 369.

corner, and a_2 in the second place in the top row or the first place in the second.

THEOREM I. *All forms PN belonging to a particular set* $T_{a_1, a_2, \ldots, a_h}$ *can be expressed as a sum*

$$\sum_{r=1}^{\lambda} A_r P_r \sigma_r N,$$

where P_r *is derived from a standard tableau,* σ_r *is a substitution which transforms* P *into* P_r*, and* A_r *is numerical.*

Let

$$P = \{b_{1,1} b_{1,2} \ldots b_{1,a_1}\} \{b_{2,1} \ldots b_{2,a_2}\} \ldots \{b_{h,1} \ldots b_{h\,a_h}\}.$$

Let the letters, as arranged in the agreed sequence, be $a_1, a_2, \ldots, a_n$. If a_1 is not in the first group of P, then

$$[1+\Sigma(a_1 b_{1,r})] PN = \frac{1}{a_1!} \{a_1 b_{1,1} b_{1,2} \ldots b_{1,a_1}\} PN = 0.*$$

Hence $$PN = -\Sigma P'(a_1 b_{1,r}) N,$$

where P' contains a_1 in its first group.

Now we may suppose that, in any group of P, the letters $b_{r,1} b_{r,2} \ldots b_{r,a_r}$ follow the fixed sequence. Let us suppose that $b_{2,s}$ precedes $b_{1,s}$; but that for values of $r < s$, $b_{1,r}$ precedes $b_{2,r}$. Then

$$\{b_{2,1} b_{2,2} \ldots b_{2,s} b_{1,s} b_{1,s+1} \ldots b_{1,a_1}\} PN = 0,$$

for the positive symmetric group on the left is of degree greater than the degree of any of the groups of P†.

Hence $$PN = -\Sigma A P' \sigma N,$$

where the first $s-1$ letters of the first group of P' are the same as in P and the s-th is one of $b_{2,1} b_{2,2} \ldots b_{2,s}$, all of which precede the other letters contained in the first two groups. The first letter in the first group of P' which can follow (in the fixed sequence) the corresponding letter of the second group is the $(s+1)$-th. Then if $b'_{2,s+1}$ precedes $b'_{1,s+1}$ we proceed again in the same way. Thus in a_2 steps at most we can express PN in the form $\Sigma A P' \sigma N$ where the letter $b'_{1,r}$ always precedes $b'_{2,r}$ in the tableau for P', according to the fixed sequence. We now apply the process to the other pairs of rows. Whenever a letter in a row precedes the corresponding letter in a higher row of the tableau, we can

* See Q.S.A. II, 370, 371.

† See Q.S.A. I, 134, 135.

perform this process of raising an earlier letter to take the place of a later one higher up. This is only possible a finite number of times, and hence ultimately we find

$$PN = \sum_{r=1}^{\lambda} A_r P_r \sigma_r N,$$

where P_r is derived from a standard tableau.

COROLLARY. The same argument is applicable to prove that

$$NP = \sum_{r=1}^{\lambda} A_r N \sigma_r P_r;$$

$$NP = \sum_{r=1}^{\lambda} A_r N_r \sigma_r P;$$

$$PN = \sum_{r=1}^{\lambda} A_r P \sigma_r N_r;$$

where in each case P_r or N_r is derived from a standard tableau, σ_r is some substitution, and A_r is numerical.

2. THEOREM II. *The number of standard tableaux belonging to* $T_{a_1, a_2, \ldots, a_h}$ *is*

$$n! \frac{\prod_{r,s} (a_r - a_s - r + s)}{\prod_r (a_r + h - r)!}.$$

We shall proceed by induction, and assume that this is true for $n-1$ letters. Consider first the case when no two a's are equal and $a_h > 1$. Let a_n be the last letter of the fixed sequence. Then it may occupy the end position in any row in a standard tableau. By removing a_n a standard tableau of the remaining $n-1$ letters is obtained. Thus, assuming the truth of the formula for $n-1$ letters, we see that the number of standard tableaux is

$$(n-1)! \frac{\prod_{r,s} (a_r - a_s - r + s)}{\prod_r (a_r + h - r)!} \left[\sum_{r=1}^{h} (a_r + h - r) \frac{\prod_{s \neq r} (a_r - 1 - a_s - r + s)}{\prod_{s \neq r} (a_r - a_s - r + s)} \right].$$

Let us call the sum in the brackets X, and let us write

$$\xi_r = a_r + h - r.$$

Then $$X = \Sigma \xi_r \frac{\prod(\xi_r - 1 - \xi_s)}{\prod(\xi_r - \xi_s)} = -\Sigma \xi_r \frac{f(\xi_r - 1)}{f'(\xi_r)},$$

where $$f(x) = (x-\xi_1)(x-\xi_2)\dots(x-\xi_h);$$

it is easy to see that no two ξ's can be equal for $\xi_1 > \xi_2 \dots > \xi_h$. Now it is well known that if the order of $\Phi(x)$ is less than $h-1$*,

$$\Sigma \frac{\Phi(\xi_r)}{f'(\xi_r)} = 0.$$

Hence $$X = -\Sigma \xi_r \frac{f(\xi_r)-f'(\xi_r)+\frac{1}{2}f''(\xi_r)}{f'(\xi_r)}$$

$$= \Sigma\xi_r - \binom{h}{2} = \Sigma a_r = n.$$

Thus the induction is proved when no two of the a's are equal and $a_h > 1$. When $a_r = a_{r+1}$ the letter a_n cannot occur in the a_r-th row of a standard tableau at all. But in this case the corresponding term in the sum X contains a factor $a_r - 1 - a_{r+1} + 1$ which is zero; so that the proof is unaffected, and we may remove the restriction that two a's must not be equal. And when $a_h = 1$, the term in X for $r = h$ is exactly that given by putting $a_h = 1$. The formula is obviously true for $n = 1, 2, 3, 4$, as can easily be seen by writing down the standard tableaux in these cases. Hence it is always true.

We shall call this number $f_{a_1, a_2, \dots, a_h}$, and drop the suffixes, when no confusion can arise; it is the number which Frobenius calls f or χ_0.

Moreover, in the series

$$1 = \Sigma A_{a_1, a_2, \dots, a_h} T'_{a_1, a_2, \dots, a_h},$$

for which the coefficients were obtained in my second paper,

$$A = \left(\frac{f}{n!}\right)^2.$$

3. The proof given for the above theorem establishes, in fact, the useful formula

$$f_{a_1-1, a_2, \dots, a_h} + f_{a_1, a_2-1, a_3, \dots, a_h} + \dots + f_{a_1, a_2, \dots, a_h-1} = f_{a_1, a_2, \dots, a_h}. \tag{1}$$

It is interesting and useful to point out that $f_{a_1, a_2, \dots, a_h}$ is the coefficient of $x_1^{a_1+h-1} x_2^{a_2+h-2} \dots x_h^{a_h}$ in the expansion of

$$E = \frac{\Delta(x_1, x_2, \dots, x_h)}{1-x_1-x_2 \dots -x_h} = \Delta\Sigma(x_1+x_2+\dots+x_h)^n,$$

* *E.g.* see Burnside and Panton, *Theory of equations*, 1 (1899), 172.

where

$$\Delta(x_1, x_2, \ldots, x_h) = (x_1-x_2)(x_1-x_3)\ldots(x_1-x_h)(x_2-x_3)\ldots(x_{h-1}-x_h)$$
$$= \{x_1 x_2 \ldots x_h\}' x_1^{h-1} x_2^{h-2} \ldots x_{h-1}.$$

If the sum $a_1+a_2+\ldots+a_h = n$, we may express the same thing by saying that f is the coefficient of $x_1^{a_1+h-1}\ldots x_h^{a_h}$ in the expansion of

$$\Delta(x_1+x_2+\ldots+x_h)^n.$$

When $n = 1$ or 2 this is easily verified. For further values the formula proves the induction from one value of n to the next.

This is proved by Frobenius (F.B.S. 1900, 522).

The formula

$$f_{a_1+1, a_2, \ldots, a_h}+f_{a_1, a_2+1, \ldots, a_h}+\ldots+f_{a_1, a_2, \ldots, a_h, 1} = (n+1) f_{a_1, a_2, \ldots, a_h} \quad (2)$$

may be proved in the same way, or else from the fact that in the expansion of

$$\Delta(x_1, x_2, \ldots, x_{h+1})(x_1+x_2+\ldots+x_{h+1})^{n+1}$$

it is the coefficient of

$$x_1^{a_1+h+1} x_2^{a_2+h-1}\ldots x_h^{a_h+1}+x_1^{a_1+h} x_2^{a_2+h} x_3^{a_3+h-2}\ldots x_h^{a_h+1}$$
$$+\ldots+x_1^{a_1+h} x_2^{a_2+h-1} x_h^{a_h+1} x_{h+1}$$
$$= x_1^{a_1+h} x_2^{a_2+h-1}\ldots x_h^{a_h+1}\{x_1+x_2+\ldots+x_{h+1}\},$$

i.e. $n+1$ times the coefficient of $x_1^{a_1+h} x_2^{a_2+h-1}\ldots x_h^{a_h+1}$ in the expansion of

$$\Delta(x_1+x_2+\ldots+x_{h+1})^n = (n+1) f_{a_1, a_2, \ldots, a_h}.$$

4. Consider the f standard tableaux $F_1, F_2, \ldots, F_f$ belonging to $T_{a_1, a_2, \ldots, a_h}$. We can arrange them in a definite sequence following the sequence of letters $a_1, a_2, \ldots, a_n$. Let the t-th letter of the s-th row of F_λ be $a_{st}^{(\lambda)}$. Then when the first $s-1$ rows of F_λ and F_μ are identical and also the first $t-1$ letters of the s-th row, but $a_{s,t}^{(\lambda)}$ precedes $a_{s,t}^{(\mu)}$ in the sequence of letters, we say that F_λ precedes F_μ in the sequence of tableaux.

We shall suppose that the suffixes of $F_1, F_2, \ldots, F_f$ are chosen so that the tableau F_r precedes F_s in this fixed sequence when $r < s$. We shall also use P_r, N_r for the products of positive or negative symmetric groups obtained from the tableau F_r.

It will be noticed that, by giving the columns of the tableaux the precedence, instead of the rows, we should obtain a different sequence; but one, indeed, equally suitable for our purpose.

THEOREM III. *If* $r < s$,

$$N_s P_r = 0.$$

Let us suppose that the first $t-1$ letters of the first row in F_r and F_s are the same, but that $a_{1,t}^{(r)}$ precedes $a_{1,t}^{(s)}$. Then, since F_s is a standard tableau, $a_{1,t}^{(s)}$ precedes $a_{1,v}^{(s)}$ unless $v < t$. Thus the letter $a_{1,t}^{(r)}$ must appear in one of the first $t-1$ columns of F_s; and N_s has a factor $\{a_{1,t}^{(r)} a_{1,v}^{(r)}\}'$ where $v < t$; and hence $N_s P_r = 0$.* The argument is the same when the first rows of F_r, F_s are identical and the distinction between them arises first in a later row.

When $r = s$, we know that $N_r P_r$ is not zero; and when $r > s$ this equation does not always hold good.

In all cases $N_f P_r = 0$, unless $r = f$, for otherwise $r < f$. Let us consider N_f, N_{f-1}, ... in the reverse order, and let N_s be the first N we come to for which there is an inequality

$$N_s P_r \neq 0,$$

when $r \neq s$. Then we know that $r > s$. Also we know that P_r contains a substitution τ_{sr} which transforms N_r into N_s†, so that

$$N_s \tau_{sr} = \tau_{sr} N_r.$$

It is possible that there is more than one value of r for which the above inequality is true. Let these be r_1, r_2, ... Let us write

$$N_s' = N_s(1-\tau_{sr_1}-\tau_{sr_2}-\ldots)$$

$$= N_s-\tau_{sr_1}N_{r_1}-\tau_{sr_2}N_{r_2}-\ldots.$$

Then $$N_s' P_r = 0$$

for all values of r other than s. For, by hypothesis, $N_t P_r$ is zero when t is less than s and not equal to r, and hence

$$N_s' P_{r_1} = N_s P_{r_1} - \tau_{sr_1} N_{r_1} P_{r_1} = 0.$$

Further $$N_s' P_s = N_s P_s.$$

We call N_s' a *prepared* form.

Proceeding, we take the next value of s for which there exists a different number r such that $N_s P_r$ is not zero. We then replace N_s by

$$N_s' = N_s - \tau_{sr_1} N_{r_1}' - \tau_{sr_2} N_{r_2}' - \ldots$$

* For P_r has a factor $\{a_{1,t}^{(r)} a_{1,v}^{(r)}\}$ and $\{a_{1,t}^{(r)} a_{1,v}^{(r)}\}'\{a_{1,t}^{(r)} a_{1,v}^{(r)}\} = 0$.

† Q.S.A. II, 363, § 3.

in the same way, only now we use the N's already prepared instead of the original ones. Then, as before,

$$N'_s P_r = 0 \quad (r \neq s)$$

and $$N'_s P_s = N_s P_s.$$

Thus we can replace each of $N_1, N_2, \ldots, N_f$ by expressions $N'_1, N'_2, \ldots, N'_f$ which have the properties

$$N'_s P_r = 0 \quad (r \neq s),$$

$$N'_s P_s = N_s P_s.$$

Moreover $N'_s = N_s[1-\tau_1-\tau_2-\ldots]$, where $\tau_1, \tau_2, \ldots$ represent certain substitutions.

Let now $\sigma_{s,r}$ be that substitution which transforms the tableau F_r into F_s. It will transform $P_r N_r$ into $P_s N_s$, so that

$$\sigma_{s,r} P_r \sigma_{s,r}^{-1} = P_s,$$

$$\sigma_{s,r} N_r \sigma_{s,r}^{-1} = N_s.$$

Then consider the f^2 expressions

$$P_r \sigma_{rs} N'_s = \sigma_{r,s} P_s N'_s$$

The product of any two is either zero or else a multiple of one of themselves. For

$$P_r \sigma_{r,s} N'_s \,.\, P_u \sigma_{u,v} N'_v$$

is zero unless $u = s$, and

$$\begin{aligned} P_r \sigma_{rs} N'_s P_s \sigma_{s,v} N'_v &= \sigma_{r,s} P_s N_s P_s \sigma_{sv} N_v[1-\tau_1-\tau_2\ldots] \\ &= \sigma_{r,s}\sigma_{sv} P_v N_v P_v N_v[1-\tau_1-\tau_2\ldots] \\ &= \frac{n!}{f}\,\sigma_{r,v} P_v N_v[1-\tau_1-\tau_2-\ldots]^* \\ &= \frac{n!}{f}\,\sigma_{r,v} P_v N'_v = \frac{n!}{f}\,P_r \sigma_{r,v} N'_v. \end{aligned}$$

None of these f^2 expressions vanishes; for otherwise

$$P_r \sigma_{r,s} N'_s P_s N_s = \sigma_{r,s} P_s N_s P_s N_s = \frac{n!}{f}\,\sigma_{r,s} P_s N_s$$

would be zero, which we know is not the case.

---

* Q.S.A. II, 366.

There can be no linear relation between them. For let us suppose that

$$\Sigma \lambda_{r,s} P_r \sigma_{r,s} N'_s = 0,$$

where $\lambda_{r,s}$ is numerical.

Multiply this equation on the right-hand side by $P_s N'_s$ and on the left-hand side by $P_r N'_r$; then every term will vanish except

$$\lambda_{r,s} \left(\frac{n!}{f}\right)^2 P_r \sigma_{r,s} N'_s,$$

and hence $\lambda_{r,s}$ vanishes for all values of r and s.

These expressions are then linearly independent. There are f^2 of them belonging to each T. The number $(f/n!)^2$ is, as we have said, the coefficient of $T_{a_1, a_2, \ldots, a_h}$ in the series

$$1 = \Sigma A_{a_1, a_2, \ldots, a_h} T_{a_1, a_2, \ldots, a_h};$$

also $$\frac{1}{n!} = \Sigma A_{a_1, a_2, \ldots, a_h}{}^*.$$

Hence $$\Sigma f^2 = n!$$

THEOREM IV. *Associated with every* $T_{a_1, a_2, \ldots, a_h}$ *there are* $f^2_{a_1, a_2, \ldots, a_h}$ *expressions* $P_r \sigma_{r,s} N'_s$ *and, taking all the possible values of* $a_1, a_2, \ldots, a_h$, *there are exactly* $n!$ *of these expressions. They are all linearly independent, and in terms of them every substitution, and every substitutional expression, can be expressed with purely numerical coefficients.*

5. Let $$T' = P_1 N'_1 + P_2 N'_2 + \ldots + P_f N'_f.$$

Then $$T' P_r \sigma_{r,s} N'_s = P_r N'_r P_r \sigma_{r,s} N'_s = \left(\frac{n!}{f}\right) P_r \sigma_{r,s} N'_s.$$

Now $$T P_r \sigma_{r,s} N'_s = T P_r N_r [1 - \tau_1 \ldots] \sigma_{r,s}$$

$$= \left(\frac{n!}{f}\right)^2 P_r \sigma_{r,s} N'_s \dagger.$$

Hence $$\left[T - \left(\frac{n!}{f}\right) T'\right] S = 0,$$

where S is any substitutional expression whatever of the n letters. We may take $S = 1$; hence

$$T = \frac{n!}{f} T'.$$

---

* Q.S.A. II, § 8, 368.

† Q.S.A. II, § 3.

THEOREM V. *The expression* $T_{a_1, a_2, \ldots, a_h}$ *is equal to*

$$\frac{n!}{f}[P_1N_1'+P_2N_2'+\ldots+P_fN_f'] = \frac{n!}{f}T',$$

that is a sum of terms, each one of which is obtained from a standard tableau. And the series may be written

$$n! = \Sigma f_{a_1, a_2, \ldots, a_h} T'_{a_1, a_2, \ldots, a_h}.$$

6. A substitutional equation may now be presented as a number of different equations each belonging to a different T, and each of the form

$$\Sigma \lambda_{r,s} P_r \sigma_{r,s} N_s' X = 0;$$

where X is the unknown function sought for. The solution desired is of the form

$$X = \Sigma \mu_{r,s} P_r \sigma_{r,s} N_s' U,$$

where U is an arbitrary function.

Substituting for X in the equation we obtain f^2 linear equations of the form

$$\sum_{s=1}^{f} \lambda_{r,s} \mu_{s,t} = 0$$

for determining the f^2 quantities $\mu_{r,s}$. Indeed, we may write substitutional expressions belonging to a particular T as matrices. They obey all the laws of matrices for addition, multiplication, &c., and, in particular, for solution of equations.

Thus we may write simply $[\lambda_{rs}]$ for $\Sigma \lambda_{r,s} P_r \sigma_{r,s} N_s'$, and consider the equation

$$[\lambda_{rs}][\mu_{rs}] = 0,$$

where λ_{rs} is known and μ_{rs} is to be found.

It is to be noticed that N_s' is never changed by any left-hand multiplication of $P_r \sigma_{r,s} N_s'$; so that no relation will connect terms $P_r \sigma_{r,s} N_s' U$ with terms having different values of s. Thus the equations we consider are really f sets of equations, there being f equations in each set, between f unknowns. The f sets of equations are identical, that is the coefficients in each set are the same; but the f unknowns in one set are distinct from those in any other set, except that they satisfy these equations. They may be called cogredient. The same remark is clear from the consideration of the matrix equations, when the unknown matrix is always on the right-hand side.

Let ρ be the rank of $[\lambda_{r,s}]$; then $(f-\rho)$ is the rank of $[\mu_{r,s}]$, and for a given value of s the equations for $\mu_{r,s}$ may be solved, the solution having $f-\rho$ undetermined quantities. Thus the expression TX has $f(f-\rho)$ undetermined quantities, and TX is a function such that the functions obtained by interchange of letters from it may be expressed linearly in terms of $f(f-\rho)$ of them. If there is more than one equation we can, by introduction of arbitrary multipliers, reduce the equations to a single one.

THEOREM VI. *The number of linearly independent values obtained from a function TX by interchange of letters is ρf where $f \geqslant \rho \geqslant 0$.*

There are two cases in which f is unity, viz. f_n and $f_{1,1,\ldots,1}$. These correspond to the symmetric and alternating functions respectively. And, as is well known, the general two-valued function is of the form

$$AT_n U + BT_{1,1,\ldots,1} U.$$

In all other cases when $n > 4$, the minimum value of f is $n-1$, which is the value of $f_{n-1,1}$ and of $f_{2,1,1,\ldots,1}$. This, of course, does not lead to an $(n-1)$-valued function in the sense required by the theory of equations; in that sense we obtain an n-valued function from a $T_{n-1,1}U = X$ which satisfies an equation $[\lambda_{r,s}]X = 0$ of rank $n-2$. But the sum of the n values is zero, and so the functions obtained can be expressed linearly in terms of $n-1$ of them. On the other hand, substitutional equations may arise when the unknown quantity appears either on the right-hand side or the left; such equations would appear if it were asked what substitutional expressions formed from the symmetric group form a group. In such cases the number of solutions TX may be any number from 0 to f^2.

What we have called T' is, when expressed as a matrix, the unit matrix; strictly speaking

$$T' = \frac{n!}{f} E_f.$$

7. In order to use the theory thus developed, it is necessary to express any substitution τ in terms of expressions $\lambda P_r \sigma_{rs} N'_s$. To do this we use the series

$$\tau = \Sigma \frac{f}{n!} \tau T',$$

and then evaluate
$$\tau T' = \sum_{r=1}^{f} \tau P_r N'_r.$$

When
$$\tau P_r \tau^{-1} = P_s,$$

where P_s is equal to the product of the positive symmetric groups formed from the rows of a standard tableau F_s, then

$$\tau = \sigma\sigma_{sr},$$

where σ is a substitution which interchanges the letters only in the same row of F_s, and hence σ belongs to P_s. Thus

$$\tau P_r N_r' = P_s \sigma_{sr} N_r';$$

and hence, if τ changes P_r into P_s, which is of standard form, the nature of $\tau P_r N_r'$ is at once determined by operating on P_r with τ.

In the case of $T_{n-1,1}$, for example, where there is only one letter in the second row of the tableau and all the rest in the first, the tableau F_r is determined by (a_{n+1-r}), the letter in the second row. If τ does not contain a_1, $\tau P_r \tau^{-1} = P_s$, which is a standard P, and hence

$$\tau P_r N_r' = P_s \sigma_{sr} N_r';$$

where, indeed, $\tau(a_{n+1-r})\,\tau^{-1} = (a_{n+1-s})$.

Here $\sigma_{sr} = (a_{n+1-s}\, a_{n-s}\, a_{n-1-s} \ldots a_{n+1-r})$ for $r > s$,

and $\sigma_{sr} = (a_{n+1-s}\, a_{n+2-s} \ldots a_{n+1-r})$ for $r < s$.

Thus, when τ does not contain a_1, $\tau T''_{n-1,1}$ is represented by a matrix obtained from the unit matrix by an interchange of rows.

If τ changes the letter in the second row of F_r into a_1, we use the equation

$$\left[1 + \sum_{t=2}^{n} (a_1 a_t)\right] \{a_2 a_3 \ldots a_n\}\, \tau N_r' = 0.$$

That is, in our case,

$$\tau P_r N_r' = -\Sigma P_{n+1-t}(a_1 a_t)\, \tau N_r'$$

$$= -\Sigma P_{n+1-t}\, \sigma_{n+1-t,\,r} N_r'$$

for $(a_1 a_t)\tau$ transforms P_r into P_{n+1-t}.

In general, for any T we have

$$\tau P_r N_r' = Q\tau N_r',$$

where Q is written to express any P belonging to T (not necessarily standard). Then, by § 1, we may express this as

$$Q\tau N_r' = \sum_{s=1}^{f} A_s P_s \varpi_s \tau N_r',$$

where $\varpi_s \tau$ transforms P_r into P_s. Then, as we have seen, $\varpi_s \tau = \sigma \sigma_{s,r}$, where $\sigma_{s\,r}$ transforms F_r into F_s, and σ belongs to P_s, and hence

$$\tau P_r N_r' = \Sigma A_s P_s \sigma_{s,r} N_r'.$$

In practice it is important to notice when applying this that, if two rows of F contain the same number of letters, the forms Q obtained by interchanging the two rows are apparently identical, but they must be regarded as distinct (at least when the number of letters in either is odd). The simplest example is the case $n = 2$, $T_{1,1}$.

Here let $\tau = (a_1 a_2)$, $Q = \{a_2\}\{a_1\}$,

$$[1+(a_1 a_2)]\,Q\tau N' = 0, \quad i.e. \quad Q\tau N' = -\{a_1\}\{a_2\}(a_1 a_2)\tau N'$$

$$= -PN'.$$

The case appears trivial, but it is of real importance in dealing with equal rows of several letters (see § 9). With this proviso we may use the tableau F_r to define any entity $P_r \sigma_{r,s} N_s$; it stands equally for f different entities, which are indistinguishable for our purposes and are independent; then $\tau F_r = \tau P_r \sigma_{r,s} N$ is simply the tableau obtained by operating on F_r with τ.

8. It is of interest to inquire what form PN will take when another letter is added. That is to express $P_r \sigma_{r,s} N_s$ for a definite set of n letters in terms of the corresponding entities when an additional letter is present. We prove

THEOREM VII. *If $T_{a_1, a_2, \ldots, a_h}$ refer to n letters, and $T'_{a_1', a_2', \ldots, a_{h'}'}$ to these n letters and one more, then $T_{a_1, \ldots, a_h} \cdot T'_{a_1', \ldots, a_{h'}'} = 0$, unless the numbers $a_1', a_2', \ldots, a_h'$ are the same as the numbers $a_1, a_2, \ldots, a_h$ except that one of them is increased by unity.*

This statement is, of course, intended to include the case when $h' = h+1$, and the suffixes of T' are $a_1, a_2, \ldots, a_h, 1$. Calling the additional letter x, we may write T' as a sum of standard entities, the sequence of letters having x as the last.

Let $T = \Sigma PN$, PN being formed from a tableau F, and let $T' = \Sigma P'N'$, $P'N'$ being formed from a tableau F'. Then in F' the letter x appears at the end of a row and also at the end of a column, for F' is a standard tableau. If, then, x be suppressed in F' we have a tableau of the original n letters, from which we may form an entity $P''N''$ belonging to $T_{\beta_1, \beta_2, \ldots, \beta_{h'}}$, where the suffixes are the suffixes $a'_1, a'_2, \ldots, a'_{h'}$ except that one of them is reduced by unity. Also $P''P' = AP'$ and $P'N'' = BN'$, where A and B are numerical and not zero. Moreover, $TP'' = 0$ or $TN'' = 0$ unless $T_{a_1} \ldots \equiv T_{\beta_1} \ldots$.

Hence, unless $\beta_1, \beta_2, \ldots$ are the same as $a_1, a_2, \ldots, TP'N' = 0$, and therefore $TT' = 0$.

This proves the theorem.

COROLLARY. *If T refers to n letters and T' to the same n letters and m more, then $TT' = 0$ unless T' is such that its tableaux can be formed from the tableaux of T by adding letters to the rows and columns. In the contrary case if PN be any entity of T, and $P'N'$ of T', then*

$$PN.P'N' = 0.$$

For example, $\begin{Bmatrix} a_1 a_2 \\ a_3 a_4 \end{Bmatrix} . T_{n-r, 1, 1, 1, \ldots, 1} = 0.$

We see the truth of this by building up the expressions one letter at a time.

9. As in the use of this calculus it is frequently necessary to use non-standard forms, it is important to be able to write down quickly the relations between them, and, in particular, to reduce them to standard forms. For this purpose we prove the following

THEOREM VIII. *A form*

$$PN \equiv \{a_1 a_2 \ldots a_r\} \{b_1 b_2 \ldots b_s\} QN$$
$$= (-)^t \Sigma \{b_1 b_2 \ldots b_t c_{t+1} c_{t+2} \ldots c_r\} \{c_1 c_2 \ldots c_t b_{t+1} b_{t+2} \ldots b_s\} Q\tau N,$$

where $c_1, c_2, \ldots, c_r$ are the letters $a_1, a_2, \ldots, a_r$ in any order, the sum extending to all terms obtainable by selecting any t of the a's for $c_1, c_2, \ldots, c_t$; and τ is that substitution which changes P into that of the particular term, and $r \geqslant s$.

When $t = 1$ the theorem gives the ordinary identity

$$\left[1 + \sum_{\rho=1}^{r} (b_1 a_\rho)\right] PN = 0,$$

and is known already.

Assume the theorem to be true up to $t = t$; then

$$0 = \{a_1 a_2 \ldots a_r b_{t+1}\} . \{b_1 b_2 \ldots b_t a_{t+1} a_{t+2} \ldots a_r\} \{a_1 a_2 \ldots a_t b_{t+1} b_{t+2} \ldots b_s\} Q\tau N;$$

hence, expanding the first positive symmetric group, we have

$$0 = \Sigma \{b_1 b_2 \dots b_t c_{t+1} c_{t+2} \dots c_r\} \{c_1 c_2 \dots c_t b_{t+1} b_{t+2} \dots b_s\} Q\tau N)$$
$$+\Sigma \{b_1 b_2 \dots b_{t+1} c_{t+2} c_{t+3} \dots c_r\} \{c_1 c_2 \dots c_{t+1} b_{t+2} \dots b_s\} Q\tau N,$$

and the theorem is true for $t+1$ if it is true for t; it is therefore always true.

COROLLARY. *When* $r = s = t$,

$$\{a_1 a_2 \dots a_r\} \{b_1 b_2 \dots b_r\} QN = (-)^r \{b_1 b_2 \dots b_r\} \{a_1 a_2 \dots a_r\} Q\tau N,$$

where τ interchanges the a's and b's; an illustration of the remark made at the end of the last paragraph.

We may extend this result thus:

THEOREM IX.

$$PN \equiv \{a_1 a_2 \dots a_r\} \{b_1 b_2 \dots b_s\} QN$$
$$= (-)^{t+1} \Sigma \{a_1 b_1 \dots b_{u-1} b_{u+1} \dots b_t c_{t+1} \dots c_r\} \{b_u c_2 \dots c_t b_{t+1} \dots b_s\} Q\tau N$$
$$+(-)^{t+1}(t-1) \Sigma \{a_1 b_1 \dots b_t c_{t+1} \dots c_r\} \{c_2 \dots c_{t+1} b_{t+1} \dots b_s\} Q\tau N,$$

where $c_2 \dots c_r$ *are the letters* $a_2 \dots a_r$ *in any order, and* u *has the values* 1, 2, ..., t, *and* $r \geqslant s$.

For

$$PN = (-)^t \Sigma \{b_1 \dots b_t a_1 c_{t+2} \dots c_r\} \{c_2 \dots c_{t+1} b_{t+1} \dots b_s\} Q\tau N$$
$$+(-)^t \Sigma \{b_1 \dots b_t c_{t+1} \dots c_r\} \{a_1 c_2 \dots c_t b_{t+1} \dots b_s\} Q\tau N$$

by the last theorem.

To every term T of the second sum we apply the result

$$T = -\sum_{u=1}^{t} (a_1 b_u) T - \sum_{u=t+1}^{r} (a_1 c_u) T,$$

and the result follows.

In the same way we obtain

$$PN = (-)^{t+2} \Big[\binom{t-1}{2} \Sigma \{a_1 a_2 b_1 \dots b_t a'_{t+3} \dots a'_r\} \{a'_3 \dots a'_{t+2} b_{t+1} \dots b_s\} Q\tau N$$
$$+(t-2) \Sigma \{a_1 a_2 b'_2 \dots b'_t a'_{t+2} \dots a'_r\} \{a'_3 \dots a'_{t+1} b'_1 b_{t+1} \dots b_s\} Q\tau N$$
$$+\Sigma \{a_1 a_2 b'_3 \dots b'_t a'_{t+1} \dots a'_r\} \{a'_3 \dots a'_t b'_1 b'_2 b_{t+1} \dots b_s\} Q\tau N \Big],$$

where $a'_3 \dots a'_{t+2}$ are selected from $a_3 a_4 \dots a_r$ in all possible ways, and b'_1 or $b'_1 b'_2$ are selected from $b_1 b_2 \dots b_t$ in all possible ways.

Further extensions are obvious.

10. The product of a positive symmetric group into a form PN was considered in Q.S.A. II, §§ 18, 19. It will be useful to detail the result here.

First consider

$$\{a_1 a_2 \dots a_\lambda\, b_1 b_2 \dots b_\mu\} \begin{pmatrix} a_1 a_2 \dots a_r \\ b_1 b_2 \dots b_s \end{pmatrix}$$

$$= [1+(b_\mu a_1)+(b_\mu a_2)+\dots+(b_\mu a_\lambda)+(b_\mu b_1)+\dots+(b_\mu b_{\mu-1})]$$

$$\times [1+(b_{\mu-1} a_1)+\dots+(b_{\mu-1} a_\lambda)+(b_{\mu-1} b_1)+\dots+(b_{\mu-1} b_{\mu-2})]$$

$$\dots [1+(b_1 a_1)+(b_1 a_2)+\dots+(b_1 a_\lambda)]\, \lambda! \begin{pmatrix} a_1 a_2 \dots a_r \\ b_1 b_2 \dots b_s \end{pmatrix}$$

$$= -[1+(b_\mu a_1)+(b_\mu a_2)+\dots+(b_\mu a_\lambda)+(b_\mu b_1)+\dots+(b_\mu b_{\mu-1})]$$

$$\dots [1+(b_2 a_1)+\dots+(b_2 a_\lambda)+(b_2 b_1)]$$

$$\times [(b_1 a_{\lambda+1})+(b_1 a_{\lambda+2})+\dots+(b_1 a_r)]\, \lambda! \begin{pmatrix} a_1 a_2 \dots a_r \\ b_1 b_2 \dots b_s \end{pmatrix}$$

$$= (-)^\mu \lambda!\, \mu!\, \Sigma \begin{pmatrix} a_1 a_2 \dots a_\lambda\, b_1 b_2 \dots b_\mu\, c_1 \dots c_{r-\lambda-\mu} \\ c_{r-\lambda-\mu+1} \dots c_{r-\lambda}\, b_{\mu+1} \dots b_s \end{pmatrix},$$

where $c_1 \dots c_{r-\lambda}$ are the letters $a_{\lambda+1} \dots a_r$ in some order, and the Σ extends to all possible ways of distributing them between the two rows. Of course, the result is zero when $\lambda+\mu > r$.

Now

$$\{a_1 a_2 \dots a_\lambda\, b_1 \dots b_\mu\} \begin{pmatrix} a_1 \dots a_r \\ b_1 \dots b_s \end{pmatrix} = \lambda!\, \mu!\, \Sigma \begin{pmatrix} d_1 \dots d_\lambda\, a_{\lambda+1} \dots a_r \\ d_{\lambda+1} \dots d_{\lambda+\mu}\, b_{\mu+1} \dots b_s \end{pmatrix},$$

where $d_1 \dots d_{\lambda+\mu}$ are the letters $a_1 a_2 \dots a_\lambda\, b_1 \dots b_\mu$ and the sign of summation extends to the different ways of distributing them between the two rows.

Thus the group $\{a_1 a_2 \dots a_\lambda\, b_1 \dots b_\mu\}$ may be said to give us the relation

$$\Sigma \begin{pmatrix} d_1 \dots d_\lambda\, a_{\lambda+1} \dots a_r \\ d_{\lambda+1} \dots d_{\lambda+\mu}\, b_{\mu+1} \dots b_s \end{pmatrix} = (-)^\mu \Sigma \begin{pmatrix} a_1 \dots a_\lambda\, b_1 \dots b_\mu\, c_1 \dots c_{r-\lambda-\mu} \\ c_{r-\lambda-\mu-1} \dots c_{r-\lambda}\, b_{\mu+1} \dots b_s \end{pmatrix}.$$

11. In applying this theory, the kind of equation considered is

$$\Sigma \lambda_r s_r X = R;$$

X being the function the properties of which it is desired to know. The way we proceed, as a rule, is to operate on this equation with a *standard* form $N'_r P_r$; where N'_r has been prepared in the manner of § 4. Thus we derive an equation of the form

$$\Sigma \mu_t N'_r \sigma_{r,t} P_t X = N'_r P_r R.$$

It will be observed that the substitutional part of every term in this equation begins with N'_r. Operation on the left-hand side with $N'_\rho \sigma_{\rho,r} P_r$ will replace N'_r by N'_ρ, leaving the rest of the equation unaltered. Thus the particular N' which appears in the equation is quite immaterial. In fact, each term is sufficiently defined by the tableau for the P_t it contains together with the numerical coefficient. Moreover, we may in general write the term $N'_r P_r sX$ of the equations down very simply by replacing $N'_r P_r s$ by $N'_r s^{-1} P_{r,1}$ (where $P_{r,1}$ is obtained from P_r by the operation s), and then the term $N'_r s^{-1} P_{r,1} X$ by the tableau for $P_{r,1}$ (which is not, in general, standard). (The sequence PN' has been here changed to $N'P$; it is, of course, just a matter of convenience which we take.)

As an example, consider the quartic invariant type*

$$(a_1 a_2)^2 (a_2 a_3)^2 (a_3 a_4)^2 (a_4 a_1)^2 = X.$$

It is subject to three equations

$$\text{(i)} \qquad [1-(\mathfrak{a}_1 \mathfrak{a}_2 \mathfrak{a}_3 \mathfrak{a}_4)] X = 0,$$

$$\text{(ii)} \qquad [1-(\mathfrak{a}_2 \mathfrak{a}_4)] X = 0,$$

$$\text{(iii)} \qquad \{\mathfrak{a}_2 \mathfrak{a}_3 \mathfrak{a}_4\} X = \tfrac{1}{2} \{\mathfrak{a}_2 \mathfrak{a}_3 \mathfrak{a}_4\} (a_1 a_2)^4 (a_3 a_4)^4.$$

There are five classes of PN or NP:

(1) T_4, we operate on each equation with $\{\mathfrak{a}_1 \mathfrak{a}_2 \mathfrak{a}_3 \mathfrak{a}_4\}$, and the only result is

$$\{\mathfrak{a}_1 \mathfrak{a}_2 \mathfrak{a}_3 \mathfrak{a}_4\} X = \tfrac{1}{2} \{\mathfrak{a}_1 \mathfrak{a}_2 \mathfrak{a}_3 \mathfrak{a}_4\} (a_1 a_2)^4 (a_3 a_4)^4.$$

(2) $T_{3,1}$, we operate in turn with $N_r P_r$ ($r = 1, 2, 3$), where

$$F_1 = \frac{a_1 a_2 a_3}{a_4}, \qquad F_2 = \frac{a_1 a_2 a_4}{a_3}, \qquad F_3 = \frac{a_1 a_3 a_4}{a_2}.$$

---

* *Proc. London Math. Soc.* (1), 30 (1898), 290.

Writing N_1P_1X in the form $\begin{pmatrix} a_1a_2a_3 \\ a_4 \end{pmatrix}$, (i) gives

$$\begin{pmatrix} a_1a_2a_3 \\ a_4 \end{pmatrix} = \begin{pmatrix} a_2a_3a_4 \\ a_1 \end{pmatrix} = \begin{pmatrix} a_3a_4a_1 \\ a_2 \end{pmatrix} = \begin{pmatrix} a_4a_1a_2 \\ a_3 \end{pmatrix} = 0;$$

because $$\begin{pmatrix} a_1a_2a_3 \\ a_4 \end{pmatrix}[1+(\mathrm{a}_1\mathrm{a}_4)+(\mathrm{a}_2\mathrm{a}_4)+(\mathrm{a}_3\mathrm{a}_4)] = 0.$$

Thus $T_{3,1}X = 0$, and we need go no further with this class.

(3) $T_{2,2}$, we operate with N_1P_1, N_2P_2, where

$$F_1 = \begin{matrix} a_1a_2 \\ a_3a_4 \end{matrix}, \qquad F_2 = \begin{matrix} a_1a_3 \\ a_2a_4 \end{matrix},$$

(i) $$\begin{pmatrix} a_1a_2 \\ a_3a_4 \end{pmatrix} = \begin{pmatrix} a_2a_3 \\ a_4a_1 \end{pmatrix} = -\begin{pmatrix} a_1a_2 \\ a_3a_4 \end{pmatrix} - \begin{pmatrix} a_1a_3 \\ a_2a_4 \end{pmatrix} \quad (\S\ 1)$$

or $$2\begin{pmatrix} a_1a_2 \\ a_3a_4 \end{pmatrix} + \begin{pmatrix} a_1a_3 \\ a_2a_4 \end{pmatrix} = 0;$$

(ii) gives the same equation again;

(iii) gives nothing, because

$$T_{2,2}\{\mathrm{a}_2\mathrm{a}_3\mathrm{a}_4\} = 0.$$

Thus in this class we have one type of form, and, since $f = 2$, we have two independent forms of this type.

(4) $T_{2,1,1}$, there are three standard tableaux,

$$\begin{matrix} a_1a_2 \\ a_3 \\ a_4 \end{matrix}, \quad \begin{matrix} a_1a_3 \\ a_2 \\ a_4 \end{matrix}, \quad \begin{matrix} a_1a_4 \\ a_2 \\ a_3 \end{matrix},$$

(i) $$\begin{pmatrix} a_1a_2 \\ a_3 \\ a_4 \end{pmatrix} = \begin{pmatrix} a_4a_1 \\ a_2 \\ a_3 \end{pmatrix} = \begin{pmatrix} a_2a_3 \\ a_4 \\ a_1 \end{pmatrix} = -\begin{pmatrix} a_1a_2 \\ a_4 \\ a_3 \end{pmatrix} - \begin{pmatrix} a_1a_3 \\ a_4 \\ a_2 \end{pmatrix},$$

and since $$-\begin{pmatrix} a_1a_2 \\ a_4 \\ a_3 \end{pmatrix} = \begin{pmatrix} a_1a_2 \\ a_3 \\ a_4 \end{pmatrix},$$

we have $$\begin{pmatrix} a_1a_3 \\ a_2 \\ a_4 \end{pmatrix} = 0;$$

(ii) $$\begin{pmatrix} a_1a_2 \\ a_3 \\ a_4 \end{pmatrix} = \begin{pmatrix} a_1a_4 \\ a_3 \\ a_2 \end{pmatrix} = -\begin{pmatrix} a_1a_4 \\ a_2 \\ a_3 \end{pmatrix} = 0,$$

on comparing with the equation obtained from (i). Thus

$$T_{2,1,1}X = 0.$$

(5) Equation (i) gives at once

$$T_{1,1,1,1}X = 0.$$

Thus the quartic invariant types of degree 4 give two irreducible forms of a single type of class (3) (*i.e.* $T_{2,2}$); and a product form of class (1) (*i.e.* T_4). The products are substitutionally one form of class (1) and a pair of a single type of class (3). The fact that there are just three products shows at once that this is the only possible arrangement.

More generally the quartic invariant type $(a_1a_2)^2(a_2a_3)^2\dots(a_na_1)^2$ is unchanged by the substitution $(\mathrm{a}_1\mathrm{a}_2\dots\mathrm{a}_n)$; and also if the sequence of the letters is reversed.

Consider the expression of the property that X is unchanged by $(a_1a_2\dots a_n)$. For $T_{n-1,1}$, we have

$$\begin{pmatrix} a_1a_2\dots a_{n-1} \\ a_n \end{pmatrix} = \begin{pmatrix} a_1a_2\dots a_{n-2}a_n \\ a_{n-1} \end{pmatrix} = \dots = \begin{pmatrix} a_2\dots a_n \\ a_1 \end{pmatrix} = 0,$$

on taking the sum and using the identity. Hence $T_{n-1,1}X = 0$.

In general, we find that we obtain a certain number of relations between the standard forms NP_rX, so that all can be expressed in terms of a smaller number. In practice we use the relations to express later standard forms in terms of earlier ones. Thus for $n = 2, 3, 4, 5, 6, 7$, when X is unchanged by $(\mathrm{a}_1\,\mathrm{a}_2\dots\mathrm{a}_n)$, we can express everything in terms of the following forms (we need only write the suffixes of the letters):

n

2; (1 2) for $\begin{pmatrix}1\\2\end{pmatrix} = 0.$

3; (1 2 3) : 0 : $\begin{pmatrix}1\\2\\3\end{pmatrix}$.

4; (1 2 3 4) : 0 : $\begin{pmatrix}1\,2\\3\,4\end{pmatrix} : \begin{pmatrix}1\,2\\3\\4\end{pmatrix} : 0.$

5; (1 2 3 4 5) : 0 : $\begin{pmatrix}1\,2\,3\\4\,5\end{pmatrix} : \begin{pmatrix}1\,2\,3\\4\\5\end{pmatrix}, \begin{pmatrix}1\,2\,4\\3\\5\end{pmatrix} : \begin{pmatrix}1\,2\\3\,4\\5\end{pmatrix} : 0 : \begin{pmatrix}1\\2\\3\\4\\5\end{pmatrix}$.

6; $(1\,2\,3\,4\,5\,6):0:\begin{pmatrix}1&2&3&4\\5&6\end{pmatrix},\begin{pmatrix}1&2&3&5\\4&6\end{pmatrix}:\begin{pmatrix}1&2&3&4\\5\\6\end{pmatrix},\begin{pmatrix}1&2&3&5\\4\\6\end{pmatrix}:\begin{pmatrix}1&2&4\\3&5&6\end{pmatrix}:$

$$\begin{pmatrix}1&2&3\\4&5\\6\end{pmatrix},\begin{pmatrix}1&2&3\\4&6\\5\end{pmatrix}:\begin{pmatrix}1&2&3\\4\\5\\6\end{pmatrix},\begin{pmatrix}1&2&4\\3\\5\\6\end{pmatrix}:$$

$$\begin{pmatrix}1&2\\3&4\\5&6\end{pmatrix},\begin{pmatrix}1&2\\3&5\\4&6\end{pmatrix}:\begin{pmatrix}1&2\\3&4\\5\\6\end{pmatrix}:\begin{pmatrix}1&2\\3\\4\\5\\6\end{pmatrix}:0.$$

7; $(1\,2\,3\,4\,5\,6\,7):0:\begin{pmatrix}1&2&3&4&5\\6&7\end{pmatrix},\begin{pmatrix}1&2&3&4&6\\5&7\end{pmatrix}:$

$$\begin{pmatrix}1&2&3&4&5\\6\\7\end{pmatrix},\begin{pmatrix}1&2&3&4&6\\5\\7\end{pmatrix},\begin{pmatrix}1&2&3&5&6\\4\\7\end{pmatrix}:\begin{pmatrix}1&2&3&4\\5&6&7\end{pmatrix},\begin{pmatrix}1&2&3&5\\4&6&7\end{pmatrix}:$$

$$\begin{pmatrix}1&2&3&4\\5&6\\7\end{pmatrix},\begin{pmatrix}1&2&3&4\\5&7\\6\end{pmatrix},\begin{pmatrix}1&2&3&5\\4&6\\7\end{pmatrix},\begin{pmatrix}1&2&3&5\\4&7\\6\end{pmatrix},\begin{pmatrix}1&2&3&6\\4&5\\7\end{pmatrix}:$$

$$\begin{pmatrix}1&2&3&4\\5\\6\\7\end{pmatrix},\begin{pmatrix}1&2&3&5\\4\\6\\7\end{pmatrix}:\begin{pmatrix}1&2&3\\4&5&6\\7\end{pmatrix},\begin{pmatrix}1&2&3\\4&5&7\\6\end{pmatrix},\begin{pmatrix}1&2&4\\3&5&6\\7\end{pmatrix}:$$

$$\begin{pmatrix}1&2&3\\4&5\\6&7\end{pmatrix},\begin{pmatrix}1&2&3\\4&6\\5&7\end{pmatrix},\begin{pmatrix}1&2&4\\3&5\\6&7\end{pmatrix}:$$

$$\begin{pmatrix}1&2&3\\4&5\\6\\7\end{pmatrix},\begin{pmatrix}1&2&3\\4&6\\5\\7\end{pmatrix},\begin{pmatrix}1&2&3\\4&7\\5\\6\end{pmatrix},\begin{pmatrix}1&2&4\\3&5\\6\\7\end{pmatrix},\begin{pmatrix}1&2&4\\3&6\\5\\7\end{pmatrix}:$$

$$\begin{pmatrix}1&2&3\\4\\5\\6\\7\end{pmatrix},\begin{pmatrix}1&2&4\\3\\5\\6\\7\end{pmatrix},\begin{pmatrix}1&2&5\\3\\4\\6\\7\end{pmatrix}:$$

$$\begin{pmatrix}1&2\\3&4\\5&6\\7\end{pmatrix},\begin{pmatrix}1&2\\3&5\\4&6\\7\end{pmatrix}:\begin{pmatrix}1&2\\3&4\\5\\6\\7\end{pmatrix},\begin{pmatrix}1&2\\3&5\\4\\6\\7\end{pmatrix}:0:\begin{pmatrix}1\\2\\3\\4\\5\\6\\7\end{pmatrix}.$$

We notice that when NP belongs to $T_{a_1, a_2, \ldots, a_h}$ it stands for $f_{a_1, a_2, \ldots, a_h}$ independent forms. Now the total number of independent forms obtained from X by interchanges of letters, when X is unaltered by $(a_1 a_2 \ldots a_n)$, is $(n-1)!$ It is easy to count up in each of the cases above that the forms given give the right number. Thus for $n = 5$ they represent $1+0+5+2 \,.\, 6+5+0+1 = 4\,!$ independent forms.

Let us call these reduced standard forms.

It is easy to see that in every case, except that of $T_{1, 1, \ldots, 1}$, 2 must occupy the second place of the first row. Otherwise, let ab be the nearest pair of letters in the first row, then, if c is a letter between them, we may introduce c into the top row by means of the ordinary identity. Then the pair ac or the pair bc is now in the top row; *i.e.* a pair nearer together than ab; we can proceed thus until we have a consecutive pair in the top row; and then, by means of the permutation $(a_1 a_2 \cdots a_n)$, we may make the consecutive pair to be $a_1 a_2$. We remark that, in our agreed sequence of tableaux, a tableau with 12 in the first row precedes all those which do not contain these digits there.

The change in the sign of every transposition is equivalent to the interchange of N's and P's, and in the case of tableaux to the interchange of columns and rows. This change leaves the substitution $(a_1 a_2 \ldots a_n)$ unaltered when n is odd, but changes its sign when n is even. Hence, when n is odd, $T_{2, 1, \ldots, 1} = 0$.

When n is even the last T is obviously zero.

12. It will be useful to give the relations which express some of the ordinary standard forms for degree 7 in terms of the reduced forms in this case. These relations are sufficiently obvious for the first four T's.

By using the group $\{4\,5\,6\,7\}$ as in § 10, we find

$$\begin{pmatrix} 1\,2\,3\,6 \\ 4\,5\,7 \end{pmatrix} = -\begin{pmatrix} 1\,2\,3\,5 \\ 4\,6\,7 \end{pmatrix} - 3\begin{pmatrix} 1\,2\,3\,4 \\ 5\,6\,7 \end{pmatrix}.$$

In the same way from $\{4\,5\,6\}\begin{pmatrix} 1\,2\,3\,6 \\ 4\,5 \\ 7 \end{pmatrix}$ we obtain

$$\begin{pmatrix} 1\,2\,3\,6 \\ 4\,7 \\ 5 \end{pmatrix} + 2\begin{pmatrix} 1\,2\,3\,6 \\ 4\,5 \\ 7 \end{pmatrix} + \begin{pmatrix} 1\,2\,3\,5 \\ 4\,7 \\ 6 \end{pmatrix}$$

$$+2\begin{pmatrix} 1\,2\,3\,5 \\ 4\,6 \\ 7 \end{pmatrix} + \begin{pmatrix} 1\,2\,3\,4 \\ 5\,7 \\ 6 \end{pmatrix} + 2\begin{pmatrix} 1\,2\,3\,4 \\ 5\,6 \\ 7 \end{pmatrix} = 0.$$

Using the ordinary identity two or three times we find

$$\left(\begin{array}{l}1\,2\,3\,6\\4\\5\\7\end{array}\right)+\left(\begin{array}{l}1\,2\,3\,5\\4\\6\\7\end{array}\right)+\left(\begin{array}{l}1\,2\,3\,4\\5\\6\\7\end{array}\right)=0.$$

For $T_{3,3,1}$, we have

$$\left(\begin{array}{l}1\,2\,3\\5\,6\,7\\4\end{array}\right)=\left(\begin{array}{l}4\,5\,6\\1\,2\,3\\7\end{array}\right)=-\left(\begin{array}{l}1\,2\,3\\4\,5\,6\\7\end{array}\right),$$

and hence

$$\left(\begin{array}{l}1\,2\,3\\4\,5\,7\\6\end{array}\right)+\left(\begin{array}{l}1\,2\,3\\4\,6\,7\\5\end{array}\right)=0.$$

Also we find

$$\left(\begin{array}{l}1\,2\,4\\3\,6\,7\\5\end{array}\right)=\left(\begin{array}{l}1\,2\,4\\3\,5\,6\\7\end{array}\right)+\left(\begin{array}{l}1\,2\,3\\4\,5\,6\\7\end{array}\right)+\left(\begin{array}{l}1\,2\,3\\4\,5\,7\\6\end{array}\right),$$

$$2\left(\begin{array}{l}1\,2\,4\\3\,5\,6\\7\end{array}\right)+\left(\begin{array}{l}1\,2\,4\\3\,5\,7\\6\end{array}\right)+\left(\begin{array}{l}1\,2\,3\\4\,5\,6\\7\end{array}\right)=0,$$

$$\left(\begin{array}{l}1\,2\,5\\3\,4\,6\\7\end{array}\right)=-\left(\begin{array}{l}1\,2\,4\\3\,6\,7\\5\end{array}\right),\quad \left(\begin{array}{l}1\,2\,5\\3\,4\,7\\6\end{array}\right)=-\left(\begin{array}{l}1\,2\,5\\3\,6\,7\\4\end{array}\right)=2\left(\begin{array}{l}1\,2\,4\\3\,5\,6\\7\end{array}\right)-\left(\begin{array}{l}1\,2\,3\\4\,5\,7\\6\end{array}\right),$$

$$\left(\begin{array}{l}1\,2\,6\\3\,4\,5\\7\end{array}\right)=\left(\begin{array}{l}1\,2\,3\\4\,5\,7\\6\end{array}\right),\quad \left(\begin{array}{l}1\,2\,6\\3\,4\,7\\5\end{array}\right)=-\left(\begin{array}{l}1\,2\,4\\3\,6\,7\\5\end{array}\right),\quad \left(\begin{array}{l}1\,2\,6\\3\,5\,7\\4\end{array}\right)=-\left(\begin{array}{l}1\,2\,4\\3\,5\,7\\6\end{array}\right).$$

For $T_{3,2,2}$:

$$\left(\begin{array}{l}1\,2\,4\\3\,6\\5\,7\end{array}\right)=-\left(\begin{array}{l}1\,2\,3\\4\,6\\5\,7\end{array}\right),\quad \left(\begin{array}{l}1\,2\,5\\3\,4\\6\,7\end{array}\right)=-\left(\begin{array}{l}1\,2\,4\\3\,5\\6\,7\end{array}\right),$$

$$\left(\begin{array}{l}1\,2\,5\\3\,6\\4\,7\end{array}\right)=3\left(\begin{array}{l}1\,2\,3\\4\,6\\5\,7\end{array}\right)+\left(\begin{array}{l}1\,2\,3\\4\,5\\6\,7\end{array}\right)-\left(\begin{array}{l}1\,2\,4\\3\,5\\6\,7\end{array}\right),$$

$$\left(\begin{array}{l}1\,2\,6\\3\,4\\5\,7\end{array}\right)=\left(\begin{array}{l}1\,2\,4\\3\,5\\6\,7\end{array}\right),\quad \left(\begin{array}{l}1\,2\,6\\3\,5\\4\,7\end{array}\right)=-\left(\begin{array}{l}1\,2\,3\\4\,6\\5\,7\end{array}\right).$$

For $T_{3,2,1,1}$:

$$\left(\begin{array}{l}1\,2\,4\\3\,7\\5\\6\end{array}\right)+\left(\begin{array}{l}1\,2\,4\\3\,6\\5\\7\end{array}\right)+\left(\begin{array}{l}1\,2\,4\\3\,5\\6\\7\end{array}\right)=-2\left(\begin{array}{l}1\,2\,3\\4\,5\\6\\7\end{array}\right),$$

$$\left(\begin{array}{l}1\,2\,5\\3\,4\\6\\7\end{array}\right)=-\left(\begin{array}{l}1\,2\,4\\3\,5\\6\\7\end{array}\right)-\left(\begin{array}{l}1\,2\,3\\4\,5\\6\\7\end{array}\right)+\left(\begin{array}{l}1\,2\,3\\4\,6\\5\\7\end{array}\right)+\left(\begin{array}{l}1\,2\,3\\4\,7\\5\\6\end{array}\right),$$

$$\left(\begin{array}{l}1\,2\,5\\3\,6\\4\\7\end{array}\right)=\left(\begin{array}{l}1\,2\,4\\3\,7\\5\\6\end{array}\right)+\left(\begin{array}{l}1\,2\,3\\4\,5\\6\\7\end{array}\right),$$

$$\left(\begin{array}{l}1\,2\,5\\3\,7\\4\\6\end{array}\right)=\left(\begin{array}{l}1\,2\,3\\4\,7\\5\\6\end{array}\right),\quad \left(\begin{array}{l}1\,2\,6\\3\,4\\5\\7\end{array}\right)=\left(\begin{array}{l}1\,2\,4\\3\,7\\5\\6\end{array}\right)-\left(\begin{array}{l}1\,2\,3\\4\,6\\5\\7\end{array}\right),$$

$$\left(\begin{array}{l}1\,2\,6\\3\,5\\4\\7\end{array}\right)=\left(\begin{array}{l}1\,2\,4\\3\,5\\6\\7\end{array}\right)-\left(\begin{array}{l}1\,2\,4\\3\,7\\5\\6\end{array}\right)+\left(\begin{array}{l}1\,2\,3\\4\,5\\6\\7\end{array}\right)-\left(\begin{array}{l}1\,2\,3\\4\,6\\5\\7\end{array}\right)-\left(\begin{array}{l}1\,2\,3\\4\,7\\5\\6\end{array}\right),$$

$$\left(\begin{array}{l}1\,2\,6\\3\,7\\4\\5\end{array}\right)=-\left(\begin{array}{l}1\,2\,4\\3\,5\\6\\7\end{array}\right)-\left(\begin{array}{l}1\,2\,3\\4\,5\\6\\7\end{array}\right)-\left(\begin{array}{l}1\,2\,3\\4\,7\\5\\6\end{array}\right).$$

The remaining forms do not call for special remark.

13. Let us write $\bar{f}_{a_1, a_2, \ldots, a_h}$ or just $\bar{f}$ as the number of reduced standard forms belonging to $T_{a_1, a_2, \ldots, a_h}$. Then $f\,.\,\bar{f}$ is the total number of linearly independent forms obtained by interchanging the letters in TX.

If we ignore the letter a_n—or rather agree that it shall always occupy the same position in X (*e.g.* the last), which we can do—then we may regard X as a function of the first $n-1$ letters only, and express it and the functions derived by interchange of letters from it in terms of the corresponding forms $N'P'_r X$. These are expressible in terms of the forms NPX involving all the n letters, and hence in terms of the reduced standard forms. Let us use f' to denote the number of standard tableaux F' corresponding to T' which involves the first $n-1$ letters only. Then the number $\Sigma f'$ is considerably greater than the number of reduced standard tableaux $\Sigma \bar{f}$, as we see at a glance of the particulars above; it is evident from the fact that the total number of linearly independent forms X is

$$\Sigma f\bar{f} = (n-1)! = \Sigma f'^2,$$

remembering that f is, in general, considerably greater than f'.

Now
$$N'_s P'_s X = \Sigma \frac{ff'}{n!(n-1)!} N'_s P'_s T N'_s P'_s X$$
$$= \Sigma \frac{ff'}{n!(n-1)!} N'_s P'_s N_s P_s N'_s P'_s X,$$

where the standard tableau F_s is obtained from the standard tableau F'_s by adding a_n, so that it occupies a position which is the end of a row and the bottom of a column.

In order to see the truth of this we observe that a_n must occupy such a position in any standard tableau F_r, and its suppression produces a standard tableau F'_r. Hence $N_r P_r = \lambda N'_r N_r P_r P'_r$, and, unless $r = s$, either

$$P'_s N'_r = 0 \quad \text{or} \quad P'_r N'_s = 0.$$

Now
$$N'_s P'_s N_s P_s N'_s P'_s X = \Sigma \lambda_{r,s} N'_s P'_s N_s \sigma_{sr} \bar{P}_r X,$$

where $\bar{P}_r$ is obtained from a reduced standard tableau. Thus

$$N'_s P'_s X = \Sigma' \Sigma \mu_{r,s} N'_s P'_s N_s \sigma_{sr} \bar{P}_r X,$$

where the first Σ' expresses the sum of the sets of terms according to the possible T''s involved, viz. those which are obtained by the increase of one of the suffixes by unity of the T' concerned on the other side. In just the same way we obtain

$$N'_s \sigma'_{st} P'_t X = \Sigma' \Sigma \mu^{(t)}_{r,s} N'_s P'_s N_s \sigma_{sr} \bar{P}_r X.$$

There are thus $f'_{a_1, a_2, \ldots, a_h}$ possible linearly independent forms with the same N'_s on the left-hand side expressed in terms of

$$\bar{f}_{a_1+1, a_2, \ldots, a_h} + \bar{f}_{a_1, a_2+1, \ldots a_h} + \ldots + \bar{f}_{a_1, a_2, \ldots, a_h, 1} = \bar{S}_f \quad \text{(say)}$$

forms on the right-hand side. Hence

$$\bar{S}_f \geqslant f'_{a_1, a_2, \ldots, a_h}.$$

Now, by § 3 (1),

$$\Sigma \bar{f}_{\beta_1, \beta_2, \ldots, \beta_h} f_{\beta_1, \beta_2, \ldots, \beta_h}$$
$$= \Sigma \bar{f}_{\beta_1, \beta_2, \ldots, \beta_h} \{ f'_{\beta_1 - 1, \beta_2, \ldots, \beta_h} + f'_{\beta_1, \beta_2 - 1, \ldots, \beta_h} + \ldots + f'_{\beta_1, \beta_2, \ldots, \beta_h - 1} \}$$
$$= \Sigma f'_{a_1, a_2, \ldots, a_h} \{ \bar{f}_{a_1+1, a_2, \ldots, a_h} + \bar{f}_{a_1, a_2+1, \ldots, a_h} + \ldots + \bar{f}_{a_1, a_2, \ldots, a_h, 1} \}$$
$$= \Sigma f'^2_{a_1, a_2, \ldots a_k}.$$

Therefore $$\Sigma f'_{a_1, a_2, \ldots, a_h} \{ \bar{S}_f - f'_{a_1, a_2, \ldots, a_h} \} = 0.$$

None of the terms on the right can be negative, as we have just seen, hence each is zero. Therefore

THEOREM X. *If* $\bar{f}_{\beta_1, \beta_2, \ldots, \beta_h}$ *be the number of reduced standard tableaux corresponding to* $T_{\beta_1, \beta_2, \ldots, \beta_h}$, *when the operand is unaltered by the substitution* $(a_1 a_2 \ldots a_n)$, *then*

$$\bar{f}_{a_1+1, a_2, \ldots, a_h} + \bar{f}_{a_1, a_2+1, \ldots, a_h} + \cdots + \bar{f}_{a_1, a_2, \ldots, a_h, 1} = f_{a_1, a_2, \ldots, a_h}.$$

It should be noticed that the proof of this theorem applies equally well to the case when X is unaltered by the substitutions of any transitive group of degree n. In fact, it applies to the following :

THEOREM XI. *If* G *be a transitive group involving* n *letters* $a_1, a_2, \ldots, a_n$; *and if* Γ *be that subgroup of* G *which leaves* a_n *unchanged, and if* $\bar{f}_{\beta_1, \beta_2, \ldots, \beta_h}$ *be the number of reduced standard tableaux corresponding to* $T_{\beta_1, \beta_2, \ldots, \beta_h}$ *when the operand is unaltered by any substitution of* G ; *and* $\bar{f}'_{a_1, a_2, \ldots, a_h}$ *the number of reduced standard tableaux corresponding to* $T'_{a_1, a_2, \ldots, a_h}$ *(involving the first* $n-1$ *letters only) when the operand is unaltered by any substitution of* Γ, *then*

$$\bar{f}_{a_1+1, a_2, \ldots, a_h} + \bar{f}_{a_1, a_2+1, \ldots, a_h} + \cdots + \bar{f}_{a_1, a_2, \ldots, a_h, 1} = \bar{f}'_{a_1, a_2, \ldots, a_h}.$$

13. When the object of discussion is not a function of the letters, but just a substitutional expression, in general the equations will involve both left- and right-hand multiplication. Thus we might seek to find those substitutional expressions which are unaltered by multiplication by $(a_1 a_2 \ldots a_n)$ on either side. We have called the number of independent forms NP belonging to T which are unchanged by multiplication on the right hand by $(a_1 a_2 \ldots a_n)$, N being supposed fixed, $\bar{f}$; and thus, taking the f different forms N, we had $f\bar{f}$ independent forms. We might equally well have used the forms PN instead of NP, and we are bound to find the same number $\bar{f}$ in this case. Now, if we are dealing with expressions also unchanged by left-hand multiplication, there are only $\bar{f}$ reduced standard forms N in NP as well as that number for P. Thus the total number of linearly independent forms of this character belonging to T is $\bar{f}^2$. And the total number altogether of such forms is $\Sigma \bar{f}^2$. These, of course, include the sum of the substitutions of any group which contains $(a_1 a_2 \ldots a_n)$.

14. The quartic invariant types are unchanged by a reversal of the sequence of letters. We then seek for standard forms further reduced, which are appropriate to the discussion of functions unchanged by the operations of the group :

$$\{(1\,2\,3\,\ldots\,n),\ (1n)(2\,\overline{n-1}),\ \ldots\},$$

These are :

n

3; $(1\,2\,3):0:0,$

4; $(1\,2\,3\,4):0:\begin{pmatrix}1&2\\3&4\end{pmatrix}:0:0,$

5; $(1\,2\,3\,4\,5):0:\begin{pmatrix}1&2&3\\4&5&\end{pmatrix}:0:\begin{pmatrix}1&2\\3&4\\5&\end{pmatrix}:0:\begin{pmatrix}1\\2\\3\\4\\5\end{pmatrix},$

6; $(1\,2\,3\,4\,5\,6):0:\begin{pmatrix}1&2&3&4\\5&6&&\end{pmatrix},\ \begin{pmatrix}1&2&3&5\\4&6&&\end{pmatrix}:0:0:\begin{pmatrix}1&2&3\\4&5&\\6&&\end{pmatrix}:\begin{pmatrix}1&2&4\\3&&\\5&&\\6&&\end{pmatrix}:$

$$\begin{pmatrix}1&2\\3&4\\5&6\end{pmatrix},\ \begin{pmatrix}1&2\\3&5\\4&6\end{pmatrix}:0:\begin{pmatrix}1&2\\3&\\4&\\5&\\6&\end{pmatrix}:0.$$

7; $(1\,2\,3\,4\,5\,6\,7):0:\begin{pmatrix}1&2&3&4&5\\6&7&&&\end{pmatrix},\ \begin{pmatrix}1&2&3&4&6\\5&7&&&\end{pmatrix}:0:\begin{pmatrix}1&2&3&4\\5&6&7&\end{pmatrix}:$

$$\begin{pmatrix}1&2&3&4\\5&6&&\\7&&&\end{pmatrix},\ \begin{pmatrix}1&2&3&5\\4&6&&\\7&&&\end{pmatrix},\ \begin{pmatrix}1&2&3&5\\4&7&&\\6&&&\end{pmatrix}:\begin{pmatrix}1&2&3&5\\4&&&\\6&&&\\7&&&\end{pmatrix}:0:$$

$$\begin{pmatrix}1&2&3\\4&5&\\6&7&\end{pmatrix},\ \begin{pmatrix}1&2&3\\4&6&\\5&7&\end{pmatrix},\ \begin{pmatrix}1&2&4\\3&5&\\6&7&\end{pmatrix}:\begin{pmatrix}1&2&3\\4&5&\\6&&\\7&&\end{pmatrix},\ \begin{pmatrix}1&2&4\\3&5&\\6&&\\7&&\end{pmatrix}:$$

$$\begin{pmatrix}1&2&3\\4&&\\5&&\\6&&\\7&&\end{pmatrix},\ \begin{pmatrix}1&2&4\\3&&\\5&&\\6&&\\7&&\end{pmatrix},\ \begin{pmatrix}1&2&5\\3&&\\4&&\\6&&\\7&&\end{pmatrix}:\begin{pmatrix}1&2\\3&4\\5&6\\7&\end{pmatrix}:0:0:0,$$

$$8;\quad (12345678):0:\left(\begin{array}{l}123456\\78\end{array}\right),\ \left(\begin{array}{l}123457\\68\end{array}\right),\ \left(\begin{array}{l}123467\\58\end{array}\right):0:$$

$$\left(\begin{array}{l}12345\\678\end{array}\right):\left(\begin{array}{l}12345\\67\\8\end{array}\right),\left(\begin{array}{l}12346\\57\\8\end{array}\right),\left(\begin{array}{l}12346\\58\\7\end{array}\right),\left(\begin{array}{l}12356\\47\\8\end{array}\right):$$

$$\left(\begin{array}{l}12346\\5\\7\\8\end{array}\right),\left(\begin{array}{l}12356\\4\\7\\8\end{array}\right):\left(\begin{array}{l}1234\\5678\end{array}\right),\ \left(\begin{array}{l}1235\\4678\end{array}\right),\ \left(\begin{array}{l}1246\\3578\end{array}\right):$$

$$\left(\begin{array}{l}1234\\567\\8\end{array}\right),\left(\begin{array}{l}1235\\467\\8\end{array}\right),\left(\begin{array}{l}1235\\468\\7\end{array}\right):\left(\begin{array}{l}1234\\56\\78\end{array}\right),\left(\begin{array}{l}1234\\57\\68\end{array}\right),$$

$$\left(\begin{array}{l}1235\\46\\78\end{array}\right),\left(\begin{array}{l}1235\\47\\68\end{array}\right),\left(\begin{array}{l}1236\\45\\78\end{array}\right),\left(\begin{array}{l}1236\\47\\58\end{array}\right),\left(\begin{array}{l}1246\\35\\78\end{array}\right):$$

$$\left(\begin{array}{l}1234\\56\\7\\8\end{array}\right),\left(\begin{array}{l}1235\\46\\7\\8\end{array}\right),\left(\begin{array}{l}1235\\47\\6\\8\end{array}\right),\left(\begin{array}{l}1246\\37\\5\\8\end{array}\right):$$

$$\left(\begin{array}{l}1234\\5\\6\\7\\8\end{array}\right),\left(\begin{array}{l}1235\\4\\6\\7\\8\end{array}\right),\left(\begin{array}{l}1236\\4\\5\\7\\8\end{array}\right),\left(\begin{array}{l}1246\\3\\5\\7\\8\end{array}\right):\left(\begin{array}{l}123\\457\\68\end{array}\right):$$

$$\left(\begin{array}{l}123\\456\\7\\8\end{array}\right),\left(\begin{array}{l}123\\457\\6\\8\end{array}\right),\left(\begin{array}{l}124\\356\\7\\8\end{array}\right),\left(\begin{array}{l}124\\357\\6\\8\end{array}\right),\left(\begin{array}{l}124\\358\\6\\7\end{array}\right):$$

$$\left(\begin{array}{l}123\\45\\67\\8\end{array}\right),\left(\begin{array}{l}123\\45\\68\\7\end{array}\right),\left(\begin{array}{l}123\\46\\57\\8\end{array}\right),\left(\begin{array}{l}124\\35\\67\\8\end{array}\right):$$

$$\left(\begin{array}{l}123\\45\\6\\7\\8\end{array}\right),\left(\begin{array}{l}123\\46\\5\\7\\8\end{array}\right),\left(\begin{array}{l}124\\35\\6\\7\\8\end{array}\right),\left(\begin{array}{l}124\\36\\5\\7\\8\end{array}\right):\left(\begin{array}{l}123\\4\\5\\6\\7\\8\end{array}\right):$$

$$\left(\begin{array}{l}12\\34\\56\\78\end{array}\right),\left(\begin{array}{l}12\\34\\57\\68\end{array}\right),\left(\begin{array}{l}12\\36\\47\\58\end{array}\right):0:\left(\begin{array}{l}12\\34\\5\\6\\7\\8\end{array}\right),\left(\begin{array}{l}12\\36\\4\\5\\7\\8\end{array}\right):0:0.$$

15. If we were to consider the corresponding relations for sextic invariant types of the form

$$(a_1 a_2)^3 (a_2 a_3)^3 \ldots (a_n a_1)^3 ;$$

we should find that, when n is odd, a reversal of the sequence of letters changes the sign. The appropriate reduced standard forms then are :

n

$$3; \quad 0:0:\begin{pmatrix}1\\2\\3\end{pmatrix}.$$

$$5; \quad 0:0:\begin{pmatrix}1\,2\,3\\4\\5\end{pmatrix}, \begin{pmatrix}1\,2\,4\\3\\5\end{pmatrix}:0:0.$$

$$7; \quad 0:0:0:\begin{pmatrix}1\,2\,3\,4\,5\\6\\7\end{pmatrix}, \begin{pmatrix}1\,2\,3\,4\,6\\5\\7\end{pmatrix}, \begin{pmatrix}1\,2\,3\,5\,6\\4\\7\end{pmatrix}:\begin{pmatrix}1\,2\,3\,5\\4\,6\,7\end{pmatrix}:$$

$$\begin{pmatrix}1\,2\,3\,4\\5\,6\\7\end{pmatrix}, \begin{pmatrix}1\,2\,3\,5\\4\,6\\7\end{pmatrix}:\begin{pmatrix}1\,2\,3\,4\\5\\6\\7\end{pmatrix}:\begin{pmatrix}1\,2\,3\\4\,5\,6\\7\end{pmatrix}, \begin{pmatrix}1\,2\,3\\4\,5\,7\\6\end{pmatrix}:\begin{pmatrix}1\,2\,4\\3\,5\,6\\7\end{pmatrix}:$$

$$0:\begin{pmatrix}1\,2\,3\\4\,5\\6\\7\end{pmatrix}, \begin{pmatrix}1\,2\,3\\4\,6\\5\\7\end{pmatrix}, \begin{pmatrix}1\,2\,4\\3\,5\\6\\7\end{pmatrix}:$$

$$0:\begin{pmatrix}1\,2\\3\,5\\4\,6\\7\end{pmatrix}:\begin{pmatrix}1\,2\\3\,4\\5\\6\\7\end{pmatrix}, \begin{pmatrix}1\,2\\3\,5\\4\\6\\7\end{pmatrix}:0:\begin{pmatrix}1\\2\\3\\4\\5\\6\\7\end{pmatrix}.$$

16. The other equations which the quartic types satisfy are of the form $\{abc\}\, I = R$, where R stands for a sum of products of invariants, and the symbol $\{abc\}$ means, in the first place, the symmetric group of any three consecutive letters; and, in the second place, a, b, c may each stand for a set of several letters, and then the symbol $\{abc\}$ is supposed to interchange the sets *en bloc* in I after the manner of the symmetric group of degree 3—(the sequence of the letters in each set being undisturbed)—it being understood that the three sets abc, as they stand,

form a consecutive set of letters in I*. A further proviso is necessary, namely that there is at least one letter in I not contained in abc. Then we see at once that, for $n = 4$,

$$I_4 = \lambda \begin{pmatrix} 1 & 2 \\ 3 & 4 \end{pmatrix} I_4 + R,$$

and there are two independent irreducible forms. For $n = 5$

$$I_5 = \lambda \begin{pmatrix} 1 & 2 \\ 3 & 4 \\ 5 & \end{pmatrix} I_5 + \mu \begin{pmatrix} 1 \\ 2 \\ 3 \\ 4 \\ 5 \end{pmatrix} I_5 + R$$

giving 6 forms.

For $n = 6$, we have to introduce a second equation, which we may write $\{1,\ 2,\ 3\ 4\}\, I = R$. If $I = [1\,2\,3\,4\,5\,6]$, this gives

$$\{1\ 2\}\,\{[1\,2\,3\,4\,5\,6] + [1\,3\,4\,2\,5\,6] + [3\,4\,1\,2\,5\,6]\} = R.$$

Now when we operate with NP on $\Sigma\lambda s\,.\,I$, the result is represented by operating on the tableau for P with s^{-1}.

Thus $\begin{pmatrix} 1 & 2 \\ 3 & 4 \\ 5 & 6 \end{pmatrix} \{1,\ 2,\ 3\ 4\}\, I = R$, gives

$$\left[\begin{pmatrix} 1 & 2 \\ 3 & 4 \\ 5 & 6 \end{pmatrix} + \begin{pmatrix} 1 & 4 \\ 2 & 3 \\ 5 & 6 \end{pmatrix} + \begin{pmatrix} 3 & 4 \\ 1 & 2 \\ 5 & 6 \end{pmatrix}\right] I = R$$

or

$$\left[\begin{pmatrix} 1 & 2 \\ 3 & 4 \\ 5 & 6 \end{pmatrix} - \begin{pmatrix} 1 & 2 \\ 3 & 5 \\ 4 & 6 \end{pmatrix}\right] I = R.$$

Similarly operating with $\begin{pmatrix} 1 & 2 \\ 3 & 5 \\ 4 & 6 \end{pmatrix}$ we have

$$\left[\begin{pmatrix} 1 & 2 \\ 3 & 5 \\ 4 & 6 \end{pmatrix} + \begin{pmatrix} 1 & 4 \\ 2 & 5 \\ 3 & 6 \end{pmatrix} + \begin{pmatrix} 3 & 4 \\ 1 & 5 \\ 2 & 6 \end{pmatrix}\right] I = R$$

or

$$\left[\left(5 \begin{pmatrix} 1 & 2 \\ 3 & 5 \\ 4 & 6 \end{pmatrix} + \begin{pmatrix} 1 & 2 \\ 3 & 5 \\ 4 & 6 \end{pmatrix}\right)\right] I = R.$$

Whence $$T_{2,2,2} I = R.$$

* *Proc. London Math. Soc.* (1), 30 (1898), 295.

Similarly we may prove that $T_{2,1,1,1,1}I = R$, though this is obvious for other reasons.

Hence
$$I_6 = \lambda \begin{pmatrix} 1 & 2 & 4 \\ 3 & & \\ 5 & & \\ 6 & & \end{pmatrix} I_6 + R,$$

and there are 10 independent forms.

For $n = 7$, we may use $\{2\,3,\ 4\,5,\ 6\,7\}\, I = R$, *i.e.*
$$\{(2\,4)(3\,5),\ (4\,6)(5\,7)\}\, I = R\,;$$

operation with $\begin{pmatrix} 1 & 2 & 4 \\ 3 & 5 & \\ 6 & 7 & \end{pmatrix}$ reduces this.

The reduction of $\begin{pmatrix} 1 & 2 & 4 \\ 3 & 5 & \\ 6 & & \\ 7 & & \end{pmatrix}$ does not come with this operation. We may, however, obtain it by operating with $\left\{\begin{matrix} 1 & 2 & 5 \\ 3 & 4 & \\ 6 & & \\ 7 & & \end{matrix}\right\}$ on
$$\{1,\ 2,\ 3\,4\}\, I = R$$
and using the equations of § 9.

The reduction of $T_{3,1,1,1,1}$ is obvious, but it may be obtained, of course, from these equations. Thus $I_7 = R$.

Similarly for I_8.

17. For the sextic invariant types mentioned above, § 15, we have equations of exactly the same form, viz.:
$$\{a\,b\,c\,d\}\, I = R,$$
where a, b, c, d each stands for a set of consecutive letters, all four sets being consecutive in I. There are no equations before degree 6. Thus
$$I_5 = \lambda \left[\begin{pmatrix} 1 & 2 & 3 \\ 4 & & \\ 5 & & \end{pmatrix} + \begin{pmatrix} 1 & 2 & 4 \\ 3 & & \\ 5 & & \end{pmatrix} \right] I_5$$
and there are 12 independent forms;
$$I_6 = \lambda \begin{pmatrix} 1 & 2 & 3 & 5 \\ 4 & 6 & & \end{pmatrix} I_6 + \ldots$$
and there are 50 forms.

Consider these sextic invariant types in general. If, in any row of P, there are 4 consecutive digits, then $NPI = R$. Further, the same is true when there are two triads of consecutive digits in any row. Let these be 1, 2, 3, and r, $r+1$, $r+2$, the result is obtained by operating with NP on the equation

$$\{1, 2, 3, 4\ 5 \ldots (r-1)\} = R.$$

When there are three separated digits in a row of P, the form may be expressed in terms of such as have the *first* pair separated only by a single digit, and of forms which have two of the digits replaced by a consecutive pair. Let P have 1, r, s in a row, but none of the intervening letters in that row, then the result is given by

$$NP\{1,\ 2\,3 \ldots (r-2),\ (r-1)\,r\,(r+1) \ldots (s-1),\ s\} = R;$$

this equation introduces also terms in which the number of digits between the first and last of the three considered is diminished, we regard this as a reduction, and use the same process to the terms so reduced, thus obtaining the result. We may regard the digits 1, r, s as systems of consecutive digits for which the first and last of any system is in the row considered, and the argument holds good. Thus we obtain the result that, if a row contains four separated digits, the form may be expressed in terms of forms in which there are four separated digits, the first two pairs being separated by a single digit, and of forms in which two of the separated digits are replaced by a consecutive pair. We treat the last three digits first, and, by what we have just proved, we may suppose that the first pair of these are separated by a single digit. Then we treat this pair and the digit between as a single digit, and the result follows.

We are now in a position to prove that a form with a row which has four separated digits can be expressed in terms of forms in which two of the digits are replaced by a consecutive pair. We have shown that we may suppose that each of the first two pairs is separated by a single digit, thus we may suppose that the digits 1, 3, 5, r lie in a row of P. Then we may include in the generic symbol R all forms in which two of these digits are replaced by a consecutive pair. Then

$$NP\{1, 2, 3, 4\}I = R$$

gives

$$NP\{2, 4\}I = R.$$

Let us write $$I = \binom{1\ \ 3\ \ 5\ \ r}{2,\ 4,\ 6,\ 7,\ \ldots,\ (r-1)},$$

the commas in the lower row indicating that any permutation of the digits in the row must be supposed to change the function. The equation

$$\{4,\ 5,\ 6\,7\ \ldots\ (r-1),\ r\}\, I = R$$

gives

$$\begin{pmatrix}1\ 3\ 5\ r \\ 2,\ 4,\ 6,\ 7,\ \ldots,\ (r-1)\end{pmatrix} + \begin{pmatrix}1\ 3\ (r-2)\ r \\ 2,\ (r-1),\ 4,\ 5,\ \ldots,\ (r-3)\end{pmatrix} = R.$$

Also by

$$\{1,\ 2,\ 3,\ 4\,5\ \ldots\ (r-3)\} \begin{pmatrix}1\ 3\ (r-2)\ r \\ 2,\ (r-1),\ 4,\ 5,\ (r-3)\end{pmatrix} = R,$$

$$\begin{pmatrix}1\ 3\ (r-2)\ r \\ 2,\ (r-1),\ 4,\ 5,\ \ldots,\ (r-3)\end{pmatrix} + \begin{pmatrix}1\ (r-4)\ (r-2)\ r \\ (r-3),\ (r-1),\ 2,\ 3,\ \ldots,\ (r-5)\end{pmatrix} = R;$$

and by

$$\{2\,3\,4\ \ldots\ (r-5),\ (r-4)\ (r-3)\ (r-2),\ (r-1),\ r\} \times \begin{pmatrix}1\ (r-4)\ (r-2)\ r \\ (r-3),\ (r-1),\ 2,\ 3,\ \ldots,\ (r-5)\end{pmatrix} = R,$$

we have

$$\begin{pmatrix}1\ (r-4)\ (r-2)\ r \\ (r-3),\ (r-1),\ 2,\ 3,\ \ldots,\ (r-5)\end{pmatrix} + \begin{pmatrix}1\ 3\ 5\ r \\ 2,\ 4,\ 6,\ 7,\ \ldots,\ (r-1)\end{pmatrix} = R.$$

Hence $$\begin{pmatrix}1\ 3\ 5\ r \\ 2,\ 4,\ 6,\ 7,\ \ldots,\ (r-1)\end{pmatrix} = R.$$

The argument holds good when one or more of the four separated digits is replaced by a consecutive pair or a consecutive triad. We may, in fact, state the result thus: in any particular row of P we may suppose that the digits are separated by not more than two gaps. Naturally we shall take the particular row to be the top one. Let us suppose that this row contains four letters. In this case we may suppose that there is not more than one gap. For, including forms of the desired nature in the symbol R, $\{1, 2, 3, 4\}$ gives

$$\begin{pmatrix}1\ 2\ 4\ r \\ 3,\ 5,\ 6,\ \ldots,\ (r-1)\end{pmatrix} + \begin{pmatrix}1\ 3\ 4\ r \\ 2,\ 5,\ 6,\ \ldots,\ (r-1)\end{pmatrix} = R,$$

$\{3,\ 4,\ 5\,6\ \ldots\ (r-1),\ r\}$ gives

$$\begin{pmatrix}1\ 3\ 4\ r \\ 2,\ 5,\ 6,\ \ldots,\ (r-1)\end{pmatrix} + \begin{pmatrix}1\ 3\ (r-1)\ r \\ 2,\ 4,\ 5,\ \ldots,\ (r-2)\end{pmatrix} = R$$

and $\{1,\ 2\,3\ \ldots\ (r-2),\ (r-1),\ r\}$ gives

$$\begin{pmatrix}1\ 3\ (r-1)\ r \\ 2,\ 4,\ 5,\ \ldots,\ (r-2)\end{pmatrix} + \begin{pmatrix}1\ 2\ 4\ r \\ 3,\ 5,\ 6,\ \ldots,\ (r-1)\end{pmatrix} = R,$$

whence the result as stated.

The same thing is true when there are five letters. The process of proof is the same, and need not be detailed; the theorem is proved first for the case when there are two pairs of consecutive digits and a single one, and then using this result for the case of a triad and two single digits.

A form with two consecutive pairs in a row can be expressed in terms of forms with a triad and a single digit, for example, if P contains 1, 2, and r, $r+1$ in a row, we obtain the result from $\{1, 2, 3\,4 \dots (r-1), r\}$.

Hence, when there are 6 digits in the top row of P we may suppose them to be a triad, a pair, and a single one; and the same process as used in the former cases may be now employed to show that we can express this in terms of forms in which there is only one gap in the top row. Now this means either two consecutive triads or a set of 4 consecutive letters, and we have already seen that either of these cases is reducible. Hence

$$T'_{\alpha_1, \alpha_2 \dots} I = R,$$

when $\alpha_1 > 5$.

Moreover, when $\alpha_1 = 4$ or 5, we need consider only forms with one gap in the top row, and, in particular, forms with a triad. For $\alpha_1 = 3$, we have already shown that, owing to the fact that I is unchanged, or changed only in sign, by the permutation $(1\,2 \dots n)$, we may suppose that the top row begins with the digits 1 2, and hence here, too, there is only one gap. Thus in all cases we need consider only forms with one gap in the top row.

Further, we may suppose that the single gap contains not more than three digits. For, writing a, b as the two consecutive sets in the top row and supposing that $a+1$, $a+2$, ..., $b-1$ are the single digits between them, we see that the equation

$$\{a\,(a+1),\ (a+2),\ (a+3)\,(a+4),\ (a+5) \dots (b-1)\,b\}\, I = R$$

gives

$$\{1+(\overline{a+2}\;\overline{a+3}\;\overline{a+4})\}\, I = R,$$

where R includes forms with a smaller gap than in I. This equation, of course, reduces I, and the reduction holds good so long as there are at least 4 digits in I.

The same process yields the result, for $b = a+4$,

$$I = -(\overline{a+1}\;\overline{a+2})\, I + R = -(\overline{a+2}\;\overline{a+3})\, I + R$$

$$= -(\overline{a+1}\;\overline{a+3})\, I + R,$$

which means that we have a reduction for a gap of three digits unless each of those digits is in a different row of P.

Let Γ_r be a negative symmetric group of r digits. Then, by Peano's theorem*, $\Gamma_r I = 0$ when $r > 7$, and $\Gamma_7 I$ has a factor Δ the determinant of the seven quantics to which the letters of Γ_7 refer. In the case of the sextic Δ is reducible. For

$$\{\text{abcdefg}\}'(ab)^3(bc)^3(ca)^3 . (de)^5(ef)(fg)^5(gd)$$

is not zero, and is therefore the invariant in question. Now it was shown in the proof of Peano's theorem, just quoted, that when this invariant is reducible for the binary n-ic, then also $\Gamma_n I$ is reducible. Hence, when there are more than 5 rows in P the form is certainly reducible. It will be noticed that this last reduction is not contained in our formal equations, for none of them reduce the invariant in question of degree 7. They would give the general result of a reduction for $\Gamma_7 I$ when the degree of I is greater than 7. Thus we can say from these equations alone, and so far as we have gone, that I is certainly reducible when the degree is greater than 30, for P cannot have more than 6 rows and more than 5 digits in a row, and with the added knowledge of the reducibility of Δ we can say that P is certainly reducible when the degree is greater than 25. Of course this system does not contain all the sextic invariant types; it does not, for instance, include the invariant of degree 15 for a single sextic. We have considered it rather as an example of substitutional equations.

18. We proceed to show briefly how group characters naturally arise in the investigation of equations of the form $\Sigma\lambda s X = R$, where s is an operation of a group H. For this purpose we consider the operations of the group as capable of quantitative expressions, as heretofore. That is, we submit them to the ordinary laws of algebra, the commutative law alone excepted. Equations composed of sums of terms involving such operations may be interpreted in two ways. In the first place a function to be operated on—an operand—may always be implicitly understood to be present, so that the sum of the operations really represents a sum of functions. Or, in the second place, we may regard the operations as capable of independent existence, as, for instance, hypercomplex units, or else as matrices, thinking of the results as referring to matrix groups.

Frobenius first considers the k different classes into which the elements of a group H may be divided, all the elements conjugate to a given

* Q.S.A., I, 141–143.

element forming one class. The number of elements belonging to a class is called h_a, and the number of elements in whole group is h, so that

$$\Sigma h_a = h.$$

If now we write S_a as the sum of the elements of the a-th class, the product $S_a S_\beta$, when expanded as a sum of elements, contains all the elements of any given class in exactly the same way, so that

$$S_a S_\beta = \sum_\gamma \frac{h_{a\beta\gamma'}}{h_\gamma} S_\gamma,$$

where γ' is the class inverse to γ, that is, if s is an operation of γ, then γ' contains that operation s^{-1} which satisfies the equation $s\,.\,s^{-1} = 1$, and $h_{a,\beta,\gamma'}$ is numerical—in fact, is defined by this equation [F.B.C. (1896), 987]. The expressions S_a obey all the laws of algebra, including the commutative law. And

$$\Sigma \lambda_a S_a \,.\, \Sigma \mu_a S_a = \Sigma \nu_a S_a,$$

whatever numerical values λ_a and μ_a may have. Moreover,

$$\Sigma \lambda_a S_a \,.\, \Sigma \mu_a S_a = \sum_\gamma \left[\sum_{a,\beta} \frac{h_{a\beta\gamma'}}{h_\gamma} \lambda_a \mu_\beta \right] S_\gamma.$$

Let us solve the equation

$$[\Sigma \lambda_a S_a]^2 = \Sigma \lambda_a S_a.$$

This is done by the aid of the numbers $\chi_a^{(\kappa)}$ called group characters. For consider $\Sigma \chi_{a'} S_a$ with the aid of the equations given by Frobenius*, quoted below:

$$[\Sigma \chi_{a'} S_a]^2 = \sum_\gamma \left[\sum_{a,\beta} \chi_{a'} \chi_{\beta'} \frac{h_{a\beta\gamma'}}{h_\gamma} \right] S_\gamma$$

$$= \sum_\gamma \left[\sum_\beta \frac{h_\beta \chi_\beta \chi_\gamma}{f} \chi_{\beta'} \right] S_\gamma \quad [\S\ 1\ (5) \text{ and } \S\ 2\ (6)]$$

$$= \sum_\gamma \frac{h \chi_{\gamma'}}{e} S_\gamma \quad [\S\ 3\ (3)].$$

Let us write $$T_\kappa = \frac{e^{(\kappa)}}{h} \Sigma \chi_{a'}^{(\kappa)} S_a,$$

then $$T_\kappa^2 = T_\kappa;$$

* F.B.C. (1896), 985 *et seq.*

and we obtain k possible solutions of the equation. Also

$$T_\kappa T_\lambda = \sum_\gamma \left[\sum_{\alpha, \beta} \frac{e^{(\kappa)} e^{(\lambda)}}{h^2} \chi_{\alpha'}^{(\kappa)} \chi_{\beta'}^{(\lambda)} \frac{h_{\alpha\beta\gamma'}}{h_\gamma} \right] S_\gamma$$

$$= \sum_\gamma \left[\sum_\beta \frac{e^{(\kappa)} e^{(\lambda)}}{h^2 f^{(\kappa)}} \chi_{\beta'}^{(\lambda)} h_\beta \chi_\beta^{(\kappa)} \chi_{\gamma'}^{(\kappa)} \right] S_\gamma$$

$$= 0 \quad [\S\ 3\ (2)].$$

When H is the symmetric group these expressions T_κ are the same as those which, in my former papers, were written $A_{a_1, a_2, \ldots, a_h} T_{a_1, a_2, \ldots, a_h}$.

The expressions S_α can be written in terms of the T_κ's; thus using [§ 3, (4)],

$$S_\alpha = \sum_\kappa \frac{h_\alpha \chi_\alpha^{(\kappa)}}{f^{(\kappa)}} T_\kappa.$$

Hence

$$\Sigma x_\alpha S_\alpha = \sum_{\alpha, \kappa} \frac{h_\alpha \chi_\alpha^{(\kappa)} x_\alpha}{f^{(\kappa)}} T_\kappa$$

$$= \xi^{(\kappa)} T_\kappa,$$

where $\xi^{(\kappa)}$ is the same as in [§ 2, 11].

Consider the specially simple form of equation

$$\Sigma x_\alpha S_\alpha X = 0,$$

which may be written

$$\Sigma \xi^{(\kappa)} T_\kappa X = 0.$$

A solution is sought of the form

$$\Sigma y_\alpha S_\alpha F = \Sigma \eta^{(\kappa)} T_\kappa F.$$

Then

$$(\Sigma \xi^{(\kappa)} T_\kappa)(\Sigma \eta^{(\kappa)} T_\kappa) F = \Sigma \xi^{(\kappa)} \eta^{(\kappa)} T_\kappa F,$$

and there is a solution other than zero when

$$\prod_\kappa \xi^{(\kappa)} = 0,$$

and only then.

In his paper *Über die Darstellung der endlichen Gruppen durch lineare Substitutionen* [B.S. (1897)], Frobenius treats functions of this kind of hypercomplex units in this way in § 6, and he further shows (p. 1010) that in general the multiplication of two such functions may be reduced to a matter of composition of matrices.

ON QUANTITATIVE SUBSTITUTIONAL ANALYSIS

Alfred Young*

1. In the *Proceedings of the London Math. Soc.* (1), vols. 33 and 34, two papers appeared with the above title, and a third has been recently communicated to the Society. This note is concerned with the application and not with the proof of the results contained in these papers.

We are concerned with a function X of n variables $a_1, a_2, \ldots, a_n$, and with the linear relations which may connect X with those functions obtained from it by a permutation of the variables. By permutation there may be $n!$ linearly independent functions, or when X belongs to a group g there may be $n!/g$ linearly independent functions. But besides belonging to a group there are other and more fundamental properties (for our purpose) which X may have. As a simple example, the symbolical product of algebra $(ab)c_x$ is a function which obeys the relation

$$\{abc\}' X = 0;$$

where we use $\{a_1 a_2 \ldots a_n\}'$ as the sum of the even permutations of the symmetric group of the letters $a_1, a_2, \ldots, a_n$ minus the sum of the odd permutations, and call it the negative symmetric group; similarly $\{a_1 a_2 \ldots a_n\}$ is written for the sum of all the permutations of the symmetric group and is called the positive symmetric group. This function $(ab)c_x$ has two linearly independent values on permutation, and we may speak of it as two-valued. It will be found that, for n variables, we can construct r-valued functions of several different types for most values of r; but, in general, there are a few values of r for which it is impossible to construct an r-valued function.

2. The variables $a_1, a_2, \ldots, a_n$ are arranged in a tableau F_r, of h rows, with α_1 letters in the first row, α_2 in the second, and so on, where $\alpha_1 \geqslant \alpha_2 \geqslant \ldots \geqslant \alpha_h$, $\Sigma\alpha = h$, and each row begins with the same column on the left. Thus

$$\begin{array}{l} a_{1,1}\, a_{1,2} \ldots a_{1,\alpha_1}, \\ a_{2,1}\, a_{2,2} \ldots a_{2,\alpha_2}, \\ \ldots \quad \ldots \quad \ldots \\ a_{h,1} \ldots a_{h,\alpha_h}. \end{array}$$

* Received 28 October, 1927; read 10 November, 1927.

The product of the positive symmetric groups of the rows, *i.e.*

$$\prod_{s=1}^{h} \{a_{s,1}\, a_{s,2} \ldots a_{s,\alpha_s}\},$$

is written P_r and the product of the negative symmetric groups of the columns is written N_r. When $\alpha_1, \alpha_2, \ldots, \alpha_h$ are kept fixed permutation of the n letters results in $n!$ tableaux F_r. We write

$$T_{\alpha_1, \alpha_2, \ldots, \alpha_h} = \Sigma P_r N_r,$$

where the Σ extends to the $n!$ terms obtained from these tableaux.

We may speak of these tableaux F_r and corresponding products $P_r N_r$ as belonging to a class, the class defined by the numbers $\alpha_1, \alpha_2, \ldots, \alpha_h$. Then the product of PN and $P'N'$, which belong to different classes, is zero. So also the product of two different T's is zero.

Next we fix the sequence of the variables $a_1, a_2, \ldots, a_n$; and define F_r to be a *standard* tableau when the letters in every row, and also those in every column, follow this fixed sequence.

Thus for five letters and class $T_{3,2}$, there are five standard tableaux

$a_1 a_2 a_3$	$a_1 a_2 a_4$	$a_1 a_2 a_5$	$a_1 a_3 a_4$	$a_1 a_3 a_5$
$a_4 a_5$	$a_3 a_5$	$a_3 a_4$	$a_2 a_5$	$a_2 a_4$
F_1	F_2	F_3	F_4	F_5.

The number of standard tableaux of a class is written $f_{\alpha_1, \alpha_2, \ldots, \alpha_h}$ or simply f; then

$$\Sigma f^2_{\alpha_1, \alpha_2, \ldots, \alpha_h} = n!^*.$$

Moreover, calling the standard tableaux $F_1, F_2, \ldots, F_f$, a certain linear function M_r of the substitutions may be obtained, such that

$$N_r M_r P_s = 0 \quad (r, s = 1, 2, \ldots, f;\ r \neq s),$$

$$N_r M_r P_r = N_r P_r.$$

In the above example, for the class $T_{3,2}$,

$$M_r = 1 \quad (r > 1), \quad M_1 = 1 - (a_1 a_5 a_3).$$

[It should be remarked that the function to be operated on is supposed to be on the right-hand side of the operator; so that $NPX = N(PX)$.]

$$* \; f = n! \frac{\prod_{r,s} (\alpha_r - \alpha_s - r + s)}{\prod_r (\alpha_r + h - r)!}.$$

Let $\sigma_{s,r}$ be that permutation which transforms F_r into F_s, then

$$\sigma_{s,r} P_r \sigma_{s,r}^{-1} = P_s, \quad \sigma_{s,r} N_r \sigma_{s,r}^{-1} = N_s.$$

There are exactly f^2 expressions $P_s \sigma_{s,r} N_r M_r$, where F_s, F_r are standard, and they possess the following properties:

(i) the product of two of them is either zero or a third,

$$P_s \sigma_{s,r} N_r M_r P_u \sigma_{u,v} N_v M_v = 0 \qquad (r \neq u),$$
$$= P_s \sigma_{s,v} N_v M_v \quad (r = u);$$

(ii) there is no linear relation between them;

(iii) taking all the classes into account, every permutation and every substitutional expression can be uniquely expressed in terms of them; for they number $\Sigma f^2 = n!$.

In particular

$$1 = \sum_{a_1, a_2, \ldots, a_h} \left[\frac{f}{n!} \sum_{r=1}^{f} P_r N_r\right] = \Sigma T'_{a_1, a_2, \ldots, a_h}.$$

Thus any substitutional expression may be written

$$S_1 = \sum_{a_1, a_2, \ldots, a_h} \left[\sum_{r,s} \lambda_{r,s} P_r \sigma_{r,s} N_s M_s\right].$$

Hence, if

$$S_2 = \sum_{a_1, \ldots, a_h} \left[\sum_{r,s} \mu_{r,s} P_r \sigma_{r,s} N_s M_s\right],$$

$$S_1 S_2 = \sum_{a_1, \ldots, a_h} \left[\sum_{r,s} \sum_{t} \lambda_{r,t} \mu_{t,s} P_r \sigma_{r,s} N_s M_s\right].$$

And the substitutional expressions may be written as matrices*, an independent matrix, of f rows and columns, for each class. They are compounded as matrices, and substitutional equations may be solved as matrix equations. If the matrix of a class γ of S_1 have the rank ρ, then in the solution of the equation

$$S_1 X = 0,$$

the matrix of class γ of X will have rank $f-\rho$. In the case of five letters, we have

$$f_5 = 1, \quad f_{4,1} = 4, \quad f_{3,2} = 5, \quad f_{3,1,1} = 6, \quad f_{2,2,1} = 5,$$
$$f_{2,1,1,1} = 4, \quad f_{1,1,1,1,1} = 1.$$

* For the general representation of a group by matrices, see Frobenius, "Über die Darstellung der endlichen Gruppen durch lineare Substitutionen", *Berliner Sitzungsberichte*, 1 (1897), 2 (1899).

3. Consider now a function X. So far as its substitutional properties are concerned, it can be defined by an equation of the form

$$X = \Sigma[\Sigma \mu_{r,s} P_r \sigma_{r,s} N_s M_s] Y.$$

The effect of permutations on this is expressed entirely by multiplication on the left-hand side by a substitutional form. This never touches the $N_s M_s$ part of a term, so that all our equations can only give relations between the coefficients $\mu_{r,s}$ which have the same second suffix (or between elements of the same column of the matrix). Thus, so far as the operand is concerned, each class is divided into f subclasses entirely independent. Moreover, the relations are the same for each value of s. Hence, if ρ be the rank of $[\mu_{r,s}]$, the number of linearly independent values that X takes on permutation is $\Sigma \rho f$, where $\rho \leqslant f$. Since P_r determines the properties of any term in X, a term may be defined completely for our purpose by the corresponding tableau F_r. And here, as we are only concerned with P_r, the letters in each row may be permuted at will.

Thus, in effect, if k be the number of classes, we have just k types of different possible functions, and these have properties defined by the tableaux. Namely: (i) the variables in each row are interchangeable; (ii) the product of a positive symmetric group of degree d which contains no letter above the r-th row in the tableau is zero when $d > a_r$.

All functions are made up linearly of functions of these types.

4. It is easy to see that an r-valued function for any value of $r \leqslant n!$ can be constructed when $n \leqslant 4$. For $n = 5$, there is no 3-valued function. But r may have any other value except 117 not greater than 120. Thus a 7-valued function is obtained by taking either the sum of

(i) one form of class (5), one of class (1, 1, 1, 1, 1), and one of class (3, 2);

(ii) by replacing the last by one of class (2, 2, 1);

(iii) one form (5) and one form (3, 1, 1);

or (iv) one form (3, 1, 1) and one form (1, 1, 1, 1, 1).

Obviously the cases r and $(n!-r)$ are reciprocal, there being the same number of possibilities in each case. For $n = 6$, the impossible values of r are 3, 4, 8, 13, 707, 712, 716, 717.

5. The relative rôles of P and N may be reversed without affecting the theory. This can simply be done by changing the sign of every transposition. A transposition of letters in one row of the tableau representing

one of the fundamental functions will now be accompanied by a change of sign.

As an illustration, let us consider

$$NPY,$$

where
$$Y = \prod_{r=1}^{n} a_{x^{(r)}}^{(r)},$$

$$a_{x^{(r)}}^{(r)} = \sum_{s=1}^{m} a_s^{(r)} x_s^{(r} ;$$

and the variables permuted are $a^{(r)}$ $(r = 1, 2, \ldots, n)$.

Let F be the tableau for NP

$$\begin{array}{l} a_{1,1}\, a_{1,2} \ldots a_{1,\alpha_1}, \\ a_{2,1} \ldots a_{2,\alpha_2}, \\ \ldots \quad \ldots \quad \ldots, \end{array}$$

where P is now formed from the columns. We suppose that in Y the variable $x_{r,s}$ is associated always with $a_{r,s}$. Then

$$PY = QY,$$

where Q is obtained from P by replacing every a by the corresponding x.

Moreover, N is permutable with Q; and NY can be written down as a mixed symbolical form. Thus, for example, when $m = 3$, $n = 6$, and

$$\begin{array}{ll} F = & a^{(1)} a^{(2)} a^{(3)} \\ & a^{(4)} a^{(5)} \\ & a^{(6)}, \end{array}$$

$$NPY = \{x^{(1)} x^{(4)} x^{(6)}\}\, \{x^{(2)} x^{(5)}\} (a^{(1)} a^{(2)} a^{(3)})(x^{(1)} x^{(2)} x^{(3)})\, a_{x^{(6)}}^{(6)} \Sigma\, [(a^{(4)} a^{(5)})(x^{(4)} x^{(5)})].$$

There are 16 standard tableaux of this class. Keeping P fixed we have 16 linearly independent forms of this nature, that is 16 forms

$$(a_1 a_2 a_3)(a_4 a_5 u)\, a_{6x}.$$

Moreover, by writing down the standard tableau, it is possible to write down at once a set in terms of which the rest may be expressed.

If, instead of replacing $x^{(1)}$, etc., by x and u, we had kept them as separate variables and taken into account the different P's we should have had 256 linearly independent forms. The symbolical product $(a_1 a_2 a_3)(a_4 a_5 u)\, a_{6x}$ is in fact, as regards permutational properties, the fundamental function NPY of class $T_{3,2,1}$; and for these purposes may always be taken to represent it. And generally each of these fundamental functions of any class may be represented as a symbolical product, linear in each of the symbols

It is well known* that any gradient can be expressed as a sum of terms of covariants. The expression may be obtained in the general case by using the symbolical notation, polarization, and then *operation* with the sum $\Sigma T'_{a_1, a_2, \ldots, a_h}$.

6. The set of standard tableaux chosen is not unique. In the case of the class $T_{3,2}$, we might take

$$\begin{matrix} a_1 a_2 a_3 & a_2 a_3 a_4 & a_3 a_4 a_5 & a_4 a_5 a_1 & a_5 a_1 a_2 \\ a_4 a_5 & a_5 a_1 & a_1 a_2 & a_2 a_3 & a_3 a_4 \end{matrix},$$

obtained by operating on the first by the permutations of a group. Such a method of selecting them appears attractive. But in many cases there is no appropriate group, *e.g.* for $T_{3,1,1}$ or $T_{3,2,1}$. It can be done for $T_{4,2}$ and for every $T_{n-1,1}$, and no doubt in many other cases.

* Elliott, *Algebra of Quantics*, 151, 152 (binary forms), 321 (ternary).

ON QUANTITATIVE SUBSTITUTIONAL ANALYSIS

By ALFRED YOUNG

(*Fourth Paper.*)

[Received 30 August, 1929.—Read 14 November, 1929.]

The third paper of this series established the theorem that any permutation Q can be expressed in terms of certain substitutional units, in the form

$$Q = \Sigma a^Q_{r,s} P_r \sigma_{rs} N_s M_s$$

(we write here $N_s M$ for what was called there N'_s). Here we show that

$$P_r \sigma_{rs} N_s M_s = \frac{f}{n!} \Sigma a^Q_{r,s} Q^{-1},$$

the point being that the numerical coefficient $a^Q_{r,s}$ is the same in both cases.

This gives a simple rule by which the group matrix may be obtained, Theorem I, § 2.

Section 3 has to do with a more detailed examination of the expression M_r.

Prof. Schur, "Über die Darstellung der symmetrischen Gruppe durch lineare homogene Substitutionen", *Berliner Sitz.* (1908), 664–678, has obtained a representation of the symmetric group by linear substitutions which is not merely equivalent to this one, but identical with it; the method is different, and so also is the form of the statement of the result. His paper has suggested Theorem II, § 4, which expresses the connection between the units used here and the *characteristic* of the irreducible linear group. The possibility suggested by Prof. Schur that the coefficients in this set of linear groups may be confined to 0, 1, -1 is shown in § 5 not to be the case. Further, § 6 establishes one of his fundamental results by these methods.

The last part of this paper proves two theorems by means of which substitutional expressions in $n-1$ letters can be transformed into expressions in n letters. Theorem III does this for a single unit, and Theorem IV for the matrix form.

The three former papers in this series which appeared in these *Proceedings* (1), 33 (1901), 97; (1), 34 (1902), 361; (2), 28 (1927), 255, are quoted as Q.S.A. I, II, III. And a paper with the same title in the *Journal*, 3 (1927), 14, is quoted as Q.S.A.J.

We are here concerned with the group matrix, *i.e.* if the matrices of a linear substitution group be A, B, C, ..., then the group matrix is the sum $\Sigma A x_A$, ...; that is, it is a matrix each of whose elements is a linear function of h variables x_A, there being one variable for each operation of the group.

Thus consider the symmetric group of degree 3, $\{abc\}$; there is a representation by means of matrices of two rows and columns, the group matrix can be written

$$\begin{bmatrix} y_1+x_3-y_2-x_2, & y_2+x_1-y_3-x_3 \\ y_3+x_1-y_2-x_2, & y_1+x_2-y_3-x_3 \end{bmatrix}.$$

This means that the several matrices of the group are the matrix coefficients of the different variables. Thus the coefficients of y_1, y_2, y_3, x_1, x_2, x_3 are

$$\begin{bmatrix} 1 & 0 \\ 0 & 1 \end{bmatrix}, \begin{bmatrix} -1 & 1 \\ -1 & 0 \end{bmatrix}, \begin{bmatrix} 0 & -1 \\ 1 & -1 \end{bmatrix}, \begin{bmatrix} 0 & 1 \\ 1 & 0 \end{bmatrix}, \begin{bmatrix} -1 & 0 \\ -1 & 1 \end{bmatrix}, \begin{bmatrix} 1 & -1 \\ 0 & -1 \end{bmatrix},$$

corresponding to 1, (abc), (acb), (bc), (ca), (ab).

See Dickson, *Modern algebraic theories* (1926), 251–270.

1. It is convenient to change slightly the notation used in the previous papers. The prepared standard forms formerly written $P_r N_r'$* are written here $P_r N_r M_r$, bringing into evidence that they are obtained by introducing a fresh factor M. Further, the numerical factor $f/n!$ will, in general, be supposed absorbed into the symbol $P_r N_r M_r$. This enables us to write

$$(P_r N_r M_r)^2 = P_r N_r M_r.$$

Moreover we use the symbol T with the definition

$$T = \sum_{r=1}^{f} P_r N_r M_r.$$

* Q.S.A. III, § 4, p. 263.

Then $$T^2 = T,$$

and we have $$1 = \Sigma T_{a_1, a_2, \dots, a_h}.$$

2. It has been shown* that any permutation Q can be expressed in the form

$$\Sigma a_{rs}^{(Q)} P_r \sigma_{rs} N_s M_s,$$

where $a_{rs}^{(Q)}$ is numerical. The matrices $[a_{rs}^{(Q)}]$ then form a group isomorphic with the symmetric group; there being one such group for each $T_{a_1, \dots, a_h}$. The degree of the matrices is f. The group matrix is $[\Sigma a_{rs}^{(Q)} x_Q]$.

Consider the operation of $\Sigma Q x_Q$ on an arbitrary function X. A set of $n!$ equations may be obtained by operating on the equation $\Sigma Q x_Q X = 0$ with each of the permutations of the symmetric group; they are linear equations satisfied by the $n!$ unknown quantities QX. The condition that there is a non-zero solution is that the numbers x_Q are such that

$$|x_{R^{-1}Q}| = 0. \tag{I}$$

Now, writing Qx_Q in terms of the units $P_r \sigma_{rs} N_s M_s$, we find the condition expressed by k conditions

$$|\Sigma a_{rs}^{(Q)} x_Q| = 0. \tag{II}$$

Thus the determinants II are the factors of I. Now Frobenius has shown that the determinant I has k irreducible factors, viz. the group determinants of the k irreducible linear substitution groups. Thus the expressions II are the group determinants of the irreducible groups. One of the determinants II cannot be a product of two of the group determinants, for the sum of the squares of the degrees is $n!$ in each case.

Thus $[\Sigma_Q a_{rs}^Q x_Q]$ is the group matrix of an irreducible representation of the symmetric group as a linear substitution group.

Now let b_{rs}^Q be the coefficient of the permutation Q in the expansion of $P_r \sigma_{rs} N_s M_s$, so that

$$P_r \sigma_{rs} N_s M_s = \sum_Q b_{rs}^Q Q. \tag{III}$$

Then, since $$Q = \sum_{\kappa, \lambda} a_{\kappa\lambda}^Q P_\kappa \sigma_{\kappa\lambda} N_\lambda M_\lambda,$$

we find that $$P_r \sigma_{rs} N_s M_s = \sum_Q \sum_{\kappa, \lambda} b_{rs}^Q a_{\kappa\lambda}^Q P_\kappa \sigma_{\kappa\lambda} N_\lambda M_\lambda.$$

Hence $$\sum_Q b_{rs}^Q a_{\kappa\lambda}^Q = 0, \tag{IV}$$

---

* Q.S.A. III, § 7, pp. 267–269.

unless $\kappa = r$ and $\lambda = s$, and $a^Q_{\kappa\lambda}$ belongs to the same T as b^Q_{rs}; also

$$\sum_Q b^Q_{rs} a^Q_{rs} = 1. \tag{V}$$

Since $[\sum_Q a^Q_{rs} x_Q]$ is an irreducible group matrix*

$$\sum_Q a^Q_{\alpha\beta} a^{Q^{-1}}_{\gamma\delta} = \frac{n!}{f} e_{\alpha\delta} e_{\gamma\beta}, \tag{VI}$$

where $[e_{rs}]$ is the unit matrix.

Keeping r, s fixed there are $n!$ equations IV and V which may be regarded as linear equations to determine the $n!$ unknowns b^Q_{rs}. These are linearly independent since the $n!$ quantities

$$Q = \sum a^Q_{\kappa\lambda} P_\kappa \sigma_{\kappa\lambda} N_\lambda M_\lambda$$

are independent. Hence the solution is unique. Consequently

$$b^Q_{rs} = \frac{f}{n!} a^{Q^{-1}}_{sr}.$$

Moreover, we see from III that $(n!/f)\, b^Q_{rs}$ is a rational integer. Thus we obtain

THEOREM I. *The group matrix of the irreducible linear substitution group of degree f corresponding to T is obtained from the matrix*

$$[m_{rs}] \equiv \left[\frac{n!}{f} P_s \sigma_{sr} N_r M_r\right]$$

by replacing every permutation Q^{-1} by the variable x_Q, the coefficients being all rational integers.

This gives a simple means of writing down a group of rational integral substitutions isomorphic with the symmetric group.

COROLLARY. $$\sum_{r=1}^{f} \frac{n!}{f} P_r N_r M_r = \sum \chi_\sigma \sigma = T' \frac{n!}{f},$$

where χ_σ is the characteristic of σ for this irreducible representation.

This follows at once from the above when we remember that the characteristic of a substitution is the sum of the elements in the leading diagonal.

---

* Schur, "Neue Begründung", *Berliner Sitzungsberichte* (1905), § 2, Theorem IV.

3. The actual form* of M_r in the expression $P_r N_r M_r$ requires a rather more explicit examination. The symbol σ_{rs} will be used here to denote that permutation which changes the tableau F_s to F_r.

Let $N_r P_s$ be not zero and let s be other than r; then we know that $s > r$. Now N_r must contain a permutation ν_r, for which $\nu_r P_s \nu_r^{-1} = P_r$†. Similarly P_s contains a permutation ϖ_s for which $\varpi_s N_r \varpi_s^{-1} = N_s$.

Hence $$\sigma_{rs} = \varpi_r \nu_r \quad \text{and} \quad \sigma_{sr} = \nu_s \varpi_s.$$

Now $\nu_r^{-1} \varpi_r \nu_r$ is a permutation of

$$\nu_r^{-1} P_r \nu_r = \nu_r^{-1} \varpi_r^{-1} P_r \varpi_r \nu_r = \sigma_{sr} P_r \sigma_{rs} = P_s.$$

Hence $$\sigma_{rs} = \nu_r \varpi_s'.$$

This permutation ϖ_s' is the inverse of ϖ_s above, since

$$\sigma_{rs} = \varpi_s^{-1} \nu_s^{-1} = \varpi_s^{-1} \nu_s^{-1} \varpi_s . \varpi_s^{-1} = \nu_r \varpi_s^{-1}.$$

Thus, slightly changing the notation, we have, in this case,

$$\sigma_{rs} = \nu_r \varpi_s, \tag{I}$$

where $\nu_r \varpi_s \nu_r^{-1}$ belongs to P_r and $\varpi_s^{-1} \nu_r \varpi_s$ to N_s. Hence when $N_r P_s$ is not zero it is equal to $\pm N_r \sigma_{rs} P_s$ according as ν_r is even or odd. And, so far as P_s alone is concerned,

$$M_r = 1 - \varpi_s.$$

It may happen that $N_s P_t \neq 0$ where $t > s$.

Then (Q.S.A. III) $$N_r M_r = N_r - \varpi_s N_s (1 - \varpi_t)$$
$$= N_r (1 - \varpi_s + \varpi_s \varpi_t).$$

In fact $$N_r \sigma_{rt} = N_r \nu_r \varpi_s \nu_s \varpi_t = N_r \nu_r' \nu_r \varpi_s \varpi_t = \pm N_r \varpi_s \varpi_t.$$

Thus, when the values of s for which

$$N_r P_s \neq 0$$

are $r_1, r_2, \ldots$; the values of s for which

$$N_{r_\lambda} P_s \neq 0$$

---

* Q.S.A. III, § 4; Q.S.A.J., p. 15.

† Q.S.A. II, 363, § 13.

are $r_{\lambda 1}$, $r_{\lambda 2}$, etc., and so on; then

$$M_r = 1 - \Sigma \varpi_{r_\lambda} + \Sigma \varpi_{r_\lambda} \varpi_{r_{\lambda\mu}} - \Sigma \varpi_{r_\lambda} \varpi_{r_{\lambda\mu}} \varpi_{r_{\lambda\mu\nu}} + \dots$$

and here $$r < r_\lambda < r_{\lambda\mu} < r_{\lambda\mu\nu} < \dots$$

It will generally be found convenient to state the law of sequence of tableaux a little differently from the statement in Q.S.A. III, § 4. We say that F_r precedes F_s when each of the letters $a_n, a_{n-1}, \dots, a_{n-\lambda+1}$ lies in the same row in both tableaux, but $a_{n-\lambda}$ lies in a lower row in F_r than in F_s. We may call this the sequence from the last letter. It is sometimes useful to use instead the sequence from the first letter, in which F_r precedes F_s when each of the letters $a_1, a_2, \dots, a_{\lambda-1}$ lies in the same row in both tableaux, but a_λ lies in a higher row in F_r than in F_s. Everything that has been said, for instance, in Q.S.A. III, § 4, applies to either of these sequences.

To find the values of r_λ when the sequence is from the last letter, we take the tableau F_r, and make a permutation in the columns only, such that the last letter which is moved goes to a higher row; the new tableau F'_r is not standard, but, if it can be made standard by a permutation ϖ in its rows only, the resulting tableau F_{r_λ} gives one of the values of r_λ; and the permutation needed for M_r in respect of it is $\varpi_{r_\lambda} = \varpi^{-1}$.

It is useful to notice that

$$M_r = 1 - \Sigma \pm \sigma_{rr_\lambda} + \Sigma \pm \sigma_{rr_{\lambda\mu}} \dots, \tag{II}$$

the signs being determined in each case by the number of column transpositions.

We shall also use the expression

$$M_r = 1 - \Sigma \varpi_{r_\lambda} M_{r_\lambda},$$

where, for each r_λ, $N_r P_{r_\lambda}$ is not zero, and where

$$\varpi_{r_\lambda}^{-1} N_r \varpi_{r_\lambda} = N_{r_\lambda}.$$

Also $$P_r N_r M_r = P_r N_r - \Sigma \pm P_r \sigma_{rr_\lambda} N_{r_\lambda} + \Sigma \pm P_r \sigma_{rr_{\lambda\mu}} N_{r_{\lambda\mu}} \dots$$

Denote the matrix $[P_r \sigma_{rs} N_s M_s]$ by I, and the matrix $[P_r \sigma_{rs} N_s]$ by J. Let S be that matrix $[a_{rs}]$, in which a_{rr} is zero and a_{rs} is zero unless $N_s P_r$ is not zero, and then a_{rs} is ∓ 1 according as

$$P_r N_r P_s N_s = \pm P_r \sigma_{rs} N_s.$$

Then $$I = J(E + S + S^2 + \dots).$$

And if a substitutional expression be $\Sigma X_{rs} P_r \sigma_{rs} N_s M_s = \Sigma Y_{rs} P_r \sigma_{rs} N_s$,

then $$[X_{rs}] = [Y_{rs}](E+S+S^2+\ldots),$$

or $$[Y_{rs}] = [X_{rs}](E-S).$$

4. We shall use the notation*

$$\Phi_{\alpha_1, \alpha_2, \ldots, \alpha_h} = \Sigma \frac{\chi_{\beta_1 \beta_2 \ldots \beta_n}}{\beta_1 ! \, \beta_2 ! \ldots \beta_n !} \left(\frac{s_1}{1}\right)^{\beta_1} \left(\frac{s_2}{2}\right)^{\beta_2} \cdots \left(\frac{s_n}{n}\right)^{\beta_n}$$

to define the characteristic of the irreducible component of the symmetric group of degree n, corresponding to $T_{\alpha_1, \alpha_2, \ldots, \alpha_h}$. Then we have

THEOREM II. *When every cycle of degree* r ($r = 1, \ldots, n$) *is replaced by the symbol* s_r:

(i) $T_{\alpha_1 \ldots \alpha_h}$ *becomes* $f_{\alpha_1 \ldots \alpha_h} \Phi_{\alpha_1 \ldots \alpha_h}$;

(ii) $P_r N_r M_r$ *becomes* $\Phi_{\alpha_1 \ldots \alpha_h}$;

(iii) $P_s \sigma_{sr} N_r M_r$ *becomes zero*;

(iv) $S . P_r N_r M_r$ *becomes* $\kappa \left(\frac{n!}{f}\right) \Phi_{\alpha_1 \ldots \alpha_h}$, *where* κ *is the coefficient of the identical permutation in* $SP_r N_r M_r$, S *being any substitutional expression, involving the* n *letters.*

The first statement follows at once from the corollary to Theorem I.

We may write $$T_{\alpha_1 \ldots \alpha_h} = \left(\frac{f}{n!}\right) \bar{\Sigma} PN,$$

where the $\bar{\Sigma}$ extends to all the $n!$ permutations of the n letters in PN. All the terms in the sum are similar, hence PN becomes $\Phi_{\alpha_1 \ldots \alpha_h}$.

The tableaux F_r, F_s being both standard, and $r < s$, there is at least one pair of letters bc in the same row in F_r and in the same column in F_s for
$$N_s P_r = 0.$$

Hence the result of interchanging b and c in $P_r \sigma_{rs} N_s$ is merely to change its sign. Thus
$$\bar{\Sigma} P_r \sigma_{rs} N_s = 0.$$

Now we have seen, (II) § 3, that
$$M_r = 1 + \Sigma \pm \sigma_{rs} \quad (s > r);$$

* Schur, *Berliner Sitzungsberichte* (1908), 665.

hence $$P_r N_r M_r = P_r N_r + \Sigma \pm P_r \sigma_{rs} N_s,$$

and $$\bar{\Sigma} P_r N_r M_r = \bar{\Sigma} P_r N_r\,;$$

which proves the second statement.

When $r=f$, s cannot be greater than r and $M_r = 1$ always; and hence

$$\bar{\Sigma} P_s \sigma_{sr} N_r M_r = 0.$$

Let us suppose that this equation has been proved for all values of r greater than a given value. Then, for this value of r,

$$P_s \sigma_{sr} N_r M_r = P_s \sigma_{sr} N_r (1 - \Sigma \varpi_{r_\lambda} M_{r_\lambda}) = P_s \sigma_{sr} N_r - \Sigma \pm P_s \sigma_{sr_\lambda} N_{r_\lambda} M_{r_\lambda},$$

and hence $$\bar{\Sigma} P_s \sigma_{sr} N_r M_r = 0,$$

unless $s = r_\lambda$; *i.e.* it is such that $N_r P_s$ is not zero. But in this case

$$\bar{\Sigma} P_{r_\lambda} \sigma_{r_\lambda r} N_r M_r = \bar{\Sigma} [P_{r_\lambda} \sigma_{r_\lambda r} N_r - P_{r_\lambda} \sigma_{r_\lambda r} N_r \varpi_{r_\lambda}],$$

since such terms as $\varpi_{r_\lambda} \varpi_{r_{\lambda\mu}} M_{r_{\lambda\mu}}$, which are introduced by M_{r_λ}, lead only to

$$\bar{\Sigma} \pm P_{r_\lambda} \sigma_{r_\lambda r_{\lambda\mu}} N_{r_{\lambda\mu}} M_{r_{\lambda\mu}},$$

which is zero by hypothesis. The second term on the right-hand side is derived from the first by the permutation ϖ_{r_λ}, which is contained in P_{r_λ}; hence this sum is zero, and the third statement is proved.

We know that $$S.P_r N_r M_r = \Sigma A_{sr} P_s \sigma_{sr} N_r M_r,$$

and hence $$\bar{\Sigma} S.P_r N_r M_r = \Sigma A_{rr} \bar{\Sigma} P_r N_r M_r.$$

The fourth statement then is the result of comparing the coefficient of s_1^n on the two sides, after replacing permutations by symbols s_r.

5. The linear group of § 2, Theorem I, is identical with that obtained by Schur*. He starts with a set of functions which are unchanged by all the operations of

$$P = \prod_{r=1}^{\lambda} \{a_{r1} \ldots a_{r\alpha_r}\} = \prod_{r=1}^{\lambda} P_\lambda,$$

and then considers the whole modul of such functions obtainable from these by the operations of the symmetric group. Amongst these he distinguishes a subdivision A obtained in effect by operating on a member of

* *Loc. cit.*

the first set by a positive symmetric group G, which contains all the letters of P_r and one more from the next lower group; A includes all those functions obtained by permutation of letters from these. He then shows that any member of M can be linearly expressed in terms of a certain set of f functions mod A. This set corresponds exactly to what I have called standard forms. The use of modulus A is in effect the application of the equations to which the forms PN have to submit. And hence the group obtained is the same. On p. 678 Schur remarks that in the case which I call $T_{a11\ldots1}$ the coefficients in the linear group are all either 0, 1, or -1; but says that he has not yet determined whether this is always so. The answer to that question is no.

Consider PN; let us write ϖ for any permutation of P and ν for any permutation of N.

Then, when $\varpi_1\nu_1 = \varpi_2\nu_2$, we must have $\varpi_1 = \varpi_2$ and $\nu_1 = \nu_2$; and hence the coefficient of every permutation in PN when expanded is ± 1. Similarly this is so in the case of $P_r\tau_{rs}N_s$. It is also true for $P_rN_rM_r$, and for $P_s\tau_{sr}N_rM_r$, in the case in which there is not more than one value of t, other than r, for which

$$N_rP_t \neq 0,$$

and no value of u, other than t, for which

$$N_tP_u \neq 0.$$

We have seen that the coefficients of the linear substitutions are the same as the coefficients of the expressions $P_r\sigma_{rs}N_sM_s$ when different permutations are represented in terms of them. In other words, they are the coefficients when a non-standard form $P\sigma N$ is expressed in terms of standard forms

$$\sum_{r=1}^{f} A_r P_r\sigma_r N.$$

Using the alphabetical sequence of letters, we find (Q.S.A. 271, Theorem IX)

$$\left\{\begin{matrix}a\,c fg\\bcd\end{matrix}\right\} = 2\left\{\begin{matrix}abcd\\efg\end{matrix}\right\} + [\,\{(bcd)\}\,\{(efg)\}\,]\left\{\begin{matrix}abce\\dfg\end{matrix}\right\}$$

(the operations included in the brackets [ ] operate on the letters of the operations which follow them; the tableaux of the P's only are here expressed, the same N being present in each case).

Thus, in $T_{4,3}$ and naturally in other cases, the coefficients are not confined to ± 1, as is seen at once by the theorem quoted.

6. Let $P_1'N_1'$ belong to $T'_{a_1,\ldots,a_h}$ with n letters, being derived from the first tableau of the set. Let g be the positive symmetric group of r more letters. And let $T_{\beta_1,\ldots,\beta_k}$ refer to the whole $n+r$ letters; we shall take the sequence of letters as beginning with the same sequence as in $P_1'N_1'$ and ending with the r additional letters. Consider

$$N_1'T_{\beta_1\ldots\beta_k}P_1'g = \Sigma N_1'PNMP_1'g.$$

The sequence of letters begins $a_1 \ldots a_{a_1}$, and, since these are the letters of the first group of P_1', if any one of them appears outside the first row of the tableau for PN, two of them must appear in its first column, and $PNMP' = 0$; hence the product

$$N_1'PNMP_1'g = 0$$

unless all the first a_1 letters lie in the first row of F. Let $b_1 b_2 \ldots b_{a_2}$ be the next a_2 letters; then N_1' contains as a factor $\{a_1 b_1\}' \ldots \{a_{a_2} b_{a_2}\}'$, and so the product is zero when any of $b_1, \ldots, b_{a_2}$ lie in the first row of F. Thus, by continuing the argument, we see that this product is zero, unless F is formed by adding the last r letters $z_1, \ldots, z_r$ to the tableau F_1' at the ends of its different rows and in a final row.

Moreover, owing to the presence of g, the number of letters β_{s+1} in the $(s+1)$-th row cannot exceed a_s, otherwise two of the z letters will be in the same column of F and the product will again be zero. Hence

$$N_1'P_1'g = \Sigma N_1'T_{a_1+\gamma_1, a_2+\gamma_2, \ldots, a_h+\gamma_h, r-\Sigma\gamma}P_1'g,$$

where $$\gamma_{s+1} \leqslant a_s - a_{s+1} \quad (s = 1, 2, \ldots, h-1).$$

Consider then a particular term, defined by certain values of $\gamma_1, \ldots, \gamma_h$. This is equal to

$$\Sigma N_1'P_sN_sM_sP_1'g,$$

where each term is derived from a tableau F_s described above. There are $r!/\{\gamma_1! \ldots \gamma_h!(r-\Sigma\gamma)!\}$ of these terms.

Consider M_s; if it contains a term σ_{st}, and any one of the first n letters lies in a lower row in F_t than in F_s, then N_tP_1' is zero. Hence, taking the sequence of tableaux from the first letter, no permutations in which the first n letters are involved need be considered in finding M_s. But, if σ_{st} involve the z letters only, one of these letters must be removed to a higher row, and, since in the column on which it lies in F_s the other letters are unchanged by σ_{st}, N_sP_t is zero; and hence M_s cannot contain permutations involving the last r letters alone. Thus

$$N_1'P_sN_sM_sP_1'g = N_1'P_sN_sP_1'g.$$

To find the coefficient of the identical permutation, consider first the part free from the last r letters. It is

$$\gamma_1!\,\gamma_2!\,\ldots\,\gamma_h!(r-\Sigma\gamma)!\,\frac{f}{(n+r)!}\,\frac{n!}{f'}\,N_1'P_1'N_1'P_1',$$

where in each $N_1'P_1'$ the constant $f'/n!$ is supposed to be absorbed,

$$f = f_{a_1+\gamma_1,\,\ldots,\,a_h+\gamma_h,\,r-\Sigma\gamma} \quad \text{and} \quad f' = f_{a_1,\,a_2,\,\ldots,\,a_h}.$$

Here too $$N_1'P_1'N_1'P_1' = N_1'P_1'.$$

Thus the coefficient of the identical permutation in $N_1'P_sN_sM_1P_1'g$ is

$$\gamma_1!\,\ldots\,\gamma_h!(r-\Sigma\gamma)!\,\frac{f}{(n+r)!};$$

and in $N_1'T_{a_1+\gamma_1,\,\ldots,\,a_h+\gamma_h,\,r-\Sigma\gamma}P_1'g$ it is $r!f/(n+r)!$.

Hence, on replacing the permutations by the symbols s_r, $N_1'P_1'g$ becomes

$$r!\,\Sigma\Phi_{a_1+\gamma_1,\,\ldots,\,a_h+\gamma_h,\,r-\Sigma\gamma},$$

where $\gamma_{s+1} \leqslant a_s - a_{s+1}$ and $r-\Sigma\gamma \leqslant a_h$. Hence

$$\Phi_{a_1,\,\ldots,\,a_h}\Phi_r = \Sigma\Phi_{a_1+\gamma_1,\,\ldots,\,a_h+\gamma_h,\,r-\Sigma\gamma},$$

and by repeated use of this we find that

$$\Phi_{a_1}\Phi_{a_2}\ldots\Phi_{a_h} = \Phi_{a_1,\,a_2,\,\ldots,\,a_h} + \Sigma c_\beta\Phi_{\beta_1,\,\ldots,\,\beta_k},$$

where the first of the differences

$$\beta_1-a_1,\quad \beta_2-a_2,\quad \ldots,$$

which does not vanish is positive, and the coefficients c_β are all positive integers*.

7. It is sometimes desirable to pass from what may be called the canonical form of a substitutional expression in a certain $T_{a_1\ldots a_h}$ to the same expression when there is one additional letter added. Before establishing the theorem by which this may be done, it is necessary to prove certain relations. For brevity, we shall understand by f the number $f_{a_1,\,a_2,\,\ldots,\,a_h}$; and by E_rf the same expression with a_r changed to a_r+1.

Let $$\phi(x) = \prod_{r=1}^{h}(x-a_r+r)$$

and $$\psi(x) = (x+h+1)\,\phi(x).$$

* See Schur, *loc. cit.*, 666.

Then
$$f = n!\,\frac{\prod\limits_{r<s}\{(a_r-r)-(a_s-s)\}}{\prod\limits_r (a_r+h-r)!},$$

and hence
$$E_\lambda f = (n+1) f\,\frac{\phi(a_\lambda+1-\lambda)}{\psi'(a_\lambda-\lambda)}$$

and
$$E_{h+1} f = f_{a_1,\ldots,a_h,1} = (n+1)\, f\,\frac{\phi\{1-(h+1)\}}{\psi'\{-(h+1)\}}.$$

Hence
$$(n+1)\frac{\phi(x+1)}{\psi(x)} f = \sum_{\lambda=1}^{h+1} \frac{E_\lambda f}{x-(a_\lambda-\lambda)}. \tag{I}$$

And, since $\phi(a_r-r) = 0$,
$$\sum_{\lambda=1}^{h+1} \frac{E_\lambda f}{(a_r-r-1)-(a_\lambda-\lambda)} = 0.$$

It is to be noticed that, in the only case when the denominator can be zero, the numerator is zero; and the term is supposed excluded before giving x a special value.

Let $x_1, x_2, \ldots, x_k$ be k different numbers, and let
$$F(x) = (x-x_1)(x-x_2)\ldots(x-x_k).$$

Then
$$\sum_{r=1}^{h+1} \frac{E_r f}{F(a_r-r)} = \sum_{r=1}^{h+1}\sum_{s=1}^{k} \frac{E_r f}{F'(x_s)(a_r-r-x_s)} = -\sum_{s=1}^{k} \frac{(n+1)\,\phi(x_s+1)}{F'(x_s)\,\psi(x_s)} f.$$

by (I).

Hence, when all the numbers x_s are roots of $\phi(x+1) = 0$,
$$\sum_{r=1}^{h+1} \frac{E_r f}{F(a_r-r)} = 0. \tag{II}$$

8. THEOREM III. *A tableau associated with $T_{a_1, a_2, \ldots, a_h}$ is $\{a_{rs}\} = F$, the first suffix defining the row, and the second the column; PN is derived from F, then*
$$PN = \sum_{\lambda=1}^{h+1} K_\lambda P_\lambda N_\lambda G_\lambda L_\lambda,$$

where $P_\lambda N_\lambda$ is formed from that tableau F_λ of $E_\lambda T_{a_1, a_2, \ldots, a_h}$ obtained from F by placing one additional final letter x at the end of the λ-th row (except that in the case when $a_\lambda = a_{\lambda-1}$ this particular term is absent or zero), where
$$K_\lambda = K_{\lambda,h} K_{\lambda,h-1} \ldots K_{\lambda,\lambda+1}$$

and
$$K_{\lambda,\mu} = 1 + \sum_{t=1}^{a_\mu} \frac{1}{a_\lambda - a_\mu + \mu - \lambda + 1} (a_{\mu t}\, x);$$

where

$$G_\lambda = \frac{1}{a_{\lambda-1}-a_\lambda}\{(a_{1,a_\lambda+1}\,a_{1,a_\lambda+2}\dots a_{1,a_{\lambda-1}})(a_{2,a_\lambda+1}\dots a_{2,a_{\lambda-1}})$$
$$\dots(a_{\lambda-1,a_\lambda+1}\dots a_{\lambda-1,a_{\lambda-1}})\};$$

and where $$L_\lambda = L_{\lambda,\lambda-2}\,L_{\lambda,\lambda-3}\dots L_{\lambda,1},$$

and $$L_{\lambda,\mu} = 1-\frac{1}{a_\mu-a_\lambda+\lambda-\mu-1}\sum_{t=a_\mu+1+1}^{a_\mu}\sum_{u=1}^{\mu}(a_{ut}x).$$

To prove this it is necessary to bear in mind the following simple facts:

(i) PN is affected by the coefficient $f/n!$, and $P_\lambda N_\lambda$ is affected by the coefficient

$$\frac{E_\lambda f}{(n+1)!} = \frac{f}{n!}\,\frac{\phi(a_\lambda+1-\lambda)}{\psi'(a_\lambda-\lambda)},$$

with the notation of the last paragraph.

(ii) $\sum_{t=1}^{a_\mu}(a_{\mu t}x)$ is permutable with any permutation of the letters $a_{\mu t}$ alone, and therefore with $\{a_{\mu_1}\dots a_{\mu a_\mu}\}$, and hence with P.

Hence K_λ is permutable with P.

Similarly both G_λ and L_λ are permutable with N.

(iii) When $s\neq u$ and $t\neq v$, and there is a letter a_{sv}, *i.e.* $v\leqslant a_s$ (which always happens when $s<u$),

$$P(a_{st}x)(a_{uv}x)\,N = 0.$$

For $$P(a_{st}x)(a_{uv}x)\,N = P(a_{st}a_{uv})(a_{st}x)\,N = (a_{st}a_{uv})\,P'(a_{st}x)\,N,$$

where P' is obtained from P by the interchange of a_{st} and a_{uv}, and hence contains the factor

$$\{a_{s1}a_{s2}\dots a_{st-1}a_{uv}a_{st+1}\dots a_{sv}\dots\};$$

the result follows at once since N has a factor

$$\{a_{1v}a_{2v}\dots a_{sv}\dots a_{uv}\dots\}'.$$

(iv) $$P(a_{st}x)(a_{sv}x)\,N = P(a_{st}x)\,N,$$
$$P(a_{st}x)(a_{ut}x)\,N = -P(a_{ut}x)\,N.$$

Now $$N_\lambda = \left[1-\sum_{t=1}^{\lambda-1}(a_{t,a_\lambda+1}x)\right]N.$$

And $$P_\lambda N_\lambda G_\lambda = P_\lambda\frac{1}{a_{\lambda-1}-a_\lambda}\sum_{\mu=a_\lambda+1}^{a_{\lambda-1}}\left[1-\sum_{t=1}^{\lambda-1}(a_{t\mu}x)\right]N = P_\lambda L_{\lambda,\lambda-1}N.$$

Also $$P_\lambda = P\left[1+\sum_{t=1}^{a_\lambda}(a_{\lambda t}\,x)\right] = PK_{\lambda\lambda}.$$

Hence from (1) and (2) the term

$$K_\lambda P_\lambda N_\lambda G_\lambda L_\lambda$$

$$= \frac{\phi(a_\lambda+1-\lambda)}{\psi'(a_\lambda-\lambda)} PK_\lambda K_{\lambda\lambda} L_{\lambda\lambda-1} L_\lambda N$$

$$= \frac{\phi(a_\lambda+1-\lambda)}{\psi'(a_\lambda-\lambda)} P\Sigma C(a_{r_1u_1}x)(a_{r_2u_2}x)\dots(a_{r_\sigma u_\sigma}x)(a_{s_1v_1}x)\dots(a_{s_\tau v_\tau}x)\,N,$$

where C is a numerical coefficient, and the number of transpositions appearing after it may be anything from 0 to h; but where

$$h \geqslant r_1 > r_2 \dots > r_\sigma \geqslant \lambda > s_t \quad (\text{where } t = 1, \dots, \tau),$$

where, of course, $u_t \leqslant a_{r_t}$, and also

$$a_\lambda < v_1 < v_2 \dots < v_\tau \leqslant a_1.$$

Owing to these inequalities no letter can appear twice in these transpositions except x. If $u_1 > u_2$ there is a letter $a_{r_1u_2}$, and hence, by (iii), the term is zero. If $u_1 = u_2$, we may use (iv) to remove $(a_{r_1u_1}x)$, changing the sign. Thus we need consider only terms

$$u_1 < u_2 < u_3 \dots < u_\sigma \leqslant a_\lambda < v_1 < v_2 \dots < v_\tau \leqslant a_1,$$

Again, when $s_{\tau-1} < s_\tau$ there is always a letter $a_{s_{\tau-1}v_\tau}$, and hence, by (iii), the term is zero. And so we need consider only the terms

$$h \geqslant r_1 > r_2 \dots > r_\sigma \geqslant \lambda > s_1 > s_2 \dots s_\tau \geqslant 1.$$

A product so reduced will be called a reduced product.

For convenience, to simplify statement and proof, the case is stated and proved for $a_1 > a_2 \dots > a_h$; the statement is true in all cases, when it is understood that the λ-th term is missing when $a_{\lambda-1} = a_\lambda$, and the proof needs alteration only in slight detail. A diagram illustrating the λ-th term is given.

. . . . o o o o o
. . . . o o o o
. . . . o o
× × × × x
× × ×
× ×
×

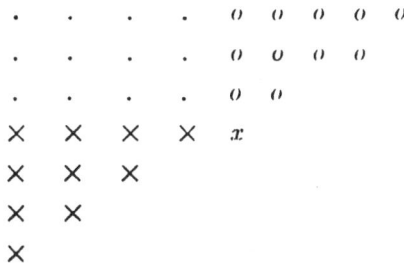

Here $\lambda = 4$ and PN belongs to $T_{9,8,6,4,3,2,1}$. The places marked with a cross indicate the letters which appear in $K_\lambda K_{\lambda\lambda}$, and those marked with an o the letters which appear in $L_{\lambda,\lambda-1} L_\lambda$.

Let us consider the coefficient necessarily attached to the transposition $(a_{rs}x)$ in a reduced product belonging to the λ-th term, where $\lambda > r$, and hence $(a_{rs}x)$ appears in $L_{\lambda,\lambda-1} L_\lambda$, and $a_r \geqslant s > a_\lambda$.

Let s lie between $a_{\rho+1}$ and a_ρ, *i.e.*

$$a_{\rho+1} < s \leqslant a_\rho.$$

Then the transposition $(a_{rs}x)$ appears in the factor $L_{\lambda,\rho}$. We have seen that any term which has a factor $(a_{\sigma\tau}x)$ to the left of this, where $r \leqslant \sigma$, either is zero or would remove the transposition $(a_{rs}x)$. We may go further and say that a transposition $(a_{\sigma\tau}x)$ does not exist on the left of $(a_{rs}x)$ for which $\sigma \leqslant \rho$, for then, by (iii), this product is zero, there being a letter $a_{\sigma s}$.

Thus any transposition $(a_{\sigma\tau}x)$ on the left of $(a_{rs}x)$ is such that $\sigma > \rho$.

Similarly a transposition $(a_{\sigma\tau}x)$ on the right of $(a_{rs}x)$, when $r > \sigma$, will still mean a zero term by (iii) unless there is no letter $a_{r\tau}$; *i.e.* unless $\tau > a_r$. But on the right of $(a_{rs}x)$ we have to remember those possible terms $(a_{r\tau}x)$ which are such that

$$P(a_{rs}x)(a_{r\tau}x)\,N = P(a_{rs}x)\,N.$$

Such terms appear in $L_{\lambda,\rho-1} L_{\lambda,\rho-2} \ldots L_{\lambda,r}$, and moreover, from what we have just seen, it is not possible for $(a_{rs}x)$ to appear in a reduced product with any other transposition from these factors. Thus associated with $(a_{rs}x)$ we have the numerical coefficient

$$\frac{-1}{a_\rho - a_\lambda + \lambda - \rho - 1}\left[1 - \frac{a_{\rho-1} - a_\rho}{a_{\rho-1} - a_\lambda + \lambda - \rho}\right]\cdots\left[1 - \frac{a_r - a_{r+1}}{a_r - a_\lambda + \lambda - r - 1}\right]$$

(the factors in this product arising severally from $L_{\lambda,\rho}, L_{\lambda,\rho-1}, \ldots, L_{\lambda,r}$)

$$= \frac{1}{(a_\lambda - \lambda) - (a_\rho - \rho - 1)}\left[\frac{(a_\lambda - \lambda) - (a_\rho - \rho - 1) - 1}{(a_\lambda - \lambda) - (a_{\rho-1} - \rho)}\right]\cdots$$
$$\left[\frac{(a_\lambda - \lambda) - (a_{r+1} - r - 2) - 1}{(a_\lambda - \lambda) - (a_r - r - 1)}\right]$$
$$= \left[1 - \frac{1}{(a_\lambda - \lambda) - (a_\rho - \rho - 1)}\right]\left[1 - \frac{1}{(a_\lambda - \lambda) - (a_{\rho-1} - \rho)}\right]\cdots$$
$$\frac{1}{(a_\lambda - \lambda) - (a_r - r - 1)}.$$

Let us turn to the case where $\lambda \leqslant r$, and the factor $(a_{rs}x)$ appears in $K_\lambda K_{\lambda\lambda}$. The argument is very similar; this transposition appears in

$K_{\lambda r}$, but there are transpositions $(a_{\sigma s}x)$ which may appear in the factors $K_{\lambda, r+1}$, $K_{\lambda, r+2}$, ..., $K_{\lambda, \rho}$, which have to be taken into account owing to the equation

$$P(a_{\sigma s}x)(a_{rs}x)N = -P(a_{rs}x)N.$$

No other transposition from the $r-\rho$ factors can appear in a reduced product. And thus the coefficient associated with the transposition is of exactly the same form as before. The value of the coefficient depends entirely on the transposition, and not on whether it appears in a K or an L factor. Consider then any reduced product. The different transpositions have coefficients which are each derived from a certain set of the factors $K_{\lambda\mu}$, $L_{\lambda\mu}$, and these sets cannot overlap, so that the actual coefficient of the whole is expressed as a sum of terms each of which is of the form

$$\frac{E_\lambda f}{(n+1)f\,F(a_\lambda-\lambda)},$$

where $F(x)$ is a function all of whose zeros are different and equal to some or all of a_r-r-1, $r=1, 2, ..., h$.

The sum of all such terms from $\lambda=1$ to $\lambda=h+1$ is zero by (II), § 7.

We have only then to calculate the numerical coefficient when x is absent. This is

$$\Sigma\frac{E_\lambda f}{(n+1)f}=1,$$

a well known result*.

Thus the theorem is proved.

9. That the factors K and L are essentially the same except for the sign of the transpositions may be seen by considering the case where $a_{\lambda-1}=a_\lambda$. The λ-th term is absent. The μ-th term ($\mu<\lambda$) has the pair of factors

$$K_{\mu,\lambda}K_{\mu,\lambda-1}=\left[1+\frac{1}{a_\mu-a_\lambda+\lambda-\mu+1}\sum_{t=1}^{a_\lambda}(a_{\lambda,t}x)\right]$$
$$\left[1+\frac{1}{a_\mu-a_{\lambda-1}+\lambda-\mu}\sum_{t=1}^{a_{\lambda-1}}(a_{\lambda-1,t}x)\right].$$

But, since $a_\lambda=a_{\lambda-1}$,

$$\sum_{t=1}^{a_\lambda}(a_{\lambda,t}x).(a_{\lambda-1,u}x)P_\mu N_\mu=(a_{\lambda-1,u}x)\sum_{t=1}^{a_\lambda}(a_{\lambda,t}a_{\lambda-1,u})P_\mu N_\mu=-(a_{\lambda-1,u}x)P_\mu N_\mu,$$

* See, for example, Q.S.A., § 3 (2).

and hence

$$K_{\mu,\lambda}K_{\mu,\lambda-1}P_\mu N_\mu$$

$$=\left[1+\frac{1}{a_\mu-a_\lambda+\lambda-\mu+1}\left\{\sum_{t=1}^{a_\lambda}(a_{\lambda,t}x)+\sum_{t=1}^{a_{\lambda-1}}(a_{\lambda-1,t}x)\right\}\right]P_\mu N_\mu$$

in formal analogy to the L factors.

10. Consider a standard tableau F_r of $T_{a_1,a_2,\ldots,a_h}$ of n letters. If the last letter a_n be removed a standard tableau $F'_{r'}$ of $n-1$ letters is obtained. Moreover, when N_rP_s is not zero and a_n lies in the same row in F_r and F_s, $N'_{r'}P'_{s'}$ is not zero. And in the same case when N_rP_s is zero, $N'_{r'}P'_{s'}$ is also zero. For otherwise the vanishing would be caused by a letter a_t which is in the same column with a_n in F_r being in the same row with it in F_s. But the change from F_r to F_s can be made by a column permutation followed by a row permutation; the column permutation would have to bring a_t down to the row of a_n, which could only be done by changing the row in which a_n lies; in fact, by moving a_n to a higher row. Thus $P_rN_rM_r$ consists of a sum of terms $\pm P_r\sigma_{r\rho}N_\rho$ such that a_n lies in a higher row in F_ρ than in F_r, together with a sum of terms which becomes $P'_{r'}N'_{r'}M'_{r'}$ when the letter a_n is dropped.

11. Consider now the expression for $P'_{r'}N'_{r'}$ in terms of members of $T_{a_1,\ldots,a_h}\equiv T$; *i.e.* the value of $T.P'_{r'}N'_{r'}$. We shall suppose the letter a_n to be added to the λ-th row, so that only one term $K_\lambda P_rN_rG_\lambda L_\lambda$ in the expression given by Theorem III for $P'_{r'}N'_{r'}$ has to be considered. So far as the rows in the matrix expression for $P'_{r'}N'_{r'}$ are concerned, these are defined by $K_\lambda P_r$. First there are terms in the row P_r, the numerical coefficient on this side being unity; in the other terms the row is defined by P_s, where a_n lies in a lower row in F_s than in F_r; *i.e.* when we define the sequence of tableaux as from the last letter, these other terms lie in a higher row in the matrix than the first. Moreover, the places in the columns are defined by $N_rG_\lambda L_\lambda$ entirely independently of $K_\lambda P_r$. Thus the effect of K_λ is really that of multiplication on the left by a matrix I of f rows and columns like the T matrix, in which all the elements in the leading diagonal are unity and all the elements on the left of and below this diagonal are zero. Moreover, the f'^2 places about the leading diagonal corresponding to that part of the T matrix occupied by terms $P_r\sigma_{rs}N_sM_s$, for which a_n lies at the end of the λ-th row both in F_r and F_s, are occupied by a unit matrix $E_{f'}$.

The terms $N_r G_\lambda L_\lambda$ may be treated in the same way, and hence also the terms

$$N_r M_r G_\lambda L_\lambda = \Sigma \pm \sigma_{rt} N_t G_\lambda L_\lambda,$$

if we remember that we have proved in the last paragraph that $N_r M_r$ differs from that expression obtained from $N'_r M'_r$ by placing a_n at the end of the λ-th row, by terms in which a_n is at the end of a higher row. Thus the effect of the factor $G_\lambda L_\lambda$ is the multiplication on the right of the T matrix expression for $P_r N_r$ by a matrix J. This matrix J has similar properties to I, but we need not dwell on them. For we see now that, if S is a substitutional expression of the first $n-1$ letters, and the matrix equivalent of $S.\Delta_\lambda T$ is $[T_\lambda]$, where $\Delta_\lambda T$ is $T_{a_1, a_2, \ldots, a_\lambda - 1, \ldots, a_h}$, then the matrix expression for ST is

$$IMJ,$$

where M is that T matrix which consists of the matrices

$$[T_1],\ [T_2],\ \ldots,\ [T_h]$$

down the leading diagonal and zero everywhere else.

Now let $S = 1$, then $[T_\lambda] = E$ and also $IMJ = E$, and therefore

$$J = I^{-1}.$$

THEOREM IV. *The matrix for $S.T$, where S is any substitutional expression involving $n-1$ letters $a_1, a_2, \ldots, a_{n-1}$ and $T = T_{a_1, a_2, \ldots, a_h}$ involves a last letter a_n, is*

$$IMI^{-1},$$

where M is that T matrix which consists of the matrices $[T_\lambda]$, $\lambda = 1, \ldots, h$, down the leading diagonal and zeros elsewhere, and $[T_\lambda]$ is the matrix for $S.\Delta_\lambda T$. The matrix I is of the form $[I_{\lambda\mu}]$, where the $I_{\lambda\mu}$ are partial matrices, $I_{\lambda\lambda}$ occupying the same place in I that $[T_\lambda]$ occupies in M. Then

$$I_{\lambda\lambda} = E, \quad I_{\lambda\mu} = 0 \quad (\mu < \lambda).$$

By applying this theorem successively we find that if S involves only the first m letters, then

$$S.T = IMI^{-1},$$

where M is a matrix which consists of the matrices corresponding to ST', where T' has to do with the first m letters only, along the leading diagonal, and zeros everywhere else. And I has the same kind of property as in the previous case.

As an example, if $n = 5$ and $T = T_{3,2}$,

$$[T_{3,2}] = I_{3,2}\begin{bmatrix}[T_{3,1}] & 0\\ 0 & [T_{22}]\end{bmatrix} I_{3,2}^{-1},$$

where $$I_{3,2} = \begin{bmatrix}E_3 & M\\ 0 & E_2\end{bmatrix} \quad \text{and} \quad M = \tfrac{1}{2}\begin{bmatrix}1 & 1\\ 1 & 0\\ 0 & 1\end{bmatrix},$$

(Here 0 stands for a zero matrix.)

12. In some cases the form of I can be readily calculated. Thus, when $h = 2$, and we consider $T_{\alpha,\beta}$, it is evident from K that

$$I_{\alpha,\beta} = \begin{bmatrix}E_{\alpha,\beta-1} & \dfrac{1}{\alpha-\beta+1}M_{\alpha,\beta}\\ 0 & E_{\alpha-1,\beta}\end{bmatrix},$$

where $E_{\alpha,\beta}$ is the unit matrix of degree $f_{\alpha,\beta}$, and $M_{\alpha,\beta}$ has $f_{\alpha,\beta-1}$ rows and $f_{\alpha-1,\beta}$ columns. To find the form of $M_{\alpha\beta}$ we notice that

$$\frac{1}{\alpha-\beta+2}M_{\alpha\beta}[T_{\alpha-1,\beta}]$$

corresponds to $[K_{1,2}-1][T_{\alpha-1,\beta}]$. In the sequence of tableaux for $T_{\alpha-1,\beta}$ we take first those, $f_{\alpha-1,\beta-1}$ in number, which have a_{n-1} in the second row, and then those, $f_{\alpha-2,\beta}$ in number, which have a_{n-1} in the first row.

In the first set we consider first those transpositions $(a_t a_n)$ of K_{12}, where $t < n-1$; the effect of these is given by the matrix $M_{\alpha,\beta-1}$, for in all of them a_{n-1} remains in the second row.

In fact we may write

$$[T_{\alpha-1,\beta}] = \begin{bmatrix}[T_{\alpha-1,\beta-1}] & [A]\\ [B] & [T_{\alpha-2,\beta}]\end{bmatrix},$$

where in $[T_{\alpha-1,\beta-1}]$ both P and N tableaux have a_{n-1} in the lower row; while in $[T_{\alpha-2,\beta}]$ in both tableaux a_{n-1} lies in the upper row.

Now let $M_{\alpha\beta} = \begin{bmatrix}m_{11} & m_{12}\\ m_{21} & m_{22}\end{bmatrix}$, so that

$$M_{\alpha\beta}[T_{\alpha-1,\beta}] = \begin{bmatrix}m_{11}[T_{\alpha-1,\beta-1}]+m_{12}[B] & m_{11}[A]+m_{12}[T_{\alpha-2,\beta}]\\ m_{21}[T_{\alpha-1,\beta-1}]+m_{22}[B] & m_{21}[A]+m_{22}[T_{\alpha-2,\beta}]\end{bmatrix}.$$

The transposition $(a_{n-1}a_n)$ gives rise to the matrix m_{21}; this transposition invariably changes a standard tableau with a_n above and a_{n-1} below into a standard tableau with a_{n-1} above and a_n below; and $m_{21} = E_{\alpha-1,\beta-1}$.

$m_{11} = M_{\alpha, \beta-1}$; $m_{12} = 0$, for none of the transpositions of K moves an a_{n-1} from the upper to the lower row; and $m_{22} = M_{\alpha-1, \beta}$.

Thus
$$M_{\alpha\beta} = \begin{bmatrix} M_{\alpha, \beta-1} & 0 \\ E_{\alpha-1, \beta-1} & M_{\alpha-1, \beta} \end{bmatrix}.$$

In particular, $I_{\alpha\alpha} = E$,
$$M_{\alpha, \alpha-1} = \begin{bmatrix} M_{\alpha, \alpha-2} \\ E_{\alpha-1, \alpha-1} \end{bmatrix},$$
and
$$M_{\alpha, 1} = [1\ 1\ \ldots\ 1],$$
a matrix of one row and $\alpha-2$ columns.

In the case of $T_{\alpha, 1, 1, \ldots, 1}$, called by Frobenius the case of unit rank,

$$K_1 PN = \left[1 + \frac{1}{\alpha+h-1}(b_h x)\right]\left[1 + \frac{1}{\alpha+h-2}(b_{h-1} x)\right]\ldots \left[1 + \frac{1}{\alpha+1}(b_2 x)\right]\begin{Bmatrix} a_1 a_2 \ldots a_\alpha x \\ b_2 \\ \vdots \\ b_h \end{Bmatrix}$$

$$= \begin{Bmatrix} a_1 a_2 \ldots a_\alpha x \\ b_2 \\ \vdots \\ b_h \end{Bmatrix} + \frac{1}{\alpha+h-1}\left[\begin{Bmatrix} a_1 a_2 \ldots a_\alpha b_h \\ b_2 \\ \vdots \\ b_{h-1} \\ x \end{Bmatrix} - \begin{Bmatrix} a_1 a_2 \ldots a_\alpha b_{h-1} \\ b_2 \\ \vdots \\ b_{h-2} \\ b_h \\ x \end{Bmatrix} + \ldots + (-)^h \begin{Bmatrix} a_1 a_2 \ldots a_\alpha b_2 \\ b_3 \\ \vdots \\ b_h \\ x \end{Bmatrix}\right].$$

Then we may designate the transforming matrix as

$$I_{(\alpha, h)} = \begin{bmatrix} E_r & \dfrac{1}{\alpha+h-1} m_{(\alpha, h)} \\ 0 & E_s \end{bmatrix},$$

where
$$r = \binom{\alpha+h-3}{h-2}, \quad s = \binom{\alpha+h-3}{h-1}.$$

Also
$$m_{(\alpha, h)} = \begin{bmatrix} -m_{(\alpha, h-1)} & 0 \\ E_t & m_{(\alpha-1, h)} \end{bmatrix},$$

where
$$t = \binom{\alpha+h-4}{h-2}.$$

ON QUANTITATIVE SUBSTITUTIONAL ANALYSIS

(*Fifth Paper*)

By ALFRED YOUNG

[Received 30 August, 1929.—Read 14 November, 1929.]

For linear substitution groups of degree two the rotation groups of the regular solids in three dimensions are of fundamental importance. It is natural to believe that the rotation groups of the regular solids in Euclidean space of any number of dimensions must occupy a prominent place in the theory of linear substitution groups of higher degrees. It is well known that in space of n dimensions when $n > 4$ there are only three regular solids, the simplex, the figure corresponding to the cube which we may call the hypercube, and that which corresponds to the octohedron, which we call here the hyper-octohedron. The simplex in n dimensions has $n+1$ vertices, the symmetric permutation group of its vertices is practically the group which includes both the rotations and the reflections of the figure into itself; the alternating group is the pure rotation group. The extended group in the general case, corresponding to Klein's extended tetrahedral group, has been worked out by Schur* as the isomorphic fractional linear group.

The other two regular solids, corresponding to the cube and the octohedron, are reciprocal and hence have the some rotation group. Thus in general there are but two types of rotation group, viz.: the symmetric and the hyper-octohedral. It is thought then worth while to discuss the hyper-octohedral group in the same way as the symmetric. Here it is shown that it lends itself to almost identical treatment by means of this analysis. The irreducible representations as a linear group are obtained both in degree and actual form; like the symmetric group they may all be

* *Journal für Math.*, 139 (1911), 155-250.

represented as linear groups with coefficients which are positive or negative integers. A generating function for the group characters is obtained which is very similar to that of Frobenius for the symmetric group.

In §§ 7–10, a self-conjugate subgroup of index 2 is considered, and its irreducible representations are obtained, again with real and integral coefficients. Finally in § 11 it is shown that the extended tetrahedral group is a subgroup of the hyper-octohedral group in four dimensions.

1. The hyper-octohedral group can be treated in exactly the same way as the symmetric group. We consider such a group of degree n, *i.e.* that related to a hyper-octohedron in Euclidean space of n dimensions. Let the pairs of vertices be A_1B_1, A_2B_2, ..., A_nB_n. Then the Abelian group $\{(A_1B_1), (A_2B_2), \ldots, (A_nB_n)\}$ generated by the transpositions (A_rB_r) $(r = 1, 2, \ldots, n)$ is a self conjugate subgroup. The factor group is the symmetric group of degree n.

The conjugate sets of operations may be compared to those of the symmetric group $\{a_1a_2\ldots a_n\}$. Corresponding to a cycle $(a_1a_2\ldots a_r)$ there are two possible permutations, viz.: $(A_1A_2\ldots A_r)(B_1B_2\ldots B_r)$, and $(A_1A_2\ldots A_rB_1\ldots B_r)$; in particular, corresponding to a cycle of one letter (a_r), there are here two possibilities $(A_r)(B_r)$, and (A_rB_r). Thus corresponding to a permutation of c cycles in the symmetric group, there must here be taken into account the different possibilities for each of these cycles separately.

The notation $\{A_1A_2\ldots A_r\}_B$ will be used to denote

$$\{(A_1A_2)(B_1B_2),\ (A_1A_3)(B_1B_3),\ \ldots,\ (A_1A_r)(B_1B_r)\},$$

that is the sum of the permutations of the group generated by these permutations, or the sum of the permutations which permute the A's in any way, and at the same time the B's in the same way. This will be called the AB positive symmetric group. In the same way $\{A_1A_2\ldots A_r\}'_B$ is used to denote the AB negative symmetric group in exact analogy to the ordinary negative symmetric group.

Then $\{A_1A_2\ldots A_r\}_{B_1}$ is used to denote

$$\frac{1}{2^r r!}\{A_1A_2\ldots A_r\}_B.\{(A_1B_1),\ (A_2B_2),\ \ldots,\ (A_rB_r)\},$$

and $\{A_1A_2\ldots A_r\}_{B_2}$ to denote

$$\frac{1}{2^r r!}\{A_1A_2\ldots A_r\}_B.\{-(A_1B_1),\ -(A_2B_2),\ \ldots,\ -(A_rB_r)\}.$$

[Here, of course,

$$\{-(A_1B_1), -(A_2B_2)\} = 1-(A_1B_1)-(A_2B_2)+(A_1B_1)(A_2B_2).]$$

Further, $\{A_1A_2\ldots A_r\}'_{B_1}$ is used to denote

$$\frac{1}{2^r r!}\{A_1A_2\ldots A_r\}'_B.\{(A_1B_1),\ (A_2B_2),\ \ldots,\ (A_rB_r)\},$$

and $\{A_1A_2\ldots A_r\}'_{B_2}$ to denote

$$\frac{1}{2^r r!}\{A_1A_2\ldots A_r\}'_B.\{-(A_1B_1),\ -(A_2B_2),\ \ldots,\ -(A_rB_r)\}.$$

Then by an obvious extension of the tableau notation we use

$$\left\{\begin{matrix} A_{11}A_{12}\ldots\ldots\ldots\ldots A_{1a_1} \\ A_{21}A_{22}\ldots\ldots\ldots A_{2a_2} \\ \ldots\ldots\ldots\ldots\ldots\ldots\ldots\ldots \\ A_{h1}A_{h2}\ldots\ldots\ldots A_{ha_h} \end{matrix}\right\}_B \equiv \{A_{rs}\}_B$$

to denote PN, where P is the product of the AB positive symmetric groups of the rows, and N is the product of the AB negative symmetric groups of the columns of the enclosed tableau.

Then we shall use $\qquad \{A_{rs}\}_{B_1}$

to denote $\qquad \dfrac{f_{a_1a_2\ldots a_h}}{2^n.n!}PNS,$

where $f_{a_1a_2\ldots a_h}$ is the constant used for symmetric groups, n is the number of letters in the tableau, S is the sum of the substitutions of the group

$$\{(A_{11}B_{11}),\ (A_{12}B_{12}),\ \ldots,\ (A_{ha_h}B_{ha_h})\},$$

and P, N have the values given above.

Similarly $\qquad \{A_{rs}\}_{B_2} = \dfrac{f_{a_1a_2\ldots a_h}}{2^n n!}PNS',$

where $\qquad S' = \{-(A_{11}B_{11}),\ -(A_{12}B_{12}),\ \ldots,\ -(A_{ha_h}B_{ha_h})\}.$

Let the letters $A_1, A_2, \ldots, A_n$ be separated into two sets of ρ and $n-\rho$ letters.

With the ρ letters we form a tableau having a_1 letters in the first row, a_2 in the second, and so on, and a_h letters in the h-th row, the last. From this tableau we form the product

$$\frac{f_{a_1\ldots a_h}}{2^\rho.\rho!}P_1N_1S.$$

With the $n-\rho$ letters we form a tableau having β_1 letters in the first row, β_2 in the second, and so on, and β_k in the last, and from this we form the product

$$\frac{f_{\beta_1\ldots\beta_k}}{2^{n-\rho}(n-\rho)!}P_2N_2S'.$$

We take the product of these two expressions (which are permutable), then we sum this product for all permutations of the letters $A_1, A_2, \ldots, A_n$ (the letters B being of course permuted simultaneously), and we call this sum

$$_{\beta_1\beta_2\ldots\beta_k}\bar{T}_{a_1a_2\ldots a_h}.$$

2. THEOREM I. *The number of conjugate sets in the group is the same as the number of expressions*

$$_{\beta_1\beta_2\ldots\beta_k}\bar{T}_{a_1a_2\ldots a_h}.$$

For consider any permutation of the group (as we have seen, it may be derived from a permutation of the symmetric group of n letters) corresponding to any cycle $(a_1a_2\ldots a_r)$ of the symmetric group; we have either $(A_1A_2\ldots A_r)(B_1B_2\ldots B_r)$ or else $(A_1A_2\ldots A_rB_1B_2\ldots B_r)$ in our group. Now the permutation of the symmetric group belongs to a conjugate set $t_{\gamma_1\gamma_2\ldots\gamma_l}$, where $\gamma_1, \gamma_2, \ldots, \gamma_l$ are the degrees of its cycles.

Then, in the same way, in our group the permutation belongs to a conjugate set

$$_{\beta_1\beta_2\ldots\beta_k}t_{a_1a_2\ldots a_h},$$

where a_r denotes the presence of the product of a pair of cycles $(A_1A_2\ldots)(B_1B_2\ldots)$ each of a_r letters, and β_r denotes the presence of a cycle of the form $(A_1A_2\ldots A_{\beta_r}B_1B_2\ldots B_{\beta_r})$.

Thus the conjugate sets and the expressions $\bar{T}$ are defined by exactly the same sets of numbers and are therefore equal in number.

THEOREM II. *The product of two different expressions $\bar{T}$ is zero.*

In the first place we observe that, in the expression P_1N_1S, if σ be any permutation of P_1N_1, then

$$\sigma S = S\sigma,$$

for S contains all the letters contained in P_1N_1, and every permutation of P_1N_1 permutes the B's in the same way as the A's. Hence

$$P_1N_1S = SP_1N_1,$$

and similarly $$P_2N_2S' = S'P_2N_2.$$

Let $$P_1N_1S_1\,.\,P_2N_2S_2'$$

belong to $\bar{T}_1$, and $P_3N_3S_3 . P_4N_4S'_4$ to $\bar{T}_2$. Then

$$P_1N_1S_1P_2N_2S'_2 . P_3N_3S_3P_4N_4S'_4 = P_1N_1P_2N_2 . S_1S'_2S_3S'_4 . P_3N_3P_4N_4.$$

Now the groups S are Abelian, and it is at once clear that this is zero unless $S_1 \equiv S_3$ and $S'_2 \equiv S'_4$. But in this case P_1N_1 and P_3N_3 involve exactly the same letters, and also P_2N_2 and P_4N_4 involve the same letters. Hence this product is equal to

$$S_1S'_2P_1N_1P_3N_3 . P_2N_2P_4N_4S_3S'_4,$$

and from the case of the symmetric group we know that this is zero, unless the tableaux for P_1N_1 and P_3N_3 have the same number of letters in each row; and also the tableaux P_2N_2 and P_4N_4. That is, this product is always zero unless the $\bar{T}$'s are the same. That is, the product of different $\bar{T}$'s is zero.

THEOREM III.

$$1 = {}_{\beta_1\beta_2\ldots\beta_k}C_{\alpha_1\alpha_2\ldots\alpha_h} \quad {}_{\beta_1\beta_2\ldots\beta_k}\bar{T}_{\alpha_1\alpha_2\ldots\alpha_h}$$

where the C's are numerical.

From the mode of formation it is manifest that

$$\bar{T} = \Sigma\lambda t,$$

where t is the sum of the members of a conjugate set, and the coefficients λ are *real* numbers.

Now the inverse of any permutation belongs to the same conjugate set as itself, and hence the square of such an expression as $(\Sigma\lambda t)$ cannot be zero when the λ's are all real. Hence the square of $\bar{T}$ is never zero.

There can be no linear relation between the expressions $\bar{T}$. For if

$$\Sigma\mu\bar{T} = 0$$

we may multiply by $\bar{T}_1$ and every term except $\mu_1\bar{T}_1^2$ vanishes, and hence $\mu_1 = 0$.

The number of expressions $\bar{T}$ is the same as the number of expressions t; hence the equations

$$\bar{T} = \Sigma\lambda t$$

can be solved in the form

$$t = \Sigma\mu\bar{T}.$$

When we take for t the identical permutation we have the required result.

8. Standard tableaux may now be introduced as in the case of the symmetric group, a standard tableau being one in which the letters follow

the prearranged sequence in every row and in every column. Further we may introduce expressions M so that PNM is a prepared form as in the former case; and we further define PNM as having absorbed the numerical factor $f/n!$

Then, when $r \neq s$, and $P_r N_r M_r$, $P_s N_s M_s$ are two different standard prepared forms involving the same letters, their product is zero, while

$$(P_r N_r M_r)^2 = P_r N_r M_r.$$

Then consider ${}_{\beta_1\beta_2\ldots\beta_k}\bar{T}_{\alpha_1\alpha_2\ldots\alpha_h}$, where

$$\alpha_1+\alpha_2+\ldots+\alpha_h = \rho,$$

$$\beta_1+\beta_2+\ldots+\beta_k = n-\rho.$$

We obtain $\binom{n}{\rho} f_{\alpha_1\alpha_2\ldots\alpha_h} f_{\beta_1\beta_2\ldots\beta_k}$ standard units

$$P_1 N_1 M_1 S_1 \,.\, P_2 N_2 M_2 S_2',$$

where first a particular set of ρ letters is chosen for the first tableau F_1, the remaining letters being assigned to the second tableau F_2', which are both arranged in standard form. The combined tableau $F_1 F_2'$ formed by placing them side by side defines the particular term.

Then we take

$${}_{\beta_1\beta_2\ldots\beta_k}T_{\alpha_1\alpha_2\ldots\alpha_h} = \Sigma P_1 N_1 M_1 S_1 \,.\, P_2 N_2 M_2 S_2',$$

where the Σ extends to forms obtained from standard tableaux only, and the factor $1/2^\rho$ is absorbed in S_1, and $1/2^{n-\rho}$ in S_2'.

THEOREM IV. *The product of* $P_1 N_1 M_1 S_1 P_2 N_2 M_2 S_2'$ by

$$P_3 N_3 M_3 S_3 P_4 N_4 M_4 S_4',$$

i.e. of two different standard forms belonging to the same T, *is zero.*

This is obvious from the case of the symmetric group when the letters in the tableau F_1 are the same as those of the tableau F_3. For then

$$P_1 N_1 M_1 P_3 N_3 M_3 = 0$$

or

$$P_2 N_2 M_2 P_4 N_4 M_4 = 0,$$

since F_1 and F_3 cannot be the same and also F_2' and F_4'.

If the letters in F_1 and F_3 are different then

$$S_1 S_4' = 0.$$

Also it is plain that

$$(P_1 N_1 M_1 S_1 P_2 N_2 M_2 S_2')^2 = P_1 N_1 M_1 S_1 P_2 N_2 M_2 S_2'.$$

Hence

$$T^2 = T.$$

Now we may write

$$T = \Sigma[\Sigma P_1 N_1 M_1 S_1][\Sigma P_2 N_2 M_2 S_2']$$

where the first Σ extends to the $\binom{n}{\rho}$ combinations of ρ letters, and the others to the standard groups for the sets of ρ and $n-\rho$ letters.

We know that $\Sigma P_1 N_1 M_1 S_1$ is a linear function of the conjugate sets of the ρ letters, *i.e.* it contains every member of a conjugate set in the same way. The same is true of $\Sigma P_2 N_2 M_2$. Also in T the letters are divided into the two sets in all possible ways, hence

$$T = \Sigma \lambda t,$$

and therefore T and $\bar{T}$ differ by a numerical factor.

Hence we have

THEOREM V. $$1 = \Sigma T$$

and $$T^2 = T.$$

4. The permutations of the hyper-octohedral group may all be written in the form $\sigma\tau$ where σ is a permutation of the A's multiplied by the same permutation of the B's, and τ is a product of transpositions $(A_r B_r)$. Remembering the numerical factor absorbed in S_1, we have

$$S_1^2 = S_1,$$

and hence $P_1 N_1 M_1 S_1 P_2 N_2 M_2 S_2' = S_1 S_2' P_1 P_2 N_1 N_2 M_1 M_2 S_1 S_2'$.

Then $\sigma\tau P_1 N_1 M_1 S_1 P_2 N_2 M_2 S_2' = \pm\sigma P_1 N_1 M_1 P_2 N_2 M_2 S_1 S_2'$.

The operation of σ on the tableau $(F_1 F_2')$ transforms it into another tableau not necessarily standard. But the process used before* may be applied to each tableau separately. Hence, if we combine the two tableaux in one $(FF')_r$, we obtain

$$\sigma\tau . (PP')_r (NMN'M')_r S_r S_r' = \Sigma \lambda_s (PP')_s \sigma_{sr} (NN'MM')_r S_r S_r'$$
$$= \Sigma \lambda_s (SS'PP')_s \sigma_{sr} (NN'MM'SS')_r .$$

Moreover the number of these *units* is

$$\Sigma\left[\binom{n}{\rho} f_{\alpha_1\alpha_2\ldots\alpha_h} f_{\beta_1\beta_2\ldots\beta_k}\right]^2 = \Sigma\binom{n}{\rho}^2 \rho!\,(n-\rho)! = 2^n . n!$$

---

* Q.S.A., III, 269.

THEOREM VI. *The $2^n n!$ operations of the hyper-octohedral group can be expressed linearly in terms of the $2^n n!$ units*

$$(SS'PP')_r \sigma_{rs} (NMN'M'SS')_s.$$

The argument of Q.S.A., IV, §2, may be applied with practically no alteration to prove

THEOREM VII. *Corresponding to each ${}_{\beta_1\beta_2\ldots\beta_k}T_{\alpha_1\alpha_2\ldots\alpha_h}$ there is a linear irreducible group isomorphic with the hyper-octohedral group. The group matrix is obtained from the matrix*

$$\left[\frac{2^n n_1!\, n_2}{f_\alpha f_\beta}(SS'PP')_s\sigma_{sr}(NMN'M'SS')_r\right] \quad (n_1 = \Sigma\alpha,\ \ n_2 = \Sigma\beta)$$

by replacing each permutation Q^{-1} by the variable x_Q. Moreover, the coefficients are all real, rational, and integral.

The *corollary*

$$\Sigma\, PP'NN'MM'SS' = \frac{f_\alpha f_\beta}{2^n n_1!\; n_2!}\,\Sigma\,\chi_\sigma\sigma$$

follows at once as in the case of the symmetric group.

5. Frobenius* has shown that for the symmetric group the character $\chi_\rho^{(\lambda)}$ is the coefficient of $x_1^{\lambda_1+n-1}x_2^{\lambda_2+n-2}\ldots x_n^{\lambda_n}$ in the expansion of

$$\Pi_r(x_1^r+x_2^r+\ldots+x_n^r)^{\rho_r}\Delta_x,$$

where Δ_x is the determinant in which the element of the r-th row and s-th column is x_s^{n-r}: the numbers $\lambda_1 \geqslant \lambda_2 \ldots \geqslant \lambda_n$ define the character (the same as is here defined by $T_{\lambda_1,\lambda_2,\ldots}$): and the conjugate set defined by ρ is that which has ρ_r cycles of r letters, $r = 1, 2, \ldots$.

A corresponding generating function for the characters of the hyper-octohedral group is given as follows:

THEOREM VIII. *The character $\chi_{\rho,\rho'}^{\alpha,\beta}$ belonging to ${}_{\beta_1\beta_2\ldots\beta_k}T_{\alpha_1\alpha_2\ldots\alpha_h}$ of a permutation of the conjugate set (ρ, ρ'), which has ρ_r pairs of cycles of the form $(A_1\ldots A_r)(B_1\ldots B_r)$ and ρ'_s cycles $(A_1\ldots A_s B_1\ldots B_s)$ (where $r, s = 1, 2, \ldots$), is the coefficient of $x_1^{\alpha_1+n-1}x_2^{\alpha_2+n-2}\ldots x_n^{\alpha_n}\,y_1^{\beta_1+n-1}y_2^{\beta_2+n-2}\ldots y_n^{\beta_n}$ in the expansion of*

$$\Pi_{r,s}(x_1^r+x_2^r+\ldots+x_n^r+y_1^r+\ldots+y_n^r)^{\rho_r}(x_1^s+x_2^s+\ldots+x_n^s-y_1^s-\ldots-y_n^s)^{\rho'_s}\Delta_x\Delta_y.$$

When $t > h$, α_t is, of course, zero.

---

* *Berliner Sitzungsberichte* (1900), 519, etc.

It is to be observed that a pair of cycles (which will be called R_1) $(A_1 \dots A_r)(B_1 \dots B_r)$ is not altered in kind by multiplication by an even number 2κ of permutations $(A_s B_s)$, where s lies between 1 and r; merely some letters A are exchanged with their corresponding B. Multiplication by an odd number of such permutations changes the kind to a single cycle R_2, of the form $(A_1 A_2 \dots A_r B_1 \dots B_r)$. Exactly the same remarks apply to the cycle R_2.

Consider a permutation made up entirely of cycles R_1. It is asserted that its characters are the coefficients in the expansion

$$\prod_r (x_1^r + x_2^r + \dots + x_n^r + y_1^r + \dots + y_n^r)^{\rho_r} \Delta_x \Delta_y.$$

The corollary to Theorem VII gives

$$\Sigma \chi_\sigma \sigma = \Sigma [\Sigma PNMS] [\Sigma P'N'M'S'] \frac{2^n n_1!\, n_2!}{f_\alpha f_\beta}$$

$$= \Sigma [2^{n_1} \Sigma \chi_{R_1} R_1 . S] [2^{n_2} \Sigma \chi_{R_1'} R_1' . S'] \qquad \text{(I)}$$

When σ is made up entirely of permutations R_1, it is formed of pairs of cycles with all A's in one cycle and all B's in its fellow, multiplied by an even number of permutations $(A_s B_s)$ of the letters of the pair. The first Σ in (I) extends to all possible selections of the letters into the two divisions. The whole of each pair of cycles must appear among the letters of one division, and as the number of permutations $(A_s B_s)$ for each pair is even, these may be selected at will and without introducing a minus sign from S or S'. The letters x are taken to refer to the first division and the letters y to the second. Thus any pair of cycles of $2c_1$ letters introduces a factor

$$(x_1^{c_1} + x_2^{c_1} + \dots + x_n^{c_1} + y_1^{c_1} + y_2^{c_1} + \dots + y_n^{c_1}),$$

and since $\chi_{R_1}^\alpha$ is the coefficient of $x_1^{\alpha_1+n-1} \dots x_n^{\alpha_n}$ in

$$\Pi (x_1^c + x_2^c + \dots + x_n^c) \Delta_x$$

it is seen that $\chi_\sigma^{\alpha, \beta}$ is the coefficient of $x_1^{\alpha_1+n-1} \dots x_n^{\alpha_n} y_2^{\beta_1+n-1} \dots y_n^{\beta_n}$ in

$$\Pi (x_1^c + \dots + x_n^c + y_1^c + \dots + y_n^c) \Delta_x \Delta_y.$$

When there is a cycle R_2 of order $2d$ this is derived from a cycle R_1 by multiplication with an odd number of permutations $(A_s B_s)$. This introduces the factor

$$x_1^d + \dots + x_n^d - y_1^d - \dots - y_n^d.$$

6. The degree of the irreducible representation is obtained as the coefficient of

$$x_1^{\alpha_1+n-1} \dots x_n^{\alpha_n} y_1^{\beta_1+n-1} \dots y_n^{\beta_n}$$

in the expansion of

$$(x_1+x_2+\ldots+x_n+y_1+y_2+\ldots+y_n)^n \, \Delta_x \Delta_y.$$

Hence we obtain

$$_{\beta_1\beta_2\ldots\beta_k}f_{\alpha_1\alpha_2\ldots\alpha_h} = \binom{n}{\Sigma\alpha} f_{\alpha_1\alpha_2\ldots\alpha_h} f_{\beta_1\beta_2\ldots\beta_k}.$$

Similarly, any character may be obtained in terms of the characters of the symmetric group. For let

$$(x+y)^p(x-y)^q = \Sigma\, C_{p,\,q}^{r,\,s} x^r y^s.$$

Then $$^{\beta_1\beta_2\ldots\beta_k}_{\sigma_1\sigma_2\ldots\sigma_n}\chi^{\alpha_1\alpha_2\ldots\alpha_h}_{\rho_1\rho_2\ldots\rho_n} = \Sigma(\Pi\, C^{\rho'_r,\,\sigma'_r}_{\rho_r,\,\sigma_r})\, \chi^{\alpha_1\alpha_2\ldots\alpha_h}_{\rho'_1\rho'_2\ldots\rho'_n}\, \chi^{\beta_1\beta_2\ldots\beta_k}_{\sigma'_1\sigma'_2\ldots\sigma'_n},$$

where $\rho'_r+\sigma'_r = \rho_r+\sigma_r$.

Thus, in the case $\Sigma\beta = 0$, we obtain

$$^{0}_{\sigma_1\sigma_2\ldots\sigma_n}\chi^{\alpha_1\alpha_2\ldots\alpha_h}_{\rho_1\rho_2\ldots\rho_n} = \chi^{\alpha_1\alpha_2\ldots\alpha_h}_{\rho_1+\sigma_1,\,\rho_2+\sigma_2,\,\ldots,\,\rho_n+\sigma_n}.$$

And in the case $\Sigma\alpha = 0$ we have

$$^{\beta_1\beta_2\ldots\beta_k}_{\sigma_1\sigma_2\ldots\sigma_n}\chi^{0}_{\rho_1\rho_2\ldots\rho_n} = (-)^{\Sigma\sigma}\chi^{\beta_1\beta_2\ldots\beta_k}_{\rho_1+\sigma_1,\,\ldots,\,\rho_n+\sigma_n}.$$

In fact, the representations $_0f_{\alpha_1\alpha_2\ldots\alpha_h}$ are just those of the symmetric group, to which the hyper-octohedral group is multiply isomorphic. The representations $_{\beta_1\beta_2\ldots\beta_k}f_0$ are these again, except for a change of sign which accompanies every transposition (A_rB_r).

The representations for which $\Sigma\beta = 1$ are nearly as simple.

The coefficient of $x^{\rho_1+\sigma_1-1}y$ in $(x+y)^{\rho_1}(x-y)^{\sigma_1}$ is $\rho_1-\sigma_1$. And hence

$$^{1}_{\sigma_1\ldots\sigma_n}\chi^{\alpha_1\ldots\alpha_h}_{\rho_1\ldots\rho_n} = (\rho_1-\sigma_1)\,\chi^{\alpha_1\ldots\alpha_h}_{\rho_1+\sigma_1-1,\,\rho_2+\sigma_2,\,\ldots,\,\rho_n+\sigma_n}.$$

7. The hyper-octohedral group has a subgroup of index 2, exactly analogous to the alternating group; its characters may be obtained from those of the alternating group by the methods used above. There is also in this case another subgroup of index 2, which, unlike that corresponding to the alternating group, has characters all real and rational; and, moreover, the linear irreducible groups isomorphic to it can all be expressed with coefficients which are all integers. If the sign of every transposition (A_rB_r) is changed, half the permutations of the hyper-octohedral group change sign; the other half, whose sign remains unaltered, form a group of index 2, which we shall call the (AB) subgroup. This subgroup we proceed to examine.

When the sign of every $(A_r B_r)$ transposition is changed, ${}_{\beta_1 \ldots \beta_k} T_{\alpha_1 \ldots \alpha_h}$ becomes ${}_{\alpha_1 \ldots \alpha_h} T_{\beta_1 \ldots \beta_k}$, and every unit of one T is changed into the corresponding unit of the other T. Such a pair of representations T of the group may be called conjugate with respect to the (AB) subgroup. Now the sum of an expression Q (a linear function of the permutations) and Q', the expression obtained by changing the sign of every (AB) transposition, is a linear function of the permutations of the (AB) subgroup. Thus, if Q_{rs} be a unit of ${}_{\beta_1 \ldots \beta_k} T_{\alpha_1 \ldots \alpha_h}$, and Q'_{rs} be the corresponding unit of ${}_{\alpha_1 \ldots \alpha_h} T_{\beta_1 \ldots \beta_k}$, then $Q_{rs}+Q'_{rs}$ will be a unit of the representation ${}_{\beta_1 \ldots \beta_k} T_{\alpha_1 \ldots \alpha_h} + {}_{\alpha_1 \ldots \alpha_h} T_{\beta_1 \ldots \beta_k}$ of the (AB) subgroup. Indeed we see that

$$(Q_{rs}+Q'_{rs})(Q_{uv}+Q'_{uv}) = 0$$

unless $u = s$, and then

$$(Q_{rs}+Q'_{rs})(Q_{st}+Q'_{st}) = Q_{rt}+Q'_{rt}.$$

Thus every pair of conjugate representations with respect to the subgroup gives us a single representation of the subgroup.

The only exception arises in the case of the self-conjugate representations ${}_{\alpha_1 \ldots \alpha_h} T_{\alpha_1 \ldots \alpha_h}$. These occur only when n is even, and then

$$\Sigma \alpha = \frac{n}{2}.$$

The number of these is the number of conjugate classes of permutations, or the number of irreducible representations of the symmetric group of degree $n/2$. From the analogy of the alternating group we should expect there to be a pair of conjugate representations of the subgroup, corresponding to each one of these self-conjugate cases; and this is what, in fact, we find.

8. Let us first consider the conjugate sets in the subgroup.

Let
$$\Gamma \equiv [\{A_1 \ldots A_n\}_{B_2}]$$

be an operation which operates on the letters of the permutations which follow it; it is the sum of all the permutations of the (AB) subgroup minus the sum of all the permutations of the hyper-octohedral group which do not belong to this subgroup.

Let s be a permutation of the (AB) subgroup, then, if all the permutations conjugate to s in the original group are also conjugate to s in the subgroup,

$$\Gamma s = 0.$$

But, if these permutations form a pair of conjugate sets in the subgroup,

$$\Gamma s = S_1 - S_2$$

and is not zero.

The permutation $(A_1 \ldots A_r)(B_1 \ldots B_r)$ belongs to the subgroup; hence

$$\Gamma(A_1 \ldots A_r B_1 \ldots B_r) = \Gamma(A_2 \ldots A_r A_1 B_2 \ldots B_r B_1) = -\Gamma(A_1 \ldots A_r B_1 \ldots B_r),$$

since Γ contains the operation $-(A_1 B_1)$. Hence, when s contains any cycle of the second kind,

$$\Gamma s = 0.$$

Moreover, this equation is true when s contains a pair of cycles of the first kind of odd degree r, for let

$$s = (A_1 A_2 \ldots A_r)(B_1 B_2 \ldots B_r),$$

then Γ contains the operation $-(A_1 B_1)(A_2 B_2) \ldots (A_r B_r)$. Hence the only conjugate sets which break up into a pair of conjugate sets for the subgroup are those which consist entirely of pairs of cycles of even degree of the first kind.

Now the permutations of the first kind all belong to the (AB) subgroup. And it has been shown that a pair of cycles of the first kind of degree δ, multiplied by an even number of $(A_r B_r)$ transpositions of the letters contained in them, is simply again a pair of cycles of the first kind of degree δ; while, when they are multiplied by an odd number of such transpositions, the result is a single cycle of the second kind. Hence the permutations of the (AB) subgroup all contain an even number of cycles of the second kind. And, *vice versa*, the condition that a permutation is not to belong to this subgroup is that it should contain an odd number of cycles of the second kind. Now the difference between the number of conjugate sets of permutations having an even number of cycles of the second kind and the number of those having an odd number of such cycles is the coefficient of x^n in the expansion of

$$\prod_{r=1}^{n}(1+x^r+x^{2r}+\ldots) \cdot \prod_{r=1}^{n}(1-x^r+x^{2r}-\ldots)$$

$$= \prod_{r=1}^{n}\frac{1}{1-x^r} \cdot \prod_{r=1}^{n}\frac{1}{1+x^r} = \prod_{r=1}^{n}(1+x^{2r}+x^{4r}+\ldots).$$

The first product relates to the cycles of the first kind, the second to those of the second kind. Hence this difference is equal to the number of conjugate sets which contain only pairs of cycles of the first kind of even degree. Let u, v be the numbers of conjugate sets having an even and odd number respectively of cycles of the second kind. Let w be the number of conjugate sets having just pairs of even cycles of the first kind. Then

$$u - v = w,$$

and the number of conjugate sets in the (AB) subgroup is $v+2w$.

Now the coefficient of x^n in the expansion of

$$\prod_{r=1}^{n} (1+x^{2r}+x^{4r}+\ldots)$$

is also the number of representations

$${}_{a_1\ldots a_h}T_{a_1\ldots a_h}.$$

The total number of representations we know to be $u+v$; since the number which are self-conjugate with respect to the (AB) subgroup is $w = u-v$, the number of conjugate pairs of representations is v.

9. Consider now a self-conjugate representation ${}_{a_1\ldots a_h}T_{a_1\ldots a_h}$.

$$\varpi = S_{r_1}S'_{r_2}P_{r_1}P_{r_2}\sigma_{rs}N_{s_1}N_{s_2}M_{s_1}M_{s_2}S_{s_1}S_{s_2}$$

is a unit. When the pair $A_1 B_1$ appears both in $S_{r_1}P_{r_1}$ and in $N_{s_1}M_{s_1}S_{s_1}$, the interchange of A_1 and B_1 in ϖ leaves it unaltered; and then

$$\Gamma\varpi = 0.$$

Thus $\Gamma\varpi$ is zero unless the letters of $S_{r_1}P_{r_1}$ are the same as those of $N_{s_2}M_{s_2}S'_{s_2}$, and similarly the letters of $S'_{r_2}P_{r_2}$ are the same as those of $N_{s_1}M_{s_1}S_{s_1}$.

We arrange the sequence of tableaux so that the tableaux $(F_{r_1}F'_{r_2})$ and $(F_{r_2}F'_{r_1})$ are at the same distance from the beginning and end respectively of the sequence. Then, when $\Gamma\varpi$ is not zero, ϖ represents an element on the secondary diagonal of the matrix (that going from the top right-hand to the bottom left-hand corner). If, therefore, W be any substitutional expression,

$$\Gamma W = \lambda m,$$

where m is the matrix whose elements on the secondary diagonal are all unity and the other elements are zero; and λ is a constant, for Γ includes the sum of all the permutations of the (AB) subgroup. Then

$$m^2 = E = T,$$

and hence the pair of new classes is $T+m$ and $T-m$. And m and T are both linear functions of the members of the (AB) subgroup only. In order to find out the appropriate units, let us write

$$[P_{r_1}N_{r_1}M_{r_1}S_{r_1}\,.\,P_{r_2}N_{r_2}M_{r_2}S'_{r_2}+P_{r_2}N_{r_2}M_{r_2}S_{r_2}\,.\,P_{r_1}N_{r_1}M_{r_1}S'_{r_1}] = p_r.$$

p_r only contains members of the (AB) subgroup. Then, writing

$$p_r(T+m) = p_{r,1}, \quad p_r(T-m) = p_{r,2},$$

and noticing that both T and m are permutable with p_r, we have two complete sets of units, in number together equal to the number of units in T, and in properties identical with those of the units in the other representations.

Let σ be any permutation of the (AB) subgroup; then

$$\sigma(\Gamma W) = (\Gamma W)\sigma.$$

But, when σ is a permutation of the hyper-octohedral group not contained in the (AB) subgroup,

$$\sigma(\Gamma W) = -(\Gamma W)\sigma.$$

Hence, if U' be what any substitutional expression U becomes when the sign of every $(A_r B_r)$ transposition is changed,

$$U(\Gamma W) = (\Gamma W)U'.$$

10. To find the value of m, we have

$$m = \frac{f}{2^n n!}\Gamma S_{r_1} S'_{r_2} P_{r_1} N_{r_1} M_{r_1} P_{r_2} N_{r_2} M_{r_2} \sigma_{r_1 r_2} P_{r_2} N_{r_2} M_{r_2} P_{r_1} N_{r_1} M_{r_1} S_{r_2} S'_{r_1}$$

$$= \frac{f}{2^n n!}\Gamma S_{r_1} S'_{r_2} P_{r_1} N_{r_1} M_{r_1} P_{r_2} N_{r_2} M_{r_2} \sigma_{r_1 r_2}.$$

Let $A_t B_t$ ($t = 1, \ldots, \frac{1}{2}n = l$) be the letters of $P_{r_1} N_{r_1} M_{r_1} S_{r_1}$ and $C_t D_t$ ($t = 1, \ldots, l$) the letters of $P_{r_2} N_{r_2} M_{r_2} S'_{r_2}$, arranged so that A_t and C_t occupy the same positions in their respective tableaux; and

$$\sigma = (A_1 C_1)(A_2 C_2) \ldots (A_l C_l)(B_1 D_1) \ldots (B_l D_l).$$

Using the brackets [ ] again to denote that the enclosed operator operates on the letters of the permutations which follow, we see that

$$[\{C_1 C_2 \ldots C_l\}_D]\, P_{r_1} N_{r_1} M_{r_1} P_{r_2} N_{r_2} M_{r_2} \sigma_{r_1 r_2}$$

$$= [\{C_1 C_2 \ldots C_l\}_D]\, P_{r_1} N_{r_1} M_{r_1} \sigma_{r_1 r_2} P_{r_1} N_{r_1} M_{r_1}$$

$$= [\{C_1 C_2 \ldots C_l\}_D]\, P_{r_1} N_{r_1} M_{r_1} P_{r_1} N_{r_1} M_{r_1} \sigma_{r_1 r_2}$$

$$= [\{C_1 C_2 \ldots C_l\}_D]\, P_{r_1} N_{r_1} M_{r_1} \sigma_{r_1 r_2}.$$

For $P_{r_1} N_{r_1} M_{r_1}$ contains permutations of the letters A alone, and the effect of permuting the C's and D's in $\sigma_{r_1 r_2}$ is the same as of permuting the A's and B's in $\sigma_{r_1 r_2}$. Moreover, Γ contains the group $\{C_1 \ldots C_l\}_D$ as a factor.

Consider, then, a cycle pair and the corresponding part of σ.

$$(A_1 A_2 \dots A_\rho)(B_1 B_2 \dots B_\rho)(A_1 C_1) \dots (A_\rho C_\rho)(B_1 D_1) \dots (B_\rho D_\rho)$$
$$= (A_1 C_1 A_2 C_2 \dots A_\rho C_\rho)(B_1 D_1 B_2 D_2 \dots B_\rho D_\rho) = \tau,$$

and $P_{r_1} N_{r_1} M_{r_1} \sigma_{r_1 r_2}$ is made up of permutations having only even pairs of cycles of the first kind.

Now the multiplication of this by an even number of transpositions $(A_t B_t)$ or $(C_t D_t)$ $(1 \leqslant t \leqslant \rho)$ produces a cycle pair of the same kind, and multiplication by an odd number of such transpositions produces a cycle of the second kind. All permutations having such a cycle of the second kind disappear under the operation of Γ.

Now τ is a pair of cycles of the first kind of even degree, and permutations made up of such belong to a conjugate set which breaks up into a pair of sets in the (AB) subgroup. Two such permutations, which differ by the interchange of an odd number of AB pairs, will belong to different conjugate sets of the pair in the subgroup. The multiplication of τ by an even number of transpositions $(A_t B_t)$, or by an even number of transpositions $(C_t D_t)$, produces a pair of cycles obtainable from τ by an even number of AB exchanges. But multiplication by one $(A_t B_t)$ and one $(C_t D_t)$, for instance by $(A_1 B_1)(C_1 D_1)$, produces a pair of cycles obtainable from τ by an odd number of AB exchanges. Hence

$$m = \frac{f}{2^n n!} \Gamma S_{r_1} S'_{r_2} P_{r_1} N_{r_1} M_{r_1} P_{r_2} N_{r_2} M_{r_2} \sigma_{r_1 r_2}$$

$$= \frac{f}{2^n n!} \Gamma S_{r_1} S'_{r_2} P_{r_1} N_{r_1} M_{r_1} \sigma_{r_1 r_2}$$

$$= \frac{f}{2^{n+1} n!} \Gamma P_{r_1} N_{r_1} M_{r_1} \sigma_{r_1 r_2}.$$

For the multiplication by $(A_t B_t)(C_u D_u)$ from SS' is accompanied by a change of sign, and the corresponding odd number of (AB) exchanges in Γ is accompanied by a change of sign, so that in the product $S_{r_1} S'_{r_2}$ odd permutations may be ignored, and even permutations with either sign replaced by $+1$. Hence $2m$ may be obtained directly from $T_{\alpha_1 \dots \alpha_h}$, where the suffixes are the same as those of one set of the representation ${}_{\alpha_1 \dots \alpha_h} T_{\alpha_1 \dots \alpha_h}$ with which we are concerned. We express $T_{\alpha_1 \dots \alpha_h}$ in terms of one set of $\frac{1}{2}n$ letters $A_1, A_2, \dots, A_l$. Then every cycle $(A_1 A_2 \dots, A_r)$ is replaced by the pair of cycles

$$(A_1 C_1 A_2 C_2 \dots A_r C_r)(B_1 D_1 B_2 D_2 \dots B_r D_r)$$

and the result is $2m$.

Thus the group characters of those conjugate sets which are the same for the (AB) subgroup as for the original group are the same as the corresponding characters in the original group in all irreducible representations; and the group characters of those pairs of conjugate sets in the (AB) subgroup which form single sets in the original group are the same for each member of the pair as the corresponding characters in the original group for all irreducible representations which are not self-conjugate, but in a self-conjugate representation ${}_{a_1 \ldots a_h}T_{a_1 \ldots a_h}$ the difference between the characters of a pair of sets is the corresponding character in the $T_{a_1 \ldots a_h}$ representation of the symmetric group of degree $\frac{1}{2}n$.

11. The extended tetrahedral group appears as a subgroup of the hyperoctohedral group of degree 4 in the following manner. The conjugate sets are :—

$$h_0 = 1,\ 1; \quad h_1 = 1,\ (A_1B_1)(A_2B_2)(A_3B_3)(A_4B_4);$$

$$h_2 = 6,\ (A_1A_2B_1B_2)(A_3A_4B_3B_4); \quad h_3 = 4,\ R^2;$$

$$h_4 = 4,\ R^4; \quad h_5 = 4,\ R^5; \quad h_6 = 4,\ R;$$

where $$R = (A_1\,B_3\,A_2\,B_1\,A_3\,B_2)(A_4\,B_4).$$

There is, in fact, a set of eight conjugate subgroups of this type; they are also conjugate in the (AB) subgroup.

Compare Frobenius, "Über die Composition der Charaktere einer Gruppe," *Berliner Sitzungsberichte* (1899), 339.

The relations between the characters of the whole group and of the subgroup (using the notation of Frobenius for the latter) are given by the equations :

$$ {}^{0}\chi^{4} = {}^{4}\chi^{0} = {}^{0}\chi^{1111} = {}^{1111}\chi^{0} = \chi^{(0)}, $$

$$ {}^{0}\chi^{22} = {}^{22}\chi^{0} = \chi^{(1)} + \chi^{(2)}, $$

$$ {}^{0}\chi^{31} = {}^{31}\chi^{0} = {}^{0}\chi^{211} = {}^{211}\chi^{0} = \chi^{(3)}, $$

$$ {}^{1}\chi^{3} = {}^{3}\chi^{1} = {}^{1}\chi^{111} = {}^{111}\chi^{1} = \chi^{(5)} + \chi^{(6)}, $$

$$ {}^{2}\chi^{2} = {}^{11}\chi^{11} = 2\chi^{(3)}, $$

$$ {}^{11}\chi^{2} = {}^{2}\chi^{11} = \chi^{(0)} + \chi^{(1)} + \chi^{(2)} + \chi^{(3)}, $$

$$ {}^{1}\chi^{21} = {}^{21}\chi^{1} = 2\chi^{(1)} + \chi^{(5)} + \chi^{(6)}. $$

The equations apply, of course, only to members of the subgroup.

ON QUANTITATIVE SUBSTITUTIONAL ANALYSIS

(*Sixth Paper*)

By Alfred Young

[Received and read 18 June, 1931.]

This paper is a continuation of the fourth paper in the series*. In the first paragraph a second proof of Theorem II in that paper is given. Section 2 gives a proof of Frobenius' generating function for the characters of the symmetric group by the methods of this analysis—a proof which I believe throws light on its nature.

In § 3 the quadratic invariant form, which forms the basis of what follows, is introduced. In § 4 this form is calculated in an elementary way for the representations of unit rank $T_{n-k, 1^k}$ (where 1^k is an abbreviation for 1 repeated k times). The results elaborated in § 5 give an indication of the nature of the more general results which follow.

In § 6 is introduced the seminormal matrix in terms of which all substitutional functions may be expressed, and Theorem II is proved; this theorem enables substitutional expressions which do not involve the last letter to be written down at once. Theorem III of § 7 gives a means of calculating the seminormal transforming matrix.

Sections 8, 9, 10, 11 lead up to Theorem IV, which gives a means of writing down the seminormal matrix for any transposition involving consecutive letters; this matrix is of a very simple form, involving only units positive or negative on the leading diagonal and quadratic matrices which need no calculation along that diagonal, the other elements being zero. Given these transpositions, all other substitutions can at once be calculated. In § 12 Theorem V the same thing is done for the orthogonal

---

* *Proc. London Math. Soc.* (2), 31 (1931), 253–288.

matrix. In § 15 a numerical function, the tableau function, is introduced, which enables the principal constants in the quadratic invariant to be calculated. In § 17 the full constants for the case $T_{n-k\,,\,k}$, where there are only two rows, are obtained. The same method could have been used for the case of unit rank, but it was felt that it was better to do this at the outset by elementary methods. I have as yet not been able to obtain the full constants for the general case.

1. The following is an alternative proof of Theorem II, Q.S.A. IV.

Let Γ be a symbol which represents the operation of taking the sum of the $n\,!$ permutations of the letters in the substitutional expression which follows it. We will consider such an expression belonging to $T_{a_1 a_2 \ldots a}$, and given by the matrix M.

Then
$$\Gamma M = \Sigma\, s M s^{-1},$$

where s is in turn each permutation of the symmetric group. We may express s as a sum of matrices belonging to the various T's, and (since the product of expressions belonging to two different T's is zero) we need only consider the matrices

$$sT_{a_1 a_2 \ldots a_h} = [\lambda_{rs}], \quad s^{-1}\, T_{a_1 a_2 \ldots a_h} = [\mu_{rs}].$$

Then
$$\Gamma M = \Sigma\, [\lambda_{rs}]\, M\, [\mu_{rs}] = \Sigma\, [\mu_{rs}]\, M\, [\lambda_{rs}].$$

Now
$$\Gamma M = \Sigma\, \lambda_\rho t_\rho,$$

where t_ρ is the sum of all the members of a conjugate set and λ_ρ is numerical. Hence

$$\Gamma M = AE,$$

where A is a constant.

Let $M_{\alpha\beta}$ be that matrix for which all the elements are zero except that in the α-th row and β-th column, which is unity. In other words

$$M_{\alpha\beta} = P_\alpha\, \sigma_{\alpha\beta}\, N_\beta\, M_\beta.$$

Then
$$[\lambda_{rs}]\, M_{\alpha\beta}\, [\mu_{rs}] = [\lambda_{r\alpha}\, \mu_{\beta s}].$$

Therefore
$$\Gamma M_{\alpha\beta} = A_{\alpha\beta}\, E = \Sigma\, [\lambda_{r\alpha}\, \mu_{\beta s}];$$

hence
$$\Sigma \lambda_{r\alpha}\mu_{\beta s}=0,$$
unless $r=s$.

But under the sign of Σ the λ and μ are interchangeable as they stand; for s and s^{-1} severally become each permutation in turn. Hence
$$\Sigma \lambda_{r\alpha}\mu_{\beta s}=0,$$
unless $\alpha=\beta$. Also
$$\Sigma\lambda_{r\alpha}\mu_{\alpha r}=\Sigma\lambda_{s\alpha}\mu_{\alpha s}=\Sigma\lambda_{r\beta}\mu_{\beta r}.$$

Thus
$$\Gamma P_\alpha N_\alpha M_\alpha=\Gamma M_{\alpha\alpha}=\Gamma M_{\beta\beta}=\frac{1}{f}\,\Gamma E=\frac{n!}{f}.$$

And
$$\Gamma P_\alpha \sigma_{\alpha\beta} N_\beta M_\beta=\Gamma M_{\alpha\beta}=0,$$
unless $\alpha=\beta$.

2. We may obtain the generating function, given by Frobenius* for the characteristics of the symmetric group, in the following manner.

Let
$$P_{\alpha\beta}=\{a_1 a_2\ldots a_\alpha\}\{a_{\alpha+1}\ldots a_{\alpha+\beta}\},$$
where $\alpha \geqslant \beta$ and the sequence of the letters is that of the suffixes.

Consider
$$P_{\alpha\beta}T_{\alpha+r,\,\beta-r}=P_{\alpha\beta}\sum_{s=1}^{f}P_sN_sM_s.$$

We express every term as the sum of standard forms
$$\Sigma B_{us}P_u\sigma_{us}N_sM_s.$$

For any term $P_sN_sM_s$, whose tableau F_s has one of the first α letters in the lower row, the sum of standard forms obtained from the product contains only terms for which u and s are different. For the others the product is always
$$\alpha!\,r!\,(\beta-r)!\,P_sN_sM_s+\sum_{s\neq u}B_{us}P_u\sigma_{us}N_sM_s;$$
moreover the number of these terms is $\binom{\beta}{r}$.

---

* *Berliner Sitzungsberichte*, 1900, 516–534.

Applying the results and method of the last paragraph, we have

$$\Gamma P_{\alpha\beta} T_{\alpha+r,\,\beta-r} = \alpha!\,\beta!\,\frac{(\alpha+\beta)!}{f_{\alpha+r,\,\beta-r}} T_{\alpha+r,\,\beta-r}.$$

Hence $$\frac{1}{(\alpha+\beta)!}\binom{\alpha+\beta}{\alpha}\Gamma P_{\alpha\beta} = \sum_{r=0}^{\beta}\frac{(\alpha+\beta)!}{f_{\alpha+r,\,\beta-r}} T_{\alpha+r,\,\beta-r}$$

$$= \frac{1}{1-S_{12}}\frac{(\alpha+\beta)!}{f_{\alpha,\beta}} T_{\alpha,\beta},$$

where S_{12} is an operator which removes a letter from the lower row to the upper one.

It is easy to see that this equation may be inverted, giving

$$\frac{(\alpha+\beta)!}{f_{\alpha,\beta}} T_{\alpha,\beta} = (1-S_{12})\frac{1}{(\alpha+\beta)!}\binom{\alpha+\beta}{\alpha}\Gamma P_{\alpha,\beta}.$$

More generally, we find in the same way that

$$\frac{1}{n!}\binom{n}{\alpha_1\,\alpha_2\ldots\alpha_h}\Gamma P_{\alpha_1\alpha_2\ldots\alpha_h} = \Sigma\,\Pi S_{rs}^{\lambda_{rs}}\frac{n!}{f_{\alpha_1\alpha_2\ldots\alpha_h}} T_{\alpha_1\alpha_2\ldots\alpha_h};$$

where $P_{\alpha_1\alpha_2\ldots\alpha_h}$ is the product of positive symmetric groups of different letters of degrees $\alpha_1, \alpha_2, \ldots, \alpha_h$ respectively; also $r < s$, and S_{rs} represents the operation of moving one letter from the s-th row up to the r-th row; and the resulting term is regarded as zero, when any row becomes less than a row below it, or when letters from the same row overlap—as, for instance, happens when $\alpha_1 = \alpha_2$ in the case $S_{13}S_{23}$.

This equation also may be inverted, and gives us

$$\frac{n!}{f_{\alpha_1\alpha_2\ldots\alpha_h}} T_{\alpha_1\alpha_2\ldots} = \Pi(1-S_{rs})\frac{1}{n!}\binom{n}{\alpha_1\,\alpha_2\ldots\alpha_h}\Gamma P_{\alpha_1\alpha_2\ldots\alpha_h}. \qquad \text{(I)}$$

Let $t_{\beta_1\beta_2\ldots\beta_k} \equiv t_\beta$ be the sum of the different members of the conjugate set of permutations, belonging to the symmetric group of n letters, which has k cycles of $\beta_1, \beta_2, \ldots, \beta_k$ letters respectively. Then it was proved (Corollary to Theorem I, Q.S.A. IV, p. 256) that

$$\frac{n!}{f_\alpha} T_\alpha = \Sigma\,\chi_\beta^\alpha t_\beta. \qquad \text{(II)}$$

Let $$W_\beta = \prod_{r=1}^{k} (x_1^{\beta_r} + x_2^{\beta_r} + \ldots + x_n^{\beta_r}) = \Sigma\, B_a^\beta x_1^{a_1} x_2^{a_2} \ldots x_h^{a_h}.$$

Then the number of ways in which a permutation of the set t_β can be formed out of the *different* groups of $P_{a_1 a_2 \ldots a_h}$ is the coefficient B_a^β of $x_1^{a_1} x_2^{a_2} \ldots x_h^{a_h}$ in the expansion of W_β; in this enumeration the letters belonging to a given group of P are regarded as all identical, but when $\beta_r = \beta_{r+1}$ the two cycles are regarded as different. We now change the notation and denote by $\tau_{\gamma_1 \gamma_2 \ldots}$ the sum of the permutations of the symmetric group of n letters which belong to that conjugate set which has γ_1 cycles of one letter, γ_2 cycles of two letters, and so on. The number of these permutations is well known to be n_γ, where

$$n_\gamma \Pi \gamma_r!\ r^{\gamma_r} = n!.$$

Consider the number $n_\gamma^{(a)}$ of the permutations of this set contained in $P_{a_1 a_2 \ldots a_h}$.

To find this we use a function W_γ, which is merely what W_β becomes on change of notation, and observe that, since the equal cycles are really equivalent, we must divide B_a^γ by $\Pi \gamma_r!$; and, since the letters of the different groups of P are really distinct, we must multiply by $\Pi a_r!$; and finally, since the letters of each cycle may be permuted cyclically without changing the cycle, we must divide by Πr^{γ_r}. Hence

$$n_\gamma^{(a)} \Pi \gamma_r!\ r^{\gamma_r} = B_a^\gamma \Pi\, a_r!;$$

and therefore $$n_\gamma B_a^\gamma = \binom{n}{a_1 a_2 \ldots a_h} n_\gamma^{(a)}.$$

Hence $$\binom{n}{a_1 a_2 \ldots a_h} \Gamma P_{a_1 a_2 \ldots a_h} = \Sigma\, n!\ B_a^\gamma t_\gamma.$$

Then, using (I) and (II), we obtain

$$\Sigma\, \chi_\gamma^a t_\gamma = \Pi (1 - S_{rs})\, \Sigma\, B_a^\gamma t_\gamma.$$

To deduce the generating function, we consider first the case $h = 2$. Then

$$\Sigma\, \chi_\gamma^{a_1 a_2} t_\gamma = (1 - S_{12})\, \Sigma\, B_{a_1 a_2}^\gamma t_\gamma$$

$$= \Sigma\, B_{a_1 a_2}^\gamma t_\gamma - \Sigma\, B_{a_1+1,\, a_2-1}^\gamma t_\gamma.$$

Hence $\chi_\gamma^{a_1 a_2}$ is the coefficient of $x_1^{a_1+1} x_2^{a_2}$ in the expansion of $(x_1 - x_2) W_\gamma$.

In the same way we see that, whatever h may be, χ_γ^a is the coefficient of

$$x_1^{n-1+a_1} x_2^{n-2+a_2} x_h^{n-h+a_h} x_{h+1}^{n-h-1} \dots$$

in the expansion of $\Delta . W_\gamma$, where Δ is the alternating function

$$\prod_{\substack{r<s\\1}}^{n} (x_r - x_s).$$

3. Corresponding to the set of forms $T_{a_1 a_2 \dots a_h}$ there is a representation of the symmetric group in f variables $x_1, x_2, \dots, x_f$.

Each variable is related to a standard tableau F, and thus we may conveniently use a notation $[F]$ for the variable, exhibiting the tableau to which it is related.

A permutation s corresponds to a transformation of the variables; this transformation may be obtained by operating on the tableau with s. Let F thus become F_1, then F_1 is not necessarily standard, but

$$sPN = P_1 sN = \Sigma \lambda_r P_r s_r N,$$

where every P_r is standard. The linear transformation corresponding to s transforms $[F]$ into $\Sigma \lambda_r [F_r]$.

We may conveniently extend the notation and use the non-standard symbol $[F_1]$ as the equivalent of the sum $\Sigma \lambda_r [F_r]$. For every real linear group there is a quadratic function of the variables, which is invariant for all the transformations of the group, and is unique when the group is irreducible*. In our case the quadratic invariant is evidently given by $\Sigma [F]^2$, where the Σ includes all tableaux whether standard or not.

4. Let us consider an example, $T_{n-k, 1^k}$. Then the tableaux, and so the variables, are completely defined by the letters in the first column $a_1 a_{\lambda_1} \dots a_{\lambda_k}$. We write the general variable as $x_{\lambda_1 \lambda_2 \dots \lambda_k}$; we may allow the suffixes to be arranged in any sequence provided that we remember that the interchange of a pair of suffixes changes the sign of the variable.

In consequence of this, in a product term of the quadratic invariant the suffixes of the two factors are the same except for one pair. It is

* Burnside, *The theory of groups* (1911), 367.

easy to calculate directly that

$$\Sigma\,(k+1)\,x^2_{\lambda_1\lambda_2\ldots\lambda_k}+2\,\Sigma\,x_{\lambda_1\lambda_2\ldots\lambda_{k-1}\mu_1}x_{\lambda_1\lambda_2\ldots\lambda_{k-1}\mu_2}$$

may be taken as the quadratic invariant. Here $1<\lambda_1<\lambda_2\ldots<\lambda_{k-1}$, but the values of μ_1, μ_2 bear no direct relation to the sequence.

THEOREM I. *The quadratic invariant for* $T_{n-k,\,1^k}$ *may be written*

$$H=\Sigma\,\frac{1}{\Pi\,(\lambda_r)_2}\,X^2_{\lambda_1\lambda_2\ldots\lambda_k},$$

where $1<\lambda_1<\lambda_2\ldots<\lambda_k$, *and*

$$X_{\lambda_1\lambda_2\ldots\lambda_k}=\Sigma\,B^{\lambda}_{\mu_1}\,B^{\lambda}_{\mu_2}\ldots\,B^{\lambda}_{\mu_k}\,x_{\mu_1\mu_2\ldots\mu_k},$$

the numerical coefficients $B^{\lambda}_{\mu_r}$ *being thus defined:*

$$B^{\lambda}_{\mu_r}=0 \quad \textit{unless} \quad \lambda_{r+1}\geqslant\mu_r\geqslant\lambda_r,$$

$$B^{\lambda}_{\mu_r}=1 \quad \textit{if} \quad \lambda_{r+1}>\mu_r>\lambda_r,$$

$$B^{\lambda}_{\mu_r}=\lambda_r \quad \textit{if} \quad \lambda_r=\mu_r,$$

$$B^{\lambda}_{\mu_r}=1-\lambda_{r+1} \quad \textit{if} \quad \lambda_{r+1}=\mu_r.$$

Since the representation of the group is irreducible, there is only one invariant quadratic. It is therefore sufficient to show that this quadratic is invariant for the group. The symmetric group of degree n is generated by the $n-1$ transpositions of consecutive letters $(a_r a_{r+1})$; therefore we need consider only these transpositions. Consider the result of operating with $(a_r a_{r+1})$ on the variable $x_{\lambda_1\lambda_2\ldots\lambda_k}$. There are three cases:

(i) Neither r nor $r+1$ appear among the suffixes of x; the variable is unchanged.

(ii) Both r and $r+1$ appear among the suffixes of x; the variable is changed in sign.

(iii) One, say r, but not $r+1$ appears amongst the suffixes; then x is changed into another variable y in which the suffix r is replaced by $r+1$.

Let us now consider the effect of this transposition on $X_{\lambda_1\lambda_2\ldots\lambda_k}$. We distinguish the same three cases:

(i) Neither r nor $r+1$ appears among the suffixes. Then, these suffixes being arranged in ascending order of magnitude, r and $r+1$ must lie between a consecutive pair, say λ_s and λ_{s+1}.

Then the coefficient of $x_{\mu_1\ldots\mu_k}$ is zero when r or $r+1$ appears amongst its suffixes, unless only one appears and that as μ_s. Moreover, when $\mu_s = r$ or $r+1$, $B^{\lambda}_{\mu_s} = 1$. Thus the linear function X is unaltered by the transformation.

(ii) When both r and $r+1$ appear among the suffixes of X, we may take $\lambda_s = r$, $\lambda_{s+1} = r+1$. Let Y be the transformed form of X.

Then the coefficient of $x_{\mu_1\ldots}$ is zero unless $\mu_s = r$ or $r+1$; and thus, when only one of r, $r+1$ appears among the suffixes of x, it is μ_s. Now for $\mu_s = r$, $B^{\lambda}_{\mu_s} = r$; and for $\mu_s = r+1$, $B^{\lambda}_{\mu_s} = -r$. Moreover, when both r and $r+1$ appear among the suffixes of x, the sign of x is changed by the transformation. Thus

$$Y = -X.$$

(iii) $$\lambda_s = r, \quad \lambda_{s+1} > r+1, \quad \lambda_{s-1} < r.$$

We shall call this for brevity X_r, and associate with it the function in which all the λ's are the same except $\lambda_s = r+1$, and call that X_{r+1}. We want to find the transformed functions Y_r and Y_{r+1}.

The coefficient of x_μ is zero in both cases when r or $r+1$ appears as any suffix except μ_s or μ_{s-1}.

When neither r nor $r+1$ appears in the suffixes of x we have a sum of terms $\Sigma A_\mu x_\mu$, the same for both X_r and X_{r+1}, and also for Y_r and Y_{r+1}. In X_r we have the following coefficients:

$$\mu_s = r, \quad B_{\mu_s} = r; \quad \mu_s = r+1, \quad B_{\mu_s} = 1; \quad \mu_{s-1} = r, \quad B_{\mu_{s-1}} = -(r-1).$$

In X_{r+1} we have the following coefficients:

$$\mu_s = r+1, \quad B_{\mu_s} = r+1; \quad \mu_{s-1} = r, \quad B_{\mu_{s-1}} = 1;$$

$$\mu_{s-1} = r+1, \quad B_{\mu_{s-1}} = -r.$$

Then, if y is a variable for which $\mu_{s-1} = r$, $\mu_s = r+1$, y will appear in X_r with coefficient $-(r-1)C$, and in X_{r+1} with coefficient $(r+1)C$. Let u_r be a variable x in which $\mu_s=r$, but $r+1$ is not among the suffixes and u_{r+1} the same when μ_s is changed to $r+1$; let v_r, v_{r+1} be similar variables when r or $r+1$ is in the position μ_{s-1}.

Then

$$X_r=\Sigma A_\mu x_\mu-(r-1)\Sigma Cy+\Sigma D(ru_r+u_{r+1})-(r-1)\Sigma Ev_r;$$

$$X_{r+1}=\Sigma A_\mu x_\mu+(r+1)\Sigma Cy+(r+1)\Sigma Du_{r+1}+\Sigma E(v_r-rv_{r+1});$$

and hence

$$X_r+Y_r=2\Sigma A_\mu x_\mu+(r+1)\Sigma D(u_r+u_{r+1})-(r-1)\Sigma E(v_r+v_{r+1})$$

$$=X_{r+1}+Y_{r+1},$$

and $$\frac{Y_r-X_r}{r-1}=2\Sigma Cy-\Sigma D(u_r-u_{r+1})+\Sigma E(v_r-v_{r+1})$$

$$=-\frac{Y_{r+1}-X_{r+1}}{r+1}.$$

Now in the sum H, X_r and X_{r+1} appear in the form

$$g\left[\frac{1}{r(r-1)}X_r^2+\frac{1}{r(r+1)}X_{r+1}^2\right]=g\left[\frac{1}{r(r-1)}Y_r^2+\frac{1}{r(r+1)}Y_{r+1}^2\right],$$

where g is a constant. Hence H is invariant for the transformation $(a_r a_{r+1})$.

We have yet to consider the transformation $(a_1 a_2)$. The only variable affected is $x_{2,\mu_2\ldots\mu_k}$, and this variable occurs only in $X_{2,\lambda_2\ldots\lambda_k}$. Thus every X_λ which has not 2 for its first suffix is unchanged by the transformation. Now $x_{2,\mu_2\ldots}$ becomes on transformation $-x_{2,\mu_2\ldots}-x_{3,\mu_2\ldots}$.

Let $X_{2,\lambda_2\ldots}$ become Y. In Y there may at first appear variables $x_{\nu_1\nu_2\ldots\nu_k}$ which have zero coefficients in X. Let us consider the possibilities, as follows:

(i) Two suffixes lie between λ_s and λ_{s+1}; this variable arises from two terms of X in which one or other of the suffixes is replaced by 2. They will appear with the same coefficient but a different sign, and their sum is zero.

(ii) Next we may have three ν's between λ_{s-1} and λ_{s+1}, one of them being λ_s; say ν_{s-1}, λ_s, ν_{s+1}.

This term arises from three terms of X, each with $\mu_1 = 2$; they are

$$\mu_{s-1} = \nu_{s-1},\ \mu_s = \lambda_s,\ \text{coefficient}\ \pm 2\lambda_s C,$$

$$\mu_{s-1} = \nu_{s-1},\ \mu_s = \nu_s,\ \text{coefficient}\ \mp 2C,$$

$$\mu_{s-1} = \lambda_s,\quad \mu_s = \nu_s,\ \text{coefficient}\ \mp 2(\lambda_s - 1)\,C,$$

and again the sum is zero.

We may treat the case of four ν's between λ_{s-2} and λ_{s+1}, viz. ν_{s-2}, λ_{s-1}, λ_s, ν_{s+1}, in the same way; the sum again is zero.

Proceeding thus we find that the coefficients of all variables which do not appear in X are zero in Y.

As regards those terms $x_{\mu_1\mu_2\ldots\mu_k}$ whose coefficients are not zero in X, we have:

(i) $\mu_1 = 2$, the term is merely changed in sign.

(ii) $\mu_1 < \lambda_2$. $B_{\mu_1} = 1$. The coefficient is C in X.

In Y it also arises from $x_{2,\mu_2\ldots\mu_k}$ and from no other term, and this gives a coefficient $-2C$; and thus finally the coefficient is $-C$.

(iii) $\mu_1 = \lambda_2$, $\mu_2 < \lambda_3$.

In Y this arises by replacing either μ_1 or μ_2 by 2, giving coefficients $-C$, $\lambda_2 C$; in X the coefficient was $-(\lambda_2 - 1)\,C$, in Y the final coefficient is $(\lambda_2 - 1)\,C$, a mere change of sign. Proceeding thus we find that

$$Y = -X.$$

Thus H is also invariant for $(a_1 a_2)$ and therefore for the whole group. It is therefore the quadratic invariant.

5. We may now take $X_{\lambda_1\ldots\lambda_k}$ as new variables; in this case the quadratic invariant is a sum of squares with numerical coefficients, the seminormal form. The transformations of the symmetric group for these variables follow a very simple law. They are derived from the transpositions of consecutive letters. We express the corresponding transformations as matrices. That for $(a_1 a_2)$ consists merely of ± 1 along the leading diagonal. That for $(a_r a_{r+1})$ is made up as follows: some of the elements in the leading diagonal are ± 1, and in these cases the row and

column on which they lie are filled elsewhere with zeros; for the rest there is a series of quadratic matrices about the remaining elements of the leading diagonal, each being

$$\begin{bmatrix} -1/r & (r-1)/r \\ (r+1)/r & 1/r \end{bmatrix}$$

for the transposition $(a_r a_{r+1})$. The rest of the elements of the matrix are zero.

It is to be observed here that the sequence of variables follows the law that $X_{\lambda_1 \dots \lambda_k}$ precedes $X_{\mu_1 \dots \mu_k}$, when the first of the differences

$$\lambda_k - \mu_k, \ \lambda_{k-1} - \mu_{k-1}, \ \dots$$

which does not vanish is positive.

Should it be desired to obtain an orthogonal representation of the group, we may take as variables*

$$\xi_{\lambda_1 \dots \lambda_k} = \frac{1}{\sqrt{\{(\lambda_1)_2 (\lambda_2)_2 \dots (\lambda_k)_2\}}} X_{\lambda_1 \lambda_2 \dots \lambda_k}.$$

It is easily seen that the matrix corresponding to $(a_r a_{r+1})$ is of the same form as in the case already given, the only difference being that the quadratic submatrices about the leading diagonal are

$$\begin{bmatrix} -1/r & \sqrt{(r^2-1)}/r \\ \sqrt{(r^2-1)}/r & 1/r \end{bmatrix}.$$

6. Let us now consider the general case $T_{a_1 a_2 \dots a_h}$. There are f standard tableaux and corresponding to them f variables which will for the general argument be written $x_1, x_2, \dots, x_f$ according to the sequence of tableaux. Then, as described, we may obtain a quadratic invariant

$$H = \Sigma A_{rr} x_r^2 + 2 \Sigma A_{rs} x_r x_s.$$

We may express H as a sum of squares. Thus we begin with x_f and the first square is

$$\frac{1}{A_{ff}} \{A_{ff} x_f + A_{ff-1} x_{f-1} + \dots + A_{f1} x_1)^2,$$

* We use $(\lambda)_k$ as an abbreviation for $\lambda(\lambda-1)\dots(\lambda-k+1)$.

the rest of H does not contain x_f; we then absorb all the terms containing x_{f-1} in the next square; the final result may be written

$$H = \Sigma\, 1/B_r X_r^2;$$

where $$X_r = \lambda_{rr} x_r + \lambda_{rr-1} x_{r-1} + \ldots + \lambda_{r1} x_1.$$

Here all the coefficients are positive or negative integers. Let us take as new variables

$$\xi_r = \frac{1}{\lambda_{rr}} X_r = x_r + \frac{\lambda_{rr-1}}{\lambda_{rr}} x_{r-1} + \ldots. \qquad \text{(III)}$$

We now turn our attention to substitutional expressions. All such have been expressed in the form

$$\Sigma\, C_{rs} P_r \sigma_{rs} N_s M_s$$

or as the matrix $[C_{rs}] = C$.

Let us now use new substitutional units derived from the old ones by the transformation (III), *i.e.*

$$\xi_{rs} = P_r \sigma_{rs} N_s M_s + \frac{\lambda_{rr-1}}{\lambda_{rr}} P_{r-1} \sigma_{r-1s} N_s M_s + \ldots.$$

Then, if the new matrix is C',

$$C = MC',$$

where

$$M = \begin{bmatrix} 1 & \frac{\lambda_{21}}{\lambda_{22}} & \frac{\lambda_{31}}{\lambda_{33}} & \ldots & \frac{\lambda_{f1}}{\lambda_{ff}} \\ 0 & 1 & \frac{\lambda_{32}}{\lambda_{33}} & \ldots & \frac{\lambda_{f2}}{\lambda_{ff}} \\ 0 & 0 & 1 & \ldots & \\ \ldots & \ldots & \ldots & & \ldots \\ 0 & 0 & 0 & \ldots & 1 \end{bmatrix}.$$

Thus $$C' = M^{-1} C.$$

At present the right-hand side $N_s M_s$ of the units has not been transformed. In order that the new matrix expression for unity may still be E, we must perform the inverse transformation at the same time on the right. The new matrix will then be

$$D = M^{-1} CM.$$

When C is the matrix form of the transformation which corresponds to the permutation s, it gives the transformation for the variables x; then that for the variables ξ is given by D.

Now this transformation with units in the leading diagonal appropriate to the seminormal quadratic invariant arranged according to the tableau sequence is definite and unique. We may take this sequence to be according to the position of the last letter, then according to the last but one, and so on. We have already obtained a transformation of the same form for passing from substitutional expressions of $n-1$ letters to those of n letters. In Q.S.A. IV, Theorem IV (p. 270), it was shown that if S is a substitutional expression of the letters $a_1 a_2 \dots a_{n-1}$, and $T_{\alpha_1 \dots \alpha_h} = T$ involves a last letter a_n, then

$$ST = IKI^{-1};$$

where K is a T matrix which consists of the matrices $[T_\lambda]$, $\lambda = h, \dots, 1$, down the leading diagonal, and $[T_\lambda]$ is the matrix for $S\Delta_\lambda T$ ($\Delta_\lambda T$ meaning T with its λ-th suffix reduced by one); and the matrix $I = [I_{\lambda\mu}]$, where $I_{\lambda\mu}$ are partial matrices, $I_{\lambda\lambda}$ occupying the same place in I as $[T_\lambda]$ does in K, and

$$I_{\lambda\lambda} = E, \quad I_{\lambda\mu} = 0 \quad (\mu < \lambda).$$

Now the product and also the inverses of matrices which have all the elements zero below the leading diagonal and all the elements of the leading diagonal unity have the same property. Let $[T_\lambda']$ be the matrix $[T_\lambda]$ transformed to its seminormal form; and let K' be the matrix obtained from K by replacing each matrix $[T_\lambda]$ by $[T_\lambda']$. Then K is transformed to K' by a matrix with the property just described. The expression S which involves only the first $n-1$ letters when expressed as a matrix in the seminormal form becomes

$$JK'J^{-1},$$

where J has the same property.

Now this transformation is unique so far as matrices of this nature are concerned. Moreover, the property that the quadratic invariant for the group contains no product terms is a property of the substitutions of the group. But the substitutions represented by K' already have this property, hence $J = 1$.

We shall refer to this representation of $T_{\alpha_1 \alpha_2 \dots \alpha_h}$ as the seminormal representation, and to the substitutional units which correspond to the elements of the seminormal matrix as the seminormal units; it will be convenient to use the notation ϖ_{rs} for these units.

The important fact for the seminormal representation, which has just been deduced, may be enunciated thus:

THEOREM II. *The seminormal matrix for $S.T$, where S is any substitutional expression involving $a_1, a_2, \ldots, a_{n-1}$, and $T = T_{a_1 a_2 \ldots}$ involves a last letter a_n, is that T matrix which consists of the matrices $[T_\lambda]$, $\lambda = h, h-1, \ldots, 1$, down the leading diagonal; where $[T_\lambda]$ is the seminormal matrix for $S.\Delta_\lambda T$.*

It should be noticed that the values of λ are given in descending order, as according to the usual sequence, tableaux with a_n in a given row precede those with a_n in a row higher up. In this particular the enunciation of Theorem IV in Q.S.A. IV is at fault.

7. For the sake of clearness the matrices hitherto used for $T_{a_1 \ldots a_h}$ corresponding to the units $P_r \sigma_{rs} N_s M_s$ will be called the natural matrices, and these units the natural units. It has been shown that there is a matrix M of special character, by means of which we can pass from the natural matrix C to the seminormal matrix D of a given expression, that in fact

$$D = M^{-1} C M.$$

We call M the seminormal transforming matrix for $T_{a_1 \ldots a_h}$. Consider now an expression S involving only the first $n-1$ letters. We have seen that, if $[T_\lambda]$ is the natural matrix and $[T_\lambda']$ the seminormal matrix for $S.\Delta_\lambda T_{a_1 \ldots a_h}$, then the natural matrix for $S.T_{a_1 \ldots a_h}$ is

$$I \begin{bmatrix} [T'_h] & 0 & 0 & \ldots \\ 0 & [T_{h-1}] & 0 & \ldots \\ 0 & 0 & [T_{h-2}] & \ldots \\ \ldots & \ldots & \ldots & \ldots \end{bmatrix} I^{-1} = IKI^{-1};$$

and the seminormal matrix is

$$\begin{bmatrix} [T_h'] & 0 & \ldots \\ 0 & [T'_{h-1}] & \ldots \\ \ldots & \ldots & \ldots \\ \ldots & \ldots & \ldots \end{bmatrix},$$

but this is equal to $J^{-1}KJ$,

where J is the matrix

$$\begin{bmatrix} M_h & 0 & 0 & \dots \\ 0 & M_{h-1} & 0 & \dots \\ 0 & 0 & M_{h-2} & \dots \\ \dots & \dots & \dots & \dots \end{bmatrix},$$

M_λ being the seminormal transforming matrix for

$$\Delta_\lambda T_{a_1 \dots a_h}.$$

We notice that S may be the perfectly general substitutional expression of the first $n-1$ letters, in which case $[T_\lambda]$ is a group matrix, and, as Schur has proved, is permutable with no other matrix except E. Then, since the matrices $[T_\lambda]$ are all different, the only matrix permutable with K is L, where L is the matrix obtained from J by replacing each M_r by $p_r E_r$, p_r being an arbitrary constant, and E_r the unit matrix of appropriate degree. Then, since

$$M^{-1} IKI^{-1} M = J^{-1} KJ,$$

K is permutable with $I^{-1} MJ^{-1}$, and

$$I^{-1} MJ^{-1} = L.$$

But both I and MJ^{-1} are matrices with units in the leading diagonal and zeros below it, hence in L every p_r is unity, so that

$$L = E,$$

and therefore $M = IJ$.

THEOREM III. *The seminormal transforming matrix for $T_{a_1 a_2 \dots a_h} = T$ being M, and that for $\Delta_\lambda T$ being M_λ, then*

$$M = IJ$$

where I is the transforming matrix corresponding to the addition of the last letter in T, and J is that matrix formed by the matrices M_λ in the leading diagonal, with zeros elsewhere.

8. Consider the matrix for $(a_{n-1}a_n)$ in the seminormal form. We first subdivide the seminormal matrix according to the positions of the last two letters. When we take the position of the last letter only we have a system of subdivision according to the sequence of tableaux for

$$\Delta_h T, \quad \Delta_{h-1} T, \quad \ldots, \quad \Delta_1 T.$$

Taking a further subdivision we use the tableaux for

$$\Delta_h^2 T, \quad \Delta_{h-1}\Delta_h T, \quad \ldots, \quad \Delta_1\Delta_h T; \qquad \Delta_h\Delta_{h-1} T, \quad \Delta_{h-1}^2 T, \quad \ldots$$

in this sequence.

By Theorem II the most general substitutional expression involving the first $n-2$ letters only is represented by the appropriate group matrices $[T_{\lambda\mu}]$ down the leading diagonal and zeros elsewhere; call this X. The transposition $(a_{n-1}a_n)$ is permutable with all such expressions. Now Schur* has shown that the only matrices permutable with X are those which are made up of quadratic submatrices of the form

$$\begin{bmatrix} p_{11}E_{\lambda\mu} & p_{12}E_{\lambda\mu} \\ p_{21}E_{\lambda\mu} & p_{22}E_{\lambda\mu} \end{bmatrix},$$

occupying the rows and columns defined by the tableaux for $\Delta_\lambda\Delta_\mu T$ and $\Delta_\mu\Delta_\lambda T$: submatrices $qE_{\lambda\lambda}$ occupying the position on the leading diagonal of $[T_{\lambda\lambda}]$ in X, and zeros everywhere else. The seminormal matrix for $(a_{n-1}a_n)$ is then of this character.

It may be noticed that there are obviously more independent matrices than those appropriate to $A+B(a_{n-1}a_n)$ of this nature. Let Γ_{n-2} be an operation which represents taking the sum of all the $(n-2)!$ permutations of the symmetric group of the first $n-2$ letters in the following expression. Then $\Gamma_{n-2}(a_1a_{n-1})$, or $\Gamma_{n-2}(a_1a_n)$, or in fact $\Gamma_{n-2}S$, where S is any substitutional expression, is of this nature.

9. Consider now the transposition $(a_{n-1}a_n)$. We first take the case where the rows of the tableaux for T are all of unequal length. Then the natural matrix for $(a_{n-1}a_n)$ consists of quadratic matrices of the form $\begin{bmatrix} O & E \\ E & O \end{bmatrix}$, and of matrices E along the leading diagonal and zero elsewhere.

* *Berliner Sitzungsberichte*, 1905, 406-432 (§3).

To make this clear we give an example, the natural matrix for $(a_5 a_6)$ in T_{42}.

The matrix has nine rows and columns, but since we are concerned only with the last two letters we can group together the first four letters. Arranging the tableaux according to their proper sequence, they will be:

$$T_4 \text{ (1 row)}, \quad T_{3,1} \text{ (3 rows)}, \quad T_{3,1} \text{ (3 rows)}, \quad T_{2,2} \text{ (2 rows)};$$

and the required matrix is

$$\begin{bmatrix} E_4 & 0 & 0 & 0 \\ 0 & 0 & E_{31} & 0 \\ 0 & E_{31} & 0 & 0 \\ 0 & 0 & 0 & E_{22} \end{bmatrix}.$$

As a second example, $(a_5 a_6)$ in T_{321} is

$$\begin{bmatrix} 0 & 0 & E_{31} & 0 & 0 & 0 \\ 0 & 0 & 0 & 0 & E_{22} & 0 \\ E_{31} & 0 & 0 & 0 & 0 & 0 \\ 0 & 0 & 0 & 0 & 0 & E_{211} \\ 0 & E_{22} & 0 & 0 & 0 & 0 \\ 0 & 0 & 0 & E_{211} & 0 & 0 \end{bmatrix}.$$

Let C be the natural and D the seminormal matrix for $(a_{n-1} a_n)$, then

$$CM = MD. \tag{IV}$$

The matrices will be subdivided as in § 8, so that

$$M = [m_{rs}].$$

Let r refer to the row defined by $\Delta_\lambda^2 T$; then in the matrix on the left of (IV) the only elements in the r-th row are $E_{\lambda\lambda} m_{rs}$, the only elements on the r-th column in the right-hand matrix are $m_{qr} D_{rr}$, and hence, equating the elements in the r-th row and column,

$$m_{rr} D_{rr} = E_{\lambda\lambda} m_{rr};$$

or $D_{rr} = E_{\lambda\lambda}$.

Let r refer to the tableau $\Delta_\lambda \Delta_\mu T$ and s to the tableau $\Delta_\mu \Delta_\lambda T$, where $\lambda < \mu$ and therefore r precedes s. Then, so far as the r-th and s-th rows are concerned, C is

$$\begin{bmatrix} 0 & E_{\lambda\mu} \\ E_{\lambda\mu} & 0 \end{bmatrix},$$

the columns being the r-th and s-th, the other elements of both these rows and columns being zeros. The same is true for these rows and columns of D except that the quadratic matrix is

$$\begin{bmatrix} p_{11} E_{\lambda\mu} & p_{12} E_{\lambda\mu} \\ p_{21} E_{\lambda\mu} & p_{22} E_{\lambda\mu} \end{bmatrix},$$

the p's being constants.

Consider the elements which occupy the position of this quadratic matrix on both sides of (IV). We have

$$\begin{bmatrix} m_{sr} \, m_{ss} \\ m_{rr} \, m_{rs} \end{bmatrix} = \begin{bmatrix} p_{11} m_{rr} + p_{21} m_{rs} & p_{12} m_{rr} + p_{22} m_{rs} \\ p_{11} m_{sr} + p_{21} m_{ss} & p_{12} m_{sr} + p_{22} m_{ss} \end{bmatrix},$$

and here $m_{sr} = 0$, for $s > r$.

Hence $$p_{11} m_{rr} = -p_{21} m_{rs}, \quad m_{rr} = p_{21} m_{ss},$$

$$m_{rs} = p_{22} m_{ss};$$

thus $$\frac{m_{rr}}{p_{21}} = \frac{m_{rs}}{-p_{11}} = m_{ss} = p_{12} m_{rr} + p_{22} m_{rs};$$

i.e. $$p_{11} = -p_{22}, \quad 1 = p_{12} p_{21} - p_{11} p_{22},$$

the matrices m_{rr}, m_{rs}, m_{ss} being all similar. To find the matrices m_{rr}, etc., we use the equation

$$M = IJ.$$

On the leading diagonal we have on the first subdivision the matrices M_λ;

and above this diagonal matrices $I_{\mu\lambda} M_\lambda$ $(\lambda < \mu)$. Along the leading diagonal of M_λ there are the matrices $M_{\mu\lambda} = M_{\lambda\mu}$; thus we see that

$$m_{rr} = m_{ss} = M_{\lambda\mu},$$

and $p_{21} = 1$.

To find m_{rs}: this comes in the matrix

$$I_{\mu\lambda} M_\lambda;$$

it lies on the rows for which, in the tableaux, a_n is in the μ-th row and a_{n-1} in the λ-th row. To find its value reference must be made to Theorem III Q.S.A. IV, from which Theorem IV was derived. We are concerned in $I_{\mu\lambda}$ with the derivation of those terms of T which have a_n in the μ-th row from the terms of $\Delta_\lambda T$. This was effected by a substitutional product applied to the left-hand side of the form $K_{\lambda,h} K_{\lambda,h-1} \ldots K_{\lambda,\lambda+1}$, where

$$K_{\lambda,\nu} = 1 + \sum_{t=1}^{a_\nu} \frac{1}{a_\lambda - a_\nu + \nu - \lambda} \quad (a_{\nu t} a_n);$$

for this purpose the length of the λ-th row is to be reckoned as $a_\lambda - 1$.

We are interested in that part of the product which, when expanded out and expressed as standard forms, gives terms with a_n in the μ-th row. Thus all those factors $K_{\lambda\nu}$ in which $\nu > \mu$ may be replaced by unity. Now the matrix $I_{\mu\lambda}$ has to be subdivided into matrices for which the rows are given by the tableaux for

$$T_{h,\mu}, \quad T_{h-1,\mu}, \quad \ldots, \quad T_{1,\mu}$$

and the columns by the tableaux for

$$T_{h,\lambda}, \quad T_{h-1,\lambda}, \quad \ldots, \quad T_{1,\lambda}.$$

We are concerned only with the row defined by $T_{\lambda\mu}$, *i.e.* which represents a_{n-1} as being finally in the λ-th row. Let $I_{\mu\lambda} = [i_{\nu\rho}]$ according to this subdivision and let $M_\lambda = [n_{\nu\rho}]$.

Then the element of $I_{\mu\lambda} M_\lambda$ which we want, *i.e.* m_{rs}, is

$$\sum_\nu i_{\lambda\nu} n_{\nu\mu}.$$

We know that $n_{\nu\mu}$ is zero unless $\nu \geqslant \mu$ and $n_{\mu\mu} = M_{\lambda\mu}$. Also $i_{\lambda\nu}$ is that part of $I_{\mu\lambda}$ which represents terms in which a_{n-1} was originally in the

ν-th row and is finally in the λ-th row. Consider how this can happen when $\nu > \mu$. The last letter a_n is lowered from the λ-th row to the μ-th row. It is useless to consider terms which have a_n in a row lower than the μ-th, for the process of expressing non-standard terms in terms of standard can never raise the last letter. Moreover, since a_{n-1} is originally in the ν-th, a row lower than the μ-th, a_{n-1} cannot be raised by this process without reducing a_n to the ν-th row; and thus, when $\nu > \mu$,

$$i_{\lambda\nu} = 0.$$

We have only then to consider

$$i_{\lambda\mu}\, n_{\mu\mu} = i_{\lambda\mu}\, M_{\lambda\mu}.$$

This can arise only from the term of the operating product K,

$$\frac{1}{a_\lambda - a_\mu + \mu - \lambda}(a_{n-1}a_n);$$

and this term is standard.

Hence
$$i_{\lambda\mu} = \frac{1}{a_\lambda - a_\mu + \mu - \lambda} E;$$

and
$$m_{rs} = \frac{1}{a_\lambda - a_\mu + \mu - \lambda} M_{\lambda\mu} = \rho M_{\lambda\mu}.$$

The quadratic matrices then which appear in the seminormal matrix for $(a_{n-1}a_n)$ are of the form

$$\begin{bmatrix} -\rho & 1-\rho^2 \\ 1 & \rho \end{bmatrix}, \qquad \text{(V)}$$

where ρ^{-1} has the value $a_\lambda - a_\mu + \mu - \lambda$.

10. It is necessary now to consider the general case in which some of the tableau rows may be of the same length. Let $a_\tau = a_{\tau+1} > a_{\tau+2}$. Then a_n cannot appear in the τ-th row of a standard tableau; the particular question that must be considered is the effect of the transposition $(a_{n-1}a_n)$ on those natural units whose P tableaux have a_{n-1} at the end of the τ-th row and a_n at the end of the $(\tau+1)$-th row. We subdivide the natural matrix $[C_{rs}] = C$ for $(a_{n-1}a_n)$ as before according to the positions of the

last two letters. And we specify the row or column defined by the positions of a_{n-1}, a_n now considered in the tableau as the t-th.

Then
$$C_{tt} = -E_{\tau\tau+1},$$
$$C_{st} = 0, \quad \text{when} \quad s > t.$$

But there will be elements C_{st} which it is impossible to give offhand when $s < t$; they arise from the equations for expressing non-standard forms in terms of standard forms.

Consider now the element in the t-th row and t-th column on both sides of equation (IV). We have to remember that there may be another pair of equal rows. For the sake of argument suppose that this is the case, and that this second pair is higher up in the tableau, and that when a_{n-1} and a_n both lie in them the tableau is specified as the u-th, where in consequence $t < u$. In this case C_{tu} will not in general be zero, but since m_{ut} is zero it will not enter our equation, which becomes

$$-E_{\tau,\tau+1}\, m_{tt} = m_{tt}\, D_{tt},$$

or
$$D_{tt} = -E_{\tau,\tau+1}.$$

It is sufficient now to notice that these unknown elements in the C matrix which arise when there are equal rows do not affect the argument of the last paragraph, because in the terms which are considered they always appear multiplied by a zero matrix of M.

Thus the seminormal matrix for $(a_{n-1} a_n)$ consists entirely of the quadratic matrices (V), unit matrices in those places on the leading diagonal defined by tableaux with a_{n-1}, a_n in the same row, and negative unit matrices in those places in the leading diagonal defined by tableaux with a_{n-1}, a_n in the same column.

11. Let us apply Theorem II to the fact just obtained. We see that the seminormal matrix for $(a_r a_{r+1})$ is known at once when we know the seminormal matrices for this transposition for all T's which refer to the first $r+1$ letters.

Now the rows and columns of the matrix for $T_{a_1 a_2 \dots a_h}$ are defined by the sequence of tableaux $F_1, F_2, \dots, F_f$. Let us specify them more particularly with respect to the positions of a_r and a_{r+1}. Let F_ϵ be a tableau in which a_r, a_{r+1} are in the same row, F_ζ one in which they are in the same column; and let F_ξ, F_η be a pair of tableaux which are identical except for the positions of a_r and a_{r+1}, F having a_r in the λ-th row and the γ-th column—or briefly in the position (λ, γ)—and a_{r+1} in the position (μ, δ), and F_η having these positions interchanged. We suppose that $\lambda < \mu$, then $\xi < \eta$. The standard condition gives $\gamma > \delta$. Then the seminormal matrix for $(a_r a_{r+1})$ has unity on the leading diagonal and zeros elsewhere on the ϵ-th row and column; -1 on the leading diagonal and zeros elsewhere on the ζ-th row and column; and the ξ-th and η-th rows and columns have the quadratic matrix

$$\begin{bmatrix} -\rho & 1-\rho^2 \\ 1 & \rho \end{bmatrix}$$

where they intersect and zeros elsewhere, where

$$\rho^{-1} = \gamma - \delta + \mu - \lambda.$$

This is the sum of the projections of the straight line joining a_r and a_{r+1} on the first row and the first column of the tableau; we call it for brevity the axial projection of $a_r a_{r+1}$. We may then enunciate the result as follows:

THEOREM IV. *The seminormal matrix for $(a_r a_{r+1})$ has zero everywhere, except (i) unity on the leading diagonal where the tableau has a_r, a_{r+1} in the same row; (ii) -1 on the leading diagonal where the tableau has a_r, a_{r+1} in the same column; (iii) a quadratic matrix*

$$\begin{bmatrix} -\rho & 1-\rho^2 \\ 1 & \rho \end{bmatrix},$$

the elements being in the positions of the intersections of a pair of rows and columns whose tableaux differ only by the exchange of a_r and a_{r+1}—the value of ρ being the reciprocal of the axial projection of $a_r a_{r+1}$ in the tableaux.

12. The seminormal matrix corresponds to the variables defined by (III), and the quadratic invariant

$$\Sigma \frac{\lambda_{rr}^2}{B_r} \xi_r^2.$$

To find an orthogonal representation of the group, new variables η_r must be chosen, for which the form $\Sigma \eta_r^2$ is invariant. This may be done by choosing

$$\eta_r = \frac{\lambda_{rr}}{\sqrt{B_r}} \xi_r = p_r \xi_r.$$

This is equivalent to transformation by a matrix which has zeros everywhere except in the leading diagonal; in other words, the r-th column of the matrices is multiplied by the same number p_r, and the r-th row is divided by p_r. In fact, we may write $\bar{D}$ as the orthogonal matrix where

$$\bar{D} = \bar{M}^{-1} D \bar{M},$$

and $\bar{M}$ has zeros everywhere except on the leading diagonal, which has the elements

$$p_1, p_2, \ldots, p_f.$$

The transformation $(a_r a_{r+1})$ for the orthogonal matrix is the same as for the seminormal matrix except that the quadratic matrices which there appear have now the form

$$\begin{bmatrix} -\rho & (1-\rho^2)\varpi^{-1} \\ \varpi & \rho \end{bmatrix}.$$

Since this matrix must be orthogonal,

$$\varpi = \sqrt{(1-\rho^2)}.$$

THEOREM V. *The orthogonal matrix for* $(a_r a_{r+1})$ *is identical with the seminormal matrix except that the quadratic matrices which appear in it are replaced by the matrices*

$$\begin{bmatrix} -\rho & \sqrt{(1-\rho^2)} \\ \sqrt{(1-\rho^2)} & \rho \end{bmatrix},$$

ρ *having the same meaning as before.*

13. We call the rows and columns which appear in the quadratic matrix in the argument of the last paragraph, the s-th and t-th, $s < t$. Then

$$\sqrt{(1-\rho^2)} = \varpi = \frac{p_s}{p_t} = \frac{\lambda_{ss}}{\lambda_{tt}} \sqrt{\left(\frac{B_t}{B_s}\right)}, \tag{VI}$$

a relation which connects the coefficients of the original quadratic invariant § 6,

$$H = \Sigma \frac{1}{B_r} X_r^2.$$

This relation will be required later.

14. Consider the product

$$(a_r a_{r+1}) \varpi_{su},$$

where ϖ_{su} is a seminormal unit, and s, t give the rows and columns of the last paragraph.

$$(a_r a_{r+1}) \varpi_{su} = -\rho \varpi_{su} + \varpi_{tu},$$

$$(a_r a_{r+1}) \varpi_{tu} = (1-\rho^2) \varpi_{su} + \rho \varpi_{tu}.$$

Now

$$\varpi_{su} = \sum_{\tau} A_{u\tau} \left\{ P_s \sigma_{s\tau} N_\tau M_\tau + \frac{\lambda_{s\,s-1}}{\lambda_{ss}} P_{s-1} \sigma_{s-1\,\tau} N_\tau M_\tau + \ldots \right\};$$

and
$$\varpi_{tu} = \sum_{\tau} A_{u\tau} \left\{ P_t \sigma_{t\tau} N_\tau M_\tau + \frac{\lambda_{t\,t-1}}{\lambda_{tt}} P_{t-1} \sigma_{t-1\,\tau} N_\tau M_\tau + \ldots \right.$$
$$\left. + \frac{\lambda_{ts}}{\lambda_{tt}} P_s \sigma_{s\tau} N_\tau M_\tau + \ldots \right\}.$$

We are not concerned with the values of the constants $A_{u\tau}$.

Consider the effect of multiplying the term beginning with P_v by $(a_r a_{r+1})$. Since $v < t$, the position of a_n in the tableau F_v is either in the same row or lower down than in the tableau F_t; if $(a_r a_{r+1}) F_v$ is non-standard, the equations expressing the resulting term in standard units cannot raise the position of a_n. Thus the only terms which can yield P_s or P_t are those terms P_v which have a_n in the same row as in F_s or F_t. We apply the same argument to each letter in turn, and conclude that the only terms commencing with P_v which, on multiplication by $(a_r a_{r+1})$ and then expressing in standard units, contain the terms P_s or P_t are those for which F_v is the same as F_s and F_t so far as the letters $a_{r+2}, a_{r+3}, \ldots, a_n$ are concerned. As regards the position of a_r, a_{r+1} in F_v we may remove these last $n-r-1$ letters, and consider only tableaux F' of fixed form of $r+1$ letters. Regarding the possible tableaux F_v we see that unless a_r, a_{r+1} terminate two consecutive equal rows $(a_r a_{r+1}) F_v$ is standard, and so the term only yields P_s or P_t when $v=s$ or t. In the case when a_r, a_{r+1} are at the ends of equal rows in F_v, the pair of equal rows must lie below the a_r row in F_s otherwise $v>t$; and thus both letters a_r, a_{r+1} lie below this row. After the transposition the standardising equations cannot raise them up to the level of that row without altering the position of later letters. Thus

$$(a_r a_{r+1})\,\varpi_{su} = \sum_{\tau} A_{u\tau} P_t \sigma_{t\tau} N_\tau M_\tau + \ldots$$

$$= -\rho \sum_{\tau} A_{u\tau} P_s \sigma_{s\tau} N_\tau M_\tau + \ldots$$

$$+ \sum_{\tau} A_{u\tau} \{P_t \sigma_{t\tau} N_\tau M_\tau + \lambda_{ts}/\lambda_{tt} P_s \sigma_{s\tau} N_\tau M_\tau + \ldots\},$$

giving $$\lambda_{ts}/\lambda_{tt} = \rho.$$

And

$$(a_r a_{r+1})\,\varpi_{tu} = \sum_{\tau} A_{u\tau} \{P_s \sigma_{s\tau} N_\tau M_\tau + \lambda_{ts}/\lambda_{tt} P_t \sigma_{t\tau} N_\tau M_\tau + \ldots\}$$

$$= (1-\rho^2) \sum_{\tau} A_{u\tau} \{P_s \sigma_{s\tau} N_\tau M_\tau + \ldots\}$$

$$+ \rho \sum_{\tau} A_{u\tau} \{P_t \sigma_{t\tau} N_\tau M_\tau + \lambda_{ts}/\lambda_{tt} P_s \sigma_{s\tau} N_\tau M_\tau + \ldots\},$$

with the same result.

Let us now consider the variables; the transformation $(a_r a_{r+1})$ becomes

$$\xi_s' = -\rho \xi_s + \xi_t,$$

$$\xi_t' = (1-\rho^2)\,\xi_s + \rho \xi_t;$$

changing to the variables X_r of § 6 and using Y_r as the transformed variable, we have

$$Y_s = -\rho X_s + \frac{\lambda_{ss}}{\lambda_{tt}} X_t,$$

$$Y_t = (1-\rho^2)\frac{\lambda_{tt}}{\lambda_{ss}} X_s + \rho X_t,$$

$$Y_s + X_s = (1-\rho)\,X_s + \frac{\lambda_{ss}}{\lambda_{tt}} X_t,$$

$$Y_t + X_t = (1-\rho^2)\frac{\lambda_{tt}}{\lambda_{ss}} X_s + (1+\rho)\,X_t.$$

If $\lambda_{ss}/\lambda_{tt} = 1+\rho$, then

$$Y_s+X_s = Y_t+X_t,$$

and
$$\frac{Y_s-X_s}{1+\rho} = -\frac{Y_t-X_t}{1-\rho},$$

exactly as was found for the case $T_{k,1^k}$.

The full justification for this ratio of λ_{ss} to λ_{tt} will be found later. Meanwhile it should be noticed that, the absolute values of λ_{rr} being still unfixed, the coefficients can all be made integral with this ratio.

15. We will now define a numerical function, the tableau function, by means of the standard tableau.

Let F be a standard tableau; it may be defined particularly as a matrix

$$F \equiv [\gamma_{r,s}];$$

$\gamma_{r,s}$ is the element in the r-th row and the s-th column; in particular, γ_{rs} is the suffix of the letter which appears in this position in the standard tableau F. The standard condition is

$$\gamma_{rs} > \gamma_{ts} \quad \text{when} \quad r > t,$$

$$\gamma_{rs} > \gamma_{rt} \quad \text{when} \quad s > t.$$

Then $a_1, a_2, \ldots, a_h$ being, as usual, the numbers of letters in the first, second, etc., rows of F, the tableau function $[F]$ is the product of $a_2+2a_3+\ldots+(h-1)a_h$ factors, which are defined thus. The element γ_{rs} gives rise to $r-1$ factors, one in relation to each row above it.

Consider the t-th row ($r > t$); we have a factor γ_{rst}. We know that $\gamma_{ts} < \gamma_{rs}$, but γ_{tu} may be either less or greater than γ_{rs} when $u > s$; it must be greater than γ_{ts}.

Let us suppose that $\gamma_{tu} < \gamma_{rs}$ and $\gamma_{tu+1} > \gamma_{rs}$ (when $u = a_t$ we shall always regard the non-existent γ_{tu+1} as greater than γ_{rs}). Then we define

$$\gamma_{rst} = u+r-s-t+1.$$

(If it is necessary to specify u more particularly we may call it u_{rst}.)

Then we define our function as

$$[F] = \Pi\, \gamma_{rst}. \qquad \text{(VII)}$$

We may take the first coefficient in the linear function X_r, *i.e.* the coefficient of the variable x_r, which corresponds to the standard tableau F_r, as the number which we have defined as $[F_r]$, the tableau function of F_r.

For consider the transposition $(a_r a_{r+1})$ which interchanges the tableaux F_s and F_t, defined as above; in F_s the position of a_r will be taken as (λ, σ), of a_{r+1} as (μ, τ); so that $\gamma_{\lambda\sigma} = r$, and $\gamma_{\mu\tau} = r+1$.

Then $$\gamma_{\mu\tau\lambda} = \sigma+\mu-\tau-\lambda+1.$$

In $[F_t]$ all the factors are the same as in $[F_s]$ with the single exception of $\gamma_{\mu\tau\lambda}$, which becomes $\sigma+\mu-\tau-\lambda$.

Now $\sigma+\mu-\tau-\lambda$ is what we have called the axial projection of $a_r a_{r+1}$ for the tableau, and have written ρ^{-1}; thus

$$[F_s]/(1+\rho) = [F_t].$$

This is the same as the ratio of the coefficients λ_{ss}, λ_{tt} given in § 14.

16. We now introduce a function which we call the second tableau function. For a given tableau it is defined by the same elements as the first tableau function, and we write it

$$[F]_2 = \Pi(\gamma_{rst}-1). \qquad \text{(VIII)}$$

Then, with the notation of § 15,

$$\frac{[F_s]_2}{[F_t]_2} = \frac{1}{1-\rho}.$$

Equation (VI) of § 13 gives

$$\frac{B_s}{B_t} = \frac{\lambda_{ss}^2}{(1-\rho^2)\lambda_{tt}^2} = \frac{1+\rho}{1-\rho} = \frac{[F_s][F_s]_2}{[F_t][F_t]_2}.$$

In the expression of § 6 for the quadratic invariant we may then put

$$B_r = [F_r][F_r]_2;$$

and $\lambda_{rr} = [F_r]$.

It will be noticed that these values agree with the results already found for $T_{n-k, 1^k}$.

Consider the coefficient of x_f^2 in H; the only term in which it appears is X_f^2; thus its coefficient is

$$[F_f]/[F_f]_2.$$

Let $\beta_1 = h$ be the number of letters in the first column of the tableaux, β_2 the number in the second, and so on; then the tableau F_f is

$$\begin{array}{llllllll} a_1 a_{\beta_1+1} & \dots & \dots & \dots & \dots & a_n \\ a_2 a_{\beta_1+2} & \dots & \dots & \dots \\ \dots \quad \dots & \dots & \dots & \dots \\ \dots \quad \dots & \dots & \dots & \dots \\ a_{\beta_2} a_{\beta_1+\beta_2} \\ \vdots \\ a_{\beta_1} \end{array}$$

Then $$\prod_{s=1}^{a_r} \gamma_{rst} = (r-t+1)^{a_r};$$

and $$\prod_{s=1}^{a_r} (\gamma_{rst}-1) = (r-t)^{a_r}.$$

Hence

$$[F_f]/[F_f]_2 = 2^{a_2} 3^{a_3} \dots h^{a_h} = \beta_1! \ \beta_2! \dots \beta_k!.$$

Let us define H exactly as the sum of the squares of all the terms derived from standard or non-standard tableaux which have different rows, a permutation in the rows only being regarded as not producing a difference; for such a permutation does not alter the corresponding function of the variables x. The number of the terms in H is then

$$\binom{n}{a_1 a_2 \dots a_h}.$$

Consider the coefficient of x_f^2 in H. It is the number of forms $P_r N_r$ which when expressed in terms of forms with a standard P contain the term P_f. Let $P_r N_r$ be such a form; then, when $a_1 > a_2$, a_n must lie in the same place in the first row of F_r; it cannot be raised by the standardizing equations. Similarly $a_{n-1}, a_{n-2}, \ldots, a_{n-\alpha_1, +\alpha_2+1}$ must all lie in the first row of F_r. In the same way when $\alpha_2 > \alpha_3$ the pair of letters $a_{n-\alpha_1+\alpha_2}, a_{n-\alpha_1+\alpha_2-1}$ must occupy places one in the first and other in the second row, for unless one is in the first row our equations will not allow either to be raised to it; and unless the other is in the second row it cannot be raised there. If both lie in the first row of F_r, then in the final result P_f does not appear at all, or appears twice, once with a + and once with a − sign. Proceeding thus we see that F_r, when the letters in each row are arranged according to the fixed sequence, is obtained from F_f by a permutation which only interchanges letters in the same column, and each such permutation will result in a linear function in which P_f appears with the coefficient ± 1; and each gives a different term of the sum H. The number of such is

$$\beta_1!\,\beta_2!\,\ldots\,\beta_k! = [F_f]/[F_f]_2.$$

THEOREM VI. *The quadratic invariant obtained from the sum of the squares obtained from all tableaux with different rows is equal to*

$$\sum_{r=1}^{f} \frac{1}{[F_r][F_r]_2} X_r^2,$$

where

$$X_r = [F_r]\,x_r + \lambda_{rr-1}\,x_{r-1} + \ldots + \lambda_{r1}\,x_1.$$

17. THEOREM VII. *The exact quadratic invariant for $T_{n-k,\,k}$ is*

$$\sum_{r=1}^{f} \frac{1}{[F_\rho]\,[F_\rho]_2} X_\rho^2,$$

where

$$X_\rho = [F_\rho]\,x_\rho + \lambda_{\rho,\rho-1}\,x_{\rho-1} + \ldots.$$

When the suffixes of the letters of the lower rows of the tableaux F_ρ, F_σ are $\lambda_1, \lambda_2, \ldots, \lambda_k$; $\mu_1, \mu_2, \ldots, \mu_k$, respectively, the values of the coefficients are

$$[F_\rho] = \lambda_1\,[\lambda_2 - 2]\,(\lambda_3 - 4] \ldots (\lambda_k - 2k + 2),$$

$$[F_\rho]_2 = (\lambda_1 - 1)\,(\lambda_2 - 3) \ldots (\lambda_k - 2k + 1),$$

$$\lambda_{\rho\sigma} = B_1^\rho\, B_2^\rho \ldots B_k^\rho;$$

where
$$B_t^\rho = 0, \quad \mu_t < \lambda_t,$$
$$B_t^\rho = \lambda_{t+u} - 2t + 2 - u, \quad \mu_t = \lambda_{t+u},$$
$$B_t^\rho = u, \quad \lambda_{t+u} > \mu_t > \lambda_{t+u-1},$$

λ_{k+1} *being reckoned as greater than* n.

In § 6 we saw that the seminormal transforming matrix M is
$$M = [m_{rs}] = [\lambda_{sr}/\lambda_{ss}],$$
where of course $\lambda_{sr} = 0$ for $r > s$.

In Theorem III it was proved that
$$M = IJ,$$
where J is a matrix which may be subdivided according to the position of a_n; when thus expressed it contains zeros everywhere except on the leading diagonal. In our case, using suffixes as for T', and thus writing $M_{n-k,k}$ for M, we have
$$J = \begin{bmatrix} M_{n-k,k-1} & 0 \\ 0 & M_{n-k-1,k} \end{bmatrix}.$$

In Q.S.A. IV, § 12, it was shown that
$$I = I_{n-k,k} = \begin{bmatrix} E_{n-k,k-1} & \dfrac{1}{n-2k+1} K_{n-k,k} \\ 0 & E_{n-k-1,k} \end{bmatrix},$$
with a necessary alteration in nomenclature. An unfortunate error appeared in the formula quoted, $1/(\alpha-\beta+2)$ being given as the coefficient of $M_{\alpha\beta}$ which should be $1/(\alpha-\beta+1)$. Further it was shown that
$$K_{n-k,k} = \begin{bmatrix} K_{n-k,k-1} & 0 \\ E_{n-k-1,k-1} & K_{n-k-1,k} \end{bmatrix}.$$

We shall use these results to establish our theorem by induction, assuming its truth for all values of n less than the one under consideration.

Let
$$K_{\alpha,\beta-\lambda} K_{\alpha-1,\beta-\lambda+1} \cdots K_{\alpha-\lambda,\beta} = K_{\alpha,\beta}^{\lambda+1}$$
$$= K_{\alpha,\beta-1}^{\lambda} K_{\alpha-\lambda,\beta}.$$

Then it easily follows by induction that

$$K^{\lambda}_{a,\beta}=K^{\lambda-1}_{a,\beta-1}K_{a-\lambda+1,\beta}$$

$$=\begin{bmatrix} K^{\lambda-1}_{a,\beta-2} & 0 \\ (\lambda-1)K^{\lambda-2}_{a-1,\beta-2} & K^{\lambda-1}_{a-1,\beta-1}\end{bmatrix}\begin{bmatrix} K_{a-\lambda+1,\beta-1} & 0 \\ E_{a-\lambda,\beta-1} & K_{a-\lambda,\beta}\end{bmatrix}$$

$$=\begin{bmatrix} K^{\lambda}_{a,\beta-1} & 0 \\ \lambda K^{\lambda-1}_{a-1,\beta-1} & K^{\lambda}_{a-1,\beta}\end{bmatrix}.$$

Let
$$M^{p}_{a,\beta}=\frac{1}{(a-\beta+1)_p}K^{p}_{a,\beta}M_{a-p,\beta},$$

then multiplication of these two quadratic matrices yields

$$M^{p}_{a,\beta}=\begin{bmatrix} \dfrac{a-\beta+2}{a-\beta+2-p}M^{p}_{a,\beta-1} & M^{p+1}_{a,\beta} \\ \dfrac{p}{a-\beta+2-p}M^{p-1}_{a-1,\beta-1} & M^{p}_{a-1,\beta}\end{bmatrix}. \tag{VIII}$$

When $p=0$,
$$M^{0}_{a,\beta}=M_{a,\beta}.$$

Consider now the matrix $M_{n-k,k}$.

$$M_{n-k,k}=IJ=\begin{bmatrix} E_{n-k,k-1} & \dfrac{1}{n-2k+1}K_{n-k,k} \\ 0 & E_{n-k-1,k}\end{bmatrix}\begin{bmatrix} M_{n-k,k-1} & 0 \\ 0 & M_{n-k-1,k}\end{bmatrix}$$

$$=\begin{bmatrix} M_{n-k,k-1} & M^{1}_{n-k,k} \\ 0 & M_{n-k-1,k}\end{bmatrix}.$$

The positions of the elements of this matrix are defined by the positions of a_n in the tableaux; in fact, in the first row they may be defined as

$$\begin{pmatrix} \\ a_n,\ a_n\end{pmatrix},\ \begin{pmatrix} & a_n \\ a_n, & \end{pmatrix} \text{ and in the second row as } \begin{pmatrix} a_n, & \\ & a_n\end{pmatrix},\ \begin{pmatrix} a_n,\ a_n \\ \end{pmatrix}.$$

The assumption that the theorem is known to be true for values of n less than the given value establishes the theorem for the whole of this matrix except the part $\begin{pmatrix} & a_n \\ a_n, & \end{pmatrix}$.

We consider then this last part of the matrix, and subdivide it according to the position of a_{n-1}. By VIII,

$$M^1_{n-k,k} = \begin{bmatrix} \frac{n-2k+2}{n-2k+1} M^1_{n-k,k-1} & M^2_{n-k,k} \\ \frac{1}{n-2k+1} M_{n-k-1,k-1} & M^1_{n-k-1,k} \end{bmatrix}.$$

Here again the whole matrix is known except the element $M^2_{n-k,k}$ in the position $\begin{pmatrix} & a_{n-1} \\ a_{n-1}, & \end{pmatrix}$, or, with reference to the original matrix, the position $\begin{pmatrix} & a_{n-1}a_n \\ a_{n-1}a_n, & \end{pmatrix}$. We keep on repeating this process, and therefore consider in detail the matrix $M^p_{n-k\ k}$ which may be defined as the element

$$\begin{pmatrix} & a_{n-p+1}a_{n-p+2}\dots a_n \\ a_{n-p+1}a_{n-p+2}\dots a_n, & \end{pmatrix}$$

of $M_{n-k,k}$.

We consider this matrix subdivided as in VIII according to the position of a_{n-p}, and write it

$$\begin{bmatrix} m_{11} & m_{12} \\ m_{21} & m_{22} \end{bmatrix}.$$

The matrix $M^p_{n-k,k-1}$ concerned with $n-1$ letters is the element

$$\begin{pmatrix} & a_{n-p}a_{n-p+1}\dots a_{n-1} \\ a_{n-p}a_{n-p+1}\dots a_{n-1}, & \end{pmatrix}$$

of the matrix $\mathrm{M}_{n-k,k-1}$.

The matrices m_{11} and $M^p_{n-k,k-1}$ are defined by the same tableaux so far as the first $n-p-1$ letters are concerned. Now the matrix

$$M = [\lambda_{sr}/\lambda_{ss}],$$

where the lower rows of the tableaux F_s, F_r are

$$\lambda_1, \lambda_2, \dots, \lambda_k;\ \mu_1, \mu_2, \dots, \mu_k,$$

respectively, the former corresponding to the column tableaux of M. Then in these two matrices $\lambda_1, \lambda_2, \dots, \lambda_{k-1};\ \mu_1, \mu_2, \dots, \mu_{k-p-1}$ are the same.

Let $$\lambda_{ss} = [F_s] = c_1 c_2 \ldots c_k,$$

and $$\lambda_{sr} = b_1 b_2 \ldots b_k,$$

the values of the factors b, c being given in the enunciation for values of n lower than the given one. The value of λ_{ss} is, in fact, that given for all values of n by Theorem VI. Then the values of $c_1, c_2, \ldots, c_{k-1}$, and $b_1, b_2, \ldots, b_{k-p-1}$, are the same according to the enunciation in corresponding elements of m_{11} and $M_{n-k,k-1}$. In the former case we should have

$$c_k = n-2k+2-p,$$

$$b_{k-p} = n-2k+2, \quad b_{k-p+1} = p, \quad b_{k-p+2} = p-1, \quad \ldots, \quad b_{k-1} = 1;$$

and in the latter case we have

$$b_{k-p} = p, \quad b_{k-p-1} = p-1, \quad \ldots, \quad b_{k-1} = 1;$$

and hence, by definition, the ratio of the corresponding elements according to the enunciation is the same as that given by (VIII), viz.:

$$m_{11} = \frac{n-2k+2}{n-2k+2-p} M^p_{n-k,k-1}.$$

Thus the induction holds for this part of the matrix. The matrix $M^{p-1}_{n-k-1,k-1}$ concerned with $n-2$ letters is the element

$$\begin{pmatrix} & a_{n-p}\, a_{n-p+1} \ldots a_{n-2} \\ a_{n-p}\, a_{n-p+1} \ldots a_{n-2}, & \end{pmatrix}$$

of $M_{n-k-1,k-1}$; and m_{21} is the element

$$\begin{pmatrix} a_{n-p}, & \\ & a_{n-p} \end{pmatrix}$$

of $M^p_{n-k,k}$. The two matrices are defined by the same tableaux so far as the first $n-p-1$ letters are concerned.

Then the values of c and b given in the enunciation which differ are, for m_{21}:

$$c_k = n-2k+2-p; \quad b_{k-p+1} = p, \quad b_{k-p+2} = p-1, \quad \ldots, \quad b_k = 1;$$

and, for $M^{p-1}_{n-k-1,k-1}$:

$$b_{k-p+1} = p-1, \quad b_{k-p+2} = p-2, \quad \ldots, \quad b_{k-1} = 1.$$

Thus the ratio of the coefficients should give

$$m_{21} = \frac{p}{n-2k+2-p} M^{p-1}_{n-k-1,k-1},$$

which we have found to be the case.

The matrix $M^{p}_{n-k-1,k}$ concerned with $n-1$ letters is the element

$$\begin{pmatrix} & a_{n-p}\,a_{n-p+1}\cdots a_{n-1} \\ a_{n-p}\,a_{n-p+1}\cdots a_{n-1}, & \end{pmatrix}$$

of $M_{n-k-1,k}$; and m_{22} is the element

$$\begin{pmatrix} a_{n-p}, & a_{n-p} \end{pmatrix}$$

of $M^{p}_{n-k,k}$. It is easy to see that the coefficients are the same according to the enunciation as in formula VIII. Thus we have proved the induction except for the element

$$m_{12} = M^{p+1}_{n-k,k}.$$

This matrix may be treated in the same way. When $p = n-2k$, $M_{n-k,k}$ is really defined by the tableaux of $T_{k,k}$ of the first $2k$ letters. But here λ_k, μ_k are both fixed, for the last letter of the lower row in a standard tableau (the two rows being equal) must be a_{2k}; and hence

$$M^{n-2k}_{n-k,k} = M^{n-2k}_{n-k,k-1}.$$

Thus the theorem is true for any value of n, provided that it is true for all lower values. It is easy to establish its truth directly for the first few values of n, and hence it is always true.

18. For some purposes it is more convenient to use the matrices appropriate to the variables X_r, instead of the seminormal matrices which have been used, namely the matrices corresponding to the variable ξ, where

$$\xi_r[F_r] = X_r.$$

We shall call these new matrices the exact seminormal matrices; they are obtained by multiplying each column F_r by $[F_r]$ and dividing the F_r row by $[F_r]$. Thus the elements on the leading diagonal are unaltered, but the seminormal element a_{rs} becomes $a_{rs}[F_s]/[F_r]$.

Consider now the quadratic matrices of Theorem IV, corresponding to the transposition $(a_r a_{r+1})$. We have seen that such a quadratic matrix occupies those elements which lie at the intersection of the s-th and t-th rows and columns, where the tableaux F_s and F_t merely differ by the interchange of a_r and a_{r+1}.

Then in § 15 it was shown that

$$[F_s] = (1+\rho)\,[F_t],$$

where ρ is the reciprocal of the axial projection of $a_r a_{r+1}$ in either tableau. Then the quadratic matrices of the Theorem IV become

$$\begin{bmatrix} -\rho & 1-\rho \\ 1+\rho & \rho \end{bmatrix}$$

in the exact seminormal matrix.

Theorem II holds good as it stands for this matrix.

BINARY FORMS WITH A VANISHING COVARIANT OF WEIGHT FOUR OR FIVE

ALFRED YOUNG*

1. This note contains merely a slight extension of what is already well known. The only covariant of the binary form f of order n, of weight 2, is the Hessian H. When this vanishes f is the n-th power of a linear form. There is only one covariant of weight 3, *i.e.* (f, H); when this vanishes, f is a quadratic raised to the power $\frac{1}{2}n$ when n is even, or the n-th power of a linear form when n is odd.

The covariants of weight 4 are

$$Af^2(f, f)^4 + BH^2,$$

where A and B are arbitrary constants.

The case when $B = 0$ and

$$(f, f)^4 = i \equiv 0$$

was shown by Brioschi† and Wedekind‡ to give three solutions (apart

* Received 27 January, 1933; read 16 February, 1933.

† *Annali di Mat.*, 8 (1876), 24.

‡ *Habil. Schrift.* (Carlsruhe), 1876. See also Gordan, *Invariantentheorie*, 204.

from the general one when f has a linear factor repeated $n-1$ times); these solutions are for $n=4$ associated with the tetrahedron, for $n=6$ with the octahedron, for $n=8$ with the icosahedron. Hilbert* obtained the condition that a form f of order mn may be the n-th power of a quantic of order m, from the equation

$$\left(y\frac{d}{dx}\right)^{m+1} f^{1/n}=0.$$

This yields the equation

$$3(2n-3)H^2-2(n-2)f^2 i\equiv 0$$

as the condition that the form f of order n may be the $\frac{1}{3}n$-th power of a cubic when n is a multiple of 3, and

$$2(3n-4)H(f, H)-(n-3)f^2(f, i)=0$$

as the condition that f may be the $\frac{1}{4}n$-th power of a quartic when n is a multiple of 4. The vanishing of the covariant $(f, f)^6$ has been discussed by Rocco†.

2. Consider the equation

$$\Phi=Af^2 i+BH^2\equiv 0.$$

Let
$$f=(a_0, a_1, \ldots, a_n \between x_1, x_2)^n.$$

In H the product $a_\lambda a_\mu$ occurs only in the coefficient of $x_1^{2n-\lambda-\mu-2}x_2^{\lambda+\mu-2}$; the numerical factor with which it is associated will be called $\binom{n}{\lambda}\binom{n}{\mu}H_{\lambda,\mu}$. Similarly we write

$$i=\Sigma\binom{n}{\lambda}\binom{n}{\mu}i_{\lambda\mu}a_\lambda a_\mu x_1^{2n-\lambda-\mu-4}x_2^{\lambda+\mu-4}.$$

The variables will be chosen with x_2 as a factor of f, so that $a_0=0$. The first coefficient of Φ, that of x_1^{4n-8}, is

$$2Aa_0^2(a_0a_4-4a_1a_3+3a_2^2)+4B(a_0a_2-a_1^2)^2;$$

when this is zero and $a_0=0$, then $a_1=0$ or $B=0$. Let us suppose that f has a factor x_2^r so that

$$a_0=0=a_1=\ldots=a_{r-1}.$$

* *Math. Annalen*, 27 (1886), 158.

† *Giornale di Mat.*, 61 (1923), 129.

Then every coefficient of Φ before that of

$$x_0^{4n-4r-4}x_2^{4r-4},$$

is zero and this coefficient is

$$\binom{n}{r}^4 (Ai_{rr}+BH_{rr}^2)\,a_r^4.$$

Hence, unless

(I) $$Ai_{rr}+BH_{rr}^2=0,$$

a_r must be zero and f has the factor x_r repeated $r+1$ times. It thus appears that, unless the ratio $A:B$ has one of a certain number of definite values, any root of f is repeated n times, and f is the n-th power of a linear form.

3. To find the values of $H_{\lambda,\mu}$ and $i_{\lambda\mu}$, we write

$$f=a_x^n=\beta_x^n.$$

Then

$$H=(a\beta)^2a_x^{n-2}\beta_x^{n-2}=2(a_1^2\beta_2^2-a_1a_2\beta_1\beta_2)\,a_x^{n-2}\beta_x^{n-2}$$

$$=2\Sigma\binom{n-2}{\lambda}\binom{n-2}{\mu}(a_\lambda a_{\mu+2}-a_{\lambda+1}a_{\mu+1})\,x_1^{2n-\lambda-\mu-4}x_2^{\lambda+\mu}$$

$$=2\Sigma\left\{\binom{n-2}{\lambda-2}\binom{n-2}{\mu}-2\binom{n-2}{\lambda-1}\binom{n-2}{\mu-1}+\binom{n-2}{\lambda}\binom{n-2}{\mu-2}\right\}$$

$$\times a_\lambda a_\mu x_1^{2n-\lambda-\mu-2}x_2^{\lambda+\mu-2}.$$

Thus

$$\tfrac{1}{2}n_2^2H_{\lambda,\mu}=\lambda_2(n-\mu)_2-2\lambda\mu(n-\lambda)(n-\mu)+\mu_2(n-\lambda)_2\quad(\lambda\neq\mu),$$

and $$\tfrac{1}{2}nn_2H_{\lambda\lambda}=-\lambda(n-\lambda).$$

Similarly,

$$i=2(a_1^4\beta_2^4-4a_1^3a_2\beta_1\beta_2^3+3a_1^2a_2^2\beta_1^2\beta_2^2)\,a_x^{n-4}\beta_x^{n-4}$$

$$=2\Sigma\binom{n-4}{\lambda}\binom{n-4}{\mu}(a_\lambda a_{\mu+4}-4a_{\lambda+1}a_{\mu+3}+3a_{\lambda+2}a_{\mu+2})\,x_1^{2n-\lambda-\mu-8}x_2^{\lambda+\mu}$$

$$=2\Sigma\left\{\binom{n-4}{\lambda}\binom{n-4}{\mu-4}-4\binom{n-4}{\lambda-1}\binom{n-4}{\mu-3}+6\binom{n-4}{\lambda-2}\binom{n-4}{\mu-2}\right.$$

$$\left.-4\binom{n-4}{\lambda-3}\binom{n-4}{\mu-1}+\binom{n-4}{\lambda-4}\binom{n-4}{\mu}\right\}a_\lambda a_\mu x_1^{2n-\lambda-\mu-4}x_2^{\lambda+\mu-4}.$$

Then

$$\tfrac{1}{2}n_4{}^2\, i_{\lambda\mu} = \lambda_4(n-\mu)_4 - 4\lambda_3\,\mu_1(n-\lambda)_1(n-\mu)_3 + 6\lambda_2\,\mu_2(n-\lambda)_2(n-\mu)_2$$
$$-4\lambda_1\,\mu_3(n-\lambda)_3(n-\mu)_1 + \mu_4(n-\lambda)_4 \quad (\lambda \neq \mu),$$

and
$$n_2\, n_4\, i_{\lambda\lambda} = 12\lambda_2(n-\lambda)_2.$$

4. The equation (I) becomes, on putting in the values found,

$$3An^2(r-1)(n-r-1) + B(n-2)(n-3)\,r(n-r) = 0.$$

We regard this as an equation for r. Unless it has a positive integral root less than n, f must be the n-th power of a linear form. When r is a root the other root is $n-r$; let $r < n-r$. Then, in any case,

$$a_0 = a_1 = \ldots = a_{r-1} = 0,$$

whatever factor of f is chosen for x_2. It may happen also that $a_r = 0$, then $a_{r+1} = a_{r+2} = \ldots = a_{n-r-1} = 0$, in which case f may be transformed into the form

$$x_1^{n-r}\, x_2{}^r.$$

Otherwise every root of f is repeated exactly r times and

$$f = \psi^r.$$

Now, in this case, $Af^2 i + BH^2$ is a covariant of ψ of weight 4; it can therefore be written

$$\psi^{4r-4}\,[A'\psi^2(\psi, \psi)^4 + B'\{(\psi, \psi)^2\}^2]\phi.$$

Since this is zero, and ψ has no repeated root, we must have $B' = 0$, and therefore $(\psi, \psi)^4 = 0$. Thus the solutions are (apart from a mere n-th power)

$$f = x_1{}^r x_2^{n-r};\; f = \psi^r;\; (\psi, \psi)^4 = 0.$$

In particular, ψ may be any cubic.

5. Covariants of weight 5 are included in the form

$$(Af^2 i + BH^2, f) = (\phi, f).$$

When x_2 is a factor of f, the coefficient of x_1^{5n-10} in (ϕ, f) is $4Ba_1{}^5$, hence every root of f is repeated unless $B = 0$.

Let
$$\phi = \phi_0 x_1^{4n-8} + \phi_1 x_1^{4n-9} x_2 + \ldots.$$

Then

$$(\phi, f) = \left\{\Sigma \binom{n}{r} r a_r x_1^{n-r} x_2^{r-1}\right\} \{\Sigma (4n-8-s)\, \phi_s x_1^{4n-9-s} x_2^{s}\}$$

$$- \left\{\Sigma \binom{n}{r} (n-r)\, a_r x_1^{n-r-1} x_2^{r}\right\} \{\Sigma s\phi_s x_1^{4n-8-s} x_2^{s-1}\}$$

$$= \Sigma \binom{n}{r} \{r(4n-8)-sn\} a_r \phi_s x_1^{5n-r-s-9} x_2^{r+s-1}.$$

Let us suppose that $a_0 = a_1 = \ldots = a_{\rho-1} = 0$, and consider the coefficient of $x_1^{5n-5\rho-5} x_2^{5\rho-5}$ in (ϕ, f). It contains products $a_r \phi_s$, where $r+s = 5\rho-4$; when $r > \rho$, $s < 4\rho-4$, and ϕ_s, which is the sum of products of the a's of degree 4 and total weight $s+4$, must be zero. Thus the only non-zero product here is $a_\rho \phi_{4\rho-4}$. Hence

$$\binom{n}{\rho}^5 (A_0 i_{\rho\rho} + BH^2_{\rho\rho}) \{\rho(4n-8)-(4\rho-4)n\} a_\rho^5 = 0.$$

There are three possibilities:

$$a_\rho = 0; \quad Ai_{\rho\rho} + BH^2_{\rho\rho} = 0; \quad \rho = \tfrac{1}{2}n.$$

The last is a new case, and the conditions are satisfied for all values of A and B when every root of f is repeated $\frac{1}{2}n$ times. It is obvious that this should be so, since f is the power of a quadratic and a quadratic has no covariant of odd weight. Let $\rho < \frac{1}{2}n$ be a solution of (I). Then every root of f is repeated at least ρ times. If a root is repeated more often, it is repeated $\frac{1}{2}n, n-\rho$, or n times. In the second and third cases, $Af^2 i + BH^2 = 0$. In the first we must have one root repeated $\frac{1}{2}n$ times, and the others ρ times; and ρ must be a factor of $\frac{1}{2}n$. Thus

$$f = \psi^m,$$

where ψ has a non-repeated root. Then

$$(\phi, f) = \psi^{5m-5}[A'\psi^2(\psi, \psi)^4 + B'\{(\psi, \psi)^2\}^2, \ \psi].$$

And, since ψ has a non-repeated root, the equation

$$A' i'_{rr} + B'(H'_{rr})^2 = 0$$

must have $r = 1$ for a solution, and hence $B' = 0$. Thus

$$\{(\psi, \psi)^4, \psi\} = 0.$$

This means that there is a functional relation between ψ of order ρ and $(\psi, \psi)^4$ of order $4\rho-8$. This requires that, unless $(\psi, \psi)^4$ is zero, $\rho = 8$ or 4. The only case needing examination is the first. Here both $(\psi, \psi)^4$ and ψ are octavics, there is a functional relation between them, and ψ has a non-repeated root. Then

$$\text{(II)} \qquad (\psi, \psi)^4 = \lambda\psi.$$

We write

$$\psi = (b_0, b_1, \ldots, b_8 \between x_1, x_2)^8$$

and choose the variables so that

$$b_0 = b_2 = 0, \quad b_1 = 1,$$

x_2 being an unrepeated factor of ψ. Then, on equating coefficients in (II), we have

$$b_3 = 0, \quad -24b_4 = 8\lambda, \quad b_5 = 0, \quad b_6 = 0,$$

$$24b_7 + 90\,b_4{}^2 = 70\lambda b_4, \quad b_8 = 0, \quad -24b_4 b_7 = 8\lambda b_7.$$

Whence

$$25b_4{}^2 = -2b_7;$$

and

$$\psi = (x_1{}^4 + 10b_4 x_1 x_2{}^3)(8x_1{}^3 x_2 - 10b_4 x_2{}^4)$$
$$= F(F, F)^2;$$

where F is a quartic for which the invariant $(F, F)^4$ is zero. The meaning of this last case was pointed out to me by Mr. Grace some years ago. Thus when a covariant of weight 5 is zero, and no covariant of lower weight vanishes, the quantic is a power of a quartic or else of an octavic of the kind just described.

NOTE ON TRANSVECTANTS

ALFRED YOUNG*

The calculation of transvectants of binary forms in terms of symbolical products frequently involves considerable arithmetical labour. The purpose of this note is to explain a method by which they can be written down at once. Consider the transvectant†

$$(a_x{}^m b_x{}^n, c_x{}^p)^\nu = \sum_{\lambda+\mu=\nu} \left\{\binom{m}{\lambda}\binom{n}{\mu}\Big/\binom{m+n}{\nu}\right\}(ac)^\lambda (bc)^\mu a_x^{m-\lambda} b_x^{n-\mu} c_x^{p-\nu}.$$

* Received 27 January, 1933; read 16 February, 1933.

† Grace and Young, *Algebra of invariants*, 48.

This may be written

$$\nu!\binom{m+n}{\nu}\binom{p}{\nu}\left(\frac{a_x{}^m}{m!}\,\frac{b_x{}^n}{n!},\ \frac{c_x{}^p}{p!}\right)^\nu=\Sigma\frac{(ac)^\lambda}{\lambda!}\,\frac{(bc)^\mu}{\mu!}\,\frac{a_x^{m-\lambda}}{(m-\lambda)!}\,\frac{b_x^{n-\mu}}{(n-\mu)!}\,\frac{c_x^{p-\nu}}{(p-\nu)!}.$$

This is a particular case of the perfectly general formula

$$\varpi!\binom{\Sigma m}{\varpi}\binom{\Sigma n}{\varpi}\left(\prod_{r=1}^{h}\frac{a_{r_x}^{m_r}}{m_r!},\ \prod_{s=1}^{k}\frac{b_{s_x}^{n_s}}{n_s!}\right)^{\varpi}=\Sigma\prod_{\substack{r=1\\ s=1}}^{\substack{s=k\\ r=h}}\frac{(a_rb_s)^{\lambda_{rs}}}{\lambda_{rs}!}\prod_{r=1}^{h}\frac{a_{r_x}^{\mu_r}}{\mu_r!}\prod_{s=1}^{k}\frac{b_{s_x}^{\nu_s}}{\nu_s!},$$

where the sum extends to all possible solutions in positive integers or zeros of the equations

$$\sum_{s=1}^{k}\lambda_{rs}+\mu_r=m_r,\quad \sum_{r=1}^{h}\lambda_{rs}+\nu_s=n_s,\quad \sum_r\sum_s\lambda_{rs}=\varpi.$$

The proof follows the ordinary process. The result is assumed true when there are h different quantics on one side of the transvectant and k on the other. We then suppose that $b_{1_x}^{n_1}$ is really the product of two quantics, and so

$$b_{1_x}^{n_1}=c_x{}^p\,d_x{}^q.$$

Hence we operate on both sides with

$$\left[c\frac{\partial}{\partial b_1}\right]^p=\left[c_1\frac{\partial}{\partial b_{11}}+c_2\frac{\partial}{\partial b_{12}}\right]^p,$$

and after operation replace the symbol b_1 by d. Then $b_{1_x}^{n_1}$ becomes, on the left,

$$(n_1!/q!)\,c_x{}^p\,d_x{}^q.$$

On the right,

$$\frac{b_{1_x}^{\nu_1}}{\nu_1!}\prod_{r=1}^{h}\frac{(a_rb_1)^{\lambda_{r1}}}{\lambda_{r1}!}$$

becomes, by using Leibniz's theorem,

$$\Sigma\binom{p}{\rho_1,\rho_{11},\rho_{21},\ldots,\rho_{h1}}c_x^{\rho_1}\frac{d_x^{\rho_2}}{\rho_2!}\prod_{r=1}^{h}(a_rc)^{\rho_{r1}}\frac{(a_rd)^{\rho_{r2}}}{\rho_{r2}!}$$

$$=p!\,\Sigma\frac{c_x^{\rho_1}}{\rho_1!}\,\frac{d_x^{\rho_2}}{\rho_2!}\prod\frac{(a_rc)^{\rho_{r1}}}{\rho_{r1}!}\,\frac{(a_rd)^{\rho_{r2}}}{\rho_{r2}!},$$

where $\rho_1+\rho_2=\nu_1$, $\rho_{r1}+\rho_{r2}=\lambda_{r1}$, $\rho_1+\Sigma\rho_{r1}=p$.

The result then follows for h and $k+1$ quantics, and so the induction is proved.

SOME GENERATING FUNCTIONS

By Alfred Young

[Received and read 10 December, 1931.]

The primary object of the following is to obtain the generating function in two variables whose coefficients are the number of concomitants of degree δ, of the weights indicated by the indices of the variables, of a single ternary n-ic. The method used is an application of quantitative substitutional analysis as developed in my third paper on that subject*.

It is proved that the generating function may be obtained from that for gradients by multiplication by factors of the form

$$(1-y)(1-z)\left(1-\frac{z}{y}\right).$$

The second part of the paper deals with generating functions of covariants of a system of binary forms of the same order; the generating function obtained is for covariants of the type

$$T_{a_1 a_2 \ldots a_h} C,$$

where the permutations interchange quantics as such. As an immediate application the generating function for combinants is given.

The third part of the paper extends the result for ternary forms to forms with any number of variables, and gives the generating function for seminvariants in terms of that for gradients.

I. *Ternary forms.*

1. Consider a gradient X of a ternary quantic

$$\Sigma \binom{n}{r,\, s} A_{r,s} x_1^{n-r-s} x_2^r x_3^s.$$

---

* *Proc. London Math. Soc.* (2), 28 (1928), 255–292.

We shall describe the coefficient $A_{r,s}$ as having three weights $n-r-s$, r, s, the first, second, and third. Thus the gradient has three weights, and the sum of the three weights is $n\delta$, where δ is the degree of X. Gradients in which the weights are in descending order of magnitude will be considered. In the symbolical notation the quantic is written $a_x{}^n$, where

$$a_x = ax_1 + a'x_2 + a''x_3,$$

using dashes for convenience as the suffixes are used to distinguish different symbolical letters. The coefficient $A_{r,s}$ is then $a^{n-r-s}a'^r a''^s$. The gradient

$$X = \Pi A_{r,s}^{\lambda_{r,s}} = \Pi a_t^{n-r_t-s_t} a_t'^{r_t} a_t''^{s_t}.$$

In order to apply the substitutional analysis we polarize this to make it linear in every symbolical letter, and so replace a_t by the n letters $a_{t1} a_{t2} \dots a_{tn}$, introducing a substitutional operator

$$\frac{1}{n!}\{a_{t1} a_{t2} \dots a_{tn}\},$$

because the letters are all equivalent.

Thus

$$X = \Gamma \Pi a_{t1} a_{t2} \dots a_{tn-r_t-s_t} a'_{tn-r_t-s_t+1} \dots a'_{tn-s_t} a''_{tn-s_t+1} \dots a''_{t,n};$$

where Γ is the product of the operators mentioned, it may be replaced by a Σ. The effect of a permutation $(\beta\gamma)$ on a product $\beta\gamma$, $\beta'\gamma'$, or $\beta''\gamma''$ is to make no change, but it interchanges the pairs of products $\beta\gamma'$, $\beta'\gamma$; $\beta\gamma''$, $\beta''\gamma$; $\beta'\gamma''$, $\beta''\gamma'$.

Thus the product above is unaltered by any permutation of the symmetric group of the w_1 undashed letters, or that of the w_2 letters with a single dash, or that of the w_3 letters with a double dash. That is, it is of the form

$$\frac{1}{w_1!\, w_2!\, w_3!} G_{w_1} G_{w_2} G_{w_3} Y = Y,$$

where G_r is a positive symmetric group of degree r. Then

$$T_{a_1 a_2 \dots a_h} Y = 0,$$

when $h > 3$, or when $a_1 < w_1$, or $a_1 + a_2 < w_1 + w_2$. Indeed, the latest TY which is not zero is $T_{w_1 w_2 w_3} Y$. The product $G_{w_1} G_{w_2} G_{w_3}$ is the product of the positive symmetric groups of the letters in each row of a tableau for this T. We may ensure that this tableau is standard by taking the letters in such a sequence that (i) no two letters of any set are separated in the sequence by a letter of another set, and (ii) those sets come first whose earliest weight is greatest. Let F_ρ be this tableau, so that

$$G_{w_1} G_{w_2} G_{w_3} = P_\rho.$$

Then $$T_{w_1 w_2 w_3} = \sum_\sigma P_\sigma N_\sigma M_\sigma$$

and* $$T_{w_1 w_2 w_3} Y = P_\rho N_\rho M_\rho P_\rho \frac{1}{w_1!\; w_2!\; w_3!} Y$$

$$= \frac{1}{f} \binom{n}{w_1 w_2 w_3} P_\rho N_\rho Y.$$

Then $N_\rho Y$ is the product of w_3 symbolical factors of the form (abc), of $w_2 - w_3$ factors of the form $(ab' - a'b) = (ab)_3$, and of $w_1 - w_2$ undashed letters, *i.e.* it is the coefficient of $u_3^{w_2 - w_3} x_1^{w_1 - w_2}$ in a concomitant

$$\Pi (abc)\, \Pi (deu)\, \Pi f_x.$$

This will be called the first term or seminvariant†.

The concomitant is completely defined by F_ρ, which in its turn is defined by the gradient X, and may be written

$$(X).$$

The argument applies equally to forms with any number of variables.

THEOREM I. *A gradient X, whose weights w_1, w_2, w_3 are in descending order of magnitude, when expressed in the form ΣTX, has for its term corresponding to $T_{w_1 w_2 w_3}$ the seminvariant*

$$(X)$$

and has no term corresponding to later T's.

2. When $T_{a_1 a_2 a_3}$ precedes $T_{w_1 w_2 w_3}$,

$$T_{a_1 a_2 a_3} X$$

is not zero, but it is a sum of terms which are the coefficients of concomitants of weights a_1, a_2, a_3. It will be more illuminating to replace

* Q.S.A., III, 264.

† It is important for our purpose to define this term thus, and not as the coefficient of $u_1^{w_2 - w_3} x_1^{w_1 - w_2}$.

the variable x by the matrix form (vu), where v and u are variables contragredient to x and cogredient with each other. Then the concomitants belonging to $T_{\alpha\beta\gamma}$ are of order $\alpha-\beta$ in (vu) and of order $\beta-\gamma$ in u. The seminvariants are the coefficients of

$$v_2^{\alpha-\beta}u_3^{\alpha-\gamma}.$$

In effect the second rows of the tableaux are filled up with $\alpha-\beta$ letters v, and the third rows with $\alpha-\gamma$ letters u, making them all of equal length.

We need the *independent* terms of a concomitant. In binary forms of order ρ with a seminvariant of weight w there is just one term of each weight from w to $w+\rho$.

In ternary forms, however, the coefficients of all the variable products

$$x_1^{\lambda_1}x_2^{\lambda_2}x_3^{\lambda_3}u_1^{\mu_1}u_2^{\mu_2}u_3^{\mu_3},$$

$$\lambda_1+\lambda_2+\lambda_3=\alpha-\beta, \quad \mu_1+\mu_2+\mu_3=\beta-\gamma,$$

have to be considered. The weights of the coefficient of this term are

$$\text{(I)} \qquad \begin{cases} w_1=\alpha-\lambda_2-\lambda_3-\mu_1=\beta+\lambda_1-\mu_1, \\ w_2=\alpha-\lambda_3-\lambda_1-\mu_2=\beta+\lambda_2-\mu_2, \\ w_3=\alpha-\lambda_1-\lambda_2-\mu_3=\beta+\lambda_3-\mu_3, \end{cases}$$

and these are not all different for the different coefficients.

But, in virtue of the equation

$$u_1x_1+u_2x_2+u_3x_3=0,$$

there is a relation between the variable products; and there is an exactly correlative relation between the coefficients of these forms. We therefore consider no terms in which the product u_2x_2 occurs; the rest will be independent with their coefficients. This is equivalent, but differently expressed, to the statement of the fact made by Grace*.

Thus to find the number of linearly independent coefficients of weights w_1, w_2, w_3 of a concomitant of type $T_{\alpha\beta\gamma}$ we consider the coefficients of those variable products for which either λ_2 or μ_2 is zero; *i.e.* either

$$\lambda_2=0, \quad \mu_2=\beta-w_2,$$

or

$$\mu_2=0, \quad \lambda_2=w_2-\beta.$$

* *Journal London Math. Soc.*, 5 (1930), 65.

3. Consider the number of gradients X of degree δ and weights w_1, w_2, w_3 which are in descending order of magnitude. Each one is expressible as a linear function of the corresponding seminvariant (X) of its own proper weights, and of terms of these weights of concomitants represented by earlier T's. Let n_{w_1, w_2, w_3} be the number of gradients X, $N_{\alpha, \beta, \gamma}$ the number of linearly independent seminvariants of weights α, β, γ; and $\begin{pmatrix} w_1 & w_2 & w_3 \\ \alpha & \beta & \gamma \end{pmatrix}$ the number of independent coefficients of weights w_1, w_2, w_3 of a concomitant of the $T_{\alpha, \beta, \gamma}$ set. Then

$$n_{w_1, w_2, w_3} = \Sigma \begin{pmatrix} w_1 & w_2 & w_3 \\ \alpha & \beta & \gamma \end{pmatrix} N_{\alpha, \beta, \gamma}.$$

And thus we proceed to obtain a relation expressing N_{w_1, w_2, w_3} in terms of $n_{\alpha, \beta, \gamma}$. When $w_3 = 0$, γ must also be zero, and we have merely the binary case

$$n_{w_1, w_2, 0} = \sum_{\rho=0}^{w_2} N_{w_1+\rho, w_2-\rho, 0}.$$

When $w_3 = 1$, consider the number of terms supplied by a concomitant

$$T_{w_1+\rho, w_2-\rho+1, 0}.$$

Unless $\rho = 0$, $w_2 - \beta$ is positive or zero and thus

$$\mu_2 = 0, \quad \lambda_2 = \rho - 1.$$

The last equation of (I) gives

$$1 = \lambda_3 + \mu_1 + \mu_2,$$

whence $\qquad \lambda_3 = 0, \ \mu_1 = 1 \quad$ or $\quad \lambda_3 = 1, \ \mu_1 = 0.$

These give two independent terms in all cases $0 < \rho < w_2 + 1$. For $\rho = 0$, we have $\lambda_2 = 0$, $\mu_2 = 1$, $\lambda_3 = 0$, $\mu_1 = 0$, and hence only one term; in the same way there is one term only for

$$T_{w_1+w_2+1, 0, 0}.$$

It is necessary to consider also forms

$$T_{w_1+\rho, w_2-\rho, 1}.$$

This may be most easily considered by noting that there is a symbolical factor (abc) of weights $(1, 1, 1)$ multiplying all the terms; the removal of this (that is the first column from the tableau) leaves us to reckon the number of independent terms of weights $(w_1-1, w_2-1, 0)$ of a form $T_{w_1-1+\rho, w_2-1-\rho, 0}$, which is of course one. Thus, collecting results and inserting the coefficients 1 or 2 according to the above several cases, we have

$$n_{w_1, w_2, 1} = N_{w_1, w_2, 1} + N_{w_1+1, w_2-1, 1} + \dots + N_{w_1+w_2-1, 1, 1}$$
$$+ N_{w_1, w_2+1, 0} + 2N_{w_1+1, w_2, 0} + \dots + 2N_{w_1+w_2, 1, 0} + N_{w_1+w_2+1, 0, 0}.$$

Hence

$$n_{w_1, w_2, 1} - n_{w_1+1, w_2-1, 1} = N_{w_1, w_2, 1} + N_{w_1, w_2+1, 0} + N_{w_1+1, w_2, 0}.$$

In the same way we find

$$n_{\alpha, \beta, 2} = N_{\alpha, \beta, 2} + N_{\alpha+1, \beta-1, 2} + \dots + N_{\alpha+\beta-2, 2, 2} + N_{\alpha, \beta+1, 1}$$
$$+ 2N_{\alpha+1, \beta, 1} + 2N_{\alpha+2, \beta-1, 1} + \dots + 2N_{\alpha+\beta-1, 2, 1} + N_{\alpha+\beta, 1, 1}$$
$$+ N_{\alpha, \beta+2, 0} + 2N_{\alpha+1, \beta+1, 0} + 3N_{\alpha+2, \beta, 0} + \dots + 3N_{\alpha+\beta, 2, 0}$$
$$+ 2N_{\alpha+\beta+1, 1, 0} + N_{\alpha+\beta+2, 0, 0}.$$

Also

$$n_{\alpha, \beta, \gamma} = N_{\alpha, \beta, \gamma} + N_{\alpha+1, \beta-1, \gamma} + \dots + N_{\alpha+\beta-\gamma, \gamma, \gamma}$$
$$+ N_{\alpha, \beta+1, \gamma-1} + 2N_{\alpha+1, \beta, \gamma-1} + \dots + \dots$$
$$+ N_{\alpha, \beta+r, \gamma-r} + 2N_{\alpha+1, \beta+r-1, \gamma-r} + \dots + (r+1) N_{\alpha+r, \beta, \gamma-r} + \dots$$
$$+ (r+1) N_{\alpha+\beta-\gamma+r, \gamma, \gamma-r} + rN_{\alpha+\beta-\gamma+r+1, \gamma-1, \gamma-r} + \dots$$
$$+ N_{\alpha+\beta-\gamma+2r, \gamma-r, \gamma-r} + \dots + \dots .$$

Then

$$n_{\alpha, \beta, \gamma} - n_{\alpha+1, \beta-1, \gamma} = N_{\alpha, \beta, \gamma} + N_{\alpha, \beta+1, \gamma-1} + N_{\alpha+1, \beta, \gamma-1}$$
$$+ N_{\alpha, \beta+2, \gamma-2} + N_{\alpha+1, \beta+1, \gamma-2} + N_{\alpha+2, \beta, \gamma-2} + \dots$$
$$+ N_{\alpha, \beta+r, \gamma-r} + N_{\alpha+1, \beta+r-1, \gamma-r} + \dots + N_{\alpha+r, \beta, \gamma-r} + \dots .$$

Hence

$$(n_{\alpha, \beta, \gamma} - n_{\alpha+1, \beta-1, \gamma}) - (n_{\alpha, \beta+1, \gamma-1} - n_{\alpha+1, \beta, \gamma-1})$$
$$- (n_{\alpha+1, \beta, \gamma-1} - n_{\alpha+2, \beta-1, \gamma-1}) + (n_{\alpha+1, \beta+1, \gamma-2} - n_{\alpha+2, \beta, \gamma-2}) = N_{\alpha, \beta, \gamma}.$$

Now let
$$\Gamma_\delta = \Sigma\, n_{\alpha, \beta, \gamma}\, y^\beta z^\gamma,$$

where δ is the degree of the seminvariants of weights α, β, γ; that is Γ_δ is the generating function for gradients of degree δ.

Then $N_{\alpha,\beta,\gamma}$ is the coefficient of $y^\beta z^\gamma$ in

$$(1-y)(1-z)\left(1-\frac{z}{y}\right)\Gamma_\delta.$$

THEOREM II. *The generating function for seminvariants of a ternary quantic is obtained from that for gradients of the same degree by multiplication by*

$$(1-y)(1-z)\left(1-\frac{z}{y}\right),$$

where the indices of y and z indicate the second and third weights.

4. To find the generating function for gradients of degree δ, we use the fact that the sum of all homogeneous powers and products of powers of $z_1, z_2, \ldots, z_m$, the roots of

$$f(z) = z^m + p_1 z^{m-1} + \ldots = 0$$

of degree δ, is*

$$\sum_{r=1}^{m} \frac{z_r^{\delta+m-1}}{f'(z_r)} = \frac{1}{\Delta}\begin{vmatrix} z_1^{\delta+m-1} & z_2^{\delta+m-1} & \ldots & z_m^{\delta+m-1} \\ z_1^{m-2} & z_2^{m-2} & \ldots & z_m^{m-2} \\ \ldots & \ldots & \ldots & \ldots \\ 1 & 1 & \ldots & 1 \end{vmatrix},$$

where

$$\Delta = \prod_{r<s} (z_r - z_s).$$

To obtain the generating function, instead of replacing the roots of $f(z)$ by the coefficients $A_{r,s}$ we replace them by the variable products $y^r z^s$. Then $z^{\delta+m-1}/f'(z)$ becomes

$$\frac{y^{r\delta} z^{s\delta}}{\prod'^{n}_{\rho+\sigma=0} (1-x^{\rho-r} y^{\sigma-s})},$$

where the product symbol has a dash affixed to indicate that one factor is omitted, viz. that for which $\rho = r$, $\sigma = s$.

THEOREM III. *The generating function for seminvariants of degree δ of the ternary n-ic is*

$$(1-y)(1-z)\left(1-\frac{z}{y}\right)\sum_{r+s=0}^{n} \frac{y^{r\delta} z^{s\delta}}{\prod'^{n}_{\rho+\sigma=0} (1-y^{\rho-r} z^{\sigma-s})}.$$

---

* A Referee has pointed out that this result is due to Jacobi; see his *Gesammelte Werke*, 3 (1884), 7.

It is evident that this form of generating function for gradients is perfectly general. Thus for quaternary forms we have

$$\sum_{r+s+t=0}^{n} \frac{y^{r\delta} z^{s\delta} u^{t\delta}}{\prod'^{\,n}_{\rho+\sigma+\tau=0} \left(1-y^{\rho-r} z^{\sigma-s} u^{\tau-t}\right)}.$$

The result may be stated in the form

$$N_{w_1, w_2, w_3} = n_{w_1, w_2, w_3} + n_{w_1+2, w_2-1, w_3-1} + n_{w_1+1, w_2+1, w_3-2}$$
$$- n_{w_1+1, w_2-1, w_3} - n_{w_1, w_2+1, w_3-1} - n_{w_1+2, w_2, w_3-2}.$$

The interchange of a pair of suffixes in the coefficients of a gradient merely interchanges the corresponding weights of the gradient, and hence

$$n_{w_1, w_2, w_3} = n_{w_1, w_3, w_2} = n_{w_2, w_1, w_3} = \ldots.$$

Thus, when the generating function G_δ for seminvariants is fully expanded, it will include terms for which the weights are not in descending order. But for this part of the expansion there is a relation to the first part of the same nature as in the case of generating functions for binary forms; but of an appropriately extended kind. In fact, from the above relations we obtain at once

$$N_{w_1, w_2, w_3} = -N_{w_2-1, w_1+1, w_3} = -N_{w_1, w_3-1, w_2+1}$$
$$= -N_{w_3-2, w_2, w_1+2} = N_{w_2-1, w_3-1, w_1+2} = N_{w_3-2, w_1+1, w_2+1}.$$

And thus

$$G_\delta = \Sigma N_{w_1, w_2, w_3} \left[x_1^{w_1} x_2^{w_2} x_3^{w_3} + x_1^{w_2-1} x_2^{w_3-1} x_3^{w_1+2} + x_1^{w_3-2} x_2^{w_1+1} x_3^{w_2+1}\right.$$
$$\left. - x_1^{w_1} x_2^{w_3-1} x_3^{w_2+1} - x_1^{w_3-2} x_2^{w_2} x_3^{w_1+2} - x_1^{w_2-1} x_2^{w_1+1} x_3^{w_3}\right],$$

where now the expansion is restricted to

$$w_1 \geqslant w_2 \geqslant w_3.$$

II. *Systems of binary forms.*

5. Consider a binary gradient type X of degree δ; that is one that includes just one coefficient from each of δ forms of order n. Then X may be obtained by polarization from a gradient of a single binary form

$$A_{r_1}^{\lambda_1} A_{r_2}^{\lambda_2} \ldots A_{r_k}^{\lambda_k} = X_1;$$

where X includes λ_1 coefficients with suffix r_1.

Permutations will be considered of the quantics themselves; then X will be unchanged by any permutation of the λ_1 quantics whose coefficient has a suffix r_1. Thus X may be regarded substitutionally as a function affected by an operator Γ which is the product of positive symmetric groups of degrees $\lambda_1, \lambda_2, \ldots, \lambda_k$, which are supposed to be in descending order of magnitude.

Operate with the series

$$1 = \Sigma T$$

on X. Then $T_{a_1 a_2 \ldots a_h} X$ is zero, unless $a_1 \geqslant \lambda_1$, $a_1 + a_2 \geqslant \lambda_1 + \lambda_2$, and so on. Consider

$$\Gamma T_{a_1 a_2 \ldots a_h} = \Gamma \sum_{r=1}^{f} P_r N_r M_r,$$

where F_r is a standard tableau. The definition of "standard" presupposes a fixed sequence of the δ letters subject to permutation. We choose this so that the λ_1 letters of the first group of Γ come first, the λ_2 letters of the second group next, and so on. Then, when a pair of letters of the same group of Γ appears in the same column of F_r, we can prove that

$$\Gamma P_r N_r M_r = \Sigma \mu_s \Gamma_s P_s \sigma N_r M_r,$$

where F_s precedes F_r, σ is a permutation which transforms F_r into F_s, and μ_s is numerical. We may suppose that $b_1 b_2 \ldots b_\lambda$ are the letters of the group in question and that part of the F_r tableau is

$$\begin{array}{l} \ldots \quad \ldots \quad \ldots \quad \ldots b_1 b_2 \ldots b_p d_1 d_2 \ldots d_h \\ c_1 c_2 \ldots c_k b_{p+1} \ldots b_q \ldots, \end{array}$$

where the letters c precede the b's, which in turn precede the d's. Then operation with the positive symmetric group

$$\{c_1 c_2 \ldots c_k b_{p+1} \ldots b_q b_1 b_2 \ldots b_p d_1 d_2 \ldots d_h\}$$

produces zero, for it is of greater degree than the length of the upper line in the tableau. On expansion we first have terms in which the letters b only are permuted; these, owing to the presence of Γ to the left of the sum, are equal to a multiple of $\Gamma P_r N_r M_r$, and all the other terms are of the form $\Gamma P_s \sigma N_r M_r$. Thus when the letters of any one group of Γ *overlap* in the rows of F_r, we can express $\Gamma P_r N_r M_r X$ in terms of earlier forms. The function X is only affected by permutations operating on the left, the permutations of PNX are all effected by a multiplication by permutations σ,

$$\sigma PNX.$$

Thus the possible forms are completely defined by the tableau of P. There are thus just the same number of linearly independent forms TX as there are standard tableaux in which the letters of the various groups of Γ do not overlap.

6. It is proposed to find how many gradients of degree δ and weight w there are which are linear in the coefficients of each of δ binary quantics of order r, and are of the substitutional type $T_{a_1 a_2 \ldots a_h} X$. The permutations considered are of the quantics themselves.

The gradient X will be said to be of the type $(\lambda_0, \lambda_1, \ldots, \lambda_r)$ when it contains λ_r coefficients with the suffix r for $r = 0, 1, \ldots, n$. Then it is to be noticed that the fact that there must be no overlapping in the tableaux of the independent forms does not depend in any way on the sequence of magnitudes of the λ groups. Thus we shall take the letters of the group λ_0 first in the sequence of letters irrespective of the magnitude of λ_0. Moreover, the respective values of $\lambda_0, \lambda_1, \ldots, \lambda_n$ are not fixed.

Let $\{a_1, a_2, \ldots, a_h;\ n\}$ be the number of independent gradients $T_{a_1 a_2 \ldots a_h} X$, that is the number of standard tableaux F_r where the letters are divided into $n+1$ groups, of any respective magnitudes $\lambda_0, \lambda_1, \ldots, \lambda_n$; the letters λ_0 come first and so on, and the letters of no group overlap. Then consider the effect of removing the letters of the last group from F_r. Since these are the last λ_n letters the result is a standard tableau F_r' of the same character but of n groups. Hence

$$\{a_1, a_2, \ldots, a_h;\ n\} = \Sigma \{\beta_1, \beta_2, \ldots, \beta_h;\ n-1\},$$

where the sum on the right includes all values of $\beta_1, \beta_2, \ldots, \beta_h$ such that

$$\beta_r \leqslant a_r, \quad \beta_r \geqslant a_{r+1},$$

for, if $\beta_r < a_{r+1}$, it would mean that, on filling up the β tableau with letters of the last group, these letters would overlap.

In the same way,

$$\{a_1, a_2, \ldots, a_h - 1;\ n\} = \Sigma \{\beta_1', \beta_2', \ldots, \beta_h';\ n-1\}.$$

The two series are the same except that in the first there are terms with $\beta_h = a_h$ not included in the second, and in the second there are terms with $\beta_{h-1} = a_h - 1$ not included in the first.

Let δ_r be an operation which diminishes the value of a_r by unity. Then

$$\text{(II)} \qquad \left[\prod_{r=1}^{h} (1-\delta_r)\right] \{a_1, a_2, \ldots, a_h;\ n\} = \{a_1, a_2, \ldots, a_h;\ n-1\}.$$

For the term on the right appears in the sum obtained for the first term only on the left, and every other term in these sums occurs an even number of times half with a positive and half with a negative sign.

7. THEOREM IV. *The generating function for covariants of type*

$$T_{a_1 a_2 \dots a_h},$$

of degree $\delta = \Sigma a$, *which are linear in the coefficients of each of* δ *quantics of order* n, *is*

$$(1-z)\left|\phi\binom{a_r+n-r+s}{n}\right|,$$

where

$$\phi\binom{\xi}{\eta} = \frac{(1-z^{\xi-\eta+1})(1-z^{\xi-\eta+2})\dots(1-z^{\xi})}{(1-z)(1-z^2)\dots(1-z^{\eta})}.$$

As usual for binary forms the generating function for covariants is obtained from that for gradients by multiplication by $1-z$; it is then only necessary to consider gradients. The well known property

$$\text{(III)} \qquad \phi\binom{\xi+1}{\eta} - z^{\eta}\phi\binom{\xi}{\eta} = \phi\binom{\xi}{\eta-1}$$

is needed.

Let $\psi(a_1, a_2, \dots, a_h;\ n)$ be the generating function for gradients of type $T_{a_1, a_2, \dots, a_h}$. Then, as in the last paragraph,

$$\psi(a_1, a_2, \dots, a_h;\ n) = \Sigma z^{n\Sigma(a-\beta)}\psi(\beta_1, \beta_2, \dots, \beta_h;\ n-1).$$

Hence, as before in (II),

$$\psi(a_1, a_2, \dots, a_h;\ n-1) = \left[\prod_{r=1}^{h}(1-z^n\delta_r)\right]\psi(a_1, a_2, \dots, a_h;\ n).$$

In virtue of the property (III),

$$\Delta = \left|\phi\binom{a_r+n-r+s}{n}\right|$$

(a determinant whose rows are given by $r = 1, 2, \dots, h$ and columns by $s = 1, 2, \dots, h$ satisfies this difference equation, for the operator $1-z^n\delta_r$ operates on the elements of the r-th row only of the determinant above. Hence, if the theorem is true for all earlier values of the arguments $a_1, a_2, \dots, a_h, n$, it is true for these.

(i) When $a_r = a_{r+1}-1$ the determinant is zero since two rows become identical; this corresponds to the fact that such a T does not exist and therefore the generating function should be zero.

(ii) When $a_h = 0$ the last row of the determinant is 0, 0, ..., 0, 1, and we reduce it to the corresponding determinant of $h-1$ rows and columns.

(iii) When $a_1 = \delta$, $a_2 = 0$, ..., $a_h = 0$, we obtain the ordinary generating function for gradients of a single quantic; this is the correct result for T_δ.

Thus the truth of the theorem follows by induction.

8. The determinant Δ can be factorized thus. In the first place, the denominators of the elements give a factor D^{-h}, where

$$D = (1-z)(1-z^2)\ldots(1-z^n).$$

The common factors of the r-th row,

$$(1-z^{a_r-r+h+1})(1-z^{a_r-r+h+2})\ldots(1-z^{a_r-r+n+1}),$$

may then be removed. We are left with the determinant

$$\Delta' = |(1-z^{a_r-r+s+1})\ldots(1-z^{a_r-r+h})(1-z^{a_r-r+n+2})\ldots(1-z^{a_r+n-r+s})|.$$

On subtracting the s-th column from the $(s+1)$-th, the r-th element of the $(s+1)$-th column becomes

$$(1-z^{a_r-r+s+2})\ldots(1-z^{a_r-r+h})(1-z^{a_r-r+n+2})\ldots(1-z^{a_r+n-r+s})(1-z^n)z^{a_r-r+s+1}.$$

Then in Δ' we subtract the last column but one from the last, then the $(h-2)$-th from the $(h-1)$-th, and so on, finally the first from the second. The result is

$$\Delta' = (1-z^n)^{h-1}z^{\binom{h+2}{2}-1}\Delta_1',$$

where the first column of Δ_1' is the same as that of Δ', but the s-th column $s > 1$ has for its r-th element

$$z^{a_r-r}(1-z^{a_r-r+s+1})\ldots(1-z^{a_r-r+h})(1-z^{a_r-r+n+2})\ldots(1-z^{a_r-r+n+s-1}).$$

Next we subtract the $(h-1)$-th column of Δ_1' from the h-th, and so on, and finally the second from the third, obtaining

$$\Delta_1' = (1-z^{n-1})^{h-2}z^{\binom{h+2}{2}-3}\Delta_2'.$$

Proceeding thus, we finally obtain

$$\Delta' = (1-z^n)^{h-1}(1-z^{n-1})^{h-2}\ldots(1-z^{n-h+2})\Delta'',$$

where $$\Delta'' = |z^{(s-1)(a_r-r+s)}(1-z^{a_r-r+s+1})\ldots(1-z^{a_r-r+h})|.$$

Now multiply the $(s-1)$-th column of Δ'' by z^{s-2} and add to the s-th; the

r-th element of the s-th column becomes

$$z^{(s-2)(a_r-r+s)}(1-z^{a_r-r+s+1})\dots(1-z^{a_r-r+h}).$$

Next we make this transformation for the last column, then for the last but one, and so on. The result is a determinant Δ_1'', whose first column is that of Δ'', and for $s>1$ the element is that just given. Then

$$\Delta''=\Delta_1''.$$

Next, in Δ_1'' we multiply the $(s-1)$-th column by z^{s-3} and add to the s-th in regular sequence from the end, as far as $s=3$; and thus obtain

$$\Delta_1''=\Delta_2''.$$

Proceeding thus, we eventually obtain

$$\Delta''=\Delta'''=|(1-z^{a_r-r+s+1})\dots(1-z^{a_r-r+h})|,$$

the last column being made up of units. Now, by means of row subtractions, we find that

$$\Delta'''=z^n\left[\prod_{r<t}(1-z^{a_r-a_t+t-r})\right]\bar{\Delta}.$$

The first row of $\bar{\Delta}$ is

$$z^{2+3+\dots+h}\ \ 0\ \ 0\ \dots\ 0.$$

In the r-th row the last $h-r$ elements are zero and the r-th element is

$$z^{(r+1)+\dots+h},$$

the last element of the last row is unity.

Thus $$\bar{\Delta}=z^{\frac{1}{3}(h+1)h(h-1)}.$$

Hence, finally,

$$\text{(IV)}\quad \left|\phi\binom{a_r+n-r+s}{n}\right|=z^{\varpi}D^{-h}(1-z^n)^{h-1}(1-z^{n-1})^{h-2}\dots(1-z^{n-h+2})$$

$$\times\prod_{r=1}^{h}(1-z^{a_r-r+h+1})\dots(1-z^{a_r-r+n+1})\prod_{r<t}(1-z^{a_r-a_t+t-r}),$$

where $$\varpi=(h-1)\,a_h+(h-2)\,a_{h-1}+\dots+a_2.$$

By writing $a_1=a_2=\dots=a_h=a$, we obtain the case of combinants of what may be called the h-th class.

THEOREM V. *The generating function for combinants of the h-th class and degree ah of the binary n-ic is*

$$z^{a\binom{h}{2}}(1-z)\,D^{-h}(1-z^{n})^{h-1}(1-z^{n-1})^{h-2}\dots(1-z^{n-h+2})$$
$$\times(1-z)^{h-1}(1-z^{2})^{h-2}\dots(1-z^{h-1})$$
$$\times(1-z^{a+1})(1-z^{a+2})^{2}\dots(1-z^{a+h})^{h}(1-z^{a+h+1})^{h}$$
$$\dots(1-z^{a+n-h+1})^{h}(1-z^{a+n-h+2})^{h-1}\dots(1-z^{a+n}),$$

when $n \geqslant 2h-1$.

9. Let $z^{a\binom{h}{2}}\psi(a, h)$ be the generating function for the combinants of class h and degree ah. Then we shall prove that

$$\psi(a, h) = \psi(a, n+1-h).$$

Let
$$n+1-h > h.$$

There are three sets of factors in the numerator of ψ; two of these do not contain a in the index of z, and we consider them first. Let these factors in $\psi(a, h)$ be P_h and let the remaining factors which have a in the index be Q_h. Then

$$P_h = (1-z^{n})^{h-1}(1-z^{n-1})^{h-2}\dots(1-z^{n-h+2})\,(1-z)^{h-1}(1-z^{2})^{h-2}\dots(1-z^{h-1}),$$

$$P_{n+1-h} = D^{n+1-2h}\,P_h,$$

$$P_h D^{-h} = P_{n+1-h}\,D^{-(n+1-h)}.$$

The factors Q_h may be written in a rectangle, the first row being those for $r=1$, the next those for $r=2$, and so on; the rectangle for Q_{n+1-h} is obtained from that for Q_h by changing columns into rows; and hence

$$Q_{n+1-h} = Q_h.$$

THEOREM VI. *The number of combinants of class h, degree ah, and weight* $w+a\binom{h}{2}$ *is the same as the number of combinants of class* $n+1-h$, *degree* $a(n+1-h)$, *and weight*

$$w+a\binom{n+1-h}{2}.$$

10. The gradient generating function (IV) should be unchanged when h is changed to $h+1$, and $a_{h+1}=0$. This can easily be verified to be

the case. In fact, ϖ is unaltered, in the first row there is an additional factor in the denominator

$$(1-z)(1-z^2)\dots(1-z^{n-h}),$$

but this appears in the numerator of the second row for $r=h+1$. The second row loses the factors of the first Π

$$\prod_{r=1}^{h}(1-z^{a_r-r+h+1}),$$

and these appear in the last Π for $t=h+1$. Thus we may take in general $h=n+1$, where it is understood that some of the a's may be zero; all the second set of factors then disappear, and the generating function takes a simpler form

$$\text{(V)} \qquad (1-z)z^{\varpi}\prod_{r=1}^{n}(1-z^r)^{r-n-1}\prod_{r<t}(1-z^{a_r-a_t+t-r}).$$

Let us now consider the generating function for the gradients

$$T_{\beta_1,\beta_2,\dots,\beta_{n+1}}X,$$

where $\beta_r=\lambda-a_{n+2-r}$, where λ is not less than a_1. It is exactly the same as (V), except for the initial factor $z^{\varpi'}$, where

$$\varpi'=\sum_{r=1}^{n+1}(r-1)\beta_r=\varpi+\lambda\binom{n+1}{2}-n\delta.$$

These two sets T_a, T_β may be called conjugate.

THEOREM VII. *The number of covariant types $T_{a_1\dots a_{n+1}}$, of degree δ and order ρ, is the same as the number of covariant types of the conjugate set of degree $(n+1)\lambda-\delta$ and of order ρ.*

11. A similar process may be applied to obtain the generating function of gradients or seminvariants of the nature

$$T_{a_1a_2\dots a_h}X$$

for forms with more than two variables. To do this we consider sets of N things, which may be represented by algebraic symbols, and such that in each set the things may be identified algebraically by what may be called their weights. These weights may be numerically additive or not. We shall first suppose that they are entirely distinct and represented by different symbols $z_1, z_2, \dots, z_N$. We consider first a homogeneous product $\bar{X}$ of symbols representing the things themselves; $\bar{X}$ being such that there is just one thing from each set. If the symbols representing the things be now replaced by the symbols representing their weight, the index of z_r is the number of things selected in $\bar{X}$ of this particular weight; let $\bar{X}$ be called X after this change.

Let now
$$\phi\binom{\delta}{N}$$

be the sum of all the products of powers of $z_1, z_2, \ldots, z_N$ of total degree δ. This is, in effect, the generating function for $\bar{X}$ when all the sets are alike. Then

(V)
$$\phi\binom{\delta+1}{N} - z_N \phi\binom{\delta}{N} = \phi\binom{\delta+1}{N-1}$$

is an obvious identity, of which indeed (III), § 7, is a particular case.

Let us now consider the expressions

$$T_{a_1 a_2 \ldots a_h} \bar{X},$$

where the permutations interchange the different sets. Let the generating function (a function of $z_1, z_2, \ldots, z_N$) of this be

$$\psi(a_1, a_2, \ldots, a_h;\ N).$$

Then, just as in §§ 6, 7, we obtain

$$\psi(a_1, a_2, \ldots, a_h;\ N) = \Sigma z_N^{\Sigma(a-\beta)} \psi(\beta_1, \beta_2, \ldots, \beta_h;\ N-1),$$

and deduce

$$\psi(a_1, a_2, \ldots, a_h;\ N-1) = \left[\prod_{r=1}^{h} (1 - z_N \delta_r)\right] \psi(a_1, a_2, \ldots, a_h;\ N).$$

That
$$\psi(a_1, a_2, \ldots, a_h;\ N) = \left| \phi\binom{a_s + s - r}{N} \right|$$

satisfies this difference equation follows at once from (V). And exactly the same considerations as were used in § 7 complete the proof by induction that this is the generating function required. It remains but to point out that, now that this form of generating function has been obtained in the general case, the result may be applied to any particular case; any relations may be introduced among the weights z_r. In fact, the sets may be the coefficients of p-ary q-ics, and the weights the appropriate sets of numbers.

THEOREM VIII. *When* $\phi\binom{\delta}{N}$ *is a generating function for gradients* X *of any quantic of degree* δ, *the quantics having* N *coefficients, the generating function for gradients* $T_{a_1 a_2 \ldots a_h} X$ *is*

$$\left|\phi\binom{a_r+s-r}{N}\right|.$$

III. *Quaternary and higher forms.*

12. The considerations used for ternary forms may be applied to higher forms, but the increase in the number of kinds of variables and their inter-relations makes the process cumbersome.

The difficulty may be overcome thus. A seminvariant of a form with h variables is given by a PN belonging to $T_{a_1 a_2 \ldots a_h}$; and this is completely defined by the corresponding tableau F. Moreover, F is completely defined by a gradient

$$X = \Pi\, A_{r_1 r_2 \ldots r_h}^{\lambda_{r_1 r_2 \ldots r_h}},$$

where in F the number of sets of n equivalent letters which have r_1 in the first row, r_2 in the second, and so on is $\lambda_{r_1 r_2 \ldots r_h}$.

Further, every seminvariant can be expressed linearly in terms of seminvariants thus defined. The weights of such a seminvariant are

$$w_1 = a_1, \quad w_2 = a_2, \quad \ldots, \quad w_h = a_h.$$

A term of the concomitant led by this seminvariant has the same substitutional properties defined by the operator PN, but its weights are different from those of the seminvariant. Consider a term whose weights are $\varpi_1, \varpi_2, \ldots, \varpi_h$. Then $w_1 \geqslant \varpi_1$, $w_1+w_2 \geqslant \varpi_1+\varpi_2$, and so on. Let δ be the degree of the concomitant, the number of letters in PN or F is then $n\delta$.

Let us use separate letters $a, b, c, \ldots$ for the moment for each of these, and consider them as ordinary symbolical letters so that

$$a_x = a_1 x_1 + a_2 x_2 + \ldots + a_h x_h$$

in the usual way.

Then of these $n\delta$ letters, in any gradient of the term considered, ϖ_1 have the suffix 1, and this gradient is unaffected by the permutations of all the letters with this suffix, *i.e.* it is unaffected by the permutations of a positive symmetric group of degree ϖ_1. Similarly, it is unaffected by the permutations of a positive symmetric group of degree ϖ_2, and so on. Thus, calling the letters with *symbolical suffix* r $\theta_{r1}, \theta_{r2}, \ldots, \theta_{r\varpi_r}$, the positive symmetric group of these letters G_r, and the product of these groups G, PN may be rewritten and supposed multiplied on the left by G. Then, as in § 5, GPN can be expressed in terms of forms $P'\sigma N$ derived from tableaux F' which are standard according to the sequence of θ sets, and in which there is no *overlapping*. Let us call the letters in the r-th row of F, which defined our concomitant,

$$B_{r1}\,B_{r2}\ldots B_{ra_r}.$$

The concomitant may be developed into actual coefficients from its PN form, by first expanding according to the symbolical letters B, and then using polarizing operators. As an example, consider the seminvariant of the Hessian of the binary cubic, defined by $A_0 A_2$ or

$$F=\begin{pmatrix} a_1\,a_2\,a_3\,b_1 \\ b_2\,b_3 \end{pmatrix}.$$

It is

$$(B_1 B_2)^2 B_1^2,$$

using B_1 for the first row letters and B_2 for the second.

We polarize with

$$\frac{1}{4!\;2!}\left(a\frac{\partial}{\partial B_1}\right)^3\left(b\frac{\partial}{\partial B_1}\right)\left(b\frac{\partial}{\partial B_2}\right)^2,$$

and obtain

$$\tfrac{1}{6}(ab)^2 a_1 b_1.$$

Consider any gradient Y of weights $\varpi_1, \varpi_2, \ldots, \varpi_h$. Then

$$T_{a_1 a_2 \ldots a_h} Y = \Sigma\lambda P' N',$$

where $P' N'$ is defined by the sets θ_r. There are then just the same number of linearly independent forms $P' N'$ as there are standard non-overlapping θ tableaux.

Next consider the correspondence between θ and B; no matter how this is arranged, $P' N'$ is unaffected by the permutations of

$$\{B_{11}\,B_{12}\ldots B_{1a_1}\}.$$

Operation with this group places each of these letters in the first row of the tableau, completely filling it. Similarly the second row can be filled with $B_{21}\ldots B_{2a_2}$. Thus for the terms of our concomitant each one of the kind $P' N'$ can be expressed in terms of the same one. Thus the number of independent terms of weights $\varpi_1, \varpi_2, \ldots, \varpi_h$ is the number of standard non-overlapping tableaux F'. Let this number be

$$\begin{bmatrix} w_1\,w_2\ldots w_h \\ \varpi_1\,\varpi_2\ldots\varpi_h \end{bmatrix};$$

this is zero when $\varpi_1 > w_1$ or $\varpi_1+\varpi_2 > w_1+w_2, \ldots$. Let $\{w_1 w_2 \ldots w_h\}$ be the number of seminvariants of these weights. Then the number of gradients of weights $\varpi_1, \varpi_2, \ldots, \varpi_h$ is

$$[\varpi_1, \varpi_2, \ldots, \varpi_h] = \sum_w \begin{bmatrix} w_1 w_2 \ldots w_h \\ \varpi_1 \varpi_2 \ldots \varpi_h \end{bmatrix} \{w_1 w_2 \ldots w_h\}.$$

Let δ_{rs} be an operator which increases the r-th row of a tableau by unity and decreases the s-th row by unity. Then $\begin{bmatrix} w_1 w_2 \ldots w_h \\ \varpi_1 \varpi_2 \ldots \varpi_h \end{bmatrix}$ is the number of terms in

$$\prod_{r<s} \delta_{rs}^{\lambda_{rs}} T_{\varpi_1 \varpi_2 \ldots \varpi_h},$$

where *overlapping* terms are excluded and the indices $\lambda_{r,s}$ are such as to make the final rows $w_1, w_2, \ldots, w_h$.

Thus, if D is the operator

$$D = \prod_{r<s} (1-\delta_{r,s}),$$

as in equation I, § 2, Q.S.A., VI*,

$$D \begin{bmatrix} w_1 w_2 \ldots w_h \\ \varpi_1 \varpi_2 \ldots \varpi_h \end{bmatrix} = 0$$

(the operators affecting the ϖ's and not the w's), except when

$$w_1 = \varpi_1, \quad w_2 = \varpi_2, \quad \ldots, \quad w_h = \varpi_h;$$

in which case the right-hand side is unity. Hence

$$D[\varpi_1, \varpi_2, \ldots, \varpi_h] = \Sigma \{w_1, w_2, \ldots, w_h\} D \begin{bmatrix} w_1 w_2 \ldots w_h \\ \varpi_1 \varpi_2 \ldots \varpi_h \end{bmatrix}$$

$$= \{\varpi_1, \varpi_2, \ldots, \varpi_h\}.$$

Then, when G is the generating function for gradients, so that

$$G = \Sigma [\varpi_1 \varpi_2 \ldots \varpi_h] x_1^{\varpi_1} x_2^{\varpi_2} \ldots x_h^{\varpi_h},$$

and Γ that for seminvariants, so that

$$\Gamma = \Sigma \{w_1, w_2, \ldots, w_h\} x_1^{w_1} x_2^{w_2} \ldots x_h^{w_h},$$

we have $$\Gamma = \left[\prod_{r<s} \left(1 - \frac{x_r}{x_s}\right)\right] G = x_1^{-h+1} x_2^{-h+2} \ldots x_{h-1}^{-1} \Delta G,$$

* *Proc. London Math. Soc.* (2), 34 (1932), 199.

where Δ is the alternating function

$$\prod_{r<s} (x_r - x_s).$$

THEOREM IX. *The generating function for the seminvariants of the q-ary p-ics is obtained from that for gradients by multiplication by*

$$x_1^{-q+1} x_2^{-q+2} \ldots x_{q-1}^{-1} \Delta,$$

where Δ is the alternating function

$$\prod_{r<s} (x_r - x_s).$$

It may be observed that the generating function for seminvariants of the q-ary p-ic (in the same way as that for binary and ternary forms) exhibits a peculiar symmetry. Thus it may be written

$$\Sigma \{w_1, w_2, \ldots, w_q\} \left[\Sigma \pm \prod_{r=1}^{q} x_r^{w_s + r - s} \right],$$

where in each product $w_1, w_2, \ldots, w_q$ appear in some sequence in the indices; and the sign is determined as positive when this sequence is obtained from that of the suffixes by a permutation belonging to the alternating group, and otherwise as negative. The whole sum is limited by the inequalities

$$w_1 \geqslant w_2 \geqslant \ldots \geqslant w_q.$$

ON QUANTITATIVE SUBSTITUTIONAL ANALYSIS

(*Seventh Paper.*)

By ALFRED YOUNG

[Received 23 May, 1932.—Read 19 May, 1932.]

The subject here is the application of quantitative substitutional analysis to the theory of invariants. In the second paper, Section V*, it was shown that a very convenient notation for the invariants of a single binary form could be obtained,

$$f(a_0, a_1, \ldots, a_n),$$

by means of the symbolical calculus and the tableaux. Equations were there obtained expressing the linear relations between invariants thus given. Two problems at once arise, the discovery of a linearly independent set of such forms in terms of which the rest can be expressed; and the expression of the product of two such forms in terms of members of the set. The notation may be used for covariants as well as for invariants, and for any concomitants of one or more quantics in any number of variables.

The first section deals with the application to concomitant types, *i.e.* concomitants linear in the coefficients of each of the ground forms concerned. And here a very simple answer is obtained in Theorem I to the first problem.

The second section develops the relations between the forms, more particularly in the binary case. The forms for this purpose are arranged according to some definite sequence, and here two sequences are introduced, (A) and (B), each of which has its own advantage.

In the third section the relations of forms given in the present tableau notation to seminvariants expressed otherwise are sought. It is shown in particular that these forms are identical with the seminvariants obtained by Elliott by means of Ω and O operators from the corresponding gradient. The

* *Proc. London Math. Soc.* (1), 34 (1902), 388.

expression for them is also obtained in terms of continued transvectants. Conversely a continued transvectant is expressed in terms of these forms. And further the actual expression of any gradient as a sum of terms of covariants is directly obtained. The fourth section gives a complete discussion of covariants of degree 4. In the fifth section the theory is illustrated by the binary cubic and quartic. In the sixth section the question of sequences is discussed. A pre-arranged sequence must underlie any enumeration of the forms. In addition to the sequences (A) and (B) already used, a sequence (C) is introduced, which is the reciprocal of (A) and (B). It is shown that this sequence (C) underlies the whole of the classical treatment of binary forms; but, that while it contains certain obvious advantages here, in that it can be said at once whether a form belongs to the linearly independent set or no, it has the disadvantage that a criterion needs to be obtained as to whether the form in question actually exists for a given finite order. A very intimate connection is here indicated between a form which belongs to the independent set according to one of the three sequences and a form which has the corresponding gradient for its leading gradient. A numerical diagram is introduced as a means by which it may be ascertained whether a seminvariant with such a leading gradient exists.

In Section VII, the invariants of degrees 5 and 6 for any binary form are obtained, with the corresponding invariants of the quintic and sextic.

In Section VIII, generating functions for types are obtained for forms in any number of variables, by means of the results of Section I.

In Section IX, these are illustrated by the discussion of the invariant types of degree 5 for binary forms.

In Section X, the independent set of seminvariants of degree 5 for a binary form is given, and exhibited as leading gradients.

In Section XI, the method is applied to obtain the irreducible invariants of a quaternary cubic.

Section XII contains some concluding remarks.

I. *Application of the analysis to algebraic forms.*

1. The general concomitant for forms with q variables is symbolically written as a product of factors of the form

$$(a_1 a_2 \dots a_r u_{r+1} \dots u_q),$$

where the original forms are $a_{1_x}^{n_1}$, $a_{2_x}^{n_2}$, ..., which may be the same or different forms:

$$a_{1_x} = a_{11} x_1 + a_{12} x_2 + \dots + a_{1q} x_q,$$

and $u_{r+1}, \ldots, u_q$ are cogredient sets of variables which are contragredient to x. For uniformity we replace x by $(u_2 u_3 \ldots u_q)$ so that the symbolical factors are all of the same type. By polarization this symbolical product can be represented as a sum of products which are linear in every symbolical letter which appears; this summation is best represented by a substitutional operator.

We take the *first term* of this concomitant, which corresponds to the seminvariant, as the coefficient of $u_{2,2}^{\lambda_2} u_{3,3}^{\lambda_3} \ldots u_{q,q}^{\lambda_q}$, where λ_2 is the number of factors which contain u_2 and is the order; λ_3 is the number of factors which contain u_3—for ternary forms the sum of the order and class—and so on.

We use a triple suffix notation where

$$a_{\lambda\mu_x} = \Sigma\, a_{\lambda\mu r} x_r,$$

and the symbolical product is in general

$$\Pi\,(a_{\lambda 1} a_{\lambda 2} \ldots a_{\lambda r_\lambda} u_{r_\lambda+1} \ldots u_q).$$

We use as letters of permutation the letters $a_{\lambda\mu}$ with two suffixes, so that, for instance, the transposition $(a_{\kappa\lambda} a_{\mu\nu})$ changes the product $a_{\kappa\lambda r} a_{\mu\nu s}$ into $a_{\mu\nu r} a_{\kappa\lambda s}$. Then the seminvariant is

$$\Pi\,\{a_{\lambda 1} a_{\lambda 2} \ldots a_{\lambda r_\lambda}\}'\, a_{\lambda 11} a_{\lambda 22} \ldots a_{\lambda r_\lambda r_\lambda} = \Pi\,\{a_{\lambda 1} a_{\lambda 2} \ldots a_{\lambda r_\lambda}\}'\, K,$$

where
$$K = G_{a_1} G_{a_2} \ldots G_{a_q} K,$$

and G_{a_r} is a positive symmetric group of degree a_r divided by its order, which contains one letter from every symbolical factor of the seminvariant which has r letters at least; for K is unaltered by the interchange of a pair of letters whose third suffix in K is the same. That is, the symbolical product is of the nature

$$NPK,$$

where NP belongs to $T_{a_1 a_2 \ldots a_q}$.

If we multiply by $\Sigma\, PN$, we can express this in terms of

$$\Sigma\, PNK.$$

The forms thus defined will all be linearly independent when the tableaux for P are standard. When, however, as is the case here, there are implicitly understood substitutional operators on the left-hand side, a further investigation is needed to determine the standard forms, since these operators usually disturb the sequence on which the standard property depends.

2. Let us consider types, that is forms which are linear in the coefficients of every ground form concerned. To start with, the ground forms are arranged in definite sequence. Let b_x^n be one of them. Then this quantic in the substitutional expressions must be represented by n letters $b_1, b_2, \ldots, b_n$; and the fact that they are all representative of the one quantic is denoted by an operator $B = \{b_1 b_2 \ldots b_n\}$ on the left of PN. Standard tableaux will include some in which two or more b's are in the same column; let us suppose that the following is part of such a tableau F:

$$\begin{matrix} \ldots & \ldots & \ldots & b_1 b_2 \ldots b_r \;\; c_1 \ldots c_k \\ d_1 \ldots d_l & b_{r+1} \ldots b_{r+s} \ldots b_n & \ldots & \ldots, \end{matrix}$$

where the letters $c_1, \ldots, c_k$ are later than b_n, and the letters $d_1, \ldots, d_l$ are earlier than b_1.

Then, if PN is given by the tableau F,

$$BPNK$$

is a sum of terms some of which are not standard. Thus one is obtained by interchange of b_1 and b_{r+s}.

Then

$$B\{d_1 \ldots d_l b_{r+1} \ldots b_{r+s} b_1 \ldots b_r c_1 \ldots c_k\} PNK = 0,$$

and hence $$\lambda BPNK = -\Sigma \mu BP' NK,$$

where F', the tableau from which P' is formed, is obtained from F by exchanging some of the letters $d_1, \ldots, d_l, b_{r+1}, \ldots, b_{r+s}$ for some of the letters $b_1, \ldots, b_r, c_1, \ldots, c_k$, and since in virtue of the presence of B interchange of letters b among themselves has no effect, at least either one of the d's is moved up or one of the c's is moved down. Thus F' is an earlier tableau than F, and also it has fewer pairs of letters b in the same column. Thus, when the tableau of a concomitant has a pair of letters belonging to the same ground form in the same column, the concomitant can be expressed in terms of others whose tableaux are earlier. We may refer to the case as overlapping, and state the result as follows:

THEOREM I. *A complete set of linearly independent concomitants linear in the coefficient of each of the ground forms* $a_{r_x}^{n_r}$ $(r = 1, 2, \ldots, \delta)$ *is given by those forms GPNK whose tableaux F are standard and do not overlap.*

Here G is that operator which expresses that the first n_1 symbolical letters refer to the first quantic, the next n_2 to the second quantic, and so on.

That these concomitants are all linearly independent follows at once from the fact that

$$GPNK = \Sigma(PNK)_\rho,$$

where every tableau F_ρ is standard. Unfortunately, the question of the reducibility of these forms is not so easily settled.

3. For the binary case the tableaux contain only two rows, and it is sufficient to define a seminvariant by giving a scheme which states how many letters belonging to each quantic lie in the second row. Thus, if the r-th quantic has λ_r letters in the second row, we may write the seminvariant as $(a_1{}^0 a_2^{\lambda_2} a_3^{\lambda_3} \dots a_\delta^{\lambda_\delta})$, or more shortly $(0\,\lambda_2 \dots \lambda_\delta)$.

The first quantic can have no letters in the second row, since that would mean overlapping. This condition further requires that

$$\lambda_2 \leqslant n_1, \quad \lambda_3 + 2\lambda_2 \leqslant n_1 + n_2, \quad \lambda_4 + 2\lambda_3 + 2\lambda_2 \leqslant n_1 + n_2 + n_3,$$

and so on. There is nothing novel about this, as a similar result is obtained at once by expressing the form as a continued transvectant

$$(\dots((f_1, f_2)^{\lambda_2} f_3)^{\lambda_3} \dots f_\delta)^{\lambda_\delta},$$

when the same inequalities must be satisfied. The similar result for forms with more variables is not so obvious.

II. *The relations between seminvariants.*

4. When two or more of the δ forms are the same we have to introduce a fresh operator on the left which disturbs the sequence of forms, and some of the seminvariants with tableaux in which there is no overlapping are expressible in terms of others with earlier tableaux.

It was proved* that all relations between such forms as PNK may be derived from relations

$$\Gamma PNK = 0,$$

where Γ is a positive symmetric group containing all the letters of one row of the tableau and one letter out of a lower row. Let PNK be the form $(a_1{}^0 a_2^{\lambda_2} \dots a_\delta^{\lambda_\delta})$ and let Γ contain all the letters of the first row and one

* Q.S.A., II, 370.

letter a_r out of the second row of the tableau. Then

$$\Gamma PNK = \sum_{s \neq r} (n_s-\lambda_s)(a_1{}^0 a_2^{\lambda_2} \dots a_s^{\lambda_s+1} \dots a_r^{\lambda_r-1} \dots a_\delta^{\lambda_\delta}) + (n_r-\lambda_r+1)(a_1{}^0 a_2^{\lambda_2} \dots a_r^{\lambda_r} \dots a_\delta^{\lambda_\delta})$$
$$= 0.$$

This may be written

(I) $$O(a_1{}^0 a_2^{\lambda_2} \dots a_r^{\lambda_r-1} \dots a_\delta^{\lambda_\delta}) = 0.$$

There is a close relation between our seminvariant $(a_1{}^0 a_2^{\lambda_2} \dots a_\delta^{\lambda_\delta})$ and the gradient

$$A_{1,0}\, A_{2,\lambda_2} \dots A_{\delta,\lambda_\delta},$$

where $A_{r,s}$ is defined by

$$a_{rx}^{n_r} = (A_{r,0}, A_{r,1}, \dots, A_{r,n_r})(x_1, x_2)^{n_r}.$$

In this respect the operator O of equation (I) appears as the well-known algebraic operator

$$O = \sum_r \sum_s (n_r-s) A_{r,s+1} \frac{\partial}{\partial A_{r,s}}.$$

In the case of seminvariants of a single form the notation may be changed with advantage to

$$(A_0^{\lambda_0} A_1^{\lambda_1} \dots A_n^{\lambda_n})$$

or simply $$(\lambda_0 \lambda_1 \dots \lambda_n),$$

where λ_r is the number of sets of n letters which have r letters in the lower row of the tableau. Then

(II) $$O(A_0^{\lambda_0} A_1^{\lambda_1} \dots A_n^{\lambda_n}) = 0,$$

where O is the ordinary algebraic operator.

5. The seminvariants of a single binary form will now be considered. Those of the same weight will be arranged in a definite sequence. For this purpose two different laws will be considered, since each possesses a certain advantage; they will be quoted as sequence (A) and sequence (B); but, since the former is that in general adopted, it will be assumed to be the one in use unless otherwise stated.

(A) The sequence from the last letter:

$$(A_0^{\lambda_0} A_1^{\lambda_1} \ldots A_n^{\lambda_n}) \quad \text{precedes} \quad (A_0^{\mu_0} A_1^{\mu_1} \ldots A_n^{\mu_n})$$

when the first of the differences

$$\lambda_n - \mu_n, \quad \lambda_{n-1} - \mu_{n-1}, \quad \ldots, \quad \lambda_0 - \mu_0$$

which does not vanish is positive.

(B) The sequence from the first letter requires that the first of the differences

$$\lambda_0 - \mu_0, \quad \lambda_1 - \mu_1, \quad \ldots, \quad \lambda_n - \mu_n$$

which does not vanish must be positive. A seminvariant which can be linearly expressed in terms of earlier seminvariants is called *reductible*, a seminvariant which cannot be so expressed is called *irreductible*.

From equation (II) it follows at once that all seminvariants for which $\lambda_1 > 0$ are reductible according to either sequence.

6. Consider now seminvariant types of degree 3. They are given by the forms

$$(a_1{}^0 a_2{}^\lambda a_3{}^\mu).$$

As in Q.S.A., II, § 28, we shall operate with the positive symmetric group of the n letters a_2; as in the place quoted this may be treated in two ways, either as a sum of permutations of the n letters a_2, in which case the effect is merely multiplication by $n!$, or as bringing up all the n letters a_2 into the first row and filling the λ places of the second row by letters a_1 and a_3 from the first row in all possible ways. By equating the two results we get

$$(-)^\lambda \binom{n}{\lambda} (a_1{}^0 a_2{}^\lambda a_3{}^\mu) = \sum_u \binom{n}{u} \binom{n-\mu}{\lambda-u} (a_1{}^u a_2{}^0 a_3^{\lambda-u+\mu}),$$

i.e.

$$(-)^\lambda \binom{n}{\lambda} \binom{n}{\mu} (a_1{}^0 a_2{}^\lambda a_3{}^\mu) = \Sigma \binom{\lambda-u+\mu}{\mu} \binom{n}{u} \binom{n}{\lambda-u+\mu} (a_1{}^u a_2{}^0 a_3^{\lambda-u+\mu}),$$

which suggests a change of notation with the definition

$$\{a_1{}^\lambda a_2{}^\mu a_3{}^\nu\} = \binom{n}{\lambda} \binom{n}{\mu} \binom{n}{\nu} (a_1{}^\lambda a_2{}^\mu a_3{}^\nu).$$

Our equation then becomes

$$(-)^\lambda \{a_1{}^0 a_2{}^\lambda a_3{}^\mu\} = \sum_v \binom{v}{\mu} \{a_1^{\lambda+\mu-v} a_2{}^0 a_3{}^v\}.$$

More generally, if the orders n_r are not necessarily equal, we obtain in the same way

$$\text{(III)}\qquad (-)^{\lambda_2}\{a_1{}^0 a_2^{\lambda_2} a_3^{\lambda_3} \dots a_s^{\lambda_s}\} = \Sigma\left[\prod_{r=3}^{s}\binom{v_r}{\lambda_r}\right]\{a_1^{w-\Sigma v} a_2{}^0 a_3^{v_3} \dots a_s^{v_s}\}.$$

7. THEOREM II. *The following relation connects seminvariant types of degree* 3:

$$(-)^{\gamma_3}\sum_{r=0}^{\gamma_1}\binom{\gamma_1+\gamma_2-r}{\gamma_2}\{a_1{}^0 a_2{}^r a_3^{w-r}\}+(-)^{\gamma_1}\sum_{r=0}^{\gamma_2}\binom{\gamma_2+\gamma_3-r}{\gamma_3}\{a_1^{w-r} a_2{}^0 a_3{}^r\}$$
$$+(-)^{\gamma_2}\sum_{r=0}^{\gamma_3}\binom{\gamma_3+\gamma_1-r}{\gamma_1}\{a_1{}^r a_2^{w-r} a_3{}^0\}=0,$$

where $\gamma_1+\gamma_2+\gamma_3=w-1.$

This is analogous to Stroh's equation*.

To prove this we use the equations

$$\{a_1^{w-r} a_2{}^0 a_3{}^r\} = (-)^{w-r}\Sigma\binom{w-s}{r}\{a_1{}^0 a_2{}^s a_3^{w-s}\},$$

$$\{a_1{}^r a_2^{w-r} a_3{}^0\} = (-)^r\Sigma\binom{s}{w-r}\{a_1{}^0 a_2{}^s a_3^{w-s}\}$$

to transform the last two sums, and we then consider the coefficient of $\{a_1{}^0 a_2{}^s a_3^{w-s}\}$.

When $s \leqslant \gamma_1$ the last sum provides no part of this coefficient. A sum is required of the form

$$\binom{\xi+\lambda}{\xi}-\eta\binom{\xi+\lambda-1}{\xi-1}+\binom{\eta}{2}\binom{\xi+\lambda-2}{\xi-2}-\dots+(-)^{\xi}\binom{\eta}{\xi}\binom{\lambda}{0}.$$

This may be regarded as a function of λ of order ξ; let us replace λ by $-\mu$, then $\binom{\xi+\lambda}{\xi}$ is

$$\frac{(\xi-\mu)(\xi-\mu-1)\dots(-\mu+1)}{\xi!}=(-)^{\xi}\binom{\mu-1}{\xi}.$$

The function becomes

$$(-)^{\xi}\left[\binom{\mu-1}{\xi}+\eta\binom{\mu-1}{\xi-1}+\dots+\binom{\eta}{\xi}\binom{\mu-1}{0}\right]=(-)^{\xi}\binom{\mu+\eta-1}{\xi},$$

---

* Grace and Young, *Algebra of invariants*, 64.

and the original sum is

(IV) $$(-)^{\xi}\binom{\eta-\lambda-1}{\xi}.$$

Thus $$\sum_{r=0}^{\gamma_3}(-)^r\binom{w-s}{r}\binom{\gamma_2+\gamma_3-r}{\gamma_3}=(-)^{\gamma_2}\binom{w-s-\gamma_3-1}{\gamma_2},$$

and the total coefficients of the terms $s\leqslant\gamma_1$ are all zero.

When s lies between γ_1 and $\gamma_1+\gamma_2+1$, then

$$0\leqslant w-s-\gamma_3-1<\gamma_2,$$

and hence the coefficient obtained from the second sum is zero, while no such terms are obtained from the other two sums.

When $s>\gamma_1+\gamma_2$ it is necessary to consider the last sum; the coefficient of the s term is

$$\sum_{r=0}^{\gamma_3}(-)^{\gamma_2+r}\binom{s}{w-r}\binom{\gamma_3+\gamma_1-r}{\gamma_1},$$

the coefficient of x^{γ_1} in the expansion of

$$(-)^{w+\gamma_2}\Big[\{1-(1+x)\}^s(1+x)^{\gamma_3+\gamma_1-w}-(1+x)^{\gamma_1+\gamma_3-w}+s(1+x)^{\gamma_1+\gamma_3-w+1}$$
$$-\ldots+(-)^{\gamma_1+\gamma_3-w}\binom{s}{w-\gamma_1-\gamma_3-1}(1+x)^{-1}\Big],$$

and this is

$$(-)^{\gamma_3}\left[\binom{\gamma_2+\gamma_1}{\gamma_2}-\binom{s}{1}\binom{\gamma_2+\gamma_1-1}{\gamma_2-1}+\ldots+(-)^{\gamma_2}\binom{s}{\gamma_2}\binom{\gamma_1}{0}\right]$$
$$=(-)^{\gamma_2+\gamma_3}\binom{s-\gamma_1-1}{\gamma_2}.$$

In this case the coefficient provided by the second sum is the coefficient of x^{γ_3} in

$$(-)^{w+\gamma_1}[1-(1+x)^{-1}]^{w-s}(1+x)^{\gamma_2+\gamma_3}=(-)^{w+\gamma_1}x^{w-s}(1+x)^{\gamma_2+\gamma_3+s-w},$$

since the index of $1+x$ is always positive. This gives

$$(-)^{\gamma_2+\gamma_3+1}\binom{s-\gamma_1-1}{s-w+\gamma_3},$$

and once more the total coefficient is zero. Thus the identity is proved. A transformation by means of equation (III) gives the following variation of the result just proved, simply and directly; we give it as a corollary because it is sometimes useful.

COROLLARY.

$$(-)^{\gamma_3}\sum_{r=0}^{\gamma_1}\binom{\gamma_1+\gamma_2-r}{\gamma_2}\{a_1^0 a_2^r a_3^{w-r}\}+(-)^{\gamma_1}\sum_{r=0}^{\gamma_2}\binom{\gamma_2+\gamma_3-r}{\gamma_3}\{a_1^{w-r}a_2^0 a_3^r\}$$
$$+(-)^{\gamma_2+\gamma_3}\sum_{r=0}^{\gamma_3}\binom{\gamma_3+\gamma_2-r}{\gamma_2}\{a_1^0 a_2^{w-r} a_3^r\}=0,$$

when $\gamma_1+\gamma_2+\gamma_3=w-1$.

8. It is to be observed that Theorem II may be applied to seminvariant types of any degree greater than 3. For by (III) we obtain in all cases

$$(-)^{\lambda_2}\{a_1^0 a_2^{\lambda_2} a_3^{\lambda_3}\dots a_\delta^{\lambda_\delta}\}=\Sigma\binom{v}{\lambda_3}\{a_1^{\lambda_2+\lambda_3-v}a_2^0 a_3^v a_4^{\lambda_4}\dots a_\delta^{\lambda_\delta}\}+K,$$

where K represents terms in which some of the indices later than that of a_3 are increased but none are diminished. Thus, in particular, we may suppose Theorem II always applicable to the first three quantics, and the result is true except for forms in which the weight is shifted to later quantics. The theorem may be applied exactly as Stroh's series or Jordan's lemma to prove that for degree 3 we may express a seminvariant type in terms of the types

$$(a_1^0 a_2^\lambda a_3^{w-\lambda}),\quad (a_1^{w-\lambda}a_2^0 a_3^\lambda),\quad (a_1^\lambda a_2^{w-\lambda}a_3^0),$$

where $$w-\lambda \geqslant 2\lambda.$$

Further, when $w=3\lambda+1$,

$$(a_1^0 a_2^\lambda a_3^{2\lambda+1})+(a_1^{2\lambda+1}a_2^0 a_3^\lambda)+(a_1^\lambda a_2^{2\lambda+1}a_3^0)$$

is expressible in terms of forms in which the highest index is increased. This is true also of the differences

$$(a_1^0 a_2^\lambda a_3^{2\lambda})-(a_1^{2\lambda}a_2^0 a_3^\lambda),$$

and so on.

When we apply these results to the seminvariants of a single form, we see that the lowest index is, of course, zero; the second must be even or by (III) the seminvariant is reductible.

Let the second lowest index be 2λ and the third λ_3. Then the form is reductible according to either sequence when $\lambda_3<4\lambda$, or when $\lambda_3=4\lambda+1$.

We have seen (§ 2) that there is no relation between types whose tableaux have no overlapping, and in which the quantics are arranged according to the fixed sequence. There are then, in general, $w+1$ independent forms $\{a_1^0 a_2^\lambda a_3^\mu\}$, and there are no relations other than those given by

Theorem II. And thus for seminvariants of degree 3 there are, in general, no others than those which we have obtained. In other words, they are really irreductible.

9. The equation (I) for seminvariant types is one in terms of which all other equations between types may be expressed. This leads to a remarkable result when the equation is expressed in terms of the new forms introduced in § 6, *i.e.*

$$\{a_1^{\lambda_1} a_2^{\lambda_2} \dots a_\delta^{\lambda_\delta}\} = \left[\prod_{r=1}^{\delta} \binom{n_r}{\lambda_r}\right] (a_1^{\lambda_1} a_2^{\lambda_2} \dots a_\delta^{\lambda_\delta}).$$

The equation (II) may now be written

$$\text{(V)} \qquad \Sigma \left(a_r \frac{\partial}{\partial a_r} a_r\right) \{a_1^{\lambda_1} a_2^{\lambda_2} \dots a_\delta^{\lambda_\delta}\} = 0;$$

the operator, of course, operates on the argument, and the numerical coefficients are brought outside. Thus the equation is

$$\Sigma(\lambda_r + 1) \{a_1^{\lambda_1} a_2^{\lambda_2} \dots a_r^{\lambda_r+1} \dots a_\delta^{\lambda_\delta}\} = 0.$$

It is to be observed that these equations are entirely independent of the order n_r of the forms. The only occurrence of the order is when the index $\lambda_r > n_r$, the coefficient $\binom{n_r}{\lambda_r}$ is zero, and the form $\{a_1^{\lambda_1} a_2^{\lambda_2} \dots\}$ is zero. Thus the relations between these types are independent of the order of the forms, so that we have

THEOREM III. *By the introduction of a numerical factor the relations between seminvariant forms may be made independent of the order of the forms and the same as for perpetuants. In particular, seminvariants of a single form are irreductible or reductible independently of its order, according to any sequence.*

10. Consider the seminvariant, of a single quantic,

$$\{a_1{}^0 a_2^{\lambda_2} \dots a_\delta^{\lambda_\delta}\},$$

where $\lambda_\delta \geqslant \lambda_{\delta-1} \geqslant \lambda_{\delta-2} \dots \geqslant \lambda_2$.

If the order n of this quantic is less than λ_δ, this seminvariant vanishes owing to the factor $\binom{n}{\lambda_\delta}$; otherwise it is reductible or irreductible independently of the value of n. Let us take $n = \lambda_\delta$.

Then it may happen that

$$w = \lambda_2 + \lambda_3 + \ldots + \lambda_\delta > \tfrac{1}{2}\delta\lambda_\delta.$$

In this case, there is no such seminvariant for $n = \lambda_\delta$, so that the form is zero. This means that for any value of n this form can be expressed in terms of forms which have one index greater than λ_δ. Hence the form is reductible according to sequence (A), but not necessarily according to sequence (B).

It has been remarked that all *types* for which λ_1 is zero are linearly independent, and also that equations (III) enable us to express all other types in terms of these, then all our relations for reductibility may be derived from (III). In fact, they are the result of taking each letter in turn as the one with zero index and then using the fact that the letters are all equivalent.

It follows from this that, if it has been proved that

$$\{a_1^0 a_2^{\lambda_2} \ldots a_r^{\lambda_r}\} \quad (r < \delta)$$

is reductible, then also must

$$\{a_1^0 a_2^{\lambda_2} \ldots a_r^{\lambda_r} \ldots a_\delta^{\lambda_\delta}\}$$

be reductible according to sequence (A), the indices being in ascending order of magnitude. For the reduction of the form of degree r might have been obtained from equations (III) [or indeed from equations (V)], and, in either case, it is applicable to the form of degree δ, it being remembered that any increase in the later indices itself constitutes a reduction.

We may state then

THEOREM IV. *A seminvariant, of a single quantic,*

$$\{a_1^0 a_2^{\lambda_2} \ldots a_\delta^{\lambda_\delta}\},$$

in which the indices are in ascending order of magnitude, is reductible according to sequence (A) *when*

$$\lambda_2 + \lambda_3 + \ldots + \lambda_r > \tfrac{1}{2} r \lambda_r,$$

where $r \leqslant \delta$.

It may be noticed that the consequence of putting $r = 3$ is the result already obtained, that for irreductibility $\lambda_3 \geqslant 2\lambda_2$.

11. The general identities proved in Q.S.A., II, 390–393, have already been referred to. They were proved for invariants, but the proof given is true also for seminvariants. The notation used differs from that used here in that the forms are defined by the numbers of letters of each set which lie in the upper row of the tableau instead of the lower row. To change the notation to that used here, each index λ must be changed to $n-\lambda$. Further, it is to be remembered that the result is given in terms of forms $(a_1^{\lambda_1} a_2^{\lambda_2} \dots a_\delta^{\lambda_\delta})$, or rather of the forms $(A_0^{\lambda_0} \dots A_n^{\lambda_n})$, and not of those denoted by $\{a_1^{\lambda_1} \dots a_\delta^{\lambda_\delta}\}$.

When the changes indicated are made we have an identity

$$\Pi\left\{x_\lambda+\mu x_{\lambda+1}+\binom{\mu}{2} x_{\lambda+2}+\dots+x_{\lambda+\mu}\right\}^{\beta_{\lambda,\mu}}$$

$$=(-)^{\varpi-\rho}\,\Pi\left\{x_\lambda+(n-\lambda-\mu)x_{\lambda+1}+\binom{n-\lambda-\mu}{2} x_{\lambda+2}+\dots+x_{n-\mu}\right\}^{\beta_{\lambda,\mu}},$$

where $\varpi-\rho=w-\Sigma\lambda\beta_{\lambda\mu}$, to which the following meaning is to be attached. Each side is expanded, but only those terms which are of weight w, when $x_0, x_1, \dots, x_n$ are regarded as the coefficients of a quantic, are retained. Then, if $\phi(\lambda_0, \lambda_1, \dots, \lambda_n)$ and $\psi(\lambda_0, \lambda_1, \dots, \lambda_n)$ are the coefficients of $x_0^{\lambda_0} x_1^{\lambda_1} \dots x_n^{\lambda_n}$ on the two sides, we have

$$\Sigma\phi(\lambda_0, \lambda_1, \dots, \lambda_n)(A_0^{\lambda_0} A_1^{\lambda_1} \dots A_n^{\lambda_n})=\Sigma\psi(\lambda_0, \lambda_1, \dots, \lambda_n)(A_0^{\lambda_0} A_1^{\lambda_1} \dots A_n^{\lambda_n}).$$

The most useful results are obtained by choosing the indices such that $\lambda\beta_{\lambda,\mu}$ is always zero.

Let us write

$$X_\mu=x_0+\mu x_1 m+\binom{\mu}{2} x_2 m^2+\dots+x_\mu m^\mu.$$

The result states that the coefficient of m^w in the equation

$$\prod_\mu X_\mu^{\beta_{0,\mu}}=(-)^w \prod_\mu X_{n-\mu}^{\beta_{0,\mu}}$$

gives an identity between seminvariants.

Now Elliott* has proved that

$$\Pi(m^{-\mu} X_\mu)^{\beta_{0,\mu}}=e^{m^{-1}\Omega}\,\Pi\, x_\mu^{\beta_{0,\mu}},$$

where $$\Omega=x_0\frac{\partial}{\partial x_1}+2x_1\frac{\partial}{\partial x_2}+\dots.$$

* *Algebra of quantics*, 111.

Here we write m^{-1} for m in Elliott's result and use our special function for his general one.

Let
$$\Sigma \mu\beta_{0,\mu} = \varpi, \quad \Sigma \beta_{0,\mu} = \delta,$$
$$n\delta - 2w = R.$$

Our equation becomes

$$m^{\varpi} e^{m^{-1}\Omega} \Pi x_{\mu}^{\beta_{0,\mu}} = (-)^w m^{n\delta-\varpi} e^{m^{-1}\Omega} \Pi x_{n-\mu}^{\beta_{0,\mu}}.$$

The result required is obtained from this by taking the coefficient of m^w on each side. On the left this is zero when $\varpi < w$, and it is $\Pi x_{\mu}^{\beta_{0,\mu}}$ when $\varpi = w$. Similarly it is zero on the right when $\varpi > n\delta - w$.

Let $\varpi = w + \rho$; then

$$n\delta - \varpi = w + R - \rho,$$

and the result of equating the coefficients of m^w gives

(VI) $$\frac{1}{\rho!} \Omega^{\rho} (A_0^{\beta_{0,0}} A_1^{\beta_{0,1}} \dots A_n^{\beta_{0,n}}) = (-)^w \left(\frac{1}{(R-\rho)!}\right) \Omega^{R-\rho} (A_0^{\beta_{0,n}} A_1^{\beta_{0,n-1}} \dots A_n^{\beta_{0,0}}),$$

where R is the order of the seminvariant. Also we obtain

(VII) $$\Omega^{\rho}(A_0^{\beta_{0,0}} A_1^{\beta_{0,1}} \dots A_n^{\beta_{0,n}}) = 0$$

when $\rho > R$.

12. For an invariant this gives

$$(A_0^{\lambda_0} A_1^{\lambda_1} \dots A_n^{\lambda_n}) = (-)^w (A_0^{\lambda_n} A_1^{\lambda_{n-1}} \dots A_n^{\lambda_0}),$$

a relation obtainable at once directly from the consideration that in the tableau for an invariant there is the same number w of letters in each of the two rows, and these rows may be interchanged, the result being multiplied by $(-)^w$.

It has been seen that, for an irreductible form, λ_1 is zero; it follows that, for an invariant, λ_{n-1} is also zero. Also, if the form is irreductible according to sequence (A), the first of the differences

$$\lambda_n - \lambda_0, \quad \lambda_{n-2} - \lambda_2, \quad \lambda_{n-3} - \lambda_3, \quad \dots$$

which does not vanish is positive; according to sequence (B) it will be negative.

Further, when all these differences vanish the form is zero unless w is even.

13. THEOREM V. *The seminvariant of order R,*

$$(A_0^{\lambda_0} A_2^{\lambda_2} A_3^{\lambda_3} \dots A_n^{\lambda_n}),$$

is always reductible, according to either sequence, when

$$\lambda_{n-1} > R.$$

From (VII)

$$\Omega^{R+1}(A_0^{\lambda_0} A_2^{\lambda_2} \dots A_{n-1}^{\lambda_{n-1}-R-1} A_n^{\lambda_n+R+1}) = 0,$$

and the result follows at once, for the form in question is the earliest term in the equation which appears.

THEOREM VI. *The seminvariant of order R,*

$$(A_0^{\lambda_0} A_2^{\lambda_2} \dots A_n^{\lambda_n}),$$

is always reductible, according to sequence (A), *when the first of the differences*

$$\lambda_0 - \lambda_n - R, \quad R - \lambda_{n-1}, \quad \lambda_2 - \lambda_{n-2}, \quad \dots$$

which does not vanish is positive; and, if all these differences vanish, when w is odd.

This follows at once from equation (VI),

$$(A_0^{\lambda_0} A_2^{\lambda_2} \dots A_n^{\lambda_n}) = (-)^w \frac{1}{R!} \Omega^R (A_0^{\lambda_n} A_1^{\lambda_{n-1}} \dots A_n^{\lambda_0}).$$

More generally, from

$$\frac{1}{\rho!} \Omega^\rho (A_0^{\lambda_0} A_2^{\lambda_2} \dots A_{n-1}^{\lambda_{n-1}-\rho} A_n^{\lambda_n+\rho}) = (-)^w \frac{1}{(R-\rho)!} \Omega^{R-\rho} (A_0^{\lambda_n+\rho} A_1^{\lambda_{n-1}-\rho} \dots A_n^{\lambda_0}),$$

we have

THEOREM VII. *The seminvariant of order R,*

$$(A_0^{\lambda_0} A_2^{\lambda_2} \dots A_n^{\lambda_n}),$$

is always reductible, according to sequence (A), *when the first of the differences*

$$\lambda_0 + \lambda_{n-1} - \lambda_n - R, \quad R - 2\lambda_{n-1}, \quad \lambda_2 - \lambda_{n-2}, \quad \dots$$

which does not vanish is positive; and also when w is odd and they all vanish.

14. In considering the reductibility of any form it will be usually written in the notation $\{a_0^{\lambda_1} a_2^{\lambda_2} \dots a_s^{\lambda}\}$, where here the indices are in

ascending order of magnitude and correspond to the suffixes in the notation used in the last few paragraphs. The irreductibility is ascertained first for the indices of lower magnitude. Thus, when λ_2 is odd, there is no need to go further; λ_1 must be zero. Then λ_3 is equal to $2\lambda_2$, or greater than $2\lambda_2+1$. For λ_4 we consider the form $\{a_1^0 a_2^{\lambda_2} a_3^{\lambda_3} a_4^{\lambda_4}\}$ as a seminvariant of degree 4 of the form of order $n = \lambda_4$. The order of this seminvariant is R_4, where

$$R_4 = 4\lambda_4 - 2(\lambda_2+\lambda_3+\lambda_4)$$

$$= 2(\lambda_4 - \lambda_2 - \lambda_3).$$

Thus $\lambda_4 \geqslant \lambda_2+\lambda_3$.

Then, writing $\lambda_2 = 2\lambda, \quad \lambda_3 = 4\lambda+\mu \quad (\mu = 0 \text{ or } > 1),$

we have $\lambda_4 = 6\lambda+\mu+\nu,$

and $R_4 = 2\nu.$

When $\lambda = 0$, it is seen from Theorem VII that the form is reductible when $\nu = 1$, for the differences are zero and w is odd. It will be seen later that, whatever λ may be, the form is reductible when $\nu = 1$; or no covariants of degree 4 and order 2 can exist, a fact easily demonstrable otherwise.

15. Another way of proceeding is as follows; operate on the seminvariant type

$$\{a_1^0 a_2^{\lambda} a_3^{\mu} a_4^{w-\lambda-\mu}\}$$

with a positive symmetric group G which contains all the n letters a_3, and $n-\rho$ of the letters a_4, the ρ letters omitted being in the second row of the tableau. To make matters clear we shall call B_r the positive symmetric group of the letters a_r, and consider the equation derived from

$$B_1 B_2 B_3 B_4 G\{a_1^0 a_2^{\lambda} a_3^{\mu} a_4^{w-\lambda-\mu}\}$$

by using G in the two different ways as in Q.S.A., II.

This yields the equation

$$\text{(VIII)} \quad \sum_u \binom{\rho+u}{\rho} \{a_1^0 a_2^{\lambda} a_3^{w-\rho-u-\lambda} a_4^{\rho+u}\}$$

$$= (-)^{w-\lambda-\rho} \sum_v \binom{\lambda+v}{\lambda} \{a_1^{w-\lambda-\rho-v} a_2^{\lambda+v} a_3^0 a_4^{\rho}\}.$$

When $\rho < \lambda$, this gives a reduction for the left-hand side according to sequence (B); that is we have a reduction for

$$\Sigma \binom{x}{\rho} \{a_1{}^0 a_2{}^\lambda a_3^{w-\lambda-x} a_4{}^x\}$$

for the values 0, 1, 2, ..., $\lambda-1$ of ρ.

Hence we deduce a reduction for

$$\Sigma \binom{x-x_1}{\rho} \{a_1{}^0 a_2{}^\lambda a_3^{w-\lambda-x} a_4{}^x\},$$

where x_1 has any selected value.

Thus any λ consecutive terms may be expressed in terms of the rest and of earlier terms in the sequence (B). Let us choose them in the centre of the series; then, when μ_3 and μ_4 are the indices of a_3 and a_4, it is possible to express all the forms in terms of those for which the difference

$$\mu_3 \sim \mu_4 \geqslant \lambda.$$

Thus all seminvariant types of degree 4 may be expressed in terms of those of the form

$$\{a_1{}^0 a_2{}^\lambda a_3^{2\lambda+\xi} a_4^{3\lambda+\xi+\eta}\},$$

where the identity of the letters is not fixed.

Exactly the same process may be applied to seminvariant types of degree δ, where the quantics $a_{\delta-1}$, a_δ of highest index are selected in place of a_3 and a_4 in the case of degree 4, and are treated in the same way. The result may be stated thus:

THEOREM VIII. *All seminvariant types of degree δ can be expressed in terms of types of the form*

$$\{a_1{}^0 a_2{}^\lambda a_3^{2\lambda+\mu_3} a_4^{3\lambda+\mu_4} \dots a_\delta^{(\delta-1)\lambda+\mu_\delta}\},$$

where

$$\mu_\delta \geqslant \mu_{\delta-1} \geqslant \mu_{\delta-2} \dots \geqslant \mu_3 \geqslant 0.$$

Thus, when using the sequence (B), it may always be asserted that the second index is equal to or less than $w \big/ \binom{\delta}{2}$.

III. *Relation of these forms to seminvariants expressed otherwise.*

16. Consider the representation of a gradient of a single binary n-ic in terms of the substitutional forms. The general gradient is

$$X = A_0^{a_1} A_1^{a_1} \dots A_n^{a_n},$$

the weight w will be supposed to be not greater than $\frac{1}{2}n\delta$, where δ is the degree.

The factor $A_r^{a_r}$ will be replaced by na_r symbolical letters, which appear for substitutional purposes as

$${}_ra_{1,1}\,{}_ra_{1,2}\cdots{}_ra_{1,n};\quad {}_ra_{2,1}\cdots{}_ra_{2,n};\quad \ldots;\quad \ldots{}_ra_{a_r,n}.$$

A symbolical letter γ appears usually in two ways, γ_1 and γ_2; for convenience they will here be written γ and γ'. A permutation $(\gamma\delta)$ will leave $\gamma\delta$ and $\gamma'\delta'$ unchanged, but it will interchange $\gamma\delta'$ and $\gamma'\delta$.

To express one factor A_r we have a product

$${}_ra'_{s,1}\,{}_ra'_{s,2}\cdots{}_ra'_{s,r}\,{}_ra_{s,r+1}\cdots{}_ra_{s,n}.$$

Then X, when written symbolically, is unchanged by any permutation of two of the undashed letters and also by any permutation of two of the dashed letters. We may therefore write it in the form

$$\frac{1}{(n\delta-w)!\,w!}\,G_{n\delta-w}\,G_w\,X,$$

where $G_{n\delta-w}$ is the positive symmetric group of all the undashed letters, and G_w is the positive symmetric group of all the dashed letters. Using the identity

$$X=\Sigma\,T_{a_1 a_2 \ldots a_h}X,$$

where $T_{a_1 a_2 \ldots a_h}$ is the substitutional expression for the $n\delta$ letters, we notice that

$$TX=0,$$

unless $h\leqslant 2$, and $a_1\geqslant n\delta-w$. The case of chief interest is

$$T_{n\delta-w,\,w}X.$$

Here $G_{n\delta-w}G_w$ is actually a P_r formed from a tableau of this set; and the sequence of letters can always be chosen so that this P_r is standard. Since

$$T=\Sigma\,P_sN_sM_s$$

and every $P_sN_sM_sP_r$ is zero unless $r=s$, and in this case (Q.S.A., III, 264)

$$P_rN_rM_rP_r=P_rN_rP_r,$$

it follows that

$$T_{n\delta-w,\,w}X=P_rN_rX.$$

That is, in the expression of a gradient X in terms of the substitutional forms, the term arising from that T which corresponds to its weight is the seminvariant (X) defined in our symbols by the gradient itself.

The argument is stated for binary forms, but every word of it applies to forms with any number of variables; there is, consequently, no need to repeat it for higher forms, and, moreover, the expression would become too cumbersome. The only difference arises from the fact that for q-ary forms there are $q-1$ weights, and these in a "seminvariant" are equal to the number of letters in each row below the first. It is from this point of view preferable to speak of q weights, and call $n\delta-w$ the first weight in the binary case. We may then state

THEOREM IX. *In the expression of a gradient X of a single quantic in any number of variables in terms of the substitutional forms, the weights of X being in descending order of magnitude, the term arising from that T which corresponds to the weights of X is the seminvariant (X) defined in substitutional symbols by the gradient itself. The terms corresponding to all later T's are zero.*

17. It has been shown that, when X represents a gradient of weight not exceeding $\frac{1}{2}n\delta$, the seminvariants (X) of a single binary form satisfy a system of linear equations which may be expressed in the form (§4)

$$O\,.\,A_r/A_{r+1}(X)=0,$$

it being understood that A_{r+1} is a factor of X. Thus, in the same notation,

$$\text{(IX)} \qquad O\Omega(X)=0.$$

Consider any seminvariant $S=\Sigma\lambda_r X_r$; when we express it in terms of these substitutional forms, we have

$$\text{(X)} \qquad S=\Sigma\lambda_r(X_r).$$

Elliott* has shown that all seminvariants can be obtained by a differential operator from gradients. He gives two equivalent forms for the seminvariant derived from the gradient X, one of which is

$$\left\{1-\frac{1}{1\,(\eta+2)}\,O\Omega+\frac{1}{1\,.\,2\,(\eta+2)(\eta+3)}\,O^2\,\Omega^2-\ldots\right\}X,$$

where η is the order of the covariant, *i.e.* $n\delta-2w$ in the notation here used.

It follows at once from (IX) and (X) that this seminvariant is (X).

---

* *Algebra of quantics* (1913), 224-225.

THEOREM X. *The seminvariant* (X) *of order* R *of a single binary form is*

$$\left\{1-\frac{1}{1\,(R+2)}\,O\Omega+\frac{1}{1\,.\,2\,(R+2)(R+3)}\,O^2\,\Omega^2-\ldots\right\}X.$$

18. The full expression for a gradient X in terms of the substitutional forms may be obtained directly by operation with the various T's. When we operate with $T_{n\delta-\varpi,\,\varpi}$, where ϖ is less than the weight w of the gradient, the result will be a sum of forms (Y) which are seminvariant forms for that particular weight. The meaning, of course, is that here we are dealing not with the seminvariant or first term of the covariant, but with a later term, in fact, the $(w-\varpi)$-th term. When (Y) is regarded as a seminvariant, then in the expression for X it must be replaced by

$$\frac{1}{(w-\varpi)\,!}\,O^{w-\varpi}(Y).$$

The expression for X may also be obtained from Theorem X as follows. It will be convenient to use the notation $(\Omega^r X)$ for the sum of the seminvariants given by the various gradients obtained by the operation, each multiplied by the corresponding numerical coefficient. By $O^s[(\Omega^r X)]$ will be meant simply the result of operating with O^s on this seminvariant.

Then, by Theorem X,

$$(\Omega X)=\left\{1-\frac{1}{1\,(R+4)}\,O\Omega+\frac{1}{1\,.\,2\,(R+4)(R+5)}\,O^2\,\Omega^2-\ldots\right\}\Omega X$$

and

(XI) $$(\Omega^s X)=\left\{1+\sum_{t=1}^{\infty}(-)^t\,\frac{1}{t\,!\;(R+2s+t+1)_t}\,O^t\,\Omega^t\right\}\Omega^s X.$$

We shall prove that

$$(X)+\frac{1}{1\,(R+2)}\,O[(\Omega X)]+\frac{1}{2\,!\;(R+4)_2}\,O^2[(\Omega^2 X)]$$
$$+\ldots+\frac{1}{(s-1)\,!\;(R+2s-2)_{s-1}}\,O^{s-1}\,[(\Omega^{s-1} X)]$$
$$=X-(-)^s\sum_{r=s}^{\infty}(-)^r\,\frac{\binom{r-1}{s-1}}{r\,!\;(R+r+s)_r}\,O^r\,\Omega^r X.$$

When $s=1$ this is Theorem X. Let us assume it to be true for any given s and use the value of $(\Omega^s X)$ in terms of $\Omega^s X$ given by (XI). We introduce the additional term on the left

$$\frac{1}{s!\,(R+2s)_s} O^s[(\Omega^s X)].$$

On the right the coefficient of

$$O^r \Omega^r X$$

becomes

$$(-)^{r-s}\frac{1}{s!\,(R+2s)_s}\,\frac{1}{(r-s)!\,(R+r+s+1)_{r-s}}-(-)^{r+s}\frac{\binom{r-1}{s-1}}{r!\,(R+r+s)_r}$$

$$=(-)^{r+s}\frac{\binom{r-1}{s}}{r!\,(R+r+s+1)_r}.$$

Thus our equation is true for $s+1$ if it is true for s, and hence it is always true. Hence, increasing s, we find

THEOREM XI. *The expression of a gradient X, for which $n\delta-2w=R$ is not negative, in terms of substitutional forms or of terms of covariants is*

$$X=\sum_{s=0}^{\infty}\frac{1}{s!\,(R+2s)_s} O^s[(\Omega^s X)].$$

19. It is useful to remember that any seminvariant given by a substitutional tableau F may be expressed in the form

$$(AB)^w A_1^{n\delta-2w}.$$

The letters A stand for all the letters in the upper row of F, and the letters B for those in the lower row. The actual expression of the seminvariant may then be obtained by repeated polarization with Aronhold operators, which replace the symbols A, B by symbols which directly represent the ground forms.

For actual calculation this method is too cumbersome to attempt. Its importance lies in the fact that it is theoretically possible; and hence the use of single symbols A, B and the form of the seminvariant written above is justifiable.

20. Let (X) be any seminvariant of order R of one or more binary forms; and let c_x^m be any other binary form. In the first place, let

$m = 1$, and consider the transvectant

$$K = \big((X), c_x\big)^1.$$

Here (X) is used for the covariant instead of the seminvariant.

As in § 2, $$(X) = GPNK,$$

where PN is given by a tableau F; this we shall write

$$\begin{array}{l} a_1 a_2 \dots a_w \dots a_{w+R} \\ b_1 b_2 \dots b_w. \end{array}$$

Then, with the numerical coefficient given in Q.S.A., IV, § 1, we have

$$(X) = \frac{f_{w+R,w}}{(2w+R)!}\{a_1 \dots a_{w+R}\}\{b_1 \dots b_w\}(a_1 b_1)(a_2 b_2)\dots(a_w b_w)\, a_{w+1_x} \dots a_{w+R_x},$$

where $$f_{w+R,R} = \frac{(2w+R)!\ (R+1)}{(w+R+1)!\ w!};$$

and

$$K = \frac{f_{w+R,w}}{(2w+R)!}\{a_1 \dots a_{w+R}\}\{b_1 \dots b_w\}(a_1 b_1)\dots(a_w b_w)(a_{w+1}c)\, a_{w+2_x} \dots a_{w+R_x}$$

$$= \frac{f_{w+R,w}}{(2w+R)!}\frac{1}{w+1}\{a_1 \dots a_{w+R}\}\{b_1 \dots b_w c\}$$

$$\times (a_1 b_1)\dots(a_w b_w)(a_{w+1}c)\, a_{w+2_x} \dots a_{w+R_x}$$

$$= \frac{R+1}{R}(XC_1).$$

For $$\frac{f_{w+R,w+1}}{(2w+R+1)!} = \frac{R}{(R+1)(w+1)}\frac{f_{w+R,w}}{(2w+R)!},$$

and C_1 stands for the coefficient of the quantic

$$c_x = (C_0, C_1 \between x_1, x_2)^1.$$

The order of (XC_1) is $R-1$, and the first transvectant of this with another linear form d_x is

$$(K, d_x)^1 = \frac{R}{R-1}\frac{R+1}{R}(XC_1 D_1),$$

and hence the second transvectant of (X) with a quadratic c_x^2 is

$$\frac{R+1}{R-1}(XC_2).$$

Proceeding thus we see that the n-th transvectant of (X) with $c_x{}^n$ is

$$\frac{R+1}{R-n+1}(XC_n).$$

Consider next the product $(X)c_x$, or the seminvariant $(X)C_0$, where the order of the C quantic is one. This is a seminvariant of order $R+1$, and therefore expressible in terms of seminvariants $T_{w+R+1,w}$. In Q.S.A., IV, Theorem III, the result of adding one more letter to a PN was obtained; since we are here concerned only with the terms of a particular T in that result (the other terms being necessarily zero), this gives us (using a notation which exhibits the tableau)

$$\begin{Bmatrix} a_1 \ldots\ldots\ldots a_{w+R} \\ b_1 \ldots b_w \end{Bmatrix} = \left[1+\frac{1}{R+2}\Sigma(cb)\right]\begin{Bmatrix} a_1 \ldots\ldots\ldots a_{w+R}\,c \\ b_1 \ldots b_w \end{Bmatrix},$$

whence $$(X)C_0=(XC_0)+\frac{1}{R+2}(C_1\Omega X).$$

Then $$(XC_n)D_0=(XC_nD_0)+\frac{1}{R-n+2}(D_1\Omega XC_n).$$

The Ω here contains the term

$$nC_{n-1}\frac{\partial}{\partial C_n},$$

and hence

$$(XC_n)D_0=(XC_nD_0)+\frac{n}{R-n+2}(XC_{n-1}D_1)+\frac{1}{R-n+2}(C_nD_1\Omega X).$$

And hence

$$\left((X),c_x^{n+1}\right)^n=\left((X),c_x{}^n\right)^n c_x$$

$$=\frac{(R+2)_2}{(R-n+2)_2}(XC_n)+\frac{R+1}{(R-n+2)_2}(C_{n+1}\Omega X).$$

Proceeding thus we obtain

THEOREM XII. *If the covariant form* (X) *is of order* R, *then the transvectant*

$$\left((X),C_x{}^n\right)^\mu=\sum_{r=0}^{n-\mu}(R+1)\binom{n-\mu}{r}\frac{(R+n-\mu+1)_{n-\mu-r}}{(R+n-2\mu+1)_{n-\mu+1}}(C_{\mu+r}\Omega^r X).$$

The truth of this result has just been proved for $n=\mu$, and for $n=\mu+1$. We assume it to be true for $n-1$. By means of the process just used one more symbolical c can be introduced, *i.e.* the order of the C quantic can be raised from $n-1$ to n. Then the coefficient of $(C_{\mu+r}\,\Omega^r X)$ on the right is

$$(R+1)\binom{n-\mu-1}{r}\frac{(R+n-\mu)_{n-\mu-r-1}}{(R+n-2\mu)_{n-\mu}}\left\{1+\frac{\mu+r}{R+n-2\mu+1}\right\}$$

$$+(R+1)\binom{n-\mu-1}{r-1}\frac{(R+n-\mu)_{n-\mu-r}}{(R+n-2\mu)_{n-\mu}}\frac{1}{R+n-2\mu+1},$$

which reduces to the coefficient given. Thus Theorem XII is proved by induction.

21. The last theorem may now be inverted with the result:

THEOREM XIII. *If (X) is a covariant form of order R, and C_μ represents a coefficient of a binary form of order n, then*

$$(XC_\mu)=\sum_{r=0}^{n-\mu}(-)^r\binom{n-\mu}{r}\frac{(R+n-2\mu+1)_{n-\mu-r+1}}{(R+n-\mu+r+1)_{n-\mu+1}}\Big((\Omega^r X),\,c_x{}^n\Big)^{\mu+r}.$$

To prove this we may put in the value of $\Big((\Omega^r X),\,C_x{}^n\Big)^{\mu+r}$ given by Theorem XII, and evaluate the coefficient of $(C_{\mu+s}\,\Omega^s X)$ in the sum on the right.

This is

$$\sum_{r=0}^{s}(-)^r\binom{n-\mu}{r}\frac{(R+n-2\mu+1)_{n-\mu-r+1}}{(R+n-\mu+r+1)_{n-\mu+1}}$$

$$\times(R+2r+1)\binom{n-\mu-r}{s-r}\frac{(R+n+r-\mu+1)_{n-\mu-s}}{(R+n-2\mu+1)_{n-\mu-r+1}}$$

$$=\binom{n-\mu}{s}\Sigma(-)^r\binom{s}{r}(R+2r+1)\frac{(R+r)!}{(R+s+r+1)!}$$

$$=\binom{n-\mu}{s}\frac{(R+1)!\;s!}{(R+2s+1)!}\Sigma(-)^r\binom{R+2s+1}{s-r}\left[\binom{R+r+1}{R+1}+\binom{R+r}{R+1}\right].$$

In (IV), §7, these sums were evaluated. We write

$$\Sigma(-)^r\binom{R+2s+1}{s-r}\binom{R+r+1}{R+1}=\Sigma(-)^{s-t}\binom{R+2s+1}{t}\binom{R+s+1-t}{R+1},$$

and, putting $\eta = R+2s+1$, $\xi = s$, $\lambda = R+1$, we obtain for the sum

$$\binom{2s-1}{s}.$$

Similarly

$$\Sigma(-)^{s-t}\binom{R+2s+1}{t}\binom{R+s-t}{R+1} = -\binom{2s-1}{s-1}.$$

Thus the coefficient required is zero provided that $s \geqslant 1$; when $s=0$, the coefficient is obviously unity. Thus the theorem is true.

22. Consider now a gradient X, of a single binary form f, for which $n\delta - 2w = R \geqslant 0$. Let A_μ be the coefficient of greatest weight which it contains; and let

$$X = YA_\mu.$$

Then
$$(X) = \lambda_0\big((Y), f\big)^\mu + \lambda_1\big((\Omega Y), f\big)^{\mu+1} + \ldots,$$

where λ_0 is not zero.

Every term, except the first, can be expressed in forms (X'), where X' is a gradient which contains a factor A_ν, where $\nu > \mu$. Thus (X) differs from a non-zero multiple of $\big((Y), f\big)^\mu$ by earlier forms according to sequence (A). This may be generalized at once. Let

$$X = A_0 A_{\mu_2} \ldots A_{\mu_s}, \quad \mu_2 \leqslant \mu_3 \ldots \leqslant \mu_s;$$

then X differs from a non-zero multiple of the continued transvectant

$$(\ldots((f, f)^{\mu_2} f)^{\mu_3} \ldots f)^{\mu_s}$$

by earlier forms according to sequence (A). Thus the representation of covariants by continued transvectants of this nature is parallel to the representation by substitutional forms. A reduction in one case that is a representation in terms of earlier forms means also a reduction in the other case.

It should be pointed out that the object of reduction here is exactly the reverse to that hitherto used in the discussion of transvectants. It has always been considered a reduction when a transvectant is expressed in terms of lower transvectants of forms obtained from the original forms by convolution; here it is a reduction when the transvectant is expressed in terms of higher transvectants of forms obtained from the original forms by *devolution* (the reverse of convolution).

THEOREM XIV. *The covariant (XA_μ) of a binary form differs from a non-zero multiple of the transvectant $\big((X), f\big)^\mu$ by higher transvectants of f with forms obtained from (X) by devolution.*

IV. *Covariant types of degree* 4.

23. The covariant types will be written

$$\{a^0 b^\lambda c^\mu d^\nu\},$$

where the indices are arranged in general in ascending order, and each indicates the number of letters of the corresponding quantic which lie in the second row of the tableau; the form is that of § 6 with the appropriate numerical coefficients attached.

Then, by § 6,

$$\{a^0 b^\lambda c^\mu d^\nu\} - (-)^\lambda \{b^0 a^\lambda c^\mu d^\nu\}$$

is reductible according to either sequence. Thus for λ even the form belongs to $\{ab\}$ and for λ odd it belongs to $\{ab\}'$ or $\left\{\begin{matrix} a \\ b \end{matrix}\right\}$.

Theorem II, § 7, provides a reduction according to either sequence when $\mu < 2\lambda$.

When $\mu = 2\lambda$, it gives a reduction for

$$\{a^0 b^\lambda c^{2\lambda} d^\nu\} - \{b^0 c^\lambda a^{2\lambda} d^\nu\}.$$

Hence, when λ is even,

$$\{a^0 b^\lambda c^{2\lambda} d^\nu\} - \{abc\}\{a^0 b^\lambda c^{2\lambda} d^\nu\}$$

is reductible, where $\{F\}$ is the (PN) corresponding to the tableau F with the proper numerical coefficient. And, when λ is odd,

$$\left[1 - \left\{\begin{matrix} a \\ b \\ c \end{matrix}\right\}\right] \{a^0 b^\lambda c^{2\lambda} d^\nu\}$$

is reductible. When $\mu = 2\lambda + 1$,

$$\{a^0 b^\lambda c^{2\lambda+1} d^\nu\} + \{b^0 c^\lambda a^{2\lambda+1} d^\nu\} + \{c^0 a^\lambda b^{2\lambda+1} d^\nu\}$$

is reductible. Thus both

$$\{abc\}\{a^0 b^\lambda c^{2\lambda+1} d^\nu\}, \qquad \left\{\begin{matrix} a \\ b \\ c \end{matrix}\right\} \{a^0 b^\lambda c^{2\lambda+1} d^\nu\}$$

are always reductible. And the form is of the nature $\begin{Bmatrix} ab \\ c \end{Bmatrix}$ when λ is even, and of the nature $\begin{Bmatrix} ab \\ c \end{Bmatrix}'$ when λ is odd, the dash indicating that the sequence PN is changed to NP. Now the number of forms of the nature $\{F\}$, where F belongs to T, is the order f of the T matrix, *i.e.* the number of independent forms into which $\{F\}$ may be changed by permutations operating on the left. Thus there is one form $\{a^0 b^\lambda c^{2\lambda}\}$, two forms $\{a^0 b^\lambda c^{2\lambda+1}\}$, and three forms $\{a^0 b^\lambda c^\mu\}$ for $\mu > 2\lambda+1$. These latter are, for λ even, one T_3 and two $T_{2,1}$; and, for λ odd, one T_{1^3} and two $T_{2,1}$.

24. In § 15 it was shown that all seminvariant types $\{a^0 b^\lambda c^\mu d^\nu\}$ can be expressed in terms of those for which

$$\mu \geqslant 2\lambda, \quad \nu \geqslant \mu+\lambda.$$

We have to consider specially the cases

$$\nu-\mu-\lambda = 0, 1.$$

As in § 15, equation (VIII) yields the result that, for $\rho < \lambda$,

$$\Sigma \binom{x-x_1}{\rho} \{a^0 b^\lambda c^x d^{\mu+\nu-x}\}$$

is expressible in terms of forms whose two lowest indices are 0, λ', where $\lambda' < \lambda$. Also

$$\Sigma \binom{x}{\lambda} \{a^0 b^\lambda c^x d^{\mu+\nu-x}\} = (-)^{\lambda+\mu+\nu} \Sigma \binom{y}{\lambda} \{c^0 d^\lambda a^y b^{\mu+\nu-y}\}.$$

Hence

$$\Sigma \binom{x-x_1}{\lambda} (a^0 b^\lambda c^{\mu+\nu-x} d^x) - (-)^{\lambda+\mu+\nu} \Sigma \binom{y-y_1}{\lambda} (c^0 d^\lambda a^{\mu+\nu-y} b^y)$$

is expressible in terms of forms whose two lowest indices are 0, $\lambda' (<\lambda)$; and here x_1 and y_1 may be chosen at will.

(i) Let $\mu+\nu = 5\lambda+2\xi$; we choose $x_1 = y_1 = 2\lambda+\xi$. We obtain a reduction which may be written in the form

$$\{a^0 b^\lambda c^{2\lambda+\xi} d^{3\lambda+\xi}\} - \{c^0 d^\lambda a^{2\lambda+\xi} b^{3\lambda+\xi}\} \equiv 0.$$

And, since

$$\{a^0 b^\lambda c^{2\lambda+\xi} d^{3\lambda+\xi}\} - (-)^\lambda \{b^0 a^\lambda c^{2\lambda+\xi} d^{3\lambda+\xi}\} \equiv 0,$$

it follows that

$$\{a^0 b^\lambda c^{2\lambda+\xi} d^{3\lambda+\xi}\} - (-)^\lambda \{a^0 b^\lambda d^{2\lambda+\xi} c^{3\lambda+\xi}\} \equiv 0.$$

Thus there are three irreductible forms, ($\xi > 1$), $\{a^0 b^\lambda c^{2\lambda+\xi} d^{3\lambda+\xi}\}$. When λ is even, they are one $\{abcd\}$ and two $\left\{\begin{matrix} ab \\ cd \end{matrix}\right\}$. When λ is odd, they are two $T_{2,2}$ and one T_{1^4}. When $\xi = 0$, there is one form, either T_4 or T_{1^4} according as λ is even or odd.

When $\xi = 1$, we have an additional equation of the form

$$\{(abc)\} = 0.$$

In this case there are only two forms which always belong to $T_{2,2}$.

(ii) Let $\mu+\nu = 5\lambda+2\xi+1$; we again choose $x_1 = y_1 = 2\lambda+\xi$. Then, writing X for $\{a^0 b^\lambda c^{2\lambda+\xi} d^{3\lambda+\xi+1}\}$, we have

$$[1+(-)^\lambda (cd)+(ac)(bd)+(-)^\lambda (acbd)]\, X \equiv 0;$$

whence $$[1+(ab)(cd)+(ac)(bd)+(ad)(bc)]\, X \equiv 0,$$

or $$[T_4+T_{2,2}+T_{1^4}]\, X \equiv 0.$$

There are nine independent forms for $\xi > 1$. When λ is even, there are six forms $T_{3,1}$ and three forms $T_{2,1^2}$, when λ is odd, there are three forms $T_{3,1}$ and six forms $T_{2,1^2}$.

When $\xi = 0$, there are three forms; for λ even they are of the nature $\left\{\begin{matrix} abc \\ d \end{matrix}\right\}$, and for λ odd they are of the nature $\left\{\begin{matrix} abc \\ d \end{matrix}\right\}'$. When $\xi = 1$, there is an additional equation

$$\{(abc)\}\, X \equiv 0;$$

there are always six forms, three $T_{3,1}$ and three $T_{2,1^2}$.

25. For a single quantic, that is a form $T_4 X$, the irreductible forms are

$$\{a^0 b^{2\lambda} c^{4\lambda+\xi} d^{6\lambda+\xi+\eta}\},$$

where $\xi = 0$ or $\xi > 1$ and $\eta = 0$ or $\eta > 1$.

Indeed, the whole result may be stated thus:

THEOREM XV. *The irreductible forms of degree 4 are included in*

$$X = \{a^0 b^\lambda c^{2\lambda+\xi} d^{3\lambda+\xi+\eta}\},$$

where the conditions are as follows:

$$T_4X \text{ and } T_{1^4}X;\quad \xi=0 \text{ or } \xi>1,\quad \eta=0 \text{ or } \eta>1,$$

and λ *even in the first and odd in the second case.*

$$T_{2,2}X;\quad \xi>0,\quad \eta=0 \quad \text{or} \quad \eta>1.$$

$$T_{3,1}X \quad \text{and} \quad T_{2,1^2}X;$$

there is one set of forms $\xi>0$, $\eta>0$ *with* $\lambda \geqslant 0$; *and there is an additional set* $\xi=0$ *or* $\xi>1$, $\eta>0$, *with* λ *even in the first and odd in the second case.*

The generating function for T_4X is

$$\frac{1}{1-x^{12}}\,\frac{1+x^6}{1-x^4}\,\frac{1+x^3}{1-x^2}=\frac{1}{(1-x^2)(1-x^3)(1-x^4)},$$

for $T_{3,1}X$, $T_{2,2}X$, $T_{2,1^2}X$, $T_{1^4}X$ the generating functions are

$$\frac{x}{(1-x)(1-x^2)(1-x^4)},\quad \frac{x^2}{(1-x^2)^2(1-x^3)},\quad \frac{x^3}{(1-x)(1-x^2)(1-x^4)},$$

$$\frac{x^6}{(1-x^2)(1-x^3)(1-x^4)}.$$

Multiplying these functions by the numbers

$$f_4,\quad f_{3,1},\quad f_{2,2},\quad f_{2,1^2},\quad f_{1^4},$$

and adding, we obtain $(1-x)^{-4}$, the generating function for all forms of degree 4, and thus verify the accuracy of the work.

V. *The cubic and quartic.*

26. This section is mainly inserted by way of illustration. For the cubic the forms $(A_0^{\lambda_0}A_2^{\lambda_2}A_3^{\lambda_3})$ have to be considered; the order of this covariant is

$$R=3(\lambda_0-\lambda_3)-\lambda_2.$$

(It was shown in § 5 that no gradients containing A_1 need be considered.)

By Theorem VII the form is reductible when

$$2(\lambda_2+\lambda_3-\lambda_0)$$

is positive, also when it is zero and the weight is odd. Thus the only forms to be considered are, for w even,

$$(A_0^{\mu_0}X_1^{\mu_1}X_2^{\mu_2}),$$

where $X_1 = A_0 A_2$, $X_2 = A_0^2 A_3^2$; and, for w odd,

$$(A_0^{\mu_0} X_1^{\mu_1} X_2^{\mu_2} X_3),$$

where $X_3 = A_0^2 A_3$.

Then (A_0) is the cubic, $(A_0 A_2)$ the Hessian, $(A_0^2 A_3)$ the cubic covariant, $(A_0^2 A_3^2)$ the invariant. The products $(A_0)^{\mu_0}(X_1)^{\mu_1}(X_2)^{\mu_2}$ are unrelated (a relation would give A_3 in terms of the other coefficients). This system is coextensive with $(A_0^{\mu_0} X_1^{\mu_1} X_2^{\mu_2})$, and hence is equivalent to it. The consideration that there are only two irreductible forms of weight 6 and degree 6, viz. $(A_0^4 A_3^2)$, $(A_0^3 A_2^3)$, leads to the syzygy

$$(A_0^2 A_3)^2 - (A_0)^2 (A_0^2 A_3^2) = \lambda (A_0 A_2)^3.$$

One is tempted to identify (XY) with $(X)(Y)$, but, though there is a connection, this is not a simple one in general. The connection, however, becomes simple and direct when the substitutional forms are transformed from the natural to the seminormal or the orthogonal forms by the transformations introduced in Q.S.A., VI; this, however, must be reserved for a later communication, but we shall return to the subject here in § 28 and in Section XII.

27. The general form for the quartic is

$$(A_0^{\lambda_0} A_2^{\lambda_2} A_3^{\lambda_3} A_4^{\lambda_4}),$$

of which the order is

$$R = 4\lambda_0 - 2\lambda_3 - 4\lambda_4.$$

By Theorem VII this is reductible when $\lambda_0 < \lambda_3 + \lambda_4$, and also when $\lambda_0 = \lambda_3 + \lambda_4$ and w is odd.

In § 10 it was seen that this form is irreductible only when $(A_0^{\lambda_0} A_2^{\lambda_2} A_3^{\lambda_3})$ is irreductible for the cubic. Thus, for w even, the irreductible forms are

$$(A_0^{\mu_0} X_1^{\mu_1} X_2^{\mu_2} X_4^{\mu_4} X_5^{\mu_5}),$$

where X_1, X_2, X_3 have the same meanings as for the cubic and

$$X_4 = A_0 A_2 A_4, \quad X_5 = A_0 A_4;$$

and, for w odd, $$(A_0^{\mu_0} X_1^{\mu_1} X_2^{\mu_2} X_4^{\mu_4} X_5^{\mu_5} X_3).$$

There is a relation between the gradients, for

$$X_1 X_5 = A_0 X_4.$$

There are only two irreductible forms of degree 4 and weight 6, viz. $(A_0^2 A_2 A_4)$ and $(A_0^2 A_3^2)$. There are also two products of forms already obtained which cannot be equal, viz. $(A_0)(A_0 A_2 A_4)$ and $(A_0 A_2)(A_0 A_4)$, so that $(A_0^2 A_3^2)$ is reducible. We see that it will be reducible for any form of order greater than 3. The syzygy obtained for the cubic will also exist for forms of order greater than 3, but its expression will be modified by expressing $(A_0^2 A_3^2)$ in terms of products of irreducible forms.

Thus the quartic forms will be

$$(A_0), \quad (A_0 A_2), \quad (A_0 A_4), \quad (A_0^2 A_3), \quad (A_0 A_2 A_4).$$

VI. *Sequences.*

28. In § 5 the seminvariant forms of a single quantic were arranged according to one or other of two sequences (A) and (B). These sequences were really sequences of gradients. Such an arrangement may be applied to the terms of a seminvariant when written out in full. According to a given sequence law, every seminvariant has a gradient which is its first term or leading gradient. Not every gradient could occupy the position of leading gradient; in fact, for any given weight and degree there must be the same number of possible leading gradients as there are linearly independent seminvariants. Thus a notation could be devised by which, for every possible leading gradient X, the seminvariant (X) is a definite seminvariant with X for leading gradient; then all other seminvariants would be linear functions of these.

The advantage of such a notation would be that a product $(X)(Y)$ would differ from (XY) by seminvariants having a later leading gradient. The main difficulty in adopting such a notation is that of determining what gradients can be leading gradients.

In all cases so far examined (and a similar statement is true for ternary and quaternary forms), the gradients X for which (X) is irreductible according to either sequence (A) or (B) are also leading gradients according to the same sequence. Although I am convinced that this is always the case, I have not yet been able to prove it in general, and hence I can state it only as an empirical law. The consequences are important, for the form (X) could be defined as a form which has X for leading gradient; with the resulting conclusions as to products. The rôle of this analysis would then be to determine which gradients can be leading gradients of a seminvariant. As has been already said in § 26, the need for this law will be removed when the seminormal forms are used. In the case of the cubic and quartic it is true not merely that there is a seminvariant which has the leading gradient X when (X) is irreductible; but, further, when (X) is also irreductible its leading gradient is X. It follows at once that there can be no linear relation between the quartic seminvariants

$$(A_0)^{\mu_0}(X_1)^{\mu_1}(X_2)^{\mu_2}(X_4)^{\mu_4}(X_5)^{\mu_5},$$

for each has a different leading gradient.

It is certainly not true in general that the gradient X is the leading gradient of the irreductible form (X); but it seems exceedingly likely to be true for the seminormal forms (X).

29. Consider a covariant C of order R, whose leading gradient X includes the factor $A_n{}^r$, but no other coefficient of weight n. Such a covariant may be obtained thus. We put $X = X'A_n{}^r$. Then* X' is a leading gradient of a covariant of a quantic of order $n-1$. This covariant can be expressed as a symbolical product ϕ, which is of order

$$R+rn-(\delta-r).$$

Then, provided that $R \geqslant \delta-r$, the transvectant

$$\psi = (\phi, f^r)^{nr}$$

gives a perfectly definite non-zero symbolical product which, on multiplication by $\delta-r$ factors a_x, b_x, ... to raise the order of the quantics represented in ϕ from $n-1$ to n, gives a symbolical product for a covariant of the nature C whose leading gradient is X according to sequence (A). Now no restriction is placed on X' except that it is a leading gradient of the $(n-1)$-ic. Hence every gradient $X'A_n{}^r$ of degree δ and order greater than or equal to $\delta-r$ is a leading gradient provided that X' is a leading gradient.

Again, when the seminvariant $(X'A_n{}^r)$ is irreductible then (X') must be irreductible. The number of irreductible forms (X') is coextensive with that of the leading gradients X. Hence, since every gradient $X'A_n{}^r$ is a leading gradient when $R \geqslant \delta-r$, it follows that every seminvariant $(X'A_n{}^r)$ is irreductible under the same conditions.

THEOREM XVI. *When (X') is irreductible or X' is a leading gradient according to sequence* (A), *then $(X'A_n{}^r)$ is irreductible or $X'A_n{}^r$ is a leading gradient, provided that the order $R \geqslant \delta-r$.*

The same argument applies to the gradient $X = A_{n-1}^s A_n{}^r X'$, when its order R is not less than $2(\delta-s-r)$. We consider then a covariant ϕ

* Elliott, *Algebra of quantics*, 222.

of the $(n-2)$-ic whose leading gradient is X', and then take the transvectant

$$(\phi, f^{r+s})^{(r+s)\,n-s}$$

whose first term is X.

This is an additional result in the case

$$\delta - r > 2(\delta - r - s),$$

i.e. $2s > \delta - r$, otherwise it gives nothing new. This is of use only when δ is large compared with n.

30. The number of seminvariants of degree δ which do not contain the coefficient A_n' is given by the generating function

$$\frac{(1-z^n)(1-z^{n+1})\dots(1-z^{n+\delta-1})}{(1-z^2)(1-z^3)\dots(1-z^\delta)}.$$

Subtracting this from the full generating function, we find

$$z^n \frac{(1-z^{n+1})(1-z^{n+2})\dots(1-z^{n+\delta-1})}{(1-z^2)\dots(1-z^{\delta-1})}$$

as the generating function for those seminvariants which actually contain the coefficient A_n'.

Consider the effect of this subtraction on the coefficients of the expansions. Let $(\delta, n, w) \equiv \{\delta, n, R\}$ be the number of seminvariants of degree δ, extent n or less, weight w, and order R, *i.e.* the coefficient of z^w in the ordinary generating function. Then the result above gives

$$(\delta, n, w) - (\delta, n-1, w) = (\delta-1, n, w-n).$$

When $R \geqslant \delta - 1$, this expresses the fact just noticed that, for every seminvariant of weight w, of extent n, and degree $\delta-1$, there is, under this condition, a seminvariant of weight $w+n$ and degree δ. When $R < \delta-1$, the coefficient of z^w in

$$\frac{(1-z^n)(1-z^{n+1})\dots(1-z^{n+\delta-1})}{(1-z^2)(1-z^3)\dots(1-z^\delta)}$$

is negative and equal to minus the coefficient of

$$z^{n\delta-\delta+1-w} = z^{w+R-\delta+1}.$$

The order of a covariant of weight $w+R-\delta+1$ of the $(n-1)$-ic is $\delta-2-R$, hence, when $R < \delta-1$,

$$\{\delta, n, R\} + \{\delta, n-1, \delta-2-R\} = \{\delta-1, n, R+n\}.$$

Thus, when (X) is an irreductible form of degree $\delta-1$ and order $R+n$, where $R<\delta-1$, (XA_n) may be reductible or irreductible, and the number of reductible forms (XA_n) is $\{\delta,\ n-1,\ \delta-2-R\}$.

31. When, for all the seminvariants (X) of a given weight, degree, and extent, which have not been reduced, it can be shown that X is a leading gradient, then it is known that every (X) is irreductible. Further, when X and Y are leading gradients, then XY is a leading gradient, and there is a strong presumption that (XY) is irreductible.

I now show how to find a seminvariant whose leading gradient is

$$X = A_0 A_{\lambda_2} A_{\lambda_3} \dots A_{\lambda_\delta}$$

when (X) is believed to be irreductible.

In § 22 a connection was established between the form (X) and the continued transvectant

$$(\dots((f, f)^{\lambda_2} f)^{\lambda_3} \dots f)^{\lambda_\delta}.$$

We look for the seminvariant in question among the symbolical product terms of this transvectant. The orthodox practice in dealing with symbolical products is to seek to express them in terms of those with the highest possible grade, the grade being the highest index to which any individual determinant factor is raised. What we need here is exactly the reverse; in fact, the lowest possible grade. The general term in this transvectant is

$$\prod_{r\neq s} (a_r a_s)^{\alpha_{rs}},$$

where r, s run through the values $1, 2, \dots, \delta$, and

$$\sum_{r=1}^{s-1} \mu_{rs} = \lambda_s.$$

Let

$$\sum_{r=s+1}^{\delta} \mu_{rs} = \mu_s - \lambda_s.$$

Then the coefficient of maximum weight in the seminvariant is A_{μ_r}, where μ_r is the greatest of the numbers μ_s. The coefficient (or cofactor) of A_{μ_r} is obtained by removing all factors containing a_r from the symbolical product. When $R = \delta\lambda_\delta - 2w \geqslant \delta-1$, it is always possible to choose the numbers so that $\mu_\delta > \mu_s$ $(s < \delta)$; and hence, when it is possible to find a symbolical product of degree $\delta-1$ with X/A_{λ_δ} for leading gradient, it is possible to find a seminvariant with X as leading gradient, as has been proved already. When $R<\delta-1$, there will be more than one of the numbers μ_r equal to λ_δ for

$$\sum_{r=1}^{\delta} \mu_r = 2w = \delta\lambda_\delta - R.$$

It is sometimes quite easy to arrange that of these the cofactor of A_{μ_δ}, *i.e.* the product obtained by removing all factors containing a_δ, has an earlier leading gradient than that for any other letter.

It is convenient to arrange the numbers μ_{rs} in a triangular table so that the value of μ_{rs} is on the r-th row of the s-th column, where $r < s$. The table fully represents the symbolical product. Thus, for example, the table for $\prod_{r=1}^{\delta} A_{2r-2} = X$ is

$$\begin{array}{ccccc} 2 & 2 & 2 & \ldots & 2 \\ & 2 & 2 & \ldots & 2 \\ & & 2 & \ldots & 2 \\ & & \ldots & & \ldots \\ & & & & 2 \end{array}$$

containing 2 in every position. The leading gradient is X, for the letters may be taken in any sequence with the same leading gradient always with a positive sign. And this is true whether the sequence is (A) or (B).

The table for $\prod_{r=1}^{\delta} A_{2r-2+\lambda_r}$ may be obtained by the superposition of the above on the table for ΠA_{λ_r}. Hence, when the latter is a leading gradient, so also is the former.

Other examples are

$$\begin{array}{cccc} 0 & 2 & 2 & 2 \\ & 1 & 2 & 2 \\ & & 1 & 2 \\ & & & 0 \end{array} \quad \text{and} \quad \begin{array}{cccc} 2 & 2 & 3 & 4 \\ & 2 & 4 & 4 \\ & & 4 & 4 \\ & & & 0 \end{array}$$

which give leading gradients

$$A_0^2 A_3 A_5 A_6 \quad \text{and} \quad A_0 A_2 A_4 A_{11} A_{12},$$

the former according to either sequence, the latter according to sequence (A) only; both give irreductible forms (X) for sequence (A), the latter is reductible for sequence (B) by Theorem VIII.

32. The sequences (A) and (B) are not the only possible sequences. In fact, the sequence underlying the whole classical treatment of invariant theory is actually the reciprocal of these sequences. The aim in the work of earlier writers has been to express the seminvariants virtually in terms of the leading gradient according to a sequence (C), which may be thus defined. A gradient $\prod_{r=1}^{\delta} A_{\lambda_r} = X_\lambda$, in which the suffixes are in ascending order of magnitude, precedes X_μ when the first of the differences

$$\lambda_\delta - \mu_\delta, \quad \lambda_{\delta-1} - \mu_{\delta-1}, \quad \ldots$$

which does not vanish is negative. From the point of view of perpetuants, this sequence has the great advantage that the necessary and sufficient condition that X may be a leading gradient is that

$$\lambda_{\delta-1} = \lambda_\delta.$$

This is in accordance with MacMahon's use of *power ending products**.

The same thing may be seen symbolically. Every perpetuant can be expressed as a symbolical product in which some given letter appears in every determinant factor. When the greatest index of these $\delta-1$ different factors is less than twice that of the next greatest index, or when this greatest index is odd, the grade can be increased. After the grade is increased, one of the letters of the factor of greatest index may be chosen to be placed in every factor; and if necessary the process may be repeated.

Thus we need only consider a product

$$\prod_{r=2}^{\delta} (a_1 a_r)^{\lambda_r},$$

where λ_δ is even and

$$\lambda_\delta \geqslant 2\lambda_{\delta-1} \geqslant 2\lambda_{\delta-2} \ldots \geqslant 2\lambda_2.$$

The leading gradient, according to sequence (C), is

$$A_{\lambda_2} \ldots A_{\lambda_{\delta-1}} A_\mu^2,$$

where $$\lambda_\delta = 2\mu;$$

this is really the method of Grace† in dealing with perpetuants.

It is easy to see that, when X is not a power ending product, then (X) is reductible according to sequence (C); for let $X = \prod A_{\lambda_r}$, where $\lambda_\delta > \lambda_{\delta-1}$;

* Elliott, *Algebra of quantics*, 241, *et seq.*

† Grace and Young, *Algebra of invariants* (1903), 326, *et seq.*

then, by equation (I), § 4,

$$O(A_{\lambda_1} A_{\lambda_2} \dots A_{\lambda_\delta - 1}) = 0$$

is an equation which reduces (X). And since the number of linearly independent seminvariants is the same as the number of power enders, every form (X) is irreductible when X is a power ender.

Thus the law which is so far but empirical for sequences (A) and (B) is an established fact for (C).

The weakness of sequence (C) lies in the fact that there is no clear indication of the order of the quantic necessary in order that (X) may exist and be independent of other forms (Y), where X and Y are power enders. In fact, like the transvectant sequence and the symbolical product sequence, the place of difficulty also is reversed in passing from sequence (C) to sequences (A) and (B).

VII. *Invariants.*

33. The discussion of invariants is facilitated by the fact that the two rows in the tableau may be exchanged with an accompanying change of sign when the weight is odd. Thus

$$(A_0 A_{\lambda_2} \dots A_{\lambda_\delta}) = (-)^w (A_0 A_{\lambda_\delta - \lambda_{\delta-1}} A_{\lambda_\delta - \lambda_{\delta-2}} \dots A_{\lambda_\delta}),$$

the order of the quantic being here taken as the weight of the last letter.

Hence for sequence (A) the form is reductible when the first of the differences

$$\lambda_\delta - \lambda_{\delta-1} - \lambda_2, \quad \lambda_\delta - \lambda_{\delta-2} - \lambda_3, \quad \dots$$

which does not vanish is positive [negative for sequence (B)], and when all these differences are zero and w is odd. The form is also reductible when λ_r is 1 or $\lambda_\delta - 1$. The form is also reductible for sequence (A) [sequence (B)] when $(A_0 A_{\lambda_2} \dots A_{\lambda_{\delta-1}})$ is reductible for sequence (A) [sequence (B)], and when $(A_0 A_{\lambda_\delta - \lambda_{\delta-1}} \dots A_{\lambda_\delta - \lambda_2})$ is reductible for sequence (B) [sequence (A)]. These conditions are the complete conditions, at least so far as degree 6 is concerned, and every other form is irreductible.

The invariants of degree 5 are then

$$(A_0 A_{2\mu} A_{4\mu+\xi} A_{6\mu+\xi+\eta} A_n),$$

where ξ, η are zero or greater than 1, and

$$3n = 24\mu + 4\xi + 2\eta.$$

We also have $$n - 6\mu - \xi - \eta \leqslant 2\mu,$$

that is $$\xi \leqslant \eta.$$

When $\xi = \eta$, we must also have w and therefore ξ even. This gives a set of forms

$$(A_0 A_{2\mu} A_{4\mu+2\xi} A_{6\mu+4\xi} A_{8\mu+4\xi}).$$

Here μ and ξ may be any positive integers or zero. Otherwise $\eta = \xi + 3\zeta$, and the form is

$$(A_0 A_{2\mu} A_{4\mu+\xi} A_{6\mu+2\xi+3\zeta} A_{8\mu+2\xi+2\zeta}).$$

This requires that $2\mu \geqslant \zeta$; also the form is reductible when ζ is odd. All possible systems of suffixes may be obtained by adding multiples of the four schemes

$$\begin{aligned} I_4 &= (0 \quad 0 \quad 2 \quad 4 \quad 4), \\ I_8 &= (0 \quad 2 \quad 4 \quad 6 \quad 8), \\ I_{12} &= (0 \quad 2 \quad 4 \quad 12 \quad 12), \\ I_{18} &= (0 \quad 2 \quad 7 \quad 18 \quad 18). \end{aligned}$$

Moreover, $$2I_{18} = 2I_{12} + 3I_4.$$

That the corresponding products are leading gradients may be seen by the diagrams

$$I_4 \begin{array}{rrrr} 0 & 1 & 1 & 2 \\ & 1 & 2 & 1 \\ & & 1 & 1 \\ & & & 0 \end{array}, \qquad I_{12} \begin{array}{rrrr} 2 & 2 & 4 & 4 \\ & 2 & 4 & 4 \\ & & 4 & 4 \\ & & & 0 \end{array}, \qquad I_{18} \begin{array}{rrrr} 2 & 4 & 6 & 6 \\ & 3 & 6 & 7 \\ & & 6 & 5 \\ & & & 0 \end{array}.$$

Moreover, these diagrams as well as the schemes are additive. Hence the former are irreductible and the gradients are leading gradients. The generating function is

$$\frac{1+x^{45}}{(1-x^{10})(1-x^{20})(1-x^{30})}.$$

The correctness of the result will again appear from the generating function.

34. The invariants of the quintic, by Hermite's law of reciprocity, should be related to those of degree 5. These are

$$(A_0^{\lambda_0} A_2^{\lambda_2} A_3^{\lambda_3} A_5^{\lambda_5}),$$

where $$\lambda_2 - \lambda_3 = 5(\lambda_5 - \lambda_0),$$

and $\lambda_5 \geqslant \lambda_0$ for sequence (A).

Moreover, when $\lambda_5=\lambda_0$, we have $\lambda_2=\lambda_3$, and w must be even. Also this form is reductible when $\lambda_2+\lambda_3>\lambda_0$ and when $\lambda_2+\lambda_3=\lambda_0$, λ_3 odd. Thus, when (X) is an irreductible invariant, X is a product of the gradients

$$X_4=A_0^2A_5^2,\quad X_8=A_0^3A_2A_3A_5^3,\quad X_{12}=A_0^4A_2^2A_3^2A_5^4,$$
$$X_{16}=A_0^5A_2^5A_5^6,\quad X_{18}=A_0^6A_2^5A_5^7.$$

Here
$$X_{18}^2=X_4X_{16}^2$$
indicates the quintic syzygy; and also
$$X_8^2=X_4X_{12}$$
indicates a syzygy or a reduction. Consider the possible invariant forms of degree 16; they must be given by the gradients

$$X_4^4,\quad X_4^2X_8,\quad X_8^2=X_4X_{12},\quad X_{16}.$$

Also it is not difficult to obtain invariants with the above leading gradients, so that the forms (X_4), (X_8), (X_{12}) are irreductible, and the forms

$$(X_4)^4,\quad (X_4)^2(X_8),\quad (X_8)^2,\quad (X_4)(X_{12}),\quad (X_{16})$$

exist. A relation between the first four would mean that (X_4) was a factor of (X_8) or of (X_{12}), which could not be the case; hence (X_{16}) is reducible—and, indeed,

$$(X_{16})=(X_4)(X_{12})-(X_8)^2.$$

For the leading gradient of the difference must be later than X_8^2 and hence cannot be either X_4^4 or $X_4^2X_8$ The generating function for the quintic invariants is then

$$\frac{1+x^{45}}{(1-x^{10})(1-x^{20})(1-x^{30})},$$

the same as for invariants of degree 5.

35. Invariants of degree 6 are given by

$$(A_0A_{2\mu}A_{4\mu+\xi}A_{6\mu+\xi+\eta}A_{8\mu+p}A_{10\mu+q}),$$

where, since the order of the form given by the first five coefficients is positive,

$$3p\geqslant 4\xi+2\eta,$$

and
$$4q=4\xi+2\eta+2p;$$

hence
$$6q\geqslant 10\xi+5\eta.$$

Also $q \leqslant p$; and, when $q = p$, $p \leqslant 2\xi + \eta$. Finally, when $q = p = 2\xi + \eta$, then

$$w = 30\mu + 6\xi + 3\eta$$

must be even.

It is, of course, a question of Diophantine equations; the solution is given by

$$\begin{aligned} I_2 &= (0\quad 0\quad 0\quad 2\quad 2\quad 2), \\ I_4 &= (0\quad 0\quad 2\quad 2\quad 4\quad 4), \\ I_6 &= (0\quad 0\quad 3\quad 3\quad 6\quad 6), \\ I_{10} &= (0\quad 2\quad 4\quad 6\quad 8\quad 10), \\ I_{12} &= (0\quad 2\quad 4\quad 6\quad 12\quad 12), \\ I_{15} &= (0\quad 2\quad 4\quad 9\quad 15\quad 15). \end{aligned}$$

The first two are invariants of the quadratic and quartic, and are known to represent leading gradients; I_{10} is the general form dealt with in § 31; that the other three are represented by leading gradients is seen from the diagrams

$$I_6 \quad \begin{array}{ccccc} 0 & 2 & 1 & 2 & 1 \\ & 1 & 2 & 1 & 2 \\ & & 0 & 2 & 1 \\ & & & 1 & 2 \\ & & & & 0 \end{array}, \qquad I_{12} \quad \begin{array}{ccccc} 2 & 2 & 2 & 3 & 3 \\ & 2 & 2 & 3 & 3 \\ & & 2 & 3 & 3 \\ & & & 3 & 3 \\ & & & & 0 \end{array}, \qquad I_{15} \quad \begin{array}{ccccc} 2 & 2 & 3 & 5 & 3 \\ & 2 & 3 & 5 & 3 \\ & & 3 & 5 & 3 \\ & & & 0 & 6 \\ & & & & 0 \end{array}.$$

The following relations connect the solutions:

$$2I_6 = 3I_4,$$

$$2I_{15} = 2I_{12} + 3I_2.$$

The generating function is

$$\frac{1+x^{45}}{(1-x^6)(1-x^{12})(1-x^{18})(1-x^{30})}.$$

36. The sextic invariants are forms

$$(A_0^{\lambda_0} A_2^{\lambda_2} A_3^{\lambda_3} A_4^{\lambda_4} A_6^{\lambda_6}),$$

where
$$\lambda_2 - \lambda_4 = 3(\lambda_6 - \lambda_0),$$
$$\lambda_2 + \lambda_3 \leqslant \lambda_0, \quad \lambda_3 + \lambda_4 \leqslant \lambda_0,$$
and, when λ_3 is odd, $\lambda_2+\lambda_3 < \lambda_0$, $\lambda_3+\lambda_4 < \lambda_0$; also $\lambda_6 \geqslant \lambda_0$, and when $\lambda_6 = \lambda_0$, then $\lambda_4 = \lambda_2$, and λ_3 is even. Then, when (X) is not reductible, X is a product of the gradients
$$X_2 = A_0 A_6, \quad X_4 = A_0 A_2 A_4 A_6, \quad X_6 = A_0^2 A_3^2 A_6^2,$$
$$X_{10} = A_0^3 A_2^3 A_6^4, \quad X_{15} = A_0^5 A_2^3 A_3 A_6^6;$$
with the single relation
$$X_{15}^2 = X_2^2 X_6 X_{10}^2$$
indicating the syzygy.

The generating function is the same as for invariants of degree 6; this is as it should be, by Hermite's law of reciprocity.

VIII. *Generating functions.*

37. In a recent paper published in these *Proceedings**, certain generating functions were obtained for invariant forms. Another set is obtained from Theorem I of the present paper as follows.

Let $A_{\delta, r}$ be the number of linearly independent types of the binary q-ic which are of degree δ and order r. Let
$$\phi(y, z) = \Sigma A_{\delta, r} z^\delta y^r.$$
Let w be the weight of the type; then
$$r = q\delta - 2w,$$
and we may write
$$A_{\delta, r} = B_{\delta, w}, \quad \phi = \Sigma B_{\delta, w} z^\delta y^{q\delta - 2w},$$
a form which puts into evidence the fact that those terms of ϕ which have z^δ in common differ by even indices of y only.

Consider the relation between coefficients of z^δ and $z^{\delta+1}$ in the light of Theorem I. We look at the tableaux of the standard forms. There must be no overlapping; the removal of the last set of letters from a standard tableau of degree $\delta+1$ must produce a standard tableau of degree δ. Thus, provided that w is not too large and that $w \geqslant q$,
$$B_{\delta+1, w} = B_{\delta, w-q} + B_{\delta, w-q+1} + \ldots + B_{\delta, w}.$$

* A. Young, "Some generating functions", *Proc. London Math. Soc.* (2), 35 (1933), 425–444.

The limitations arise when $r = q\delta - 2w$ is negative, in which case $B_{\delta, w}$ is zero, and when $w < q$. The last equation may be written

$$A_{\delta+1, r} = A_{\delta, r+q} + A_{\delta, r+q-2} + \ldots + A_{\delta, r-q}.$$

In the limiting cases we have

$$A_{\delta+1, 0} = A_{\delta, q},$$

$$A_{\delta+1, 1} = A_{\delta, q+1} + A_{\delta, q-1},$$

$$\ldots \quad \ldots \quad \ldots \quad \ldots \quad \ldots \quad \ldots$$

$$A_{\delta+1, r} = A_{\delta, q+r} + A_{\delta, q+r-2} + \ldots + A_{\delta, q-r} \quad (r < q).$$

Consider now

$$[1 - (y^q + y^{q-2} + \ldots + y^{-q}) z] \phi(y, z).$$

In the coefficient of $z^{\delta+1}$ all the terms whose y index is greater than q at once disappear. For the rest we have

$$-A_{\delta, 0}(y^{q-2} + \ldots + y^{-q}) - A_{\delta, 1}(y^{q-3} + \ldots + y^{-q+1})$$
$$- \ldots - A_{\delta, r}(y^{q-r-2} + \ldots + y^{-q+r}) - \ldots - A_{\delta, q-1} y^{-1}.$$

Every alternate A is zero, but this does not affect the argument. The first term in ϕ is zy^q, Thus

$$[1 - (y^q + y^{q-2} + \ldots + y^{-q}) z] y\phi(y, z)$$
$$= zy^{q+1} - \sum_{\delta=2}^{\infty} \sum_{r=0}^{q-1} A_{\delta, r}(y^{q-r-1} + y^{q-r-3} + \ldots + y^{-q+r+1}).$$

Replace y by y^{-1} and subtract, then

$$y\phi(y, z) - y^{-1}\phi(y^{-1}, z) = z(y^{q+1} - y^{-q-1}) \left\{1 - z\frac{y^{q+1} - y^{-q-1}}{y - y^{-1}}\right\}^{-1}$$
$$= \sum_{\delta=1}^{\infty} z^{\delta}(y^{q+1} - y^{-q-1})^{\delta}(y - y^{-1})^{-\delta+1}.$$

THEOREM XVII. *The number of linearly independent covariant types of the binary q-ic of degree δ and order r is the coefficient of y^{r+1} in the expansion of*

$$(y^{q+1} - y^{-q-1})^{\delta}(y - y^{-1})^{-\delta+1}.$$

Exactly the same process may be applied to the case when the orders of the ground forms are not all the same. We thus obtain

THEOREM XVIII. *The number of linearly independent covariants of order r, which are of unit degree in the coefficients of each of δ different binary* *quantics whose orders are* $q_1, q_2, \ldots, q_\delta$, *is the coefficient of* y^{r+1} *in the expansion of*

$$(y-y^{-1})^{-\delta+1} \prod_{s=1}^{\delta} (y^{q_s+1}-y^{-q_s-1}).$$

38. Just the same process can be applied to ternary and higher forms. Let A_{δ, r_1, r_2} be the number of concomitant types of the ternary q-ic which are of order r_1 in x, and r_2 in u, and of degree δ. Let

$$\phi(z, y_1, y_2) = \Sigma A_{\delta, r_1, r_2} z^\delta y_1^{r_1} y_2^{r_2}.$$

Then r_1 is the difference between the lengths of the first two rows in the tableau, and r_2 is the difference between the lengths of the second and third rows. As before, we consider the relation between the coefficients of $z^{\delta+1}$ and of z^δ. Provided that neither r_1 nor r_2 is less than q, we have

$$\begin{aligned} A_{\delta+1, r_1, r_2} = \; & A_{\delta, r_1, r_2+q} + A_{\delta, r_1+1, r_2+q-2} + \ldots + A_{\delta, r_1+q, r_2-q} \\ & + A_{\delta, r_1-1, r_2+q-1} + A_{\delta, r_1, r_2+q-3} + \ldots + A_{\delta\ r_1+q-2, r_2-q+1} \\ & + \quad \ldots \quad \ldots \quad \ldots \quad \ldots \quad \ldots \quad \ldots \\ & \ldots \quad \ldots \quad \ldots \quad \ldots \quad \ldots \quad \ldots \quad \ldots \\ & + A_{\delta, r_1-q, r_2}. \end{aligned}$$

Thus we multiply ϕ, in the same way as before, by

$$1 - Zz,$$

where

$$\begin{aligned} Z = \; & y_2^{-q} + y_1^{-1} y_2^{-q+2} + \ldots + y_1^{-q} y_2^{q} \\ & + y_1 y_2^{-q+1} + y_2^{-q+3} + \ldots + y_1^{-q+2} y_2^{q-1} \\ & + \quad \ldots \quad \ldots \quad \ldots \quad \ldots \\ & \ldots \quad \ldots \quad \ldots \quad \ldots \quad \ldots \\ & + y_1^{q} \\ = \; & \Delta_q / \Delta_0, \end{aligned}$$

where

$$\Delta_q = \begin{vmatrix} y_1^{q+2} & y_1^{-q-2} y_2^{q+2} & y_2^{-q-2} \\ y_1 & y_1^{-1} y_2 & y_2^{-1} \\ 1 & 1 & 1 \end{vmatrix}.$$

We must now transform the variables y_1, y_2 to ξ, η, ζ, where

$$\xi = y_1, \quad \eta = y_1^{-1} y_2, \quad \zeta = y_2^{-1},$$

i.e. y_1 is replaced by ξ and y_2 is replaced by $\xi\eta$ in ϕ. Then, when (uv) is used for x, the index of ξ represents the order in u, and the index of η the order in v. Expressed in terms of ξ, η, ζ, the quotient Δ_q/Δ_0 is the sum of homogeneous products of degree q.

The coefficient of $z^{\delta+1}$ in ϕ is

$$\sum_{r,s} A_{\delta+1,r,s} y_1^r y_2^s = \sum_{\rho,\sigma} A_{\delta,\rho,\sigma} y_1^\rho y_2^\sigma \left\{ \sum_{r,s} y_1^{r-\rho} y_2^{s-\sigma} \right\};$$

the last sum is given by the relations which express $A_{\delta+1,r,s}$ in terms of $A_{\delta,\rho,\sigma}$ and which are obtained by the removal of the last set of letters from the tableaux. Let this last set occupy λ, μ, ν places respectively in the first, second, and third rows of the tableau for a form of order r and class s; and, when this set has been removed, let ρ be the order and σ the class of the resulting form. Then

$$\xi^\lambda \eta^\mu \zeta^\nu = y_1^{r-\rho} y_2^{s-\sigma}, \quad \lambda+\mu+\nu = q.$$

This is how the term $y_1^{r-\rho} y_2^{s-\sigma}$ arises in the last sum above. In general this sum is Z, but it is necessary to take care that only such products are retained as represent not more than r removals from the first row and not more than s removals from the second row; otherwise a non-standard tableau has been introduced. It need not trouble us if a term is included which represents the removal of more letters from a row in a particular case than that row contains, for the corresponding $A_{\delta,\rho,\sigma}$ is then necessarily zero. In the coefficients of $-z^{\delta+1}$ in

$$(1 - z\Delta_q/\Delta_0)\,\phi(z, y_1, y_2),$$

we replace each $A_{\delta+1,r,s}$ by its value in terms of $\Sigma A_{\delta,\rho,\sigma}$. Then $A_{\delta,\rho,\sigma}$ appears multiplied by a sum of products $\xi^\lambda \eta^\mu \zeta^\nu$, of total degree q, which are just those products which do not appear in the corresponding

$$A_{\delta+1,\,\rho+\lambda-\mu,\,\sigma+\mu-\nu}$$

terms. That is

$$\lambda > \rho+\lambda-\mu \quad \text{or} \quad \mu > \sigma+\mu-\nu.$$

Hence

$$\rho < \mu \quad \text{or} \quad \sigma < \nu.$$

Thus

$$(1-\Delta_q/\Delta_0)\,\phi(z, \xi, \xi\eta) = z\xi^q - \sum_{\delta=1}^{\infty} z^{\delta+1} \Sigma A_{\delta,\rho,\sigma} \Sigma \xi^{\rho+\sigma+\lambda} \eta^{\sigma+\mu} \zeta^\nu.$$

The second Σ extends to all values of ρ and σ which are not both greater than $q-1$, and the third sum extends to those values of λ, μ, ν for which $\lambda+\mu+\nu$ is q, and μ exceeds ρ, or ν exceeds σ.

Multiply both sides of this equation by $\xi^2\eta$ and then operate with $\{\xi\eta\zeta\}'$. Consider first the terms of the sum on the right for which $\nu > \sigma$;

they may be arranged in pairs, the sum of each pair being annihilated by the operator. A pair is

$$\lambda,\ \mu,\ \nu > \sigma\ :\ \lambda' = \lambda,\ \mu' = \nu - \sigma - 1,\ \nu' = \mu + \sigma + 1.$$

We are left with terms for which $\nu \leqslant \sigma$, $\mu > \rho$; these again can be arranged in annihilated pairs, thus:

$$\lambda,\ \mu > \rho,\ \nu\ :\ \lambda' = \mu - \rho - 1,\ \mu' = \lambda + \rho + 1,\ \nu' = \nu.$$

Thus the sum on the right disappears, and we have

$$\{\xi\eta\zeta\}'\,\xi^2\eta(1 - z\Delta_q/\Delta_0)\,\phi(z,\ \xi,\ \xi\eta) = z\Delta_q.$$

And hence $$\{\xi\eta\zeta\}'\,\xi^2\eta\phi(z,\ \xi,\ \xi\eta) = \Sigma z^\delta \Delta_0^{-\delta+1} \Delta_q^{\ \delta}.$$

The fact that the process is just as applicable to the case when the orders of the δ ground forms are not all the same, as in the case of binary forms, is obvious. We may therefore make the following statement:

THEOREM XIX. *The number of linearly independent concomitants of order ρ and class κ, linear in the coefficients of each of δ ternary quantics of orders $q_1, q_2, \ldots, q_\delta$, is the coefficient of $y_1^{\rho+1} y_2^{\kappa+1}$ in the expansion of*

$$\Delta_0^{-\delta+1} \prod_{s=1}^{\delta} \Delta_{q_s},$$

where $$\Delta_q = \{\xi\eta\zeta\}'\,\xi^{q+2}\eta,$$

and $$\xi = y_1,\quad \eta = y_1^{-1} y_2,\quad \zeta = y_2^{-1}.$$

It is perhaps necessary to add a further remark. We started with the generating function

$$\phi(z,\ y_1,\ y_2) = \phi(z,\ \xi,\ \xi\eta),$$

in which the coefficient of $z^\delta y_1^{\ \rho} y_2^{\ \kappa}$ is the number of independent types of degree δ, order ρ, and class κ, there being no terms where no types correspond, *e.g.* for negative values of ρ and κ. The generating function actually obtained is

$$\{\xi\eta\zeta\}'\,\xi^2\eta\phi(z,\ \xi,\ \xi\eta).$$

This is a sum of six functions like $\xi^2\eta\phi$. The reason why the method is effective is that the terms arising from these six functions fall into six mutually exclusive compartments. Thus, consider a term of $\xi^2\eta\phi$,

$$z^\delta \xi^\lambda \eta^\mu = z^\delta y_1^{\lambda-\mu} y_2^{\ \mu} = z^\delta y_1^{\rho+1} y_2^{\kappa+1}.$$

Then $$\lambda > \mu > 0.$$

This term in the final sum gives rise to six terms, one belonging to each function; the corresponding values of ρ, κ may be written ρ_r, κ_r, where

$$r = 1, 2, \ldots, 6 : \rho_1 = \rho, \ \kappa_1 = \kappa.$$

These values are

$$\rho, \kappa : -\rho-\kappa-3, \ \rho : \kappa, \ -\rho-\kappa-3,$$

$$\rho+\kappa+1, \ -\kappa-2 : -\kappa-2, \ -\rho-2 : -\rho-2, \ \rho+\kappa+1.$$

and thus no confusion arises.

39. The same arguments may be applied when there are p sets of variables. We begin with the function

$$\phi(z, y_1, y_2, \ldots, y_{p-1}) = \Sigma z^\delta A_{\delta, r_1, r_2, \ldots, r_{p-1}} y_1^{r_1} y_2^{r_2} \ldots y_{p-1}^{r_{p-1}},$$

where $A_{\delta, r_1, r_2, \ldots, r_{p-1}}$ is the number of types of degree δ and of orders $r_1, r_2, \ldots, r_{p-1}$ in the different kinds of variables. The concomitants are represented by tableaux of $T_{a_1, a_2, \ldots, a_p}$, where

$$a_1 - a_2 = r_1, \quad a_2 - a_3 = r_2, \quad \ldots, \quad a_{p-1} - a_p = r_{p-1}.$$

Then the variables $y_1, y_2, \ldots$ are changed by the transformation

$$\xi_1 = y_1, \quad \xi_2 = y_1^{-1} y_2, \quad \xi_3 = y_2^{-1} y_3, \quad \ldots, \quad \xi_p = y_{p-1}^{-1},$$

or

$$y_1 = \xi_1, \quad y_2 = \xi_1 \xi_2, \quad \ldots, \quad y_{p-1} = \xi_1 \xi_2 \ldots \xi_{p-1}, \quad 1 = \xi_1 \xi_2 \ldots \xi_p.$$

We write

$$\Delta_q = \{\xi_1 \xi_2 \ldots \xi_p\}' \xi_1^{p+q-1} \xi_2^{p-2} \xi_3^{p-3} \ldots \xi_{p-1}.$$

Then $A_{\delta, r_1, r_2, \ldots, r_{p-1}}$ is the coefficient of $y_1^{r_1+1} y_2^{r_2+1} \ldots y_{p-1}^{r_{p-1}+1}$ in the expansion of

$$\Delta_0^{-\delta+1} \Delta_q^{\ \delta}.$$

And similarly the generating function is obtained for the case when the orders of the ground forms are not all the same.

40. The result of § 37 may be used to obtain another form of generating function. We consider, first, types of even degree 2δ of a binary quantic. Let $A_{q,r}$ be the number of types of order r, when the order of the ground form is q; and let

$$\phi(x, y) = \Sigma A_{q,r} x^q y^r.$$

Then, by Theorem XVII,

$$y\phi(x, y)-y^{-1}\phi(x, y^{-1})$$

$$=\sum_{q=0}^{\infty}(y^{q+1}-y^{-q-1})^{2\delta}(y-y^{-1})^{-2\delta+1}x^q$$

$$=(y-y^{-1})^{-2\delta+1}\sum_{q=0}^{\infty}\left[\sum_{s=0}^{\delta-1}(-)^s\binom{2\delta}{s}\{y^{2(q+1)(\delta-s)}+y^{-2(q+1)(\delta-s)}\}+(-)^\delta\binom{2\delta}{\delta}\right]x^q$$

$$=(y-y^{-1})^{-2\delta+1}\left[\sum_{s=0}^{\delta-1}(-)^s\binom{2\delta}{s}\{y^{2(\delta-s)}(1-xy^{2(\delta-s)})^{-1}+y^{-2(\delta-s)}(1-xy^{-2(\delta-s)})^{-1}\}\right.$$

$$\left.+(-)^\delta\binom{2\delta}{\delta}(1-x)^{-1}\right].$$

Now $y\phi(x, y)$ contains only positive powers of y, and $y^{-1}\phi(x, y^{-1})$ only negative powers. Hence

(XII) $(y-y^{-1})^{2\delta-1}y\phi(x, y)$

$$=\sum_{r=0}^{\delta-1}(-)^r\binom{2\delta}{r}y^{2(\delta-r)}(1-xy^{2(\delta-r)})^{-1}+(-)^\delta\tfrac{1}{2}\binom{2\delta}{\delta}(1-x)^{-1}+\chi(x, y).$$

Here $\chi(x, y)$ is a function with the following properties:

$$\chi(x, y)+\chi(x, y^{-1})=0;$$

it only contains even powers of y; it cannot contain a term in which the index of y is less than $-2\delta+2$. Hence

$$\chi(x, y)=A_1(y^2-y^{-2})+A_2(y^4-y^{-4})+\ldots+A_{\delta-1}(y^{2\delta-2}-y^{-2\delta+2}).$$

The coefficients A are functions of x. Since the expression on the left of (XII) has the factor y^2-1 repeated $2\delta-1$ times, each of the expressions obtained by differentiating the right-hand side of (XII) $0, 1, 2, \ldots, 2\delta-2$ times with respect to y^2 and then putting $y=1$ must vanish; we thus obtain $2\delta-1$ equations satisfied by $A_1, A_2, \ldots, A_{\delta-1}$.

The method is too cumbersome for calculation, but it yields at once the important fact that the denominator of $\phi(x, y)$ is

$$\prod_{r=1}^{\delta}(1-xy^{2r})(1-x)^{2\delta-2},$$

the numerator being a function $\psi(x, y)$, integral and algebraic in x, y, which contains no power of x greater than $3\delta-4$, since the coefficient of $x^{3\delta-3}$ is always zero.

41. When the degree $2\delta+1$ is odd the process is the same, but the form of the result is slightly different; for we then have

$$y\phi(x, y)-y^{-1}\phi(x, y^{-1})$$
$$=(y-y^{-1})^{-2\delta}\sum_{s=0}^{\delta}(-)^s\binom{2\delta+1}{s}\{y^{2\delta+1-\overset{2}{\kappa}s}(1-xy^{2\delta+1-\overset{2}{\kappa}s})^{-1}$$
$$-y^{-2\delta-1+\overset{2}{\kappa}s}(1-xy^{-2\delta-1+\overset{2}{\kappa}s})^{-1}\}.$$

Hence

$$(y-y^{-1})^{2\delta}y\phi(x, y)=\sum_{r=0}^{\delta}(-)^r\binom{2\delta+1}{r}y^{2\delta+1-\overset{2}{\kappa}r}(1-xy^{2\delta+1-\overset{2}{\kappa}r})^{-1}+\chi(x, y),$$

where

$$\chi(x, y)=A_0+A_1(y+y^{-1})+A_2(y^2+y^{-2})+\ldots+A_{2\delta-1}(y^{2\delta-1}+y^{-2\delta+1}).$$

The denominator of ϕ is

$$(1-x^2)^{2\delta-1}\prod_{r=0}^{\delta}(1-xy^{2r+1}).$$

The numerator of ϕ is $\psi(x, y)$ in which the highest index of x is $5\delta-3$, for the coefficient of $x^{5\delta-2}$ is always zero.

42. The following are the values of $\phi(x, y)$ for the first six values of δ:

$$\delta=1,\quad \phi=\frac{1}{1-xy};$$

$$\delta=2,\quad \phi=\frac{1}{(1-x)(1-xy^2)}.$$

For higher values of δ it is sufficient to give the numerator ψ.

$$\delta=3,\quad \psi=1+xy+x^2y^2;$$

$$\delta=4,\quad \psi=(1+xy^2)^2;$$

$$\delta=5,\quad \psi=1+3x^2+x^4+y(4x+2x^3-x^5)$$
$$+y^2(10x^3-5x^4)+y^3(3x+4x^3-7x^5)$$
$$-y^4(-7x^2+4x^4+3x^6)-y^5(-5x^3+10x^5)$$
$$-y^6(-x^2+2x^4+4x^6)-y^7(x^3+3x^5+x^7);$$

$$\delta=6,\quad \psi=1+x+x^2+y^2(8x-x^2-x^3)+y^4(4x+10x^2-11x^3)$$
$$-y^6(-11x^2+10x^3+4x^4)+y^8(x^2+x^3-8x^4)-y^{10}(x^3+x^4+x^5).$$

There is a curious form of symmetry possessed by these functions, namely that the terms appear in pairs with the same coefficients, but in the last two cases with opposite signs, the products of the variables of every pair having a common value. Thus for degree 5 the common value is $x^7 y^7$, and for degree 6 it is $x^5 y^{10}$.

43. The generating function of Theorem XVII, § 37, may be put in another form. The coefficient of y^{r+1} in that function is the number of independent covariant types of order r. Now

$$r = q\delta - 2w.$$

If then we multiply by $y^{-q\delta-1}$ and replace y^{-2} by η, we shall have a function of η such that the coefficient of η^w is the number of independent covariant types of degree δ and of weight w. This generating function is

$$(1-\eta)^{-\delta+1}(1-\eta^{q+1})^{\delta}.$$

When q is infinite we obtain at once the well-known generating function for perpetuant types,

$$(1-\eta)^{-\delta+1}.$$

44. A similar transformation may be applied to the ternary generating function of Theorem XIX. Here we have the number of types of order r_1 and class r_2 expressed as the coefficient of $y_1^{r_1+1} y_2^{r_2+1}$. Now let the weights be w_1, w_2, w_3, *i.e.* the numbers of letters in the three rows of the tableau representing the form. Then

$$r_1 = q\delta - 2w_2 - w_3, \quad r_2 = w_2 - w_3.$$

We multiply the function obtained by $y_1^{-q\delta-1} y_2^{-1}$ and then make the transformation

$$\eta_2 = y_1^{-2} y_2, \quad \eta_3 = y_1^{-1} y_2^{-1}.$$

Hence, in the expansion of

$$\eta_2^{-1} D_0^{-\delta+1} D_q^{\delta},$$

where

$$D_q = \begin{vmatrix} 1 & \eta_2^{q+2} & \eta_3^{q+2} \\ 1 & \eta_2 & \eta_3 \\ 1 & 1 & 1 \end{vmatrix},$$

the coefficient of $\eta_2^{w_2} \eta_3^{w_3}$ is the number of types of degree δ, and of weights $\delta q - w_2 - w_3$, w_2, w_3.

This function may at once be applied to forms of infinite order and we obtain

THEOREM XX. *The number of linearly independent concomitant types, of a ternary form of infinite order, of degree* δ, *having second and third weights* w_2 *and* w_3, *is the coefficient of* $\eta_2^{w_2}\eta_3^{w_3}$ *in the expansion of*

$$(1-\eta_3\eta_2^{-1})(1-\eta_2)^{-\delta+1}(1-\eta_3)^{-\delta+1}.$$

Exactly the same process is applicable to forms with any number of variables. Thus the generating function for quaternary perpetuant types is

$$(1-\eta_3\eta_2^{-1})(1-\eta_4\eta_2^{-1})(1-\eta_4\eta_3^{-1})(1-\eta_2)^{-\delta+1}(1-\eta_3)^{-\delta+1}(1-\eta_4)^{-\delta+1}.$$

IX. *Invariant types of degree* 5.

45. A change in notation helps to bring out the essential points of the argument.

A type of degree δ has been written

$$\{a_1{}^0 a_2^{\lambda_2} \dots a_\delta^{\lambda_\delta}\}.$$

When the identity of the different quantics is immaterial, this may be written

$$\{0\ \lambda_2\ \lambda_3\ \dots\ \lambda_\delta\}.$$

Such a form will be used even with permutations; thus

$$\{a_r a_s\}\{0\ \lambda_2\ \lambda_3\ \dots\ \lambda_\delta\}$$

will indicate the sum of the type and that obtained by interchanging the quantics a_r and a_s, these being those which appear with the indices λ_r and λ_s.

The symbol $\bar{I}$ indicates the complete inversion of the sequence of quantics, thus

$$\bar{I}\{a_1{}^0 a_2^{\lambda_2} \dots a_\delta^{\lambda_\delta}\} = \{a_\delta{}^0 a_{\delta-1}^2 \dots a_1^{\lambda_\delta}\}.$$

The peculiarity of invariants lies in the fact that the two rows of the tableau may be interchanged; such interchange is accompanied with a change of sign when w, the number of letters in a row, is odd.

Thus, for invariants,

$$\{0\ \lambda_2\ \dots\ \lambda_\delta\} = (-)^w\bar{I}\{0\ \lambda_\delta-\lambda_{\delta-1}\ \lambda_\delta-\lambda_{\delta-2}\ \dots\ \lambda_\delta\},$$

the numbers $\lambda_2, \ldots, \lambda_\delta$ being in ascending order of magnitude. From our point of view the essential thing is the set of differences between the consecutive numbers $0, \lambda_2, \ldots, \lambda_\delta$. We then write

$$i_r = \lambda_{r+1} - \lambda_r,$$

and define the form in terms of these *intervals*,

$$[i_1, i_2, i_3, \ldots, i_{\delta-1}].$$

It is convenient to use Sylvester's term *excess* of a gradient for

$$\delta\lambda - 2w,$$

where δ is its degree, λ its extent (*i.e.* the maximum weight of a coefficient contained), and w its weight. Then the excess of

$$\{0 \ \lambda_2 \ \ldots \ \lambda_\delta\}$$

is
$$E_\delta = \delta\lambda_\delta - 2w.$$

We shall require also the excess for part of the form, and so shall use E_r for the excess of $\{0 \ \lambda_2 \ \ldots \ \lambda_r\}$. Then it is not difficult to show that the excess in terms of the intervals is

$$E_\delta = (\delta-2)(i_{\delta-1} - i_1) + (\delta-4)(i_{\delta-2} - i_2) + \ldots.$$

The law of reductibility of types is taken here to be the same as for covariants for a single form according to sequence (A), Then a reduction is a relation

$$[i_1, i_2, \ldots, i_{\delta-1}] = \Sigma\, Bs\, [j_1, j_2, \ldots, j_{\delta-1}],$$

where s is a permutation of quantics, B is numerical, and the first of the differences $i_{\delta-1} - j_{\delta-1}$, $i_{\delta-2} - j_{\delta-2}$, ... which does not vanish is positive.

46. Consider then

$$[i, i_2, i_3, i_4],$$

where
$$E_5 = 3(i_4 - i) + i_3 - i_2 = 0.$$

Then there is always a reduction when E_r $(r < \delta)$ is negative. Thus

$$\tfrac{1}{2}E_4 = i_3 - i = \eta \geqslant 0,$$

$$E_3 = i_2 - i = \xi \geqslant 0.$$

For invariants,

(XIII) $$[i, i_2, i_3, i_4] = (-)^w \bar{I}[i_4, i_3, i_2, i].$$

Thus the form is reductible when the first of the differences i_4-i, i_3-i_2 which does not vanish is positive.

Also, when the dexter is reductible according to sequence (B), the sinister is reductible sequence (A). Hence, by Theorem VIII, there is a reduction when any i_r is less than i_4. This does not, however, yield any fact not obtained otherwise for degree 5, but it is useful for higher degrees.

When there is a relation of the form

$$\text{(XIV)} \qquad \Sigma\, Bs\,[i_1, i_2, \ldots, i_{\delta-1}] = R,$$

where R is a linear function of earlier forms, either (i) according to sequence (A) or (ii) according to sequence (B), then a similar relation is satisfied by

$$\text{(XV)} \qquad [i_1+\mu, i_2+\mu, \ldots, i_{\delta-1}+\mu],$$

the sign of every transposition in s being altered when μ is odd.

This follows at once from the equations of § IV, for $\delta < 5$; that it is true for invariants of degree 5 follows from equation (XIII), which is all that is required here. I am convinced that the above statement is true for all values of δ, but cannot give a formal proof. When the intervals given refer to the leading gradient instead of to the substitutional form, its truth is clear. For, if we take K to represent the symbolical product form of (XIV), then that of (XV) may be taken as

$$K \,.\, \prod_{r>s} (a_r a_s)^{\mu},$$

whether the sequence is (A) or (B).

We need then consider only the cases

$$i_4 = 0, \quad i_3 = 4i+\xi, \quad i_2 = i+\xi,$$

that is the invariant

$$C = [i \quad i+\xi \quad 4i+\xi \quad 0].$$

When $i > 0$, $\xi > 1$, the only relations of reduction for C are (see § IV)

$$[1-(-)^i(a_1\, a_2)]\, C = R, \quad [1-(a_4\, a_5)]\, C = R.$$

That is, among the 5! functions obtained from C by permutation of quantics thirty are independent. Taking the permutations according to this analysis, we may represent the functions as

$$(i \text{ even}) \qquad 1\,.\,T_5 + 2\,.\,T_{4,1} + 2\,.\,T_{3,2} + 1\,.\,T_{3,1^2} + 1\,.\,T_{2^2,1},$$

$$(i \text{ odd}) \qquad 1\,.\,T_{4,1} + 1\,.\,T_{3,2} + 2\,.\,T_{3,1^2} + 1\,.\,T_{2^2,1} + 1\,.\,T_{2,1^3}.$$

It will be remembered that $\lambda T'_{\alpha_1 \dots}$ stands for $\lambda f_{\alpha_1 \dots}$ forms belonging to this T*.

The form is $\{0\ i\ 2i+\xi\ 6i+2\xi\ 6i+2\xi\}$, and the diagram

$$\begin{matrix} i & i+\xi & 2i+\xi & 2i \\ & i & 2i+\xi & 2i+\xi \\ & & 2i & 2i+\xi \\ & & & 0 \end{matrix}$$

shows that there exists an invariant, with a corresponding leading gradient, which exactly fulfils the conditions.

When $i > 0$, $\xi = 0$:

i even, C belongs to $\{a_1 a_2 a_3\}$ and to $\{a_4 a_5\}$; substitutionally it contains

$$1 . T_5 + 1 . T_{4,1} + 1 . T_{3,2}.$$

i odd, C belongs to $\{a_1 a_2 a_3\}'$ and to $\{a_4 a_5\}$; substitutionally it contains

$$1 . T_{3,1^2} + 1 . T_{2,1^3}.$$

The same diagram may be used as before to establish the existence of the invariant.

When $i > 0$, $\xi = 1$, C obeys the same relations as

$$\begin{Bmatrix} a_1 a_2 \\ a_3 \end{Bmatrix} \quad \{a_4 a_5\},$$

and the forms are

$$1 . T_{4,1} + 1 . T_{3,2} + 1 . T_{3,1^2} + 1 . T_{2^2,1}.$$

When $i = 0$, $\xi > 1$.

$$[1 - (-)^w (a_1 a_5)(a_2 a_4)]\, C = 0,$$

the forms are

w even, $\quad 1 . T_5 + 1 . T_{4,1} + 1 . T_{3,2} + 1 . T_{2^2,1}$;

and w odd, $\quad 1 . T_{4,1} + 1 . T_{3,2} + 1 . T_{3,1^2}$.

Finally, when $i = 0$, $\xi = 1$, there is one form

$$(0 \quad 0 \quad 1 \quad 2 \quad 2)$$

---

* See Q.S.A., III, § 6.

of the nature $T_{3,1^2}$; for this we may take the diagram

$$\begin{matrix} 0 & 1 & 1 & 0 \\ & 0 & 1 & 1 \\ & & 0 & 1 \\ & & & 0. \end{matrix}$$

47. The results now may easily be verified with the generating functions. The invariant is of the form

$$[i+\mu,\quad i+\mu+\xi,\quad 4i+\mu+\xi,\quad \mu],$$

or, in the other notation,

$$\{0 \quad i+\mu \quad 2i+2\mu+\xi \quad 6i+3\mu+2\xi \quad 6i+4\mu+2\xi\}.$$

The generating function is $\Sigma A_r x^r$, where $r = 6i+4\mu+2\xi$; following Elliott (*loc. cit.*, 138), we call r the *extent* of the invariant. Its form corresponds to the common denominator of the functions

$$D = (1-x^4)(1-x^6)(1-x^8);$$

the indices corresponding to the coefficients of μ and ξ are double their coefficients, owing to the change of sign of transpositions when μ is odd, and the peculiarity of the case $\xi = 1$.

When the forms corresponding to the various T's in the last paragraph are picked out, the following generating functions are obtained; the numerator alone is given in each case, the common denominator being D.

$$\begin{array}{llll} T_5 & : \; 1-x^6+x^{12}. & T_{4,1} & : \; x^4(1+x^2)(1+x^4). \\ T_{3,2} & : \; x^4(1+x^2+2x^4+x^8), & T_{3,1^2} & : \; x^2(1+x^4+x^8)(1+x^4). \\ T_{2^2,1} & : \; x^4(1+2x^4+x^6+x^8). & T_{2,1^3} & : \; x^6(1+x^2)(1+x^4). \\ T_{1^5} & : \; x^4(1-x^6+x^{12}). & & \end{array}$$

If G and G' are the generating functions for two T's obtained from each other by interchanging rows and columns (*e.g.* $T_{3,2}$ and $T_{2^2,1}$), then

$$xG(x) = x^{-1}G'(x^{-1}).$$

The sum $$\Sigma f_{a_1,\ldots} G_{a_1,\ldots} = (1-x^2)^{-3}(1+3x^2+x^4),$$

as it should be by § 42.

This shows that no reduction has been omitted, that the forms included are irreductible, and that the generating functions obtained are correct;

this also follows from the fact that leading gradients correspond to the forms obtained.

X. *Covariants of degree* 5, *of a single quantic.*

48. As has been proved (§ 29), the covariants of order greater than $\delta-2$ and of degree δ can be written down at once when those of lower degree are known. It is only necessary then to find those whose order is not greater than $\delta-2$. For degree 5 I have shown that all covariants are reductible other than those given, but for reasons of space I do not give the proofs here; the diagrams, by which it is seen that leading gradients correspond to the forms, are given.

Covariants of order 1:

$$C_5^{1}=\{0\quad 0\quad 2\quad 5\quad 5\};\qquad\qquad C_7^{1}=\{0\quad 0\quad 3\quad 7\quad 7\};$$

$$\begin{array}{c} 0\quad 1\quad 2\quad 2 \\ 1\quad 2\quad 2 \\ 1\quad 1 \\ 0 \end{array}\qquad\qquad\qquad\qquad \begin{array}{c} 0\quad 2\quad 2\quad 3 \\ 1\quad 3\quad 2 \\ 2\quad 2 \\ 0 \end{array}$$

$$C_{11}^{1}=\{0\quad 2\quad 4\quad 10\quad 11\};\qquad\qquad C_{13}^{1}=\{0\quad 2\quad 4\quad 13\quad 13\};$$

$$\begin{array}{c} 2\quad 2\quad 4\quad 3 \\ 2\quad 3\quad 4 \\ 3\quad 4 \\ 0 \end{array}\qquad\qquad\qquad\qquad \begin{array}{c} 2\quad 2\quad 5\quad 4 \\ 2\quad 4\quad 5 \\ 4\quad 4 \\ 0 \end{array}$$

$$C_{17}^{1}=\{0\quad 2\quad 7\quad 16\quad 17\}.$$

$$\begin{array}{c} 2\quad 4\quad 5\quad 6 \\ 3\quad 6\quad 6 \\ 5\quad 5 \\ 0 \end{array}$$

The remaining forms are obtained by the addition of the schemes for the invariants I_4, I_8, I_{12} (§ 33) to those of one or other C^1. Incidentally, we see that

$$I_4+C_{13}^{1}=I_{12}+C_5^{1}.$$

Also the addition of I_{18} only produces forms obtainable otherwise.

The generating function is

$$x^5\{(1-x^2)(1-x^6)(1-x^8)\}^{-1}.$$

Covariants of order 2:

$$C_2^2=\{0\ \ 0\ \ 0\ \ 2\ \ 2\};\qquad C_6^2=\{0\ \ 0\ \ 3\ \ 5\ \ 6\};$$

$$\begin{array}{cccc} 0 & 0 & 1 & 1 \\ & 0 & 1 & 1 \\ & & 0 & 0 \\ & & & 0 \end{array}\qquad\qquad \begin{array}{cccc} 0 & 2 & 2 & 2 \\ & 1 & 2 & 2 \\ & & 1 & 2 \\ & & & 0 \end{array}$$

$$C_8^2=\{0\ \ 0\ \ 3\ \ 8\ \ 8\};\qquad C_{12}^2=\{0\ \ 2\ \ 4\ \ 11\ \ 12\};$$

$$\begin{array}{cccc} 0 & 2 & 3 & 2 \\ & 1 & 3 & 3 \\ & & 2 & 3 \\ & & & 0 \end{array}\qquad\qquad \begin{array}{cccc} 2 & 2 & 4 & 4 \\ & 2 & 4 & 4 \\ & & 3 & 4 \\ & & & 0 \end{array}$$

$$C_{12t-2}^2=\{0\ \ 0\ \ 8t-2\ \ 10t-2\ \ 12t-2\};$$

$$\begin{array}{cccc} 0 & 4t-1 & 4t-1 & 4t-1 \\ & 4t-1 & 4t-1 & 4t-1 \\ & & 2t & 2t \\ & & & 2t \end{array}$$

$$C_{12t+16}^2=\{0\ \ 0\ \ 8t+9\ \ 10t+14\ \ 12t+16\}$$

$$\begin{array}{cccc} 0 & 4t+5 & 4t+5 & 4t+5 \\ & 4t+4 & 4t+5 & 4t+6 \\ & & 2t+4 & 2t+3 \\ & & & 2t+2 \end{array}$$

As before, the other forms are obtained from I_4, I_8, I_{12} by additions.

The number sets are connected by the relations

$$I_{12}+C_{12t-2}^2=4I_4+I_8+C_{12(t-1)-2}^2,$$
$$I_{12}+C_{12t+16}^2=4I_4+I_8+C_{12(t-1)+16}^2.$$

The generating function is

$$(x^2+x^6+x^{10})\{(1-x^4)(1-x^6)(1-x^8)\}^{-1}.$$

Covariants of order 3. There is just one condition. The irreductible covariants of order 3 are those forms

$$\{0 \quad \lambda_2 \quad \lambda_3 \quad \lambda_4 \quad \lambda_5\}$$

of order 3 for which $2\lambda_4 - \lambda_3 - \lambda_5 \geqslant 0$, and which are irreductible for degree 4.

In every case a leading gradient corresponds.

The generating function is

$$x^3\{(1-x^2)(1-x^4)(1-x^6)\}^{-1}.$$

The forms may be given by addition of I_8 to

$$(0 \quad 0 \quad 4\sigma+\rho-3 \quad 5\sigma+2\rho-3 \quad 6\sigma+2\rho-3),$$

$$(0 \quad 2\lambda \quad 4\lambda+\rho \quad 12\lambda+2\rho+3 \quad 12\lambda+2\rho+3),$$

$$(0 \quad 2\lambda \quad 4\lambda+\rho \quad 12\lambda+2\rho \quad 12\lambda+2\rho+1).$$

It is unnecessary to give the corresponding diagrams, which may easily be written down.

XI. *The invariants of the quaternary cubic.*

49. A simple example of the general method is furnished by the invariants of a single quaternary cubic. Here the tableau has four rows of equal length. Let δ be the degree of the invariant; then the tableau contains 3δ letters and δ must be a multiple of 4. The letters occur in sets of three.

Let us suppose that there are a_1 sets which have all three letters in the first row, and that there is another set $aa'a''$ which has two letters in the first row and the other a in the second row. Then, if G is the positive symmetric group of the $3(a_1+1)$ letters of these sets, and PNX is the substitutional form obtained from the tableau*,

$$GPNX = \Sigma\, P'N'X,$$

where in $P'N'X$ all the letters of G are in the first row, and the other rows are the same as in PNX, except for the single letter a.

The left-hand side of this equation is simply a sum of terms, which differs from PNX only by an interchange of equivalent sets. Thus, when two letters of a set are in one row and the third in another row, the form

* Q.S.A., III, § 10.

may be expressed in terms of forms which have more sets completely in that row. This is regarded as a reduction. By repeated application of this reduction, we may express any invariant in terms of those whose tableaux only contain sets of two kinds:

(i) Sets with all three letters in the same row.

(ii) Sets with all three letters in different rows.

Let a_r be the number of sets in the r-th row; and let β_r be the number of sets of the second kind, which contain no letter in the r-th row. Now let G be the positive symmetric group of all the letters of the $\sum_{r=1}^{4} \beta_r$ sets of the second kind. We use the equation

$$GPNX - \Sigma P'N'X,$$

where all the letters of G are moved into the first row in every term on the right. When there are more letters in G than in a row the right-hand side is zero.

Consider the left-hand side. It is the sum of terms obtained by interchange of all the letters of G in PNX. In some terms, some of these sets of G are no longer of the second kind, and these either have a greater number of sets of the first kind or are equal to forms with a greater number of sets of the first kind. An increase in the number of sets of the first kind is regarded as a reduction. In the remaining terms the sets of G are all of the second kind, and, the number of places in each row being unaltered, it is easy to see that the numbers β_1, β_2, β_3, β_4 are necessarily unchanged; that is the terms differ from PNX by an interchange of equivalent sets.

On the right there are $\Sigma \beta_r$ sets with all three letters in the first row, and the a_2, the a_3, and the a_4 sets are unchanged; hence there is a reduction when

$$\Sigma \beta_r > a_1.$$

Again, in the equation used, the sign* of the terms on the right is plus or minus according as the number of letters moved up into the first row is even or odd. Thus there is still a reduction when

$$\Sigma \beta_r = a_1,$$

and this sign is minus, that is $3\beta_1+2\beta_2+2\beta_3+2\beta_4$ is odd; that is β_1 must be odd.

---

* *Loc. cit.*

Any other row might be taken instead of the first, with a similar result. Thus, for an irreductible form,

$$\alpha_s \geqslant \Sigma\beta,$$

and if $\alpha_s = \Sigma\beta$, β_s is even.

The number of letters in the first row is

$$3\alpha_1+\beta_2+\beta_3+\beta_4 = \tfrac{3}{4}\delta.$$

Hence

$$3\alpha_1-\beta_1 = 3\alpha_2-\beta_2 = 3\alpha_3-\beta_3 = 3\alpha_4-\beta_4.$$

When $\frac{1}{4}\delta$ is odd, the interchange of a pair of rows changes the sign; if then $\alpha_1 = \alpha_2$ in this case, the first two rows are identical and the form is zero. Thus, when $\frac{1}{4}\delta$ is odd, no two of the numbers α can be equal. These facts give the whole theory.

We write an invariant in the notation

$$\begin{Bmatrix} \alpha_1 & \alpha_2 & \alpha_3 & \alpha_4 \\ \beta_1 & \beta_2 & \beta_3 & \beta_4 \end{Bmatrix},$$

and we confine ourselves to forms whose schemes cannot be expressed as the sum of the schemes of forms already obtained. Then we have

$$I_8 = \begin{Bmatrix} 2 & 2 & 2 & 2 \\ 0 & 0 & 0 & 0 \end{Bmatrix},$$

$$I_{16} = \begin{Bmatrix} 4 & 3 & 3 & 3 \\ 3 & 0 & 0 & 0 \end{Bmatrix},$$

$$I_{24} = \begin{Bmatrix} 5 & 5 & 5 & 5 \\ 1 & 1 & 1 & 1 \end{Bmatrix},$$

$$I_{32} = \begin{Bmatrix} 7 & 7 & 6 & 6 \\ 3 & 3 & 0 & 0 \end{Bmatrix},$$

$$I_{40} = \begin{Bmatrix} 8 & 8 & 8 & 8 \\ 2 & 2 & 2 & 2 \end{Bmatrix},$$

$$I_{48} = \begin{Bmatrix} 10 & 10 & 10 & 9 \\ 3 & 3 & 3 & 0 \end{Bmatrix},$$

$$I_{100} = \begin{Bmatrix} 22 & 21 & 20 & 19 \\ 9 & 6 & 3 & 0 \end{Bmatrix}.$$

In adding schemes, a convention about the identity of rows is necessary; this is effected by agreeing that $\alpha_1 \geqslant \alpha_2 \geqslant \alpha_3 \geqslant \alpha_4$.

It is a simple arithmetical problem to show that all schemes not reductible according to the rules above can be obtained as sums from the above seven. Moreover, we have

$$2I_{100} = 2I_{48} + 2I_{32} + 2I_{16} + I_8,$$

which indicates a syzygy for I_{100}^2.

50. Wakeford* has shown that the quantic of order n in any number of variables x_r, in which the coefficients of all terms

$$x_r^{n-1} x_s$$

are zero, is canonical.

We consider then a quaternary cubic of this form, which exactly corresponds to the substitutional schemes used here.

Consider the complex

$$C_{4,p}^6 = \begin{Bmatrix} 2 & 2 & 0 & 0 \\ 0 & 0 & 0 & 0 \end{Bmatrix}$$

(which resembles the invariant of a binary cubic).

Let $$D_{p,q} = \Sigma \frac{\partial^2}{\partial p_r \partial q_r}.$$

Then $$D_{p,q}^6 C_{4,p}^6 C_{4,q}^6$$

is an invariant of the form I_8, and its leading gradient is that of the scheme for I_8, the sequence of gradients following the convention about sequence of rows in a scheme. Let

$$E_{xu} = \Sigma \frac{\partial^2}{\partial x_r \partial u_r}.$$

Also let f be the cubic and let

$$\Delta_{4u} = (abcu)(bcdu)(cdau)(dabu).$$

Then the leading gradient of

$$E_{xu}^{12} f^4 (\Delta_{4u})^3$$

is given by I_{16}.

Consider the complex

$$C_{12,p}^{14} = \begin{Bmatrix} 5 & 5 & 0 & 0 \\ 1 & 1 & 0 & 0 \end{Bmatrix}.$$

* *Proc. London Math. Soc.* (2), 18 (1920), 403–410 (408).

Then $$D^{14}_{p,q}\,C^{14}_{12,p}\,C^{14}_{12,q}$$
is an invariant whose leading gradient is given by I_{24}. For I_{32} we use in the same way the complexes

$$C^{18}_{20,p}=\begin{Bmatrix}7 & 7 & 0 & 0\\ 3 & 3 & 0 & 0\end{Bmatrix},$$

$$C^{18}_{12,p}=\begin{Bmatrix}6 & 6 & 0 & 0\\ 0 & 0 & 0 & 0\end{Bmatrix},$$

and take the invariant

$$D^{18}_{p,q}\,C^{18}_{20,p}\,C^{18}_{12,q}.$$

For I_{40} we use $$\begin{Bmatrix}8 & 8 & 0 & 0\\ 2 & 2 & 0 & 0\end{Bmatrix}.$$

Up to this point we are assured that the invariants actually exist, and that no reduction has been missed. Later the difficulties increase. But for I_{48} an indirect proof can be given. By addition of the schemes,

$$2I_{24}=I_8+I_{40}.$$

In other words, using the invariants in the sense of those given by the leading gradients,

$$I_{24}^2-I_8\,.\,I_{40}$$

is an invariant with a later leading gradient.

It cannot be zero, for then I_8 must be reducible or a factor of I_{24}, neither of which is possible. It must then be I_{48}, which consequently exists and is reducible. The direct verification of the existence of I_{100} I have not carried out. But we conclude that there are six irreducible invariants of the quaternary cubic. That of the highest degree is *gauche* in the sense used for the invariant of degree 18 of the binary quantic, since, in that case, its square is expressible in terms of the other invariants. There is thus one syzygy.

XII. *Concluding remarks.*

51. Certain conclusions of a general nature may be mentioned, as the leading results of the foregoing work. And in connection with them, I would call attention to certain results which, as yet, I have been unable to prove; but they have been verified in so many cases that I feel justified in recording them as empirical theorems. In the first place, substitutional analysis and the tableau notation provide a means of

writing any concomitant of quantics in any number of variables as a function of a certain number of arguments; each argument is a positive integer and the number of arguments is, in the first place, that of the coefficients of the ground forms. The concomitants thus defined are not all independent, and, indeed, the set of functions may be limited to those in which certain of the arguments are zero. There is a (1, 1) correspondence between arguments and coefficients, and it would appear that there is a connection between coefficients which may be omitted in a canonical form and arguments which may be omitted as zero, at least in the case of invariants.

There are still relations among the functions when they are limited to those which have these particular arguments zero. And the first object has been to define exactly the arguments as sets of integers, so that they give a linearly independent set of concomitants. This has been done completely for covariants of a binary form up to degree 5, and in certain other cases.

When the arguments are such a set, the form is called irreductible, otherwise it is called reductible.

Now, in all cases considered, the sets are found to be additive. That is, the addition of arguments of two irreductible forms gives an irreductible form. This, indeed, must necessarily follow from the fact that the relations which fix the sets are Diophantine inequalities. The arguments may be said to belong to a compound additive " field " in the sense of the definition of Dickson*.

When it is remembered that such a form obtained by addition of arguments from two others has the same characteristics, such as degree, order, and so on, as the product of these other forms, one is tempted to look upon it as actually the product, and to regard the set of functions as providing a kind of logarithmic theory of forms, in the sense that products are represented by a sum of arguments. But investigation shows that this is not the case in general for the substitutional forms as introduced. However, in the sixth paper on this subject, the substitutional units have been transformed into normal and into semi-normal units. And I have established a theorem (not yet published) which shows that, at least in the binary case, the suggested result is actually true when normal or semi-normal units are used as the basis of the covariant functions instead of the natural units. I know of no reason to anticipate insurmountable difficulty in extending the theorem to the general case. This must be left for a future communication.

---

* *Linear groups* (1901), 5.

The introduction of the notion of the leading gradient of a concomitant provides another way of proceeding to a like result.

It is to be remembered that both leading gradient and field of arguments presuppose the definition of an agreed sequence, and this sequence is naturally taken to be the same in both cases.

EMPIRICAL THEOREM I. *In general, the complete set of leading gradients is defined in the same way as the complete set of irreductible forms.*

I believe that it is possible to construct a sequence law for which this is not the case, but that such a law is exceptional. The theorem has been proved true for binary forms with sequence (C). It has stood all tests that have been applied in all cases so far. It is certainly true for sequence (A) when the degree is not greater than 5; also I have proved it for degree 6 except for covariants of order 4 where the work is not yet complete. It is true also for sequence (A), for degree δ when the order is greater than $\delta-2$, provided that it is true for degree $\delta-1$. This suggests the definition of the forms by their leading gradients, but as different forms have a common leading gradient something more is needed, in addition to the discovery of what gradients are to be included.

I believe that the semi-normal form which follows naturally from sequence (A) actually has the leading gradient defined by its arguments.

Another advantage of the semi-normal forms arises from the fact that the product of a permutation which affects no letter later than a_r with a semi-normal unit Q is expressed in terms of units which are the same as Q so far as letters later than a_r are concerned. It follows that the expression of the binary semi-normal form as continued transvectants, as in Theorem XIII, contains just one term; *i.e.* it is the continued transvectant whose transvection indices are the arguments of the form.

52. A point of less general interest is touched on in §46. It may be stated thus:

EMPIRICAL THEOREM II. *A permutational equation of reduction for a covariant type of a binary form, defined as in* §45 *by the intervals* $i_1, i_2, \ldots,$ —*when the sequence is* (A) *or* (B)—*still holds good when all the intervals are increased by the same arbitrary number; when, however, this number is odd the sign of every transposition in the equation must be changed.*

This theorem is true, as was shown, for the leading gradients, and hence follows from the first theorem, provided that that theorem is true.

53. An invariant of a binary quantic is a sum of products of the differences of the roots; a covariant is a sum of products of the differences of the roots and of differences of the variable and the roots. This leads to a substitutional expression for invariants and covariants of a nature exactly similar to that already developed. If the quantic is of order n and the covariant of degree δ, there are $n\delta$ letters in the substitutional form, as before, but now they are grouped into n sets of δ letters each, each set representing one of the n roots. Thus we have two different expressions for a covariant, one by means of the symbolical calculus and the coefficients, in which the $n\delta$ letters appear in δ equivalent sets, each set containing n letters which represent one letter of that calculus. In the other expression, the $n\delta$ letters appear in n sets of δ letters. each set representing one root of the form. In both cases, the form belongs to the same substitutional set $T_{n\delta-w,\,w}$. This, of course, leads at once to Hermite's law of reciprocity. According to the first empirical theorem, the irreductible form—sequence (A)—

$$\{0\ \lambda_2\ \ldots\ \lambda_\delta\}$$

has for its leading gradient

$$\Pi\, A_{\lambda_r}.$$

This may be expressed in terms of a sum of products of the roots, which sum may be taken to begin with a term which we write

$$\begin{array}{cccccc} a_1 & a_2 & \cdots & \cdots & \cdots & a_{\lambda_\delta} \\ a_1 & a_2 & \cdots & \cdots & a_{\lambda_{\delta-1}} & \\ \cdots & \cdots & \cdots & \cdots & & \\ a_1 & a_2 & \cdots & a_{\lambda_2}, & & \end{array}$$

where a_1, a_2, ... are the roots in a definite sequence. The rows of this tableau indicate the leading gradient. The columns give us the term in the roots

$$a_1^{\delta-1}\, a_2^{\delta-1} \ldots a_{\lambda_2}^{\delta-1}\, a_{\lambda_2+1}^{\delta-2} \ldots.$$

This is the first term according to a sequence (D) of a covariant form in the differences of the roots. The sequence (D) is the reciprocal of the sequence (A), and is defined by the law that $\{\lambda_1 \lambda_2 \ldots \lambda_\delta\}$ precedes $\{\mu_1 \mu_2 \ldots \mu_\delta\}$ when the first of the differences

$$\lambda_1 - \mu_1, \quad \lambda_2 - \mu_2, \quad \ldots$$

which does not vanish is negative.

The sequence (C) is the reciprocal of the sequence (B).

When a covariant is expressed in terms of the coefficients as K, and in terms of the roots as L, we must look for the relations between K and L to be developed in such a way that they follow reciprocal sequences; either (A) and (D) or (B) and (C). The actual forms irreductible for sequence (D) are the same as those for sequence (C), the argument being the same, and, further, the forms represent leading gradients in both cases. The above is merely an indication of what is to be expected. Development on these lines should follow directly from the use of semi-normal units. Obviously, the expression for the discriminant in terms of the roots is, sequence (A) or (B),

$$\{0\ 2\ 4\ \dots\ 2n-2\}.$$

In terms of the coefficients this is the catalecticant of the $(2n-2)$-ic.

Reciprocally the expression, sequence (C) or (D), is

$$\{1,\ 1,\ \ 2,\ 2,\ \ \dots,\ \ n-1,\ n-1\}.$$

From the well-known determinant expression for the discriminant it is easy to see that this gives its leading gradient according to sequences (C) or (D). Two reciprocal expressions, one in the coefficients and one in the roots, are not in general equivalent, at least for natural units. They will differ by earlier forms of one or other sequence.

ON QUANTITATIVE SUBSTITUTIONAL ANALYSIS

(Eighth Paper.)

By Alfred Young

[Received 25 January, 1933.—Read 19 January, 1933.]

The first part of this paper, Sections I–VII, continues the general theory from the sixth paper. The last part, Sections VIII, IX, has to do with applications to the algebra of invariants in continuation of the seventh paper. The theory in the first part is developed mainly with a view to the applications made later; but I believe it to be of interest in itself.

The semi-normal group matrix introduced in Q.S.A., VI, presupposed a definite sequence of tableaux, taken to be that which was defined as the sequence from the last letter. In Section I it is shown that other sequences will do equally well, and lead to the same seminormal units. In fact, a whole class of sequences is defined as *principal sequences* for which this is true.

In Section II very simple expressions are found for those seminormal matrices which are unchanged on multiplication on the right or on the left by the permutations of the symmetric group of a set of consecutive letters; or else are changed in sign by the odd permutations. It is interesting to find that the matrix elements depend only on the tableau functions. And, further, that the numerical average of all the elements, in either case, is simply expressible in terms of the tableau functions of the first and last tableaux of the whole set.

It is important for applications to invariants, and I think of general interest, to find the expression for any seminormal unit of a set of consecutive letters in terms of the units of the whole. This problem is attacked in Section III, but it has only been solved in special cases. It leads at once to the notion of a distorted tableau, where the upper rows are bodily shifted to the right. In Section IV the special case when the tableaux contain two rows only is discussed and solved. This is the case needed for binary covariants. In Section V there are introduced certain simple algebraic forms which represent the seminormal units, in that for a given unit we can put down a form which has the substitutional properties of the unit for left-hand multiplication and no others. Theorem XV, which connects the first term of such a form directly with the tableau, is of importance for its applications. Section VI obtains the matrix which transforms the regular group matrix of a symmetric group into its exact seminormal irreducible components.

Frobenius introduced a double group matrix $[x_{PQ^{-1}}+y_{Q^{-1}P}]$, and showed that it possessed the same number of irreducible components as the ordinary group matrix, but that the order of each irreducible matrix was f^2 instead of f. In Section VII the transformation effecting this reduction is discussed, and it is shown that the double group matrix has a remarkably simple form in a limited number of cases.

Section VIII deals with the expression of concomitants by means of seminormal units; it is shown that for concomitants of a single quantic the sets of numbers defining the concomitant and the leading gradient are identical. This establishes what was merely put forward as an empirical theorem in Q.S.A., VII. This may be regarded as the fundamental theorem of this presentation of the subject. One immediate result of it is that the exhibition of a form with a given leading gradient is proof of the irreducibility of the corresponding concomitant, whatever principal sequence is chosen. A second result is that a form whose leading gradient is the product of two leading gradients may be regarded as reducible. Section IX deals with applications to binary forms.

I. *Principal sequences.*

1. The seminormal matrix for $T_{\alpha_1\alpha_2\ldots\alpha_h}$ was obtained by means of the quadratic invariant H of the symmetric group. The variables were written $x_1, x_2, \ldots, x_f$, the numbers referring to the tableaux $F_1, F_2, \ldots, F_f$; and the variables were connected to the natural units P_r, N_r, M_r defined by these tableaux. The first step was taken in § 6 of the sixth paper by expressing H as a sum of squares. This was done by first collecting all the terms which contained x_f and so obtaining

$$H = H' + A_f X_f^2,$$

where H' does not contain x_f.

Thus the actual form of the sum of squares, and hence also the seminormal matrix, depends on the sequence of tableaux. In general, when a different sequence is taken from that adopted, namely that which has been styled the sequence from the last letter, a different seminormal matrix will be obtained; and one, in fact, for which the very simple form for the transpositions of consecutive letters, obtained in the paper referred to, no longer holds good.

It may happen that an alteration in sequence does not alter the matrix except by interchanges of rows, and at the same time of columns. We write

$$H = \sum_{r=1}^{f} A_r X_r^2;$$

then, if X_{r+1} does not contain the variable x_r, the simple interchange of F_r and F_{r+1} in the sequence does not change the form of H, it only produces an interchange of a pair of rows and a pair of columns in the seminormal matrix, and the same change in the seminormal transforming matrix; and in this latter matrix the element $m_{r,r+1}$ is necessarily zero, like $m_{r+1,r}$ and all the elements below the leading diagonal; the seminormal units corresponding to the different tableaux are also unchanged. In order that a new sequence law may produce the same matrix except for the sequence of rows and columns, it is a necessary and sufficient condition that the element m_{rs} of the old seminormal matrix is zero when F_s precedes F_r according to the new law. It will be shown that certain other sequences may be adopted equally as well as that from the last letter to produce the same seminormal units. Thus, for instance, the sequence from the first letter is a case in point. As an example, for $T_{3,2}$,

$$H = \tfrac{1}{72} X_1^2 + \tfrac{1}{36} X_2^2 + \tfrac{1}{12} X_3^2 + \tfrac{1}{12} X_4^2 + \tfrac{1}{4} X_5^2,$$

where (see Q.S.A., VI, § 17)

$$X_1 = 12x_1, \quad X_2 = 9x_2 + 3x_1, \quad X_3 = 6x_3 + 3x_2 + 3x_1,$$

$$X_4 = 6x_4 + 3x_2 + 3x_1, \quad X_5 = 4x_5 + 2x_4 + 2x_3 + x_2 + 3x_1.$$

Here the sequence is taken from the last letter and the tableaux may be written

$$\begin{matrix} 123 \\ 45 \end{matrix}, \quad \begin{matrix} 124 \\ 35 \end{matrix}, \quad \begin{matrix} 134 \\ 25 \end{matrix}, \quad \begin{matrix} 125 \\ 34 \end{matrix}, \quad \begin{matrix} 135 \\ 24 \end{matrix}.$$

But, since X_4 does not contain x_3, the form of H is unaltered by the interchange of F_3 and F_4 in the tableau sequence; and this is the same as the sequence from the first letter.

2. It is possible to give many different laws of sequence of standard tableaux; among them there is a class of sequences which have important properties in common. We therefore define this class as follows:

Definition. A sequence of standard tableaux is called a *principal sequence* when it possesses the property that F_r precedes F_s whenever the permutation σ_{rs} which transforms F_s into F_r can be expressed as a product of transpositions $\tau_1, \tau_2, \ldots, \tau_k$, which is such that, when the permutation on F_s is made a transposition at a time, at each stage either the transposition raises the earlier letter in the tableau operated on and depresses the later letter, or else it merely interchanges two letters in the same row.

The sequence from the last letter and the sequence from the first letter are examples of a principal sequence. Other examples will be introduced later.

In the case of T_{32} the relative positions of all the tableaux are fixed for a principal sequence, except that of F_3 and F_4.

3. For the present purpose, and also for future use, it is necessary to consider more exactly the natural matrix C corresponding to an interchange of a pair of consecutive letters. In this case the seminormal matrix D is known, and the equation (Q.S.A., VI, § 6)

$$CM = MD$$

determines the seminormal transforming matrix M.

Let b, b' be the consecutive pair of letters. In a particular tableau F_r there are three ways in which these letters may appear:

(i) In the same row; if n_{rr} is the corresponding natural unit,

$$(bb')\, n_{rr} = n_{rr}.$$

(ii) In the same column:

$$(bb')\, n_{rr} = -n_{rr} + \ldots,$$

and the terms on the right need examination.

(iii) Obliquely, then there is a second standard tableau F_s obtained from F_r by the interchange of b and b', so that

$$(bb')\, n_{rr} = n_{sr}.$$

It is the second case that we have to examine.

The natural unit

$$n_{rs} = P_r \sigma_{rs} N_s M_s.$$

At the moment, the attention is confined to permutations applied to the left-hand side only of the units. The equations then are of the form

$$\tau n_{rs} = \Sigma \lambda_t n_{ts};$$

and the second suffix is unchanged, the equations being the same whatever its value provided that it is kept unaltered.

We require $(bb')\, n_{rs}$, where a pair of consecutive rows of F_r have the form

$$\begin{array}{l} \;\;\;\;\;\;\;\;\;\;\;\; \ldots\ldots\, b c_1 c_2 \ldots c_k \\ a_1 \ldots a_l b' \ldots \end{array}$$

the sequence of the letters exposed being

$$a_1 a_2 \ldots a_l b b' c_1 c_2 \ldots c_k.$$

After the transposition (bb') the standardizing equations must be used, and first of all

$$G\,[(bb')\, n_{rs}] = 0,$$

where $$G = \{a_1 a_2 \ldots a_l b b' c_1 \ldots c_k\}.$$

The result is $$\{bb'\}\, n_{rs} = -\Sigma q.$$

The terms q are those obtained from the different tableaux resulting from the operations of G on F_r. They are not necessarily all standard; but so far as the two rows at the moment considered are concerned, they are standard. Moreover, each q tableau is obtained from F_r by a permutation of the character described in the definition of the last paragraph. Hence those which are standard are earlier than F_r in any principal sequence.

Let us now consider a q tableau Φ which is not standard. Then a pair of consecutive rows exists (necessarily one of them is one of the rows already disturbed) for which at some point the letter in the upper row is later than that in the lower row. Starting from the left-hand side, we select the first pair of letters for which this is so, and apply the standardizing equation for this pair. As before,

$$q = -\Sigma q',$$

where the tableau for each q' is obtained from Φ by a permutation of the character described in the definition of the last paragraph. Then, if this tableau is standard, it precedes F_r in any principal sequence. If not, the standardizing equations must be used again. That the process is finite we know; the argument is the same at each step; and hence we ultimately arrive at

THEOREM I. *When the letters a_p, a_{p+1} lie in the same column of the tableau F_r,*

$$\{a_p a_{p+1}\} n_{ru} = \Sigma \lambda_s n_{su},$$

where F_s precedes F_r in any principal sequence. Further, a_p, a_{p+1} lie in the same row of F_s, or else obliquely, in which case $\lambda_t = \lambda_s$ when

$$(a_p a_{p+1}) n_{su} = n_{tu};$$

but they never lie in the same column.

The equality of λ_s and λ_t in the last part of this statement is obvious when we operate on the equation with $\{a_p a_{p+1}\}$. For then the left-hand side is merely multiplied by two, and hence the same must be true for the right-hand side, since no linear relation exists between the units.

4. The sequences used so far have been that from the last letter and that from the first letter. It will be found that a principal sequence, which will be called the sequence from the last row, has advantages for certain purposes, in that the matrices thus arranged assume a regularity and symmetry which is lacking with other sequences. In the sequence from the last row the tableau F_r precedes the tableau F_s when the latest letter in the last row of F_r is later than the latest letter of the last row of F_s, or, when these are the same, the next latest letter in the last row of F_r precedes the next latest in the last row of F_s, and so on. When the last rows of F_r and F_s are the same, we consider the latest row which differs in the two tableaux and determine the sequence from that.

The corresponding sequence from the first row is also a principal sequence and appears to enjoy a like advantage. These sequences may be illustrated by giving the relative places of a set of six tableaux which are identical except for the positions of three consecutive letters *a*, *b*, *c*. The places occupied by these letters are supposed to be such that no two are in the same row or column. We write them in a row, the first in the row occupying that position which is highest in the tableaux and most to the right, and the last that which is lowest and most to the left. The sequence from the last letter is

$$\begin{matrix} abc & bac & acb & bca & cab & cba \\ F_1, & F_2, & F_3, & F_4, & F_5, & F_6. \end{matrix}$$

That from the first letter is

$$F_1, \quad F_3, \quad F_2, \quad F_5, \quad F_4, \quad F_6.$$

That from the last row is

$$F_1, \quad F_2, \quad F_3, \quad F_5, \quad F_4, \quad F_6.$$

That from the first row is

$$F_1, \quad F_3, \quad F_2, \quad F_4, \quad F_5, \quad F_6.$$

Another form of sequence sometimes useful is obtained by grouping the n letters permuted into two consecutive sets. The first m are called the letters a, the last $n-m$ the letters b. Then, when the letters b are erased from a standard tableau F_r, we are left with a tableau F_r' of the a's only which is still standard. The tableau F_r' belongs to a representation T_r' of the symmetric group of $a_1 a_2 \ldots a_m$. Then we may define F_r to precede F_s whenever T_r' precedes T_s'. When F_r' and F_s' belong to the same T', we may define the precedence of F_r and F_s by that of F_r' and F_s' according to any principal sequence. When $F_r' \equiv F_s'$, the sequence is defined by the letters b according to some principal sequence. A sequence thus defined is a principal sequence.

A moment's reflection will show the truth of

THEOREM II. *In every principal sequence the first tableau is the same, and also the last tableau.*

5. THEOREM III. *The element $m_{\xi\eta}$ of the seminormal transforming matrix M is zero, when F_η precedes F_ξ according to any one of the principal sequences.*

This is known to be the case when the sequence is from the last letter; we consider any other principal sequence S; and proceed by a double induction. We take the tableaux according to S, and notice that, the first tableau F_1 being the same in all the principal sequences, the theorem is true when $\eta = 1$. Then the theorem is assumed for every value of η prior to the given one in S, and the tableaux F_ξ are considered beginning with the last F_f which is the same in all principal sequences. The theorem then is known to be true when $\xi = f$; it is then assumed for all values of ξ

later in S than the given one and the particular η in question. and also for all earlier values of η in S. Consider a particular F_η; except in the trivial cases T_n and T_{1^n}, a consecutive pair of letters can be found in F_η which lie obliquely.

Indeed, it is possible to pass from F_1 to F_f by successive transpositions of such pairs of consecutive letters.

Let a, a' be such a consecutive pair of letters. Let C be the natural matrix for (aa'), and D the seminormal matrix; let M be the seminormal transforming matrix. Then the results of Theorem I must be applied to the equation

$$\text{(I)} \qquad CM = MD.$$

Let F_p have a, a' in one row, let F_r have them in one column, and let $F_s = (aa')F_t$ have them obliquely, where a' lies lower in F_s than in F_t. In this case F_s precedes F_t in every principal sequence.

Then
$$\{aa'\} n_{rv} \doteq \Sigma \lambda_{ru} n_{uv},$$

where F_u precedes F_r in every principal sequence. In the columns F_p, F_r of D every element is zero except that in the primary diagonal, which is ± 1. In the pair of columns F_s, F_t of D the elements are zero except for the primary diagonal matrix

$$\begin{bmatrix} -\rho & 1-\rho^2 \\ 1 & \rho \end{bmatrix}.$$

Consider the element (ξ, η) in (I).

On the left we have four cases $\xi = p, r, s, t$:

$$m_{p\eta} + \Sigma \lambda_{\varpi p} m_{\varpi\eta}, \qquad -m_{r\eta} + \Sigma \lambda_{\varpi r} m_{\varpi\eta},$$
$$m_{t\eta} + \Sigma \lambda_{\varpi s} m_{\varpi\eta}, \qquad m_{s\eta} + \Sigma \lambda_{\varpi t} m_{\varpi\eta}.$$

Here F_ϖ has a, a' in the same column, and $\lambda_{\varpi\xi}$ is zero unless F_ξ precedes F_ϖ in every principal sequence. Hence, on the assumption of the induction, $\lambda_{\varpi\xi} m_{\varpi\eta}$ is zero in every case; for we are considering the case in which F_η precedes F_ξ according to S. Thus on the left-hand side we have

$$m_{p\eta}, \quad -m_{r\eta}, \quad m_{t\eta}, \quad m_{s\eta},$$

in the four cases.

In F_η the letters a, a' lie obliquely, we write $(aa')F_\eta = F_\zeta$ and suppose that F_ζ precedes F_η. Then on the right we have

$$(1-\rho^2) m_{\xi\zeta} + \rho m_{\xi\eta}.$$

Then, for $\xi = p$,

$$m_{p\eta} = (1-\rho^2)\,m_{p\zeta} + \rho m_{p\zeta} = (1+\rho)\,m_{p\zeta} = 0,$$

for $\xi = r$, $$m_{r\eta} = -(1-\rho)\,m_{r\zeta} = 0,$$

for $\xi = s$, $$m_{t\eta} - \rho m_{s\eta} = (1-\rho^2)\,m_{s\zeta} = 0,$$

for $\xi = t$, $$m_{s\eta} - \rho m_{t\eta} = (1-\rho^2)\,m_{t\zeta} = 0,$$

and hence $$m_{s\eta} = m_{t\eta} = 0.$$

It has been assumed that F_η is the later of the pair of F_η and F_ζ. The consecutive pair a, a' can always be chosen so that F_η is later than its fellow F_ζ unless $\eta = 1$. For consider any tableau F_η, and each letter in order in connection with its consecutive. If a_2 is in the second row, let a_{r+1} be the first letter in the second column, then a_r, a_{r+1} will be a suitable pair to take. If a_2 is in the first row, let a_r be the first letter not in the first row, then let a_s be the letter which follows a_{r-1} in the first row, then a_{s-1}, a_s will be a suitable pair. The only exception to the success of this is when the first row letters are all consecutive; we then proceed to the second row, and obtain a suitable pair unless these are consecutive, and so on.

Thus the induction is proved, and the theorem is true. As an immediate result we obtain

THEOREM IV. *The seminormal matrix is the same except for row and column interchanges whichever of the principal sequences originally is adopted; and the seminormal units are the same in each case.*

II. *Positive and negative symmetric groups of a set of consecutive letters.*

6. Consider an expression S given in terms of the *exact* seminormal units $\varpi_{u,v}$; we are to be concerned with permutations applied on the left-hand side, and so are not concerned with the second suffix of ϖ which is supposed always the same, *i.e.* we write

$$S = \Sigma \lambda_u \varpi_u,$$

in effect an exact seminormal matrix consisting of a single column and zeros elsewhere. In the exact seminormal matrix, for a transposition of a pair of consecutive letters, the diagonal quadratic matrices take the form

$$\begin{bmatrix} -\rho & 1-\rho \\ 1+\rho & \rho \end{bmatrix}.$$

Multiply S on the left by the transposition (aa'). Then, using the notation of the last paragraph,

$$\lambda_p \varpi_p \quad \text{becomes} \quad \lambda_p \varpi_p,$$

$$\lambda_r \varpi_r \quad \text{becomes} \quad -\lambda_r \varpi_r,$$

$$\lambda_s \varpi_s + \lambda_t \varpi_t \quad \text{becomes} \quad \{-\rho\lambda_s + (1-\rho)\lambda_t\}\varpi_s + \{(1+\rho)\lambda_s + \rho\lambda_t\}\varpi_t.$$

When S is unchanged, $\qquad\qquad \lambda_r = 0,$

$$(1+\rho)\lambda_s = (1-\rho)\lambda_t.$$

Hence (Q.S.A., VI, § 16)

$$\lambda_s [F_s]_1 [F_s]_2 = \lambda_t [F_t]_1 [F_t]_2;$$

$[F]_1$, $[F]_2$ being the first and second tableau functions.

Thus when S is unchanged by the permutations of the symmetric group of m consecutive letters $b_1, b_2, ..., b_m$, then S is a sum of sets of terms $\Sigma \lambda_u \varpi_u$, a set being such that the letters b collectively occupy the same position in the tableaux F_u, no two b's lie in the same column, and the coefficient

$$\lambda_u = \frac{A}{[F_u]_1 [F_u]_2},$$

where A is a constant for the set.

We can now prove that the sum of the coefficients λ_u is $AN/[F_1]_1 [F_l]_2$, where N is their number, F_1 is the first and F_l the last tableau of the set.

The proof is by induction; the theorem is assumed true for the case of m consecutive letters $b_1, ..., b_m$, and then it is established for the symmetric group of $b_1, ..., b_{m+1}$. For definiteness the case is first taken for which each of the letters b is in a different row in the tableaux F_u. The first and the last tableaux will be called F_1 and F_l. For the group of the first m letters there are just $m+1$ sets, and the first and the last tableaux of the r-th set will be denoted by $F_{r,1}$, $F_{r,l}$. In F_1 the axial projections of $b_1 b_2, b_2 b_3, ..., b_m b_{m+1}$ will be denoted by $a_1, a_2, ..., a_m$. Then, of course, that of $b_1 b_3$ is $a_1 + a_2$, and so on. Then, by Q.S.A., VI, § 15,

$$[F_1]_1 = [F_{1,1}]_1 = \{1 + 1/a_m\} [F_{2,1}]_1 = \{1 + 1/a_{m-1}\}\{1 + 1/(a_{m-1} + a_m)\} [F_{3,1}]_1$$

$$= ... = \{1 + 1/a_1\}\{1 + 1/(a_1 + a_2)\} ... \{1 + 1/(a_1 + a_2 + ... + a_m)\}[F_{m+1,1}]_1;$$

$$[F_l]_2 = [F_{m+1,l}]_2 = \{1 - 1/a_1\} [F_{m,l}]_2 = \{1 - 1/a_2\}\{1 - 1/(a_2 + a_1)\} [F_{m-1,l}]_2$$

$$= ... = \{1 - 1/a_m\}\{1 - 1/(a_m + a_{m-1})\} ... \{1 - 1/(a_m + ... + a_2 + a_1)\} [F_{1,l}]_2.$$

Then, by hypothesis,

$$\sum_{r=1}^{l} \frac{A}{[F_r]_1\,[F_r]_2} = \sum_{r=1}^{m+1} \frac{A\,.\,m!}{[F_{r,1}]_1\,[F_{r,l}]_2} = \frac{Am!}{[F_1]_1\,[F_l]_2}\,X,$$

where

$$X = \sum_{r=1}^{m+1} \{1+1/a_{m+2-r}\}\{1+1/(a_{m+2-r}+a_{m+3-r})\}\ldots$$
$$\times\{1-1/a_{m+1-r}\}\{1-1/(a_{m+1-r}+a_{m-r})\}\ldots.$$

To calculate X we first make a_r very small; there are two terms in which the very large factor $1/a_r$ appears. The coefficient of $1/a_r$ is seen to be (on neglecting a_r)

$$-\{1-1/a_{r-1}\}\{1-1/(a_{r-1}+a_{r-2})\}\ldots\{1-1/(a_{r-1}+\ldots+a_1)\}$$
$$\times\{1+1/a_{r+1}\}\{1+1/(a_{r+1}+a_{r+2})\}\ldots\{1+1/(a_{r+1}+\ldots+a_{m+1})\}$$
$$+\{1+1/a_{r+1}\}\ldots\{1+1/(a_{r+1}+\ldots+a_{m+1})\}$$
$$\times\{1-1/a_{r-1}\}\ldots\{1-1/(a_{r-1}+\ldots+a_1)\},$$

that is, zero. Hence, when the expression $X-m-1$ is brought to a common denominator D, so that

$$X-m-1 = Y/D,$$

then a_r is a factor of Y.

Next let us make a_r+a_{r-1} very small; then the coefficient of the large factor $1/(a_r+a_{r-1})$ becomes

$$-\{1-1/a_r\}\{1-1/a_{r-2}\}\ldots\{1-1/(a_{r-2}+\ldots+a_1)\}$$
$$\times\{1+1/a_{r+1}\}\ldots\{1+1/(a_{r+1}+\ldots+a_{m+1})\}$$
$$+\{1+1/a_{r-1}\}\{1+1/a_{r+1}\}\ldots\{1+1/(a_{r+1}+\ldots+a_{m+1})\}$$
$$\times\{1-1/a_{r-2}\}\ldots\{1-1/(a_{r-2}+\ldots+a_1)\}$$
$$= 0,$$

since $$a_{r-1} = -a_r.$$

Hence a_r+a_{r-1} is a factor of Y.

In the same way we find that every factor of D is a factor of Y, and hence, since Y is of lower order than D,

$$Y = 0.$$

Therefore $$X = m+1,$$

and the induction is proved; for it is quite simple to prove the result for $m = 1$ or 2.

To prove the more general case, we suppose that the letters $b_1, \ldots, b_{m+1}$ occupy r_1 places in one row, r_2 in a lower, and so on, ending with r_k in the lowest row touched. Then

$$[F_1]_1 = [F_{1,1}]_1$$
$$= \{1+1/(a_{k-1}+r_k-1)\}\{1+1/(a_{k-1}+r_k-2)\}\ldots\{1+1/a_{k-1}\}[F_{2,1}]_1;$$

here a_{k-1} is the axial projection of a pair of letters occupying the extreme right-hand positions in the last two rows. Then

$$[F_1]_1 = \{1+r_k/a_{k-1}\}[F_{2,1}]_1 = \{1+r_{k-1}/a_{k-2}\}\{1+r_k/(a_{k-2}+a_{k-1})\}[F_{3,1}]_1 = \ldots$$

and

$$[F_l]_2 = [F_{k,l}]_2 = \{1-r_1/a_1\}[F_{k-1,l}]_2 = \{1-r_2/a_2\}\{1-r_1/(a_2+a_1)\}[F_{k-2,l}]_2 = \ldots.$$

The total number of tableaux is

$$N = (m+1)!/\Pi r!,$$

and the number of tableaux when b_{m+1} is in the s-th row is

$$N.r_s/(m+1) = N_{k+1-s}.$$

We have to sum $$\sum_{s=1}^{k} N_s/[F_{s,1}]_1[F_{s,l}]_2.$$

The result follows exactly as before.

THEOREM V. *A substitutional expression S, which is unaltered by left-hand multiplication with the permutations of the positive symmetric group of m consecutive letters, may be expressed in exact seminormal units as a sum of sets of terms*

$$\Sigma A\varpi_u/([F_u]_1[F_u]_2),$$

where A is constant for each set, and the tableaux of a set are those tableaux obtained from each other by operations of the group, no two of the letters of the group lying on the same column. Further, when F_1, F_l are the first and last tableaux of the set, the average of the coefficients is

$$A/([F_1]_1[F_l]_2).$$

Multiplication by transpositions of consecutive letters on the right shows that for the symmetric group of m letters the columns corresponding to the tableaux F_u of a set are all identical. Consider then the matrix representation of $G/m!$, where G is the positive symmetric group of m consecutive letters.

The form is determined by Theorem V, and the fact that the u columns are all identical. The constant A is given by the fact that $G/m!$ multiplied by itself simply reproduces itself.

But this multiplication by itself simply reproduces each element multiplied by the sum of the elements. This sum is

$$NA/\{[F_1]_1\,[F_t]_2\} = 1,$$

which gives A.

7. In exactly the same way it is found that, for multiplication of S on the left by a transposition, S is changed in sign when

$$\lambda_p = 0, \quad \lambda_s = -\lambda_t.$$

For multiplication on the right we have to consider a row in the matrix or

$$S = \sum_v \lambda_v\,\varpi_{uv};$$

and then S is changed in sign provided that

$$\lambda_p = 0, \quad (1-\rho)\lambda_s = -(1+\rho)\lambda_t,$$

i.e.

$$\lambda_s/\{[F_s]_1\,[F_s]_2\} = -\lambda_t/\{[F_t]_1\,[F_t]_2\}.$$

Thus we obtain

THEOREM VI. *A substitutional expression S, which is changed in sign by right-hand multiplication with any odd permutation of the symmetric group of m consecutive letters, may be expressed in exact seminormal units as a sum of sets of terms*

$$\sum_v \pm A[F_v]_1\,[F_v]_2\,\varpi_{uv},$$

where A is constant for each set, the tableaux F_v of a set are those obtained by permuting the m letters in one of them, no two of the letters being in the same row. Further, when F_1, F_t are the first and last tableaux of the set, the average numerical value of the coefficients is

$$A\,[F_t]_1\,[F_1]_2.$$

The proof of the last fact follows exactly the same lines as that of the corresponding fact of Theorem V. The expression for $\Gamma/m!$, where Γ is the negative symmetric group of m consecutive letters, is obtained as before; it is necessary to remember in this case that the signs are alternately positive and negative both in the rows and columns of the matrix.

III. *The expression for a unit belonging to a consecutive set of letters.*

8. It is necessary for certain applications of this analysis to express the units $\Pi_{r,s}$ of a representation $T_{\beta_1\beta_2\ldots}$ of the symmetric group of $b_1 b_2 \ldots b_n$ in terms of the units $\varpi_{\rho\sigma}$ of a representation $T_{\gamma_1\gamma_2\ldots}$ of the symmetric group of $a_1 a_2 \ldots a_m b_1 \ldots b_n$. Exact seminormal units will be used, so that a transposition of consecutive letters will take the form of quadratic matrices

$$\begin{bmatrix} -\rho & 1-\rho \\ 1+\rho & \rho \end{bmatrix},$$

or linear matrices [1] or [−1] on the leading diagonal and zero elsewhere. The letter sequence will be taken as that written above, so that all the letters b are consecutive.

Consider any tableau F_ρ of T_γ. Let all the letters b be erased and there will remain a tableau of the a's alone; this will be standard and we shall call it F_{ρ_1}. Now erase the letters a in F_ρ, there remains a tableau of the b's alone, which, however, differs in general from the ordinary tableau in that some of the lower rows begin on an earlier column than the higher rows. It is a tableau distorted by moving some of the upper rows to the right. It must be standard as it is, but will not, in general, be standard when the rows are moved into the regular position. We call the distorted b tableau F'_{ρ_2}.

Then we use, instead of F_ρ, the double suffix notation F_{ρ_1,ρ_2}.

The quadratic matrices which appear in the representation of a transposition of consecutive letters have elements whose row and column tableaux differ only by this transposition. Hence any substitutional expression S_b in the b's can be expressed in the form

$$S_b = \sum_{\rho_1} \Big[\sum_{\rho_2,\sigma_2} \lambda_{\rho_2\sigma_2} \varpi_{\rho_1\rho_2:\rho_1\sigma_2} \Big]. \quad \text{(II)}$$

That is, each set of terms with the tableau F_{ρ_1} in common is really a separate representation so far as the letters b are concerned.

Similarly we find

$$S_a = \sum_{\rho_2} \left[\sum_{\rho_1, \sigma_1} \lambda_{\rho_1 \sigma_1} \varpi_{\rho_1 \rho_2 : \sigma_1 \rho_2} \right].$$

When S_a belongs to the representation $T_{a_1 a_2 \ldots}$, each of the tableaux F_{ρ_1} belongs to this representation, and, in fact,

$$S_a = \Sigma \lambda_{\rho_1 \sigma_1} \varpi_{\rho_1 \sigma_1}.$$

When all the rows of F_{ρ_1} are equal, and in number equal to that of the rows in the whole F_{ρ_1, ρ_2} tableau, F'_{ρ_2} is no longer distorted. The expression for S_b is obtained then directly from the b matrices. Hence we have

THEOREM VII. *Let the representation τ_{a^h} (which has tableaux with equal rows) have units ϖ_{rs} and let*

$$S_1 = \Sigma \lambda_{rs} \varpi_{rs};$$

let $T_{\beta_1 \beta_2 \ldots \beta_k}$, a representation of a set of independent letters, have units ϖ_{uv} and let

$$S_2 = \Sigma \mu_{uv} \varpi_{uv}.$$

Then $$S_1 S_2 T_{a+\beta_1, a+\beta_2, \ldots} = \Sigma \lambda_{rs} \mu_{uv} \varpi_{ru:sv},$$

h being equal to or greater than k, and the sequence of letters being such that those of S_1 precede those of S_2.

9. Consider now equation (II) when the tableau F'_{ρ_2} is distorted.

Those terms for which the first tableau F_{ρ_1} are the same form a set. An operation, on the left or right, with any permutation of the b's on a unit of this set gives a linear function of units of the same set. Thus the matrix formed by the coefficients of the set is a representation of the symmetric group of $b_1 b_2 \ldots b_n$. This representation corresponds to the distorted tableaux F'_{ρ_2}; let g be its order. Then, in general, this is not an irreducible representation and so a matrix Λ exists such that $\Lambda^{-1}[\lambda_{\rho_2 \sigma_2}]\Lambda$ is a matrix with partial square matrices on the leading diagonal (which are irreducible group matrices of the b's), and zeros elsewhere. Now let $S_b = [\mu_{rs}]$ be the group matrix of the b's for the representation $T_{\beta_1 \beta_2 \ldots \beta_k}$. Then the corresponding matrix $[\lambda_{\rho_2 \sigma_2}]$ for the particular set defined by F_{ρ_1} is transformable into a matrix which has $[\mu_{rs}]$ for a diagonal matrix, possibly repeated, and zeros elsewhere.

Thus

$$\text{(III)} \qquad [\lambda_{\rho_2\sigma_2}]\Lambda = \Lambda \begin{bmatrix} [\mu_{rs}] & 0 & \dots & 0 \\ 0 & \dots & \dots & \dots \\ \vdots & \dots & \dots & \dots \\ 0 & \dots & \dots & \dots \end{bmatrix}.$$

Let f be the order of $[\mu_{rs}]$; then, if we select the f columns of Λ which correspond to the position of $[\mu_{rs}]$ in the last matrix (the first f columns as it is written), and delete the rest, we shall obtain a matrix Λ_1 of g rows and f columns such that

$$[\lambda_{\rho_2\sigma_2}]\Lambda_1 = \Lambda_1[\mu_{rs}].$$

The value of a transposition of consecutive letters is known in terms of both λ and μ matrices, and the equations so obtained give the means of calculating Λ_1. When the solution is unique, except for a proportional factor, the matrix appears but once in the matrix on the right of (III). When it is repeated there, the solution for Λ_1 is not unique.

In the same way we may find a matrix Λ_1' of f rows and g columns, which is part of the matrix Λ^{-1}, such that

$$\Lambda_1' [\lambda_{\rho_2\sigma_2}] = [\mu_{rs}]\Lambda_1'.$$

And from the relation of Λ_1, Λ_1' to Λ, Λ^{-1} we obtain

$$\Lambda_1'\Lambda_1 = E_f.$$

10. Consider, then, the equation

$$\Lambda D = \Delta\Lambda.$$

A transposition τ of consecutive letters is represented in D and Δ on a given row (or column) by $+1$ or -1 on the diagonal, or else on a pair of rows (or columns), by a quadratic matrix of the form

$$\begin{bmatrix} -p & 1-p \\ 1+p & p \end{bmatrix} \equiv \{p\}_{r,s},$$

where the suffixes denote the rows and columns concerned. Then the above equation indicates the equality of corresponding elements; it is convenient to group the elements according to the elementary diagonal matrices. There are thus nine cases.

(i) As regards D, τ may have the value $+1$ or -1 on the r-th place on the diagonal or the matrix $\{p\}_{r,s}$ where the r-th and s-th rows and columns intersect.

(ii) Similarly for Δ, there are three cases

$$\{1\}_\rho, \quad \{-1\}_\rho, \quad \{q\}_{\rho,\sigma}.$$

The simultaneous consideration of $\{1\}_r$ and $\{1\}_\rho$, or of $\{-1\}_r$ and $\{-1\}_\rho$, gives no information about $\lambda_{\rho r}$.

There are six cases of interest:

(I) $$\Lambda_\rho\{1\}_r=\{-1\}_\rho\Lambda_r, \quad \text{or} \quad \Lambda_\rho\{-1\}_r=\{1\}_\rho\Lambda_r,$$

gives $$\lambda_{\rho r}=0.$$

(II) $$\Lambda_{\rho,\sigma}\{1\}_r-\{q\}_{\rho,\sigma}\Lambda_r$$

gives $$(1+q)\lambda_{\rho r}=(1-q)\lambda_{\sigma r}.$$

(III) $$\Lambda_{\rho,\sigma}\{-1\}_r=\{q\}_{\rho,\sigma}\Lambda_r$$

gives $$\lambda_{\rho r}=-\lambda_{\sigma r}.$$

(IV) $$\Lambda_\rho\{p\}_{r,s}=\{1\}_\rho\Lambda_{r,s}$$

gives $$\lambda_{\rho r}=\lambda_{\rho s}.$$

(V) $$\Lambda_\rho\{p\}_{r,s}=\{-1\}_\rho\Lambda_{r,s}$$

gives $$(1-p)\lambda_{\rho r}=-(1+p)\lambda_{\rho s}.$$

(VI) $$\Lambda_{\rho\sigma}\{p\}_{r,s}=\{q\}_{\rho,\sigma}\Lambda_{r,s}$$

gives four equations, each one of which omits one of the four quantities concerned, giving what is, in general, a unique relation between the other three; for only two of the equations are independent. The relation

$$\lambda_{\rho r}+\lambda_{\sigma r}=\lambda_{\rho s}+\lambda_{\sigma s}$$

may be deduced from them.

For our purposes, the equations of most importance are those obtained when it is known otherwise that one of the four elements is zero. These are:

(A) $\lambda_{\rho r}=0$: $$\lambda_{\rho s}/(1-q)=\lambda_{\sigma r}/(1+p)=\lambda_{\sigma s}/(p+q).$$

(B) $\lambda_{\rho s}=0$: $$\lambda_{\rho r}/(1-q)=\lambda_{\sigma s}/(1-p)=\lambda_{\sigma r}/(q-p).$$

(C) $\lambda_{\sigma r}=0$: $$\lambda_{\rho r}/(1+p)=\lambda_{\sigma s}/(1+q)=\lambda_{\rho s}/(p-q).$$

(D) $\lambda_{\sigma s}=0$: $$\lambda_{\rho s}/(1-p)=\lambda_{\sigma r}/(1+q)=-\lambda_{\rho s}/(p+q).$$

(VI a) The special case $p=q$ yields

$$\lambda_{\rho s}/(1-p)=\lambda_{\sigma r}/(1+p)=(\lambda_{\sigma s}-\lambda_{\rho r})/2p.$$

11. Let D be a matrix of order f of $T_{\beta_1\beta_2\ldots\beta_k}$ and Δ the corresponding matrix of order g which results by distorting the tableau, the number of letters in each row remaining the same. Then Λ is a matrix with g rows and f columns. Consider the first row; this corresponds to the distorted tableau F_1' in which the first β_1 letters lie in the first row, and so on. For the transposition of any consecutive pair of letters from the first β_1 the diagonal matrix for the first row is $\{1\}_1$. Let F_r be a tableau with b_2 below the first row, then a diagonal matrix of D for $(b_1 b_2)$ is $\{-1\}_r$.

Hence, by (I), $\lambda_{1r}=0.$

Then, from $(b_2 b_3)$ and (IV),

$$\lambda_{1s}=0,$$

where $(b_2 b_3)\,F_r=F_s.$

In the same way we see that, when F_s has any one of the letters $b_2, b_3, \ldots, b_{\beta_1}$ in the first place of the second row,

$$\lambda_{1s}=0.$$

Now b_{β_1+1} is the latest letter which can occupy this position, and when it does so the first row is necessarily the same as in F_1. We next observe that, when the first row in F_r is as in F_1 but b_{β_1+2} is in the third row, then

$$\lambda_{1r}=0.$$

Proceeding as before, we find that

$$\lambda_{1r}=0$$

unless the first two rows of F_r are the same as in F_1. The next row can be then considered, and we finally see that

$$\lambda_{1r}=0$$

unless $r=1$.

The tableaux F_r of D are arranged according to some principal sequence, and the same sequence is used for the tableaux F_ρ'. Two tableaux F_r, F_ρ' are said to correspond when the rows of the two tableaux are the same, and this will be written

$$F_\rho'=\bar{F}_r.$$

In general, F_{r+1} and $F'_{\rho+1}$ do not correspond, since between $\bar{F}_r$ and $\bar{F}_{r+1}$ there may be distorted tableaux which become non-standard when the rows are shifted back into the usual position. It will now be proved that $\lambda_{\rho r}$ is zero when $\bar{F}_r$ is a tableau later than F_ρ' in any principal sequence. This has been proved for $\rho = 1$; it will be assumed for all values of ρ less than the given one. Since $\rho > 1$, there is some letter b_u which is in a lower row in F_ρ' than b_{u+1}.

Then
$$(b_u b_{u+1}) F_\rho' = F_\pi',$$
where $\pi < \rho$.

Let $\bar{F}_r$ be a tableau later than F_ρ'; then, by hypothesis.
$$\lambda_{\pi r} = 0;$$
and, unless b_u, b_{u+1} lie obliquely in F_r, it follows at once that
$$\lambda_{\rho r} = 0.$$

When b_u, b_{u+1} lie obliquely in F_r, let
$$(b_u b_{u+1}) F_r = F_s.$$

When $s > r$,
$$\lambda_{\pi s} = 0;$$
and hence again
$$\lambda_{\rho r} = 0.$$

When $s < r$, there is always some principal sequence according to which F_π' precedes $\bar{F}_s$. For it is possible to arrange a principal sequence that regards b_u and b_{u+1} as identical letters in all cases except that of the determination of the relative position of a pair of tableaux which are identical except for the interchange of these letters. Thus, when F_ρ' precedes $\bar{F}_r$ according to any principal sequence, then a principal sequence can be found for which F_π' precedes $\bar{F}_s$.

Hence, by hypothesis,
$$\lambda_{\pi r} = 0, \quad \lambda_{\pi s} = 0.$$

When two of the four elements in (VI) are zero, the other two are zero, and hence
$$\lambda_{\rho r} = 0, \quad \lambda_{\rho s} = 0,$$
which proves the statement.

Let
$$F_\rho' = \bar{F}_r, \quad F_\sigma' = \bar{F}_s$$

be tableaux such that $r > s$ and

$$\tau F_r = F_s;$$

where τ is a transposition of consecutive letters. Then $\lambda_{\rho s} = 0$, and hence, by (VI), (B),

$$\lambda_{\rho r}/(1-q) = \lambda_{\sigma s}/(1-p) = \lambda_{\sigma r}/(p-q);$$

whence
$$\frac{\lambda_{\rho r}}{\lambda_{\sigma s}} = \frac{[F_r]_2}{[F_\rho']_2} \Big/ \frac{[F_s]_2}{[F_\sigma']_2}.$$

Thus the "skin" elements, those dividing the zero part of the matrix from the rest on the rows whose tableaux correspond to standard undistorted tableaux, have the form

$$\lambda_{\rho r} = \lambda\, [F_r]_2/[F_\rho']_2.$$

In exactly the same way it may be shown for the matrix Λ', for which

$$D\Lambda' = \Lambda' \Delta,$$

that $\lambda'_{r\rho} = 0$, when $\bar{F}_r$ is later than F_ρ' and that, for the skin elements $\lambda'_{r\rho}$,

$$\lambda'_{r\rho} = \lambda' [F_\rho']_1/[F_r]_1.$$

THEOREM VIII. *The group matrix D of order f for the representation $T_{\beta_1\beta_2\ldots\beta_k}$ is represented as a matrix Δ of order g obtained by lateral distortion of the tableaux, the sizes of the rows being unchanged. Then a matrix Λ of g rows and f columns exists, which is unique but for a proportional factor, such that*

$$\Lambda D = \Delta\Lambda.$$

The element $\lambda_{\rho r}$ of Λ is zero when the distorted tableau F_ρ' precedes $\bar{F}_r$ in any principal sequence, and the skin elements are

$$\lambda_{\rho r} = \lambda [F_r]_2/[F_\rho']_2.$$

Also there is a unique matrix Λ' of f rows and g columns, for which

$$D\Lambda' = \Lambda' \Delta.$$

The element $\lambda'_{r\rho}$ of Λ' is zero when F_ρ' precedes $\bar{F}_r$ in any principal sequence, *and the skin elements are*

$$\lambda'_{r\rho} = \lambda'[F_\rho']_1/[F_r]_1.$$

Also $$\Lambda'\Lambda = E_f.$$

The only statement in the enunciation not yet proved is that the matrices Λ, Λ' are unique.

If Λ is not unique, let M be a second matrix fulfilling the conditions, then

$$(\Lambda - M)D = \Delta(\Lambda - M).$$

The proportional factors may be chosen so that in

$$\Lambda - M = [\mu_{\rho r}]$$

any definite element is zero.

This will be taken to be μ_{11}. Then $[\mu_{\rho r}]$ satisfies all the equations obtained for $[\lambda_{\rho r}]$, and hence the whole of the first row is zero. But the equations of § 10 obtain all the elements of a row defined by F_ρ' in terms of those of any row defined by F_σ' for which

$$F_\sigma' = \tau F_\rho',$$

where τ is a transposition of consecutive letters. The elements of any row, by repeated application of this, can then be expressed in terms of those of the first row; and thus all the rows are zero.

The matrix Λ is then unique, and also Λ'. This principle may be used in all cases to which the equations of § 10 apply; and the general result is that the number of undetermined constants in every row is the same. That is, the number of times a given group matrix appears in a distorted tableau representation is the same as the number of independent elements in any row of Λ as given by the equations of § 10. The group matrix is repeated only when the sizes of the rows are changed.

12. The last three paragraphs may be illustrated by the representation T_{32}. Let us make a distortion by shifting the second row one place to the left. There are nine distorted tableaux with the sequence from the last letter; their second rows are:

	45,	35,	25,	15,	34,	24,	14,	23,	13
$\rho =$	1	2	3	4	5	6	7	8	9.

Then
$$F_r' = \bar{F}_r \quad (r = 1, 2, 3).$$
$$F_5' = \bar{F}_4, \quad \bar{F}_6' = \bar{F}_5.$$

Also
$$36\Lambda = \begin{bmatrix} 18 & 0 & 0 & 0 & 0 \\ -2 & 16 & 0 & 0 & 0 \\ -4 & -4 & 12 & 0 & 0 \\ -12 & -12 & -12 & 0 & 0 \\ -4 & -4 & 0 & 12 & 0 \\ -8 & 1 & -3 & -3 & 9 \\ -24 & 3 & 3 & -9 & -9 \\ 0 & -9 & -9 & -9 & -9 \\ 0 & -27 & 9 & -27 & 9 \end{bmatrix}$$

and
$$36\Lambda' = \begin{bmatrix} 60 & -4 & -4 & -4 & -4 & -4 & -4 & 0 & 0 \\ 0 & 64 & -8 & -8 & -8 & 1 & 1 & -3 & -3 \\ 0 & 0 & 72 & -24 & 0 & -9 & 3 & -9 & 3 \\ 0 & 0 & 0 & 0 & 72 & -9 & -9 & -9 & -9 \\ 0 & 0 & 0 & 0 & 0 & 81 & -27 & -27 & 9 \end{bmatrix}$$

13. Let ϖ_{uv} be a unit of $T_{\alpha_1\alpha_2\ldots\alpha_h}$ of the letters $a_1, a_2, \ldots, a_m$, and Π_{rs} a unit of $T_{\beta_1\beta_2\ldots\beta_k}$ of $b_1, b_2, \ldots, b_n$. The product $\varpi_{uv}\Pi_{rs}$, expressed in terms of units of $a_1, a_2, \ldots, a_m, b_1, \ldots, b_n$, has terms which are units from many different representations $T_{\gamma_1\gamma_2\ldots}$. For our purposes the terms which belong to $T_{\alpha_1+\beta_1, \alpha_2+\beta_2, \ldots}$ have a special importance; they are
$$\varpi_{uv}\Pi_{rs}T_{\alpha_1+\beta_1, \alpha_2+\beta_2, \ldots}.$$

In § 8 it was seen that this is
$$\Sigma A_{\rho,\sigma}\varpi_{u,\rho;v,\sigma},$$

where the tableau $F_{u,\rho}$ becomes the "a" tableau F_u on removal of the b's, and the distorted "b" tableau F_ρ' on the removal of the a's.

In § 9 it was seen that a non-singular matrix L exists such that
$$L^{-1}[A_{\rho,\sigma}]L = K,$$

a matrix which has matrices D_1, D_2, ... on the leading diagonal and zeros elsewhere, where D_1, D_2, ... are the particular values for the case in point of the irreducible group matrices of the b's. Here every D is zero except those which arise from the representation $T_{\beta_1\beta_2\ldots\beta_k}$. These D's are, of course, all equal, and, indeed, in the case before us have the element $d_{rs}=1$ and all the other elements zero. They may be placed in any desired position on the diagonal.

Theorem VIII tells us that there is just one D, which we suppose to occupy the intersections of the first rows and columns in K, the rest of the matrix being zero. Moreover, it gives us the first f columns of L, and the first f rows of L^{-1}, thus

$$[A_{\rho\sigma}]=LKL^{-1}.$$

Hence $$A_{\rho\sigma}=\lambda_{\rho r}\lambda'_{s\sigma}.$$

This is zero when F_ρ' precedes $\bar{F}_r$ or F_σ' precedes $\bar{F}_s$ in any principal sequence. It is not zero when F_ρ' corresponds to F_r, and F_σ' to F_s. This result is important when considering products of functions.

THEOREM IX. *The seminormal unit ϖ_{uv} belongs to $T_{\alpha_1\ldots\alpha_h}$ of the letters a; the unit Π_{rs} belongs to $T_{\beta_1\ldots\beta_k}$ of the letters b. Then*

$$\varpi_{uv}\Pi_{rs}T_{\alpha_1+\beta_1,\,\alpha_2+\beta_2,\ldots}=\Sigma A_{\rho\sigma}\varpi_{u,\rho;\,v,\sigma},$$

where $A_{\rho\sigma}$ is zero when F_ρ' precedes $\bar{F}_r$ or F_σ' precedes $\bar{F}_s$ in any principal sequence, and is not zero when F_ρ' corresponds to F_r, and F_σ' to F_s.

IV. *The case when the tableaux have two rows only.*

14. When the tableaux contain only two rows the matrices Λ, Λ' may be calculated as follows. We consider the most general case of this, that is, the product of a matrix belonging to $T_{n-k,\,k}$ of $b_1, b_2, \ldots, b_n$ by a matrix belonging to $T_{\alpha+\gamma,\,\alpha}$ of $a_1, a_2, \ldots, a_m$ $(m=2\alpha+\gamma)$ expressed in terms of a matrix belonging to $T_{n-k+\alpha+\gamma-l,\,k+\alpha+l}$. Obviously there will be no such matrix when

$$l>n-2k \quad \text{or} \quad l>\gamma.$$

As before, we have to consider a matrix D belonging to $T_{n-k,\,k}$ and a matrix Δ corresponding to a distortion of γ places to the left in

$$T_{n-k-l,\,k+l}.$$

The equations for λ, λ' are simplified by writing

$$\lambda_{\rho r}[F_\rho']_1\,[F_\rho']_2 = \mu_{\rho r},$$

$$\lambda'_{r\rho}[F_r]_1\,[F_r]_2 = \mu'_{r\rho}.$$

Then, from the equations of § 10, it follows that, when a_t, a_{t+1} lie in the same row of F_ρ', but their interchange changes F_r into F_s,

$$\mu_{\rho r} = \mu_{\rho s}, \quad \mu'_{r\rho} = \mu'_{s\rho}.$$

When $(a_t a_{t+1})$ changes F_ρ' into F_σ' but leaves F_r unchanged,

$$\mu_{\rho r} = \mu_{\sigma r}, \quad \mu'_{r\rho} = \mu'_{r\sigma}.$$

When a_t, a_{t+1} lie in the same row of F_ρ' and the same column of F_r,

$$\mu_{\rho r} = 0 = \mu'_{r\rho}.$$

The tableaux will be defined by their second rows:

$$F_\rho' : b_{x_1} b_{x_2} \dots b_{x_{k+l}}; \quad x_1 < x_2 \dots < x_{k+l}.$$

$$F_r : b_{y_1} b_{y_2} \dots b_{y_k}; \quad y_1 < y_2 \dots < y_k.$$

Let F_r be a tableau in which $y_{u-1} < y_u - 1$, so that b_{y_u-1} lies in the upper row. Let its place be the ξ-th; then

$$\xi + u = y_u,$$

and the axial projection of $b_{y_u-1} b_{y_u}$ is

$$\xi - u + 1 = y_u - 2u + 1.$$

In the case of the distorted tableau F_ρ' this becomes

$$x_u - 2u + 1 + \gamma.$$

Thus
$$[F_r]_1 = \prod_{u=1}^{k} (y_u - 2u + 2);$$

$$[F_\rho']_2 = \prod_{u=1}^{k+l} (x_u - 2u + 1 + \gamma).$$

When $y_1 = 2$ and $x_1 > 2$,

$$\lambda_{\rho r} = 0 = \lambda'_{r\rho}.$$

When $x_1 > 3$, and $(b_2 b_3)$ changes F_r to F_s,

$$\lambda_{\rho s} = 0 = \lambda'_{s\rho}.$$

Thus, when $y_1 < x_1$, $\lambda_{\rho r}$ and $\lambda'_{r\rho}$ are zero.

Let us assume that, provided that $u < v$, $\lambda_{\rho r}$ and $\lambda'_{r\rho}$ are zero when $y_u < x_u$. Let

$$y_v = 2v < x_v, \quad y_{v-1} = 2(v-1) \geqslant x_{v-1}.$$

Then $\lambda_{\rho r}$, $\lambda'_{r\rho}$ are zero, for b_{2v}, b_{2v-1} lie in the same row of F_ρ' and the same column of F_r; and hence $\lambda_{\rho s}$ and $\lambda'_{s\rho}$ are also zero when

$$y_{v-1} = 2(v-1) \geqslant x_{v-1}, \quad y_v < x_v.$$

Taking $x_{v-1} < y_{v-1}$, we may use the transpositions $(b_{2v-2} b_{2v-1})$, etc., to show that the pair of elements is zero when $y_v < x_v$ and

$$x_{v-1} < 2(v-1) < y_{v-1}.$$

When $x_{v-1} = y_{v-1}$, we use $(b_{2v-2} b_{2v-1}) = \tau_1$, but now it is case (VI) of § 10 involving four elements

$$\lambda_{\rho s}, \quad \lambda_{\rho t}, \quad \lambda_{\sigma s}, \quad \lambda_{\sigma t},$$

where

$$\tau F_\rho' = F_\sigma', \quad \tau F_s = F_t.$$

In $\lambda_{\sigma s}$,

$$y_{v-1} = 2(v-1) < x_{v-1},$$

and, by hypothesis, $\lambda_{\sigma s}$ is zero; also $\lambda_{\rho s}$ is zero, hence $\lambda_{\rho t}$ and λ_{st} are zero. Proceeding thus, we see that $\lambda_{\rho r}$ is zero when $y_v < x_v$, provided that neither x_{v-2} nor y_{v-2} is greater than $2v-3$. This last proviso can be removed step by step as before. Hence, in all cases, $\lambda_{\rho r}$ and $\lambda'_{r\rho}$ vanish when $y_u < x_u$. Let

$$t+1 = y_u = x_{u+v}, \quad y_u > y_{u-1}+1, \quad x_{u+v} > x_{u+v-1}+1, \quad \tau = (b_t b_{t+1}).$$

$$\tau F_r = F_s, \quad \tau F_\rho' = F_\sigma'.$$

Then τ gives the equations

$$\Lambda_{\rho,\sigma}\{(t-2u+2)^{-1}\}_{r,s} = \{(t-2u-2v+2+\gamma)^{-1}\}_{\rho,\sigma}\Lambda_{r,s},$$

$$\{(t-2u+2)^{-1}\}_{r,s}\Lambda'_{\rho,\sigma} = \Lambda'_{r,s}\{(t-2u-2v+2+\gamma)^{-1}\}_{\rho,\sigma}.$$

As was seen in § 10, each equation gives two independent equations for the

λ elements. These equations are satisfied when

$$\mu_{\rho r}=(t-2u-v+2)(t-2u-v+3+\gamma)=\mu'_{r\rho},$$

$$\mu_{\sigma s}=(t-2u-v+1)(t-2u-v+2+\gamma)=\mu'_{s\sigma},$$

$$\mu_{\rho s}=-v(\gamma-v+1)=\mu'_{s\rho},$$

$$\mu_{\sigma r}=-(v+1)(\gamma-v)=\mu'_{r\sigma}.$$

When $v=0$, $\mu_{\rho s}=0$, as is already known, for here $y_u<x_u$; and in this case the ratios of the other three elements are uniquely given by the equations. We take then

$$\mu_{\rho r}=\mu'_{r\rho}=\prod_{u=1}^{k} B_u^{\rho r},$$

where $$B_u=(y_u-2u-v+1)(y_u-2u-v+2+\gamma),$$

when $$y_u=x_{u+v};$$

$$B_u=-(v+1)(\gamma-v),$$

when $$x_{u+v+1}>y_u>x_{u+v};$$

$$B_u=0,$$

when $$y_u<x_u;$$

and we observe that the result has been proved for $v=0$. Let

$$x_{u+v+1}=t+2, \quad y_{u+1}>t+2,$$

$$(b_{t+1}\,b_{t+2})\,F_r=F_w.$$

Then $$\mu_{\rho r}=\mu_{\rho w}.$$

But in $\mu_{\rho w}$, $$y_u=t+2=x_{u+v+1};$$

the factor B_u, as given above, is unchanged in value, hence if the value is true for v it is true for $v+1$, as regards the case $y_u=x_{u+v}$. The other case follows from the fact that, as we have seen, the set of values for v verifies the equations of (VI), § 10.

The equations are thus uniquely solved. Thus, in general, the group representation by the matrix for the tableaux of $T_{n-k-l,\,k+l}$ distorted by a lateral shift of the lower row γ places to the left is made up of each irreducible representation $T_{n-g,g}$, $g = 0, 1, \ldots, k+l$, or

$$g = k+l-\gamma,\ k+l-\gamma+1,\ \ldots\ k+l$$

when $\gamma < k+l$, once and once only. This may be further verified by noticing that the order of the reducible representation is equal to the sum of the orders of the irreducible representations. In the case when $l = 0$, the matrices Λ, Λ' thus obtained satisfy the equation

$$\Lambda' \Lambda = E.$$

The equations have determined the matrices except for a proportional factor. It is then sufficient to consider a single element of the product to determine them completely. We take

$$S = \sum_\rho \lambda'_{1\rho} \lambda_{\rho 1} = \{[F_1]_1 [F_1]_2\}^{-1} \sum_\rho \{[F_\rho']_1 [F_\rho']_2\}^{-1} \prod_u [B_u^{\rho, 1}]^{\lambda}.$$

In the first place the value of γ will be supposed greater than k; the modifications for smaller values of γ will be indicated later.

The tableau F_1 has

$$y_1 = n-k+1,\quad y_2 = n-k+2,\quad \ldots,\quad y_k = n.$$

Let F_ρ' have

$$x_{k-r} < n-k+1,\quad x_{k-r+1} \geqslant n-k+1.$$

Then

$$\prod_u B_u^{\rho, 1} = (-)^{k-r}(k-r)!\, \gamma_{k-r} (n-2k+r)_r (\gamma+n-2k+r+1)_r.$$

This is independent of the particular values of $x_1, x_2, \ldots, x_{k-r}$, provided that all are less than $n-k+1$, and of those of $x_{k-r+1}, \ldots, x_k$, provided that each is greater than $n-k$. The factors of $[F_\rho']_1$ due to the first $k-r$ suffixes $x_1, \ldots$ will be denoted by $[F'_{\rho_1}]_1$, and those due to the last r suffixes by $[F'_{\rho_2}]_1$. Thus

$$[F_\rho']_1 = [F'_{\rho_1}]_1 [F'_{\rho_2}]_1, \qquad [F_\rho']_2 = [F'_{\rho_1}]_2 [F'_{\rho_2}]_2.$$

The variation of each set of factors is independent of the other set. Hence

$$\sum_\rho \{[F_\rho']_1 [F_\rho']_2\}^{-1} = \left[\sum_{\rho_1} \{[F'_{\rho_1}]_1 [F'_{\rho_1}]_2\}^{-1}\right] \left[\sum_{\rho_2} \{[F'_{\rho_2}]_1 [F'_{\rho_2}]_2\}^{-1}\right].$$

Each of these sums is given by Theorem V, and thus their product is

$$\frac{\binom{n-k}{k-r}}{(\gamma+n-2k+r+1)_{k-r}\gamma_{k-r}} \cdot \frac{\binom{k}{r}}{(\gamma+n-2k+r+1)_r (\gamma+n-3k+2r)_r}.$$

Thus

$$(n-k+1)_k\,(n-k)_k\,S$$

$$= \sum_{r=0}^{k} \frac{(n-k)!\;k!\;\gamma_{k-r}\,(n-2k+r)_r\,(\gamma+n-3k+2r+1)}{(n-2k)!\;r!\;(\gamma+n-2k+1)_{k-r+1}}$$

$$=(n-k)_k \sum_{r=0}^{k} \frac{k!\;\gamma!\;(\gamma+p-k+r)!}{(\gamma-k+r)!\;(\gamma+p+1)!}\binom{p+r}{r}(\gamma+p-k+2r+1),$$

where $$p=n-2k.$$

It will now be shown that

$$\sum_{r=0}^{k}(\gamma+p-k+2r+1)\binom{p+r}{p}\binom{\gamma+p-k+r}{p}$$

$$=(p+1)\sum_{r=0}^{k}\left[\binom{p+r}{p}\binom{\gamma+p-k+r}{p+1}+\binom{p+r+1}{p+1}\binom{\gamma+p-k+r}{p}\right]$$

$$=(p+1)\Sigma=(p+1)\binom{\gamma+p+1}{\gamma}\binom{p+k+1}{k}.$$

This is equivalent to $S=1$.

The result for Σ is true for $k=0$ and for $k=1$, irrespective of the values of γ and p.

It will be assumed to be true for a particular value of k. Let Σ become Σ' when k is replaced by $k+1$, and at the same time γ by $\gamma+1$; then

$$\Sigma'-\Sigma=\binom{p+k+1}{p}\binom{\gamma+p+1}{p+1}+\binom{p+k+2}{p+1}\binom{\gamma+p+1}{p}$$

$$=\binom{\gamma+p+2}{\gamma+1}\binom{p+k+2}{k+1}-\binom{\gamma+p+1}{\gamma}\binom{p+k+1}{k}.$$

Thus the result is true in general. It is to be noticed that the number of terms in Σ is finite, and the whole can be expressed as an algebraic identity of degree $p+1$ in γ for positive integral values of p and k. When $\gamma \leqslant k$ the argument as given breaks down, for when $\gamma < k-r$ the number of terms ρ_1 is diminished, and indeed Theorem V is no longer applicable to these terms. But in this case the general result has a vanishing factor affecting these terms, and thus the general result is still true.

Let us now consider the general case when the tableau F_ρ' has $k+l$ letters in the second row. When

$$x_{k+l-r+1} \geqslant n-k+1 > x_{k+l-r},$$

$$\prod_{u=1}^{k} B_u^{\rho,1}=(-)^{k-r}(k+l-r)_{k-r}\,\gamma_{k-r}\,(n-2k-l+r)_r\,(\gamma+n-2k-l+r+1)_r.$$

Also, for all those tableaux F_ρ' for which these inequalities are true,

$$\Sigma\{[F_\rho']_1[F_\rho']_2\}^{-1} = \frac{\binom{n-k}{\gamma+l-r}}{(\gamma+n-2k-l+r+1)_{k+l-r}\,\gamma_{k+l-r}} \times \frac{\binom{k}{r}}{(\gamma+n-2k-2l+r+1)_r(\gamma+n-3k-2l+2r)_r}.$$

Now

$$(k+l-r)_{k-r}\,(n-2k-l+r)_r\binom{n-k}{k+l-r} = (n-k)_k\binom{n-2k}{l}.$$

Also

$$(\gamma+n-2k-l+1)_{k+l-2r}\,(\gamma+n-2k-2l+r+1)_r\,(\gamma+n-3k-2l+2r)_r$$
$$= (\gamma+n-2k-l+1)_l\,(\gamma+n-2k-2l+r+1)_{k+1}(\gamma+n-3k-2l+2r+1)^{-1}.$$

Let $$S = \sum_\rho \lambda'_{1\rho}\lambda_{\rho 1} = \Sigma\,[B_u^{\rho;\,1}]^2\{[F_\rho']_1[F_\rho']_2[F_1]_1[F_1]_2\}^{-1}.$$

Then

$$S(n-k+1)_k\,(\gamma+n-2k-l+1)_l\,\gamma_l\left\{\binom{n-2k}{l}\right\}^{-1}$$
$$= \sum_{r=0}^{k}\binom{k}{r}\frac{(k+l-r)_{k-r}\,(\gamma-l)_{k-r}\,(n-2k-l+r)_r \times (\gamma+n-2k-l+r+1)_r(\gamma+n-3k-2l+2r+1)}{(\gamma+n-2k-2l+r+1)_{k+1}}$$
$$= \frac{k!\,l!\,(p-l)!\,(\gamma-l)!}{(\gamma+p-l+1)!}\,Q,$$

where $p = n-2k$, and

$$Q = \phi(\gamma, p, l, k)$$
$$= \sum_{r=0}^{k}\binom{k+l-r}{l}\binom{\gamma+p-l+r+1}{l}\left[\binom{p-l+r+1}{p-l+1}\binom{\gamma+p-k-2l+r}{p-l} + \binom{p-l+r}{p-l}\binom{\gamma+p-k-2l+r}{p-l+1}\right].$$

It will now be shown that

$$S = \binom{n-2k}{l}\Big/\{(\gamma+n-2k-l+1)_l\,\gamma_l\};$$

this gives unity when $l = 0$, a result already established. It requires

$$Q = \frac{(\gamma+p-l+1)!}{(p-l)!\,(\gamma-l)!\,l!}\binom{p+k+1}{k}.$$

Now

$$\binom{k+l-r+1}{l}\binom{\gamma+p-l+r+2}{l}$$

$$=\binom{k+l-r}{l}\binom{\gamma+p-l+r+1}{l}+\frac{\gamma+p+k-l+3}{l}\binom{k+l-r}{l-1}\binom{\gamma+p-l+r+1}{l-1}.$$

Hence

$$\phi(\gamma+1, p, l, k+1)-\phi(\gamma, p, l, k)$$
$$=l^{-1}(\gamma+p+k-l+3)\phi(\gamma, p-1, l-1, k+1).$$

It is a simple matter to verify that the proposed result satisfies this equation. If then the result is true up to a given value of k, also for $k+1$ and $l-1$, irrespective of γ and p, then it is true for $k+1$ and l, and hence for $k+1$ and $l+1$, etc. Now it is true for $l=0$ and any k. Hence it is true for $k+1$ if it is true for k. It is easy to see that it is true for $k=0$, 1, or 2, and any l. Thus the result is true universally.

THEOREM X. *The exact seminormal unit Π_{uv} belongs to $T_{\alpha+\gamma,\alpha}$ of $a_1, a_2, \ldots, a_m$; the unit π_{rs} belongs to $T_{n-k,k}$ of $b_1, b_2, \ldots, b_n$. Then*

$$\Pi_{u,v}\pi_{r,s}T_{n-k+\alpha+\gamma-1,\alpha+k+l}$$
$$=\gamma_l(\gamma+n-2k-l+1)_l\left\{\binom{n-2k}{l}\right\}^{-1}\sum_{\rho,\sigma}\lambda_{\rho r}\lambda'_{s\sigma}\varpi_{u\rho;\,v\sigma},$$

where ϖ is the unit whose row tableau is formed by the combination of F_u and F_ρ' and column tableau by the combination F_v and F_σ'; where the second row letters defining the tableaux are

$$F_\rho':\quad b_{x_1}b_{x_2}\ldots b_{x_{k+l}};\quad x_1<x_2\ldots<x_{k+l};$$

$$F_r:\quad b_{y_1}b_{y_2}\ldots b_{y_k};\quad y_1<y_2\ldots<y_k;$$

$$\lambda_{\rho r}[F_\rho']_1[F_\rho']_2=\lambda'_{r\rho}[F_r]_1[F_r]_2=\prod_{t=0}^{k}B_t^{\rho,r};$$

and where $\qquad B_t^{\rho,r}=0,\quad y_t<x_t,$

$$B_t^{\rho,r}=(y_t-2t-w+1)(\gamma+y_t-2t-w+2),\quad y_t=x_{t+w},$$

$$B_t^{\rho,r}=-(w+1)(\gamma-w),\quad x_{t+w+1}>y_t>x_{t+w}.$$

15. Two results are given here without proof. They are, I think, of sufficient interest and value to be put on record; for they show the

manner in which the numerical values of the matrix elements depend on the tableaux, and particularly on the tableau functions; and they are of use in calculation. On the other hand, once the form of these results is known, the verification is merely a matter of repeated application of processes already given.

In Q.S.A., VI, § 17, the seminormal transforming matrix M for the representation on $T_{n-k,k}$ was obtained. When D is the seminormal and C the natural matrix,

$$D = M^{-1} CM.$$

The following theorem gives the reciprocal matrix M^{-1}. It has been demonstrated by induction in a manner which is practically a repetition of the argument of § 17 just referred to. Probably the methods of this paper would verify its truth more quickly.

THEOREM XI. *The reciprocal M^{-1} of the seminormal transforming matrix for the representation $T_{n-k,k}$ is*

$$M^{-1} = [m_{\lambda\mu}] = [p_{\lambda\mu}/p_{\lambda\lambda}],$$

where the rows and columns are defined by the tableaux F_λ, F_μ, with second row suffixes

$$\lambda_1, \lambda_2, \ldots, \lambda_k; \quad \mu_1, \mu_2, \ldots, \mu_k;$$

and where

$$p_{\lambda\lambda} = [F_\lambda]_2 = \prod_{r=1}^{k} (\lambda_r - 2r + 1), \quad p_{\lambda\mu} = \prod_{r=1}^{k} B_r^{\lambda\mu},$$

$$B_r^{\lambda\mu} = 0, \quad \mu_r > \lambda_r.$$

$$B_r^{\lambda\mu} = \lambda_{r-s} - 2r + s + 1, \quad \mu_r = \lambda_{r-s},$$

$$B_r^{\lambda\mu} = -s, \quad \lambda_{r-s+1} > \mu_r > \lambda_{r-s}.$$

The second result may be verified by the same method as the proof of Theorem X given here. It gives the expression for the permutation

$$(a_1 a_n)(a_2 a_{n-1}) \ldots (a_r a_{n+1-r}) \ldots,$$

which inverts the sequence of letters and may be called the inverting permutation; the result is as follows.

THEOREM XII. *The exact seminormal matrix*

$$K = [\kappa_{xy}]$$

is the expression for the inverting permutation in the representation $T_{n-k,k}$, *where*

$$[F_x]_1 [F_x]_2 \kappa_{xy} = \varpi_{xy},$$

$$\varpi_{xy} = \prod_{r=1}^{k} B_r^{xy},$$

$$B_r^{xy} = 0, \quad y_r < n+1-x_{k-r+1},$$

$$B_r^{xy} = (x_{k-r+1-s} - 2k+2r-1+s)(n-2r+2-s-x_{k-r+1-s}),$$
$$y_r = n+1-x_{k+1-r-s},$$

$$B_r^{xy} = -s(n-2k+s+1), \quad x_{k-r+1-s} < n+1-y_r < x_{k-r+2-s}.$$

V. *Algebraic equivalents of seminormal units.*

16. The relations between the seminormal units and the natural units are, of course, inherent in the equation

$$MDM^{-1} = C$$

connecting the seminormal matrix D to the natural matrix C. In a few cases the relations between the units may be very simply expressed.

THEOREM XIII. *The seminormal units* ϖ_{11}, ϖ_{ff}, ϖ_{1f}, ϖ_{f1} *are given by*

$$\varpi_{11} = f/(n!\, g) P_1 N_1 P_1,$$

$$\varpi_{ff} = f/(n!\, \gamma) N_f P_f N_f,$$

$$\varpi_{1f} = p_{1f} = f/n!\; P_1 \sigma_{1f} N_f,$$

$$\varpi_{f1} = 1/g\gamma\, N_f \sigma_{f1} P_1,$$

where g is the product of the orders of the groups of P, γ is that of the orders of the groups of N and p_{1f} is the natural unit.

In every case the matrices M, M^{-1} have the property that their elements m_{rs}, m'_{rs} are zero for $r > s$, and unity for $r = s$.

To find the seminormal unit ϖ_{11} in terms of the natural units, we put for D that matrix which has $d_{11} = 1$, and all other elements zero. Then

$$\varpi_{11} = \Sigma m'_{1r} p_{1r}.$$

Hence $$\varpi_{11} P_1 = p_{11} P_1 = f/n!\, P_1 N_1 P_1.$$

Now every permutation of P_1, expressed as a seminormal matrix, has zero for every element of the first row and column, except for that on the leading diagonal which is unity. Hence

$$\varpi_{11} P_1 = g\varpi_{11};$$

this establishes the value of ϖ_{11}. The value of ϖ_{ff} is obtained in the same way; that of ϖ_{1f} comes even more directly from the matrix equation.

Finally,

$$N_f \varpi_{f1} P_1 = g\gamma\varpi_{f1}.$$

Hence $$\varpi_{f1} = \lambda N_f \sigma_{f1} P_1.$$

The value of λ is obtained at once from the product

$$\varpi_{f1}\varpi_{1f} = \varpi_{ff}.$$

The *exact* units are the same for ϖ_{11} and ϖ_{ff}, but for ϖ_{1f} we must multiply by $[F_f]_1/[F_1]_1$ and for ϖ_{f1} by $[F_1]_1/[F_f]_1$: see Q.S.A., VI, §18.

COROLLARY. *It follows from the above reasoning that*

$$TP_1 = g\varpi_{11}, \quad TN_f = \gamma\varpi_{ff}.$$

17. A representation $T_{\alpha_1\alpha_2\ldots\alpha_h}$ is defined by the lengths of the rows of its tableaux, it might equally well be defined by the lengths of the columns $\beta_1 = h, \beta_2, \ldots, \beta_k$. Let us suppose the letters $a_1, a_2, \ldots, a_n$ which are the subject of permutations to be really h-ary variables. Then with any column $b_1 b_2 \ldots b_\beta$ in the tableau may be associated a determinant

$$\{b_1 b_2 \ldots b_\beta\}' b_{11} b_{22} \ldots b_{\beta\beta} = (b_1 b_2 \ldots b_\beta),$$

and with the whole tableau the product of these determinant factors.

THEOREM XIV. *The product*

$$X = (a_1 a_2 \ldots a_{\beta_1})(a_{\beta_1+1} a_{\beta_1+2} \ldots a_{\beta_1+\beta_2}) \ldots$$

has the substitutional properties of ϖ_{ff} *and no others.*

It is seen at once that

$$N_f X = \gamma X.$$

Hence $T'X$ is zero when the first of the differences $\beta_1 - \beta_1', \beta_2 - \beta_2', \ldots$ which does not vanish is positive. Also when $\beta_1' > \beta_1$,

$$N'X = 0,$$

for the variables contain no more than β_1 possible second suffixes. Similarly, this equation is true when the first of the differences

$$\beta_1'-\beta_1,\ \beta_2'-\beta_2,\ \ldots$$

which does not vanish is positive. Thus it is true unless $T'=T$. Hence

$$T'X=0,$$

unless $$T'=T.$$

Hence $$X=TX=1/\gamma\, TN_f X=\varpi_{ff}X,$$

which proves the statement.

COROLLARY. *The form, in single non-homogeneous variables,*

$$Y=\Delta_{1,2\ \ldots,\beta_1}\Delta_{\beta_1+1,\beta_1+2,\ldots,\beta_1+\beta_2}\cdots\Delta_{\beta_1+\beta_2+\ldots+\beta_{k-1}+1,\ldots,\beta_1+\beta_2+\ldots+\beta_k},$$

where $$\Delta_{1,2,\ldots,\beta_1}=\{a_1a_2\ldots a_{\beta_1}\}'\,a_1^{\beta_1-1}a_2^{\beta_1-2}\ldots a_{\beta_1-1},$$

has the same property.

18. By operating with different permutations on the function X of the last paragraph, all kinds of products like X may be obtained; their substitutional form is given by the operations of these permutations on ϖ_{ff}. Thus there may be obtained a linear function X_s of such products corresponding to every unit ϖ_{sf}; this will be called the symbolical product equivalent of the tableau F_s. In exactly the same way we may obtain from Y a determinantal equivalent of the tableau F_s.

The symbolical products may be multiplied out, in which case X_s becomes a sum of products $\prod_{r=1}^{n} a_{r,u}$ with numerical coefficients. These terms may be arranged in a sequence just as the sequence of tableaux, the second suffix u_r defining the row of the letter a_r.

THEOREM XV. *The first term of the symbolical product equivalent of any tableau F_s according to any principal sequence is that defined by the tableau itself.*

Here a principal sequence is one defined as such in §2. The term defined by the tableau itself is that obtained by writing for the second suffix u_r the number of the row in which the letter a_r lies in the tableau.

The theorem is obviously true for the tableau F_f. We assume its truth for all tableaux F_τ when $\tau \geqslant \sigma$. Then, when $\sigma < f$, there exists a transposition $(a_r a_{r+1})$ which changes F_σ into F_τ where $\tau > \sigma$.

Then $$(a_r a_{r+1}) X_\tau = (1-\rho) X_\sigma + \rho X_\tau,$$

where ρ is the inverse of the axial projection of $a_r a_{r+1}$ in either tableau. The earliest term of X_τ is S_τ, that given by the tableau F_τ. Hence the term S_σ given by the tableau F_σ, which is obtained by the transposition $(a_r a_{r+1})$ from S_τ, is certainly a term of $(a_r a_{r+1}) X_\tau$, while it is not a term of X_τ, hence it is a term of X_σ. Moreover, if possible, let S be an earlier term of X_σ than that defined by F_σ. Then $(a_r a_{r+1}) S = S'$ is a term of F_τ.

By hypothesis, S' is a later term than S according to any principal sequence. Taking the sequence from the first letter, if S' differs from S_τ in the first $r-1$ letters, then S' (and also S) is later than S_τ and also S_σ in this sequence. In the same way, taking the sequence from the last letter, if S' differs from S by a permutation affecting one of the last $n-r-1$ letters, it is later than S_τ in this sequence, and hence also S is later than S_σ in this sequence. Thus S cannot exist. And, in all cases, the term defined by the tableau F_s is the earliest term of the function X_s.

COROLLARY. *The function Y_s defined by the tableau F_s in the last paragraph has for its first term the term defined by the tableau.*

The above reasoning applies to this case.

VI. *The regular group matrix.*

19. The regular group matrix R of the symmetric group of n letters contains $n!$ rows and columns. It is usually defined as the matrix $[x_{PQ^{-1}}]$, where P is the same for each row, Q for each column, and P and Q are respectively each operation of the group in turn; x_P then represents a set of $n!$ arbitrary variables. Then there is a non-singular matrix Λ for which

$$\Lambda^{-1} R \Lambda = D,$$

where D is a matrix which consists of Σf smaller matrices about the principal diagonal and zeros elsewhere. These smaller matrices may be made to consist of a succession of sets each set containing f identical group matrices of order f. Further, we may suppose Λ chosen so that these group matrices are the exact seminormal group matrices. The matrix Λ

consists of a series of bands of columns. Corresponding to a particular group matrix of D of order f, there is a band of f columns which occupy the same position in Λ as the columns on which this matrix lies in D. And further, corresponding to the f identical group matrices in D, there will be f bands, of f columns each, which, however, are not identical since Λ is non-singular. In the same way there are bands of rows in Λ^{-1}. Schur has shown that the only matrices permutable with D are of the form K, where in K the sub-matrix defined by the f^2 rows and columns, occupied by a set of f identical group matrices of order f in D, is

$$[B_{rs} E_f],$$

where B_{rs} are f^2 arbitrary constants, and all the other elements are zero.

Then Λ may be replaced by ΛK, as the most general form of Λ. This, in effect, replaces the f bands corresponding to a particular representation by any f linear functions of them which are independent.

Let us now consider a transposition $\varpi = (a_\alpha a_{\alpha+1})$ of a pair of consecutive letters in R; the coefficient of x_ϖ is represented by $\frac{1}{2}n!$ quadratic matrices about the leading diagonal of the form

$$\begin{bmatrix} 0 & 1 \\ 1 & 0 \end{bmatrix}.$$

In D we must consider each group matrix separately, and the coefficient of x_ϖ is either ± 1 on the leading diagonal or else a quadratic matrix of the form

$$\begin{bmatrix} -\rho & 1-\rho \\ 1+\rho & \rho \end{bmatrix}$$

about this diagonal, and zeros elsewhere. Thus the equation

$$R\Lambda = \Lambda D$$

gives us a series of equations, of one of the two forms

(i) $$\begin{bmatrix} 0 & 1 \\ 1 & 0 \end{bmatrix}\begin{bmatrix} \lambda_{r\sigma} \\ \lambda_{s\sigma} \end{bmatrix} = \pm\begin{bmatrix} \lambda_{r\sigma} \\ \lambda_{s\sigma} \end{bmatrix},$$

whence $$\lambda_{r\sigma} = \pm\lambda_{s\sigma}.$$

(ii) $$\begin{bmatrix} 0 & 1 \\ 1 & 0 \end{bmatrix}\begin{bmatrix} \lambda_{r\sigma} & \lambda_{r\tau} \\ \lambda_{s\sigma} & \lambda_{s\tau} \end{bmatrix} = \begin{bmatrix} \lambda_{r\sigma} & \lambda_{r\tau} \\ \lambda_{s\sigma} & \lambda_{s\tau} \end{bmatrix}\begin{bmatrix} -\rho & 1-\rho \\ 1+\rho & \rho \end{bmatrix},$$

whence

$$\left.\begin{aligned}(1-\rho)\lambda_{s\sigma} &= -\rho\lambda_{s\tau}+\lambda_{\tau\tau},\\ (1-\rho)\lambda_{\tau\sigma} &= -\rho\lambda_{\tau\tau}+\lambda_{s\tau}.\end{aligned}\right\}, \qquad \left.\begin{aligned}\lambda_{\tau\sigma} &= -\rho\lambda_{s\sigma}+(1+\rho)\lambda_{s\tau},\\ \lambda_{s\sigma} &= -\rho\lambda_{\tau\sigma}+(1+\rho)\lambda_{\tau\tau}.\end{aligned}\right\}.$$

Similarly, if

$$\Lambda^{-1}=[\mu_{rs}],$$

$$(1+\rho)\mu_{\sigma\tau} = -\rho\mu_{\tau\tau}+\mu_{\tau s},$$

$$(1+\rho)\mu_{\sigma s} = -\rho\mu_{\tau s}+\mu_{\tau\tau}.$$

Now the rows and columns of R may be supposed to be defined by Φ, which is nothing more than the n letters $a_1, a_2, \ldots, a_n$ arranged in some particular sequence. Then, in the above, if r is the numerical equivalent of Φ, s is that of $(a_\alpha a_{\alpha+1})\Phi$. Consider the last column of a band in Λ, the remaining $f-1$ columns of this band are obtained from it by the equations (ii) above.

The elements of this last column will be written λ_{rf}, where r has the values 1, 2, ..., $n!$, corresponding to the $n!$ sequences Φ. Corresponding to $r=1$, the letters in Φ are arranged in order of the suffixes. An arbitrary function of n variables has, in general, $n!$ different values corresponding to the possible $n!$ permutations of the variables. We may then suppose λ_{rf} to be such a function ϕ_r of

$$A_1, A_2, \ldots, A_n,$$

derived from a function $\lambda_{1f}=\phi_1$ by permutations. But we define the connection between ϕ_r and ϕ_s thus. Let $(a_\alpha a_{\alpha+1})=\varpi$ change Φ_r to Φ_s, let a_α occupy the p-th place in Φ_r and the q-th place in Φ_s, then

(IV) $$\phi_s=(A_p A_q)\phi_r.$$

Corresponding to any permutation ϵ_r for which

$$\Phi_r=\epsilon_r\Phi_1,$$

there exists a permutation $\bar{\epsilon}_r$ for which

$$\phi_r=\bar{\epsilon}_r\phi_1.$$

Then, if ϵ_r changes a_α into a_{u_α}, $\bar{\epsilon}_r$ will change A_{u_α} into A_α, as can be easily proved by induction, so that

$$\bar{\epsilon}_r=\epsilon_r^{-1}.$$

Let $\beta_1, \beta_2, \ldots$ be the column lengths in the tableaux of the representations considered; then the equations (i) above show that ϕ_1 is a function which changes sign for every odd permutation of the groups

$$\{A_1 A_2 \ldots A_{\beta_1}\}, \quad \{A_{\beta_1+1} \ldots A_{\beta_1+\beta_2}\}, \quad \ldots.$$

Also when S is a substitutional expression which satisfies the equation

$$TS = 0,$$

the sum
$$[S\Phi, f] = 0,$$

where
$$[\Phi_r, f] \equiv \lambda_{rf},$$

and S is supposed written out with the appropriate numerical coefficients. This follows from the equation

$$R\Lambda = \Lambda D$$

by giving the variables x_Q in R the values corresponding to the coefficients of S, and then considering the f column in the product; this column vanishes in D with TS. Let $\bar{S}$ be the result of replacing the permutations of a in S by the inverse permutations of A, then

$$\bar{S}\lambda_{rf} = 0.$$

Hence
$$T'\lambda_{rf} = 0,$$

where T' is any representation other than T, and hence $\lambda_{1f} = \phi_1$ is of the nature ϖ_{ff}. The appropriate form of ϕ_1 is then that defined in § 18 as Y_f.

The remaining columns of this band are given by the equations (ii). It is to be observed that certain substitutional equations are involved.

That these are actually satisfied may be seen as follows. Let us assume that

$$\lambda_{r\tau} = \varpi_{\tau\tau}\lambda_{r\tau},$$

and that
$$\lambda_{s\tau} = (a_\alpha a_{\alpha+1})\,\varpi_{\tau\tau}\lambda_{r\tau}.$$

Then
$$\varpi_{\sigma\sigma}(a_\alpha a_{\alpha+1}) = \begin{bmatrix} 1 & 0 \\ 0 & 0 \end{bmatrix}\begin{bmatrix} -\rho & 1-\rho \\ 1+\rho & \rho \end{bmatrix}$$

$$= -\rho\varpi_{\sigma\sigma} + (1-\rho)\,\varpi_{\sigma\tau},$$

and
$$\varpi_{\tau\tau}(a_\alpha a_{\alpha+1}) = (1+\rho)\,\varpi_{\tau\sigma} + \rho\varpi_{\tau\tau}.$$

Now, by equations (ii),

$$(1-\rho)\lambda_{r\sigma} = -\rho\lambda_{r\tau}+\lambda_{s\tau},$$

whence $$(1-\rho)\varpi_{\sigma\sigma}\lambda_{r\sigma} = \varpi_{\sigma\sigma}\lambda_{s\tau} = (1-\rho)\varpi_{\sigma\tau}\lambda_{r\tau};$$

and similarly $$\varpi_{\tau\tau}\lambda_{r\sigma} = 0.$$

In the same way it may be shown that every

$$\varpi_{\xi\xi}\lambda_{r\sigma} = 0, \quad \xi \neq \sigma;$$

hence $$\varpi_{\sigma\sigma}\lambda_{r\sigma} = \lambda_{r\sigma}.$$

It is to be remarked that it does not signify that $\lambda_{r\tau}$ is not a function of a but of A; all that matters is that a permutation of the a's on $\lambda_{r\tau}$ has a definite meaning. There are f independent bands connected with T, and all others are expressible as a linear function of such f bands. We conclude that it is possible to choose f sets of quantities

$$A_{r,1}, A_{r,2}, \ldots, A_{r,n} \quad (r = 1, 2, \ldots, f);$$

such that, for any other set,

$$Y_f = B_1 Y_f^{(1)} + B_2 Y_f^{(2)} + \ldots + B_f Y_f^{(f)}.$$

Next consider $\Lambda^{-1} = [\mu_{rs}]$. There are f bands of rows connected with T. The last row of a band must also have the form Y_f; but equations (ii) show that the $[\mu]$ rows differ from the $[\lambda]$ columns in that they are changed by the transposition $(a_\alpha a_{\alpha+1})$ in the ratio

$$\lambda_{-\sigma} : \mu_{\sigma-} = 1+\rho : 1-\rho = [F_\sigma]_1 [F_\sigma]_2 : [F_\tau]_1 [F_\tau]_2.$$

THEOREM XVI. *The matrix Λ of degree $n!$ which transforms the regular group matrix of the symmetric group of degree n into its irreducible seminormal components is made up of a set of f bands of f columns for each representation T. The elements of a column are obtained by permutation from a function Y_σ of n variables, corresponding to the tableau F_σ which defines the column. A similar result is true for the rows of the inverse matrix Λ^{-1}.*

20. LEMMA. *In the orthogonal group matrix, if every permutation is replaced by its inverse the unit ϖ_{rs} becomes ϖ_{sr}.*

It was shown (Q.S.A., VI, Theorem V) that in the orthogonal matrix for a transposition of consecutive letters, the quadratic matrices which appear have the form

$$\begin{bmatrix} -\rho & \sqrt{(1-\rho^2)} \\ \sqrt{(1-\rho^2)} & \rho \end{bmatrix};$$

hence any transposition of consecutive letters is of the form

$$\Sigma \lambda_r \varpi_{rr} + \Sigma \lambda_{rs}(\varpi_{rs} + \varpi_{sr}).$$

If any permutation is

$$\Sigma \mu_r \varpi_{rr} + \Sigma \mu_{rs} \varpi_{rs} + \Sigma \mu_{sr} \varpi_{sr},$$

it may be obtained as a product of consecutive transpositions; its inverse is obtained by taking this product in the reverse order and is therefore

$$\Sigma \mu_r \varpi_{rr} + \Sigma \mu_{rs} \varpi_{sr} + \Sigma \mu_{sr} \varpi_{rs}.$$

Hence, also, the result of replacing each permutation by its inverse is to change ϖ_{rs} into ϖ_{sr} and to leave ϖ_{rr} unchanged.

The change from the orthogonal units to the seminormal units is effected by replacing each unit ϖ_{rs} in an expression given in orthogonal units by

$$\varpi_{rs} \left\{ \frac{[F_s]_1 [F_s]_2}{[F_r]_1 [F_r]_2} \right\}^{\frac{1}{2}}.$$

Thus when each permutation is changed into its inverse the seminormal unit ϖ_{rr} is unaltered, but

$$[F_s]_1 [F_s]_2 \varpi_{rs} \quad \text{is changed into} \quad [F_r]_1 [F_r]_2 \varpi_{sr}.$$

21. Consider a product of two functions

$$\phi(x_1, x_2, \ldots, x_n) \psi(y_1, y_2, \ldots, y_n).$$

Let G be the simultaneous symmetric group of x and y. That is when x_r and x_s are interchanged, then y_r and y_s are also interchanged, and G is the sum of these $n!$ permutations. Let k be a permutation of the x's and κ the same permutation of the y's, so that $k\kappa$ is contained in G. Also let p_{rr} be a seminormal unit of the x's, and π_{rr} the same unit of the y's. Then

$$G\phi\, \kappa^{-1} \psi = G k \kappa \phi\, \kappa^{-1} \psi = G k \phi\, \psi;$$

and hence, by the last paragraph,

$$G\phi\, \varpi_{rr} \psi = G p_{rr} \phi\, \psi.$$

Hence $$Gp_{rr}\phi\,\varpi_{ss}\psi = 0,$$

unless the units not only belong to the same representation, but are identical.

Hence the sum of products of corresponding elements of two different columns of Λ is zero unless the two columns are defined by the same tableau of the same representation. Again, the second pair of equation (ii) § 19, give, on squaring and adding,

$$(\lambda_{r\sigma}^2+\lambda_{s\sigma}^2)(1-\rho^2)-(1+\rho)^2(\lambda_{r\tau}^2+\lambda_{s\tau}^2) = -2(1+\rho)\{\lambda_{s\sigma}\lambda_{s\tau}+\lambda_{r\sigma}\lambda_{r\tau}\}.$$

Now sum for the whole length of the columns; the sum on the right is zero and we have

$$(1-\rho)\sum_r \lambda_{r\sigma}^2 = (1+\rho)\sum_r \lambda_{r\tau}^2,$$

whence $\sum_r \lambda_{r\sigma}^2$ is proportional to $[F_\sigma]_1\,[F_\sigma]_2$.

THEOREM XVII. *The functions Y_σ of $A_1, A_2, \ldots, A_n$ and Z_σ of $B_1, B_2, \ldots, B_n$ are formed from the tableau F_σ. Then*

$$\Sigma\, Y_\sigma Z_\tau = 0 \quad (\sigma \neq \tau);$$

$$\Sigma\, Y_\sigma Z_\sigma = \Phi\,[F_\sigma]_1\,[F_\sigma]_2;$$

where the sum extends to all simultaneous permutations of A and B, and Φ is a function which depends only on the particular representation T.

This corresponds to the fact that the elements of Λ and Λ^{-1} can each be put into a form Y_σ while the product of the two matrices is E.

VII. *The double group matrix.*

22. The regular group matrix $[x_{PQ^{-1}}]$ is reducible to Σf irreducible matrices which are identical in k sets of f matrices each. Frobenius* showed that the matrix $[x_{PQ^{-1}}+y_{Q^{-1}P}]$ is reducible to k irreducible matrices of order f^2, each of which reduces to f identical matrices in x or y when the other set of variables is suppressed. It follows from the results of Schur† that, if D_1 is the double regular group matrix transformed so that its k components occupy positions as principal minors, the only matrix permutable with D_1 is that which has the place of these minors

* *Berliner Sitzungsberichte*, 1896, 1343–1382 (1372).

† *Berliner Sitzungsberichte*, 1905, 406–432, § 3.

filled by matrices BE_{f^2} and zeros elsewhere. Thus it has k arbitrary constants, one for each irreducible representation of the symmetric group. Then for the double regular group matrix there is a matrix Λ_1 such that

$$R_1\Lambda_1 = \Lambda_1 D_1.$$

This matrix Λ_1 is obtained by giving particular values to the constants in the transforming matrix Λ already discussed. It consists of k bands of f^2 columns in a band, and each band as a whole is unique except for one arbitrary constant which multiplies every element of the band.

Now $[y_{Q^{-1}P}]$ may be looked upon as a group matrix, indeed, if $y_P = z_{P^{-1}}$,

$$[y_{Q^{-1}P}] = [z_{P^{-1}Q}].$$

This is the same as the matrix $[x_{PQ^{-1}}]$ except for a rearrangement of the rows and columns. The regular matrix R is orthogonal in the sense that the coefficient of every variable x_P is an orthogonal matrix, and hence it gives an orthogonal representation of the group. When the matrix Λ is orthogonal the matrix D is also orthogonal. In this case Λ^{-1} is obtained from Λ by changing columns into rows and *vice versa*. Thus, when in D_1 the irreducible minors of order f^2 are orthogonal so far as the x's alone are concerned, they are also so as regards the y's alone. Does it follow that when this minor appears as f separate minors in x about the leading diagonal, the variables y being suppressed, the same property will be preserved for the y's when the variables x are suppressed? The most obvious arangement is as follows. Consider the principal minor of D_1 of f^2 rows and columns corresponding to a particular representation T. The rows and columns are made to correspond to a double set of tableaux FF' one of which refers to the x units and the other to the y units. The elements follow the rule that that element whose row and column tableaux are $F_r F'_{r'}$, $F_c F'_{c'}$ respectively contains x, only when $F'_{r'}$ and $F'_{c'}$ are the same, and then as the function given by the unit ω_{rc}; and similarly it contains y, only when F_r and F_c are the same, and then as the function given by $\omega'_{r'c'}$, it being remembered that the y variables are associated with the permutations in the inverse manner to the x variables. Thus the elements in the leading diagonal are functions of both x and y, and are unique in this. It will be shown that for all representations of unit rank the double group matrix can be transformed into this form, and also in some other cases, but that it is not possible in general.

Consider then the elements of Λ_1 in that column which corresponds to the double tableau $F_f F_f'$, supposing that this arrangement is possible. Having regard to the transformation of the x's alone, each element must have the form Y_f, just as in the seminormal case; for the distinction between the seminormal and the orthogonal cases arises only when passing from one row to another. As regards the variables y, we suppose that the elements λ_{rf} can be alternatively expressed as functions of a set of quantities A_1', A_2', ..., A_n'. Then, whereas in the former case the rule (IV), § 19, was obtained, here we must have

$$(A'_a A'_{a+1})\lambda'_{rf} = \lambda'_{sf},$$

where λ'_{rf} is λ_{rf} differently expressed.

The first element of the column λ_{1f} is that function Y_f derived directly from F_f by writing the variables A_a in the tableau; and λ'_{1f} is the result of replacing A_a by A_a'. But this relation between λ_{rf} and its equivalent λ'_{rf} does not hold in general. Consider a representation $T_{n-k+1,1^{k-1}}$,

$$\lambda_{1f} = Y_f = \Delta_{1,2,\dots,h} = \Delta'_{1,2,\dots,h}.$$

The suffixes denote the variables present. The function $\Delta'_{1,2,\dots,h-1,h+1}$ is the value of λ'_{rf} when r corresponds to even permutations ϵ of

$$\{a_1 a_2 \dots a_{h-1} a_{h+1}\},$$

where $$\Phi_r = \epsilon\Phi_{r_1} = \epsilon(a_h a_{h+1})\Phi_1;$$

and of $-\lambda'_{rf}$ for odd permutations ϵ. On the other hand, the value of λ_{rf}, while governed by the same rule of signs, has each of the variables $A_1, A_2, \dots, A_{h-1}$ omitted in turn in this range. Hence, in order that the two forms λ, λ' of the element may co-exist, we obtain the equations (after division by a common factor)

$$\frac{\Delta_{2,3,\dots,h}}{A_1-A_{h+1}} = \frac{-\Delta_{1,3,\dots,h}}{A_2-A_{h+1}} = \dots.$$

Whence

$$\sum_{a=1}^{h} A_a = hA_{h+1}, \quad \sum_{a=1}^{h} A_a^2 = hA_{h+1}^2, \quad \dots, \quad \sum_{a=1}^{h-1} A_a^{h-1} = hA_{h+1}^{h-1}.$$

And therefore $A_1, A_2, \dots, A_h$ are the roots of

$$(x-A_{h+1})^h = (-A_{h+1})^h + B,$$

Let ϖ be a primitive h-th root of unity, we may write

$$A_a = A_{h+1} + \varpi^a \xi.$$

In the same way A_{h+2} may be treated with the same result, so that

$$A_{h+1} = A_{h+2} = \dots = A_n.$$

Since the functions are functions of the differences only, they are unaffected by the subtraction of A_{h+1} from each variable; thus we have

$$A_a = \varpi^a \xi \quad (a = 1, 2, \ldots, h),$$

$$A_a = 0 \quad (a > h).$$

For orthogonal transformations the relations between different columns are given by

$$\begin{bmatrix} 0 & 1 \\ 1 & 0 \end{bmatrix} \begin{bmatrix} \lambda_{r\sigma} & \lambda_{r\tau} \\ \lambda_{s\sigma} & \lambda_{s\tau} \end{bmatrix} = \begin{bmatrix} \lambda_{r\sigma} & \lambda_{r\tau} \\ \lambda_{s\sigma} & \lambda_{s\tau} \end{bmatrix} \begin{bmatrix} -\rho & \sqrt{(1-\rho^2)} \\ \sqrt{(1-\rho^2)} & \rho \end{bmatrix},$$

whence

$$\sqrt{(1-\rho^2)}\,\lambda_{r\sigma} = -\rho\lambda_{r\tau} + \lambda_{s\tau}.$$

The remaining f^2-1 columns may be calculated by means of this equation from that already determined.

23. As an example, we consider the double regular group matrix of degree 3. The sequence of arrangements Φ is taken to be

$$\begin{matrix} abc, & bac, & acb, & bca, & cab, & cba \\ 1 & 2 & 3 & 4 & 5 & 6 \end{matrix}$$

The suffix r of each variable represents the permutation which changes abc into the corresponding arrangement Φ_r. Then

$$R_1 = \begin{bmatrix} x_1+y_1 & x_2+y_2 & x_3+y_3 & x_5+y_5 & x_4+y_4 & x_6+y_6 \\ x_2+y_2 & x_1+y_1 & x_4+y_5 & x_6+y_3 & x_3+y_6 & x_5+y_4 \\ x_3+y_3 & x_5+y_4 & x_1+y_1 & x_2+y_6 & x_6+y_2 & x_4+y_5 \\ x_4+y_4 & x_6+y_3 & x_2+y_6 & x_1+y_1 & x_5+y_5 & x_3+y_2 \\ x_5+y_5 & x_3+y_6 & x_6+y_2 & x_4+y_4 & x_1+y_1 & x_2+y_3 \\ x_6+y_6 & x_4+y_5 & x_5+y_4 & x_3+y_2 & x_2+y_3 & x_1+y_1 \end{bmatrix},$$

$$\Lambda_1 = \frac{1}{\sqrt{(12)}} \begin{bmatrix} \sqrt{2} & 2 & 0 & 0 & 2 & \sqrt{2} \\ \sqrt{2} & 2 & 0 & 0 & -2 & -\sqrt{2} \\ \sqrt{2} & -1 & \sqrt{3} & \sqrt{3} & 1 & -\sqrt{2} \\ \sqrt{2} & -1 & -\sqrt{3} & \sqrt{3} & -1 & \sqrt{2} \\ \sqrt{2} & -1 & \sqrt{3} & -\sqrt{3} & -1 & \sqrt{2} \\ \sqrt{2} & -1 & -\sqrt{3} & -\sqrt{3} & 1 & -\sqrt{2} \end{bmatrix}.$$

The orthogonal double group matrix T_{21} is then

$$\begin{bmatrix} X_{11}+Y_{11} & X_{12} & Y_{12} & 0 \\ X_{21} & X_{22}+Y_{11} & 0 & Y_{12} \\ Y_{21} & 0 & X_{11}+Y_{22} & X_{12} \\ 0 & Y_{21} & X_{21} & X_{22}+Y_{22} \end{bmatrix},$$

where
$$X_{11}=x_1+x_2-\tfrac{1}{2}(x_3+x_4+x_5+x_6),$$
$$X_{22}=x_1-x_2-\tfrac{1}{2}(x_4+x_5-x_3-x_6);$$

and Y_{11}, Y_{22} are the same functions of y:

$$X_{12}=\tfrac{1}{2}\sqrt{3}(x_4-x_5+x_3-x_6),\quad X_{21}=\tfrac{1}{2}\sqrt{3}(x_5-x_4+x_3-x_6),$$
$$Y_{12}=\tfrac{1}{2}\sqrt{3}(y_5-y_4+y_3-y_6),\quad Y_{21}=\tfrac{1}{2}\sqrt{3}(y_4-y_5+y_3-y_6).$$

It is necessarily irreducible.

24. Consider next the representation T_{22}. Here

$$\lambda_{1f}=(A_1-A_2)(A_3-A_4)=(A_1'-A_2')(A_3'-A_4').$$

There are four values of r for which λ'_{rf}, and four for which $-\lambda'_{rf}$, is

$$(A_1'-A_3')(A_2'-A_4').$$

In the first four cases, λ_{rf} is $(A_1-A_3)(A_2-A_4)$, in the other four λ_{rf} is $(A_1-A_4)(A_2-A_3)$; hence

$$(A_1-A_3)(A_2-A_4)+(A_1-A_4)(A_2-A_3)=0.$$

That is, if A_1, A_2, A_3, A_4 represent points on a straight line, the pairs A_1A_2 and A_3A_4 must be harmonically conjugate. This will, of course, give virtually the same double group matrix as in the former case; there are, indeed, four times as many variables, but they correspond in sets of four to each of the former variables.

Similarly, for the representation T_{33} (which again is practically the same as T_{32}), the pairs A_1A_2, A_3A_4, A_5A_6, on which λ_{rf} depends, must represent pairs of points mutually harmonic two and two. Such pairs are known to exist and hence the double group matrix can be transformed into the required form. But for a representation T_{kk}, where $k>3$, since it is not possible to find more than three pairs of mutually harmonic points, it is not possible to express the group matrix in the form suggested.

It is not difficult to show that, apart from the representations of unit rank, the only representations for which this is possible are $T_{n-2,2}$, T_{33}, and the corresponding representations when rows and columns are interchanged in the tableaux, *i.e.* $T_{2^2 1^k}$ and T_{2^3}.

VIII. *Concomitants given by seminormal units.*

25. In what follows the results which have been obtained here are applied to the theory of algebraic invariants. The general method of the applications of this analysis to invariants has been explained (Q.S.A., VII). In the first place, with the use of natural units n_{rs}, concomitant types take the form

$$Gn_{rs}X,$$

where G is a product of positive symmetric groups, expressing the facts that the first n_1 letters are equivalent as all representing the first quantic, the next n_2 letters are equivalent and so on. Here we are dealing with the seminormal units ϖ_{rs} and concomitant types of the form

$$G\varpi_{rs}X.$$

In both cases the second suffix s of the unit may be always taken as the same; in such applications we are only concerned with one column of the matrix. Now the operator G really means taking a sum of different terms, or else a number of elements in the column. These are obtained, from the one element expressed, by permuting the letters belonging to each quantic among themselves in all possible ways; the actual numerical coefficients attached to the different elements and units are given by Theorem V; and we see that in F_r no two elements belonging to any quantic may lie in the same column. It is not contemplated to write out the expression in full, but the power to do so gives complete definiteness to the abbreviated expression.

26. A concomitant of a single quantic requires another substitutional operator Γ in either set of units. This represents the fact that all the quantics concerned are the same, or that each set of n letters is interchangeable with any other set. Thus the full expression is $\Gamma Gn_{rs}X$, or $\Gamma G\varpi_{rs}X$. Like G, the operator Γ represents a sum. Any one of the units occurring in this sum might be taken for n_{rs} (or ϖ_{rs}) to give the expression, but for the sake of the developments of the theory the earliest unit according to the sequence of tableaux selected is chosen. This unit or its tableau defines also a gradient, a product of the coefficients of the quantic; and, conversely, the gradient defines the concomitant. Every gradient defines a concomitant in this way, some of these concomitants are zero, and some can be equally well expressed by earlier gradients. A

concomitant as thus defined by the earliest possible gradient is called irreductible; naturally the term implies also a definite sequence. There is a distinction to be remembered between the natural and seminormal units; in the natural units the letters of permutation are all on the same footing and their sequence is of no particular consequence until the sequence of tableaux has to be determined; in the seminormal units the sequence of the letters is of the fundamental texture of the theory. The terms of the sum given by Γ in this latter case are to be supposed calculated as from the values of the transpositions of consecutive letters, they are not arrived at by mere permutation. Further, in the natural units any pre-arranged sequence of tableaux might be used, and thus any sequence of gradients might be conceived as the foundation for the representation of concomitants. In seminormal units a practical limitation in this matter is laid upon us by the sequence used in the transformation from natural to seminormal units. There is still a considerable choice, as is shown by Theorem IV, but in seminormal units the sequence of tableaux must be a principal sequence, such as is defined in § 2.

27. Consider the transformation from natural to seminormal units. The natural matrix C becomes the seminormal matrix D, where

$$D = M^{-1}CM,$$

and M, M^{-1} are matrices with nothing but zero elements below leading diagonals of positive units. Thus the natural unit n_{rs} becomes

$$\sum_{uv} m'_{ur} m_{sv} \varpi_{uv},$$

where m, m' are elements of M, M^{-1}, respectively, and $\varpi_{u,v}$ are the seminormal units. The coefficient of ϖ_{uv} is zero unless $u \leqslant r$, $v \geqslant s$; also the coefficient of ϖ_{rs} is unity. Hence an algebraic function given in natural units in the form

$$n_{rs} X$$

becomes
$$\sum_{u=1}^{r} \lambda_u \varpi_{uv} X',$$

where $\lambda_r = 1$, when we change to seminormal units. As has been observed, we are only really concerned with one column of the matrix; the transformation leads to several columns of the seminormal matrix, but they are all proportional to each other, and they may be absorbed into one by the change from X to X'; thus the coefficient λ_u only really relates to the first suffix of the unit. We deduce at once from the above:

THEOREM XVIII. *The difference between the algebraic forms or concomitants given by the same given tableau by natural and by seminormal units is a linear function of forms given by earlier tableaux in any principal sequence in either set of units.*

The result as to concomitants follows at once when n_{rs} is taken as the earliest unit in the concomitant $\Gamma G n_{rs} X$.

28. It is necessary to be clear as to what constitutes a principal sequence. This last theorem shows that, in the case of concomitants of a single form, the sequence, and consequently the determining tableau F_r, may be selected for the natural units first, and then the sequence of letters of permutation determined accordingly as a necessary preliminary to the transformation to seminormal units. Thus, for binary forms under certain conditions, a gradient

$$A_0^{\lambda_0} A_2^{\lambda_2} \dots A_\delta^{\lambda_\delta}$$

has been found to represent an irreductible seminvariant. This is first polarized $\delta - 1$ times in order to obtain δ different sets of letters; and a particular term in the result is selected. In the sequence of letters of permutation the letters corresponding to a coefficient suffix zero are taken first, then those of another, and so on, until these are exhausted, then those of suffix 2, and so on. With such a method both the sequences A and B are principal sequences. Quite another sequence is used for the invariants of the quaternary cubic (Q.S.A., VII, section XI), but it is not difficult to see that the sequence of the letters of permutation may be selected, so that this also is a principal sequence. The sequences C and D would require the separation of the letters of the different sets; the sequence of letters of permutation being begun with one letter from each set, then a second, and so on. They do not appear to be suitable sequences for use with seminormal units.

29. An empirical theorem has been stated (Q.S.A., VII, section XII, §51). No proof had been found, but very strong experimental evidence of its truth had been obtained. This theorem will now be demonstrated. With the demonstration its scope is greatly extended, and accurately defined. It may be regarded as the fundamental theorem of this approach to the

algebra of invariants. It may be enunciated thus:

THEOREM XIX. *With any principal sequence, the leading gradient of an irreductible concomitant given in seminormal units is that which defines the concomitant. And the same gradients define the irréductible concomitants in natural units.*

The last part of the enunciation is an immediate consequence of Theorem XVIII. The first part follows from Theorem XV, § 18, when it is remembered that the tableau taken as representative of the whole form is that of the earliest unit in the sum. The theorem quoted applies to every term of the sum, and hence the leading gradient is the leading gradient of the first term. The second empirical theorem of the last paper follows at once from this, as was pointed out there. It thus becomes a demonstrated theorem and is no longer empirical. It is of much less importance.

30. The product of two concomitants has for its leading gradient the product of their leading gradients. And thus it follows that a concomitant given in seminormal units by the product of two leading gradients differs from the product of the corresponding concomitants by concomitants which are given by *later* gradients. This fact is given by Theorem IX, independently of the theorem just proved.

In looking for the complete irreducible system of concomitants of a quantic, all leading gradients have to be considered; but those which are products of other leading gradients may be rejected at once as reducible.

In binary forms it is often easy to show that a certain gradient is a leading gradient, by a method explained in the last paper. When the number of leading gradients found in this way is equal to the full number of linearly independent forms, as given by generating functions, then, as these are also the gradients giving irreductible covariants, the covariant given by any other gradient is reductible.

31. It was pointed out in § 51 (Q.S.A., VII) that the covariant type given by a binary gradient

$$A_0^{(1)} A_{\lambda_2}^{(2)} A_{\lambda_3}^{(3)} \dots A_{\lambda_\delta}^{(\delta)}$$

in seminormal units is the continued transvectant

$$(\dots((f_1, f_2)^{\lambda_2} f_3)^{\lambda_3} \dots f_\delta)^{\lambda_\delta};$$

and hence also the covariant of a single quintic given by the irreductible form

$$A_0 A_{\lambda_2} \dots A_{\lambda_s}$$

is

$$(\dots((f, f)^{\lambda_2} f)^{\lambda_3} \dots f)^{\lambda_s}.$$

Hence we deduce:

THEOREM XX. *The continued transvectant*

$$(\dots((f, f)^{\lambda_2} f)^{\lambda_3} \dots f)^{\lambda_s}$$

has the leading gradient

$$A_0 A_{\lambda_2} A_{\lambda_3} \dots A_{\lambda_s}$$

according to a principal sequence, when and only when it is irreductible according to that sequence.

In the case of ternary forms the coefficients of the ground form require two suffixes for their expression. They are connected with the tableau by regarding the first suffix as giving the number of letters of the corresponding set which lie in the second row, while the second suffix gives the number in the third row. The addition of the last quantic in a concomitant type may be described as a two-indexed transvectant (in a manner already known). And the result of this transvectant can be obtained by differential operators. In the same way for quaternary forms, a three-indexed transvectant is to be used. Using the word transvectant in this extended sense, we see that an irreductible concomitant given by a gradient is equal to the corresponding continued transvectant, no matter how many variables there are. Further, Theorem XX is true for all cases when the multiple indexes of the transvectants and the corresponding multiple suffixes of the coefficients are inserted, and not merely for binary forms alone.

IX. *Binary forms.*

32. For binary forms the product of two seminvariants is given by Theorem X. It is the case of the product of the form

$$T_{\alpha_1 \alpha_2} T_{\beta_1 \beta_2} T_{\alpha_1+\beta_1, \alpha_2+\beta_2}.$$

It does not follow that all the terms in the product represent separate seminvariants; some of them will unite under the operation of ΓG. But when a leading gradient may be expressed as a product of two leading gradients the difference of the products will give us at once a reduction for a later gradient. The case of the product of the form

$$T_{\alpha_1\alpha_2}\, T_{\beta_1\beta_2}\, T_{\alpha_1+\beta_1-\kappa,\ \alpha_2+\beta_2+\kappa},$$

dealt with by the same theorem, gives us the transvectant

$$(C_\alpha, C_\beta)^\kappa.$$

Incidentally, we learn at once from the result what is the leading gradient of this transvectant.

33. That exactly the same analysis may be applied to seminvariants as functions of differences of the roots has been already pointed out. (Q.S.A., VII, § 53.) The results obtained here, so far as binary forms are concerned, may be interpreted this way. A direct connection between the seminvariants expressed in terms of the coefficients and the seminvariants expressed in terms of the roots, which was suggested in the place quoted, will now be established. A leading gradient, which therefore represents an irreducible seminvariant, is considered, $A_0 A_{\lambda_2} \ldots A_{\lambda_\delta}$; where, as always, $\lambda_\delta \geqslant \lambda_{\delta-1} \geqslant \lambda_{\delta-2} \ldots$. A tableau of $\delta-1$ rows all beginning with the same column is written down, the first row being $a_1 a_2 \ldots a_{\lambda_\delta}$, the second $a_1 a_2 \ldots a_{\lambda_{\delta-1}}$, and so on. Then the gradient may be looked upon as the symbolical product

$$a_1{}^0 a_2^{\lambda_2} \ldots a_\delta^{\lambda_\delta}.$$

The tableau as constructed yields a *root* product

$$a_1^{\delta-1} \ldots,$$

where each index is the length of the corresponding column. The two products will be called reciprocal; one is always given by the rows of a tableau, the other by the columns of the same tableau. The symbolical products are arranged in a definite sequence, this determines a definite sequence for the reciprocal products, which will be called the reciprocal sequence.

34. In a seminvariant expressed by substitutional forms, the letters of permutation are regarded as the symbolical letters and they appear in sets of n (the order of the ground form), each set representing one symbol; the equivalence of members of a set gives rise to the operator G, the equivalence of sets as a whole to the operator Γ. If the seminvariant is regarded as a function of the roots, the ground form is written as a product of n linear factors, each factor giving rise to one letter of permutation for each degree of the seminvariant. The letters again appear in sets of n

and the operator G is still required because for each coefficient of the ground form in the seminvariant the roots appear symmetrically. The seminvariant expressed in the roots is, in fact, of the same form as its expression in the coefficients, except that a new operator H must be introduced. This is the product of the symmetric groups (divided by the order $\delta!$ in each case) of the letters which represent each individual root separately, there being one group for each letter. Thus the seminvariant

$$\Gamma G \varpi_{u,v} X$$

becomes the function of the differences of the roots

$$H \Gamma G \varpi_{u,v} X.$$

Conversely, we may start with a function of the differences of the roots $\Gamma' G' \varpi_{u,v} X$, and obtain the seminvariant in the coefficients

$$H' \Gamma' G' \varpi_{u,v} X.$$

Here G' represents the fact that the letters really represent a function of degree δ in the coefficients of each of the linear factors; Γ' represents the fact that the function is symmetrical in the different roots; and, finally, H' puts the letters in sets of n proper to the symbolical representation.

The two forms of expression of a seminvariant are thus entirely complementary. To a leading gradient on one hand,

$$A_0 A_{\lambda_2} \dots A_{\lambda_\delta},$$

corresponds a first term on the other,

$$a_2^{\lambda_2} \dots a_\delta^{\lambda_\delta},$$

where $a_1, a_2, \dots, a_\delta$ are the roots of a binary δ-ic. Anything which can be proved about the one can be interpreted in the other case. With this in view, the term leading gradient may be used with either meaning.

35. THEOREM XXI. *The reciprocal of a leading gradient for any principal sequence is a leading gradient in the reciprocal sequence.*

Consider a leading gradient

$$g = A_0 A_{\lambda_2} \dots A_{\lambda_\delta},$$

the irreductible covariant

$$(\dots((f, f)^{\lambda_2} f)^{\lambda_3} \dots f)^{\lambda_\delta}$$

corresponds to it. For n, the order of f, we have

$$n \geqslant \lambda_\delta \geqslant \lambda_{\delta-1} \dots \geqslant \lambda_2.$$

The covariant exists for $n = \lambda_\delta$, and hence in this case there is a function ϕ_{λ_δ} of the differences of the λ_δ roots, which is such that $\Sigma \phi_{\lambda_\delta}$ is equal to the seminvariant in question. For every irreductible covariant of f of this degree and weight there is such a function of the differences of the λ_δ roots. And the several functions $\Sigma \phi_{\lambda_\delta}$ are all non-zero and linearly independent. When the order of f is increased, the seminvaraints corresponding to the leading gradients g remain irreductible. The functions $\Sigma \phi_{\lambda_\delta}$, the same in form but with extension in summation, remain also non-zero and linearly independent. Hence a definite non-zero covariant may be obtained from g in the following way. The factors of f are arranged in fixed sequence $a_{1_x} a_{2_x} \dots a_{n_x}$. The product of the first μ factors is written f_μ, that of the next $\nu-\mu$ factors is written $f_{\nu, -\mu}$; then the sum for all permutations of the roots

$$\Sigma (f_{\lambda_2}, f_{\lambda_2})^{\lambda_2} f^2_{\lambda_3, -\lambda_2}, \quad f_{\lambda_3})^{\lambda_3} f^3_{\lambda_4, -\lambda_3}, \quad f_{\lambda_4})^{\lambda_4} \dots, \quad f_{\lambda_\delta})^{\lambda_\delta} \quad f^\delta_{n, -\lambda_\delta}$$

is a definite non-zero covariant.

The first term of its seminvariant considered as a function of the roots, according to the reciprocal sequence, is the reciprocal of g. And hence the theorem is proved.

By an obvious application of the argument used in § 22 (Q.S.A., VII) we see that this seminvariant expressed in terms of the roots whose first term is g', the reciprocal of g, differs from the seminvariant defined by g as a continued transvectant by seminvariants defined by earlier gradients than g in the principal sequence used, and hence in any principal sequence. Thus

THEOREM XXII. *The seminvariant defined by a leading gradient differs from that defined as above by a product of differences of the roots with the first term reciprocal to this gradient by seminvariants defined by earlier leading gradients in any principal sequence.*

36. We now see that a covariant may be defined either by a leading gradient in a principal sequence, or by one in a reciprocal sequence in terms of the coefficients or else in terms of the roots. Thus every leading gradient stands for two covariants, one of degree δ of an n-ic, the other of degree n of a δ-ic. When a complete system of leading gradients in a principal or a reciprocal sequence can be obtained, that in the other can be written down. The difficulty is to obtain the law by which it may be

known at once whether a given gradient is a leading one or not. It has been shown (Q.S.A., VII, § 32) that the condition in a reciprocal sequence is that the coefficient of highest index appears with an index equal to 2 at least: the proof, which was there given for a particular reciprocal sequence, applies equally to all. But this statement loses sight of the fact that such a gradient is only a leading gradient when the order of the ground form exceeds some lower limit, and that no statement has yet been given as to what this limit is. Reciprocally, if g is any gradient whatever which does not contain A_1, there is some number γ for which $A_0{}^{\gamma} g$ is a leading gradient for a given principal sequence and $A_0^{\gamma-1} g$ is not one; γ may be called the *index* of g for this sequence.

37. The leading gradients for sequence A of degree 5 have been obtained (Q.S.A., VII, invariants, Section VII; covariants, Section X). By taking the reciprocals we have the leading gradients for the quintic, for the sequence D reciprocal to A.

The invariants

$$I_4 = A_3{}^2 A_2{}^2, \quad I_8 = A_4{}^2 A_3{}^2 A_2{}^2 A_1{}^2, \quad I_{12} = A_4{}^2 A_3{}^2 A_2{}^8,$$

$$I_{18} = A_4{}^2 A_3{}^5 A_2^{11}.$$

Moreover, the relation

$$I_{18}^2 = I_{12}^2 I_4{}^3$$

indicates the syzygy.

The reciprocal of a form obtained by adding the arguments of two forms sequence A is evidently the product of the reciprocal forms.

Covariants of order 1.

$$C_5{}^1 = A_3{}^2 A_2{}^3, \quad C_7{}^1 = A_3{}^3 A_2{}^4, \quad C_{11}^1 = A_4{}^2 A_3{}^2 A_2{}^6 A_1,$$

$$C_{13}^1 = A_4{}^2 A_3{}^2 A_2{}^9, \quad C_{17}^1 = A_4{}^2 A_3{}^5 A_2{}^9 A_1.$$

Here $$I_{12}\, C_5{}^1 - I_4\, C_{13}^1 = 0,$$

so far as their representative gradients are concerned, but the left-hand side cannot be accurately zero; it is, therefore, a covariant of order one given by a later gradient sequence D. It can only be C_{17}^1 which is therefore reducible. We thus have the four well-known linear covariants.

Covariants of order 2.

$$C_2^2 = A_2^2, \quad C_6^2 = A_3^3 A_2^2 A_1, \quad C_8^2 = A_3^3 A_2^5, \quad C_{12}^2 = A_4^2 A_3^2 A_2^7 A_1,$$

$$C_{12t-2}^2 = A_3^{8t-2} A_2^{2t} A_1^{2t}, \quad C_{12t+16}^2 = A_3^{8t+9} A_2^{2t+5} A_1^{2t+2}.$$

A reduction for C_{12}^2 arises from

$$I_4\, C_8^2 - C_5^1\, C_7^1,$$

which cannot be accurately zero, and C_{12}^2 is the only form which it can be. The reductions of C_{12t-2}^2, C_{12t+16}^2 must follow in the same way.

Covariants of order 3.

$$C_{6\sigma+2\rho-3}^3 = A_3^{4\sigma+\rho-3} A_2^{\sigma+\rho} A_1^{\sigma},$$

$$C_{12\sigma+2\rho+3}^3 = A_4^{2\sigma} A_3^{2\sigma+\rho} A_2^{8\sigma+\rho+3} \quad (\rho \neq 1),$$

$$C_{12\sigma+2\rho+1}^3 = A_4^{2\sigma} A_3^{2\sigma+\rho} A_2^{8\sigma+\rho} A_1 \quad (\rho \neq 1).$$

Covariants of order greater than 3.

$$A_4^{2\lambda} A_3^{2\lambda+\xi} A_2^{2\lambda+\eta} A_1^{\zeta},$$

where $\xi \neq 1, \quad \eta \neq 1, \quad 3\zeta - 6\lambda - \xi + \eta > 3.$

III—The Application of Substitutional Analysis to Invariants

By ALFRED YOUNG, F.R.S.

(Received June 13, 1934)

In a previous paper* it was proved that the generating function for any class of ternary concomitants might be obtained from the corresponding generating function for gradients (coefficient products) by multiplication by $(1 - x)(1 - y)(x - y)$. A generating function for ternary gradients was given in Theorem III of that paper, but it is of such a character that it is useless for purposes of calculation. In this paper a new system of generating functions is obtained applicable to perpetuants or to forms of finite order, and also to binary, ternary, or any forms.

In Section I a class of polynomial function $f_\alpha(z)$ is discussed which appeared in the paper just quoted in connection with the generating function for binary gradients of particular substitutional form in the perpetuant case. In Section II a one to one correspondence between binary perpetuants of particular substitutional form and the terms of the corresponding generating function is obtained by means of the tableau notation; and this is used to give a very simple extension of GRACE's Theorem on irreducible perpetuants to the case of perpetuants of particular substitutional form. Section III deals with the properties of the functions for forms of finite order which correspond to the functions $f_\alpha(z)$ for perpetuants.

The generating functions for the different substitutional classes must by addition give the generating function for types. This in some cases has been obtained independently. Thus there arise certain algebraic identities. In Section IV a general theorem is established covering all these identities. It is obtained by means of the Characteristic Function of SCHUR. The same method is then used to express the binary generating functions in a new form.

In Section V, the Compound Symmetric group is discussed, which is formed by the simultaneous permutation of two sets of letters. It is shown that certain results due to FROBENIUS are applicable to the problem in hand. The results of this section are developed with a view to their application to the problem of ternary generating functions.

In Section VI, the functions $f_\alpha(x, y)$ are defined and discussed; they are the extension to two variables of the function $f_\alpha(z)$ of Section I, and they play a corresponding role in ternary generating functions. The theorems of Section (IV)

* YOUNG, 'Proc. London Math. Soc.,' vol. 35, p. 425 (1933).

[Published February 12, 1935.

are then extended to these functions. The generating functions for ternary perpetuants thus obtained contain two sets of terms, only one of which is applicable to the problem in hand. The task of separating the function into two, one of which is the actual generating function required is carried out in some of the simpler cases.

In Section VII the generating function for ternary perpetuants is definitely established ; and in the simplest cases it is compared to the corresponding symbolical products.

In Section VIII the generating function for ternary forms of finite order is obtained.

I—The Functions $f_{\alpha_1, \alpha_2, \dots \alpha_h}(z)$

§ 1. It has been proved* that the generating function for binary covariant types of degree δ, and substitutional form $T_{\alpha_1, \alpha_2, \dots \alpha_h}$ $(\Sigma\alpha = \delta)$, of quantics of order n is

$$(1 - z)\, z^{\varpi} D^{-h} (1 - z^{n})^{h-1} (1 - z^{n-1})^{h-2} \dots (1 - z^{n-h+2})$$

$$. \prod_{r=1}^{h} (1 - z^{\alpha_r - r + h + 1}) \dots (1 - z^{\alpha_r - r + n + 1}) \prod_{r<t} (1 - z^{\alpha_r - \alpha_t + t - r}),$$

where

$$D = (1 - z)(1 - z^2) \dots (1 - z^n),$$

and

$$\varpi = (h - 1)\,\alpha_h + (h - 2)\,\alpha_{h-1} + \dots + \alpha_2.$$

We will use the notation

$$[m] = 1 - z^m$$

$$[m]\,! = (1 - z)(1 - z^2) \dots (1 - z^m)$$

$$[m]_r = (1 - z^m)(1 - z^{m-1}) \dots (1 - z^{m-r+1}).$$

Then this generating function becomes

$$\text{(I)} \qquad z^{\varpi}\, [1] \prod_{r=1}^{h} \frac{[\alpha_r + n + 1 - r]_{\alpha_r}}{[\alpha_r + h - r]\,!} \,.\, \prod_{r<t} [\alpha_r - \alpha_t + t - r].$$

When n is increased indefinitely, we obtain the perpetuant generating function

$$\text{(II)} \qquad z^{\varpi} \frac{[1] \prod_{r<t} [\alpha_r - \alpha_t + t - r]}{\prod [\alpha_r + h - r]\,!} = \frac{z^{\varpi}\,[1]}{[\delta]\,!} [f_{\alpha_1, \alpha_2, \dots \alpha_h}],$$

the last symbol being introduced in formal analogy to the new notation, for there is a factor $[k]$ corresponding to each factor k in f. In the case of f_δ, the perpetuants of a single form, this generating function becomes the familiar function

$$\frac{[1]}{[\delta]\,!}.$$

It is more convenient to write $[f_{\alpha_1, \alpha_2, \dots \alpha_h}]$ in the form $f_{\alpha_1, \alpha_2, \dots \alpha_h}(z)$.

* Young, 'Proc. London Math. Soc.,' vol. 35, p. 437 (1933).

§ 2. It follows from the substitutional identity

$$1 = \Sigma \mathrm{T}$$

that every covariant can be expressed as a sum of covariant types, each term of the sum being a type of some definite substitutional form $\mathrm{T}_{\alpha_1, \alpha_2, \dots \alpha_h}$. Hence the generating function for covariant types in general must be the sum of the generating functions for the types of each particular substitutional form. We apply this to binary perpetuant types and obtain

$$\text{(III)} \qquad \frac{1}{(1-z)^{\delta-1}} = \Sigma \frac{1}{[\delta]\,!} z^{\varpi} f_{\alpha_1, \alpha_2, \dots \alpha_h} \cdot f_{\alpha_1, \alpha_2, \dots \alpha_h} (z)\,;$$

for there will be f different covariants for each type of substitutional form T obtained by permutation of quantics. Thus for small values of δ,

$$(1-z)^{-1} = (1-z^2)^{-1} \{f_2 \cdot f_2 (z) + z f_{1^2} \cdot f_{1^2} (z)\} = (1-z^2)^{-1} (1+z)$$

$$(1-z)^{-2} = \{(1-z^2)(1-z^3)\}^{-1} \{1 + 2z(1+z) + z^3\}$$

$$(1-z)^{-3} = \{(1-z^2)(1-z^3)(1-z^4)\}^{-1} \{1 + 3z(1+z+z^2) + 2z^2(1+z^2) + 3z^3(1+z+z^2) + z^6\}.$$

Equation (III) may be written

$$[\delta]\,!/[1]^{\delta} = \Sigma z^{\varpi} f_{\alpha_1, \alpha_2, \dots \alpha_h} \cdot f_{\alpha_1, \alpha_2, \dots \alpha_h} (z) = (1+z)(1+z+z^2) \dots (1+z+z^2+\dots+z^{\delta-1}).$$

The sum of the coefficients in the expansion of the last expression is $\delta\,! = \Sigma f^2$; thus we are led to expect that $f(z)$ is always an integral function of z, and that the sum of its coefficients is f.

§ 3. The value of f was originally obtained by FROBENIUS* in the form

$$f_{\alpha_1, \alpha_2, \dots \alpha_h} = \delta\,! \left| \frac{1}{(\alpha_r - r + s)\,!} \right|,$$

(where r defines the row and s the column in the determinant) as an immediate deduction from the fact that $f_{\alpha_1, \alpha_2, \dots \alpha_h}$ is the coefficient of $x_1^{\alpha_1+h-1} x_2^{\alpha_2+h-2} \dots x_h^{\alpha_h}$, in

$$(x_1 + x_2 + \dots + x_h)^{\delta} \Delta,$$

where

$$\Delta = \{x_1 x_2 \dots x_h\}' x_1^{h-1} x_2^{h-2} \dots x_{h-1},$$

is the alternant of the h variables x.

To make the next paragraph more clear, we deduce the ordinary form of f.

Now

$$f \,.\, \Pi (\alpha_r + h - r)\,! = \phi\, \delta\,!,$$

where ϕ is integral.

* 'S. B. berl. math. Ges.,' p. 522 (1900).

We regard the determinant as a function of h unknown quantities α_r. Then ϕ is zero when

$$\alpha_r - r = \alpha_t - t.$$

Thus ϕ has a factor

$$\prod_{r<t} (\alpha_r - \alpha_t - r + t).$$

The total order of ϕ in the α's is $\binom{\delta}{2}$ showing that the remaining factor is numerical. Comparison of coefficients of suitable products of powers of the α's shows that this factor is unity. Hence we obtain the usual form

$$f = \delta\,! \frac{\prod_{r<t} (\alpha_r - \alpha_t - r + t)}{\prod (\alpha_r - r + h)\,!}.$$

§ 4. THEOREM I—*The function* $f_{\alpha_1, \alpha_2, \dots \alpha_h}(z)$ *may be written as a determinant*

$$[\delta]\,! \left| \frac{1}{[\alpha_r - r + s]\,!} \right|,$$

where r defines the row, and s the column of the element.

Let Δ be the value of this product, then

$$\Delta \prod [\alpha_r + h - r]\,! = [\delta]\,! \,\phi,$$

where ϕ is an integral function of z.

Let $\psi(z)$ be an integral function of z with integral coefficients, then if

$$\psi(z) = 0,$$

when z is any root of $z^n = 1$, $\psi(z)$ has a factor $z^n - 1$. Now ϕ is zero when $\alpha_r - r = \alpha_t - t$, and $\alpha_r - r$, $\alpha_t - t$ only appear as indices of z, thus ϕ is zero whenever $z^{\alpha_r - r} = z^{\alpha_t - t}$, *i.e.*, ϕ has a factor

$$1 - z^{\alpha_r - \alpha_t + t - r}.$$

Thus ϕ is a function which contains all the factors of

$$\prod_{r<t} [\alpha_r - \alpha_t + t - r].$$

In view of the fact that $\alpha_1, \alpha_2, \dots \alpha_h$ may be looked on at present as arbitrary numbers, the only necessarily repeated factor in this is $(1 - z)^{\binom{h}{2}}$.

Now when z approaches the value 1, we may take $1 - z = \zeta$ a small quantity. Then when ζ is small

$$[k] = k\zeta\,; \text{ and}$$

$$\Delta = \delta\,! \left| \frac{1}{(\alpha_r - r + s)\,!} \right| = f_{\alpha_1, \alpha_2, \dots \alpha_h}.$$

Thus Δ has no factor $1 - z$. The number of factors $1 - z$ in $\Pi\, [\alpha_r + h - r]\,!$ is

$$\Sigma\,(\alpha_r + h - r) = \delta + \binom{h}{2},$$

hence ϕ has the factor $(1 - z)^{\binom{h}{2}}$, and therefore the factor $\prod\limits_{r>t} [\alpha_r - \alpha_t - r + t]$.

No other factor is possible for this is of the same order as ϕ, and the numerical factor to be attached is easily seen to be unity. Hence, according to definition

$$\Delta = f_{\alpha_1, \alpha_2, \ldots \alpha_h}(z).$$

§ 5. Theorem II—*The function $f_{\alpha_1, \alpha_2, \ldots \alpha}(z)$ is an integral function of z.*

The fact that $f_{\alpha_1, \alpha_2, \ldots \alpha_h}$ is always an integer is not sufficient; for instance, the function $\frac{[4]\,[1]}{[2]\,[2]}$ is not integral.

We notice first that when $\Sigma\alpha = \delta$,

$$Q = \frac{[\delta]\,!}{[\alpha_1]\,!\,[\alpha_2]\,! \ldots [\alpha_h]\,!}$$

is integral. For let

$$P_{r,\alpha} = [r + 1]\,[r + 2] \ldots [r + \alpha],$$

then

$$\frac{P_{r,\alpha}}{[\alpha]\,!} = \frac{P_{r,\alpha-1}}{[\alpha - 1]\,!} + z^{\alpha}\frac{P_{r-1,\alpha}}{[\alpha]\,!}.$$

Hence, noticing that $P_{2,1}/[1]$, and $P_{0,\alpha}/[\alpha]!$ are always integral, it is easy to see by induction that $P_{r,\alpha}/[\alpha]!$ is integral. It follows at once that Q is integral. Now the determinant form of $f\,(z)$ given in Theorem I is a sum of expressions such as Q, which are now seen to be integral, hence $f\,(z)$ is necessarily integral.

Corollary—*The sum of the coefficients of the powers of z in $f\,(z)$ is f.*

This follows at once since we proved in the last paragraph that $\lim\limits_{z \to 1} \Delta = f$.

§ 6. Theorem III—*If the tableaux for the representations* $T_{\alpha_1, \ldots \alpha_h}$, $T_{\beta_1, \ldots \beta_k}$, *are obtained from each other by the change of rows into columns and vice versa*, i.e., *if they are conjugate representations, then*

$$f_{\alpha_1, \alpha_2 \ldots \alpha_h}(z) = f_{\beta_1, \beta_2, \ldots \beta_k}(z).$$

Consider the case $h = 3$,

$$f_{\alpha_1, \alpha_2, \alpha_3}(z) = [\delta]\,!\,\frac{[\alpha_1 - \alpha_2 + 1]\,[\alpha_1 - \alpha_3 + 2]\,[\alpha_2 - \alpha_3 + 1]}{[\alpha_1 + 2]\,!\,[\alpha_2 + 1]\,!\,[\alpha_3]\,!} = [\delta]\,!\,A/B.$$

Let

$$f_{\beta_1, \beta_2, \ldots \beta_k}(z) = [\delta]\,!\,A'/B'.$$

It has to be shown that

$$AB' = A'B.$$

The values of β_r and $\beta_r + k - r$ are

$$\begin{array}{llllllll} r = & 1, & 2, \ldots & \alpha_3 & , \alpha_3+1 & , \ldots & \alpha_2 & , \alpha_2+1, \ldots \alpha_1, \\ \beta_r = & 3, & 3, \ldots & 3 & , 2 & , \ldots & 2 & , 1, \ldots 1, \\ \beta_r+k-r = & \alpha_1+2, & \alpha_1+1, \ldots & \alpha_1-\alpha_3+3, & \alpha_1-\alpha_3+1, & \ldots & \alpha_1-\alpha_2+2, & \alpha_1-\alpha_2, \ldots 1; \end{array}$$

that is, $\beta_r + k - r$ has all the values from 1 up to $\alpha_1 + 2$, in turn excepting only $\alpha_1 - \alpha_2 + 1, \alpha_1 - \alpha_3 + 2$. In the general case it will have all the values from 1 up to $\alpha_1 + h - 1$, in turn with the exception of $\alpha_1 - \alpha_2 + 1$, $\alpha_1 - \alpha_3 + 2, \ldots$ $\alpha_1 - \alpha_h + h - 1$, the numbers containing α_1, which appear in A.

Further, A′ is the product of those factors which are defined by the differences of the various values of $\beta_r + k - r$, there being one factor for each difference.

We insert now the missing numbers, and so obtain the complete series of numbers from 1 up to $\alpha_1 + 2$. The product of factors defined by these differences is

$$[\alpha_1 + 1]\,!\,[\alpha_1]\,!\,[\alpha_1 - 1]\,!\ldots[1]\,!,$$

which may be written $[\alpha_1 + 1]\,!!$.

Let P_1 be the product of those factors which are defined by the differences between $\alpha_1 - \alpha_2 + 1$ and the other numbers, P_2 the product defined by the differences between $\alpha_1 - \alpha_3 + 2$ and the other numbers, and P' the function $[\alpha_2 - \alpha_3 + 1]$ defined by the difference between these two numbers, there being only one such factor for $h = 3$.

Then

$$P'.\,[\alpha_1 + 1]\,!! = P_1 P_2 A',$$

where

$$P' = [\alpha_2 - \alpha_3 + 1],\ P_1 = [\alpha_1 - \alpha_2]\,!\,[\alpha_2 + 1]\,!,\ P_2 = [\alpha_1 - \alpha_3 + 1]\,!\,[\alpha_3]\,!\,.$$

Also

$$[\alpha_1 - \alpha_2 + 1]\,!\,[\alpha_1 - \alpha_3 + 2]\,!\,B' = [\alpha_1 + 2]\,!!.$$

Hence

$$AB' = \frac{[\alpha_1 + 2]\,!!\,[\alpha_2 - \alpha_3 + 1]}{[\alpha_1 - \alpha_2]\,!\,[\alpha_1 - \alpha_3 + 1]\,!} = A'B,$$

the required result. The method of proof is applicable to any value of h.

It will be seen from equation (II), § 1, that the highest power π of z in $f_{\alpha_1, \alpha_2, \ldots \alpha_h}(z)$ is

$$\pi = \binom{\delta + 1}{2} + \sum_{r<t} (\alpha_r - \alpha_t + t - r) - \Sigma \binom{\alpha_r + h - r + 1}{2} = \binom{\delta}{2} - \Sigma \binom{\alpha_r}{2} - \varpi.$$

Now from the values for r and β_r given it is easy to see that

$$\varpi' = \sum_{r=1}^{k} (r - 1)\,\beta_r = \Sigma \binom{\alpha_r}{2},$$

hence

(IV) $$\pi + \varpi + \varpi' = \binom{\delta}{2}.$$

Consider then the series

$$(1-z)^{-\delta+1}=\frac{[1]}{[\delta]\,!}\{1+\ldots+z^{\varpi}f.f(z)+\ldots+z^{\varpi'}f'.f'(z)+\ldots+z^{\binom{\delta}{2}}\},$$

where f and f' are conjugate. Then write z^{-1} for z, and we have

$$z^{\delta-1}(1-z)^{-\delta+1}=\frac{z^{\binom{\delta+1}{2}-1}[1]}{[\delta]\,!}\{1+\ldots+z^{-\pi-\varpi}f.f(z)+\ldots+z^{-\pi'-\varpi'}f'.f'(z)$$
$$+\ldots+z^{-\binom{\delta}{2}}\},$$

and in view of (IV) and the equality of $f(z)$ and $f'(z)$, we see that the generating functions for conjugate representations merely change places in the series after this transformation.

§ 7. The numbers f satisfy the relation*

$$(\delta+1)f_{\alpha_1,\alpha_2,\ldots\alpha_h}=\sum_{r=1}^{h+1}E_r f_{\alpha_1,\alpha_2,\ldots\alpha_h},$$

where E_r is an operator which increases α_r by unity. This relation has its counterpart for the functions $f(z)$, as we proceed to show.

THEOREM IV—*The functions $f(z)$ for types of degree δ are connected with those for types of degree $\delta+1$, by the equations* :—

$$z^{\Sigma(s-1)\alpha_s}f_{\alpha_1,\alpha_2,\ldots\alpha}(z)\,[\delta+1]=\sum_{r=1}^{h+1}[1]\,E_r\{z^{\Sigma(s-1)\alpha_s}f_{\alpha_1,\alpha_2,\ldots\alpha_h}(z)\}.$$

For

$$\frac{[1]\,E_r\,z^{\Sigma(s-1)\alpha_s}f_{\alpha_1,\alpha_2,\ldots\alpha_h}(z)}{[\delta+1]\,z^{\Sigma(s-1)\alpha_s}f_{\alpha_1,\alpha_2,\ldots\alpha_h}(z)}$$
$$=z^{r-1}\frac{[1]}{[\alpha_r+1-r+h]}\prod_{s<r}\frac{[\alpha_s-\alpha_r-s+r-1]}{[\alpha_s-\alpha_r-s+r]}\prod_{s>r}\frac{[\alpha_r-\alpha_s-r+s+1]}{[\alpha_r-\alpha_s-r+s]}$$
$$=\left(\prod_{s\neq r}\frac{z^{\alpha_r+1+h-r}-z^{\alpha_s+h-s}}{z^{\alpha_r+h-r}-z^{\alpha_s+h-s}}\right)\frac{1-z^{-1}}{z^{\alpha_r+h-r}-z^{-1}}.$$

When $r=h+1$, we have $\prod_{s=1}^{h}\frac{1-z^{\alpha_s+h-s}}{z^{-1}-z^{\alpha_s+h-s}}$.

Consider the two functions

$$\phi(y)=y(y-z^{-1})\prod_{r=1}^{h}(y-z^{\alpha_s+h-s}),$$

whose roots are $y_1, y_2, \ldots y_{h+2}$, where

$$y_r=z^{\alpha_r+h-r},\ (r<h+2)\ ;\ y_{h+2}=0\ ;$$

and

$$\psi(y)=\prod_{s=1}^{h}(yz-z^{\alpha_s+h-s}),$$

* YOUNG, 'Proc. London Math. Soc.,' vol. 28, p. 262 (1928).

whose order is less by 2 than that of $\phi(y)$. Then by a well-known theorem

$$\sum_{r=1}^{h+2} \frac{\psi(y_r)}{\phi'(y_r)} = 0.$$

Now

$$\frac{\psi(y_{h+2})}{\phi'(y_{h+2})} = -z, \quad \frac{\psi(y_{h+1})}{\phi'(y_{h+1})} = z \prod_{s=1}^{h} \frac{1 - z^{a_s+h-s}}{z^{-1} - z^{a_s+h-s}},$$

and

$$\frac{\psi(y_r)}{\phi'(y_r)} = \frac{1}{z^{a_r+h-r}} \cdot \frac{z^{a_r+1+h-r} - z^{a_r+h-r}}{z^{a_r+h-r} - z^{-1}} \prod_{s \neq r} \frac{z^{a_r+1+h-r} - z^{a_s+h-s}}{z^{a_r+h-r} - z^{a_s+h-s}}$$

$$= z \frac{[1]}{[\delta+1]} \cdot \frac{E_r\, z^{\Sigma(s-1)a_s} f_{a_1, a_2, \dots a_h}(z)}{z^{\Sigma(s-1)a} f_{a_1, a_2, \dots a_h}(z)}.$$

On putting in these values of the terms in the equation just obtained, and dividing by z, we obtain at once the result required.

One hoped to obtain an analogue to the equation*

$$f_{a_1, a_2, \dots a_h} = \Sigma\, \delta_r f_{a_1, a_2, \dots a_h},$$

where δ_r indicates the decrease of α_r by unity; but the attempt failed. The nearest approach obtained to such an analogue was

$$f_{\alpha, 2}(z) = z f_{\alpha-1, 2}(z) + z^2 f_{\alpha, 1}(z^2),$$

but this relation is accidental rather than general.

II—Binary Perpetuants

§ 8. A one to one correspondence between terms in the generating function with the covariants themselves, for the case of perpetuants, can be obtained as follows. In the first place, when the substitutional form is T_δ, the covariants are covariants of a single form. It has been proved (Q.S.A. VII)† that all such forms can be expressed linearly in terms of the forms

$$(a_1 a_2)^{\lambda_1} (a_2 a_3)^{\lambda_2} \dots (a_{\delta-2} a_{\delta-1})^{\lambda_{\delta-2}} (a_{\delta-1} a_\delta)^{2\lambda_{\delta-1}},$$

where

$$\lambda_{\delta-1} \geqq \lambda_{\delta-2} \geqq \lambda_{\delta-3} \dots \geqq \lambda_1;$$

of which the first term sequence C is

$$A_{\lambda_1} A_{\lambda_2} \dots A_{\lambda_{\delta-2}} A^2_{\lambda_{\delta-1}}.$$

The generating function for T_δ is

$$\frac{1}{(1-z^2)(1-z^3)\dots(1-z^\delta)} = \Sigma\, z^{\Sigma r \mu_r},$$

where r is one of the numbers 2, 3, ... δ.

* Young, 'Proc. London Math. Soc.,' vol. 28, p. 261 (1928).
† Young, 'Proc. London Math. Soc.,' vol. 36, p. 339 (1934).

The correspondence is established by means of a diagram of rows and columns of dots, all rows begin with the same column on the left, and all columns with the same upper row like the tableaux. We represent $\delta\mu_\delta$ first by μ_δ columns of δ dots, then immediately to the right we put $\mu_{\delta-1}$ columns of $\delta - 1$ dots and so on, finally we have on the extreme right μ_2 columns of two dots. Thus this is a perfectly general diagram of dots except that there are no columns with only one dot, and there are no columns with more than δ dots. Then there are δ rows, and calling the rows beginning with the first $\lambda_\delta, \lambda_{\delta-1} \dots \lambda_1$, we have

$$\lambda_\delta = \lambda_{\delta-1} \geqq \lambda_{\delta-2} \geqq \dots \geqq \lambda_1.$$

Thus the symbolical product representation, or the leading gradients sequence C, is made to correspond to the generating function terms by means of the rows and columns of a diagram.

§ 9. Let PN belong to $T_{a_1, a_2, \dots a_h}$, then any covariant belonging to this representation may be written in the form PNC. This will be looked on as a symbolical product. Now a symbolical letter a appears in two kinds, a_1, a_2, the letters of the first kind will be replaced here by 1, so that the symbolical letters appear only in one kind, and the symbolical product is not homogeneous. Let $T_\delta D$ be a symbolical product sum representing a covariant of a single form. Then

$$\text{PNC} \ . \ T_\delta D$$

is also a symbolical product sum, and it is of the substitutional form $T_{a_1, a_2, \dots a_h}$. Moreover, the leading gradient of this new form is given at once by the product of the symbolical expressions of the leading gradients of its two component parts. Thus, when we have obtained covariants of this substitutional form and of weights given by the generating function $z^\varpi f_{a_1, a_2, \dots a_h}(z)$, $f_{a_1, a_2, \dots a_h}$ in number, we can write down the rest corresponding to the full generating function

$$z^\varpi f_{a_1, a_2, \dots a_h}(z)\ [1]/[\delta]\ !$$

by multiplication of symbolical forms ; or by addition of suffixes in leading gradients ; or by addition of indices in the symbolical form given in the last paragraph.

§ 10. Any symbolical product representing a perpetuant type of degree δ can be expressed in terms of the products

$$(a_1a_2)^{\lambda_1}\,(a_1a_3)^{\lambda_2} \dots (a_1a_\delta)^{\lambda_\delta},$$

with a pre-selected letter a_1 in each symbolical factor. And the generating function $(1 - x)^{-\delta+1}$ simply represents all possible types of this form, they are linearly independent. By the use of the symbolical identities and STROH's Lemma all such forms can be expressed in terms of forms obtainable by permutation from

$$(a_1a_2)^{\lambda_1}\,(a_2a_3)^{\lambda_2} \dots (a_{\delta-2}a_{\delta-1})^{\lambda_{\delta-2}}\,(a_{\delta-1}a_\delta)^{\lambda_{\delta-1}+\lambda_\delta},$$

where

(V) $$\lambda_{\delta-1} \geqq \lambda_{\delta-2} \geqq \dots \geqq \lambda_1,$$

and $\lambda_\delta - \lambda_{\delta-1} = 0$ or 1. The leading gradient sequence C of this form is

$$A_{1,\lambda_1} A_{2,\lambda_2} \dots A_{\delta-1,\lambda_{\delta-1}} A_{\delta,\lambda_\delta} \equiv X,$$

where the first suffix defines the quantic and the second the weight.

Then all possible leading gradients can be expressed in terms of the gradients X, and those obtained by permutation of quantics from X.

If we operate on X with $\{A_{r_1} A_{r_2} \dots A_{r_\rho}\}'$, the result is zero unless $\lambda_{r_1}, \lambda_{r_2} \dots \lambda_{r_\rho}$ are all different. Consider the operation of $P_\rho N_\rho$ on X, where $P_\rho N_\rho$ is derived from a tableau F_ρ of $T_{a_1, a_2, \dots a_h}$. Let Λ_ρ be the tableau obtained from F_ρ by replacing each letter A_r by its corresponding weight suffix λ_r in X; then $P_\rho N_\rho X$ is zero unless the numbers in every column of Λ_ρ are all different. We shall call Λ_ρ the weight tableau of X.

Definition—*A normal leading gradient* X_ρ *corresponding to a standard tableau* F_ρ *is one in which the weights satisfy the conditions* (V) *the quantics being arranged in proper sequence, and for which the weight tableau* Λ_ρ *is such that* $\lambda_{r+1} > \lambda_r$ *when* λ_{r+1} *is on a lower row than* λ_r. *And also* $\lambda_\delta = \lambda_{\delta-1}$ *when* λ_δ *does not lie on a lower row than* $\lambda_{\delta-1}$, *and* $\lambda_\delta = \lambda_{\delta-1} + 1$ *when* λ_δ *lies on a lower row.*

For a normal leading gradient X_ρ it is obvious at once that $P_\rho N_\rho X_\rho$ is not zero.

THEOREM V—*For any leading gradient* X, *and any standard tableau* F_ρ, *either*

$$\text{(i)}\ P_\rho N_\rho X = 0,$$

or

(ii) X *is a normal leading gradient corresponding to* F_ρ,

or

$$\text{(iii)}\ P_\rho N_\rho X = \Sigma\, \beta s P_\sigma N_\sigma Y_\sigma,$$

where Y_σ *is a normal leading gradient corresponding to the standard tableau* F_σ *from which* Y_σ *is derived; s is some permutation and* β *is numerical.*

The sequence of quantics in X will be first supposed to be the fixed pre-arranged sequence, so that according to conditions (V) $\lambda_{r+1} \geqq \lambda_r$ always. Consider the weights in order commencing with λ_1, if $\lambda_2 > \lambda_1$, $\lambda_3 > \lambda_2$ and so on, the conditions that X may be normal are being fulfilled. Let the sign of equality first appear in the case $\lambda_{r+1} = \lambda_r$, then unless A_{r+1} lies on a lower row than A_r, X is still normal for F_ρ. Let us then suppose that A_{r+1} lies in a lower row than A_r, but not in the same column, otherwise

$$P_\rho N_\rho X = 0.$$

Then, since

$$\lambda_{r+1} = \lambda_r, \qquad X = (A_r A_{r+1})\, X,$$

and

$$P_\rho N_\rho X = (A_r A_{r+1})\, P_\sigma N_\sigma X,$$

where F_σ is a standard tableau derived from F_ρ by the interchange of A_r and A_{r+1}. Now A_{r+1} lies in a higher row than A_r in F_σ and thus X is normal for F_σ up to this point. Proceeding thus step by step we find a standard tableau F_σ which is such that

$$P_\rho N_\rho X = s P_\sigma N_\sigma X,$$

where s is some permutation, and X is a normal leading gradient for F_σ at least so far as the letter $A_{\delta-1}$ is concerned ; and, indeed, as far as the last letter is concerned, unless A_δ does not lie in a lower row than $A_{\delta-1}$ and $\lambda_\delta = \lambda_{\delta-1} + 1$.

In this case the last condition of the definition is not fulfilled. For this case it is necessary to consider the symbolical form Z from which the leading gradient was obtained.

Since $\lambda_\delta + \lambda_{\delta-1}$ is odd

$$\{A_{\delta-1}A_\delta\}\, Z = \Sigma\, Z',$$

where Z′ is a form which has an increased value of $\lambda_\delta + \lambda_{\delta-1}$; the theorem will be assumed true for such forms when the total weight does not exceed that of X. Then

$$\begin{aligned} P_\sigma N_\sigma X &= -\, P_\sigma N_\sigma\, (A_{\delta-1}A_\delta)\, X + \Sigma\, \beta s P_\tau N_\tau Y_\tau \\ &= -\, (A_{\delta-1}A_\delta)\, P_\varpi N_\varpi X + \Sigma\, \beta s P_\tau N_\tau Y_\tau, \end{aligned}$$

here F_ϖ is derived from F_σ by the interchange of $A_{\delta-1}$ and A_δ. Thus, $A_{\delta-1}$ is raised in the tableau in changing from F_σ to F_ϖ, and X is normal for F_ϖ.

The theorem is thus true when the quantics in X are arranged according to the fixed sequence ; when this is not the case

$$X = sY,$$

where the quantics in Y are arranged in the fixed sequence.

Then

$$P_\rho N_\rho X = P_\rho N_\rho s Y = \Sigma \beta s' P_\sigma N_\sigma Y,$$

from the former case. Thus the theorem is always true.

§ 11. THEOREM VI—*Every leading gradient can be linearly expressed in terms of forms obtained by permutation from* $P_\rho N_\rho X_\rho$; *where* X_ρ *is a normal leading gradient for the tableau* F_ρ *from which* $P_\rho N_\rho$ *is derived.*

Let X be a leading gradient ; then

$$X = \Sigma\, TX = \Sigma\, \Sigma\, PNMX = \Sigma\, \beta s P_\rho N_\rho X_\rho,$$

by the last theorem.

The number of different forms obtainable from $P_\rho N_\rho X_\rho$ by permutation, *i.e.*, by left-hand multiplication with a permutation, is $f_{a_1, a_2, \ldots a_h}$, when the tableau F_ρ belongs to $T_{a_1, a_2, \ldots a_h}$.

Amongst the normal leading gradients corresponding to F_μ, there is one which is of special importance, the normal leading gradient of minimum weight. This is obtained by the rule that when A_{r+1} is in a lower row of F_ρ than A_r

$$\lambda_{r+1} = \lambda_r + 1, \quad \text{and otherwise} \quad \lambda_{r+1} = \lambda_r, \quad \text{while} \quad \lambda_1 = 0.$$

For example, the standard tableaux of T_{32} are

$$\begin{pmatrix} A_1 & A_2 & A_3 \\ A_4 & A_5 & \end{pmatrix}, \begin{pmatrix} A_1 & A_2 & A_4 \\ A_3 & A_5 & \end{pmatrix}, \begin{pmatrix} A_1 & A_2 & A_5 \\ A_3 & A_4 & \end{pmatrix}, \begin{pmatrix} A_1 & A_3 & A_4 \\ A_2 & A_5 & \end{pmatrix}, \begin{pmatrix} A_1 & A_3 & A_5 \\ A_2 & A_4 & \end{pmatrix};$$

the corresponding weight tableaux for normal leading gradients of minimum weight are

$$\begin{pmatrix} 0 & 0 & 0 \\ 1 & 1 & \end{pmatrix}, \quad \begin{pmatrix} 0 & 0 & 1 \\ 1 & 2 & \end{pmatrix}, \quad \begin{pmatrix} 0 & 0 & 1 \\ 1 & 1 & \end{pmatrix}, \quad \begin{pmatrix} 0 & 1 & 1 \\ 1 & 2 & \end{pmatrix}, \quad \begin{pmatrix} 0 & 1 & 2 \\ 1 & 2 & \end{pmatrix}.$$

It will be seen at once that these correspond to the terms of the generating function $z^2 f_{3,2}(z)$. It will be found that in every case the normal leading gradients of minimum weight have $z^{\varpi} f_{\alpha_1, \alpha_2, \ldots \alpha_h}(z)$ for generating function.

§ 12. Let

$$X = \prod_{r=1}^{\delta} A_{r, \lambda_r}, \quad Y = \prod_{r=1}^{\delta} A_{r, \mu_r},$$

be both normal leading gradients corresponding to the standard tableau F, Y being the minimum leading gradient.

Then $\lambda_{r+1} - \lambda_r \geqq \mu_{r+1} - \mu_r$, and therefore $\lambda_{r+1} - \mu_{r+1} \geqq \lambda_r - \mu_r$; while $\lambda_\delta - \mu_\delta = \lambda_{\delta-1} - \mu_{\delta-1}$. Let $\lambda_r - \mu_r = v_r$. Then $Z = \prod_{r=1}^{\delta} A_{r, v_r}$ is a normal leading gradient corresponding to the tableau for T_δ which has only one row. Moreover, $T_\delta Z_\delta$ is the general leading gradient for a single quantic; and these forms are enumerated by the generating function $[1]\{[\delta]!\}^{-1}$ as described in §8.

Thus the normal leading gradients for $T_{\alpha_1, \alpha_2, \ldots \alpha_h}$ are enumerated by the generating function $g\,[1]\{[\delta]!\}^{-1}$, where g is the generating function for the minimum normal leading gradient. But this we have found to be (§9)

$$z^{\varpi} f_{\alpha_1, \alpha_2, \ldots \alpha_h}(z)\,[1]\{[\delta]!\}^{-1},$$

hence

$$g = z^{\varpi} f_{\alpha_1, \alpha_2, \ldots \alpha_h}.$$

THEOREM VII—*The generating function for the minimum normal leading gradients corresponding to* $T_{\alpha_1, \alpha_2, \ldots \alpha_h}$ *is*

$$z^{\varpi} f_{\alpha_1, \alpha_2, \ldots \alpha_h}[z].$$

It is easy to see that the lowest weight of the minimum normal leading gradients corresponding to $T_{\alpha_1, \alpha_2, \ldots \alpha_h}$ is given by the first tableau, that in which the first α_1 letters lie in the first row, the next α_2 in the second row and so on. In the corresponding weight tableau every element of the rth row is $r - 1$, and the total weight is

$$\alpha_2 + 2\alpha_3 + \ldots + (h-1)\,\alpha_h = \varpi.$$

Thus the leading gradients are all identified in correspondence with the terms of the generating function.

§ 13. There is no difficulty in extending these results to irreducible perpetuants. GRACE* proved that all perpetuant types can be expressed in terms of those of the form

(VI) $$(a_1 a_2)^{\lambda_1} (a_2 a_3)^{\lambda_2} \ldots (a_{\delta-2} a_{\delta-1})^{\lambda_{\delta-2}} (a_{\delta-1} a_\delta)^{\lambda_{\delta-1}},$$

* 'Proc. London Math. Soc.,' vol. 35, p. 107 (1902); and GRACE and YOUNG, "Algebra of Invariants," pp. 327-330 and Appendix IV (1903).

where

$$\lambda_1 \geqq 1, \qquad \lambda_2 \geqq 2, \ldots \qquad \lambda_{\delta-2} \geqq 2^{\delta-3}, \qquad \lambda_{\delta-1} \geqq 2^{\delta-2},$$

and the sequence of letters is fixed; and of products of forms. The generating function for the forms (VI) is $\dfrac{x^{2^{\delta-1}-1}}{(1-x)^{\delta-1}}$.

Wood* demonstrated that this result is exact, or, in other words, that there can be no linear relation between the forms (VI) and products of forms.

Grace† proved further that for a single form where the letters are interchangeable the index conditions for the forms (VI) may be written

$$\lambda_1 \geqq 1, \quad \lambda_2 \geqq 1+\lambda_1, \quad \lambda_3 \geqq 2+\lambda_2, \ldots \quad \lambda_{\delta-2} \geqq 2^{\delta-4}+\lambda_{\delta-3}, \quad \lambda_{\delta-1} \geqq 2\lambda_{\delta-2};$$

the generating function being

$$\frac{x^{2^{\delta-1}-1}}{(1-x^2)(1-x^3)\ldots(1-x^\delta)}.$$

The argument used establishes at once that all perpetuant types may be expressed linearly in terms of products, and of forms

$$(a_1a_2)^{\lambda_1}(a_2a_3)^{\lambda_2}\ldots(a_{\delta-2}a_{\delta-1})^{\lambda_{\delta-2}}(a_{\delta-1}a_\delta)^{\lambda_{\delta-1}+\lambda_\delta},$$

where the quantics are arranged in fixed sequence, and

$$\lambda_1 \geqq 1, \qquad \lambda_2 \geqq 1+\lambda_1, \ldots \qquad \lambda_{\delta-2} \geqq 2^{\delta-4}+\lambda_{\delta-3}, \qquad \lambda_{\delta-1} \geqq 2^{\delta-3}+\lambda_{\delta-2}$$

and $\lambda_\delta - \lambda_{\delta-1} = 0$ or 1; and of forms obtained from these by permutation. The leading gradient sequence C of this form is

$$A_{1,\lambda_1}A_{2,\lambda_2}\ldots A_{\delta,\lambda_\delta}.$$

The whole of the preceding argument may be applied to this result, the generating function for perpetuant types being multiplied in every case by $x^{2^{\delta-1}-1}$ to give the corresponding generating function for irreducible perpetuant types. The normal leading gradient for an irreducible perpetuant corresponding to a given tableau is the same in form as that in the former case except that each weight suffix λ_r is increased by 2^{r-1}, except in the case of λ_δ which is increased by $2^{\delta-2}$.

III—Application to Binary Forms of Finite Order

§ 14. The generating function for covariant types of substitutional form $T_{\alpha_1,\alpha_2,\ldots\alpha_h}$ of the binary n—*ic* has been given in (I), § 1. This will be written

$$z^{\varpi}f^{n}_{\alpha_1,\alpha_2,\ldots\alpha_h}(z);$$

* 'Proc. London Math. Soc.,' vol. 1, p. 480 (1904).

† 'Proc. London Math. Soc.,' vol. 35, p. 319 (1903) and Grace and Young, *op. cit.*

it is an integral function of z, for it was obtained in the first place in the paper quoted in § 1 in the form

$$(1-z)\left|\phi\binom{\alpha_r+n-r+s}{n}\right|,$$

in which all the elements of the determinant are integral functions of z; hence the function itself is an integral function of z. Comparison with $f_{\alpha_1, \alpha_2, \dots \alpha_h}(z)$ gives the relation

$$\text{(VII)} \qquad [\delta]!\, f^n_{\alpha_1, \alpha_2, \dots \alpha_h}(z) = [1]\, f_{\alpha_1, \alpha_2, \dots \alpha_h}(z) \prod_{r=1}^{h} [\alpha_r + n + 1 - r]_{\alpha_r}.$$

The highest index of z in $f^n_{\alpha_1, \alpha_2, \dots \alpha_h}(z)$ must then be

$$1+\pi+\Sigma\binom{\alpha_r+1}{2}+\Sigma\,\alpha_r(n+1-r)-\binom{\delta+1}{2}=1+\delta n-2\varpi=0,$$

on using the notation and results of § 6.

Replace z by z^{-1} in (VII) and multiply the result by $z^{\binom{\delta+1}{2}+\theta}$ the right-hand side is simply multiplied by $(-)^{\delta+1}$, hence

$$f^n_{\alpha_1, \alpha_2, \dots \alpha_h}(z) = \Sigma\, B_r (z^r - z^{\theta-r}).$$

From the generating function for covariants linear in the coefficients of each of δ quantics of order n, we obtain—in the same way as for perpetuants—

$$\frac{[n+1]^\delta}{[1]^{\delta-1}} = \Sigma\, f_{\alpha_1, \alpha_2, \dots \alpha_h} z^{\varpi} f^n_{\alpha_1, \alpha_2, \dots \alpha_h}(z).$$

When $T_{\alpha_1, \alpha_2, \dots \alpha_h}$, $T_{\beta_1, \beta_2, \dots \beta_k}$ are conjugate, and their tableaux differ by interchange of rows and columns, the generating functions are not quite so simply related as in the case of perpetuants. There is a relation, as we proceed to show. As in § 6 we take $h = 3$, and use the value of β_r as found there. Then

$$\begin{aligned}\Pi\,[\beta_r+n+1-r]_{\beta_r} &= [n+3]_3\,[n+2]_3 \dots [n+4-\alpha_3]_3\\ &\quad [n+2-\alpha_3]_2\,[n+1-\alpha_3]_2 \dots [n+3-\alpha_2]_2\\ &\quad [n+1-\alpha_2]\,[n-\alpha_2] \dots [n+2-\alpha_1]\\ &= [n+3]_{\alpha_3}\,[n+2]_{\alpha_2}\,[n+1]_{\alpha_1}.\end{aligned}$$

Now

$$\begin{aligned}[n+r]_{\alpha_r} &= \prod_{s=1}^{\alpha_r}[n+r-\alpha_r+s]\\ &= (-)^{\alpha_r} z^{-(m+1-r)\alpha_r - \binom{\alpha_r+1}{2}}\,[m+1+\alpha_r-r]_{\alpha_r},\end{aligned}$$

where

$$n = -m-2.$$

Hence

$$\Pi\,[n+r]_{\alpha_r} = (-)^\delta z^\chi\, \Pi\,[m+1+\alpha_r-r]_{\alpha_r},$$

where

$$\chi = -(m+1)\delta + \varpi - \varpi'.$$

The other part of $f^n_{\alpha_1, \alpha_2, \dots \alpha_h}(z)$ does not contain n, and is the same as for the conjugate set as was seen in § 6. Thus the transformation of n into $-m-2$ which changes $\frac{[n+1]^\delta}{[1]^{\delta-1}}$ into $(-)^\delta z^{-(m+1)\delta} \cdot \frac{[m+1]^\delta}{[1]^{\delta-1}}$ at the same time interchanges conjugate generating functions.

IV—The Characteristic Function

§ 15. In the fourth paper on Quantitative Substitutional Analysis* the characteristic function $\Phi_{\alpha_1, \alpha_2, \dots \alpha_h}$ of the symmetric group corresponding to the representation $T_{\alpha_1, \alpha_2, \dots \alpha_h}$ was introduced. The conception is due to Schur. The characteristic function is defined as the function

$$\Phi_{\alpha_1, \alpha_2, \dots \alpha_h} = \Sigma \frac{\chi_{\beta_1, \beta_2, \dots \beta_n}}{\beta_1! \, \beta_2! \dots \beta_n!} \left(\frac{s_1}{1}\right)^{\beta_1} \left(\frac{s_2}{2}\right)^{\beta_2} \dots \left(\frac{s_n}{n}\right)^{\beta_n},$$

where $\chi_{\beta_1, \beta_2, \dots \beta_n}$ is the characteristic of the permutation which has β_r cycles of r letters each, for the various values of r. It was shown in the paper referred to that when every cycle of degree r is replaced by the symbol s_r then $T_{\alpha_1, \alpha_2, \dots \alpha_h}$ becomes $f_{\alpha_1, \alpha_2, \dots \alpha_h} \Phi_{\alpha_1, \alpha_2, \dots \alpha_h}$, $P_r N_r M_r$ becomes $\Phi_{\alpha_1, \alpha_2, \dots \alpha_h}$ and $P_r \sigma_{rs} N_s M_s$ becomes zero when r and s are different.

Theorem VIII—*The characteristic function* $\Phi_{\alpha_1, \alpha_2, \dots \alpha_h}$ *is equal to* $|\Phi_{\alpha_r+s-r}|$, *where* r, s *define the row and column respectively in the determinant; and* Φ_λ *is the characteristic function of the representation* T_λ *of the symmetric group of* λ *letters.*

This result follows at once from the equation proved in Q.S.A. VI†

$$\text{(VIII)} \qquad \frac{n!}{f_{\alpha_1, \alpha_2, \dots \alpha_h}} T_{\alpha_1, \alpha_2, \dots \alpha_h} = \prod_{r<s} (1 - S_{rs}) \frac{1}{n!} \binom{n}{\alpha_1 \alpha_2 \dots \alpha_h} \Gamma P_{\alpha_1, \alpha_2, \dots \alpha_h},$$

where $P_{\alpha_1, \alpha_2, \dots \alpha_h}$ is the product of symmetric groups of degrees $\alpha_1, \alpha_2, \dots \alpha_h$ respectively; Γ indicates that the sum of all the $n!$ permutations of the n letters are to be taken, S_{rs} represents the operation of moving one letter from the sth row of the tableau up to the rth row, and the resulting term is taken to be zero when any row becomes less than a row below it or when letters from the same row overlap. In fact, when cycles of degree r are replaced by the symbol s_r this equation becomes

$$\Phi_{\alpha_1, \alpha_2, \dots \alpha_h} = \prod_{r<s} (1 - S_{r,s}) \Phi_{\alpha_1} \Phi_{\alpha_2} \dots \Phi_{\alpha_h},$$

which gives the result stated, provided the restriction as to zero terms is observed. It will be noticed that the suffix of the element Φ_{α_r+s-r} in the determinant may become zero or negative. In the former case the value of the element is unity, and in the latter case it is zero.

* Young, 'Proc. London Math. Soc.,' vol. 31, p. 259 (1930).
† Young, 'Proc. London Math. Soc.,' vol. 34, p. 199 (1932).

COROLLARY—*The coefficient in* $\Phi_{\alpha_1, \alpha_2, \ldots \alpha_h}$ *of any permutation which contains a cycle of degree greater than* $\alpha_1 + h - 1$ *is zero. And hence the characteristic of such a permutation is zero.*

We may use equation (VIII) to put Theorem VIII in another useful form. The symbol G_α is used for the positive symmetric group of α letters divided by its order, then

THEOREM IX—

$$T_{\alpha_1, \alpha_2, \ldots \alpha_h} = \frac{f_{\alpha_1, \alpha_2, \ldots \alpha_h}}{n\,!} \Gamma \,|\, G_{\alpha_r + s - r} \,|,$$

where r, s define the rows and columns of the determinant; and in its expansion every letter appears in one and only one factor of each term.

§ 16. From the equation

$$1 = \Sigma\, T_{\alpha_1, \alpha_2, \ldots \alpha_h},$$

we deduce at once the equation

$$1 = \Sigma\, f_{\alpha_1, \alpha_2, \ldots \alpha_h}\, \Phi_{\alpha_1, \alpha_2, \ldots \alpha_h},$$

and hence by Theorem VIII the equation

$$1 = \Sigma\, f_{\alpha_1, \alpha_2, \ldots \alpha_h} \,|\, \Phi_{\alpha_r + s - r} \,|.$$

Now

$$\Phi_1, \Phi_2, \ldots \Phi_n$$

are a series of functions each of which contains a new independent variable which does not appear in those which precede it; thus s_r first appears in Φ_r. The functions may then be regarded as completely arbitrary and independent. Hence

THEOREM X—*The quantities* Φ_r *for positive integral values of r are arbitrary, for negative values of r they are zero, and* $\Phi_0 = 1$, *then*

$$\Phi_1{}^n = \sum_{\Sigma \alpha = n} f_{\alpha_1, \alpha_2, \ldots \alpha_h} \,|\, \Phi_{\alpha_r + s - r} \,|.$$

That this equation is true when Φ is a generating function is an immediate deduction from Theorem VIII in Some Generating Functions,* and hence it is true for a wide range of functions. It was well to determine, whether it expressed an inherent property of generating functions as such, we see that it does not, that, in fact, it is not a property of Φ at all, but only of the numbers $f_{\alpha_1, \alpha_2, \ldots \alpha_h}$. That the value of the left-hand side of the equation is $\Phi_1{}^n$ is seen by considering the last term of the series

$$f_{1^n} \,|\, \Phi_{1 + s - r} \,|,$$

which alone of all the terms contains $\Phi_1{}^n$, the only term on the right-hand side not containing one of the other Φ's.

§ 17. LEMMA I.

$$\sum_{r=1}^{\delta} \frac{1}{[r] \,.\, [\delta - r]\,!} = \frac{\delta}{[\delta]\,!}.$$

* YOUNG, 'Proc. London Math. Soc.,' vol. 35, p. 440 (1933).

It is quite easy to verify this for small values of δ, so its truth will be assumed for all values of δ less than that one under consideration.

Multiply each side by $[\delta]!$ then it is required to prove that

$$\phi_\delta(z) = \sum_{r=1}^{\delta} \frac{[\delta]_r}{[r]} - \delta = 0.$$

Then

$$\begin{aligned}\phi_\delta(z) - \phi_{\delta-1}(z) &= z^{\delta-1} + [\delta-1]z^{\delta-2} + [\delta-1]_2 z^{\delta-3} + \ldots + [\delta-1]_{\delta-2} z + [\delta-1]! - 1\\ &= [\delta-1]\{z^{\delta-2} + [\delta-2]z^{\delta-3} + [\delta-2]_2 z^{\delta-4} + \ldots + [\delta-2]! - 1\}\\ &= [\delta-1]\{\phi_{\delta-1}(z) - \phi_{\delta-2}(z)\}.\end{aligned}$$

The truth of the lemma is then at once apparent by induction.

§ 18. THEOREM XI—*When in the generalized form Φ_δ of the symmetric group the symbol s_r for a cycle of degree r is replaced by $(1 - z^r)^{-1}$, the replacement being made for all cycles,*

$$\Phi_\delta = \frac{\delta\,!}{[\delta]\,!}.$$

This is verified without difficulty for $\delta = 1, 2, 3$; it will be assumed true for all degrees less than δ. In the symmetric group we have first those terms in which the last letter b is not permuted, these form the symmetric group of degree $\delta - 1$, then those terms in which b occurs in a cycle of two letters and so on. Let P_r denote the sum of all terms in which b occurs in a cycle of r letters. Then the sum of all terms is $\sum_{r=1}^{\delta} P_r$.

Consider P_r when the $r - 1$ letters associated with b are fixed, we may have the remaining $\delta - r$ letters permuted by any permutation of the symmetric group of degree $\delta - r$. Also the $r - 1$ letters may be selected in $\binom{\delta - 1}{r - 1}$ ways and also may have $(r - 1)\,!$ different positions in the cycle relative to b. Then when the replacement of cycles is made

$$P_r = \frac{(\delta - 1)\,!}{[r]\,[\delta - r]\,!}.$$

And

$$\Phi_\delta = \sum_{r=1}^{\delta} \frac{(\delta-1)\,!}{[r]\,[\delta - r]\,!} = \frac{\delta\,!}{[\delta]\,!},$$

by the lemma, which proves the theorem.

§ 19. THEOREM XII—

$$\delta\,!\, z^{\varpi_\alpha} f_\alpha(z) = \sum_\beta h_\beta \chi_\beta^{(\alpha)} \frac{[\delta]\,!}{[\beta_1]\,[\beta_2] \ldots [\beta_k]}.$$

Here α, β are generic symbols, and β stands for $\beta_1, \beta_2, \ldots \beta_k$, where

$$\beta_1 + \beta_2 + \ldots + \beta_k = \delta\,;$$

and h_β is the number of members of the β conjugate class of the symmetric group which consists of operations having cycles of degrees $\beta_1, \beta_2, \ldots \beta_k$; $\chi_\beta^{(\alpha)}$ is the characteristic of this class in the representation T_α.

By Theorem VIII the characteristic function for this representation is

$$\Phi_\alpha \equiv \Phi_{\alpha_1, \alpha_2, \ldots \alpha_h} = |\Phi_{\alpha_r + s - r}|.$$

Here the cycles are represented by variables $s_1, s_2, \ldots s_\delta$, and the characteristic function is at once given by this equation with the ordinary rules of algebra. If we write $(1 - x^r)^{-1}$ for s_r we obtain

$$\Phi_{\alpha_r + s - r} = \frac{1}{[\alpha_r + s - r]\,!}.$$

But by Theorem I, § 4,

$$z^{\varpi} f_{\alpha_1, \alpha_2, \ldots \alpha_h}(z) \equiv [\delta]\,! \left|\frac{1}{[\alpha_r + s - r]\,!}\right| = [\delta]\,!\,\Phi_\alpha = \frac{1}{\delta\,!}[\delta]\,! \sum_\beta h_\beta \chi_\beta^{(\alpha)} s_{\beta_1} s_{\beta_2} \ldots s_{\beta_k}$$

$$= \frac{1}{\delta\,!} \Sigma\, h_\beta \chi_\beta^{(\alpha)} \frac{[\delta]\,!}{[\beta_1]\,[\beta_2]\ldots[\beta_k]};$$

which was to be proved.

THEOREM XIII—

$$\frac{[\delta]\,!}{[\beta_1]\,[\beta_2]\ldots[\beta_k]} = \sum_\alpha \chi_\beta^{(\alpha)} z^{\varpi_\alpha} f_\alpha(z).$$

To prove this we use the well-known results

$$\sum_\alpha \chi_\gamma^{(\alpha)} \chi_\beta^{(\alpha)} = 0, \qquad \beta \neq \gamma; \qquad \sum_\alpha (\chi_\beta^{(\alpha)})^2 = \delta\,!/h_\beta.$$

Multiply the identity of the last theorem by $\chi_\gamma^{(\alpha)}$ and then sum the result for all representations α, we obtain the desired result at once.

Equation (III), § 2, is a particular case of this theorem.

§ 20. The generating functions for binary forms of finite order may be treated in the same way. We begin with a lemma as before.

LEMMA II.

$$\phi(n, \delta) \equiv \sum_{r=1}^{\delta} \frac{[\delta]_r}{[r]} [rn + r]\,[n + \delta - r]_{\delta - r} - \delta\,[n + \delta]_\delta = 0.$$

This identity is easy to prove directly for the values 1, 2, 3 of δ, and all values of n; it is also obviously true for $n = 0$, and all values of δ. We proceed to establish a double induction.

Now

$$[\delta]_r = [\delta - 1]_r + z^{\delta - r}[\delta - 1]_{r-1}\,[r];$$

and

$$[rn + r] = [rn + 2r] - z^{rn + r}\,[r];$$

hence

$$\phi(n,\delta) - [n+1]\,\phi(n+1,\delta-1) = \sum_{r=1}^{\delta}\{-z^{nr+r}[\delta-1]\,[n+\delta-r]_{\delta-r}$$
$$+ z^{\delta-r}[\delta-1]_{r-1}[rn+r]\,[n+\delta-r]_{\delta-r}\} - [n+\delta]_{\delta}.$$

The conventions are adopted here that

$$[\delta-1]_{\delta}=0, \qquad [\delta-1]_{-1}=0.$$

Then

$$z^{\delta-r}[\delta-1]_{r-1}[rn+r] - z^{nr+r}[\delta-1]_r = [\delta-1]_{r-1}[rn+r] - [\delta-1]_r$$

and hence

$$\phi(n,\delta) - [n+1]\,\phi(n+1,\delta-1)$$
$$= \sum_{r=0}^{\delta}\{-[\delta-1]_r[n+\delta-r]_{\delta-r} + [\delta-1]_{r-1}[rn+r]\,[n+\delta-r]_{\delta-r}\}.$$

Again

$$-[\delta-1]_r[n+\delta-r]_{\delta-r} + [\delta-1]_{r-1}[rn+r]\,[n+\delta-r]_{\delta-r}$$
$$+ z^{n+1}[\delta-1]_r[n+\delta-r]_{\delta-r} - z^{n+1}[\delta-1]_{r-1}[rn+r-n-1]\,[n+\delta-r]_{\delta-r}$$
$$= [n+1]\{-[\delta-1]_r[n+\delta-r]_{\delta-r} + [\delta-1]_{r-1}[n+\delta-r]_{\delta-r}\}$$
$$= [n+1]\,z^{\delta-r}[\delta-1]_{r-1}[n+\delta-r]_{\delta-r}.$$

Hence

$$\phi(n,\delta) - [n+1]\,\phi(n+1,\delta-1) - z^{n+1}[\delta-1]\{\phi(n,\delta-1) - [n+1]\,\phi(n+1,\delta-2)\}$$
$$= [n+1]\{\sum_{r=1}^{\delta} z^{\delta-r}[\delta-1]_{r-1}[n+\delta-r]_{\delta-r} - [n+\delta]_{\delta-1}\}.$$

Further

$$z^{\delta-r}[n+\delta-r]_{\delta-r}[\delta-1]_{r-1} = z^{\delta-r}[n-1+\delta-r]_{\delta-r}[\delta-1]_{r-1}$$
$$+ z^{n+1}[\delta-1]\,.\,z^{\delta-r-1}[n+\delta-1-r]_{\delta-1-r}[\delta-2]_{r-1};$$

and

$$[n+\delta]_{\delta-1} = [n-1+\delta]_{\delta-1} + z^{n+1}[\delta-1]\,[n+\delta-1]_{\delta-2}.$$

We deduce at once the identity

$$\phi(n,\delta) - [n+1]\,\phi(n+1,\delta-1) - z^{n+1}[\delta-1]\{\phi(n,\delta-1) - [n+1]\,\phi(n+1,\delta-2)\}$$
$$= \frac{[n+1]}{[n]}(\phi(n-1,\delta) - [n]\,\phi(n,\delta-1)$$
$$- z^{n}[\delta-1]\{\phi(n-1,\delta-1) - [n]\,\phi(n,\delta-2)\})$$
$$+ z^{n+1}[\delta-1](\phi(n,\delta-1) - [n+1]\,\phi(n+1,\delta-2)$$
$$- z^{n+1}[\delta-2]\{\phi(n,\delta-2) - [n+1]\,\phi(n+1,\delta-2)\}).$$

This equation enables us to deduce that $\phi(n,\delta)=0$, provided this is true for all lower values of δ and any value of n and also for the given value of δ and values of

n, less than that considered. The induction is thus complete for we already know the truth of the result when $n = 0$ and δ has any value, and also when $\delta = 1, 2,$ or 3 and n has any value.

§ 21. THEOREM XIV—*When in the generalized form* Φ_δ *of the symmetric group the symbol* s_r *for a cycle of degree* r *is replaced by*

$$\frac{[(n+1)\,r]}{[r]},$$

then

$$\Phi_\delta = \frac{\delta\,!\;[n+\delta]_\delta}{[\delta]\,!}.$$

This is easily verified for $\delta = 1, 2, 3$ and we proceed by induction.

Let P_r be that part of the expression which arises from those operations of the group which contain a particular letter b in a cycle of degree r. Then

$$P_r = (\delta - 1)\,!\;\frac{[n+\delta-r]_{\delta-r}\,[(n+1)\,r]}{[r]\,[\delta-r]\,!}.$$

Hence

$$\Phi_\delta = \sum_{r=1}^{\delta} P_r = \delta\,!\;\frac{[n+\delta]_\delta}{[\delta]\,!},$$

on using the lemma.

The arguments of § 19 may now be repeated to establish the following :—

THEOREM XV—

$$\delta\,!\; z^{\varpi_a} f_a^{\,n}(z) = \sum_\beta h_\beta\,\chi_\beta^{\,a}\,[\delta]\,!\;\prod_{r=1}^{k}\frac{[(n+1)\,\beta_r]}{[\beta_r]}$$

and

$$[\delta]\,!\;\prod_{r=1}^{k}\frac{[(n+1)\,\beta_r]}{[\beta_r]} = \sum_a \chi_\beta^{(a)}\,z^{\varpi_a} f_a^{\,n}(z).$$

V—THE COMPOUND SYMMETRIC GROUP

§ 22. The group generated by the permutations

$$(a, a_{r+1})\,(b, b_{r+1}), \qquad r = 1, 2, \ldots \delta - 1$$

will be called the compound symmetric group of degree δ. It possesses the same number of irreducible representations as the ordinary symmetric group and they are of identically the same form ; we may write them

$$T^{(ab)}_{a_1, a_2, \ldots a_h}.$$

Further corresponding to such a representation there are semi-normal units, which will be written

$$\Pi_{\rho,\sigma}.$$

A unit $\Pi_{\rho,\sigma}$ is a linear function of the permutations of the compound symmetric group. Every such permutation is a product of a permutation of the a's by a permutation of the b's. Let $\varpi_{r,s}$, $\varpi'_{u,v}$ be semi-normal units of any representations of the a's and of the b's respectively. Then

$$\Pi_{\rho,\sigma} = \Sigma\, \lambda \varpi_{r,s} \varpi'_{u,v},$$

where λ is numerical.

For our purposes, it is important to calculate this expression for $\Pi_{\rho,\sigma}$ or at least for

$$T^{(ab)} = \sum_{\rho=1}^{f} \Pi_{\rho,\rho}\,.$$

In the first place consider the representation

$$T_{\delta}^{(ab)} = \{A_1 A_2 \ldots A_\delta\}\,;$$

that is, the sum of all the permutations of the compound symmetric group divided by their number. In Q.S.A., VIII*, § 20, it was shown that

$$\{A_1 A_2 \ldots A_\delta\}\, \varpi_{r,s}\, \varpi'_{u,v} = 0,$$

unless the units $\varpi_{r,s}$, $\varpi'_{u,v}$ are identical, except for the different letters employed. Hence

(IX)
$$\begin{aligned}\{A_1\, A_2\, .\, .\, .\, A_\delta\} &= \Sigma \lambda \varpi_{r,s}\, \varpi'_{u,v}\\ &= \Sigma \lambda\, \{A_1\, A_2\, .\, .\, .\, A_\delta\}\, \varpi_{r,s}\, \varpi'_{u,v}\\ &= \Sigma \lambda_{r,s}\, \varpi_{r,s}\, \varpi'_{r,s}\,.\end{aligned}$$

Now multiply each side by

$$(A_k A_{k+1}) = (a_k a_{k+1})\, (b_k b_{k+1})$$

on the left, and let us suppose that the tableau F_r is changed by $(a_k a_{k+1})$ into the later tableau F_t. Then (using orthogonal units)

$$\begin{aligned}\Sigma \lambda_{rs} \varpi_{rs} \varpi'_{rs} = \Sigma \lambda_{rs}\, &[-\rho^{-1}\varpi_{rs} + \rho^{-1}\sqrt{(\rho^2-1)}\,\varpi_{ts}]\,[-\rho^{-1}\varpi'_{rs} + \rho^{-1}\sqrt{(\rho^2-1)}\,.\,\varpi'_{ts}]\\ &+ \Sigma \lambda_{ts}\,[\rho^{-1}\sqrt{(\rho^2-1)}\,\varpi_{rs} + \rho^{-1}\varpi_{ts}]\,[\rho^{-1}\sqrt{(\rho^2-1)}\,\varpi'_{rs} + \rho^{-1}\varpi'_{ts}]\,;\end{aligned}$$

hence $\lambda_{rs} = \lambda_{ts}$. Similarly, by right-hand multiplication we find $\lambda_{sr} = \lambda_{st}$. Thus all the coefficients are the same, for any particular representation $T_{\alpha_1 . \alpha_2 \ldots \alpha_h}$ to which $\varpi_{r,s}$ and $\varpi'_{r,s}$ respectively belong.

Consider a particular representation T then since

$$\varpi_{r,s}\varpi'_{r,s}\,.\,\varpi_{u,v}\varpi'_{u,v} = 0,$$

unless $s = u$, and

$$\varpi_{r,s}\varpi'_{r,s}\,.\,\varpi_{s,v}\varpi'_{s,v} = \varpi_{r,v}\varpi'_{r,v},$$

the sum $\Sigma \lambda_{rs} \varpi_{r,s} \varpi'_{r,s}$ can be expressed as the matrix $[\lambda_{rs}]$ of order f for this representation. In our case all the elements λ_{rs} have the same value.

* Young, 'Proc. London Math. Soc.,' vol. 37, p. 441 (1934).

Now take the square of each side of equation (IX), we obtain

$$\{A_1A_2 \ldots A_\delta\} = \Sigma Q^2 = \Sigma Q,$$

where Q is represented by the matrix $[\lambda_{rs}]$; and hence

$$Q^2 = f\lambda_{rs}Q.$$

Hence

$$f\lambda_{rs} = 1,$$

and

(X) $$\{A_1A_2 \ldots A_\delta\} = \Sigma f^{-1}\varpi_{rs}\varpi'_{rs}.$$

Here the Σ extends to every unit of every representation.

§ 23. There are certain fairly obvious restrictions to the coefficients in the equation

$$\Pi_{\rho,\sigma} = \Sigma\lambda\varpi_{r,s}\varpi'_{u,v}.$$

We take orthogonal units.

(i) $\Pi_{\rho,\sigma}$ is a linear function of the form

$$\Sigma\mu\tau\tau',$$

where μ is numerical, τ is a permutation of the a's, and τ' the same permutation of the b's. We deduce at once

$$\Pi_{\rho,\sigma} = \Sigma\lambda\,(\varpi_{r,s}\varpi'_{u,v} + \varpi_{u,v}\varpi'_{r,s}).$$

(ii) $$\Pi_{\sigma,\rho} = \Sigma\lambda\,(\varpi_{s,r}\varpi'_{v,u} + \varpi_{v,u}\varpi'_{s,r}).$$

(iii) $$\Pi_{\rho,\rho} = \Sigma\lambda\,(\varpi_{r,s}\varpi'_{u,v} + \varpi_{u,v}\varpi'_{r,s} + \varpi_{s,r}\varpi'_{v,u} + \varpi_{v,u}\varpi'_{s,r}).$$

(iv) When every transposition both of the a's and b's is changed in sign, the permutations of the compound group are all unchanged; hence, if the suffix letters of the units be also taken to define the corresponding tableaux, and $F_{r'}$ be the tableau obtained by interchanging rows and columns in F_r,

$$\Pi_{\rho,\sigma} = \Sigma\lambda\,(\varpi_{r,s}\varpi'_{u,v} + \varpi_{u,v}\varpi'_{r,s} + \varpi_{r',s'}\varpi'_{u',v'} + \varpi_{u',v'}\varpi'_{r',s'}).$$

(v) When every transposition of the a's is changed in sign, but the transpositions of the b's are unchanged, we obtain

$$\Pi_{\rho'\sigma'} = \Sigma\lambda\,(\varpi_{r',s'}\varpi'_{u,v} + \varpi_{u',v'}\varpi'_{r,s} + \varpi_{r,s}\varpi'_{u',v'} + \varpi_{u,v}\varpi_{r',s'}).$$

The last consideration enables us to write down without further enquiry the value

$$T_{1\delta}^{(ab)} = \Sigma f^{-1}\varpi_{rs}\varpi'_{r',s'}.$$

It is to be noticed that the orders f, f' of conjugate representations are the same.

§ 24. Let $\Gamma^{(a)}$ represent as before the operation of taking the sum of the δ ! terms obtained by permuting the letters a in the operand. Then

(XI)
$$\Gamma^{(a)}T^{(ab)}_{a_1, a_2 \ldots a_h} = \sum_{\beta} \mu_\beta t_\beta^{(a)} t_\beta^{(b)}$$
$$= \sum_{\epsilon, \zeta} \nu_{\epsilon, \zeta} T_\epsilon^{(a)} T_\zeta^{(b)},$$

where μ_β, $\nu_{\epsilon, \zeta}$ are numerical, t_β is the sum of the members of a particular conjugate set of permutations of the letters concerned, ϵ, ζ define certain definite irreducible representations, and $\nu_{\epsilon, \zeta} = \nu_{\zeta, \epsilon}$. Hence, when

$$T_a^{(ab)} = \Sigma \lambda_{rs, uv} \varpi_{r, s} \varpi'_{u, v},$$

we find

$$\Gamma^{(a)} \Sigma \lambda_{rs, uv} \varpi_{r, s} \varpi'_{u, v} = \sum_{\epsilon, \zeta} \nu_{\epsilon, \zeta} T_\epsilon^{(a)} T_\zeta^{(b)}.$$

The equation (XI) will be written

(XII)
$$\Gamma^{(a)} T_a^{(ab)} = \Sigma f_{\alpha\beta\gamma} T_\beta^{(a)} T_\gamma^{(b)}.$$

§ 25. Frobenius* discussed a closely allied problem, and his results come in very usefully here. Consider two linear substitution groups, which are representations of the same abstract group. Let

$$u_\alpha = \sum_{\beta} a_{\alpha\beta} v_\beta \qquad (\alpha, \beta = 1, 2, \ldots f)$$

and

$$u'_\gamma = \sum_{\delta} a'_{\gamma\delta} v'_\delta \qquad (\gamma, \delta = 1, 2, \ldots f')$$

be corresponding substitutions in the two groups. A third group is constructed by compounding the substitutions of the first two, in this the substitution corresponding to those written above is

$$u_\alpha u'_\gamma = \sum_{\beta, \delta} a_{\alpha\beta} a'_{\gamma, \delta} v_\beta v'_\delta.$$

This group is a representation of the same abstract group as a linear substitution group with ff' variables. In general, this third group is reducible ; let Φ be its group determinant, then the reduction may be expressed by the equation

$$\Phi = \prod_{\mu} \Phi_{\mu'}^{f_{\kappa\lambda\mu}},$$

where the suffixes κ, λ, μ define irreducible representations of the group, μ and μ' being conjugate imaginary representations (they are the same when the representation is real), and κ, λ define the representations by the two groups from which we started.

* 'S. B. berl. math. Ges.' (volumes are not numbered), p. 330 (1899).

FROBENIUS proved that $f_{\kappa\lambda\mu}$ is a positive integer or zero unchanged in value by any permutation of the suffixes. Its value is given by

$$hf_{\kappa\lambda\mu} = \sum_{R} \chi^{(\kappa)}(R)\, \chi^{(\lambda)}(R)\, \chi^{(\mu)}(R)\,;$$

where h is the order of the group, and $\chi^{(\kappa)}(R)$ is the characteristic of the group element R in the representation κ.

Also

$$f_{\kappa} f_{\lambda} = \sum_{\mu} f_{\kappa\lambda\mu} f_{\mu},$$

$$\Sigma f_{\kappa\lambda\mu}{}^2 = \sum_{\rho} h/h_{\rho},$$

where h_{ρ} is the number of operations in the ρth conjugate class.

§ 26. The problem of FROBENIUS is not quite identical with that considered here, and so it is not permissible merely to quote his result without further enquiry. Let $\chi^{a}(R)$ be the characteristic of the permutation R in the representation T_a of the symmetric group of degree δ. When the permuted letters have to be expressed, we write $R^{(a)}$ or $R^{(b)}$ as the case may be, and for the compound group

$$R^{(ab)} = R^{(a)} R^{(b)}.$$

Then from equation (XII)

$$\frac{\delta\,!}{h_R} \chi^{(a)}(R) = \sum_{\beta,\gamma} f_{a\beta\gamma}\, \chi^{(\beta)}(R)\, \chi^{(\gamma)}(R),$$

and

$$0 = \sum_{\beta,\gamma} f_{a\beta\gamma}\, \chi^{(\beta)}(R)\, \chi^{(\gamma)}(S),$$

when R and S are not conjugate. Moreover, these equations are the necessary and sufficient conditions for the truth of (XII). FROBENIUS found the equation

$$\text{(XIII)} \qquad \chi^{\kappa}(R)\, \chi^{\lambda}(R) = \sum_{\mu} f_{\kappa\lambda\mu'}\, \chi^{(\mu)}(R'),$$

and from which he derived the value of $f_{\kappa\lambda\mu}$. Multiply (XIII) by $\chi^{(\lambda)}(S)$, and sum the equation for all values of λ, we have when S is not conjugate to R

$$0 = \sum_{\lambda\mu} f_{\kappa\lambda\mu'}\, \chi^{(\lambda)}(S)\, \chi^{(\mu)}(R)\,;$$

and when S = R

$$\frac{h}{h_R} \chi^{(\kappa)}(R) = \sum_{\lambda\mu} f_{\kappa\lambda\mu'}\, \chi^{\lambda}(R)\, \chi^{\mu}(R).$$

These equations are identical with (XII), when it is remembered that the representation μ is the same as the conjugate imaginary representation for the symmetric group. Hence the quantity $f_{a\beta\gamma}$ of the present discussion is the same as that introduced by FROBENIUS. We therefore may use his results.

VI—The Function $f_\delta(x, y)$

§27. We define this function in terms of the functions of Section I, thus:

$$f_\delta(x, y) = \Sigma\, x^{\varpi} y^{\varpi} f_{\alpha_1, \alpha_2, \dots \alpha_h}(x)\, f_{\alpha_1, \alpha_2, \dots \alpha_h}(y),$$

the summation extends to all the representations of the symmetric group of δ letters.

The importance here of this function lies in the fact that the generating function for ternary perpetuants of degree δ for a single form may be written

$$\frac{(1-x)(1-y)(x-y)\, f_\delta(x, y)}{x \,.\, \{[\delta]\, !\}_x \{[\delta]\, !\}_y},$$

where the suffix after the bracket indicates the particular variable in the function.

The attempt is made to express this in the form

$$\frac{\phi(x, xy)}{\psi(x, xy)} - \frac{y\phi(y, xy)}{x\psi(y, xy)},$$

where ϕ, ψ are rational integral functions; thus the function would be separated into two parts of which obviously the first only concerns the perpetuant problem in hand. The attempt is successful up to degree 5; and indications are obtained as to what may be expected in general.

Theorem XVI—$\delta\, !\, f_\delta(x, y) = \sum\limits_{\beta} h_\beta X_\beta Y_\beta$, *where* β *stands for a partition* $\beta_1, \beta_2, \dots \beta_k$ *of* δ; $X_\beta = \left\{\dfrac{[\delta]\, !}{[\beta_1]\, [\beta_2] \dots [\beta_k]}\right\}_x$, Y_β *is the same function of* y, *and* h_β *is the number of members of the symmetric group of degree* δ *which belong to the conjugate class defined by* β.

Consider the effect of replacing each "a" cycle of degree r by $(1-x^r)^{-1}$ and each "b" cycle of degree r by $(1-y^r)^{-1}$. The compound symmetric group of degree δ, $\delta\, !\, T_\delta$, becomes

$$\Sigma \frac{h_\beta X_\beta Y_\beta}{\{[\delta]\, !\}_x \{[\delta]\, !\}_y}.$$

Also as in §§ 15, 19, we see that $\varpi_{r,s}$ becomes zero unless $r = s$, and then it has the value $x^{\varpi_\alpha} f_\alpha(x)/\{[\delta]\, !\}_x$, where T_α is the representation to which the unit $\omega_{r,r}$ belongs. Thus, from equation (X), § 22, T_δ becomes, after making this replacement, $f_\delta(x, y)/(\{[\delta]\, !\}_x \{[\delta]\, !\}_y)$. The theorem is proved by equating the two values obtained for T_δ.

§ 28. A whole set of functions may now be defined analogous to the functions $f_{\alpha_1, \alpha_2, \dots \alpha_h}(x)$, as follows:—

$$f_{\alpha_1, \alpha_2, \dots \alpha_h}(x, y) = \sum_{\beta, \gamma} f_{\alpha\beta\gamma}\, x^{\varpi_\beta} y^{\varpi_\gamma} f_\beta(x) f_\gamma(y).$$

That $f_\delta(x, y)$ is a particular case will be seen at once.

THEOREM XVII—

$$\delta\,!\,f_\alpha(x,y) = \sum_\beta h_\beta \chi_\beta^{(\alpha)} X_\beta Y_\beta.$$

And

$$X_\beta Y_\beta = \sum_\alpha \chi_\beta^{(\alpha)} f_\alpha(x,y).$$

The proof is practically a repetition of §§ 19 and 27.

§ 29. The same processes may be extended. Thus one or both of the sets of letters a, b may be supposed to represent compound groups—or multiply compound groups. Thus we may have m sets of letters

$$a_{r,s}\qquad r = 1, 2, \dots m,\qquad s = 1, 2, \dots \delta\,;$$

and consider the symmetric group of degree δ formed by permuting all the sets simultaneously in exactly the same way. We obtain functions

$$f_{a_1, a_2, \dots a_h}(x_1, x_2, \dots x_m).$$

These may be obtained in the first place by grouping the m sets into two groups and using the above arguments ; and then we can break these groups up further. It is to be noticed that the same final result must be obtained however the m sets are thus divided. Or else following FROBENIUS we may use his numerical function

$$f_{\alpha\beta\gamma\dots},$$

with $m + 1$ suffixes, and write

$$f_\alpha(x_1, x_2, \dots x_m) = \sum_{\beta,\gamma} f_{\alpha\beta\gamma\dots}\, x_1^{\varpi_\beta} x_2^{\varpi_\gamma} \dots f_\beta(x_1) f_\gamma(x_2)\dots.$$

FROBENIUS* proved that

$$f_{\alpha\beta\gamma\epsilon\dots} = \sum_R \chi^{(\alpha)}(R)\,\chi^{(\beta)}(R)\,\chi^{(\gamma)}(R)\,\chi^{(\epsilon)}(R)\dots.$$

Then, as before, we obtain

THEOREM XVIII—

$$\delta\,!\,f_\alpha(x_1, x_2, \dots x_m) = \sum_\beta h_\beta \chi_\beta^{(\alpha)} X_\beta^{(1)} X_\beta^{(2)} \dots X_\beta^{(m)},$$

And

$$X_\beta^{(1)} X_\beta^{(2)} \dots X_\beta^{(m)} = \sum_\alpha \chi_\beta^{(\alpha)} f_\alpha(x_1, x_2, \dots x_m).$$

THEOREM XIX—*When* T_α, T_α', *are conjugate representations*

$$f_{\alpha'}(x,y) = y^{\binom{\delta}{2}} f_\alpha(x, y^{-1}) = x^{\binom{\delta}{2}} f_\alpha(x^{-1}, y).$$

It was shown § 6 that

$$z^{\varpi'} f_{\alpha'}(z) = z^{\binom{\delta}{2}}\,[z^{-\varpi} f_\alpha(z^{-1})].$$

Also it follows from (v), § 23, that

$$f_{\alpha\beta\gamma} = f_{\alpha'\beta\gamma'} = f_{\alpha'\beta'\gamma} = f_{\alpha\beta'\gamma'}.$$

* 'S. B. berl. math. Ges.,' p. 330 (1899).

The result follows at once on replacing y by y^{-1} in the equation

$$f_\alpha(x, y) = \sum_{\beta\gamma} f_{\alpha\beta\gamma} x^{\varpi_\beta} y^{\varpi_\gamma} f_\beta(x) f_\gamma(y).$$

§30. The next problem is to separate the function

$$\frac{(1-x)(1-y)(x-y) f_\delta(x, y)}{x\{[\delta]!\}_x\{[\delta]!\}_y} = \frac{\phi(x, xy)}{\psi(x, xy)} - \frac{y\phi(y, xy)}{x\psi(y, xy)}.$$

For $\delta = 2, 3, 4$ the problem is solved quite easily. We use the negative symmetric group $\{xy\}'$ thus :—

$$(x-y) f_2(x, y) = \{xy\}'(x + x^2y) = \{xy\}' x(1-y^2);$$

hence

$$\frac{(x-y) f_2(x, y)}{x(1-x^2)(1-y^2)} = \frac{1}{1-x^2} - \frac{y}{x(1-y^2)}.$$

For $\delta = 3$,

$$(x-y) f_3(x, y)(1-x^2y^2) = \{xy\}'(1-y^2)(1-y^3)(x+yx^5),$$

hence

$$\frac{(x-y) f_3(x, y)}{x(1-x^2)(1-x^3)(1-y^2)(1-y^3)} = \frac{1+yx^4}{(1-x^2)(1-x^3)(1-x^2y^2)} - \frac{y+xy^5}{x(1-y^2)(1-y^3)(1-x^2y^2)}.$$

For $\delta = 4$,

$$\begin{aligned}(x-y) f_4(x, y)&(1-x^2y^2)(1-x^3y^3)(1-x^4y^4)\\ &= \{xy\}'(1-y^2)(1-y^3)(1-y^4)[x + y(x^5+x^6+x^7-x^9) + y^2(x^4+x^6)\\ &\quad - y^3(x^4-x^5-x^6-x^7+x^9) + y^4(x^7-x^9-x^{10}-x^{11}+x^{12})\\ &\quad - y^5(x^{11}+x^{12}) + y^6(x^7-x^9-x^{10}-x^{11}) - y^7x^{15}].\end{aligned}$$

There is no need to put down the second part of the function

$$x^{-1}(1-x)(1-y)(x-y) f_4(x, y)/\{[4]!\}_x\{[4]!\}_y,$$

the first part is

$$\begin{aligned}&\{(1-x^2)(1-x^3)(1-x^4)(1-x^2y^2)(1-x^3y^3)(1-x^4y^4)\}^{-1}\\ &\quad\times[1 + yx^4(1+x+x^2-x^4) + y^2x^3(1+x) - y^3x^3(1-x-x^2-x^3+x^5)\\ &\quad + y^4x^6(1-x^2-x^3-x^4+x^5) - y^5x^{10}(1+x)\\ &\quad + y^6x^6(1-x^2-x^3-x^4) - y^7x^{14}].\end{aligned}$$

§31. The difficulty of calculating the function increases rapidly with δ. It appears to be easiest to use the form of $f_\delta(x, y)$ obtained in § 27 when $\delta > 4$; for instead of a single function $f_\delta(x, y)$ to be discussed, there are several much simpler functions which can be dealt with separately. There is the additional advantage that the solution of the problem for $f_\delta(x,y)$ obtained in this form can be at once applied to $f_\alpha(x,y)$.

Theorem XVI gives us :—

$$f_3(x,y) = \tfrac{1}{6}(1+x)(1+x+x^2)(1+y)(1+y+y^2) + \tfrac{1}{3}(1-x)(1-x^2)(1-y)(1-y^2) + \tfrac{1}{2}(1-x^3)(1-y^3)$$

$$f_4(x,y) = \tfrac{1}{24}X_{1^4}Y_{1^4} + \tfrac{1}{4}X_4Y_4 + \tfrac{1}{8}X_{2^2}Y_{2^2} + \tfrac{1}{3}X_{3,1}Y_{3,1} + \tfrac{1}{4}X_{2,1^2}Y_{2,1^2}$$

$$f_5(x,y) = \tfrac{1}{120}X_{1^5}Y_{1^5} + \tfrac{1}{5}X_5Y_5 + \tfrac{1}{4}X_{4,1}Y_{4,1} + \tfrac{1}{6}X_{3,2}Y_{3,2} + \tfrac{1}{6}X_{3,1^2}Y_{3,1^2} + \tfrac{1}{8}X_{2^2,1}Y_{2^2,1} + \tfrac{1}{12}X_{2,1^3}Y_{2,1^3}.$$

In the first place, we have to consider

$$\frac{x-y}{(1-x)^{\delta-1}(1-y)^{\delta-1}}.$$

Let us write z for xy.
Then

$$(1-z) = x(1-y) + y(1-x) + (1-x)(1-y),$$

and

$$x - y = x(1-y) - y(1-x).$$

Let us write

$$(x-y)(1-z)^{2\delta-5} = (1-y)^{\delta-1}xP_\delta(x,z) - (1-x)^{\delta-1}yP_\delta(y,z);$$

for the index $2\delta - 5$ of z is both necessary and sufficient in order that an integral function $P_\delta(x,z)$ may be found which satisfies the condition. Then also

$$(x-y)(1-z)^{2\delta-5} = (1-z)^2[(1-y)^{\delta-2}xP_{\delta-1}(x,z) - (1-x)^{\delta-2}yP_{\delta-1}(y,z)],$$

hence

$$(1-y)^{\delta-2}[(1-z)^2xP_{\delta-1}(x,z) - (x-z)P_\delta(x,z)]$$
$$= (1-x)^{\delta-2}[(1-z)^2yP_{\delta-1}(y,z) - (y-z)P_\delta(y,z)]$$
$$= (1-x)^{\delta-2}(1-y)^{\delta-2}\Phi_\delta(z);$$

for it is a function of degree $\delta - 2$ in x and in y and has a factor $(1-x)^{\delta-2}$, and also $(1-y)^{\delta-2}$.
Hence

$$(1-z)^2xP_{\delta-1}(x,z) - (x-z)P_\delta(x,z) = (1-x)^{\delta-2}\Phi_\delta(z).$$

Now put $x = z$,

$$\Phi_\delta(z) = z(1-z)^{-\delta+4}P_{\delta-1}(z,z).$$

Thus

$$(x-z)P_\delta(x,z) = (1-z)^2xP_{\delta-1}(x,z) - z(1-x)^{\delta-2}(1-z)^{-\delta+4}P_{\delta-1}(z,z),$$

gives a scale of relation to calculate P_δ.

The following results are obtained :—

$$P_3(x, z) = 1,\ P_4(x, z) = 1 - xz,$$
$$P_5(x, z) = 1 + z - 4xz + zx^2 + z^2x^2.$$

§ 32. The second term in $f_5(x, y)$, due to X_5Y_5, is

$$\frac{(1-x)(1-y)(x-y)}{x(1-x^5)(1-y^5)}.$$

Then

$$\{xy\}'\, x(1-x)(1-y)(1-z^5) = \{xy\}'(1-y^5)\, x\,[1 - x + z(1-x^3) + z^2x^2(1-x)].$$

And

$$\frac{(1-x)(1-y)(x-y)}{x(1-x^5)(1-y^5)} = \frac{1 - x + z(1-x^3) + z^2x^2(1-x)}{(1-x^5)(1-z^5)} - y\,\frac{1 - y + z(1-y^3) + z^2y^2(1-y)}{x(1-y^5)(1-z^5)}.$$

The third term $X_{4,1}$, $Y_{4,1}$, belongs to a very simple class, viz. :—

$$\frac{x-y}{x(1-x^n)(1-y^n)}.$$

Here we may use

$$1 - z^n = x^n(1-y^n) + y^n(1-x^n) + (1-x^n)(1-y^n).$$

In our case $n = 4$ and the result is

$$\frac{1-zx^2}{(1-x^4)(1-z^4)} - y\,\frac{1-zy^2}{x(1-y^4)(1-z^4)}.$$

The fourth and fifth terms may be taken together, for

$$(1-x)(1-y)(x-y)(X_{3,2}Y_{3,2} + X_{3,1^2}Y_{3,1^2})/(x\{[5]\,!\}_x\{[5]\,!\}_y)$$
$$= 2\,\frac{(x-y)(1+z)}{x(1-x^2)(1-x^3)(1-y^2)(1-y^3)}.$$

It is necessary to introduce the factors $(1-z^3)(1-z^6)$ and the x, z function is

$$2\{(1-x^2)(1-x^3)(1-z^3)(1-z^6)\}^{-1} \times [1 - zx(1-x^2) + z^3x(1-x) - z^5(1-x^2) - z^6x^3].$$

The sixth term $X_{2^2,1}\,Y_{2^2,1}$ is

$$\frac{x-y}{x(1-x^2)^2(1-y^2)^2},$$

and the corresponding x, z function is

$$\frac{1-z}{(1-x^2)^2(1-z^2)^2} - \frac{z(1-z)^2}{(1-x^2)(1-z^2)^3}.$$

Lastly, we have $X_{2,\,1^3}\,Y_{2,\,1^3}$, and

$$\frac{x-y}{x\,(1-x)^2\,(1-x^2)\,(1-y)^2\,(1-y^2)},$$

with the x, z function

$$\frac{x^2}{(1-x)^2\,(1-x^2)\,(1-z)\,(1-z^2)}+\frac{2x\,(1-x)+z\,(1-x^2)}{(1-x)^2\,(1-x^2)\,(1-z)\,(1-z^2)^2}$$
$$+\frac{1-z}{(1-x^2)\,(1-z^2)^3}.$$

§33. Before giving the final result for the case $\delta=5$. There is a general remark that can now be made clear about these functions. The form of result that is sought is as in §27 :—

$$\frac{(1-x)\,(1-y)\,(x-y)\,f_\delta\,(x,y)}{x\,\{[\delta]!\}_x\,\{[\delta]\,!\}_y}=\frac{(1-x)\,\phi\,(x,z)}{\{[\delta]\,!\}_x\,\psi\,(z)}-\frac{(1-y)\,y\,\phi\,(y,z)}{x\{[\delta]\,!\}_y\,\psi\,(z)}.$$

Write x^{-1} for x, and y^{-1} for y, then z becomes z^{-1}; also

$$x^{\binom{\delta}{2}}\,y^{\binom{\delta}{2}}\,f_\delta\,(x^{-1},y^{-1})=f_\delta\,(x,y),$$

by Theorem III, §6. Then

$$(-)^{\delta-1}\left\{\frac{x^{\binom{\delta+1}{2}-1}\,(1-x)\,\phi\,(x^{-1},z^{-1})}{\{[\delta]\,!\}_x\,\psi\,(z^{-1})}-\frac{y^{\binom{\delta+1}{2}-2}\,x\,(1-y)\,\phi\,(y^{-1},z^{-1})}{\{[\delta]\,!\}_y\,\psi\,(z^{-1})}\right\}$$
$$=-\frac{x^{\delta-1}\,y^{\delta-2}\,(1-x)\,(1-y)\,(x-y)\,f_\delta\,(x,y)}{\{[\delta]\,!\}_x\ \ \{[\delta]\,!\}_y}$$
$$=-\,x^\delta\,y^{\delta-2}\left\{\frac{(1-x)\,\phi\,(x,z)}{\{[\delta]\,!\}_x\,\psi\,(z)}-\frac{y\,(1-y)\,\phi\,(y,z)}{x\,\{[\delta]\,!\}_y\,\psi\,(z)}\right\},$$

whence

$$x^{\binom{\delta+1}{2}-3}\,\phi\,(x^{-1},z^{-1})\,\psi\,(z)=(-)^\delta\,z^{\delta-2}\,\phi\,(x,z)\,\psi\,(z^{-1}).$$

Now ϕ always contains a term independent of x and thus the degree of ϕ in x is exactly $\binom{\delta+1}{2}-3$. Owing to the equation

$$1-z^n=y^n\,(1-x^n)+x^n\,(1-y^n)+(1-x^n)\,(1-y^n),$$

it is clear that it is always possible to choose ψ as a product of binomial factors $1-z^n$, just as has been done for the cases discussed. Let p be the number of binomial factors in ψ and q its total degree, then

$$\psi\,(z)=(-)^p\,z^q\,\psi\,(z^{-1}).$$

The degree of ϕ in z is $q-\delta+2$, and, in fact,

(XIV) $$\phi\,(x,z)=(-)^{\delta+p}\,x^{\binom{\delta+1}{2}-3}\,z^{q-\delta+2}\phi\,(x^{-1},z^{-1}).$$

§34. For $\delta = 5$ we insert the numerical coefficients in the results of §§31, 32 to obtain the x, z part of

$$\frac{(1-x)(1-y)(x-y)f_5(x,y)}{x\{[5]!\}_x\ \{[5]!\}_y},$$

viz.:—

$$\frac{1+z(1-4x+x^2)+z^2x^2}{120(1-x)^4(1-z)^5}+\frac{1-x+z(1-x^3)+z^2x^2(1-x)}{5(1-x^5)(1-z^5)}$$

$$+\frac{1-zx^2}{4(1-x^4)(1-z^4)}+\frac{1-zx(1-x^2)+z^3x(1-x)-z^5(1-x^2)-z^6x^3}{3(1-x^2)(1-x^3)(1-z^3)(1-z^6)}$$

$$+\frac{1-z}{8(1-x^2)^2(1-z^2)^2}-\frac{z(1-z)^2}{8(1-x^2)(1-z^2)^3}$$

$$+\frac{x^2}{12(1-x)^2(1-x^2)(1-z)(1-z^2)}$$

$$+\frac{2x(1-x)+z(1-x^2)}{12(1-x)(1-x^2)(1-z)(1-z^2)^2}+\frac{1-z}{12(1-x^2)(1-z^2)^3}.$$

In the final result the denominator is

$$(1-x^2)(1-x^3)(1-x^4)(1-x^5)(1-z^2)(1-z^3)(1-z^4)(1-z^5)(1-z^6),$$

and extraneous numerical factors must disappear—a useful check on the arithmetic. The numerator we call $\phi(x, z)$ and for this equation (XIV) gives

$$\phi(x, z) = x^{12}z^{17}\phi(x^{-1}, z^{-1}).$$

In consequence of this equation it is only necessary to give the terms up to z^8 the rest can be written down at once; then

$$\begin{aligned}\phi(x,z) = 1 &+ z(x^3+x^4+2x^5+x^6-x^8-x^9-x^{10}+x^{12})\\
&+z^2(x+2x^2+x^3+x^4)\\
&+z^3(-1+2x+2x^2+3x^3+2x^4-x^6-x^7-x^8+x^{10})\\
&+z^4(1+2x+3x^2+2x^3+x^4-2x^5-2x^6-2x^7+x^9+x^{10}+x^{11}-x^{12})\\
&+z^5(-1+2x+3x^2+4x^3+4x^4+x^5-4x^6-5x^7-3x^8-2x^9+x^{10}+2x^{11}+x^{12})\\
&+z^6(3+3x+3x^2-2x^4-5x^5-5x^6-3x^7-x^8+x^9+2x^{10}+2x^{11}-x^{12})\\
&+z^7(1+2x+3x^2+3x^3-x^4-3x^5-6x^6-7x^7-5x^8-x^9+x^{10}+2x^{11}+3x^{12})\\
&+z^8(4+2x+2x^2-x^3-4x^4-6x^5-7x^6-5x^7-2x^8+x^9+x^{10}+3x^{11}+x^{12})\\
&+\dots+z^{17}x^{12}.\end{aligned}$$

VII—The Generating Function for Ternary Perpetuants of a Single Form

§ 35. It has been proved* that if G_δ be a generating function for ternary forms of any particular type, and Γ_δ be the corresponding generating function for gradients (*i.e.*, coefficient products),

$$G_\delta = (1 - x)\,(1 - y)\,(1 - y/x)\,\Gamma_\delta,$$

where the indices of x and y respectively refer as usual to the second and third weights of the form (or of the coefficients).

We therefore proceed to consider the generating function for gradients of degree δ of a single ternary form of infinite order. A gradient is a product

$$A_{p_1 q_1} \ldots A_{p_\delta q_\delta},$$

each factor being a coefficient of the form, and the two suffices of the coefficient are its second and third weights.

For gradient types we name the coefficients further, say, with a prefix, to define which of the δ quantics has supplied the particular coefficients. For a single quantic, this fact is substitutionally expressed by the operator $\{{}_1A\ {}_2A \ldots {}_\delta A\}$, the suffices not being here expressed as the prefixes alone are permuted. For binary forms, where there is only one suffix, the fact is equally well and more conveniently expressed by a permutation of the suffices. In ternary forms where there are two suffixes the permutation $({}_rA\ {}_sA)$ is expressed by the compound permutation $(p_r\, p_s)\ (q_r\, q_s)$ of the suffixes. Thus in the ternary case we have to do with the compound symmetric group. We have seen § 22 equation (X) that the compound symmetric group

$$T_\delta^{(pq)} = \Sigma\, \frac{1}{f}\, \varpi_{rs}\, \varpi'_{rs}.$$

Let

$$\omega_{rs}\, P$$

be any gradient of the binary coefficients ${}_rA_{pq}$, where ϖ_{rs} is a unit of $T^{(p)}{}_{\alpha_1 \alpha_2 \ldots \alpha_h}$, then by operation with ϖ_{tr} we may obtain a gradient

$$\omega_{ts}\, P,$$

and thus f distinct gradients of this type. The number of distinct gradients for this representation $T_{\alpha_1 \alpha_2 \ldots \alpha_h}$ has been found to be given by the generating function

$$\{[\delta]\,!\}^{-1} f_{\alpha_1 \ldots \alpha_h} f_{\alpha_1 \ldots \alpha_h}(x)\,.\,x^{\varpi}.$$

Hence the generating function for ternary gradients of a single form of infinite order is

$$\Sigma\ (\{[\delta]\,!\}^{-1})_x\,(\{[\delta]\,!\}^{-1})_y\, f_{\alpha_1 \ldots \alpha_h}(x)\, f_{\alpha_1 \ldots \alpha_h}(y)\ x^{\varpi} y^{\varpi}, = \{[\delta]\,!\}_x^{-1}\,\{[\delta]\,!\}_y^{-1}\, f_\delta(x, y).$$

* Young, 'Proc. Lond. Math. Soc.,' vol. 35, p. 431 (1933).

THEOREM XX—*The generating function for ternary perpetuants of degree* δ *for a single form is*

$$\frac{(1-x)(1-y)(x-y)f_\delta(x,y)}{x\{[\delta]!\}_x\{[\delta]!\}_y}.$$

§36. Let us now consider the results of the last section in respect to the generating function. The expansion is an infinite series in both x and y. The terms $x^r y^s$ where $s > r$ do not really concern us, as the seminvariants, with which only we are concerned, never have the third weight greater than the second. In fact, the generating function proper is the part called above

$$G_\delta = \frac{(1-x)\,\phi(x,z)}{([\delta]!)_x\,\psi(z)} = \Sigma\, B_{r,\rho}\, x^r z^\rho.$$

Consider the expression of the perpetuant as a symbolical product, it essentially contains two kinds of factors (abu), (abc); then the index of x in the generating function is the number of factors (abu), and the index of z is the number of factors (abc).

When $\delta = 2$,

$$G_2 = \frac{1}{1-x^2},$$

giving the obviously correct result.

When $\delta = 3$

$$G_3 = \frac{1+zx^3}{(1-x^2)(1-x^3)(1-z^2)}.$$

Here as in the general case the x factors in the denominator refer to (abu) terms which may be taken exactly as in the binary case, viz. :—

$$(a_1a_2u)^{\lambda_1}(a_2a_3u)^{\lambda_2}\ldots(a_{\delta-2}a_{\delta-1}u)^{\lambda_{\delta-2}}(a_{\delta-1}a_\delta u)^{2\lambda_{\delta-1}},$$

where

$$\lambda_{\delta-1} \geqq \lambda_{\delta-2} \ldots \geqq \lambda_1,$$

see §8.

The denominator factor $1-z^2$ here refers to a factor $(a_1\,a_2\,a_3)^{2\mu}$; and the numerator term zx^3 corresponds to the form

(XV) $$(a_1\,a_2\,a_3)(a_1\,a_2\,u)(a_2\,a_3\,u)(a_3\,a_1\,u);$$

there are thus two sets of perpetuants of degree 3, one

$$(a_1\,a_2\,a_3)^{2\mu}(a_1\,a_2\,u)^{\lambda}(a_2\,a_3\,u)^{2\lambda+2\rho};$$

and the other this set multiplied by the form (XV) given by zx^3.

When $\delta = 4$:

$$G_4 = \{1 + z(x^3+x^4+x^5-x^7) + z^2(x+x^2) - z^3(1-x-x^2-x^3+x^6) + z^4(x^2-x^4-x^5-x^6+x^7) - z^5(x^5+x^6) + z^6(1-x^2-x^3-x^4) - z^7x^7\} \times \{(1-x^2)(1-x^3)(1-x^4)(1-z^2)(1-z^3)(1-z^4)\}^{-1}.$$

The terms independent of x are

$$\frac{1+z^9}{(1-z^2)(1-z^4)(1-z^6)}.$$

It is easy to see that we may take the perpetuants corresponding to the denominator factors to be

$$(a_1\,a_2\,a_3)^2,\ (a_1\,a_2\,a_3)^2\,(a_1\,a_2\,a_4)^2,$$

$$(a_1\,a_2\,a_3)^3\,(a_1\,a_2\,a_4)^3.$$

And that corresponding to the numerator term z^9 to be

$$(a_1\,a_2\,a_3)^6\,(a_1\,a_2\,a_4)^2\,(a_1\,a_3\,a_4).$$

The generating function assures us that all perpetuants of degree four, such that the second and third weights are equal, may be expressed as a sum of terms

$$(a_1a_2a_3)^{\lambda_1}\,(a_1a_2a_4)^{\lambda_2}\,(a_1a_3a_4)^{\lambda_3};$$

where

$$\lambda_1 = 2\mu_1 + 2\mu_2 + 3\mu_3 + 6\varepsilon,\quad \lambda_2 = 2\mu_2 + 3\mu_3 + 2\varepsilon,\quad \lambda_3 = \varepsilon,\quad \varepsilon = 0 \text{ or } 1.$$

It is useless to say anything about the other terms without a careful investigation of them, which has not yet been undertaken.

VIII—Generating Functions for Ternary Forms of Finite Order

§ 37. We concern ourselves with the generating function for gradients.

Theorem XXI—*The generating function $\Phi_\delta^{(n)}$ for gradients of degree δ of the ternary n-ic is given by*

$$\delta\,!\,\Phi_\delta^{(n)} = \sum_\beta h_\beta \prod_{r=1}^{k} \mathrm{X}_{\beta_r}^{(n)};$$

where β defines the partition $\beta_1, \beta_2, \dots \beta_k$ of δ, h_β is the number of members of the corresponding class of conjugate operations of the symmetric group of degree δ, and

$$\mathrm{X}_\beta^{(n)} \begin{vmatrix} 1 & x^{2\beta} & y^{2\beta} \\ 1 & x^{\beta} & y^{\beta} \\ 1 & 1 & 1 \end{vmatrix} = \begin{vmatrix} 1 & x^{(n+2)\beta} & y^{(n+2)\beta} \\ 1 & x^{\beta} & y^{\beta} \\ 1 & 1 & 1 \end{vmatrix}.$$

When $n = 0$,

$$\mathrm{X}_\beta^{(0)} = 1,\qquad \Phi_\delta^{(0)} = 1,$$

and the theorem is a truism.

We assume it true for all ternary quantics of order less than n.

Consider a gradient $\prod_{r=1}^{\delta} \mathrm{A}_{\lambda_r,\mu_r}$, the sum of the suffixes of each coefficient must be equal to or less than n. Let each gradient be divided into two factors, let the first factor contain all the coefficients the sum of whose suffixes is n, and the second all the coefficients the sum of whose suffices in less than n. Let P_r be the generating function for those gradients in which r of the coefficients have the sum of the coefficients equal to n. Then

$$P_r = Q_r \cdot \Phi_{\delta-r}^{(n-1)},$$

where Q_r is obtained at once from the generating function for the gradients of degree r of the binary n-ic, *i.e.*,

$$Q_r = \frac{(x^{n+1}-y^{n+1})(x^{n+2}-y^{n+2})\dots(x^{n+r}-y^{n+r})}{(x-y)(x^2-y^2)\dots(x^r-y^r)} = \sum_\gamma \frac{h_\gamma^{(r)}}{r!}\prod_{s=1}^{l}\frac{x^{(n+1)\gamma_s}-y^{(n+1)\gamma_s}}{x^{\gamma_s}-y^{\gamma_s}}$$

$$= \sum_\gamma \frac{h_\gamma^{(r)}}{r!}\prod_{s=1}^{l}(X_{\gamma_s}^{(n)} - X_{\gamma_s}^{(n-1)}),$$

by Theorem XV. Here $h_\gamma^{(r)}$ is the number of members of the γ conjugate set in the symmetric group of degree r.

Thus we obtain

$$\Phi_\delta^{(n)} = \sum_{r=0}^{\delta} P_r = \sum_{r=0}^{\delta}\left[\sum_\gamma \frac{h_\gamma^{(r)}}{r!}\prod_{s=1}^{l}(X_{\gamma_s}^{(n)} - X_{\gamma_s}^{(n-1)})\right]\left[\sum_{\gamma'}\frac{h_{\gamma'}^{(\delta-r)}}{(\delta-r)!}\prod_{s=1}^{l}X_{\gamma'_s}^{(n-1)}\right].$$

The generating function for binary forms can be treated in exactly the same way, and the same equation is obtained, where now

$$X_\beta^n = \frac{[(n+1)\beta]}{[\beta]};$$

by § 21 the result is

$$\sum_\beta h_\beta \prod_{r=1}^{k} X_{\beta_r}^{(n)}.$$

It is evident that there can be no linear relation with constant coefficients between the products $\prod_{r=1}^{k} X_{\beta_r}^{(n)}$; and hence the result is due to the values of the coefficients $h_\gamma^{(r)}$, $h_\gamma^{(\delta-r)}$, and is quite independent of the form assigned to the functions $X_\beta^{(n)}$; hence for ternary forms

$$\Phi_\delta^{(n)} = \sum_{r=0}^{\delta} h_\beta \prod_{r=1}^{k} X_{\beta_r}^{(n)}$$

as was to be proved.

§ 38. It will be noticed that the generating function just obtained is the result of putting for every cycle of degree r in the symmetric group of degree δ the function $X_r^{(n)}$ and then dividing by δ !. The same methods give us the generating function for gradient types of degree δ, and of any particular substitutional form. The result is

THEOREM XXII—*The generating function* $\Phi_\alpha^{(n)}$ *for gradient types, of degree* δ *of ternary n-ics, of substitutional form* T_α *is*

$$\delta\,!\;\Phi_\alpha^{(n)} = \sum_\beta h_\beta\, \chi_\beta^{(\alpha)} \prod_{r=1}^{k} X_{\beta_r}.$$

These results may be extended at once to quaternary and higher forms. For quaternary forms we take

$$X_{\beta}\begin{vmatrix} 1 & x^{3\beta} & y^{3\beta} & z^{3\beta} \\ 1 & x^{2\beta} & y^{2\beta} & z^{2\beta} \\ 1 & x^{\beta} & y^{\beta} & z^{\beta} \\ 1 & 1 & 1 & 1 \end{vmatrix} = \begin{vmatrix} 1 & x^{(n+3)\beta} & y^{(n+3)\beta} & z^{(n+3)\beta} \\ 1 & x^{2\beta} & y^{2\beta} & z^{2\beta} \\ 1 & x^{\beta} & y^{\beta} & z^{\beta} \\ 1 & 1 & 1 & 1 \end{vmatrix}.$$

§ 39. As a verification of the results we may deduce from them the generating functions for types and compare them with those obtained in Q.S.A., VII,* Section VIII. Thus for ternary forms the generating functions for gradient types of the ternary *n-ic* is

$$\sum_{\alpha} f_{\alpha}\, \Phi_{\alpha}^{(n)} = \{X_1^{(n)}\}^{\delta}$$

the same result as that obtained by other means in § 44 of the paper just quoted.

Summary

The particular application, that is the main object of this paper, is the determination of a generating function for concomitants of ternary forms, of a nature more amenable to calculation than that obtained in a previous paper entitled "Some Generating Functions."† This object is achieved and a generating function for concomitants of any particular substitutional form, of quantics of finite or of infinite order, and with any number of variables is obtained. Amongst other results certain polynominal functions which appeared in the generating functions for binary forms in the paper just quoted have properties of interest which are discussed ; and some of these properties are extended to the corresponding functions with two or more variables. A one to one correspondence between the generating function for binary perpetuants of particular substitutional form and the perpetuants themselves is obtained by means of the tableaux and this leads at once to the corresponding extension of Grace's Theorem on irreducible perpetuant types.

Incidentally, certain properties of the Characteristic Function of Schur are considered, and a curious identity is obtained involving a series of determinants.

* Young, 'Proc. London Math. Soc.,' vol. 36, p. 304 (1934).

† 'Proc. London Math. Soc.,' vol. 35, p. 425 (1933).

ON QUANTITATIVE SUBSTITUTIONAL ANALYSIS

(*Ninth Paper*)

By ALFRED YOUNG

[Received 29 May, 1951.—Read 14 June, 1951.]

Introduction—By G. de B. Robinson

(i) The present paper was found in typescript form amongst the very considerable body of MSS. left by Alfred Young. Young's habits of work were systematic and consistent throughout his life. Unable to devote his whole time to mathematics, he could lay down a piece of work and pick it up again after a lapse of time with little apparent loss of continuity. He worked at different aspects of what he called "Substitutional analysis" and filled numerous folders designated A-Z, A^2, AB, ..., [Aux A], B^2, BC. As a subject developed he would write a paper on it including material, it may be, from different folders, but destroying the final draft and typescript after the paper appeared in print. It appears, however, that he kept his folders intact throughout the greater part of 50 years, though he only dated work done in the last few years of his life in any consistent manner. Young also appears to have kept the rough work attached to the folders.

In accord with the above observations the material contained in the earlier folders, at least, had been well sifted through and incorporated in his published papers. Shortly before his election to the Royal Society in 1934 Young began the writing of Q.S.A. IX; much of the material seems, however, to have been used in the paper on Invariants published in the Philosophical Transactions of the Royal Society in 1935. About this time it would appear to have occurred to him to utilize the canonical form of a ground form and relate this to the symbolic method and his own analysis. The preliminary drafts seem to have been destroyed but the typescript mentioned above was evidently considered by Young as almost ready for printing. The formulae had been carefully filled in and it shows signs of considerable revision, though an introduction is lacking.

It is perhaps worth noting briefly those folders which appear to have occupied Young during the last two years of his life.

[Aux A] begins with the sentence "In the paper under preparation Q.S.A. IX . . ." and consists of an extension and further application of the methods described and illustrated in the typescript. This MS. is dated and appears to have been worked on from October 1939 to January 1940.

[B²] appears to have been started earlier but taken up again in February 1940. The subject matter is different from that of the typescript and [Aux A] and does not seem to have reached a definitive form. It is chiefly notable for an increased emphasis on the actual matrices of representations of the symmetric group. Again dates appear and we can say that the folder was worked on up to, and for a short time after, October 8, 1940.

[BC]. This folder too had been started earlier and deals with group representations. On December 2, Young returned to this more abstract part of his work and began what was to have been a systematic study of the representation theory of subgroups of the Symmetric Group in relation to his own analysis. He died on December 15, 1940.

(ii) In Q.S.A. IX Young is concerned with his theory of reductibility as defined in Q.S.A. VII* and in making clear the extent to which reductibility differs from the familiar concept of reducibility. Part I extends the reduction formula for binary forms developed in Theorem II of Q.S.A. VII and its corollary (p. 313). In Part II these ideas are further developed and applied to invariants and covariants of orders 1, 2. Part III corrects the statements of two important theorems in Q.S.A. VIII†. (The ideas involved here will be the subject of comment in (iii) below.) Parts IV, V and VI contain essentially new material and show how the classical theory of canonical forms fits into and makes more precise the criteria of reductibility already obtained. In Part VII these criteria are analysed in the general case and in Part VIII they are applied to obtain a list of irreductible forms for the quintic. In the last sentence of the paper Young promises a discussion of reducibility, but this seems never to have been attempted.

(iii) In concluding this Introduction it has seemed worth while to comment briefly on Young's later work in so far as it has contact with that of his successors, in particular, D. E. Littlewood, J. A. Todd and the Editor‡.

Young's notion of *reductibility* which dominates his published papers Q.S.A. VII, VIII and the present paper, and [Aux A] as well, is defined in Q.S.A. VII with reference to a particular *sequence* of the standard tableaux

* *Proc. London Math. Soc.* (2), 36 (1934), 304.

† *Ibid.*, 37 (1934), 441.

‡ Littlewood, *Phil. Trans.* (A), 239 (1944), 305; Todd, *Proc. London Math. Soc.* (2), 52 (1948), 271; Robinson, *Canadian Journal Math.*, 2 (1950), 334.

associated with a given irreducible representation of the symmetric group. Those sequences which interested him particularly were the following: (A) according to the last letter; (B) according to the first letter; (C) the reciprocal of (A); (D) the reciprocal of (B); (a) according to the last row; and (b) according to the first row. It is to be noted that the correction involved in Part III of the present paper has to do with this notion of *sequence*.

What is the reason for the significance of *sequence* and in what respect do the sequences differ in significance? Consider the *restriction* of S_{m+n} to the direct product $S_m \times S_n$. The essential point of Young's seminormal or orthogonal form of the representation $[\alpha]$ of S_{m+n} is that the representations of the subgroups S_m on the symbols 1, 2, ..., m and S_n on $m+1$, $m+2$, ..., $m+n$ do not "mix up"; *i.e.*, the representation of $S_m \times S_n$ thus obtained is a direct sum of *Kronecker products*. This fact appears most clearly using sequences (a) and (b); Young noted the point in Q.S.A. VIII, p. 446.

On the other hand sequences (A) and (B) are significant because of their association with the *inducing* process. Consider, for instance, the representation $[2].[2].[2]$ of S_6 induced by the Kronecker product representation $[2]\times[2]\times[2]$ of the subgroup $S_2 \times S_2 \times S_2$. We may obtain the reduction into irreducible components by the Littlewood-Richardson Rule:

$$\begin{array}{l}aa\\bb\\cc\end{array} = aabbcc \dotplus \left(\begin{array}{l}aabcc\\b\end{array}\dotplus\begin{array}{l}aabbc\\c\end{array}\right)\dotplus\left(\begin{array}{l}aabb\\cc\end{array}\dotplus\begin{array}{l}aabc\\bc\end{array}\dotplus\begin{array}{l}aacc\\bb\end{array}\right)\dotplus\begin{array}{l}aabc\\b\\c\end{array}\dotplus\begin{array}{l}aab\\bcc\end{array}$$

$$\dotplus\left(\begin{array}{l}aab\\bc\\c\end{array}\dotplus\begin{array}{l}aac\\bb\\c\end{array}\right)\dotplus\begin{array}{l}aa\\bb,\\cc\end{array}$$

i.e. $[2].[2].[2] = [6]\dotplus 2[5, 1]\dotplus 3[4, 2]\dotplus [4, 1^2]\dotplus [3^2]\dotplus 2[3, 2, 1]\dotplus [2^3]$.

Now Young considered the three diagrams [4, 2] with regard to their *sequence* only and he remarks that the sequences (C) and (D) which "split up the symbols" are unsuitable to his purpose (Q.S.A. VIII, p. 488). As usual his intuition was unerring, even though he had not the formal machinery to explain his preference of sequence (A) or (B) in this connection. The significance of these diagrams as *representations* is the new element in the situation, and it is the Editor's belief that the ideas contained in Young's later papers Q.S.A. VII-IX when taken in conjunction with subsequent group theoretical developments will prove of importance for the problem of Invariants.

I. *Complete sets of types which are linearly independent*

1. The expression $(a_1^{\lambda_1} a_2^{\lambda_2} \dots a_\delta^{\lambda_\delta})$ was introduced in Q.S.A. VII, p. 310, to denote the covariant obtained from the natural substitutional unit which has λ_1 of the n_1 letters a_1, λ_2 of the n_2 letters a_2 and so on, in the second row; and then this was modified to $\{a_1^{\lambda_1} \dots\}$ by the relation

$$\{a_1^{\lambda_1} \dots a_\delta^{\lambda_\delta}\} = \left[\Pi \binom{n_r}{\lambda_r}\right] (a_1^{\lambda_1} \dots a_\delta^{\lambda_\delta}).$$

The relation

$$(-)^{\lambda_2} \{a_1^0 a_2^{\lambda_2} \dots a_\delta^{\lambda_\delta}\} = \Sigma \left[\Pi \binom{v_r}{\lambda_r}\right] \{a_1^{w-\Sigma v} a_2^0 a_3^{v_3} \dots a_\delta^{v_\delta}\}$$

was then obtained. We will use $\binom{m}{\lambda}$ as the coefficient of x^λ in $(1+x)^m$ whether m is positive or negative. Then

$$\{a_1^{w-\Sigma r} a_2^{r_2} a_3^{r_3} \dots a_{\delta-1}^{r_{\delta-1}} a_\delta^0\} = (-)^w \Sigma \left[\prod_{t=2}^{\delta-1} (-)^{r_t} \binom{s_t}{r_t}\right] \{a_1^0 a_2^{s_2} a_3^{s_3} \dots a_{\delta-1}^{s_{\delta-1}} a_\delta^{w-\Sigma s}\};$$

and hence

$$\text{(I)} \quad \left[\prod_{t=2}^{\delta-1} \sum_{r_t=0}^{s_t} \binom{\xi_t+\vartheta_t-r_t}{\xi_t}\right] \{a_1^{w-\Sigma r} a_2^{r_2} \dots a_{\delta-1}^{r_{\delta-1}} a_\delta^0\}$$

$$= (-)^w \Sigma \left[\prod_{t=2}^{\delta-1} \Sigma_{s_t} (-)^{s_t} \binom{s_t-\xi_t-1}{\vartheta_t}\right] \{a_1^0 a_2^{s_2} \dots a_{\delta-1}^{s_{\delta-1}} a_\delta^{w-\Sigma s}\};$$

since

$$\sum_{r_t=0}^{s_t} (-)^{r_t} \binom{\xi_t+\vartheta_t-r_t}{\xi_t} \binom{s_t}{r_t} = \Sigma_{u_t} (-)^{s_t-u_t} \binom{\xi_t+u_t}{u_t} \binom{s_t}{\vartheta_t-u_t}$$

$$= \left((x^{\vartheta_t})\right) . (1-x)^{-\xi_t-1} (1-x)^{s_t} = (-)^{\vartheta_t} \binom{s_t-\xi_t-1}{\vartheta_t}.$$

Let us write $\kappa_t = \xi_t+\vartheta_t+1$, $t = 2, 3, \dots, \delta-1$; $\kappa_1 = w-\Sigma\vartheta$, $\kappa_\delta = w-\Sigma\xi$. Then the terms on the left belong to a set marked by the fact that the index of a_1 is equal to or greater than κ_1. In the terms on the right either $s_t \geqslant \kappa_t$, or else $s_t \leqslant \xi_t$; thus in each of the terms on the right for some t the index of a_t is equal to or greater than κ_t, where $1 < t < \delta$ or else the index of a_δ is equal to or greater than κ_δ. Now

$$\sum_{t=1}^{\delta} \kappa_t = 2w+\delta-2,$$

and the numbers ϑ can be chosen at will; hence we may select δ quantities κ such that

$$\text{(II)} \qquad \sum_{t=1}^{\delta} \kappa_t = 2w+\delta-1,$$

where κ_t defines a set of covariant types K_t, in each of which the index of a_t is equal to or greater than κ_t. Of course some types will belong to two or more of such sets. Then the equation (I) may be regarded as an equation for expressing a type which does not belong to any one of the δ sets, but in which the index of a_1 is p in terms of types which have a_1^{p+1} and of members of the sets K_t. Successive applications of equation (I) enable us to express all types of weight w in terms of the δ sets, provided only equation (II) is satisfied.

Theorem I. *All covariants of weight w, linear in the coefficients of each of δ quantics $a_{t_x}^{n_t}$ $(t = 1, 2, \ldots, \delta)$ can be linearly expressed in terms of covariants*

$$\{a_1^{\lambda_1} a_2^{\lambda_2} \ldots a_\delta^{\lambda_\delta}\}$$

for which, for some value of t, λ_t is equal to or greater than κ_t; where $\kappa_1, \kappa_2, \ldots, \kappa_\delta$ are δ positive integers, selected beforehand at will which satisfy the equation

$$\sum_{t=1}^{\delta} \kappa_t = 2w+\delta-1.$$

2. To calculate the number N of different members of these δ sets it is necessary to add the numbers for each set separately, to subtract the number of those which belong to two sets, then to add that of those which belong to three sets and so on.

Consider the expression

$$X = \binom{w+\delta-2}{\delta-2} - \sum_s \binom{w-\kappa_s+\delta-2}{\delta-2} + \sum_{s,t} \binom{w-\kappa\;-\kappa_t+\delta-2}{\delta-2} - \ldots$$

$$+(-)^s \binom{w-\Sigma\kappa+\delta-2}{\delta-2}$$

$$= \left((x^{\delta-2})\right)\left[(1+x)^{w+\delta-2} - \sum_s (1+x)^{w-\kappa_s+\delta-2} + \ldots\right]$$

$$= \left((x^{\delta-2})\right)\left[\prod_{t\;1}^{\delta} \{1-(1+x)^{-\kappa_t}\}\right](1+x)^{w+\delta-2} = 0.$$

Now $w+\lambda-\Sigma\kappa+\delta-2 = -(w-\lambda)-1$, and hence

$$\binom{w+\lambda-\Sigma\kappa+\delta-2}{\delta-2} = (-)^{\delta-2}\binom{w-\lambda+\delta-2}{\delta-2}.$$

Thus the terms of X occur in pairs, and the members of a pair are equal in value and have the same sign; the two members (apart from sign) have the form $\binom{m}{\delta-2}$, $\binom{-m'}{\delta-2}$ where m, m' are positive integers, and the first properly belongs to N and the second does not. Hence

$$X = 2\binom{w+\delta-2}{\delta-2} - 2N = 0.$$

Thus the number of different members of the δ sets is equal to the whole number of independent perpetuant types. Thus when each n_t exceeds w in Theorem I, the number of covariants in the δ sets is exactly equal to the number of linearly independent covariants; and since all covariants can be expressed in terms of the members of these sets, there can be no linear relation between these members. Also, as it was remarked (Q.S.A. VII, p. 314), the relations between the covariant forms used are independent of the orders of the ground forms, and hence

THEOREM II. *The members of the δ sets are all linearly independent, and their total number is independent of the particular selection of the values of κ_t, provided only they satisfy equation* (II).

The identity for degree three, or rather its corollary, Q.S.A. VII, §7, is just a particular case of (I).

3. In general the only terms of importance of equation (I) for our applications are those in which the index of a_t in the set K_t is exactly κ_t; and those in which this index is greater than κ_t will generally be ignored. Then equation (I) becomes

$$\text{(III)} \quad (-)^{w}\{a_1^{\kappa_1} a_2^{\vartheta_2} a_3^{\vartheta_3} \ldots a_{\delta-1}^{\vartheta_{\delta-1}} a_\delta^{0}\}$$

$$= \{a_1^{0} a_2^{\xi_2} a_3^{\xi_3} \ldots a_{\delta-1}^{\xi_{\delta-1}} a_\delta^{\kappa_\delta}\}$$

$$+(-)^{\vartheta_{\delta-1}} \Sigma \binom{\vartheta_2+\rho_2}{\vartheta_2} \cdots \binom{\vartheta_{\delta-2}+\rho_{\delta-2}}{\vartheta_{\delta-2}} \{a_1^{0} a_2^{\xi_2-\rho_2} \ldots a_{\delta-2}^{\xi_{\delta-2}-\rho_{\delta-2}} a_{\delta-1}^{\kappa_{\delta-1}} a_\delta^{\kappa_\delta-\vartheta_{\delta-1}-1+\Sigma\rho}\}$$

$$+\ldots$$

$$+(-)^{\vartheta_2} \Sigma \binom{\vartheta_3+\rho_3}{\vartheta_3} \cdots \binom{\vartheta_{\delta-1}+\rho_{\delta-1}}{\vartheta_{\delta-1}} \{a_1^{0} a_2^{\kappa_2} a_3^{\xi_3-\rho_3} \ldots a_{\delta-1}^{\xi_{\delta-1}-\rho_{\delta-1}} a_\delta^{\kappa_\delta-\vartheta_2+\Sigma\rho-1}\}$$

$$+(-)^{\vartheta_{\delta-2}+\vartheta_{\delta-1}} \Sigma \binom{\vartheta_2+\rho_2}{\vartheta_2} \cdots \binom{\vartheta_{\delta-3}+\rho_{\delta-3}}{\vartheta_{\delta-3}}$$

$$\times \{a_1^{0} a_2^{\xi_2-\rho_2} \ldots a_{\delta-2}^{\kappa_{\delta-2}} a_{\delta-1}^{\kappa_{\delta-1}} a_\delta^{\kappa_\delta-\vartheta_{\delta-2}-\vartheta_{\delta-1}+\Sigma\rho-2}\}$$

$$+\ldots.$$

So long as there are forms on the right which belong to only one of the δ sets, we shall generally ignore the others.

4. The following transformation is sometimes useful. Consider the K terms in (III), we write $\xi_t - \rho_t = r_t$ and use equation (I) as for degree $\delta - 1$, *i.e.* for all the quantics except a_2: then the result is true except for certain terms in which the index of a_2 is increased, these terms for which the index of a_2 is greater than κ_2 being already amongst those ignored. Further take $\kappa_1', \kappa_3', \ldots, \kappa_\delta'$ for the values of κ for this application; and $\kappa_t' = \kappa_t$ for $2 < t < \delta$, and also $\xi_t' = \vartheta_t$, $\vartheta_t' = \xi_t$. Then

$$\kappa_\delta' = w - \kappa_2 = \sum_{t=3}^{\delta-1} \vartheta_t' = (w - \kappa_2) - (w - \kappa_\delta - \xi_2) = \kappa_\delta + \xi_2 - \kappa_2,$$

$$\kappa_1' = w - \kappa_2 - \Sigma \xi_t' = \kappa_1 + \vartheta_2 - \kappa_2.$$

For here the positions of a_1 and a_δ are exchanged. Hence

$$\text{(IV)} \quad \Sigma \binom{\vartheta_3 + \rho_3}{\vartheta_3} \cdots \binom{\vartheta_{\delta-1} + \rho_{\delta-1}}{\vartheta_{\delta-1}} \{a_1^0 a_2^{\kappa_2} a_3^{\xi_3 - \rho_3} \ldots a_{\delta-1}^{\xi_{\delta-1} - \rho_{\delta-1}} a_\delta^{\kappa_\delta - \vartheta_2 + \Sigma \rho}\}$$

$$= \Sigma \binom{\kappa_3 - 1 - r_3}{\vartheta_3} \cdots \binom{\kappa_{\delta-1} - 1 - r_{\delta-1}}{\vartheta_{\delta-1}} \{a_1^0 a_2^{\kappa_2} a_3^{r_3} \ldots a_{\delta-1}^{r_{\delta-1}} a_\delta^{w - \kappa_2 - \Sigma r}\}$$

$$= (-)^{w - \kappa_2} \{a_1^{\kappa_1 + \vartheta_2 - \kappa_2} a_2^{\kappa_2} a_3^{\vartheta_3} \ldots a_{\delta-1}^{\vartheta_{\delta-1}} a_\delta^0\}$$

+ terms which belong to two of the original sets and terms with a higher index of a_1.

5. The following extension is found useful. Consider equation (I) (§1); the factor $\binom{s_t - \xi_t - 1}{\vartheta_t}$ on the right-hand side limits the value of s_t to two ranges of values, the one from 0 to ξ_t, and the other equal to or greater than $\xi_t + \vartheta_t + 1$, *i.e.* equal to or greater than κ_t. At the moment all values of s_t which are greater than κ_t will be ignored. On the left-hand side of (I) r_t is limited to the range from 0 to ϑ_t. We introduce on the left the term $r_t = \kappa_t$ and the coefficient $\binom{\xi_t + \vartheta_t - r_t}{\xi_t}$ becomes $(-)^{\xi_t}$; on the right this will bring a coefficient with the factor

$$(-)^{\kappa_t + \xi_t} \binom{s_t}{\kappa_t} = (-)^{\vartheta_t + 1} \binom{s_t}{\kappa_t},$$

the other factors being as before. Here s_t is of necessity equal to or greater than κ_t and hence the only term considered now is $s_t = \kappa_t$. Thus the term introduced on the left is equal to in value and opposite in sign to that already on the right; and on addition the appearance of the new κ_t term on the left removes the old one on the right.

Thus equation (III) becomes

$$
\begin{aligned}
\text{(V)}\quad (-)^{w}\Big[& \{a_1^{\kappa_1} a_2^{\vartheta_2} a_3^{\vartheta_3} \dots a_{\delta-1}^{\vartheta_{\delta-1}} a_\delta^0\} \\
& + (-)^{\xi_2} \Sigma \binom{\xi_3+\sigma_3}{\xi_3} \dots \binom{\xi_{\delta-1}+\sigma_{\delta-1}}{\xi_{\delta-1}} \{a_1^{\kappa_1-\xi_2-1+\Sigma\sigma} a_2^{\kappa_2} a_3^{\vartheta_3-\sigma_3} \dots a_\delta^0\} \\
& + \dots \\
& + (-)^{\xi_t} \Sigma \binom{\xi_2+\sigma_2}{\xi_2} \dots \binom{\xi_{t-1}+\sigma_{t-1}}{\xi_{t-1}} \binom{\xi_{t+1}+\sigma_{t+1}}{\xi_{t+1}} \dots \binom{\xi_{\delta-1}+\sigma_{\delta-1}}{\xi_{\delta-1}} \\
& \qquad \{a_1^{\kappa_1-\xi_t-1+\Sigma\sigma} a_2^{\vartheta_2-\sigma_2} \dots a_{t-1}^{\vartheta_{t-1}-\sigma_{t-1}} a_t^{\kappa_t} a_{t+1}^{\vartheta_{t+1}-\sigma_{t+1}} \dots a_\delta^0\}\Big] \\
-\Big[& \{a_1^0 a_2^{\xi_2} \dots a_{\delta-1}^{\xi_{\delta-1}} a_\delta^{\kappa_\delta}\} \\
& + (-)^{\vartheta_{\delta-1}} \Sigma \binom{\vartheta_2+\rho_2}{\vartheta_2} \dots \binom{\vartheta_{\delta-2}+\rho_{\delta-2}}{\vartheta_{\delta-2}} \\
& \qquad \{a_1^0 a_2^{\xi_2-\rho_2} \dots a_{\delta-2}^{\xi_{\delta-2}-\rho_{\delta-2}} a_{\delta-1}^{\kappa_{\delta-1}} a_\delta^{\kappa_\delta-\vartheta_{\delta-1}+\Sigma\rho-1}\} \\
& + \dots \\
& + (-)^{\vartheta_{t+1}} \Sigma \binom{\vartheta_2+\rho_2}{\vartheta_2} \dots \binom{\vartheta_t+\rho_t}{\vartheta_t} \binom{\vartheta_{t+2}+\rho_{t+2}}{\vartheta_{t+2}} \dots \binom{\vartheta_{\delta-1}+\rho_{\delta-1}}{\vartheta_{\delta-1}} \\
& \qquad \{a_1^0 a_2^{\xi_2-\rho_2} \dots a_t^{\xi_t-\rho_t} a_{t+1}^{\kappa_{t+1}} a_{t+2}^{\xi_{t+2}-\rho_{t+2}} \dots a_{\delta-1}^{\xi_{\delta-1}-\rho_{\delta-1}} a_\delta^{\kappa_\delta-\vartheta_{t+1}-1+\Sigma\rho}\}\Big].
\end{aligned}
$$

= terms which belong to two or more of the original sets + terms which contain a letter $a_r^{p_r}$ for some r with $p_r > \kappa_r$.

It will be noticed that equation (IV) leads to the particular case of (V) where $t = 1$. Also that in the result the two sets of terms, *i.e.* the ϑ and ξ terms, are interchangeable as a whole as they should be.

6. It is useful to extend equation (I) somewhat. There may be other letters present, with which the result is not primarily concerned, they will introduce other terms which are sometimes needed. It is only necessary to introduce one such letter α, and the results for any number are at once apparent. Then as in §1,

$$
\{a_1^{w-\Sigma r} a_2^{r_2} \dots a_{\delta-1}^{r_{\delta-1}} a_\delta^0 \alpha^u\}
= (-)^w \Sigma \left[\binom{u+\varpi}{u} \Pi (-)^{r_t} \binom{s_t}{r_t}\right] \{a_1^0 a_2^{s_2} \dots a_{\delta-1}^{s_{\delta-1}} a_\delta^{w-\varpi-\Sigma s} \alpha^{u+\varpi}\};
$$

and hence

$$\text{(VI)}\quad \left[\prod_{t=2}^{\delta-1}\sum_{r=0}^{\vartheta_t}\binom{\xi_t+\vartheta_t-r_t}{\xi_t}\right]\{a_1^{w-\Sigma r}a_2^{r_2}\dots a_{\delta-1}^{r_{\delta-1}}a_\delta^0 a^u\}$$

$$=(-)^w\Sigma\binom{u+\varpi}{\varpi}\prod_{s_t}(-)^{\vartheta_t}\binom{s_t-\xi_t-1}{\vartheta_t}\{a_1^0a_2^{s_2}\dots a_{\delta-1}^{s_{\delta-1}}a_\delta^{w-\Sigma s-\varpi}a^{u+\varpi}\}$$

$$=(-)^w\left[\Sigma\prod_{s_t}(-)^{\vartheta_t}\binom{s_t-\xi_t-1}{\vartheta_t}\right]\{a_1^0a_2^{s_2}\dots a_{\delta-1}^{s_{\delta-1}}a_\delta^{w-\Sigma s}a^u\}$$

$$+(u+1)\Pi\Sigma\binom{\xi_t+\vartheta_t-r_t}{\xi_t}\{a_1^{w-\Sigma r-1}a_2^{r_2}\dots a_{\delta-1}^{r_{\delta-1}}a_\delta^0a^{u+1}\}$$

$+$terms in a^{u+2}, etc.

II. *First applications of the identity of* §1.

7. Let us consider the covariant types of binary forms of order n, and of degree δ. They may be written in the form

$$\{a_1^{\kappa_1}a_2^{\vartheta_2}\dots a_\delta^0\}=C,$$

and it is supposed here that the indices are in descending order of magnitude. Then when $\kappa_1+\kappa_2+\dots+\kappa_\delta\geqslant 2w+\delta-1$ we have δ sets of forms defined by the presence of $a_r^{\kappa_r}$, $r=1,\dots,\delta$, in the defining gradient, such that no relation is possible between members of different sets. Let

$$\kappa_1=\kappa_2=\dots=\kappa_\delta=n,$$

then provided $$n\delta-2w=R\geqslant\delta-1,$$

where R is the order of C, C is irreductible sequence A provided $\{a_2^{\vartheta_2}\dots a_\delta^0\}$ is an irreductible type sequence A of degree $\delta-1$. This of course is already well known.

When $R<\delta-1$, we may take $\kappa_1=\kappa_2=\dots=\kappa_{R+2}=n$, and

$$\kappa_{R+3}=\kappa_{R+4}=\dots=\kappa_\delta=n+1:$$

then $$\Sigma\kappa=2w+\delta-2,$$

and equation III gives a relation between the first $R+2$ sets.

Now this relation together with the reductions for types of lower degree gives all the relations for these types as is shown by Theorems I and II.

Let us consider the results in the simplest cases.

8. For *invariants* $R=0$, and the result is

$$\text{(VI)}\qquad (-)^w\{a_1^na_2^{\vartheta_2}\dots a_{\delta-1}^{\vartheta_{\delta-1}}a^0\}=\{a_1^0a_2^{\xi_2}\dots a_{\delta-1}^{\xi_{\delta-1}}a_\delta^n\}$$

where $n = \vartheta_r + \xi_2$. This is known already and is obtained at once by the interchange of the upper and lower rows which are of the same length, in the substitutional form. But no proof had been obtained that no other relation independent of this and of the relations for degree $\delta-1$ could exist. It is to be noticed that when the ϑ form is irreductible sequence (A) then the ξ form is irreductible sequence (B) and vice versa, for any reduction of the ϑ form sequence (A) leads to a reduction of the ξ form sequence (B) and the other way round. Thus we know* that all semi-invariant types irreductible sequence (B) can be expressed in the form

$$\{a_1^0 a_2^\lambda a_3^{2\lambda+\mu_3} a_4^{3\lambda+\mu_4} \ldots a_\delta^{(\delta-1)\lambda+\mu_\delta}\}, \quad \text{where} \quad \mu_\delta \geqslant \mu_{\delta-1} \geqslant \ldots \geqslant \mu_3 \geqslant 0.$$

Hence when the right-hand side of (VI) is irreductible sequence (A), then $\xi_{\delta-1} = n$, or every difference $\xi_{r+1} - \xi_r \geqslant n - \xi_{\delta-1}$. In fact in all cases the differences

$$0, \quad 2\xi_{\delta-1} - \xi_{\delta-2} - n, \quad 3\xi_{\delta-1} - \xi_{\delta-3} - 2n, \quad 4\xi_{\delta-1} - \xi_{\delta-4} - 3n, \quad \ldots, \quad \delta\xi_{\delta-1} - (\delta-1)n$$

are in ascending order of magnitude.

THEOREM III. *The necessary and sufficient conditions that an invariant type of degree δ of binary forms of order n may be irreductible are given by equation* (VI) *together with the reductions for covariants of degree $\delta-1$.*

It may happen that $\vartheta_r = \xi_{\delta+1-r}$, $r = 2, 3, \ldots, \delta-1$. Then the invariant

$$I_\xi = \{a_1^0 a_2^{\xi_2} \ldots a_{\delta-1}^{\xi_{\delta-1}} a_\delta^n\}$$

has permutational properties defined by the substitutional form into which it may be put, *i.e.* $\Sigma \lambda_r P_r N_r I$.

The various terms are defined by the tableaux F_r, which give $P_r N_r$. When the letter a_δ is dropped F_r becomes F_r', a tableau appropriate to the covariant given by the first $\delta-1$ letters of I. In this case the letter a_δ must be added to the tableau F_r' in a position consistent with the fact that

$$[1 - (-)^w (a_1 a_\delta)(a_2 a_{\delta-1}) \ldots] I = 0.$$

The equation VI itself always introduces limitations on the introduction of a_δ into F_r'. Thus when $\vartheta_{\delta-1}$ is even I_ϑ is unchanged by the operation $(a_\delta a_{\delta-1})$, and when $\vartheta_{\delta-1}$ is odd this operation changes the sign of I_ϑ and therefore also of I_ξ. Other results involving $a_{\delta-2}$ etc. are obtained in the same way.

These remarks will be illustrated by considering in particular some values of δ later on.

* Q.S.A. VII, p. 320, Theorem VIII.

9. *Covariants of order one.*

The appropriate equation is

$$\text{(VII)}\quad (-)^{w}\{a_1^n a_2^{\vartheta_2} a_3^{\vartheta_3} \dots a_{\delta-1}^{\vartheta_{\delta-1}} a_\delta^0\} = C_\vartheta$$

$$= \{a_1^0 a_2^{\xi_2} a_3^{\xi_3} \dots a_{\delta-1}^{\xi_{\delta-1}} a_\delta^n\}$$

$$+(-)^{\vartheta_{\delta-1}} \Sigma \binom{\vartheta_2+\rho_2}{\vartheta_2} \dots \binom{\vartheta_{\delta-2}+\rho_{\delta-2}}{\vartheta_{\delta-2}} \{a_1^0 a_2^{\xi_2-\rho_2} \dots a_{\delta-1}^n a_\delta^{n-\vartheta_{\delta-1}-1+\Sigma\rho}\}$$

where $n = \vartheta_{\delta-1}+\xi_{\delta-1}+1 = \vartheta_r+\xi_r$, $r = 2, 3, \dots, \delta-2$. There is a reduction sequence (A) of C_ϑ when the first of the differences $\xi_{\delta-1}-\vartheta_2$, $\xi_{\delta-2}-\vartheta_3$, ... which does not vanish is positive. Now $\xi_{\delta-1}-\vartheta_2 = n-\vartheta_{\delta-1}-\vartheta_2-1$. Thus the condition series is $n-\vartheta_{\delta-1}-\vartheta_2-1$, $n-\vartheta_{\delta-2}-\vartheta_3$, ..., $n-\vartheta_2-\vartheta_{\delta-1}$ and the differences cannot all be zero for if the first is zero the last is $+1$.

The covariant C_ξ is subject to an equation which reduces

$$[1+(-)^{\vartheta_{\delta-1}}(a_{\delta-1} a_\delta)]\, C_\xi$$

when the first of the differences $\vartheta_2-\xi_{\delta-1}$, $\vartheta_3-\xi_{\delta-2}$, ... which does not vanish is positive.

And here the condition series is

$$n-\xi_2-\xi_{\delta-1},\ n-\xi_3-\xi_{\delta-2},\ \dots.$$

Theorem IV. *Conditions for the reduction of the covariant type of order one*

$$\{a_1^0 a_2^{\xi_2} \dots a_{\delta-1}^{\xi_{\delta-1}} a_\delta^n\} = C_\xi$$

are given by the reductions for degree $\delta-1$ together with the fact that C_ξ is reductible, when the first of the differences

$$n-\xi_2-\xi_{\delta-1}-1,\ n-\xi_3-\xi_{\delta-2},\ \dots$$

which does not vanish is positive. And $[1+(-)^{\xi_{\delta-1}}(a_{\delta-1} a_\delta)]C_\xi$ is reductible when the first of the differences

$$n-\xi_2-\xi_{\delta-1},\ n-\xi_3-\xi_{\delta-2},\ \dots$$

which does not vanish is positive.

It should be remarked that for a covariant of odd order both n and δ must be odd, and hence

$$(-)^{\vartheta_{\delta-1}} = (-)^{n-1-\xi_{\delta-1}} = (-)^{\xi_{\delta-1}}.$$

A reduction is obtained at once for $[1+(-)^{\xi_{\delta-1}}(a_{\delta-1}a_\delta)]\,C_\xi$ when C_ϑ is reductible sequence (B) for then each term on the left when C_ϑ is reduced will give by the use of (VII) only terms which are earlier than C_ξ. The multiplicity of terms on the right of (VII) prevents the similar conclusion being made for C_ϑ; in fact it is easy to give instances where it is plain that C_ϑ is irreductible sequence (A), while the latest C_ξ on the right is reductible sequence (B).

10. *Covariants of order two.*

The natural course here is to take $\kappa_1 = \kappa_\delta = \kappa_{\delta-1} = \kappa_{\delta-2} = n$ in equation (III), every other κ having the value $n+1$. The terms of the set $K_{\delta-2}$ introduce the difficulty that the latest term is no longer

$$\{a_1^0 a_2^{\xi_2} \dots a_{\delta-2}^n a_{\delta-1}^{\xi_{\delta-1}} a_\delta^{\xi_{\delta-2}}\}$$

when $n > \xi_{\delta-1} > \xi_{\delta-2}$. For there are terms with gradient factor $a_{\delta-1}^{\xi_{\delta-1}-\rho} a_\delta^{\xi_{\delta-2}+\rho}$ which may be later sequence (A). In this case we multiply the equation (III) by a numerical factor and sum as follows:

$$\text{(VIII)} \quad (-)^w \sum_{r=0}^{\gamma_1} \binom{\gamma_1+\gamma_2-r}{\gamma_2} \{a_1^n a_2^{\vartheta_2} \dots a_{\delta-3}^{\vartheta_{\delta-3}} a_{\delta-2}^{\vartheta-r} a_{\delta-1}^r a_\delta^0\}$$

$$- \sum_{r=0}^{\gamma_1} \binom{\gamma_1+\gamma_2-r}{\gamma_2} [1+(-)^r(a_{\delta-1}a_\delta)] \{a_1^0 a_2^{\xi_2} \dots a_{\delta-3}^{\xi_{\delta-3}} a_{\delta-2}^{n-1-\vartheta+r} a_{\delta-1}^{n-1-r} a_\delta^n\}$$

$$= (-)^\vartheta \Sigma (-)^r \binom{\gamma_1+\gamma_2-r}{\gamma_2} \binom{r+\sigma}{r} \binom{\vartheta_2+\rho_2}{\vartheta_2} \dots \binom{\vartheta_{\delta-3}+\rho_{\delta-3}}{\vartheta_{\delta-3}}$$

$$\times \{a_1^0 a_2^{\xi_2-\rho_2} \dots a_{\delta-2}^n a_{\delta-1}^{n-1-r-\sigma} a_\delta^{n-1-\vartheta+r+\sigma+\Sigma\rho}\} + \text{earlier forms}$$

$$= (-)^\vartheta \{a_1^0 a_2^{\xi_2} \dots a_{\delta-3}^{\xi_{\delta-3}} a_{\delta-2}^n a_{\delta-1}^{n-1-\gamma_2} a_\delta^{n-1-\vartheta+\gamma_2}\}$$

$$+ (-)^{\vartheta+\gamma_1} \{a_1^0 a_2^{\xi_2} \dots a_{\delta-3}^{\xi_{\delta-3}} a_{\delta-2}^n a_{\delta-1}^{n-1-\vartheta+\gamma_3} a_\delta^{n-1-\gamma_3}\} + \text{earlier forms}$$

where
$$\gamma_1+\gamma_2+\gamma_3 = \vartheta - 1;$$

for
$$\sum_{r=1}^{\gamma_1} (-)^r \binom{\gamma_1+\gamma_2-r}{r} \binom{p}{r} = (-)^{\gamma_1} \binom{p-\gamma_2-1}{\gamma_1}.$$

There is a reduction here for that form or set of forms which has the second highest index with the lowest value. The possibilities are $n-\gamma_1-1$, $n-\gamma_2-1$, $n-\gamma_3-1$, $\vartheta_2 = n-\xi_2$.

The latest term on the right, which is the term to be reduced, will in general be written

$$C_\xi = (0, \xi_2, \xi_3, \ldots, \xi_{\delta-1}, n).$$

The conditions to be fulfilled that this may be later than any term on the left of (VII) are called the conditions (a_1): these are (in general) $n > \xi_2 + \xi_{\delta-1}$ or $n = \xi_2 + \xi_{\delta-1} > \xi_3 + \xi_{\delta-2}$, and so on.

(*A*) $\gamma_3 \leqslant \gamma_2 - 1,\ \gamma_1 \leqslant \gamma_2 - 1,\ \vartheta - 1 = \gamma_1 + \gamma_2 + \gamma_3 \leqslant 3\gamma_2 - 2,\ \xi_{\delta-1} = n - \gamma_2 - 1.$

gives a complete reduction for C_ξ. Then

$$\vartheta = 2n - 2 - \xi_{\delta-1} - \xi_{\delta-2} \leqslant 3(n - \xi_{\delta-1} - 1) - 1;$$

i.e. $n + \xi_{\delta-2} \geqslant 2\xi_{\delta-1} + 2$, is the condition for this reduction in terms of the C_ξ coefficients.

(*B*) $\gamma_2 = \gamma_3 = \gamma_1 + 1 = \frac{1}{3}\vartheta$, or $\gamma_2 = \gamma_1 = \gamma_3 + 1 = \frac{1}{3}\vartheta$,

in either case $n + \xi_{\delta-2} = 2\xi_{\delta-1} + 1$, reductions are obtained for the two forms

$$[1 + (-)^{n+\xi_{\delta-1}}(a_{\delta-2}\, a_{\delta-1})]\, C_\xi,$$

$$[1 - (-)^{n+\xi_{\delta-1}}(a_{\delta-1}\, a_\delta) - (-)^{n+\xi_{\delta-1}}(a_{\delta-2}\, a_\delta)]\, C_\xi.$$

(*C*) $\gamma_1 = \gamma_2 = \gamma_3 = \frac{1}{3}(\vartheta - 1),$

i.e. $n + \xi_{\delta-2} = 2\xi_{\delta-1}$ and the reduction is for

$$[1 - (-)^{n+\xi_{\delta-1}}(a_{\delta-1}\, a_\delta)][1 + (-)^{n+\xi_{\delta-1}}(a_{\delta-2}\, a_\delta)]\, C_\xi.$$

11. Equation (V) gives a simpler reduction

$$(-)^w[(n, \vartheta_2, \ldots, \vartheta_{\delta-1}, 0) + (-)^{\xi_2}(\vartheta_2, n, \vartheta_3, \ldots, 0)]$$
$$= (0, \xi_2, \ldots, \xi_{\delta-1}, n) + (-)^{\vartheta_{\delta-1}}(0, \xi_2, \ldots, n, \xi_{\delta-1}),$$

where $n = \kappa_1 = \kappa_2 = \kappa_{\delta-1} = \kappa_\delta$ and the other κ's have the value $n+1$.

There is a set of conditions (a_2) that the terms on the right may be later than those on the left; they may be expressed by the rule that the first of the differences

$$n - \xi_{\delta-1} - \xi_2 - 1,\quad n - \xi_{\delta-2} - \xi_3,\quad \ldots$$

which does not vanish is positive. When (a_2) is satisfied there is a reduction for

$$[1 + (-)^{n-1-\xi_{\delta-1}}(a_{\delta-1}\, a_\delta)]\, C_\xi.$$

Further, when the conditions (b_2) which require the vanishing of all the differences in (a_2) are satisfied, there is a reduction for

$$[1-(-)^w I][1+(-)^{n-1-\xi_{\delta-1}}(a_{\delta-1}a_\delta)]\,C_\xi,$$

where I is the permutation $(a_1 a_\delta)(a_2 a_{\delta-1})\ldots$ which we call the inverter.

It is to be noticed that the conditions (a_1) are always satisfied when either (a_2) or (b_2) are satisfied. Thus, when (a_2) is satisfied and

$$n+\xi_{\delta-1}=2\xi_{\delta-2}+1,$$

C_ξ is completely reduced by (B). And using (C) when (a_2) is satisfied C_ξ is seen to have the substitutional properties of $\{a_{\delta-2}a_{\delta-1}a_\delta\}$ or of $\{a_{\delta-2}a_{\delta-1}a_\delta\}'$ according as $n+\xi_{\delta-1}$ is even or odd.

12. There is a further reduction when $n+\xi_{\delta-2}=2\xi_{\delta-1}-1$, and (a_2) is satisfied. To obtain this, the complete reduction for

$$C_\xi'=(0,\,\xi_2,\,\ldots,\,\xi_{\delta-3},\,\xi_{\delta-2}+1,\,\xi_{\delta-1}-1,\,n)$$

is obtained as a form of class (A) §10.

Then the reduction for $[1+(-)^{n-\xi_{\delta-1}}(a_{\delta-1}a_\delta)]\,C_\xi'$ is obtained from this; and it is also obtained by the reduction of §11. By elimination of C_ξ' a reduction is obtained for the next latest form. The reduction in §10 is obtained from (VII) by writing $\gamma_1=n-\xi_{\delta-1}$, $\gamma_2=n-\xi_{\delta-1}-1=\gamma_3$; *i.e.*

$$-[1+(-)^{n-\xi_{\delta-1}}(a_{\delta-1}a_\delta)]\,C_\xi'-(n-\xi_{\delta-1})[1-(-)^{n-\xi_{\delta-1}}(a_{\delta-1}a_\delta)]\,C_\xi$$
$$-(-)^{n-\xi_{\delta-1}}(n-\xi_{\delta-2}-1)(a_{\delta-1}a_\delta)\,C_\xi$$
$$=(-)^{n-\xi_{\delta-1}-1}(a_{\delta-2}a_\delta)\,C_\xi-(a_{\delta-1}a_\delta)(a_{\delta-2}a_\delta)\,C_\xi+\text{forms earlier than } C_\xi.$$

The reduction in §11 gives

$$[1+(-)^{n-\xi_{\delta-1}}(a_{\delta-1}a_\delta)]\,C_\xi'+(-)^{n-\xi_{\delta-1}}(n-\xi_{\delta-2})(a_{\delta-1}a_\delta)\,C_\xi=\text{earlier forms};$$

and also $$[1-(-)^{n-\xi_{\delta-1}}(a_{\delta-1}a_\delta)]\,C_\xi=\text{earlier forms}.$$

Whence, $$(a_{\delta-1}a_\delta)\,C_\xi=-(a_{\delta-2}a_\delta)\,C_\xi-(a_{\delta-2}a_{\delta-1})\,C_\xi$$

or $$[1+(a_{\delta-2}a_{\delta-1}a_\delta)+(a_{\delta-1}a_{\delta-2}a_\delta)]\,C_\xi,$$

is reduced. Therefore C_ξ is of the form

$$\{a_\delta a_{\delta-1}\}\{a_\delta a_{\delta-2}\}' \quad\text{or}\quad \{a_\delta a_{\delta-1}\}'\{a_\delta a_{\delta-2}\}$$

according as $n+\xi_{\delta-1}$ is even or odd.

13. To sum up, the following cases may be distinguished:

(α) A reduction of C_ξ when (a_2) is satisfied and also $n-2\xi_{\delta-1}+\xi_{\delta-2}\geqslant 1$; also when (a_1) alone is satisfied and $n-2\xi_{\delta-1}+\xi_{\delta-2}\geqslant 2$.

For the rest C_ξ is reduced except for certain representations in the last three letters.

(β) C_ξ must belong to T_3 or T_{1^3} according as $n+\xi_{\delta-1}$ is even or odd when (a_2) is satisfied and $n-2\xi_{\delta-1}+\xi_{\delta-2}=0$.

(γ) C_ξ is of one representation T_{21} (the general function has f_α representations T_α) when (a_2) is satisfied and $n-2\xi_{\delta-1}+\xi_{\delta-2}=-1$, or when only (a_1) is satisfied and $n-2\xi_{\delta-1}+\xi_{\delta-2}=1$.

(δ) The representation of C_ξ is T_3+T_{21} or $T_{21}+T_{1^3}$ according as $n+\xi_{\delta-1}$ is even or odd when (a_2) is satisfied and $n-2\xi_{\delta-1}+\xi_{\delta-2}\leqslant -2$.

(ϵ) C_ξ belongs to $T_3+T_{21}+T_{1^3}$ when (a_1) only is satisfied and

$$n-2\xi_{\delta-1}+\xi_{\delta-2}=0.$$

The reductions when the conditions (b_2) are satisfied are far more difficult to analyse, involving as they do the inverter permutation.

(ζ) A reduction of C_ξ when $n>\xi_2+\xi_{\delta-1}+1$, and $2n\geqslant \xi_3+2\xi_{\delta-1}+1$. And also when $n=\xi_2+\xi_{\delta-1}+1>\xi_3+\xi_{\delta-2}+1$, and $\xi_3-2\xi_2=0$ or 1.

III. *A correction.*

14. It is necessary to point out that Theorem XIX, Q.S.A. VIII, p. 489 is not true as it stands; it requires restatement. This also applies to Theorem XX. It is true as was pointed out in §31, p. 489, that the covariant type given by a binary gradient $A_0^{(1)}A_{\lambda_2}^{(2)}\dots A_{\lambda_\delta}^{(\delta)}$ in semi-normal units is the continued transvectant

$$\left(\dots\left((f_1, f_2)^{\lambda_2}f_3\right)^{\lambda_3}\dots f_\delta\right)^{\lambda_\delta}.$$

No proof was given in the place there referred to. It follows at once from the fact that in semi-normal units, a permutation of some of the first r letters, has no effect on the later letters; and in the result the position of the later letters is unaffected. Thus any property which can be expressed by permutational operations on the semi-normal form

$$C_{\delta-1}=\{A_0^{(1)}A_{\lambda_2}^{(2)}\dots A_{\lambda_{\delta-1}}^{(\delta-1)}\}$$

holds in the same way for the form

$$C_\delta = \{A_0^{(1)} A_{\lambda_2}^{(2)} \ldots A_{\lambda_{\delta-1}}^{(\delta-1)} A_{\lambda_\delta}^{(\delta)}\}.$$

That is C_δ is a covariant of $C_{\delta-1}$ and f_δ and hence must be a numerical multiple of $(C_{\delta-1}, f_\delta)^{\lambda_\delta}$.

15. With a given defining gradient G for types, the natural form $\{G\}$, the semi-normal form $\{G\}$ and its equivalent continued transvectant all differ from each other by forms given by an earlier defining gradient*.

Hence the natural forms $\{G\}$ in equations (III) or (V) may be replaced by the semi-normal forms or by the continued transvectants.

Consider now the types of binary forms of a definite order n for which the defining gradient $a_1^0 a_2^{\lambda_2} \ldots a_\delta^{\lambda_\delta} = A_0^{(1)} A_{\lambda_2}^{(2)} \ldots A_{\lambda_\delta}^{(\delta)}$ has at least one factor a_δ^n. It will be supposed that $\lambda_\delta \geqslant \lambda_{\delta-1} \ldots \geqslant \lambda_2 \geqslant 0$. Then a modified continued transvectant will be introduced; we may take this to be

$$\left(\ldots\left(\left((a_{1_x}^{\lambda_2}, a_{2_x}^{\lambda_2})^{\lambda_2} a_{1_x}^{\lambda_3-\lambda_2} a_{2_x}^{\lambda_3-\lambda_2}, a_{3_x}^{\lambda_3}\right)^{\lambda_3} a_{1_x}^{\lambda_4-\lambda_3} a_{2_x}^{\lambda_4-\lambda_3} a_{3_x}^{\lambda_4-\lambda_3}, a_{4_x}^{\lambda_4}\right)^{\lambda_4} \ldots a_{\delta_x}^{\lambda_\delta}\right)^{\lambda_\delta}.$$

THEOREM V. *The leading gradients sequence* (A), *and the defining gradients for irreductible types of degree* δ *of forms of order* n *are the same.*

It is well known to be true when $\delta \leqslant 4$. It will be assumed to be true for all degrees up to $\delta-1$. Consider the covariant type

$$C = \{a_1^0 a_2^{\lambda_2} \ldots a_{\delta-1}^{\lambda_{\delta-1}} a_\delta^{\lambda_\delta}\} = \{G_{\delta-1} a_\delta^{\lambda_\delta}\},$$

$$\lambda_\delta \geqslant \lambda_{\delta-1} \ldots \geqslant \lambda_2 \geqslant 0.$$

Then in place of C we may consider the transvectant $(P_{\delta-1}, a_{\delta_x}^{\lambda_\delta})^{\lambda_\delta}$, where the leading gradient of $P_{\delta-1}$ is $G_{\delta-1}$. When R the order of C is greater than $\delta-2$, and $\lambda_\delta > \lambda_{\delta-1}$ the modified transvectant

$$(\Pi_{\delta-1} a_{1_x}^{\lambda_\delta-\lambda_{\delta-1}-1} \ldots a_{(\delta-1)_x}^{\lambda_\delta-\lambda_{\delta-1}-1}, a_{\delta_x}^{\lambda_\delta})^{\lambda_\delta}$$

may be used, where $\Pi_{\delta-1}$ is the covariant of the form of order $\lambda_{\delta-1}$ for which $P_{\delta-1}$ is leading gradient. Then a_δ is the only letter which can receive the weight λ_δ and thus the leading gradient is the same as the defining gradient. When $\lambda_\delta = \lambda_{\delta-1} > \lambda_{\delta-2}$ the modified transvectant

$$(\Pi_{\delta-2} a_{1_x}^{\lambda_\delta-\lambda_{\delta-2}-1} \ldots a_{(\delta-2)_x}^{\lambda_\delta-\lambda_{\delta-2}-1}, a_{(\delta-1)_x}^{\lambda_\delta} a_{\delta_x}^{\lambda_\delta})^{2\lambda_\delta}$$

---

* Q.S.A. VIII, Theorem XVIII. See also Q.S.A. VII, p. 328, §22.

may be used in the same way, provided $R \geqslant \delta-2$. Consider now the covariant types $0 \leqslant R \leqslant \delta-2$ which have defining gradients in which the highest index is λ_δ. They may be regarded as types of forms of order λ_δ. Thus the presence of a higher index is impossible: such a gradient is zero, for $a_r^{\lambda_\delta+1}=0$.

The leading gradient is alike in this and further its highest index is λ_δ, otherwise it would correspond to a covariant type of forms of order $\lambda_{\delta-1}$, *i.e.* a covariant of the negative order $R-\delta$, which is impossible. The leading gradient is then $G_{\delta-1} a_\delta^{\lambda_\delta}$, where $G_{\delta-1}$ is a leading gradient of degree $\delta-1$. Now in equation (III) the defining gradients on the left are given by the sequence $a_1, a_2, \ldots, a_\delta$ on the right the first set is given by the sequence beginning $a_\delta, a_{\delta-1}, \ldots,$ the second by the sequence $a_{\delta-1}, a_\delta, a_{\delta-2}, \ldots$ and so on. In each case the leading gradient and the defining gradient are the same for the sequence used. The result in either case is to express the covariant types in terms of the earliest possible forms according to the sequence (A), which depends on the indices and not on the letters. And the result must be the same, and in each case will involve certain substitutional restrictions, such as $T_a C$ is or is not reductible, where T_a refers to permutations of quantics.

This completes the proof.

IV. *Preliminary remarks.*

16. When the indices of the form on the left in equation (III) are each increased to $\vartheta_1', \vartheta_2', \ldots,$ so that $\vartheta_r'=\vartheta_r+2(\delta-r)$, $\kappa_r'=\kappa_r+2(\delta-1)$, then $\xi_r'=\xi_r+2(r-1)$ and the corresponding indices ϑ_r, $\xi_{\delta-r+1}$ are such that

$$\vartheta_r'-\xi_{\delta-r+1}'=\vartheta_r-\xi_{\delta-r+1};$$

it follows that the reductions effected are the same in the two cases. When Π is a seminvariant with leading gradient $a_1^0 a_2^{\lambda_2} \ldots a_\delta^{\lambda_\delta}$ and Δ is the product of the squared differences of the symbolical letters, a seminvariant with the leading gradient $a_1^0 a_2^{\lambda_2+2} a_3^{\lambda_3+4} \ldots a_\delta^{\lambda_\delta+2\delta-2}$ and the same substitutional properties as Π is obtained by multiplication of the symbolical form of Π by Δ.

A similar result is obtained when the differences between consecutive indices in the leading gradient are in every case increased by unity, provided that the sign of every transposition of the accompanying substitutional expression is changed. Thus for a single quantic the form $T_\delta C$ is really considered, with the implied assumption that

$$\lambda_\delta \geqslant \lambda_{\delta-1} \geqslant \lambda_{\delta-2} \cdots \geqslant 0.$$

Changing the sign of all transpositions gives $T_{1^\delta} C$ with the implied assumption

$$\lambda_\delta \geqslant \lambda_{\delta-1}+1 \geqslant \lambda_{\delta-2}+2 \ldots \geqslant 0+\delta-1.$$

And in any form PNC there is a definite appropriate assumption obtained from F the tableau of PN, given by the rule that when λ_{r+1} is on a lower row in F than λ_r, $\lambda_{r+1} \geqslant \lambda_r+1$ otherwise, $\lambda_{r+1} \geqslant \lambda_r$.

THEOREM VI. *Gradients C_1, C_2 are of degree δ and the successive index differences in C_2 are increased by a constant quantity d from those in C_1; then both $(PN)_a C_1$, and $(PN)_a C_2$ are reductible or both irreductible when d is even; and the same is true for $(PN)_a C_1$ and $(PN)_{a'} C_2$ when d is odd: the tableau $F_{a'}$ being obtained from F_a by the interchange of rows and columns. Always provided that C_1 is an appropriate form for F_a.*

17. This theorem may be applied in the following way.

COROLLARY. *The discussion of the reductibility of a covariant*

$$C = \{a_1^0 a_2^{\lambda_2} \ldots a_\delta^{\lambda_\delta}\}$$

may be limited to those cases in which one pair of consecutive indices differ by 0 or 1; and this is true whether the covariants are those of a single form, i.e. $T_\delta C$; or belong to any substitutional representation.

For the theorem may be used to limit the discussion to such cases.

18. Consider a seminvariant of a single quantic of order n which is defined as usual by a substitional form $\{P\}$. Let P have the factor $A_r^\lambda A_{r+1}^\mu$. This means that there are λ sets of n letters each which have $n-r$ in the upper row of the tableau and r letters in the lower row; and also μ sets of n letters each which have $n-r-1$ letters in the upper row, and $r+1$ in the lower. Let g be the positive symmetric group of all these $(\lambda+\mu)n$ letters; let the covariant be actually $K = HGPNX$; consider the form $HGgPNX$. This differs from a multiple of K by earlier forms according to either sequence (A) or (B). For any term which differs from K has either an increase in λ and a new letter A_s, $s > r+1$; or an increase in μ and a new letter A_t, $t < r$; or a new letter A_s and a new letter A_t, in the defining gradient. In fact if we replace the product $A_r^\lambda A_{r+1}^\mu$ by a new coefficient $B_{(\lambda+\mu)r+\mu}$ which is the coefficient of a new quantic $\phi = f^{\lambda+\mu}$, in the gradient P; and call the new gradient Q; then (Q) differs from (P) by earlier forms according to either of the sequences (A) and (B). This form of replacement is the basis of most of what follows.

From the last paragraph we need only consider forms sequence (A) in which at least one such replacement has been made.

It is possible that (P) may be expressed with compound coefficients in two or more ways; thus the product $A_r^{\lambda} A_{r+1}^{\mu} A_{r+2}^{\nu}$ may be written $(A_r^{\lambda} A_{r+1}^{\mu}) A_{r+2}^{\nu}$ or $A_r^{\lambda}(A_{r+1}^{\mu} A_{r+2}^{\nu})$. Let (P) be expressed thus as (Q) or (Q') and let the degrees of the coefficients in Q be $d_1 \geqslant d_2 \geqslant d_3$. Then (Q) will be regarded as reductible when the first of the differences $d_1 - d_1', d_2 - d_2', \ldots$ which does not vanish is negative. When they are all zero, and the weight of the coefficient of degree d_r in (Q) is w_r, (Q) will be regarded as reductible when the first of the differences $w_1 - w_1'$, $w_2 - w_2'$, ... which does not vanish is positive. Thus the combination $(A_r^{\lambda} A_{r+1}^{\mu}) A_{r+2}^{\lambda}$ is preferred to the combination $A_r^{\lambda} (A_{r+1}^{\mu} A_{r+2}^{\lambda})$.

19. For a single quantic of even order the only possible leading gradients are of the form $A_0^{\lambda_0} A_2^{\lambda_2} A_3^{\lambda_3} \ldots A_{2n}^{\lambda_{2n}}$. From what has just been said we may consider instead the gradient $b_1^0 b_2^{2\lambda_2+3\lambda_3} \ldots b_{n+1}^{2n\lambda_{2n}}$, where b_1^0 is the first coefficient of the form f^{λ_0}, $b_2^{2\lambda_2+3\lambda_3}$ the coefficient of $x_2^{2\lambda_2+3\lambda_3}$ in $f^{\lambda_2+\lambda_3}$ and so on.

Symbolical letters referring to each of these new quantics may be used, and then Theorem VI is applicable. The result is that $\{b_1^0 b_2^{2\lambda_2+3\lambda_3} \ldots b_{n+1}^{2n\lambda_{2n}}\}$ and $\{c_1^0 c_2^{2(\lambda_2+1)+3\lambda_3} c_3^{4(\lambda_4+1)+5\lambda_5} \ldots c_{n+1}^{2n(\lambda_{2n}+1)}\}$ are both reductible or both irreductible.

The letters c are used because c_r is a letter referring to $f^{\lambda_{2r-2}+\lambda_{2r-1}+1}$, while b_2 refers to $f^{\lambda_{2r-2}+\lambda_{2r-1}}$. And as covariants of a single quantic are being considered it follows that the discussion may be limited to gradients which have one of the indices $\lambda_0, \lambda_2, \lambda_4, \ldots, \lambda_{2n}$ zero.

THEOREM VII. *The covariants of a quantic of even order may be expressed in terms of those which have the catalecticant as a factor; and those for which the defining gradient lacks one or more of the coefficients* $A_2, A_4, \ldots, A_{2n}$.

V. *The binary canonical forms for quantics of odd order.*

20. The well-known theorem that a binary form of odd order $2n+1$ can always be expressed as a sum of the $(2n+1)$-th powers of $n+1$ linear forms, *i.e.* as $\sum_{r=1}^{n+1} a_{r_x}^{2n+1}$, can be, and has been, applied as a basis for the expression of the covariant system as simultaneous covariants of these linear forms. In such an expression there is a sum of products of invariant factors $(a_r a_s)$ and of variable factors a_{r_x}, as in the symbolical calculus. As there too each letter appears a regular number of times—but here it is any multiple of $2n+1$; the letters are also interchangeable. In fact any symbolical product really stands for the sum of the products obtained by inter-

changing the $n+1$ letters in all possible ways. Substitutional analysis may be applied to this form of expression in the same way as to the symbolical form.

Now the ordinary substitutional expression for a covariant of this quantic may be written

$$C=(A_0^{\lambda_0}A_2^{\lambda_2}A_3^{\lambda_3}\dots A_{2n}^{\lambda_{2n}}A_{2n+1}^{\lambda_{2n+1}})$$

where λ_r is the number of sets of $2n+1$ letters which have r letters in the lower row and $2n+1-r$ in the upper row of the tableau.

Actually C is an abbreviation for $HGPNX$ where H is an operator implying the equivalence of the δ sets, G is the product of the positive symmetric groups of the δ sets, PN has the ordinary substitutional meaning, and X is the algebraic operand. Let g_r be the positive symmetric group of all the letters of the $\lambda_{2r}+\lambda_{2r+1}$ sets which provide the factors in C, $A_{2r}^{\lambda_{2r}}A_{2r+1}^{\lambda_{2r+1}}$. Then consider $K=HGg_rPNX$, this will consist of a sum of terms, first there will be a numerical multiple of C. Then some of the sets may have fewer letters in the lower row of the tableau, in which case others must have an increased number, in other words K differs from a multiple of C by covariants C' in which one or more of the indices $\lambda_{2r+1}, \lambda_{2r+2}, \dots$ is increased; it may happen that λ_{2r+1} is decreased but then one or more of λ_{2r+2} is increased. Hence K differs from a multiple of C by covariants C' which are earlier than C sequence (A) and also, as is seen in the same way, sequence (B). Thus C differs from

(IX) $$(a_1^0 a_2^{2\lambda_2+3\lambda_3}\dots a_{n+1}^{2n\lambda_{2n}+(2n+1)\lambda_{2n+1}})$$

by earlier covariants, where a_r represents a quantic of order

$$n_r(2n+1)=(\lambda_{2r-2}+\lambda_{2r-1})(2n+1),$$

a form of expression in entire correspondence with that of the canonical form.

21. THEOREM VIII. *The covariant* $C=(A_0^{\lambda_0}A_2^{\lambda_2}A_3^{\lambda_3}\dots A_{2n+1}^{\lambda_{2n+1}})$ *of order* R *of a binary form of order* $2n+1$ *is reductible (sequence* A*) unless*

$$R \geqslant \lambda_{2n}+\nu+\tau,$$

where
$$\nu_r=\lambda_{2r-2}+\lambda_{2r-1}-\lambda_{2n+1},$$
$$\nu=\Sigma\nu_r,$$

the sum being confined to the values of r *for which* ν_r *is positive; and* τ *is the number of values of* r *for which* $\nu_r>\lambda_{2n}$.

We consider the form (IX) for C, and apply Theorem (I). There is a reduction if C can be expressed in terms of similar covariants in which the index of a_r is equal to or greater* than κ_r

$$= 2n\,.\,n_r + \lambda_{2n+1}, \qquad n_r > n_{n+1},$$

$$= (2n+1)n_r - \nu_r$$

or

$$= 2n\,.\,n_r + \lambda_{2n+1} + 1, \qquad n_{n+1} \geqslant n_r \geqslant \lambda_{2n+1},$$

$$= (2n+1)n_r - \nu_r + 1$$

or

$$= (2n+1)n_r + 1, \qquad \lambda_{2n+1} > n_r;$$

for these covariants precede C; in fact the last set only exist for quantics of higher order.

Then
$$\sum_{r=1}^{n} \kappa_r = (2n+1)(\delta - n_{n+1}) - \nu + (n - \tau),$$

for
$$\sum_{r=1}^{n+1} n_r = \delta, \quad \text{where } \delta \text{ is the degree of } C.$$

We take
$$\kappa_{n+1} = (2n+1)n_{n+1} - \lambda_{2n} + 1.$$

Then C can be expressed in terms of the $n+1$ sets defined by κ_r, unless

$$(2n+1)\delta - \nu - \tau + n - \lambda_{2n} + 1 \geqslant 2w + n + 1$$

i.e.
$$R \geqslant \lambda_{2n} + \nu + \tau.$$

It is quite clear that when the form in $a_1, a_2, \ldots, a_{n+1}$ is irreductible then the form in $a_1, a_2, \ldots, a_{2n+1}$ is irreductible. For if we take the canonical form all the irreductible forms a give actual covariants. And thus as each irreductible covariant can be expressed in terms of a separate irreductible form in the a's there must be an actual one to one correspondence between irreductible forms in the two modes of expression.

22. The theorem just proved is not quite complete; it gives a necessary condition; but it needs a special examination of the cases $n_r = n_{n+1}$ before the sufficient condition can be given. This and further aspects of the case may be illustrated by the cubic and quintic.

---

* These three choices correspond to $\nu_r > \lambda_{2n}$, $\lambda_{2n} \geqslant \nu_r \geqslant 0$ and $\nu_r < 0$. Note further that ν_r is summed from 1 to n to yield ν; also that $R = (2n+1)\delta - 2n$. (*Ed.*)

For the cubic the order of $(A_0^{\lambda_0} A_2^{\lambda_2} A_3^{\lambda_3})$ is

$$R = 3(\lambda_0+\lambda_2+\lambda_3)-2(2\lambda_2+3\lambda_3) = 3(\lambda_0-\lambda_3)-\lambda_2 \geqslant \lambda_2+\nu+\tau$$

and hence $\lambda_0 > \lambda_3$; since $\nu = \lambda_0-\lambda_3$, the inequality becomes

$$2(\lambda_0-\lambda_2-\lambda_3) \geqslant \tau.$$

This is always satisfied when $n_1 > n_2$, *i.e.* when $\lambda_0 > \lambda_2+\lambda_3$, for then $\tau = 1$. When $n_1 = n_2$, *i.e.* when $\lambda_0 = \lambda_2+\lambda_3$, the form may be written $(a_1 a_2)^w$ where a_1, a_2 represent the same quantic, this is zero when w is odd, but a covariant when w is even.

Thus, w even, $\quad \nu \leqslant \lambda_2,\ \tau = 0;\quad \nu > \lambda_2,\ \tau = 1;$

w odd, $\quad \nu < \lambda_2,\ \tau = 0;\quad \nu \geqslant \lambda_2,\ \tau = 1.$

For the quintic the forms $(a_1^0 a_2^r a_3^{w-r})$ are considered: they are subject to the equation*

$$\sum_{r=0}^{\gamma_1} \binom{\gamma_1+\gamma_2-r}{\gamma_2} \{a_1^0 a_2^r a_3^{w-r}\}+(-)^{\gamma_2} \sum_{r=0}^{\gamma_1} \binom{\gamma_3+\gamma_2-r}{\gamma_2} (a_1^0 a_2^{w-r} a_3^r)$$

$$+(-)^{\gamma_1+\gamma_3} \sum_{r=0}^{\gamma_2} \binom{\gamma_2+\gamma_3-r}{\gamma_3} (a_1^{w-r} a_2^0 a_3^r) = 0,$$

where $$\gamma_1+\gamma_2+\gamma_3 = w-1.$$

When $\gamma_2 = \gamma_3$ and w is odd, there is a relation

$$\{a_1 a_2\} \sum_{r=0}^{\gamma_2} \binom{2\gamma_2-r}{\gamma_2} (a_1^{w-r} a_2^0 a_3^r) = (-)^{\gamma_2} \sum_{r=0}^{\gamma_1} \binom{\gamma_1+\gamma_2-r}{\gamma_2} (a_1^0 a_2^r a_3^{w-r})$$

which gives a reduction for $(a_1^{w-\gamma_2} a_2^0 a_3^{\gamma_2})$ in terms of earlier forms provided $\gamma_2 > \gamma_1$; this relation only holds good when $n_1 = n_2$ otherwise the operation $\{a_1 a_2\}$ does not permute equivalent quantics. Moreover it does not hold for w even.

Thus when $\nu_1 = \lambda_4$ and w is odd then $\tau_1 = 1$.

Similarly when $\nu_2 = \lambda_4$ and w is odd, $\tau_2 = 1$.

Thus again for the quintic, the same rule is found, the necessary and sufficient condition that the form is irreductible is

$$R \geqslant \lambda_4+\nu_1+\nu_2+\tau_1+\tau_2,$$

* Q.S.A. VII, Theorem II, Corollary, p. 313.

where ν_1, ν_2 are as defined above and

$$\text{when } w \text{ is even} \quad \tau_2 = 0,\ \nu_2 \leqslant \lambda_4; \quad \tau_2 = 1,\ \nu_2 > \lambda_1;$$

$$\text{when } w \text{ is odd} \quad \tau_2 = 0,\ \nu_2 < \lambda_4; \quad \tau_2 = 1,\ \nu_2 \geqslant \lambda_1.$$

Other cases must be examined further.

VI. *The binary canonical forms, quantics of even order.*

23. In the ordinary theory the canonical form of a binary form f of order $2n$ is

$$f = a_{1_x}^{2n} + a_{2_x}^{2n} + \ldots + a_{n_x}^{2n} + p a_{1_x}^{2} a_{2_x}^{2} \ldots a_{n_x}^{2};$$

and the condition that the form can be expressed as the sum of n $2n$-th powers is that the catalecticant invariant

$$K_n = \begin{vmatrix} A_0 & A_1 & \ldots & A_n \\ A_1 & A_2 & \ldots & A_{n+1} \\ \ldots & \ldots & \ldots & \ldots \\ \ldots & \ldots & \ldots & \ldots \\ A_n & A_{n+1} & \ldots & A_{2n} \end{vmatrix}$$

vanishes, so that p vanishes with K_n.

Now it has been shown (Theorem VII, §19) that the covariants of f may be expressed in terms of those which have a factor K_n and those for which the defining gradient lacks one of the coefficients $A_2, A_4, \ldots, A_{2n}$. When A_{2n} is absent the defining gradient gives a covariant of the quantic of order $2n-1$, and all these are supposed known. When A_{2r} is missing, the covariant differs from a form

$$(a_0^{\lambda_0} a_2^{2\lambda_2+3\lambda_3} \ldots a_r^{(2r-2)\lambda_{2r-2}+(2r-1)\lambda_{2r-1}} a_{r+1}^{(2r+1)\lambda_{2r+1}+(2r+2)\lambda_{2r+2}} \ldots a_n^{(2n-1)\lambda_{2n-1}+2n\lambda_{2n}})$$

by covariants with earlier leading gradients; just as in the last case. This brings the substitutional theory exactly into line with the ordinary theory in the matter of canonical forms.

Now the argument of Theorem VIII may be applied word for word here, it does not really matter in the application which letter A_{2r} is absent; the real point is there are not more than n different letters a, there may indeed be fewer.

Theorem IX. *The covariant* $C = (A_0^{\lambda_0} A_2^{\lambda_2} \ldots A_{2n}^{\lambda_{2n}})$ *of order* R *of the binary form of order* $2n$, *is a product of* K_n *and another covariant unless one of the indices* $\lambda_2, \lambda_4, \ldots, \lambda_{2n}$ *is zero. And further when* $\lambda_{2r} = 0$ *it is reductible* *unless* $R \geqslant \lambda_{2n-1}+\nu+\tau$, *where*

$$r \leqslant \rho, \quad \nu_r = \lambda_{2r-2}+\lambda_{2r-1}-\lambda_{2n}$$
$$r > \rho, \quad \nu_r = \lambda_{2r-1}+\lambda_{2r}-\lambda_{2n}$$
$$\nu = \Sigma\, \nu_r$$

the sum being confined to the values of r *for which* ν_r *is positive; and* τ *is the number of values of* r *for which* $\nu_r > \lambda_{2n-1}$.

24. It will be noticed at once that the matter of chief consequence is the number of letters a that have to be used. Up to the quartic only two such letters have to be used: and the conditions are very simple. For a quantic of any order n, such forms exist, for instance the covariants

$$(A_0^{\lambda_0} A_{n-1}^{\lambda_{n-1}} A_n^{\lambda_n}),$$

and the condition that this is irreductible is always

$$w \text{ even}, \quad \lambda_0 \geqslant \lambda_{n-1}+\lambda_n,$$
$$w \text{ odd}, \quad \lambda_0 > \lambda_{n-1}+\lambda_n.$$

In the same way when the covariants can be expressed in a form with three letters a, as can always be done up to the sextic, the necessary and sufficient condition that the form is irreductible is that

$$R \geqslant \lambda_{n-1}+\nu+\tau$$

where $\tau = \Sigma\tau_r$, and

$$w \text{ even} \quad \tau_r = 1, \text{ when } \nu_r > \lambda_{n-1}; \quad \tau_r = 0 \text{ when } \nu_r \leqslant \lambda_{n-1};$$
$$w \text{ odd} \quad \tau_r = 1, \text{ when } \nu_r \geqslant \lambda_{n-1}; \quad \tau_r = 0 \text{ when } \nu_r < \lambda_{n-1};$$

where n is the order of the quantic considered.

VII. *Further results from and extensions of Theorem VIII.*

25. When the sets K_r in Equation III, that is the sets whose defining gradients have one of the factors $a_r^{\kappa_r}$, $r = 2, 3, \ldots, \delta-1$, represent earlier forms in the sequence chosen, than the sets K_1, K_δ this equation becomes

$$\text{(X)} \qquad (-)^w(a_1^{\kappa_1} a_2^{\vartheta_2} \ldots a_{\delta-1}^{\vartheta_{\delta-1}} a_\delta^0) = (a_1^0 a_2^{\xi_2} \ldots a_{\delta-1}^{\xi_{\delta-1}} a_\delta^{\xi_\delta}) + \text{earlier terms.}$$

Consider the case in which each letter represents a quantic which is some power of the original quantic as in Section V. Then let us define κ_r, $r = 2, \ldots, n-1$ as in §21 writing $n+1$ instead of δ as the number of letters in use, and keeping δ as the degree in the coefficients of the original quantic. The form with K_{n+1} is taken to be C of §21; and further the case $n_1 = n_{n+1}$ is to be considered, and in particular it will be assumed that $\kappa_1 = \kappa_{n+1} = 2n \,.\, n_1 + \lambda_{2n+1}$, or $(2n+1)\delta - \nu - \tau + n - \lambda_{2n} = 2w + n - 1$, whence $R = \lambda_{2n} + \nu + \tau$. Then the equation (X) is true for this case. There is a reduction for $C_\xi = C$, always when $a_2^{\vartheta_2}$ represents an earlier form than $a_{\delta-1}^{\xi_\delta-1}$.

Now $a_{\delta-1}^{\xi_\delta-1}$ represents $A_{2n-2}^{\lambda_{2n-2}} A_{2n-1}^{\lambda_{2n-1}}$; while $\vartheta_2 = \kappa_2 - 1 - \xi_2$. If $n_2 > n_{n+1}$,

$$\kappa_2 = 2n \,.\, n_2 + \lambda_{2n+1}, \quad n_2 = \lambda_2 + \lambda_3, \quad \xi_2 = 2\lambda_2 + 3\lambda_3,$$

$$\vartheta_2 = (2n-2)\, n_2 - 1 + \lambda_{2n+1} - \lambda_3,$$

this means $$a_2^{\vartheta_2} = A_{2n-2}^{\lambda_2 + 2\lambda_3 + 1 - \lambda_{2n+1}} A_{2n-1}^{\lambda_{2n+1} - \lambda_3 - 1};$$

or when the index of A_{2n-1} is negative it means

$$A_{2n-3}^{\lambda_3 + 1 - \lambda_{2n+1}} A_{2n-2}^{\lambda_2 + \lambda_{2n+1} - 1}.$$

There is a reduction when $\lambda_{2n-1} + \lambda_3 + 1 < \lambda_{2n+1}$. In the case of equality there is a reduction when $n_2 > n_n$, and not when $n_2 < n_n$. If

$$n_{n+1} \geqslant n_2 \geqslant \lambda_{2n+1}, \; \kappa_2 = 2n \,.\, n_2 + \lambda_{2n+1} + 1,$$

and there is a reduction when $\lambda_{2n-1} + \lambda_3 < \lambda_{2n+1}$. If

$$\lambda_{2n+1} > n_2, \; \kappa_2 = (2n+1)\, n_2 + 1,$$

there is a reduction when $\lambda_{2n-1} < \lambda_2$.

Cases of equality have to be considered step by step. Ultimately when the two forms are identical as covariants of a single quantic the question of reductibility or not rests on w: if it is even there is no reduction, if it is odd there is a reduction. When $n = 2$, $\vartheta_2 = \xi_2$ and as we have seen the sign $(-)^w$ decides the matter.

In such cases the reduction is accounted for by putting $\tau_1 = 1$ when $n_1 = n+1$ and w is odd; and $\tau_1 = 0$ when w is even.

It is possible that there is a reduction not accounted for already, due to the reductibility of C_9 for degree n, omitting the last letter.

26. When there are no cases of equality $n_r = n_{n+1}$; and

$$R \geqslant \lambda_{2n} + \nu + \tau,$$

then the values given for κ_r in §21, except that

$$\kappa_{n+1} = 2n \,.\, n_{n+1} + \lambda_{2n+1},$$

are such that $\sum_{r=1}^{n+1} \kappa_2 = 2w + (n+1-1)$

and hence by Theorem II, the members of the different sets are entirely independent of each other.

THEOREM X. *When there is no equality of the nature $n_r = n_{n+1}$ between the degrees of the different compound coefficients and the degree of the last; the necessary condition $R \geqslant \lambda_{2n} + \nu + \tau$ for irreductibility in Theorem VIII is both necessary and sufficient, the earlier conditions being satisfied.*

This is also true for the extended cases of Theorem IX.

27. It is clear now that the general condition for irreductibility for covariants of a single form of order n is

$$R \geqslant \lambda_{n-1} + \nu + \tau.$$

The definition of $\nu = \Sigma \nu_r$ is also quite definite, the definition of τ is also clear, viz. $\tau = \Sigma \tau_r$ where τ_r 0 or 1. $\tau_r = 0$, $\nu_r < \lambda_{n-1}$, $\tau_r = 1$, $\nu_r > \lambda_{n-1}$. But when $\nu_r = \lambda_{n-1}$, and w is odd, the condition requires examination in general.

It is well known that when $R \geqslant \delta - 1$ the irreductibility of the form is unaltered by the introduction of the last letter. When $\lambda_n = 1$ and $\lambda_{n-1} = 0$ there can be constructed forms for which the condition $R \geqslant \delta - 1$ is necessary as well as sufficient. For instance in the case of $A_0^{\lambda_0} A_r^{\lambda_r} A_s^{\lambda_s} A_t^{\lambda_t} A_n$ where $\lambda_0 > 1$, $\lambda_r > 1$, $\lambda_s > 1$, $\lambda_t > 1$, each τ is 1 and $\nu + \tau = \delta - 1$, and so the condition $R \geqslant \delta - 1$ is necessary. This will always be the case when $\lambda_n = 1$, $\lambda_{n-1} = 0$ unless there is at least one isolated coefficient $A_r^{\lambda_r}$ with $\lambda_r = 1$; for then the corresponding τ_r may be zero.

When $\lambda_n = 2$, $\lambda_{n-1} = 0$, the part due to $A_r^{\lambda_r} A_{r+1}^{\lambda_{r+1}} = a_t^{\mu_t}$ where

$$n_t = \lambda_r + \lambda_{r+1}, \; \mu_t = r\lambda_r + (r+1)\lambda_{r+1},$$

is $\nu_t = n_t - 2$ and $\tau_2 = 1$ for $n_t > 2$; in this case $\nu_t + \tau_t = n_t - 1$. Thus for each compound letter $\nu_t + \tau_t \leqslant n_t - 1$. When there are m compound letters $\nu + \tau \leqslant \delta - 2 - (m-1)$. When λ_{n-1} is not zero

$$\delta = \Sigma n_t + \lambda_{n-1} + \lambda_n, \; \lambda_{n-1} + \nu + \tau \leqslant \delta - 2 - (m-1)$$

and thus when $\lambda_n = 2$ and $R \geqslant \delta - 2 - (m-1)$ the covariant is irreductible, and the reductibility question for lower values of R depends entirely on the first $m-1$ compound letters.

More generally when $\lambda_{n-1} = 0$, and

$$n_t > \lambda_n, \; \nu_t = n_t - \lambda_n, \; \tau_t = 1, \; \nu_t + \tau_t = n_t - \lambda_n + 1.$$

The maximum value of $\nu + \tau$ is then $\delta - 2\lambda_n + 1$, and when $\lambda_{n-1} > 0$ this is the maximum value of $\lambda_{n-1} + \nu + \tau$; and it only occurs when $m = 2$.

THEOREM XI. *For a single binary form of order n, a covariant whose defining gradient G ends with $A_n^{\lambda_n}$ and whose order is*

$$R \geqslant \delta - 2\lambda_n + 1$$

is irreductible, provided that the covariant whose defining gradient is obtained by removing the factors in A_{n-1} and A_n from G is also irreductible. This limit for R can be lowered in general when G contains more than 2 compound coefficients.

This is a much lower limit for R than that found previously*, viz. $R \geqslant \delta - \lambda_n$.

28. In general the covariants of a single binary form are each expressed by means of compound coefficients, so that the number of these is as small as possible.

When there is only one compound coefficient, the only possible form is A_0^r, *i.e.* a power of the quantic itself.

When there are two compound coefficients, the leading gradients are made up of powers and products of the gradients

$$A_0 A_{2r+2},\ A_0^2 A_{2r+3},\ A_0^2 A_{2r+3}^2,\ A_0.$$

29. The following is a sketch of the case of three compound coefficients,

$$(A_0^{\lambda_0} A_r^{\lambda_r} A_{r+1}^{\lambda_{r+1}} A_{n-1}^{\lambda_{n-1}} A_n^{\lambda_n}).$$

The first two compound coefficients must satisfy the conditions for two. Hence $\lambda_0 > \lambda_r + \lambda_{r+1}$ or $\lambda_0 = \lambda_r + \lambda_{r+1}$ and $r\lambda_r + (r+1)\lambda_{r+1}$ even.

(i) The most important case is $\lambda_0 = \lambda_r + \lambda_{r+1} = \lambda_{n-1} + \lambda_n$, and this is obtained from the case of $(a^0 \beta^\gamma \gamma^{w-\gamma})$, where a, β, γ are equivalent symbols. There is a reduction unless γ is even and $w - \gamma = 2\gamma$, or $w - \gamma > 2\gamma + 1$. Hence

$$(n-1)\lambda_{n-1} + n\lambda_n = 2r\lambda_r + 2(r+1)\lambda_{r+1} + \xi,\ \xi = 0 \text{ or } > 1;$$

i.e.

$$(n-2r)\lambda_0 = \lambda_{n-1} + 2\lambda_{r+1} + \xi.$$

When $n = 2r$, $\lambda_{n-1} = 0 = \lambda_{r+1} = \xi$. The solutions are all powers of $(A_0 A_r A_{2r})$, r even and of $(A_0^2 A_r^2 A_{2r}^2)$ when r is odd.

When $n = 2r+1$, $\lambda_0 = \lambda_{n-1} + 2\lambda_{r+1} + \xi$, $\lambda_r = \lambda_{n-1} + \lambda_{r+1} + \xi$, $\lambda_n = \lambda_{2r+1} + \xi$, and the form is $(A_0^{2\lambda_{r+1}+\lambda_{n-1}+\xi} A_r^{\lambda_{r+1}+\lambda_{n-1}+\xi} A_{r+1}^{\lambda_{r+1}} A_{2r}^{\lambda_{n-1}} A_{2r+1}^{2\lambda_{r+1}+\xi})$.

* Q.S.A. VII, p. 335, Theorem XVI.

r even: λ_{r+1} must be even, and there is a factor $(A_0^4 A_r^2 A_{r+1}^2 A_{2r+1}^4)$ unless $\lambda_{r+1} = 0$. The other forms are $(A_0 A_r A_{2r})$ when $\lambda_{n-1} > 0$, $(A_0^2 A_r^2 A_{2r+1}^2)$, $(A_0^3 A_r^3 A_{2r+1}^3)$.

r odd: λ_r must be even, there is a factor $(A_0^4 A_r^2 A_{r+1}^2 A_{2r+1}^4)$ unless $\lambda_{r+1} < 2$; a factor $(A_0^3 A_r^2 A_{r+1} A_{2r} A_{2r+1}^2)$ when $\lambda_{r+1} > 0$, $\lambda_{n-1} > 0$. A factor $(A_0^2 A_r^2 A_{2r}^2)$ unless $\lambda_{n-1} < 2$. A factor $(A_0^2 A_r^2 A_{2r+1}^2)$, or $(A_0^3 A_r^3 A_{2r+1}^3)$ unless $\xi = 0$.

When $n = 2r+2$, $2\lambda_r = \lambda_{n-1} + \xi$, $\lambda_r + \lambda_n = \lambda_{r+1} + \xi$. And the form is $(A_0^{\lambda_r+\lambda_{r+1}} A_r^{\lambda_r} A_{r+1}^{\lambda_{r+1}} A_{2r+1}^{2\lambda_r-\xi} A_{2r+2}^{\lambda_{r+1}-\lambda_r+\xi})$.

r even: λ_{r+1} is even. There is a factor

$$(A_0^2 A_r^2 A_{2r+1}^2),\ (A_0^3 A_r^3 A_{2r+1}^3) \text{ or } (A_0 A_r A_{2r+2}),$$

unless $\lambda_r = 0$ or $\lambda_r = 1$, $\xi = 0$. A factor $(A_0^2 A_{r+1}^2 A_{2r+2}^2)$ unless $\lambda_{r+1} = 0$, or $\lambda_{r+1} - \lambda_r + \xi < 2$; now when $\lambda_r = 0$ then $\xi = 0$, but when $\lambda_r = 1$, $\xi = 1$ it is necessary to include $(A_0^3 A_r A_{r+1}^2 A_{2r+1}^2 A_{2r+2})$.

r odd: λ_{r+1} is even: factor $(A_0 A_{r+1} A_{2r+2})$ unless $\lambda_{r+1} = \lambda_r - \xi$, or $\lambda_{r+1} = 0$. When $\lambda_{r+1} = \lambda_r - \xi$, the case is reduced to $n = 2r+1$. Thus the only other forms are $(A_0^2 A_r^2 A_{2r+2}^2)$, $(A_0^2 A_r^2 A_{2r+1} A_{2r+2})$.

When $n = 2r+s$, $s > 2$ there is nothing of interest unless $\xi \geqslant 2$. Let us write $s = \sigma+3$, then $\sigma\lambda_0 + \lambda_n + 2\lambda_r = \xi$, and the form is

$$(A_0^{\lambda_r+\lambda_{r+1}} A_r^{\lambda_r} A_{r+1}^{\lambda_{r+1}} A_{2r+\sigma+2}^{\lambda_{n-1}} A_{2r+\sigma+3}^{\lambda_n}).$$

r even: λ_{r+1} is even. We assume $\lambda_n > 0$. There is a factor $(A_0 A_r A_{2r+\sigma+3})$ unless $\lambda_r = 0$. When $\sigma > 0$ there is a factor $(A_0^2 A_{r+1}^2 A_{2r+\sigma+2}^2)$ or a factor $(A_0^2 A_{r+1}^2 A_{2r+\sigma+2} A_{2r+\sigma+3})$ when $\lambda_{r+1} > 0$. When $\sigma = 0$, $\lambda_r = 0$, $\lambda_n > 1$ for $\xi \neq 1$, and there is a factor $(A_0^2 A_{r+1}^2 A_{2r+3}^2)$.

r odd: λ_2 even, $\sigma > 0$ there is a factor $(A_0^2 A_r^2 A_{2r+\sigma+2} A_{2r+\sigma+3})$ or $(A_0 A_{r+1} A_{2r+\sigma+3})$ or $(A_0^2 A_r^2 A_{2r+\sigma+2} A_{2r+\sigma+3})$. When $\sigma = 0$, there is one of the factors $(A_0 A_{r+1} A_{2r+2})$ $(A_0^2 A_{r+1}^2 A_{2r+3}^2)$, $(A_0^3 A_{r+1}^3 A_{2r+3}^3)$, $(A_0^2 A_r^2 A_{2r+3}^2)$, $(A_0^2 A_r^2 A_{2r+2}^2)$, $(A_0^2 A_r^2 A_{2r+2} A_{2r+3})$.

Thus summing up for this class, the forms which are not obvious products are included in the following:

$$(A_0 A_{2r} A_{4r+\xi}),\ \xi = 0 \text{ or } > 1:\quad (A_0^2 A_{2r}^2 A_{4r+1}^2),\quad (A_0^3 A_{2r}^3 A_{4r+1}^3),$$

$$(A_0^2 A_{2r+3}^2 A_{4r+6+s}^2),\ (A_0^4 A_{r+2}^2 A_{r+3}^2 A_{2r+5}^4),\ (A_0^2 A_{r+2}^2 A_{2r+s+5} A_{2r+s+6}),$$

$$(A_0^3 A_{2r+3}^2 A_{2r+4} A_{4r+6}^2 A_{4r+7}),\ (A_0^3 A_{2r+2} A_{2r+3}^2 A_{4r+5}^2 A_{4r+6}).$$

30. The other forms may be classified as follows:

(ii) $\lambda_n \geqslant \lambda_0 \geqslant \lambda_r + \lambda_{r+1}$.

Here $\nu = 0$; and $\tau = 0$ unless $\lambda_n = \lambda_0 > \lambda_r + \lambda_{r+1}$, $\lambda_{n-1} = 0$, and w is odd, when $\tau = 1$; (the case of (i) is excluded).

It will be seen that there is no form unless $n > 2r$.

Here either

$$(a) \quad R = n(\lambda_0 - \lambda_n) - (n-2)\lambda_{n-1} + (n-2r)\lambda_r + (n-2r-2)\lambda_{r+1} \geqslant \lambda_{n-1},$$

$$i.e. \quad (n-2r)(\lambda_r + \lambda_{r+1}) \geqslant n(\lambda_n + \lambda_{n-1} - \lambda_0) + 2\lambda_{r+1} - \lambda_{n-1}.$$

(b) Or else w is odd, and $\lambda_n = \lambda_0 > \lambda_r + \lambda_{r+1}$, $\lambda_{n-1} = 0$, $R \geqslant 1$;

$$i.e. \quad (n-2r)(\lambda_r + \lambda_{r+1}) \geqslant 2\lambda_{r+1} + 1.$$

(iii) $\lambda_0 > \lambda_n \geqslant \lambda_r + \lambda_{r+1}$, $R \geqslant \lambda_{n-1} + \lambda_0 - \lambda_n + \tau_1 + \tau_2$,

$$i.e. \quad (n-1)(\lambda_0 - \lambda_{n-1} - \lambda_n) + (n-2r)(\lambda_r + \lambda_{r+1}) \geqslant 2\lambda_{r+1} + \tau_1 + \tau_2.$$

(a) $\lambda_0 < \lambda_{n-1} + \lambda_n$, $\tau_1 + \tau_2 = 0$,

$$(n-2r)(\lambda_r + \lambda_{r+1}) \geqslant (n-1)(\lambda_{n-1} + \lambda_n - \lambda_0) + 2\lambda_{r+1}.$$

There is no form here unless $n > 2r$.

(b) $\lambda_0 = \lambda_{n-1} + \lambda_n$, and then $\lambda_{n-1} > 0$, $\tau_2 = 0$,

$$(n-2r)(\lambda_r + \lambda_{r+1}) \geqslant 2\lambda_{r+1} + \epsilon,$$

where $\epsilon = 0$, w even; and $\epsilon = 1$, w odd. There is no form unless $n > 2r$, or $n = 2r$, $\lambda_{r+1} = 0$ and w even.

(c) $\lambda_0 > \lambda_{n-1} + \lambda_n$; $(n-1)(\lambda_0 - \lambda_{n-1} - \lambda_n)$

$$+ (n-2r)(\lambda_r + \lambda_{r+1}) \geqslant 2\lambda_{r+1} + 1 + \epsilon,$$

where $\epsilon = 0$ unless $\lambda_n = \lambda_r + \lambda_{r+1}$, $\lambda_{n-1} = 0$, w is odd and then $\epsilon = 1$.

(iv) $\lambda_0 \geqslant \lambda_r + \lambda_{r+1} > \lambda_n$, $R \geqslant \lambda_{n-1} + \lambda_0 + \lambda_r + \lambda_{r+1} - 2\lambda_n + \tau_1 + \tau_2$;

$$i.e. \quad (n-1)(\lambda_0 - \lambda_{n-1} - \lambda_n) + (n-2r-1)(\lambda_r + \lambda_{r+1}) + \lambda_n \geqslant 2\lambda_{r+1} + \tau_1 + \tau_2.$$

The results for these last cases will be worked out for the quintic*.

There is no restriction when $n > 2r+2$, beyond $\lambda_0 \geqslant \lambda_r + \lambda_{r+1} > \lambda_n$.

* Young added the words "and the sextic", but this latter was not undertaken. (*Ed.*)

VIII. *Covariants of the binary quintic sequence* (*A*).

31. There are first those whose defining gradients have not more than two compound coefficients.

$$(A_0)=f,\ (A_0A_2)=H,\ (A_0A_4)=i,\ (A_0^2A_3)=t,\ (A_0^2A_3^2),\ (A_0^2A_5)=q,$$
$$(A_0^2A_5^2)=I_4.$$

With three compound coefficients the forms of class (i) may be written down from §29. They are

$$(A_0A_2A_4)=j,\quad (A_0^2A_2^2A_5^2)=(jj)^2=\tau,\quad (A_0^3A_2^3A_5^3)=(j\tau),$$
$$(A_0^4A_2^2A_3^2A_5^4)=I_{12}.$$

The other classes need calculation.

(ii) $\lambda_5 \geqslant \lambda_0 \geqslant \lambda_2+\lambda_3$.

(*a*) $$\lambda_5=\lambda_0+\xi_2,\quad \lambda_0=\lambda_2+\lambda_3+\xi_3,$$
$$\lambda_2=5(\lambda_5-\lambda_0)+4\lambda_4+\lambda_3+\xi_1=4\lambda_4+\lambda_3+\xi_1+5\xi_2.$$

There is a limitation that those cases for which $\xi_2=0$, $\lambda_4=0$, w odd must be excluded. Without this limitation the solutions are products of powers of $(A_0^4A_2^4A_4A_5^4)$, $(A_0^5A_2^5A_5^6)$; (A_0A_5), $(A_0A_2A_5)$, $(A_0^2A_2A_3A_5^2)$. These forms are obtained by giving the value 1 to each of the following quantities in turn while the rest are zero: λ_4, ξ_2, ξ_3, ξ_1, λ_3. The first two are actual forms, the others are not but may produce actual forms as products. Thus all the forms here are powers and products of the forms

$$(A_0^4A_2^4A_4A_5^4),\quad (A_0^5A_2^4A_4A_5^5),\quad (A_0^5A_2^5A_4A_5^5),\quad (A_0^5A_2^5A_5^6),$$
$$(A_0^6A_2^5A_5^7),\quad (A_0^6A_2^6A_5^7),\quad (A_0^2A_2A_5^2),\quad (A_0^3A_2A_3A_5^3).$$

(*b*) $\lambda_5=\lambda_0=\lambda_2+\lambda_3+\xi_2+1$, $\lambda_4=0$, $\lambda_2=\lambda_3+\xi_1+1$, $\lambda_3+\xi_1+\xi_2$ odd; hence $\lambda_5=2\lambda_3+\xi_1+\xi_2+2=\lambda_0$. The forms are those products of $(A_0^2A_2A_5^2)$ with $(A_0^2A_2A_3A_5^2)$, $(A_0A_2A_5)$, (A_0A_5) and their products and powers which have an odd number of these factors; any even number of factors is already included in (*a*). Thus we have to record here the three forms

$$(A_0^3A_2A_5^3),\quad (A_0^3A_2^2A_5^3),\quad (A_0^4A_2^2A_3A_5^4).$$

(iii) (*a*) $\lambda_4+\lambda_5=\lambda_0+\xi_1+1$, $\lambda_5=\lambda_2+\lambda_3+\xi_2$, $\lambda_2=\lambda_3+4\xi_1+\xi_3+4$,
$$\lambda_0=\lambda_5+\xi_4+1.$$

Hence

$$\lambda_4 = \xi_1+\xi_4+2, \quad \lambda_5 = 2\lambda_3+4\xi_1+\xi_2+\xi_3+4, \quad \lambda_0 = 2\lambda_3+4\xi_1+\xi_2+\xi_3+\xi_4+5.$$

The forms are the product of $(A_0^5 A_2^4 A_4^2 A_5^4)$ with any powers and products of $(A_0^2 A_2 A_3 A_5^2)$, $(A_0^4 A_2^4 A_4 A_5^4)$, $(A_0 A_5)$, $(A_0 A_2 A_5)$, $(A_0 A_4)$. Now the basic form $(A_0^5 A_2^4 A_4^2 A_5^4)$ is itself the product

$$(A_0 A_4)(A_0^4 A_2^4 A_4 A_5^4)$$

and hence the only covariant to be recorded here is $(A_0^7 A_2^5 A_3 A_4^2 A_5^6)$.

(*b*) $\lambda_0 = \lambda_4+\lambda_5$, $\lambda_5 = \lambda_2+\lambda_3+\xi_1$, $\lambda_2 = \lambda_3+\xi_2$, where $\xi_1 > 0$ when w is odd; $\lambda_4 > 0$ since $\lambda_0 = \lambda_5$ is included in (ii). The covariants are

$$(A_0 A_4)^{\lambda_4} (A_0^2 A_2 A_3 A_5^2)^{\lambda_3} (A_0 A_5)^{\xi_1} (A_0 A_2 A_5)^{\xi_2}.$$

The only new covariant is $(A_0^2 A_2 A_4 A_5)$.

(*c*) $\lambda_0 = \lambda_4+\lambda_5+\xi_1+1$, $\lambda_5 = \lambda_2+\lambda_3+\xi_2$, $4\xi_1+3-\epsilon+\lambda_2 = \lambda_3+\xi_3$ where $\epsilon = 0$, unless $\lambda_4 = 0 = \xi_2$ and λ_2 is odd, when $\epsilon = 1$. The basic forms are $(A_0^4 A_3^3 A_5^3)$, $(A_0^3 A_3^2 A_5^2)$, $(A_0^2 A_3 A_5)$, (A_0); the first is excluded for odd weights and $\lambda_4 = 0 = \xi_2$ and in place of it $(A_0^8 A_3^6 A_5^6)$ should be included.

The additional factors are obtained as the solutions of

$$\lambda_0 = \lambda_4+\lambda_5+\xi_1, \quad \lambda_5 = \lambda_2+\lambda_3+\xi_2, \quad 4\xi_1+\lambda_2 = \lambda_3+\xi_3.$$

The last equation is solved by

$$\lambda_3 = y_1+4y_3+3y_4+2y_5+y_6, \quad \xi_3 = y_2+y_4+2y_5+3y_6+4y_7.$$

$$\lambda_2 = y_1+y_2, \quad \xi_1 = y_3+y_4+y_5+y_6+y_7.$$

The factors are

$$A_0, \quad A_0 A_4, \quad A_0^2 A_3 A_5, \quad A_0^3 A_3^2 A_5^2, \quad A_0^4 A_3^3 A_5^3;$$

$$A_0 A_5, \quad A_0 A_2 A_5, \quad A_0^2 A_2 A_3 A_5^2, \quad A_0^5 A_3^4 A_5^4.$$

The first five factors are leading gradients, and need not be discussed. The squares of the next three factors are also leading gradients, as also is the product of $A_0 A_5$ and either of the other two. Thus the following covariants have to be recorded:

$$(A_0^2 A_3 A_5), \quad (A_0^3 A_3^2 A_5^2), \quad (A_0^4 A_3^3 A_5^3), \quad (A_0^3 A_3 A_5^2), \quad (A_0^4 A_3^2 A_5^3),$$

$$(A_0^5 A_3^3 A_5^4), (A_0^2 A_2 A_5), (A_0^3 A_2 A_3 A_5^2), (A_0^4 A_2 A_3^2 A_5^3), (A_0^5 A_2 A_3^3 A_5^4),$$

$$(A_0^{5k+4} A_3^{4k+3} A_5^{4k+3}), \quad (A_0^{5k+5} A_3^{4k+3} A_5^{4k+4}), \quad (A_0^{5k+5} A_2 A_3^{4k+3} A_5^{4k+4}).$$

This excludes those covariants whose leading gradients are the direct products of two leading gradients of lower degree.

(iv) $$\lambda_0 = \lambda_2 + \lambda_3 + \xi_1, \quad \lambda_2 + \lambda_3 = \lambda_5 + \xi_2 + 1,$$
$$4(\lambda_0 - \lambda_4 - \lambda_5) + \lambda_5 - 2\lambda_3 = \xi_3 + \tau_1 + \tau_2.$$

There are certain factors, not all leading gradients that enable us to obtain the solutions here with ease. In each case the factor is supposed to be an actual factor of the defining gradient of the covariant considered, and the corresponding co-factor is a leading gradient: where necessary the condition for this is stated. Factors which are leading gradients are numbered (a_r), those which are not (β_r).

(a_1) $(A_0 A_2 A_4)$ always when $\lambda_2 + \lambda_3 > \lambda_5$

(β_1) $(A_0 A_2 A_5)$ when $\xi_3 > 0$, $\lambda_4 = 0$; and some other cases.

(a_2) $(A_0 A_4)$ always when $\xi_3 > 0$ and $\lambda_2 = 0$; and some other cases.

(β_2) $(A_0{}^6 A_3{}^6 A_5{}^4)$ always when $\lambda_2 + \lambda_3 > \lambda_5 + 2$.

(β_3) $(A_0{}^5 A_3{}^4 A_4{}^4)$ always when $\lambda_0 > \mu_0$ where μ_0 is the lowest index of A_0 for which $(A_0^{\mu_0} A_2^{\lambda_2} A_3^{\lambda_3} A_4^{\lambda_4})$ is a quartic covariant.

Thus all gradients in which both A_2 and A_4 are present are excluded by (a_1). Then all gradients which have either A_2 or A_4 present are excluded unless $\xi_3 = 0$.

The (β_3) factor always is present when the indices are high enough unless $\lambda_0 = \mu_0$ and so this equality will first be assumed.

With this condition the only forms to be considered for which $\lambda_2 + \lambda_3 = \lambda_5 + 1$ are:

$$(A_0^{2\mu_3+2\epsilon+\lambda_4} A_3^{2\mu_3+\epsilon} A_4^{\lambda_4} A_5^{2\mu_3+\epsilon-1}), \quad \text{and} \quad (A_0^{2\mu_3+2\epsilon+\lambda_2} A_2^{\lambda_2} A_3^{2\mu_3+\epsilon} A_5^{\lambda_2+2\mu_3+\epsilon-1}),$$

where $\epsilon = 0$ or 1.

The existence conditions are $3\epsilon + 2 = 2\mu_3 + \xi_3 + \tau_2$, $\lambda_2 + 3\epsilon + 1 = 2\mu_3 + \xi_3$. When $\lambda_2 > 1$ there is a factor $(A_0{}^4 A_2{}^2 A_3{}^2 A_5{}^4) = I_{12}$; otherwise $\lambda_3 < 6$ in either case. Thus in all cases there is a factor (β_2) or I_{12} when the indices are great enough. It remains to consider the lower indices.

(a) $$\lambda_5 \geqslant 4, \quad \lambda_3 < 6.$$

There is a factor I_{12} unless $\lambda_2 < 2$ or $\lambda_3 < 2$. When $\lambda_2 = 1$, $\lambda_3 \geqslant \lambda_5 \geqslant 4$. There is then a factor (β_3) unless $\lambda_0 = \mu_0$. There are no other covariants $\lambda_2 = 1$, except direct products.

For $\lambda_2=0$, $\lambda_3>\lambda_5$ and hence $\lambda_3=5$, $\lambda_5=4$, $\lambda_0=\lambda_4+6$; and the only covariants are $(A_0^6 A_3^5 A_5^4)$, and the direct products of this and $(A_0 A_4)=i$.

$$(b)\ \lambda_5=3: \quad \lambda_3=2\mu_3+\epsilon, \quad \lambda_0=\lambda_0'+\lambda_2+2\mu_3+2\epsilon, \quad \lambda_4=0;$$

$$4(\lambda_0'+\lambda_2+\mu_3)+6\epsilon=\xi_3+11.$$

There is a factor $(A_0 A_2 A_5)$ unless $\xi_3=0$ and this is impossible. Otherwise $\lambda_2=0$, $\lambda_0=\lambda'_0+\lambda_4+2\mu_3+\epsilon$, $4(\lambda_0'+\mu_3)+6\epsilon=\xi_3+\tau+9$, $\mu_3\geqslant 2$. There is a factor $(A_0 A_4)$ unless $\xi_3=0$ or $\lambda_4=0$. The conditions cannot be satisfied with $\xi_3=0$. Then the covariants to be recorded are only $(A_0^5 A_3^4 A_5^3)$, $(A_0^6 A_3^6 A_5^3)$.

$$(c)\ \lambda_5=2: \quad 4(\lambda_0'+\lambda_2+\mu_3)+6\epsilon=\xi_3+8, \quad \lambda_2+\lambda_3\geqslant\lambda_5+1=3;$$

hence when $\xi_3=0$, $\lambda_2=1$, $\lambda_3=2$, and the covariant is $(A_0^3 A_2 A_3^2 A_5^2)$. Otherwise $4(\lambda_0'+\mu_3)+6\epsilon=\xi_3+\tau+6$ and $\lambda_3\geqslant 3$; the only forms to be recorded are $(A_0^4 A_3^4 A_5^2)$, $(A_0^4 A_3^3 A_5^2)$.

$$(d)\ \lambda_5=1: \quad 4(\lambda_0'+\lambda_2+\mu_3)+6\epsilon=\xi_3+5.$$

There can be no solution $\xi_3=0$ so $\lambda_2=0$ only need be considered. Then $4(\lambda_0'+\mu_3)+6\epsilon=\xi_3+\tau_2+4$, also $\lambda_3\geqslant 2$. There is a solution $\lambda_3=2$, $\xi_3=0$, $\tau_2=0$; it is $(A_0^4 A_3^2 A_4^2 A_5)$. Otherwise $\lambda_4=0$ and the solutions $(A_0^3 A_3^2 A_5)$, $(A_0^4 A_3^4 A_5)$ must be recorded.

The covariants which have one of the β factors must now be considered. As a preliminary, we make a list of the covariants already obtained in (iv) attaching to each in a bracket the value of $\Lambda_1=4(\lambda_0-\lambda_4-\lambda_5)+\lambda_5-2\lambda_3$;

$$(A_0^6 A_3^5 A_5^4)\ (2),\quad (A_0^5 A_3^4 A_5^3)\ (3),\quad (A_0^6 A_3^6 A_5^3)\ (3),\quad (A_0^3 A_2 A_3^2 A_5^2)\ (2),$$

$$(A_0^4 A_3^4 A_5^2)\ (2),\quad (A_0^4 A_3^3 A_5^2)\ (4),\quad (A_0^4 A_3^2 A_4^2 A_5)\ (1),\quad (A_0^3 A_3^2 A_5)\ (5),$$

$$(A_0^4 A_3^4 A_5)\ (5).$$

The factor (β_1) does not change the class under which the covariants are calculated and hence the product of $(A_0 A_2 A_4)$ and the 9 covariants above and the covariants of Class (i) for which $\lambda_2+\lambda_3>\lambda_5$ alone need be considered. These yield

$$(A_0^4 A_2^2 A_3^2 A_5^3)\ (3),\quad (A_0^5 A_2 A_3^4 A_5^3)\ (3),\quad (A_0^2 A_2^2 A_5)\ (5),$$

$$(A_0^3 A_2 A_3^2 A_5)\ (5).$$

The remaining factors (β_2), (β_3) provide forms with a factor

$$A_0^{5k+6l} A_3^{4k+6l} A_5^{4k+4l}$$

for which $\Lambda_1 = 0$, and a second factor which may be any covariant for which $\Lambda_1 \geqslant 2$. In certain cases, for instance when $l = 0$, covariants with a lower value of Λ_1 may be considered: this will be done later. The reason of this limitation in Λ_1 is that when $\tau = 2$ it is the condition of existence. The possible covariants to be considered are all those as yet obtained; a complete list is therefore given together with the values of Λ. It is

(i) (A_0) (4), (A_0A_2) (4), (A_0A_4) (0), $(A_0^2A_3)$ (6), $(A_0^2A_3^2)$ (4), $(A_0^2A_5)$ (5), $(A_0^2A_5^2)$ (2), $(A_0A_2A_4)$ (0), $(A_0^2A_2^2A_5^2)$ (2), $(A_0^3A_2^3A_5^3)$ (3), $(A_0^4A_2^2A_3^2A_5^4)$ (0).

(ii) $(A_0^4A_2^4A_4A_5^4)$ (0), $(A_0^5A_2^4A_4A_5^5)$ (1), $(A_0^5A_2^5A_4A_5^5)$ (1), $(A_0^5A_2^5A_5^6)$ (2), $(A_0^6A_2^5A_5^7)$ (3), $(A_0^6A_2^6A_5^7)$ (3), $(A_0^2A_2A_5^2)$ (2), $(A_0^3A_2A_3A_5^3)$ (1), $(A_0^3A_2A_5^3)$ (3), $(A_0^3A_2^2A_5^3)$ (3), $(A_0^4A_2^2A_3A_5^4)$ (2).

(iii) $(A_0^7A_2^5A_3A_4^2A_5^6)$ (0), $(A_0^2A_2A_4A_5)$ (1), $(A_0^2A_3A_5)$ (3), $(A_0^3A_3^2A_5^2)$ (2), $(A_0^4A_3^3A_5^3)$ (1), $(A_0^3A_3A_5^2)$ (4), $(A_0^4A_3^2A_5^3)$ (3), $(A_0^5A_3^3A_5^4)$ (2), $(A_0^2A_2A_5)$ (5), $(A_0^3A_2A_3A_5^2)$ (4), $(A_0^4A_2A_3^2A_5^3)$ (3), $(A_0^5A_2A_3^3A_5^4)$ (2), $(A_0^{10}A_2A_3^7A_5^8)$ (2).

The covariants (iv) have already been given.

The covariants for which $\lambda_2+\lambda_3 \leqslant \lambda_5$ having been previously dealt with, no forms with low values of k and l need be here considered unless in the product $\lambda_2+\lambda_3 > \lambda_5$.

All forms listed give covariants when $\Lambda_1 \geqslant 2$ also the product of any two covariants for each of which $\Lambda_1 = 1$. Those forms which give direct products may be omitted. In finding those which give direct products it is useful to observe that any leading gradient which multiplied by $A_0^2A_3^2A_5^2$ gives a leading gradient may be omitted as yielding a direct product; for $A_0^6A_3^6A_5^4 = A_0^2A_3^2A_5^2 . A_0^4A_3^4A_5^2$, $A_0^5A_3^4A_5^4 = A_0^2A_3^2A_5^2 . A_0^3A_3^2A_5^2$.

The following are the only leading gradients $(\Lambda_1 > 1)$ which yield a covariant but not a direct product

(i) None.

(ii) $(A_0^2A_2A_5^2)$.

(iii) $(A_0^3A_3^2A_5^2)$, $(A_0^5A_3^3A_5^4)$, $(A_0^4A_2A_3^2A_5^3)$, $(A_0^5A_2A_3^3A_5^4)$, $(A_0^8A_3^6A_5^6)$.

(iv) $(A_0^6A_3^5A_5^4)$, $(A_0^6A_3^6A_5^3)$, $(A_0^{10}A_2A_3^7A_5^8)$, $(A_0^3A_2A_3^2A_5^2)$, $(A_0^4A_3^4A_5^2)$, $(A_0^5A_2A_3^4A_5^3)$; and also $(A_0^5A_3^4A_5^3)$ when $k=0$

32. The following is a complete list of the leading gradients, other than direct products. The ordinary nomenclature* is given also of the corresponding covariant when it is irreducible.

$A_0 = f$,	$A_0^2 A_2 A_5 = (jf)$,	$A_0^2 A_3 A_5 = (jf)^2 = p$,	$A_0^3 A_2 A_3^2 A_5$,
$A_0 A_2 = H$,	$A_0^2 A_2^2 A_5 = (Hj)$,	$A_0^3 A_3^2 A_5$,	$A_0^3 A_2 A_3 A_5^2 = (pj)$,
$A_0 A_4 = i$,	$A_0^2 A_2 A_5^2 = \alpha$,	$A_0^3 A_3 A_5^2 = (pi)$,	$A_0^3 A_2 A_3^2 A_5^2$,
$A_0 A_2 A_4 = j$,	$A_0^2 A_2^2 A_5^2 = \tau = (jj)^2$,	$A_0^3 A_3^2 A_5^2$,	$A_0^3 A_2 A_3 A_5^3 = I_8$,
$A_0^2 A_3 = t$,	$A_0^3 A_2 A_5^3 = \beta$,	$A_0^4 A_3^4 A_5$,	$A_0^4 A_2 A_3^2 A_5^3$,
$A_0^2 A_3^2$,	$A_0^3 A_2^2 A_5^3 = (i\tau) = \vartheta$,	$A_0^4 A_3^3 A_5^2$,	$A_0^4 A_2^2 A_3^2 A_5^3$,
$A_0^2 A_5 = (if)$,	$A_0^3 A_2^3 A_5^3 = (j\tau)$,	$A_0^4 A_3^4 A_5^2$,	$A_0^4 A_2^2 A_3 A_5^4 = \gamma$,
$A_0^2 A_5^2 = I_4$,	$A_0^5 A_2^5 A_5^6$,	$A_0^4 A_3^2 A_5^3$,	$A_0^4 A_2^2 A_3^2 A_5^4 = I_{12}$,
	$A_0^6 A_2^5 A_5^7 = I_{18}$,	$A_0^4 A_3^3 A_5^3$,	$A_0^5 A_2 A_3^4 A_5^3$,
	$A_0^6 A_2^6 A_5^7$,	$A_0^5 A_3^4 A_5^3$,	$A_0^5 A_2 A_3^3 A_5^4$,
	$A_0^2 A_2 A_4 A_5 = (ij)$,	$A_0^5 A_3^3 A_5^4$,	$A_0^4 A_3^2 A_4^2 A_5$,
	$A_0^4 A_2^4 A_4 A_5^4 = \delta$,	$A_0^6 A_3^6 A_5^3$,	$A_0^7 A_2^5 A_3 A_4^2 A_5^6$.
	$A_0^5 A_2^4 A_4 A_5^5$,	$A_0^6 A_3^5 A_5^4$,	
	$A_0^5 A_2^5 A_4 A_5^5$,		

And also those products of $A_0^{5k+6l} A_3^{4k+6l} A_5^{4k+4l}$ and X for which $\lambda_2 + \lambda_3 > 5$, where X is one of the gradients

$$A_0^2 A_2 A_5^2, \quad A_0^5 A_3^3 A_5^4, \quad A_0^8 A_3^6 A_5^6, \quad A_0^5 A_2 A_3^4 A_5^3,$$
$$A_0^3 A_3^2 A_5^2, \quad A_0^6 A_3^6 A_5^3, \quad A_0^3 A_2 A_3^2 A_5^2, \quad A_0^5 A_2 A_3^3 A_5^4,$$
$$A_0^4 A_3^4 A_5^2, \quad A_0^6 A_3^5 A_5^4, \quad A_0^4 A_2 A_3^2 A_5^3;$$

and also the products of $A_0^{5k} A_3^{4k} A_5^{4k}$ and one of the gradients $A_0^4 A_3^3 A_5^3$, $A_0^5 A_3^3 A_5^4$, $A_0^5 A_2 A_3^3 A_5^4$; and finally the gradients $A_0^{6l+5} A_3^{6l+4} A_5^{4l+3}$.

There are thus 47 leading gradients and 15 sets of gradients, which are not direct products. The question of reducibility will be discussed later.

* Grace and Young, *Algebra of Invariants* (Cambridge, 1903), 132. See also Q.S.A. VII, p. 342.